Single-Channel Recording

Second Edition

Single-Channel Recording

Second Edition

Edited by

Bert Sakmann

Max-Planck-Institut für medizinische Forschung
Heidelberg, Germany

and

Erwin Neher

Max-Planck-Institut für biophysikalische Chemie
Göttingen, Germany

PLENUM PRESS • NEW YORK AND LONDON

Library of Congress Cataloging-in-Publication Data

Single-channel recording / edited by Bert Sakmann and Erwin Neher. --
2nd ed.
p. cm.
Includes bibliographical references and index.
ISBN 0-306-44870-X
1. Ion channels--Research--Methodology. I. Sakmann, Bert, 1942-
II. Neher, Erwin, 1944-
QH603.I54S56 1995
574.87'5--dc20 95-3364
CIP

Cover illustrations by Dr. Claudia Racca (neuron) and Dr. Nelson Spruston (trace).

ISBN 0-306-44870-X

A Division of Plenum Publishing Corporation
233 Spring Street, New York, N. Y. 10013

10 9 8 7 6 5 4 3

Printed in the United States of America

Contributors

Klaus Benndorf Department of Physiology, University of Cologne, D-50931 Cologne, Germany

Robert H. Chow Department of Membrane Biophysics, Max-Planck-Insitute for Biophysical Chemistry, Am Fassberg, D-37077 Göttingen-Nikolausberg, Germany.

David Colquhoun Department of Pharmacology, University College London, London WC1E 6BT, England

Jens Eilers International Physiological Institute, University of Saarland, D-66421 Homburg/Saar, Germany

Kevin D. Gillis Department of Membrane Biophysics, Max-Planck-Institute for Biophysical Chemistry, Am Fassberg, D-37077 Göttingen, Germany

Walter Häberle IBM Physics Group Munich, D-80799 Munich, Germany

Owen P. Hamill Department of Physiology and Biophysics, The University of Texas Medical Branch, Galveston, Texas 77555-0641

Alan G. Hawkes European Business Management School, University of Wales Swansea, Swansea SA2 8PP, Wales.

Rainer Hedrich Institute for Biophysics, University of Hannover, D-30419 Hannover, Germany

Stefan H. Heinemann Max-Planck Society, Research Unit, "Molecular and Cellular Biophysics," D-07747 Jena, Germany

Donald W. Hilgemann Department of Physiology, University of Texas Southwestern, Medical Center at Dallas, Dallas, Texas 75235-9040

J. K. Heinrich Hörber Department of Cell Biophysics, European Molecular Biology Laboratory (EMBL), D-69117 Heidelberg, Germany

Peter Jonas Department of Cell Physiology, Max-Planck-Institute for Medical Research, D-69120 Heidelberg, Germany

Arthur Konnerth First Physiological Institute, University of Saarland, D-66421 Homburg/Saar, Germany

P. Läuger Department of Biology, University of Konstanz, D-78434, Konstanz, Germany. *Deceased:* direct correspondence to H.-J. Apell at this address.

Alain Marty Neurobiology Laboratory, Teacher's Training College, F-75005 Paris France; *Present address:* Cellular Neurobiology Workgroup, Max-Planck-Insitutute for Biophysical Chemistry Am Fassberg, D-37077 Göttingen, Germany

Don W. McBride, Jr. Department of Physiology and Biophysics, The University of Texas Medical Branch, Galveston, Texas 77555-0641

Hannah Monyer Center for Molecular Biology (ZMBH), Im Neuenheimer Feld 282, D-69120 Heidelberg, Germany

Johannes Mosbacher Department of Cell Physiology, Max-Planck-Institute for Medical Research, D-69120 Heidelberg, Germany

Erwin Neher Department of Membrane Biophysics, Max-Planck-Institute for Biophysical Chemistry, Am Fassberg, D-37077 Göttingen, Germany

Anant B. Parekh Max-Planck-Institute for Biophysical Chemistry, Am Fassberg, D-37077 Göttingen, Germany

Reinhold Penner Department of Membrane Biophysics, Max-Planck-Institute for Biophysical Chemistry, Am Fassberg, D-37077 Göttingen, Germany

Bert Sakmann Department of Cell Physiology, Max-Planck-Institute for Medical Research, D-69120 Heidelberg, Germany

Ralf Schneggenburger International Physiological Institute, University of Saarland, D-66421 Homburg/Saar, Germany. *Present address:* Neurobiology Laboratory, Teacher's Training College, 46 Rue d'Ulm, F-75005 Paris, France.

F. J. Sigworth Department of Cellular and Molecular Physiology, Yale University School of Medicine, New Haven, Connecticut 06510

Greg Stuart Department of Cell Physiology, Max-Planck-Institute for Medical Research, D-69120 Heidelberg, Germany

Walter Stühmer Max-Planck-Institute for Experimental Medicine, D-37075 Göttingen, Germany

Ludolf von Rüden Department of Membrane Biophysics, Max-Planck-Institute for Biophysical Chemistry, Am Fassberg, D-37077 Göttingen-Nikolausberg, Germany. *Present address:* Howard Hughes Medical Institute, Department of Cellular Physiology, Stanford University, Stanford, CA 94305-5428

Preface

The single-channel recording technique has reached the status of a routine method, and the view that conductance changes in biological membranes are caused by the openings and closings of ion channels is now almost universally accepted. The most convincing early evidence for channels mediating flow of ions across biological membranes was provided in 1972 by Bernard Katz and Ricardo Miledi through the observation of membrane noise and their estimate of the size of the underlying "elementary event." The patch-clamp method has confirmed their view directly. In 1993, the work of Nigel Unwin permitted a first visual glance through an ion channel in a biological membrane.

In the sense we use it in this book, the concept and the word *Kanal* was used first by the Austrian physiologist Ernst Brücke in 1843 to describe his view of the mechanism of transport of solutes (via water-filled capillary tubes) through a biological membrane separating two fluids. Patch-clamp recording and molecular cloning of channel genes have revealed an enormous diversity of ion channels. As has been found for other proteins, ion channels fall into different families that share common functional properties. Bertil Hille speculated that the very diverse channel subtypes may have evolved from prototypical channels (or even from an "*Ur-Kanal*"). Neither the ion selectivity nor the gating mechanism nor amino acid sequence motifs of the putative channel ancestors are known. Also, the evolutionary relationships between channel families are just emerging. On the other hand, the evolution of the word channel is fairly straightforward to derive (Fig. 1).

The words *channel* (English), *canal* (French), *Kanal* (German), or *canale* (Italian) derive from Latin *canalis* meaning a small water-filled tube or pipe. It is derived from the Greek word καννα, which also means cane, tube, or pipe. Καννα is a loan-word that was adopted by the Greeks from the Semitic word *qanû*, used likewise by the Phoenicians. This root is preserved, for example, in biblical Hebrew as **קנה**. The word *qanû* is of Assyro-Babylonian origin and means, among other things, a pipe made from reed. The Sumerian equivalent of *qanû* is gi, which designates the common and the giant reed growing in Mesopotamia and in the Near East and which is, in systematic botany, referred to as *Phragmites australis* and *Arundo donax L.* In early Sumerian cuneiform writing the shape of the reed (Fig. 2) is preserved. It is interesting to note that the same word καννα is also the root of such common words as canon or canonical, meaning a set of rules that have reached the status of official truth.

Figure 1. Evolution of the word channel from Sumerogram gi.

Figure 2. Photograph of giant reed (*Arundo donax* L).

Presumably in ancient times tubes made from this reed were used to suck or expel water and solutes between different reservoirs. Until the present day, *Arundo donax* tubes have been used for making water pipes for houses in the Near East.

Canalis was originally the word for devices used to direct the flow of water and solutes and was only later replaced in western Europe by the French words pipe and pipette. This book is about both channels and pipettes, since the essence of patch clamping is the attempt to join a channel to a pipette in order to measure the flow of solutes through both.

Patch pipettes have become more useful than originally thought; i.e., they are useful not only for measuring flow of ions through channels. In this new edition of *Single-Channel Recording,* we include a number of new chapters that describe techniques that rely on the use of pipettes to study cellular mechanisms that are only indirectly related to single ion channels.

In our opinion some important new applications have developed since the first edition of the "blue book":

- Capacitance measurements allowing the detection of single fusion events of secretory vesicles.
- Single-cell PCR measurements allowing detection of mRNA molecules in single cells by combining patch-clamp methods with molecular biology methods.
- Whole-cell recording from neurons in brain slices in combination with imaging techniques.
- Atomic force microscopy of cells and membranes attached to glass pipettes in the hope of allowing the detection of the structure of molecules in membranes.

The new edition therefore includes new chapters that give accounts of these wider applications. Also, three introductory chapters were added, which are intended to introduce the newcomer to patch clamping and to provide access to the vast literature on patch-clamp technology that has accumulated in the meantime. We do not try to cover all aspects of the technique, since quite recent reviews handling the different areas are available, such as *Methods in Enzymology,* Vol. 207, *The Plymouth Workshop Handbook* (D. C. Ogden, ed., Academic Press) and the *Axon Guide* (distributed by Axon Instruments).

We would like to thank our colleagues who contributed chapters to the new edition and also Prof. Waetzoldt and Dr. Kramer of Heidelberg University and Prof. Dani Dagan from the Technion for their help in tracing the origin of the word channel.

Bert Sakmann
Erwin Neher

Contents

Part I. INTRODUCTION TO PATCH CLAMPING

Chapter 1
A Practical Guide to Patch Clamping

Reinhold Penner

Chapter 2
Tight-Seal Whole-Cell Recording

Alain Marty and Erwin Neher

Chapter 3
Guide to Data Acquisition and Analysis

Stefan H. Heinemann

Chapter 8

Patch-Pipette Recordings from the Soma, Dendrites, and Axon of Neurons in Brain Slices

Bert Sakmann and Greg Stuart

Chapter 9
Patch Clamp and Calcium Imaging in Brain Slices

Jens Eilers, Ralf Schneggenburger, and Arthur Konnerth

Chapter 10
Fast Application of Agonists to Isolated Membrane Patches

Peter Jonas

Chapter 13

The Giant Membrane Patch

Donald W. Hilgemann

Chapter 14

A Fast Pressure-Clamp Technique for Studying Mechanogated Channels

Don W. McBride, Jr., and Owen P. Hamill

Chapter 15

Electrophysiological Recordings from *Xenopus* Oocytes

Walter Stühmer and Anant B. Parekh

Part III. ANALYSIS

Chapter 18
The Principles of the Stochastic Interpretation of Ion-Channel Mechanisms

David Colquhoun and Alan G. Hawkes

Chapter 19
Fitting and Statistical Analysis of Single-Channel Records

David Colquhoun and F. J. Sigworth

Chapter 20

A Q-Matrix Cookbook: How to Write Only One Program to Calculate the Single-Channel and Macroscopic Predictions for Any Kinetic Mechanism

David Colquhoun and Alan G. Hawkes

Part IV. "CLASSICS"

Chapter 21
Geometric Parameters of Pipettes and Membrane Patches

Bert Sakmann and Erwin Neher

Chapter 22
Conformational Transitions of Ionic Channels

P. Läuger

Color Plates

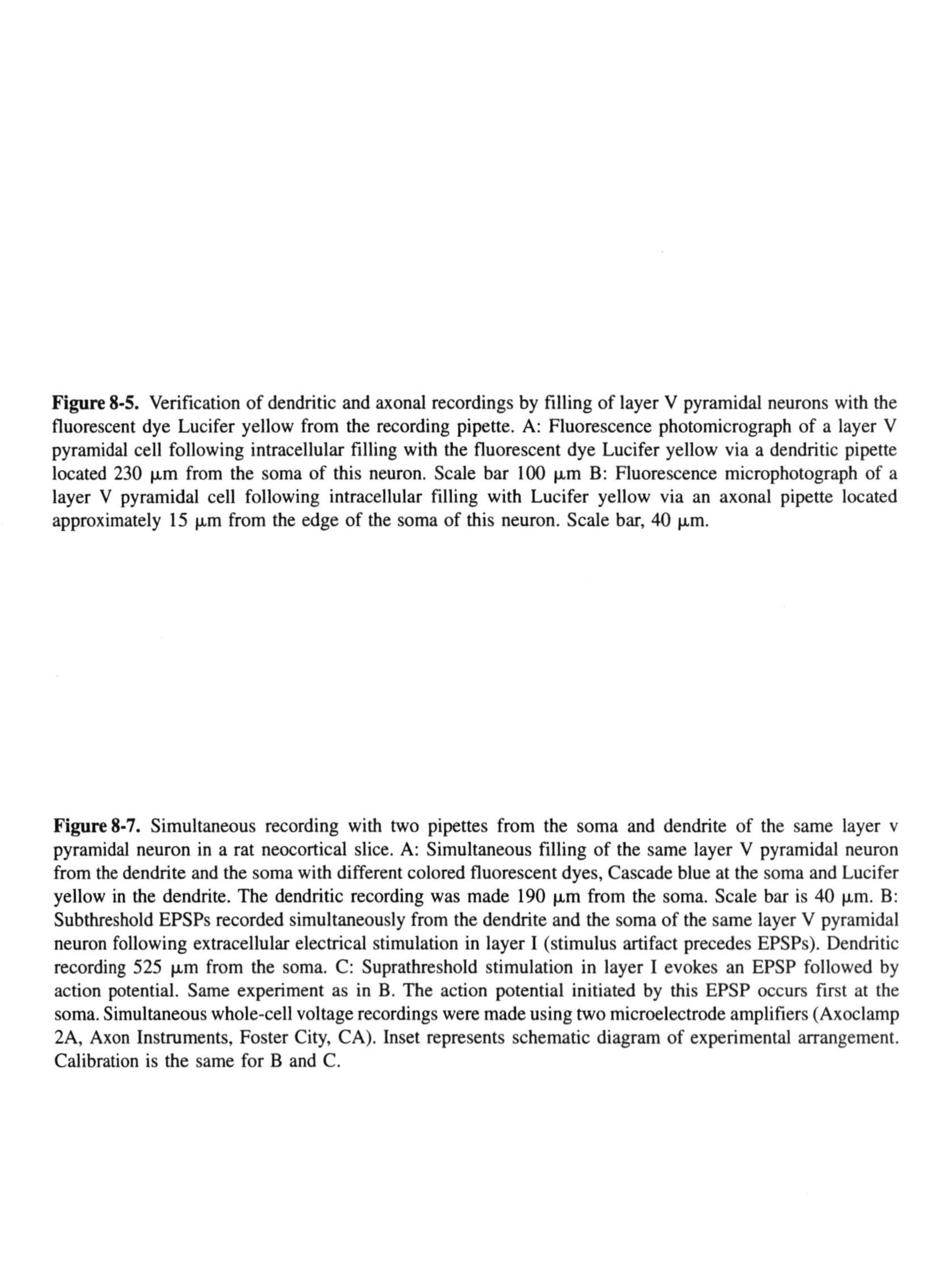

Figure 8-5. Verification of dendritic and axonal recordings by filling of layer V pyramidal neurons with the fluorescent dye Lucifer yellow from the recording pipette. A: Fluorescence photomicrograph of a layer V pyramidal cell following intracellular filling with the fluorescent dye Lucifer yellow via a dendritic pipette located 230 μm from the soma of this neuron. Scale bar 100 μm B: Fluorescence microphotograph of a layer V pyramidal cell following intracellular filling with Lucifer yellow via an axonal pipette located approximately 15 μm from the edge of the soma of this neuron. Scale bar, 40 μm.

Figure 8-7. Simultaneous recording with two pipettes from the soma and dendrite of the same layer v pyramidal neuron in a rat neocortical slice. A: Simultaneous filling of the same layer V pyramidal neuron from the dendrite and the soma with different colored fluorescent dyes, Cascade blue at the soma and Lucifer yellow in the dendrite. The dendritic recording was made 190 μm from the soma. Scale bar is 40 μm. B: Subthreshold EPSPs recorded simultaneously from the dendrite and the soma of the same layer V pyramidal neuron following extracellular electrical stimulation in layer I (stimulus artifact precedes EPSPs). Dendritic recording 525 μm from the soma. C: Suprathreshold stimulation in layer I evokes an EPSP followed by action potential. Same experiment as in B. The action potential initiated by this EPSP occurs first at the soma. Simultaneous whole-cell voltage recordings were made using two microelectrode amplifiers (Axoclamp 2A, Axon Instruments, Foster City, CA). Inset represents schematic diagram of experimental arrangement. Calibration is the same for B and C.

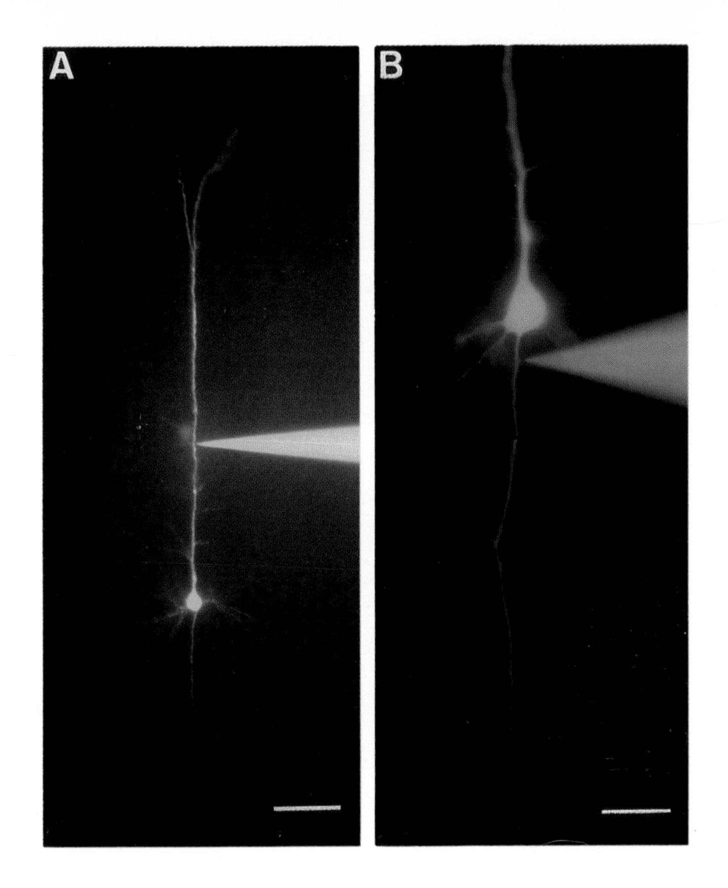
A
B

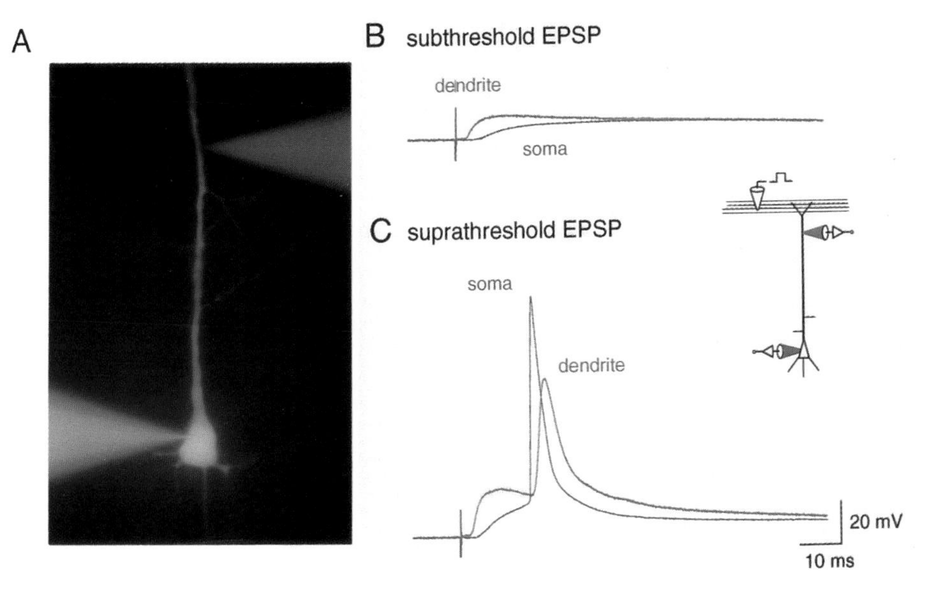
A
B subthreshold EPSP
dendrite
soma
C suprathreshold EPSP
soma
dendrite
20 mV
10 ms

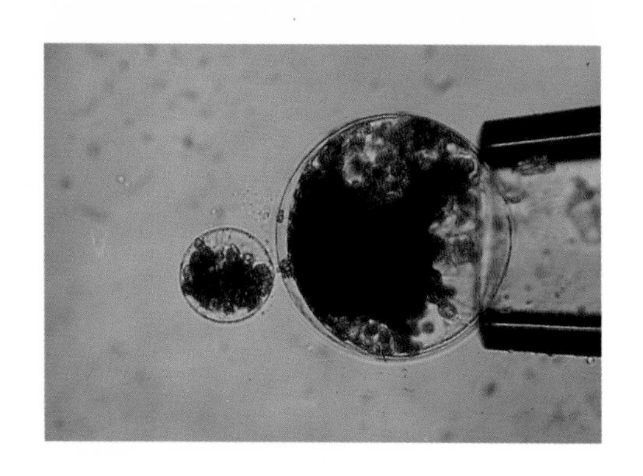

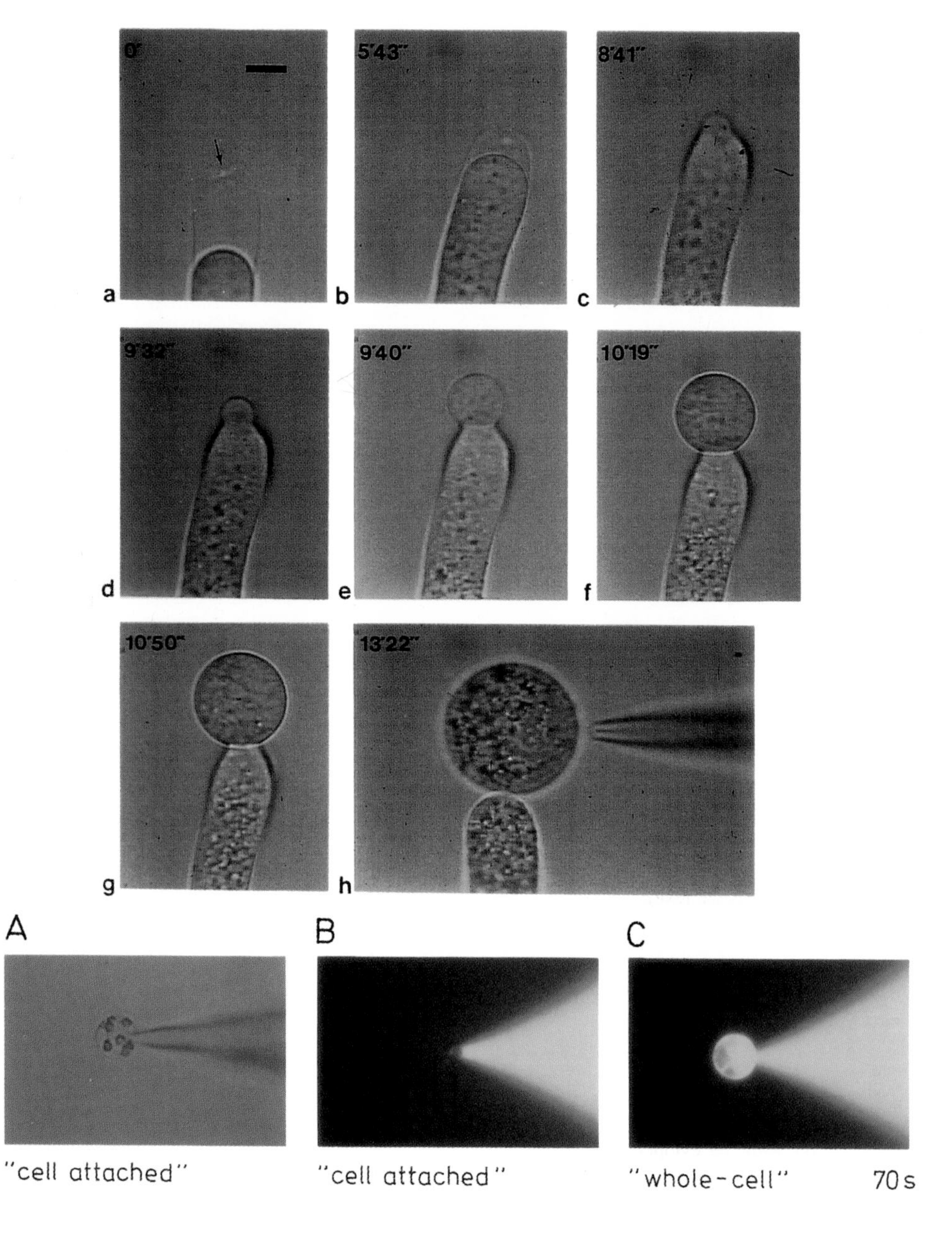

0'
5'43"
8'41"
a
b
c
9'32"
9'40"
10'19"
d
e
f
10'50"
13'22"
g
h
A
B
C
"cell attached"
"cell attached"
"whole-cell"
70s

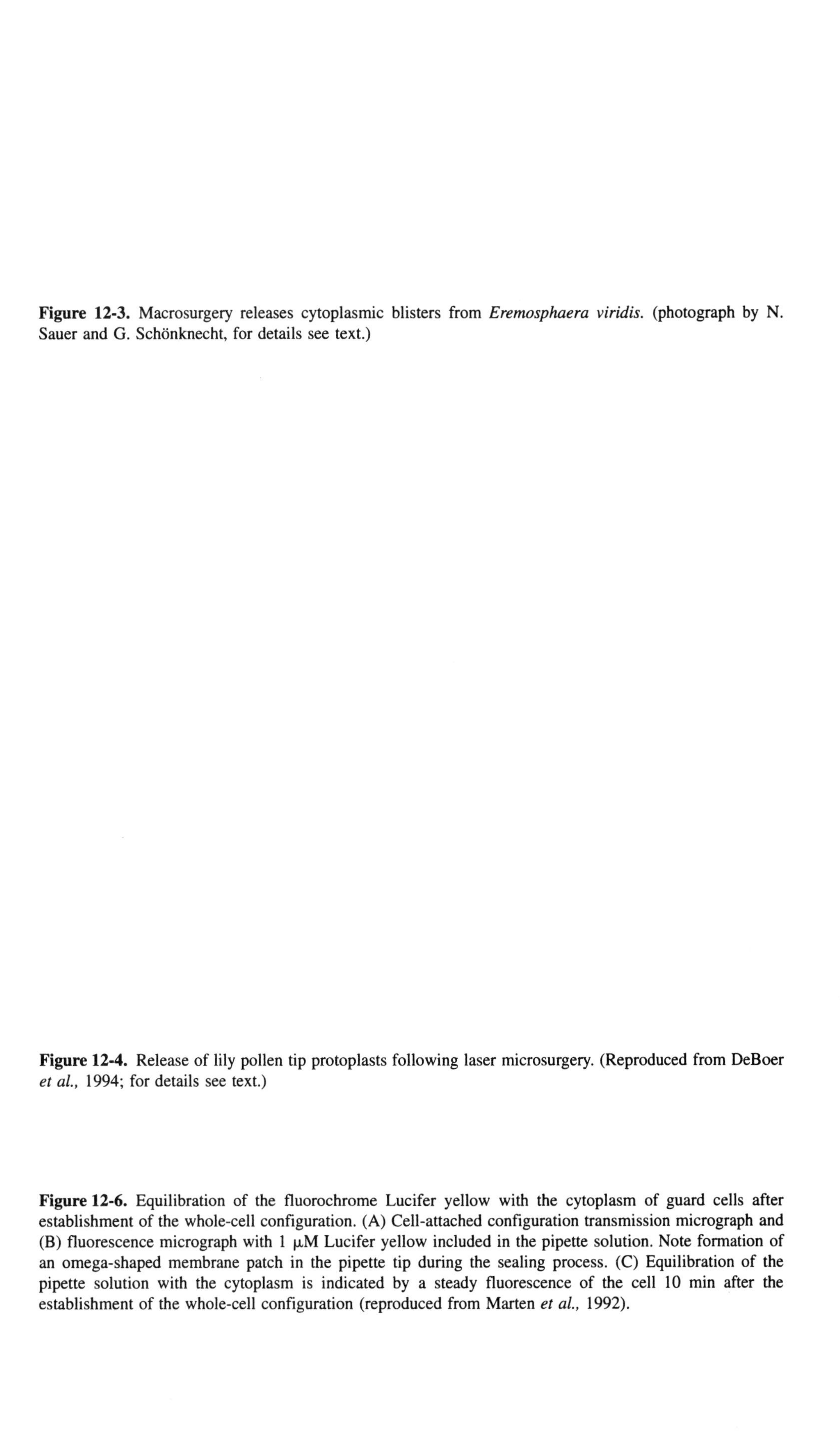

Figure 12-3. Macrosurgery releases cytoplasmic blisters from *Eremosphaera viridis.* (photograph by N. Sauer and G. Schönknecht, for details see text.)

Figure 12-4. Release of lily pollen tip protoplasts following laser microsurgery. (Reproduced from DeBoer *et al.,* 1994; for details see text.)

Figure 12-6. Equilibration of the fluorochrome Lucifer yellow with the cytoplasm of guard cells after establishment of the whole-cell configuration. (A) Cell-attached configuration transmission micrograph and (B) fluorescence micrograph with 1 μM Lucifer yellow included in the pipette solution. Note formation of an omega-shaped membrane patch in the pipette tip during the sealing process. (C) Equilibration of the pipette solution with the cytoplasm is indicated by a steady fluorescence of the cell 10 min after the establishment of the whole-cell configuration (reproduced from Marten *et al.,* 1992).

Part I

INTRODUCTION TO PATCH CLAMPING

Chapter 1

A Practical Guide to Patch Clamping

REINHOLD PENNER

1. Introduction

The patch-clamp technique is an extremely powerful and versatile method for studying electrophysiological properties of biological membranes. Soon after its development by Erwin Neher and Bert Sakmann, it was adopted by numerous laboratories and subsequently caused a revolutionary advancement of many research areas in both cellular and molecular biology. Not surprisingly, the developers of this technique were awarded the highest scientific recognition. The Nobel Assembly in Stockholm issued the following press release on 7 October 1991:

> The Nobel Assembly at the Karolinska Institute has today decided to award the Nobel Prize in Physiology or Medicine for 1991 jointly to Erwin Neher and Bert Sakmann for their discoveries concerning "The Function of Single Ion Channels in Cells."
>
> Each living cell is surrounded by a membrane which separates the world within the cell from its exterior. In this membrane there are channels, through which the cell communicates with its surroundings. These channels consist of single molecules or complexes of molecules and have the ability to allow passage of charged atoms, that is, ions. The regulation of ion channels influences the life of the cell and its functions under normal and pathological conditions. The Nobel Prize in Physiology or Medicine for 1991 is awarded for the discoveries of the function of ion channels. The two German cell physiologists Erwin Neher and Bert Sakmann have together developed a technique that allows the registration of the incredibly small electrical currents (amounting to a picoampere—10^{-12} A) that passes through a single ion channel. The technique is unique in that it records how a single channel molecule alters its shape and in that way controls the flow of current within a time frame of a few millionths of a second.
>
> Neher and Sakmann conclusively established with their technique that ion channels do exist and how they function. They have demonstrated what happens during the opening or closure of an ion channel with a diameter corresponding to that of a single sodium or chloride ion. Several ion channels are regulated by a receptor localized to one part of the channel molecule which upon activation alters its shape. Neher and Sakmann have shown which parts of the molecule constitute the "sensor" and the interior wall of the channel. They also showed how the channel regulates the passage of positively or negatively charged ions. This new knowledge and this new analytical tool has during the past ten years revolutionized modern biology, facilitated research, and contributed to the understanding of the cellular mechanisms underlying several diseases, including diabetes and cystic fibrosis.

Since the first demonstration of single channels in a biological membrane using this methodology (Neher and Sakmann, 1976), several key improvements have refined its use

REINHOLD PENNER • Department of Membrane Biophysics, Max-Planck-Institute for Biophysical Chemistry, Am Fassberg D-37077 Göttingen, Germany.

Single-Channel Recording, Second Edition, edited by Bert Sakmann and Erwin Neher. Plenum Press, New York, 1995.

and applicability to virtually all biological preparations (including animal and plant cells, bacteria, yeast, and cell organelles). The development of the "gigaseal" (Sigworth and Neher, 1980; Neher, 1982) and the establishment of the various recording configurations ("cell-attached"/"inside-out"/"outside-out"/"whole-cell") allowed patch recording from the cell surface or cell-free membrane patches as well as intracellular recordings (Hamill *et al.,* 1981). In recent years, even more technical and experimental variations of the patch-clamp method have emerged and further expanded its power to address previously unapproachable questions in cell biology. Today, the patch clamp is the method of choice when it comes to investigating cellular and molecular aspects of electrophysiology. At present, more than 1000 scientific articles employing the patch-clamp technique are published each year. Paired with cell and molecular biological approaches (e.g., protein chemistry, cloning and expression techniques, microfluorimetry), the patch-clamp technique constitutes an indispensible pillar of modern cell biology. Comprehensive reviews on the development of the patch-clamp technique and its applications can be found in the Nobel lectures given by Neher and Sakmann (Neher, 1992b; Sakmann, 1992) and other review articles (Sakmann and Neher, 1984; Sigworth, 1986; Neher, 1988; Neher and Sakmann, 1992).

Given the rapid pace at which molecular biology advances and the wealth and complexity of cell proteins interacting with each other to bring about cellular function, the potential use of the patch-clamp technique is still growing and attracting newcomers into either collaborating with patch-clampers or establishing this area of research in their own laboratories. This chapter is designed to be a practical guide to patch-clamping for newcomers and a starting point for students entering a patch-clamp laboratory. It discusses the very basic features of the patch-clamp technique, how a patch-clamp setup functions, what type of equipment is needed, some basic experimental procedures, and what the potential problems with this technique are. This chapter is not meant to discuss these topics fully, as there are numerous well-written and detailed descriptions of each of these points available in the literature. The reader who is interested in introductory reading material relevant to various aspects of modern electrophysiology is directed to some excellent books on the biology of ion channels (Hille, 1992) and on patch-clamp methodology (Kettenmann and Grantyn, 1992; Rudy and Iverson, 1992; Sherman-Gold, 1993). Some of the topics introduced in this chapter are discussed in more detail in individual chapters of this volume.

2. Patch-Clamp Techniques

Originally, the patch-clamp technique referred to voltage-clamp of a small membrane patch, but it now generally refers to both voltage-clamp and current-clamp measurements using "patch-clamp"-type micropipettes. The patch-clamp technique is an electrophysiological method that allows the recording of macroscopic whole-cell or microscopic single-channel currents flowing across biological membranes through ion channels. Active transporters may also be studied in cases where they produce measurable electrical currents, that is, if the transport is not electroneutral (e.g., Na^{+}–Ca^{2+} exchanger, amino acid transporters). The technique allows one to experimentally control and manipulate the voltage of membrane patches or the whole cell (voltage clamp), thus allowing the study of the voltage dependence of ion channels. Alternatively, one may monitor the changes in membrane potential in response to currents flowing across ion channels (current clamp), which constitute the physiological response of a cell (e.g., action potentials). Thus, the main targets of patch-clamp investigations are membrane-contained ion channels, including voltage-dependent ion channels (e.g., Na^{+},

K^+, Ca^{2+}, Cl^- channels), receptor-activated channels (e.g., those activated by neurotransmitters, hormones, mechanical or osmotic stress, exogenous chemical mediators), and second-messenger-activated channels (e.g., those activated by $[Ca^{2+}]_i$, cAMP, cGMP, IP_3, G proteins, or kinases). Indeed, the gating mechanisms or the presence in certain cell types of many of these channels has been discovered as a direct consequence of using the patch-clamp technique in previously inaccessible preparations.

Other electrical parameters may be monitored as well, most notably the cell membrane capacitance, which is indicative of the plasma membrane surface area. The quantification of membrane area not only allows the determination of current densities but the time-resolved monitoring of cell capacitance may also be used to assess exocytotic and endocytotic activity of single secretory cells (Neher and Marty, 1982; Lindau and Neher, 1988; see also Chapter 7, this volume).

With the opportunity afforded by the whole-cell configuration to selectively perfuse and dialyze cells intracellularly with any desired biological or pharmacological probe while monitoring its effects on cell function, the possible applications of the technique are limited only by the ingenuity of the experimental design. In combination with additional techniques (e.g., microfluorimetry, amperometry), as discussed in Chapters 9 and 11 (this volume), one can correlate the measurements to events not amenable to electrophysiological techniques. Furthermore, with the development of powerful molecular biological tools (e.g., polymerase chain reaction, mRNA amplification) it is even feasible to obtain relevant genetic information from the very same cell that was characterized electrophysiologically by extracting the cytosol into the patch pipette and analyzing it later on as described in Chapter 16 (this volume).

2.1. The Patch-Clamp Configurations

The basic approach to measure small ionic currents in the picoampere range through single channels requires a low-noise recording technique. This is achieved by tightly sealing a glass microelectrode onto the plasma membrane of an intact cell, thereby isolating a small patch. The currents flowing through ion channels enclosed by the pipette tip within that patch are measured by means of a connected patch-clamp amplifier. This so-called "cell-attached" configuration is the precursor to all other variants of the patch-clamp technique. The resistance between pipette and plasma membrane is critical for determining the electrical background noise from which the channel currents need to be separated. The seal resistance should typically be in excess of 10^9 Ω ("gigaseal").

The cell-attached configuration may be used (as such) to record single-channel activity, or one may proceed to isolate the patch from its environment by withdrawing the pipette from the cell. This usually retains the integrity of the gigaseal pipette–patch assembly and allows one to study ion channels in the excised patch configuration. This configuration is called "inside-out" because the cytosolic side of the patch now faces the outside bath solution.

As an alternative to excising the patch, one can simply break the patch by applying a pulse of suction through the patch pipette, thereby creating a hole in the plasma membrane and gaining access to the cell interior. Amazingly, this maneuver does not compromise the gigaseal between pipette and plasma membrane. The tightness of the gigaseal both prevents leak currents flowing between the pipette and the reference electrode and prevents flooding of the cell with the constituents of the bath solution. This configuration is characterized by a low-resistance access to the cell interior through the pipette tip (typically a few megohms with tip diameters of about 1 μm and appropriate pipette solutions), allowing one to voltage-clamp the whole cell ("whole-cell" configuration).

From the whole-cell configuration, one may proceed further by withdrawing the pipette from the cell. This will generally result in resealing of both the plasma membrane of the cell and the patch at the pipette tip. This time, however, the geometric orientation of the patch results in the outside of the membrane facing the bath solution ("outside-out" patch).

2.2. Applications, Advantages, Problems

In the very early days of patch clamp, most studies concentrated on the classification of ion channel types in different cells and the characterization of their biophysical properties in terms of conductance, voltage dependence, selectivity, open probability, and pharmacological profile. It was soon realized, however, that cells had a large variety of ion channels (even for the same ion species), and complementary whole-cell measurements were necessary to reveal the relative importance and the physiological role of a given ion channel for the entire cell. Today, whole-cell recordings are the most popular patch-clamp configuration, and single-channel measurements are performed to complement the whole-cell results.

Which patch-clamp configuration is chosen as the experimental paradigm depends on the type of question to be addressed and the kind of ion channel under study. Each of the configurations has its peculiarities, advantages, and disadvantages.

2.2.1. Cell-Attached Recording

This is mainly used when the channel type in question requires unknown cytosolic factors for gating and these would be lost following patch excision. Also, because this configuration is noninvasive, leaving the ion channel in its physiological environment, it may be used to test for possible alterations of channel properties after patch excision. Another important application for cell-attached recording is to determine whether a particular ion channel is gated by a cytosolic diffusible second messenger. In this type of experiment, the ion channel enclosed in the cell-attached patch is isolated from the bath (by the pipette) and cannot be gated directly by the receptor agonist. If ion channel activity in the patch changes subsequent to addition of an agonist to the bath, then it is clear that some intracellular messenger must have been generated and diffused in the cytosol to gate the ion channel in the patch. Generally, the main disadvantages of the cell-attached configuration are the lack of knowledge of the cell membrane resting potential (which adds to the applied pipette potential) and the inability to effectively control and change the ionic composition of the solutions on both sides of the patch during the measurement.

2.2.2. Inside-Out Recording

This configuration enables one easily to change the cytosolic side of the patch. It is therefore the method of choice to study the gating of second-messenger-activated channels at the single-channel level. Because most channels are modulated in one way or another by intracellular processes, the effects of cytosolic signaling molecules or enzymatic activity on channel behavior can be studied using this configuration. Commonly, the main problems with inside-out recordings arise from the loss of key cytosolic factors controlling the behavior of some ion channels. Also, more often than with outside-out patches, there is the chance of obtaining vesicles in the pipette tip rather than planar patches.

2.2.3. Outside-Out Recording

This configuration allows one easily to change the extracellular side of the patch. It is therefore often used to study receptor-operated ion channels. As for the inside-out configuration, the cytosolic environment of the channels is lost on patch excision. Furthermore, high-quality and stable outside-out recordings are more difficult to obtain, because more steps are required to reach the outside-out configuration (from cell-attached to whole-cell to patch excision). In order for the patch to be excised successfully, cells need to adhere very well to the bottom of the recording chamber.

2.2.4. Whole-Cell Recording

This configuration is employed when ion currents of the entire cell are recorded. This method is essentially like an outside out recording with the advantage of recording an average response of all channels in the cell membrane. The whole-cell configuration is also suitable to measure exocytotic activity of secretory cells by measuring cell membrane capacitance. The main disadvantages again are the possible loss of cytosolic factors and the inability to change the cytosolic solution easily without pipette perfusion.

2.3. Special Techniques

Several modifications of the above configurations have been developed in order to overcome some of their limitations or to perform certain experiments that cannot readily be accomplished with the standard patch-clamp configurations.

2.3.1. Perforated Patch Recording

One of the major problems in patch clamp is the washout of cytosolic constituents following patch excision or dialysis of cells during long-lasting whole-cell measurements (Pusch and Neher, 1988). Several methods may be used to alleviate these problems (see Horn and Korn, 1992, for review). The perforated patch technique aims at retaining the cytosolic constituents by selectively perforating the membrane patch by including channel-forming substances in the pipette solution, e.g., ATP (Lindau and Fernandez, 1986), nystatin (Horn and Marty, 1988), and amphotericin B (Rae *et al.,* 1991). Although the channels formed in the patch essentially allow a low-resistance access to the cell comparable to that in the standard whole-cell configuration, the small size of these channels allows passage of only small ions, thus preventing the washout of cytosolic factors. A variation of this technique is provided by the perforated vesicle configuration (Levitan and Kramer, 1990), in which the pipette is withdrawn from the cell to obtain a small vesicle retaining many of the cytosolic constituents. The vesicle membrane facing the pipette interior is perforated and allows recording of single channels from the outside-out patch facing the bath solution. The main tradeoffs of this technique are the long time required to obtain low-resistance access to the cell, the larger noise associated with the recordings, possible osmotic effects, and the inability to effectively control the cytosolic environment.

2.3.2. Double Patch Recording

This method is used to study gap-junction channels in intercellular communication (Neyton and Trautmann, 1985; Veenstra and DeHaan, 1986). The method is basically a paired whole-cell measurement in two connected cells in which synchronously occurring signals correspond to channel activity of junctional channels between the two cells (for review, see Kolb, 1992).

2.3.3. Loose Patch Recording

Focal recordings with large-diameter pipettes (Strickholm, 1961; Neher and Lux, 1969) may be considered a predecessor of the tight-seal patch-clamp technique, comparable to a cell-attached recording with a large-diameter pipette tip (normally 5–20 μm), where the seal resistance is only a few megohms. The so-called loose patch technique is an improved variant of this method in which special electronic enhancements such as leak compensation or the use of special pipettes with two concentric patch tips allow large currents to be measured (for reviews see Roberts and Almers, 1992; Stühmer, 1992). This approach is mainly used to map the distribution of ion channels and current densities of large cells (e.g., muscle cells, giant axons).

2.3.4. Giant Patch Recording

Normally, gigaseal formation becomes increasingly more difficult as the size of the pipette tip is increased. However, when special hydrocarbon mixtures are applied to the rims of large-tipped patch pipettes (tip diameters of 10–40 μm), giant patches with gigohm seal resistances can be obtained (Hilgemann, 1990; see also Chapter 13, this volume). Much like the loose patch technique (yet with high seal resistance), this method is basically a cell-attached configuration and may be used to study macroscopic currents through ion channels or transporters.

2.3.5. Detector-Patch Recording

Ion channels can be exquisitely sensitive detectors of neurotransmitters or second messengers, and this method takes advantage of such ion channels. For example, excising an outside-out patch containing a neurotransmitter detector channel and placing it close to a synaptic structure releasing the appropriate transmitter may be used to monitor synaptic activity (Hume *et al.*, 1983; Young and Poo, 1983; for review see Young and Poo, 1992). Another application is to excise an inside-out patch containing Ca^{2+}-activated K channels and "cramming" the patch into a large cell such as an oocyte to monitor changes in $[Ca^{2+}]_i$ (Kramer, 1990).

2.3.6. Pipette Perfusion

One of the major problems in any patch-clamp configuration is the difficulty of exchanging the pipette solution. Several methods have been designed to achieve an effective pipette perfusion (e.g., Soejima and Noma, 1984; Lapointe and Szabo, 1987; Neher and Eckert,

1988; Tang *et al.,* 1990). They all share the basic principle of having one or more capillary tubes (polyethylene or quartz) inserted into the patch pipette and placed as close as possible to the pipette tip. By means of gentle pressure, the desired solution within the capillaries is expelled and replaces the original pipette solution within seconds to minutes.

2.3.7. Tip-Dip Bilayer Recording

This method is basically a modification of the lipid bilayer technique used to study reconstituted ion channels in a small patch environment. The pipette tip is repetitively immersed in a monolayer lipid film to produce a lipid bilayer at the pipette tip (Coronado and Latorre, 1983; Suarez-Isla *et al.,* 1983). The main advantage consists of minimizing capacitative artifacts of conventional bilayer experiments.

3. The Patch-Clamp Setup

The diversity of experimental preparations and types of experiment that can be studied with the patch-clamp technique is reflected by the variety of patch-clamp setups used in various laboratories. The expansion of the patch-clamp methodology has been paralleled by a growth of the commercial supply industry and availability of patch-clamp equipment. Although there are some basic features of instrumentation common to all functional patch-clamp workstations, in practice, there is no such thing as the "standard" patch-clamp setup. Every laboratory has its own equipment preferences, quite often supplemented by custom-built devices and/or modified commercial instruments.

As desirable as it might seem to the newcomer, it would be impossible to list, describe, or even evaluate all of the available instrumentation for patch clamp in this chapter. Also, the pace at which introduction and modification of available hardware and software occurs would make this attempt useless at the time of publication of this volume. However, the author is willing to compile a fairly complete and continuously updated list of the available patch-clamp equipment. This list is currently being compiled and will be made accessible in electronic form on a fileserver on the internet. The equipment list may be retrieved via anonymous FTP from "ftp.gwdg.de" in the directory "pub/patchclamp." The files will contain information in the form of data sheets grouped into categories corresponding to the subheadings below.

In its simplest form, a patch-clamp setup may consist of a microscope (for cell visualization) placed on a vibration isolation table within a Faraday cage, a patch-clamp amplifier and pulse generator for voltage-clamping the cells, a micromanipulator holding the amplifier probe for positioning the attached patch pipette, and data-recording devices (e.g., ocilloscope, computer, chart recorder). In addition, some instruments for pipette fabrication are required (i.e., a pipette puller and a microforge). The addition of other instruments to the setup can extend the range of possible applications, increase the efficiency, or simply make certain tedious tasks more convenient. An example of a patch-clamp setup (as used in our laboratory) is shown in Fig. 1. There are many variants for all of the devices described below, and there is no strict objective criterion that may be applied to make the right choice. The only advice that can be given here is to collect from colleagues as much information as possible about experiences with the equipment in question. Some detailed discussion of advantages and disadvantages of certain types of equipment and instrumentation are also found in the literature

Figure 1. An example of a patch-clamp setup. The two panels show the instrument rack (left panel) and the patch-clamp workstation inside a Faraday cage (right panel). This particular setup is used in our laboratory and is equipped with the following instruments: (1) vibration isolation table (Physik Instrumente T-250); (2) supertable with Faraday cage (custom made); (3) inverted microscope (Zeiss Axiovert 100, fluorescence equipped) and lamp power supply (4); (5) hydraulic pump for bath perfusion; (6) motorized micromanipulator for recording pipette (Eppendorf 5171) with joystick (7) and controller units for manipulator and motorized microscope stage (8); (9) hydraulic micromanipulator for drug application pipette (Narishige WR 88); (10) hydraulic micromanipulator for second patch pipette (Newport MX630); (11) video camera (Kappa CF-6) and video monitor (12); (13) photomultiplier tube (Seefelder Messtechnik SMT ME 930); (14) patch-clamp amplifier (HEKA EPC-9); (15) oscilloscope (Philips PM 3335); (16) computer (Macintosh Quadra 800) with two monitors (17), magnetooptical disk drives (18), keyboard (19), and laser printer (20); (21) multipressure control unit (custom made); (22) function generator (Wavetek Model 19); and (23) various control instruments for dual-wavelength fluorescence excitation of fura-2.

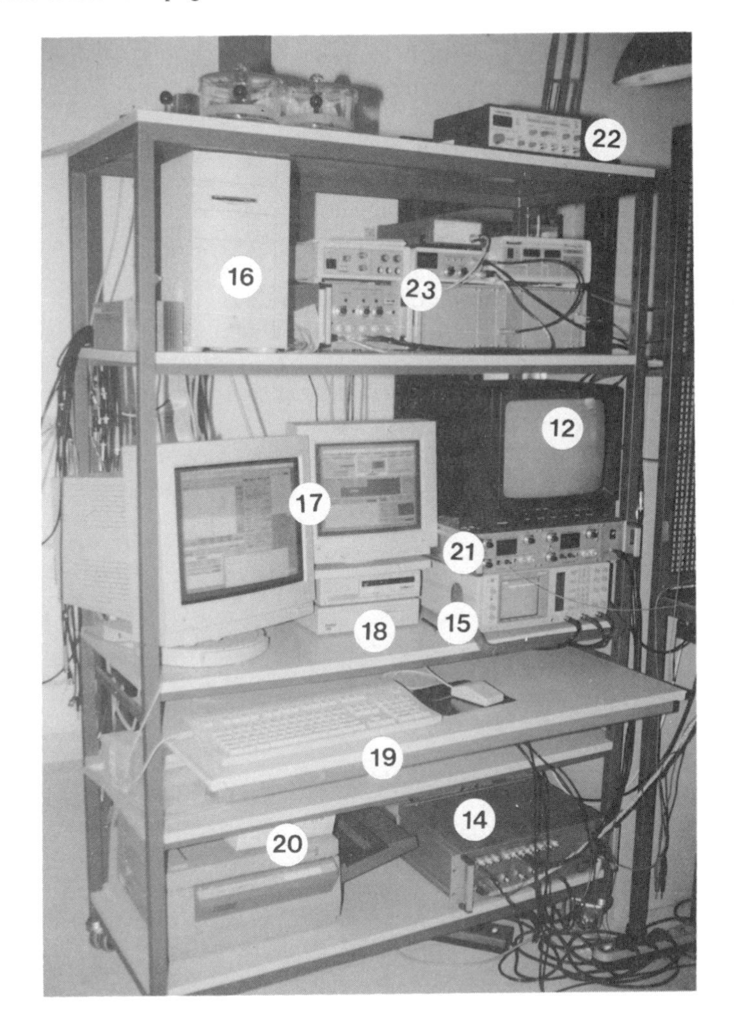

Figure 1. *Continued.*

(see, e.g., Levis and Rae, 1992; Sherman-Gold, 1993). The following section deals in short with the instrumentation of basic and advanced patch-clamp setups.

3.1. Mechanics

In most patch-clamp experiments (particularly whole-cell or cell-attached recordings from small cells), mechanical stability of all setup components must be considered crucial, as even the slightest vibrations or relative movements of the pipette/cell assembly are detrimental to stable recordings.

3.1.1. Vibration Isolation Table

Microscopic movements and vibrations are present to different degrees in all buildings and must be damped out by an appropriate vibration isolation table. Most low-end air-suspension tables available from optical companies will reduce vibrations beyond a few Hertz, which is usually sufficient for the purpose of patch clamping. It is a good idea to place this table in a corner of a room or close to a wall, where vibrations are generally smaller. In some bad cases (e.g., plastered or wooden floors), one might consider supporting the table by replacing the elastic floor with a concrete base that connects to the support structure of the building to further reduce vibration pickup. Many vibration isolation tables preferentially damp vertical movements and are less effective at horizontal displacements. If horizontal vibrations are severe, one may have to resort to more expensive tables with isotropic properties and/or active feedback mechanisms.

In order to avoid accidental touching of the air table during experiments, it is a good idea to surround the air table (without touching it) by a superstructure that basically represents a table slightly larger than the air table itself. A small area of tabletop should be cut out such that the microscope, which is placed on the air table, emerges through it. This "supertable" could be made of galvanized steel and be used to support a Faraday cage placed on top of it.

3.1.2. Faraday Cage

Most patch-clamp setups have a Faraday cage surrounding it. Its main purpose is to shield the sensitive patch-clamp preamplifier from electrical noise. Although this is not an essential requirement (if proper electrical shielding can be accomplished otherwise), it is nevertheless useful to have a cage of some sort around the setup. The inside walls of the Faraday cage can have shelves or mounting brackets to hold solution bottles, peristaltic pumps, etc. If patch-clamp experiments are performed with light-sensitive cells or substances, or if patch clamp is combined with optical measurements that require light shielding, the outside walls of the cage can be draped with cloth or carton to reduce disturbance from ambient light.

3.1.3. Racks

Depending on the needs, a significant number of electronic instruments must be placed outside the Faraday cage, reasonably close to the recording stage. There are commercially available racks that hold standard 19-inch chassis instruments and have provisions for stacking nonstandard instruments on shelves. However, because most commercial racks do not provide for arrangement of computer keyboards, and, in practice, patch-clamp instruments rarely remain fixed in the rack, it is best to have a custom-made rack to hold the equipment. This can be constructed to provide the best ergonomic placement and allow for easy rearrangement of instruments (see Fig. 1-1).

3.2. Optics

The vast majority of patch-clamp studies are carried out on small single cells. Microscopic observation of the cells during the measurement and, even more importantly, the approach to the cell by the patch pipette for seal formation generally require a good optical visualization of the preparation.

3.2.1. Microscopes

In principle, any microscope that will allow observation of the desired cell at micrometer resolution is suitable for patch clamping as long as it allows access of the cell by a patch pipette. Most investigators use inverted microscopes for studying acutely dissociated cells or cultured cell lines because this arrangement allows both good visualization of the cells and unhindered access of patch pipettes from the top. Inverted microscopes also tend to be mechanically more stable. Because focusing is usually accomplished by moving the objective rather than the microscope stage, the bath chamber as well as the micromanipulators can be fixed to the stage for good mechanical stability. Upright microscopes are mainly used for studying cells in sliced tissues, and for this, one has to use objectives with long working distance in order to be able to place the patch pipette underneath the objective. The main disadvantage of conventional upright microscopes is that focusing is accomplished by moving the microscope stage and leaving the objective fixed (although recently, some optical companies have introduced special versions of upright microscopes with fixed stages). The conventional stages sometimes lack the mechanical stability and rigidity to support the weight of some micromanipulators. A remedy to this problem is to detach the stage from the microscope and fix it to a mechanically rigid superstructure that also holds the micromanipulator. The microscope itself is placed on a mounting plate that can be moved vertically to allow focusing, while the stage movement is controlled by X–Y translators. An alternative is to leave the microscope fixed and have the stage moving in all three axes (including the focus axis), in which case the manipulator has to be fixed to the stage.

3.2.2. Video

Video cameras can be attached to most microscopes (inverted or upright). Since affordable video technology is available these days, it is highly recommended to supplement a microscope with video monitoring. Most low-end video cameras and appropriate monitors are suitable for simple observation of cells. Of course, more dedicated video equipment is needed if fluorescence imaging, video-enhanced microscopy, or time-lapse video is an integral part of the experiments. The advantages of having a video camera to monitor the preparation during seal formation and throughout the experiment are numerous. One can form seals much more easily, as one can simultaneously observe the approach of the patch pipette and the change in pipette resistance on the oscilloscope or computer screen when the pipette touches the cell membrane. During an experiment one can monitor any morphological changes of the cell under investigation (e.g., swelling or shrinking, blebbing, contraction).

3.3. Micromanipulation

In order to place patch pipettes on cells as small as a few micrometers, it is essential to be able to precisely control the movement of the patch pipette in the submicrometer range. Another important requirement is that the position of the pipette be free of drift after seal formation to maintain stable recordings for several minutes. This is accomplished by micromanipulators, of which there exist a large variety.

The types of commercially available manipulators include mechanical, hydraulic, motorized, and piezoelectric drives. In principle, all of these varieties can be used for patch-clamp experiments. The ideal manipulator should be reasonably small and mechanically rigid. It should allow long travel distances at fast speed and smooth submicron movements in at least three axes; it should be exceptionally stable and drift-free and ideally allow for remote control. Motorized manipulators are probably closest to such ideal manipulators, followed by hydraulic manipulators (although the latter can sometimes drift considerably). Some of the criteria mentioned may also be fulfilled by combining coarse and fine manipulators, e.g., by using long-range mechanical manipulators with attached fine-positioning devices such as piezoelectric drives for the final approach.

In typical setups, the amplifier probe is mounted directly on the micromanipulator. The probes are normally supplied with a plastic mounting plate that can be fixed on a flat surface of the manipulator. Some manipulators have clamps for holding the probe. The headstage should be fixed tightly to the manipulator, but care must be taken that the metal enclosure of the probe should never be in contact with metallic parts of the manipulator. The arrangement of the manipulator and the attached headstage should also allow for easy access of the pipette holder for exchange of pipettes. This may be accomplished by mounting the manipulator or the headstage on a rotatable or tiltable platform that makes it possible to swing out the headstage for pipette exchange and return it to its fixed position for experiments.

3.4. Amplifiers

A number of commercial patch-clamp amplifiers capable of recording single channels as well as whole-cell currents and operating in voltage- or current-clamp mode are available. There is no general advice that can be given in favor of or against a certain model. One might point out that there are two basic modes in which current-to-voltage conversion is implemented: resistive and capacitive feedback. Resistive feedback is the classical mode and suitable for all types of patch-clamp recordings, whereas capacitive feedback is currently superior only for ultra-low-noise single-channel recordings (provided all other noise sources are meticulously eliminated). Some amplifiers allow switching between resistive and capacitive feedback mode. Others provide for total digital control of the amplifier with automatic series resistance and capacitance compensation.

3.5. Stimulators

Patch-clamp amplifiers are usually capable of applying a steady command voltage to the pipette, and sometimes a test-pulse generator is provided. However, in order to apply complex stimulus protocols in the form of square, ramp, or even more complex voltage pulses, there is a need for more sophisticated pulse generators.

There are basically two options to consider: stand-alone analogue or digital stimulators and computer-based stimulators that are integrated into a data acquisition program. The latter option is favored by most investigators because it allows convenient stimulation using complex stimulus patterns (including leak-pulse protocols) and proper processing of acquired data (see Chapter 3, this volume). Analogue stimulators (often digitally controlled) are sometimes required when very fast repetitive or very long-lasting pulses need to be applied. An analogue lock-in amplifier may be considered a special type of stimulator. It is used for measuring membrane capacitance. It features a sine-wave generator (used to stimulate the cell) and special circuitry that analyzes the membrane currents in response to the applied sine-wave stimulus at different phase angles (see Chapter 7, this volume). An alternative to analogue lock-in amplifiers is provided by the implementation of a software lock-in featured by some commercial or public-domain data acquisition packages.

3.6. Data Acquisition and Analysis

The registration and documentation of patch-clamp recordings requires equipment for data acquisition, storage, and data analysis.

3.6.1. Oscilloscopes

Most computerized data acquisition systems feature both data acquisition and display of acquired data, often in a leak-corrected or otherwise processed form. However, for test purposes, it is still a good idea to have an oscilloscope connected to the amplifier. It is often convenient to observe the voltage and current monitor signals on the oscilloscope rather than from the computer-processed screen display (which might often be quite sluggish and more difficult to scale appropriately). In many situations it is easier to observe the signals on the oscilloscope (maybe at an increased resolution or less heavily filtered), as some fine details may become apparent that might be missed by the data acquisition software. Furthermore, a look at the oscilloscope and comparison with the digitized recordings can increase the experimenter's confidence that the data are being recorded and processed correctly by the computer.

3.6.2. Chart Recorders

Quite often it is of interest to keep track of the entire time course of an experiment, monitoring simultaneously various additional parameters relevant for the experimental results. Multichannel chart recorders offer this possibility by recording any voltage-encoded signal onto a chart paper. These can monitor, e.g., development of the holding current, temperature changes, pH, solution changes, stimulation protocols, or other parameters of interest. Annotations can be made by simply scribbling remarks or marking special experimental procedures on the chart paper. In recent times, classical paper chart recorders must compete with more versatile and convenient computer-based charting programs, which offer more flexibility and certainly a more effective way of analyzing the acquired data.

3.6.3. Tape Recorders

Continuous high-resolution acquisition of single-channel data from a single patch can easily fill even very large computer hard disks within a few minutes. It is therefore often unavoidable to resort to analogue or digital recordings on tape recorders. Analogue acquisition of electrical signals on frequency-modulated magnetic tape recorders is rarely used these days. Two main systems are currently favored for acquisition of patch-clamp data: VCR/PCM combinations in which analogue signals are converted into pulse-code modulated signals recorded on a video tape, and DAT (digital audio tape) recorders, which use a more convenient recording medium.

3.6.4. Filters

Most patch-clamp amplifiers have built-in filters, often complemented by the capability of offline digital filtering of the acquired data. In some instances, particularly when there is need for accurate filtering of single-channel data or when performing noise analysis studies, a more sophisticated analogue filter with different filter characteristics may be required.

3.6.5. Computerized Systems

Most current recordings, be they single-channel or whole-cell data, published in patch-clamp studies rely on powerful data acquisition and analysis software available on different computer platforms (see Chapter 3, this volume). The data acquisition and analysis of the early days of patch clamp was largely performed on DEC (Digital Equipment Corporation) computers running self-programmed software developed in the leading laboratories. The rapid pace at which personal computers grew in performance and affordability promoted the development of dedicated software for these computers on a commercial basis. Today most of the commercial data acquisition packages are running either on personal computers under the DOS operating system or on the Macintosh line of computers. Some laboratories still use software tailored to their specific needs and often are willing to share it with colleagues who ask for it.

With the rate at which computer hardware turns over these days, it is not so much a question of which computer platform to use but rather which software to obtain. As for any of the other components of a patch-clamp setup, this is a difficult question that cannot be answered by naming a particular software package. In general, most of the commercial acquisition software is quite adequate to perform basic patch-clamp experiments by providing stimulation output and acquisition of data at high speed. However, there are differences in performance in terms of user interface, ease of operation, flexibility, and special features between different software packages. It is worth looking into the idiosyncrasies of the data acquisition software, as software performance will often be the limiting factor in what type of experiments can be done and how effective or time-consuming data acquisition and analysis will be.

3.7. Grounding the Setup

Because of the extreme sensitivity of the headstage, special care must be taken in grounding all surfaces that will be near the probe input in order to minimize line-frequency interference. Even 1 mV of AC on a nearby surface, which can easily arise from a ground loop, can result in significant 50- or 60-Hz noise. A high-quality ground is available at the terminal of the probe; this is internally connected through the probe's cable directly to the signal ground in the main amplifier unit. The ground terminal on the probe is best used for the bath electrode and perhaps for grounding nearby objects such as the microscope.

All other metallic surfaces (e.g., the air table, manipulators, Faraday cage) should be grounded by low-resistance ground cables at a central point, usually on the amplifier's signal ground. It is a good idea to have a brass or copper rod inside the Faraday cage to which all grounds are connected. This grounding rod is then connected by a high quality ground wire to the signal ground on the amplifier. It is best to have this ground wire run parallel to the probe's cable in order to avoid magnetic pickup and ground loop effects. Besides 50- or 60-

Hz magnetic pickup, there may be some 35-kHz pickup from the magnetic deflection of the computer monitor. This pickup becomes visible only when the filters are set to high frequencies; it can usually be nulled by changing the orientation or spacing of the ground wire from the probe cable.

In most cases, the patch clamp is used in conjunction with a microscope; it and its stage typically constitute the conducting surfaces nearest the pipette and holder. In a well-grounded setup, the microscope can provide most of the shielding. It should be made sure that there is electrical continuity between the various parts of the microscope, especially between the microscope frame and the stage and condenser, which are usually the large parts nearest the pipette.

Electrically floating surfaces can act as "antennas," picking up line-frequency signals and coupling them to the pipette. It is important that the lamp housing also be well grounded. It is usually not necessary to supply DC power to the lamp provided that the cable to the lamp is shielded and that this shield is grounded at the microscope.

4. Pipette Fabrication

Procedures for fabricating pipettes are presented in some detail elsewhere (Cavalié *et al.*, 1992; Rae and Levis, 1992a), and the shapes and properties of such patch pipettes are described in Chapter 21 (this volume). The basic equipment required, a summary of the procedures, and some tips that might be helpful are presented in this section. Depending on which patch-clamp configuration is used and the abundance of ion channels in a given preparation, an optimal adjustment of the size, shape, glass type, and coating of the patch pipette is required. The main steps in pipette fabrication involve pulling of appropriately shaped pipettes from glass capillary tubes, coating the pipette with a suitable insulation to reduce the background noise, and fire-polishing the tip of the pipette to allow gigaseals to be formed without damaging the cell membrane. Various instruments can contribute to obtaining the best possible results when fabricating patch pipettes.

4.1. Pipette Pullers

These are used to pull patch pipettes from glass capillary tubes. In its simplest form, a patch-pipette puller passes large currents through a metal filament made of tungsten or platinum and uses gravitation to pull the glass apart as the heat starts to melt the glass. Such vertical pullers usually employ a two-step pull mechanism in which the first pull softens the glass and pulls it a short distance to thin the capillary, after which the second pull (usually with lower heat) separates the capillary, yielding two pipettes with large-diameter tips. Other types of pipette pullers operate in a horizontal arrangement and apply elastic or motorized force to pull the glass in one or multiple steps. These latter pullers are also suitable for fabricating standard intracellular electrodes. They are often microprocessor controlled, and some even use laser technology, thus allowing one to pull quartz glass. Some pullers feature heat polishing while pulling, which is only useful when coating of pipettes is not necessary.

4.2. Pipette Microforges

Once pipettes have been pulled, they are often further processed. One objective is to reduce the pipette capacitance by coating the tapered shank of the pipette up to a few micrometers of the pipette tip with a hydrophobic material (e.g., Sylgard®), which will prevent liquid films creeping up the pipette. A second purpose is to optimize the success rate of seal formation and to obtain stable seals for longer periods of time by smoothing the pipette tip (fire polishing). Pipette microforges have been designed to ease these manipulations. They basically consist of a microscope with a low-magnification objective for controlling the coating step and a high-magnification objective for monitoring the fire-polishing step. In addition, there must be a means by which a jet of hot air is directed to the pipette tip (used for curing of the Sylgard® coating) and a heated platinum wire for melting the pipette tip.

4.3. Glass Capillaries

Pipettes can be made from many different types of glass (for review, see Rae and Levis, 1992a). It has been found that different types of glass work better on different cell types. Glass capillaries are available from soft (soda glass, flint glass) or hard glasses (borosilicate, aluminosilicate, quartz). Soft-glass pipettes have a lower melting point (800°C vs. 1200°C), are easily polished, and can be pulled to have a resistance of 1–2 MΩ. They are often used for whole-cell recording, where series resistance rather than noise is the limiting criterion. The large dielectric relaxation in soft glass sometimes results in additional capacitive transient components that interfere with good capacitance compensation. Hard-glass pipettes often have a narrow shank after pulling and consequently a higher resistance. Hard glass tends to have better noise and relaxation properties; however, the important parameter here is the dielectric loss parameter, which describes the AC conductivity of the glass. Although the DC conductivity of most types of glass is very low, soft glasses in particular have some conductivity around 1 kHz; that is sufficiently high to become the major source of thermal noise in a patch-clamp recording (see Chapter 5, this volume). Borosilicate and aluminosilicate glasses have lower dielectric loss and produce less noise. Quartz glass may be used for exceptionally low-noise recordings (Rae and Levis, 1992b) but requires a laser-driven puller for pipette fabrication.

4.4. Pulling

Depending on the puller used, pipettes are pulled in two or more stages: the first to thin the glass to 200–400 μm at the narrowest point over a 7- to 10-mm region, and the next to pull the two halves apart, leaving clean, symmetrical breaks. Both halves can be used. The length of the first pull and the heat of the last pull are the main determinants of the tip diameter of the final pipette.

4.5. Coating

The capacitance between the pipette interior and the bath, and also the noise from dielectric loss in the glass, can be reduced by coating the pipette with an insulating agent such as Sylgard.® Sylgard® is precured by mixing the resin and catalyst oil and allowing it to sit at room temperature for several hours (or in an oven at 50°C for 20 min) until it begins to thicken. It can then be stored at −18°C for many weeks until use. The Sylgard®

is applied around the lower few millimeters of the electrode to within 10–20 μm of the tip and then rapidly cured by a hot-air jet or by heat from a coil. Coating should be done before the final heat polishing of the pipette, so that the heat can evaporate or burn off any residue left from the coating process.

4.6. Heat Polishing

Heat polishing is used to smooth the edges of the pipette tip and remove any contaminants left on the tip from coating. It is done in a microforge or similar setup in which the pipette tip can be observed at a magnification of 400–800×. The heat source is typically a platinum or platinum–iridium wire. To avoid metal evaporation onto the pipette, the filament should be coated with glass at the point where the pipette will approach it. This is done by simply pressing a noncoated patch pipette onto the glowing filament until it melts and forms a drop of liquid glass covering the bare metal. To produce a steep temperature gradient near the filament (which helps make the pipette tip sharply convergent), an air stream can be directed at the filament. The amount of current to pass through the filament must be determined empirically for each type of glass, but a good place to start is with sufficient current to get the filament barely glowing. The typical practice is to turn on the filament current and move either the filament or the pipette (whichever is movable) into close proximity of the other until the pipette tip starts to melt and the desired tip size is reached. Because the opening in the pipette tip is usually at the limit of resolution of viewing, one might not see the change in shape at the tip but instead only a darkening of the tip. One can tell whether the tip was melted closed, and also get an idea of the tip diameter, by blowing air bubbles in methanol with air pressure supplied to the back of the pipette by a small syringe.

4.7. Use of Pipettes

Pipettes should be used within 5–8 hr after fabrication, even if stored in a covered container; small dust particles from the air stick readily to the glass and can prevent sealing. However, with some easy-sealing cells, experience has been that pipettes may even be used the next day. It is very important to filter the filling solutions (e.g., using a 0.2-μm syringe filter). Pipettes can be filled by sucking up a small amount of solution through the tip. This can be done by capillary force (simply dipping the tip for a few seconds into a beaker containing the pipette solution) or by applying negative pressure to the back of the pipette (e.g., using a 5-ml syringe). Thereafter, the pipette is back-filled, and any bubbles left in the pipette can be removed by tapping the side of the pipette. Overfilling the pipette has disastrous consequences for background noise because the solution can spill into the holder, wetting its internal surfaces with films that introduce thermal noise. Therefore, the pipette should only be partially filled, just far enough to make reasonable contact with the electrode wire (the pipette holder is not filled with solution but is left dry). However, one might still want to fill the pipette high enough such that hydrostatic pressure outweighs the capillary suction, thus causing outflow of solution when the tip enters the bath. If this produces untolerable noise, one may reduce the filling level and apply slight positive pressure to the pipette to obtain outflow of solution.

4.8. Pipette Holders

Commercial pipette holders come in different varieties for accommodating different sizes of pipettes. Many laboratories with decent workshops have their holders custom made to meet their needs, sometimes modified to allow pipette perfusion. Holders are usually made from Teflon® or polycarbonate, both having low dielectric loss. The pipette electrode is simply a thin silver wire that is soldered onto the pin that plugs into the probe's connector. The chloride coating on the wire gets scratched when pipettes are exchanged, but this does not degrade the stability very much; the wire does need to be rechlorided occasionally, perhaps once per month or whenever a significant drift in pipette potential occurs. A wire for the standard electrode holder should be about 4.5 cm long; after it is chlorided along its entire length, an O-ring is slipped onto it, and the wire is inserted into the holder. A good alternative for a bare silver wire is one that is coated by a Teflon® insulation, where only a few millimeters at the tip of the wire are stripped and chlorided; this reduces the scratching of the Ag–AgCl coating during pipette exchanges and capacitative noise. Chloriding can be done by passing current (e.g., 1 mA) between the wire and another silver or platinum wire in a Cl^- containing solution (e.g., 100 mM KCl, or physiological saline). Current is passed in the direction that attracts Cl^- ions to the electrode wire; this produces a gray coating.

The noise level of a holder can be tested by mounting it (with the electrode wire installed but dry) on the probe input and measuring the noise using the noise test facility of the patch-clamp amplifier. The probe should be in a shielded enclosure so that no line-frequency pickup is visible on an oscilloscope connected to the current monitor output at a bandwidth of 3 kHz or less. A good holder increases the rms noise of the headstage alone by only about 20%. The final noise level relevant for the actual recording, which includes all noise sources other than the cell itself, can be estimated by measuring the noise with a filled pipette just above the bath surface.

For low-noise recording, the electrode holder should be cleaned before each experiment with a methanol flush, followed by drying with a nitrogen jet. Before inserting a pipette into the holder, it is a good idea to touch a metal surface of the setup to discharge any static electricity that one may have picked up. The holder should be tightened firmly enough that the pipette does not move (on a scale of 1 μm) when suction is applied.

4.9. Reference Electrodes

The main requirements for a bath electrode are stability, reversibility, and reproducibility of the electrode potential (for review see Alvarez-Leefmans, 1992). A bare, chlorided silver wire makes a good bath electrode unless the cell type under investigation shows intolerable sensitivity to Ag^+ ions. A good alternative is an electrode incorporating an agar salt bridge, in which case the silver wire is either embedded in an agar-filled tube inserted directly into the bath or, as another option, the chlorided silver wire is immersed in saline kept outside the bath but in contact with it through an agar-filled bridge made of a U-shaped capillary tube. The agar should be made up in a solution that can be a typical bath solution or something similar, such as 150 mM NaCl. More concentrated salt solutions are not necessary, and they can leak out, changing the composition of the bath solution. The technical problems related to reference electrodes are discussed in Chapter 6 (this volume).

5. Experimental Procedures

The technical aspects of patch-clamp experiments are fairly simple, provided some basic precautions are taken. In the following, a brief description of the techniques for establishing a seal and recording from either a membrane patch or from an entire small cell are given. This is complemented by some practical tips as well as notes on possible problems that may be helpful when conducting these experiments. The reader is referred to more detailed reviews on recording techniques and data analysis in this volume (Chapters 2, 3, 5, 18–20) and other publications covering data acquisition in general (French and Wonderlin, 1992) as well as practical and theoretical considerations of single-channel data analysis (Jackson, 1992; Magleby, 1992; Sigworth and Zhou, 1992).

5.1. Preparing Experiments

Usually, the aim is to obtain as many experiments as possible from a given preparation (including test and control experiments). Often, experiments have to be carried out at a rapid pace, as some preparations have a short lifetime. To avoid delays and interruptions, it is therefore a good idea to get the technical aspects organized before an experimental session. It is helpful to have all the required solutions made ready, have a reasonable number of patch pipettes prepared, have the stimulation protocols programmed, have the tools for pipette filling and bath exchange at hand, etc.

Often-used extracellular solutions may be kept in stock (liter quantities) in a refrigerator. Intracellular solutions may be frozen in smaller stocks (5–10 ml) and thawed before experiments. If solutions need be prepared during an experimental session, they can be mixed from appropriately concentrated stock solutions. In any case, the pH and osmolarity of the solutions should be adjusted appropriately. Solutions should always be filtered unless a "sticky" substance is included; such substances should be added after the filtering of the normal saline. Supplements (e.g., ATP, GTP, fluorescent dyes, second messengers) are added from frozen stocks to a small volume (100–500 μl) of the pipette-filling solution in an Eppendorf tube as needed shortly before the experiment. The pipette-filling solutions might be kept in a rack on top of a container filled with ice to prevent degradation of labile ingredients. The patch pipettes are filled as described earlier.

The choice of solutions is probably one of the most crucial determinants for any patch-clamp experiment. The formulation of individual bath and pipette solutions depends on many factors. In general, the bath solution should mimic the natural extracellular environment of the cell, while the pipette solution should substitute for the cytosol. On the other hand, one often wants to study a certain ion current in isolation or increase its amplitude. In such cases one needs to alter the ion composition of external or internal or both solutions in order to abolish the masking effect of currents interfering with the conductance of interest (or increase the latter). Some general guidelines on choosing saline compositions are given elsewhere (Swandulla and Chow, 1992). Furthermore, it is a good idea to screen the available literature on the cell type under study and the particular composition of solutions used to study the ion current in question.

5.2. Forming a Seal

The process of seal formation is monitored by observing the pipette currents on an oscilloscope while applying voltage pulses to the pipette. A convenient pulse amplitude is

2 mV, which can be obtained from a pulse generator. Before the pipette is inserted into the bath, the current trace should be flat except for very small capacitive transients caused by the stray capacitance of the pipette and holder. When the pipette enters the bath, the 2-mV pulses will cause 1 nA to flow in a 2-MΩ pipette. The approach to the cell membrane and the formation of a gigaseal will cause the resistance to increase, reducing the currents. For observation of the current pulses, it is convenient to pick a gain setting and oscilloscope sensitivity such that the current through the open pipette is reasonably sized.

5.2.1. Entering the Bath

The surface of the bath solution is relatively "dirty," even if (as is strongly recommended) one aspirates some solution from the surface to suck off dust and contaminants. For this reason it is important always to have solution flowing out of the pipette until the pipette is in contact with the cell (either by applying a small amount of positive pressure to the pipette or by filling it appropriately). Also, one should avoid going through the air–water interface more than once before forming a seal. When moving the pipette tip into the bath, the current trace may go off scale; in that case, one needs to reduce the amplifier gain until the trace reappears. Then, one needs to cancel any offset potentials between pipette and reference electrode; this is done by setting the holding potential of the patch-clamp amplifier to 0 mV and adjusting the pipette offset control such that the DC pipette currents are close to zero. For a detailed discussion of offset compensation procedures, see Chapter 6 (this volume). From the size of the current response to the test pulses, the pipette resistance can be calculated (good data acquisition software usually provides for this).

5.2.2. Forming a Gigaseal

After the pipette has entered the bath, one should proceed as fast as possible to obtain a gigaseal because the success rate of sealing is inversely proportional to the time the pipette tip is exposed to the bath solution (presumably from an increased probability of picking up floating particles). It is even more important to obtain a seal rapidly when the pipette solution contains peptides or proteins, as they tend to cover the pipette tip and interfere with seal formation. When the pipette is pushed against a cell, the current pulses will become slightly smaller, reflecting an increase in resistance; when the positive pressure is released from the pipette, the resistance usually increases further. Some cell types require more "push" from the pipette than others, but a 50% increase in resistance (i.e., a reduction in the current pulse amplitude by this value) is typical. Application of gentle suction should increase the resistance further and result (sometimes gradually, over maybe 30 sec; sometimes suddenly) in the formation of a gigaseal, which is characterized by the current trace becoming essentially flat again (hyperpolarizing the pipette to −40 to −90 mV often helps to obtain or speed the seal formation). To verify gigaseal formation, one may increase the amplifier gain; the trace should still appear essentially flat except for capacitive spikes at the start and end of the voltage pulse.

5.3. Patch Recording

It is a good idea to start out with the holding potential set to zero (this will leave the patch at the cell's resting potential); if a whole-cell experiment is planned, an alternative is to start with the desired holding potential (e.g., −70 mV) in order not to depolarize the cell when breaking the patch. When the test pulse is applied, the fast capacitive spikes recorded arise mainly from the pipette itself and the enclosed membrane patch. One should now compensate the capacitance by adjusting the amplitude and time constant of the fast capacitance neutralization controls of the patch-clamp amplifier to minimize the size of these spikes (some amplifiers offer an automatic compensation of this capacitance). Transient cancellation will be essential if one will be giving voltage pulses in the experiment; if not, the test pulse should be discontinued to avoid introducing artifacts.

In single-channel recordings, the gain should be set to at least 50 mV/pA or above for lower noise (most patch-clamp amplifiers use a high feedback resistor only for these high-gain settings); in whole-cell configuration this setting will depend largely on the size of currents to be recorded. The gain setting should be calculated by dividing the output voltage of the current monitor (e.g., ±10 V) by the gain setting (e.g., 10 mV/pA). This will record currents up to ±1 nA without saturation. It should be kept in mind, however, that the output voltage needs to be sampled by an AD converter, which will resolve the input voltage of ±10 V with its intrinsic accuracy (typically 12 or 16 bits, equivalent to 4,096 or 65,536 discrete levels, amounting to a digitized resolution of a ± 10 V signal of 5 and 0.3 mV, respectively). It is therefore important that the expected current amplitudes be recorded with an appropriate gain setting, thus providing for the best possible resolution of currents.

If one is applying voltage pulses to the patch membrane, it is important to try to cancel the capacitive transients as well as possible in order to avoid saturating any amplifiers, the recording medium, or the AD converter. It is a good idea to set the fast capacitance neutralization controls while observing the signal without any filtering beyond the 10-kHz filtering. Then, during the recording, one should observe to see if the clipping indicator of the amplifier flashes. If it does, it means that internal amplifiers are about to saturate and/or that the current monitor output voltage is going above 10–15 V on the peaks of the transients, and one should readjust the transient cancellation controls. Otherwise, it is likely that the recording will be nonlinear, and subtraction will not work correctly.

The fast transient cancellation is not sufficient to cancel all of the capacitive transients in a patch recording. This is partly because the pipette capacitance is distributed along the length of the pipette; therefore, each element of capacitance has a different amount of resistance in series with it, so that a single value of the time constant of the fast capacitance will not provide perfect cancellation. The time course of the transients also reflects dielectric relaxation in the material of the pipette holder and in the pipette glass. These relaxations are not simple exponentials but occur on time scales of about 1 msec or longer. If one is using pipette glass with low dielectric loss (e.g., aluminosilicate glass), or if one is careful to coat the pipette with a thick coating and near to the tip, the relaxations will be smaller. Remaining transients can be canceled by subtracting control traces without channel openings from the traces containing the channels of interest.

For cell-attached or inside-out patch configurations, positive pipette voltages correspond to a hyperpolarization of the patch membrane, and inward membrane currents appear as positive signals at the current monitor outputs. Some data acquisition programs compensate for this by inverting digital stimulus and sampled values in these recording configurations such that the stimulation protocols, holding voltages, and displays of current records in the oscilloscope all follow the standard electrophysiological convention. In this convention, outward currents are positive, and positive voltages are depolarizing. However, even if the data acquisition software conveniently processes and displays the signals in apparently "physiological" polarity, the analogue current and voltage monitor outputs are not inverted in these recording modes. One should always make sure that one really understands what exact data processing is in effect during data acquisition, and the software should allow the user to reconstruct off line what the experimental settings were during data acquisition.

5.4. Whole-Cell Recording

5.4.1. Breaking the Patch

After a gigaseal is formed, the patch membrane can be broken by additional suction or, in some cells, by high-voltage pulses (these need to be established empirically for the particular cell type; 600–800 mV for 200–500 μsec is a good starting point). Electrical access to the cell's interior is indicated by a sudden increase in the capacitive transients from the test pulse and, depending on the cell's input resistance, a shift in the current level or background noise. Additional suction pulses sometimes lower the access resistance, causing the capacitive transients to become larger in amplitude but shorter in duration. Low values of the access (series) resistance (R_s) are desirable, and when R_s compensation is in use, it is important that the resistance be stable as well. A high level of Ca-buffering capacity in the pipette solution (e.g., with 10 mM EGTA) helps prevent spontaneous increases in the access resistance as a result of partial resealing of the patch membrane, which is favored by high intracellular Ca^{2+} concentrations.

5.4.2. Capacitive Transient Cancellation

If the fast capacitance cancellation was adjusted (as described above) before breaking the patch, then all of the additional capacitance transient will be attributable to the cell capacitance. Canceling this transient using the *C*-slow and *R*-series controls on the amplifier will then give estimates of the membrane capacitance and the series resistance. With small round cells, it should be possible to reduce the transient to only a few percent of its original amplitude. However, if the cell has an unfavorable shape (for example, a long cylindrical cell or one with long processes), the cell capacitance transient will not be a single exponential, and the cancellation will not be as complete.

5.4.3. Series Resistance Compensation

In whole-cell voltage-clamp recording, the membrane potential of the cell is controlled by the potential applied to the pipette electrode. This control of potential is not complete but depends on the size of the access resistance between the pipette and the cell interior and on the size of the currents that must flow through this resistance. This access resistance is called the series resistance (R_s) because it constitutes a resistance in series with the membrane. Part of the series resistance arises from the pipette itself, but normally the major part arises from the residual resistance of the broken patch membrane, which provides the electrical access to the cell interior. In practice, the series resistance usually cannot be reduced below a value about two times the resistance of the pipette alone.

Series resistance has several detrimental effects in practical recording situations. First, it slows the charging of the cell membrane capacitance because it impedes the flow of the capacitive charging currents when a voltage step is applied to the pipette electrode. The time constant of charging is given by $\tau_u = R_s \cdot C_m$, where C_m is the membrane capacitance. For typical values of $R_s = 5\ \mathrm{M\Omega}$ and $C_m = 20$ pF, the time constant is 100 μsec. This time constant is excessively long for studying rapid voltage-activated currents such as Na^+ currents in morphologically convoluted cells such as neurons, especially because several time constants are required for the membrane potential to settle at its new value after a step change. Series resistance and cell capacitance also impose limitations on the recording bandwidth of the acquired currents, which are filtered by the combination of R_s and C_m. For an *RC* filter, the corner frequency (−3 dB) is calculated from $f = 1/(2\pi R_s C_m)$, and in the above example we arrive at a limiting bandwidth of 1.6 kHz. This bandwidth would be reduced with larger capacitance and/or series resistance. It should be realized that although series resistance can be compensated for electronically (R_s compensation control of the patch-clamp amplifier), the capacitance cannot (capacitance neutralization controls of patch-clamp amplifiers do not increase the bandwidth of the recording, nor do they speed up the charging of the capacitance). Another detrimental effect of series resistance is that it yields errors in membrane potential when large membrane currents flow. In the case of $R_s = 5\ \mathrm{M\Omega}$, a current of 2 nA will give rise to a voltage error of 10 mV, which is a fairly large error.

To use R_s compensation in practice, one first has to adjust the transient-cancellation controls (including *C*-fast and τ-fast if necessary) to provide the best cancellation. Then one activates the R_s compensation control by turning it up to provide the desired percentage of compensation. Most patch-clamp amplifiers use this setting to determine the amount of positive feedback being applied for compensation. It should be adjusted with some care, because too high a setting causes overcompensation (the amplifier will think that R_s is larger than it is); this can cause oscillation and damage to the cell under observation.

Optimal settings of the R_s compensation controls depend on the approximate value of the uncompensated membrane-charging time constant τ_u, which can be calculated as the product of the *C*-slow and *R*-series settings (for example, suppose *C*-slow is 20 pF and *R*-series is 10 MΩ; the time constant τ_u is then $20\ \mathrm{pF} \cdot 10\ \mathrm{M\Omega} = 200\ \mu\mathrm{sec}$). The speed of the R_s compensation circuitry can also be adjusted on most amplifiers. If τ_u is smaller than about 500 μsec one should use a fast setting of the R_s compensation circuitry to provide the necessary rapid compensation. The slower settings, on the other hand, will provide compensation that is less prone to high-frequency oscillations from misadjustment of the controls. How much compensation one can apply is also determined by τ_u. If τ_u is larger than about 100 μsec, one can use any degree up to the maximum of 90% compensation without serious overshoot or ringing in the voltage-clamp response. For smaller values of τ_u the R_s compensation setting should be kept below the point at which ringing appears in the current trace.

As in the case for patch recording, there is rarely need to use the full bandwidth of the amplifier in whole-cell recording. This is because typical membrane charging time constants (even after R_s compensation) are considerably longer than 16 μsec, which is the time constant corresponding to a 10-kHz bandwidth. Thus, the current monitor signal is expected to contain no useful information beyond this bandwidth.

In whole-cell recording, the voltage and current monitor signals follow the usual convention, with outward currents being positive. This is because the pipette has electrical access to the cell interior.

6. Caveats and Sources of Artifacts

For newcomers to the patch-clamp technique it is usually not so difficult to master the technical aspects of the methodology. It is rather easy to gather a large amount of data, but the problems emerge in analyzing and evaluating the acquired data. As in any other method there are pitfalls and artifacts, too many to address here. Nevertheless, some problems in using the technique and how to avoid them are briefly discussed in the following paragraphs.

6.1. Solutions

Many problems arise from the composition of extracellular and intracellular solutions. The following are some points to be considered.

Differences in osmolarity and pH between bath and pipette solutions can seriously affect all sorts of ion currents and should be avoided (unless one specifically wants to study their effects). For example, hypo- or hyperosmotic solutions cause cell swelling or shrinking accompanied by modulation of volume-regulatory conductances such as Cl^-, K^+, and cation channels (Sarkadi and Parker, 1991; Hoffmann, 1992), and protons are known to modify the properties of many ion channels (for review see Moody, 1984; Chesler and Kaila, 1992).

Divalent ions in the bath can screen surface membrane charges, thereby affecting the voltage dependence of ion channels. Shifts in the activation and inactivation curves of virtually all voltage-gated ion channels by divalents have been described and are reviewed in more detail elsewhere (Green and Andersen, 1991; Latorre *et al.*, 1992). Complete removal of certain ions can alter the properties of ion channels. The most dramatic effects are observed following removal of Ca^{2+} from the bath. Among other effects, this may cause Ca^{2+} channels to lose selectivity and become permeant to monovalent ions (Almers and McCleskey, 1984; Hess and Tsien, 1984), it will shift the activation curve of Na^+ channels to the left (Campbell and Hille, 1976), enhance currents through inward rectifying K^+ channels (Biermans *et al.*, 1987), and long exposure to Ca^{2+}-free bath solutions will eventually cause nonspecific leaks in the plasma membrane.

Another problem might arise from the precipitation of divalent ions in the saline when sulfate or carbonate ions are present, yielding erroneous estimates of the effective concentration of these ions.

Many organic compounds are not easily soluble in aqueous solutions and need to be dissolved in organic solvents such as ethanol or dimethylsulfoxide, whose final concentration should not exceed 0.1%. In any case, appropriate control experiments of the vehicle should be carried out.

The chloride ions are the primary charge transfer ions between the aqueous phases and the silver wires serving as electrodes. Therefore, these electrodes must be immersed in a solution that contains at least some Cl^- ions (at least 10 mM). A complete removal of Cl^- ions in the bath or pipette-filling solution is not feasible unless an agar bridge is used.

In all patch-clamp configurations a number of offsets have to be taken into account. These include amplifier offsets ($\pm$30 mV), electrode potentials ($\pm$200 mV, depending on Cl^- concentration of pipette and reference electrode), liquid junction potentials, and potentials of membrane(s) in series with the membrane under study. Some of these offsets are fixed during an experiment (such as amplifier and electrode offsets); some are variable. It is standard practice to take care of voltage offsets by performing a reference measurement at the beginning of an experiment. An adjustable amplifier offset is then set for zero pipette current. Thereafter, the command potential of the amplifier will be equal in magnitude to the membrane potential if no changes in offset potentials occur.

The polarity of the command potential will be that of the membrane for whole-cell and outside-out configurations but will be inverted in the cell-attached and inside-out configurations. In cell-attached configurations an additional offset is present because of the resting potential of the cell under study. Liquid-junction potentials may appear or disappear during the measurement when solution changes are performed or in cases in which the pipette solution is different from the bath solution (Barry and Lynch, 1991; Neher, 1992a). A detailed discussion of liquid junction potentials and how to correct for them is found in Chapter 6 (this volume).

In order to maintain cells viable to receptor-activated signal transduction or to avoid rundown of certain ion currents, it is a good idea and common practice to include ATP, GTP, and other nucleotides in the pipette-filling solution. Thus, the cell is unavoidably exposed to the pipette solution during the approach for seal formation. At least for ATP a note of caution is appropriate, because many cells possess purinergic receptors. Activation of these receptors and the resulting signal transduction events may take place before a recording is started. A similar problem might arise when glutamate is used as the internal anion when studying cells that are sensitive to this neurotransmitter. In order to realize and consider the effects caused by the intracellular solution prior to starting an actual recording, it might be useful to let a cell recover by waiting some time after seal formation (during which the bath solution should be exchanged), then establish the whole-cell recording configuration and apply the internal solution extracellularly.

Contaminations of the solutions with foreign substances that might affect ion channels are very difficult to eliminate completely, because containers, syringes, tubings, needles, or filters may release small amounts of leachable substances or detergents into the solution. Some ion channels are extremely sensitive to such contaminations. Therefore, solutions should always be prepared from chemicals of the highest purity, and the possible sources of contaminants should be thoroughly cleaned and rinsed.

6.2. Electrodes

The reference electrode in the bath and the test electrode in the pipette holder are usually silver wires coated with AgCl. This coating gets scratched during multiple exchanges of pipettes and may also degrade with time when large currents are passed (effectively dissolving the AgCl coating as the Cl^- ions are released into the saline). If the electrodes are not regularly chlorided, shifts in the electrode potential may become so severe that voltage drifts become noticeable in the course of an experiment, making the measurements inaccurate. Stability of the electrodes should be verified occasionally by monitoring the currents of an open pipette in the bath at zero-current potential over a few minutes or by comparing the zero-current potential before and after an experiment. If the current is not stable, and one needs to adjust the holding potential by more than 1–2 mV to return to the zero-current potential, then rechloriding is necessary.

Another potential problem arises from shunts or high resistances in the current path between saline and electrodes, which usually becomes apparent on entering the bath with a patch pipette. If there should be no or only a small current flow in response to a test pulse, there might be an open circuit, for example, (1) a bubble in the pipette, (2) a faulty connection to the probe input, or (3) a missing connection to the bath electrode. If large currents or erratic noise appears, there might be problems with the pipette or bath electrode, for example, (1) the pipette tip is broken, (2) the reference electrode is short-circuited to grounded parts of the chamber holder or microscope (e.g., through spilled bath solution), (3) the pipette holder has spilled, or (4) the setup is not well grounded.

6.3. Data Acquisition

The main problems in this respect originate from leak subtraction procedures, choice of holding potential and stimulation protocols, or inappropriate sampling and filtering.

Leak subtraction is in common use for voltage-activated ion currents to compensate and cancel linear leak and capacitive currents. Typically, a variable number of small voltage pulses is applied in a voltage range that does not recruit voltage-dependent ion currents (leak pulses). The size of each of the leak pulses is calculated from the test-pulse amplitude (P) divided by number of leak pulses (n), hence the term P/n leak correction (Bezanilla and Armstrong, 1977). The currents recorded during the leak pulses are summed, and the resulting leak current is subtracted from the actual test pulse. There are many variations of the P/n protocol, including scaling procedures, leak pulses with alternating polarity, etc. (see Chapter 3, this volume). Considerable artifacts may be introduced inadvertedly by using an inappropriate leak-subtraction protocol. It is advisable to inspect the leak pulses to ascertain that no nonlinear current components are subtracted.

Voltage-activated currents are often subject to steady-state inactivation. This can be exploited to dissect currents, as has been shown for Ca^{2+} channels (Tsien *et al.,* 1988). However, problems may arise, e.g., when long-lasting leak protocols at more negative potentials are applied, since some channels (that were inactivated at the holding potential) might recover from inactivation. In addition, frequency-dependent phenomena (e.g., rundown or facilitation of ion currents) should be taken into account when designing pulse protocols.

Sampling frequency and filtering of data should be appropriate for the signals to be recorded. Otherwise, aliasing effects might occur, or the kinetic information is not accurate. Chapter 3 (this volume) addresses the details of sampling and filtering procedures.

The list of problems mentioned above is certainly not complete; there are many more possible complications when conducting patch-clamp experiments, and some have to be actually experienced in order to be fully appreciated. However, this should not deter anybody from moving into the field, since the reward of being able to observe biological processes at the cellular and molecular level in real time as well as the esthetically pleasing records of ion currents offer more than sufficient reward to make the effort worthwhile.

References

Almers, W., and McCleskey, E. W., 1984, Non-selective conductance in calcium channels of frog muscle: Calcium selectivity in a single-file pore, *J. Physiol.* **353:**585–608.

Alvarez-Leefmans, F. J., 1992, Extracellular reference electrodes, in: *Practical Electrophysiological Methods* (H. Kettenmann and R. Grantyn, eds.), pp. 171–182, Wiley-Liss, New York.

Barry, P. H., and Lynch, J. W., 1991, Liquid junction potentials and small cell effects in patch-clamp analysis, *J. Membr. Biol.* **121:**101–117.

Bezanilla, F., and Armstrong, C. M., 1977, Inactivation of the sodium channel. I. Sodium current experiments, *J. Gen. Physiol.* **70:**549–566.

Biermans, G., Vereecke, J., and Carmeliet, E., 1987, The mechanism of the inactivation of the inward-rectifying K current during hyperpolarizing steps in guinea-pig ventricular myocytes, *Pflügers Arch.* **410:**604–613.

Campbell, D., and Hille, B., 1976, Kinetic and pharmacological properties of the sodium channel of frog skeletal muscle, *J. Gen. Physiol.* **67:**309–323.

Cavalié, A., Grantyn, R., and Lux, H.-D., 1992, Fabrication of patch clamp pipettes, in: *Practical Electrophysiological Methods* (H. Kettenmann and R. Grantyn, eds.), pp. 235–240, Wiley-Liss, New York.

Chesler, M., and Kaila, K., 1992, Modulation of pH by neuronal activity, *Trends Neurosci.* **15:**396–402.

Coronado, R., and Latorre, R., 1983, Phospholipid bilayers made from monolayers on patch-clamp pipettes, *Biophys. J.* **43:**231–236.

French, R. J., and Wonderlin, W. F., 1992, Software for acquisition and analysis of ion channel data: Choices, tasks, and strategies, *Methods Enzymol.* **207:**711–728.
Green, W. N., and Andersen, O. S., 1991, Surface charges and ion channel function, *Annu. Rev. Physiol.* **53:**341–359.
Hamill, O. P., Marty, A., Neher, E., Sakmann, B., and Sigworth, F. J., 1981, Improved patch-clamp techniques for high-resolution current recording from cells and cell-free membrane patches, *Pflügers Arch.* **391:**85–100.
Hess, P., and Tsien, R. W., 1984, Mechanism of ion permeation through calcium channels, *Nature* **309:**453–456.
Hilgemann, D. W., 1990, Regulation and deregulation of cardiac Na–Ca exchange in giant excised sarcolemmal membrane patches, *Nature* **344:**242–245.
Hille, B., 1992, *Ionic Channels of Excitable Membranes,* 2nd ed., Sinauer Associates, Sunderland, MA.
Hoffmann, E. K., 1992, Cell swelling and volume regulation, *Can. J. Physiol. Pharmacol.* **70:**310–313.
Horn, R., and Korn, S. J., 1992, Prevention of rundown in electrophysiological recording, *Methods Enzymol.* **207:**149–155.
Horn, R., and Marty, A., 1988, Muscarinic activation of ionic currents measured by a new whole-cell recording method, *J. Gen. Physiol.* **92:**145–159.
Hume, R. I., Role, L. W., and Fischbach, G. D., 1983, Acetylcholine release from growth cones detected with patches of acetylcholine-rich membranes, *Nature* **305:**632–634.
Jackson, M. B., 1992, Stationary single-channel analysis, *Methods Enzymol.* **207:**729–746.
Kettenmann, H., and Grantyn, R. (eds.), 1992, *Practical Electrophysiological Methods,* Wiley-Liss, New York.
Kolb, H. A., 1992, Double whole-cell patch clamp technique, in: *Practical Electrophysiological Methods* (H. Kettenmann and R. Grantyn, eds.), pp. 289–295, Wiley-Liss, New York.
Kramer, R. H., 1990, Patch cramming: Monitoring intracellular messengers in intact cells with membrane patches containing detector ion channels, *Neuron* **4:**335–341.
Lapointe, J. Y., and Szabo, G., 1987, A novel holder allowing internal perfusion of patch-clamp pipettes, *Pflügers Arch.* **410:**212–216.
Latorre, R., Labarca, P., and Naranjo, D., 1992, Surface charge effects on ion conduction in ion channels, *Methods Enzymol.* **207:**471–501.
Levis, R. A., and Rae, J. L., 1992, Constructing a patch clamp setup, *Methods Enzymol.* **207:**14–66.
Levitan, E. S., and Kramer, R. H., 1990, Neuropeptide modulation of single calcium and potassium channels detected with a new patch clamp configuration, *Nature* **348:**545–547.
Lindau, M., and Fernandez, J. M., 1986, IgE-mediated degranulation of mast cells does not require opening of ion channels, *Nature* **319:**150–153.
Lindau, M., and Neher, E., 1988, Patch-clamp techniques for time-resolved capacitance measurements in single cells, *Pflügers Arch.* **411:**137–146.
Magleby, K. L., 1992, Preventing artifacts and reducing errors in single-channel analysis, *Methods Enzymol.* **207:**763–791.
Moody, W. J., 1984, Effects of intracellular H^+ on the electrical properties of excitable cells, *Annu. Rev. Neurosci.* **7:**257–278.
Neher, E., 1982, Unit conductance studies in biological membranes, in: *Techniques in Cellular Physiology* (P. F. Baker, ed.), pp. 1–16, Elsevier, Amsterdam.
Neher, E., 1988, Exploring secretion control by patch-clamp techniques, in: *Fidia Research Foundation Neuroscience Award Lecture,* pp. 37–53, Raven Press, New York.
Neher, E., 1992a, Correction for liquid junction potentials in patch clamp experiments, *Methods Enzymol.* **207:**123–131.
Neher, E., 1992b, Ion channels for communication between and within cells, *Science* **256:**498–502.
Neher, E., and Eckert, R., 1988, Fast patch-pipette internal perfusion with minimum solution flow, in: *Calcium and Ion Channel Modulation* (A. D. Grinnell, D. Armstrong, and M. B. Jackson, eds.), pp. 371–377, Plenum Press, New York.
Neher, E., and Lux, H. D., 1969, Voltage clamp on *Helix pomatia* neuronal membrane: Current measurement over a limited area of the soma surface, *Pflügers Arch.* **311:**272–277.
Neher, E., and Marty, A., 1982, Discrete changes of cell membrane capacitance observed under conditions of enhanced secretion in bovine adrenal chromaffin cells, *Proc. Natl. Acad. Sci. USA* **79:**6712–6716.
Neher, E., and Sakmann, B., 1976, Single-channel currents recorded from membrane of denervated frog muscle fibres, *Nature* **260:**799–802.
Neher, E., and Sakmann, B., 1992, The patch clamp technique, *Sci. Am.* **266:**28–35.

Neyton, J., and Trautmann, A., 1985, Single-channel currents of an intercellular junction, *Nature* **317**:331–335.
Pusch, M., and Neher, E., 1988, Rates of diffusional exchange between small cells and a measuring patch pipette, *Pflügers Arch.* **411**:204–211.
Rae, J. L., and Levis, R. A., 1992a, Glass technology for patch clamp electrodes, *Methods Enzymol.* **207**:66–92.
Rae, J. L., and Levis, R. A., 1992b, A method for exceptionally low noise single channel recordings, *Pflügers Arch.* **420**:618–620.
Rae, J., Cooper, K., Gates, P., and Watsky, M., 1991, Low access resistance perforated patch recordings using amphotericin B, *J. Neurosci. Methods* **37**:15–26.
Roberts, W. M., and Almers, W., 1992, Patch voltage clamping with low-resistance seals: loose patch clamp, *Methods Enzymol.* **207**:155–176.
Rudy, B., and Iverson, L. E. (eds.), 1992, *Methods in Enzymology,* Vol. 207, *Ion Channels,* Academic Press, San Diego.
Sakmann, B., 1992, Elementary steps in synaptic transmission revealed by currents through single ion channels, *Science* **256**:503–512.
Sakmann, B., and Neher, E., 1984, Patch clamp techniques for studying ionic channels in excitable membranes, *Annu. Rev. Physiol.* **46**:455–472.
Sarkadi, B., and Parker, J. C., 1991, Activation of ion transport pathways by changes in cell volume, *Biochim. Biophys. Acta* **1071**:407–427.
Sherman-Gold, R. (eds.), 1993, *The Axon Guide,* Axon Instruments, Foster City, CA.
Sigworth, F. J., 1986,, The patch clamp is more useful than anyone had expected, *Fed. Proc.* **45**:2673–2677.
Sigworth, F. J., and Neher, E., 1980, Single Na^+ channel currents observed in cultured rat muscle cells, *Nature* **287**:447–449.
Sigworth, F. J., and Zhou, J., 1992, Analysis of nonstationary single-channel currents, *Methods Enzymol.* **207**:746–762.
Soejima, M., and Noma, A., 1984, Mode of regulation of the ACh-sensitive K-channel by the muscarinic receptor in rabbit atrial cells, *Pflügers Arch.* **400**:424–431.
Strickholm, A., 1961, Impedance of a small electrically isolated area of the muscle cell surface, *J. Gen. Physiol.* **44**:1073–1088.
Stühmer, W., 1992, Loose patch recording, in: *Practical Electrophysiological Methods* (H. Kettenmann and R. Grantyn, eds.), pp. 271–273, Wiley-Liss, New York.
Suarez-Isla, B. A., Wan, K., Lindstrom, J., and Montal, M., 1983, Single-channel recordings from purified acetylcholine receptors reconstituted in bilayers formed at the tip of patch pipets, *Biochemistry* **22**:2319–2323.
Swandulla, D., and Chow, R. H., 1992, Recording solutions for isolating specific ionic channel currents, in: *Practical Electrophysiological Methods* (H. Kettenmann and R. Grantyn, eds.), pp. 164–168 Wiley-Liss, New York.
Tang, J. M., Wang, J., Quandt, F. N., and Eisenberg, R. S., 1990, Perfusing pipettes, *Pflügers Arch.* **416**:347–350.
Tsien, R. W., Lipscombe, D., Madison, D. V., Bley, K. R., and Fox, A. P., 1988, Multiple types of neuronal calcium channels and their selective modulation, *Trends Neurosci.* **11**:431–438.
Veenstra, R. D., and DeHaan, R. L., 1986, Measurement of single channel currents from cardiac gap junctions, *Science* **233**:972–974.
Young, S. H., and Poo, M. M., 1983, Spontaneous release of transmitter from growth cones of embryonic neurones, *Nature* **305**:634–637.
Young, S. H., and Poo, M.-M., 1992, Moving patch method for measurement of transmitter release, in: *Practical Electrophysiological Methods* (H. Kettenmann and R. Grantyn, eds.), pp. 354–357, Wiley-Liss, New York.

Chapter 2

Tight-Seal Whole-Cell Recording

ALAIN MARTY and ERWIN NEHER

1. Introduction

The tight-seal whole-cell recording method, often abbreviated as "whole-cell recording" (WCR), allows one to record from cells and modify their internal environment by using a patch-clamp pipette. This has become the most commonly used configuration of the patch-clamp technique. In the present chapter, we first describe the basic experimental procedures used to obtain whole-cell recordings. We then discuss the pipette–cell interactions during whole-cell recording, first from an electrical point of view and then from a chemical point of view. We finally compare the tight-seal whole-cell recording with other methods for studying electrical properties of cells.

2. Basic Procedures

To perform whole-cell recordings, patch-clamp pipettes are fabricated and filled with an appropriate low-Ca^{2+} solution. The pipette is pressed onto the cell membrane to establish a "gigaseal" at the contact area. The pipette potential is then changed to a negative voltage (such as 70 mV below the bath potential), and repetitive voltage steps of a few millivolts amplitude are given. At this stage, the fast capacitance compensation is adjusted to cancel the transient caused by the capacitance of the pipette holder and pipette wall (Fig. 1). Pulses of suction are applied to the pipette interior until a sudden increase in the size of the capacitive transients is observed (Fig. 1). This additional current reflects the contribution of the cell membrane to the pipette input capacitance following the destruction of the patch membrane. An alternative but rarely used method to break the patch membrane is to apply to the pipette voltage pulses of large amplitude to induce membrane breakdown ("zapping"). The pulses are very short (e.g., 10–500 μsec) such that once the patch is broken, the cell capacitance does not have the time to be loaded to an appreciable fraction of the applied potential, thus protecting the cell from total dielectric breakdown.

ALAIN MARTY • Neurobiology Laboratory, Teacher's Training College, École Normale Supérieure, F-75005, Paris, France. ERWIN NEHER • Department of Membrane Biophysics, Max-Planck-Institute for Biophysical Chemistry, Am Fassberg, D-37077 Göttingen, Germany. *Present adress For A.M.:* Cellular Neurobiology Workgroup, Max-Planck Institute for Biophysical Chemistry, Am Fassberg, D-37077 Göttingen, Germany

Single-Channel Recording, Second Edition, edited by Bert Sakmann and Erwin Neher. Plenum Press, New York, 1995.

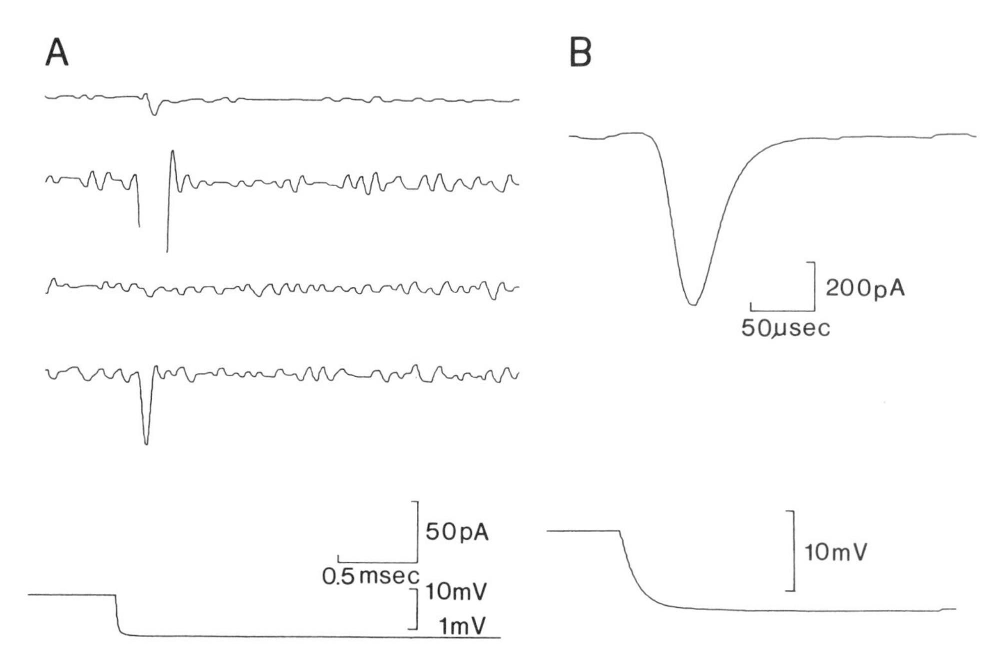

Figure 1. Capacitive currents observed at different stages of WCR. A: The top trace shows the response to a 10-mV voltage pulse after formation of a gigaseal and after cancellation of pipette capacitance, before rupture of the patch. The second trace shows the response to the same stimulus immediately after rupture. The capacitive artifact is off scale; note increase in background noise. The third trace shows the same response after cancellation of cell membrane capacitance (dial settings: $C = 3.56$ pF; $\tau = 20$ μsec). The fourth trace shows again a response without cell capacitance cancellation, employing a tenfold attenuated stimulus. The lowest trace shows the time course of the voltage pulses. B: Response to a 10-mV hyperpolarizing voltage pulse on a faster time scale (same experiment, WCR without capacitance cancellation).

Once the WCR mode is established, it is often helpful to lift the pipette somewhat to relieve the strain imposed on the cell when the initial seal was established. In some cases, the cell may be lifted further to detach it from the bottom of the recording chamber. In such cases, the solution surrounding the cell can be changed rapidly without impairing the stability of the recording.

Whole-cell recordings can be performed for at least 1 hr without electrical signs of deterioration. (However, as is discussed below, some currents may subside during such long recordings.) Provided that the cell under study is firmly attached to the culture dish, it is then possible to form an outside-out patch simply by pulling away the recording pipette (Hamill *et al.,* 1981). After this maneuver, the cell membrane reseals, leaving the cell essentially intact except that its internal solution has been replaced with the pipette solution. This property can be exploited in several ways. (1) The method may be used to alter the internal solution of a chosen cell with minimal damage to its membrane. (2) Successive WCRs with different internal solutions can be performed on a given cell by changing the recording pipette. (3) Both macroscopic and single-channel data may be obtained in one experiment. It should be noted, however, that the electrical stability of outside-out patches formed after a long WCR is often not as good as that of patches obtained after a few minutes of WCR.

3. The Whole-Cell Recording Configuration from an Electrical Point of View

3.1. The Equivalent Circuit for a Simple Cell

We assume here a reasonably small cell (longest dimension $\leq$ 100 μm) with a resistance at rest R on the order of 1 GΩ and an input capacitance C on the order of 10–100 pF. This corresponds to the situation found for a large number of animal cells. This cell is studied using a patch-clamp pipette that has a resistance R_S *during recording* (R_S is different from the pipette input resistance measured before contacting the cell; see below). The equivalent circuit is illustrated in Fig. 2. If V_P, the pipette potential, follows an imposed square impulse of amplitude ΔV, the current I through such a circuit is described by a simple exponential function of time, as illustrated in Fig. 2B. If one assumes that $R_S << R$ (typical values are 10 MΩ and 1 GΩ, respectively) one obtains:

$$I_{in} = \Delta V / R_S \tag{1}$$

$$I_{SS} = \Delta V / R \tag{2}$$

and

$$\tau = R_S C \tag{3}$$

Where I_{in} is the "instantaneous current" obtained just after the jump, τ the time constant of the current relaxation, and I_{ss} the steady-state current. I_{in}, I_{ss}, and τ can be all measured from the current record, and equations 1–3 can then be used to calculate R, R_S, and C. Thus, a simple analysis of the current changes recorded in response to a square impulse yields all the parameters of the electrical circuit of the cell but one (E_R, the resting potential of the cell). Note that R_S, as calculated from equation 1, is different from the value found for the resistance of the pipette before making the seal, R_{in}. Typical ratios of R_S/R_{in} are 2–5. If several pulses of suction are applied to the pipette interior, this value can be brought down to 2–3 but not below. This is probably because cytosolic elements impose a lower conductivity to the part of the cell that is sucked into the pipette compared to the conductivity of the bath solution.

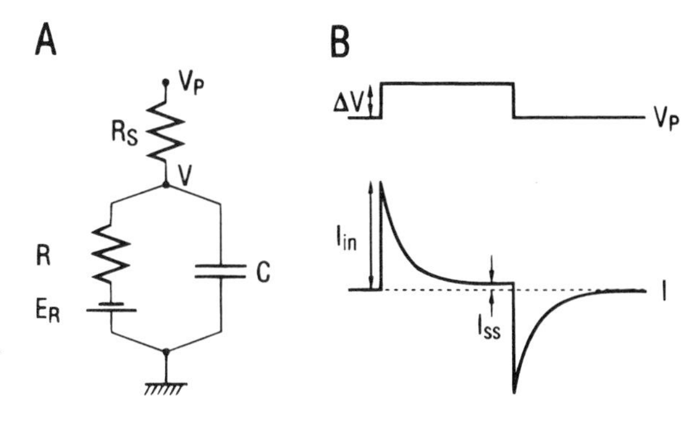

Figure 2. Passive electrical properties for a simple cell. A: Equivalent electrical circuit. E_R is the cell resting potential. R_S, R, and C represent the pipette resistance, the cell resistance, and the cell capacitance. B: Current response to a ΔV displacement of the pipette potential.

3.2. Series Resistance Errors

To illustrate some of the problems associated with series resistance errors, let us assume that we want to measure a Na^+ current in a cell clamped with $R_S = 10\ M\Omega$. The "ideal" I–V curve is as drawn in Fig. 3A. This is the curve that would be recorded if R_S were zero. For each current value actually recorded, the flow of current along the pipette tip results in a discrepancy between the pipette potential V_P and the cell potential V. According to Ohm's law $V - V_P = R_S I$. Using this equation, it is possible to predict the distortion brought about by series resistance errors in the I–V curve. A graphic method can be employed to go from the ideal $I(V)$ curve to the experimental $I(V_P)$ curve (or vice versa) as shown in Fig. 3B. As a result of the series resistance error, the $I(V_P)$ curve has a very abrupt take-off near -40 mV, and it reaches its maximum within a few millivolts. In fact, if the maximal slope of the original $I(V)$ curve exceeds 0.1 nA/mV (corresponding to the ratio $1/R_S$), the $I(V_P)$ curve has a jump with infinite slope. The discontinuity occurs as the voltage-clamp system is unable to prevent the cytosolic potential V from firing an action potential. This happens in practice if the maximum I_{Na} current is on the order of 3 nA, a rather modest value. Thus, series resistance errors can severely distort I–V curves. In regions of the curves where the derivative dI/dV is negative, total voltage clamp failure can occur.

A second, potentially serious consequence of the presence of R_S is the fact that, even for large values of R, V does not follow V_P immediately. Thus, if V_P follows a square impulse, V is rounded off with a time constant τ as given by equation 3. With $R_S = 10\ M\Omega$ and $C = 100$ pF, we obtain $\tau = 1$ msec. Changes in channel-opening probability can occur much faster than this, for example, during deactivation of Ca^{2+} channels on returning from a depolarized test potential to the holding potential. If, because

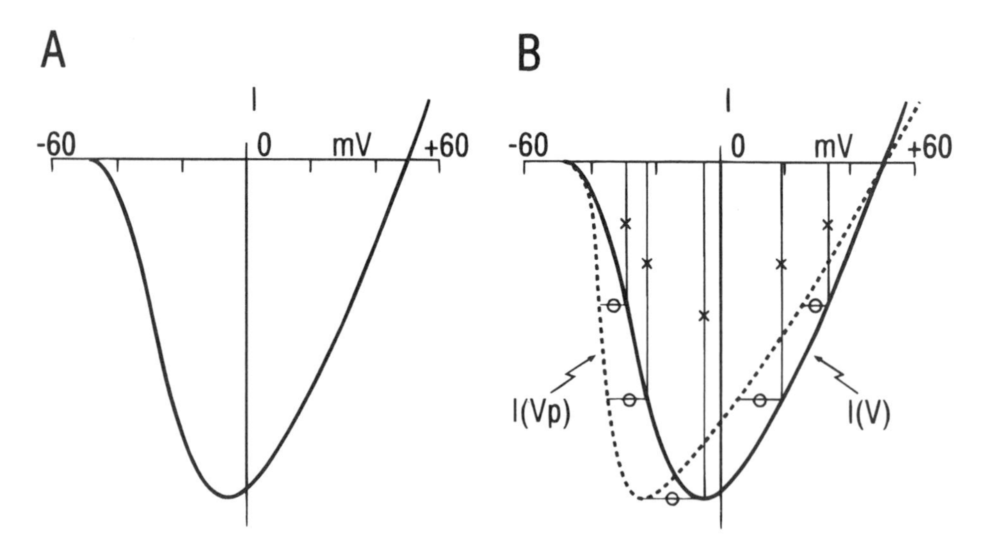

Figure 3. Series resistance errors. A: Assumed ideal I–V curve for peak Na current measurements. B: Because of series resistance errors, the measured curve $I(V_P)$ differs from the ideal curve $I(V)$. It is possible to transform one curve into the other by a graphic method. Here the $I(V) \rightarrow I(V_P)$ transform is illustrated. For all pairs of points the values of the ratio x/o are the same. Note that the $I(V_P)$ curve is much steeper than the $I(V)$ curve in the -40– to -20-mV voltage range.

of the R_S error, V actually changes with a time constant of 1 msec, such fast processes will have to wait for the changes in V to have occurred, resulting in artifactual changes in the current trace.

Series resistance errors can be dealt with in various ways. To some extent, they can be reduced by series resistance compensation (see below). Residual errors on steady-state currents can be appreciated using the type of transform illustrated in Fig. 3, and I–V curves can be corrected accordingly if R_S is known. Alternative strategies include the use of a single-electrode switching clamp instead of the usual continuous voltage-clamp design.

3.3. Cell Capacitance Cancellation

This should not be confused with series resistance compensation. Capacitance cancellation is performed (1) to avoid saturation of the amplifiers of the recording circuit during the sudden high-amplitude currents generated at the onset of voltage changes and (2) to eliminate error signals caused by the pipette and holder capacitance. In a first step, the small capacitor representing the pipette and pipette holder is canceled at the cell-attached stage. This is called "fast" capacitance cancellation because the time constant of the corresponding current is very fast, in the 0.5- to 5-μsec range. This first step is essential if one wishes to estimate R_S by using equation 1 once the WCR mode has been established. In WCR, a second cancellation is performed to remove the cell membrane capacitive current. At this stage, most commercial amplifiers display the values of the cell capacitance and access resistance that correspond to the second (slow) capacitive current. Capacitive current cancellation consists of supplying the current needed to charge the capacitor directly to the summing input of the headstage amplifier through a small capacitor C_h (Fig. 4A). To achieve this, an exponentially shaped voltage command is applied to C_h. By adjusting the amplitude and time constant of this voltage V_{shaped}, I_{cap} is made to flow through C_h and not through R_f. It is therefore not recorded. To perform simultaneous cancellation of fast (holder and pipette) and slow (cell) capacitive currents, the sum of two exponentials is fed to $V_{shaped.}$

The problem of capacitance cancellation is treated in greater detail in Chapters 4 and 7 (this volume).

3.4. Series Resistance Compensation

The target here is to limit errors in I–V curves and to improve the effective speed of the voltage clamp in order to be able to record fast current changes. The basic idea is to take the current output, scale it with an adjustable gain α, and add the result to the command voltage (Fig. 4B). During voltage pulses, this results in an overshoot of the effectively applied potential (Fig. 5). Also, the system automatically corrects V_P if a sizable ionic current starts to flow. This correction is in essence the inverse of the graphic method used to go from the $I(V)$ curve to the $I(V_P)$ curve in Fig. 3.

One problem with series resistance compensation is that, because it employs a positive feedback, it can escape control ("ringing"). For this reason it is impossible to reach 100% compensation. Also, the compensation only works if the value of R_S has been determined correctly. R_S may be evaluated by analyzing capacitive current transients

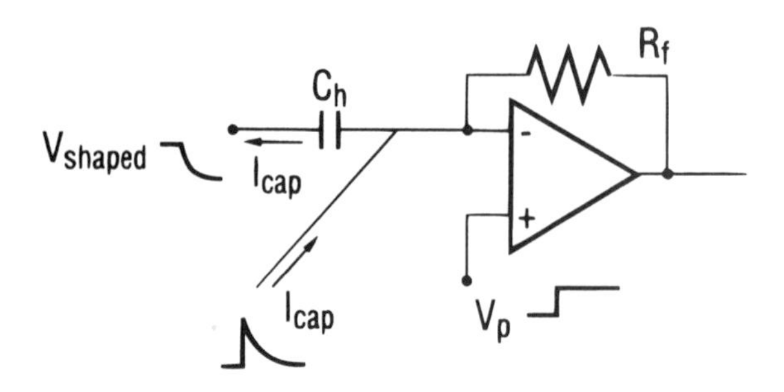

capacitive current cancellation

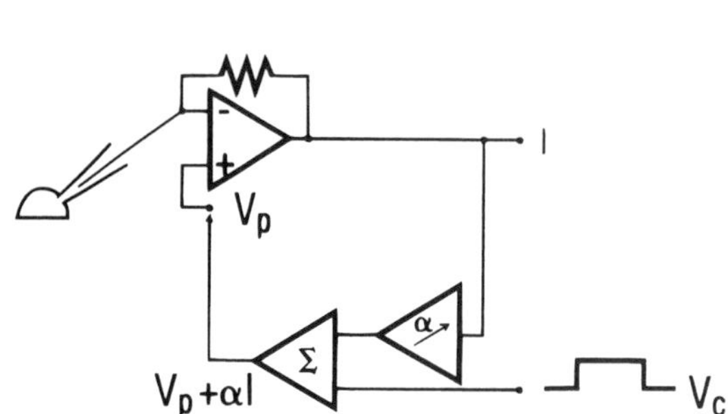

series resistance compensation

Figure 4. Cell capacitance cancellation and series resistance compensation. A: Cell capacitance cancellation. At the onset of voltage jumps, an adjustable signal V_{shaped} is fed through a small capacitor C_h to the negative input of the headstage amplifier. The corresponding current flow cancels the capacitive current I_{cap}, such that no net current is recorded by the amplifier. B: Series resistance compensation. The command potential V_P is the sum of the command potential V_c and a signal proportional to the measured current (αI).

using equation 1 or by reading the setting of an appropriately designed capacitance compensation network. Unfortunately, the value of R_S may change spontaneously during recording, thus making later R_S compensations inaccurate. The only way out of this problem is to determine R_S often during the course of the experiment. If R_S increases markedly, it is usually possible to decrease it again, at least transiently, by applying a new pulse of suction to the pipette interior. Computer-controlled patch-clamp amplifiers make it possible to estimate and compensate R_S automatically before each depolarizing voltage pulse.

As an alternative or complement to series resistance compensation, the effective onset of applied potential steps can be accelerated by adding a square pulse of large amplitude and short duration (~ 5 μsec) at the leading edge of command potential steps. The amplitude of the pulse is adjusted empirically in order to minimize the capacitive current transient ("supercharging": Armstrong and Chow, 1987).

Series resistance compensation is usually performed after full cancellation of pipette, holder, and cell membrane capacitive currents (Fig. 5). However, there is no real obligation to proceed that way. Provided that R_S is set at the right value, most commercial amplifiers will perform a correct series resistance compensation (at least for small voltage pulses) in the absence of cell membrane capacitance cancellation. If this is done, however,

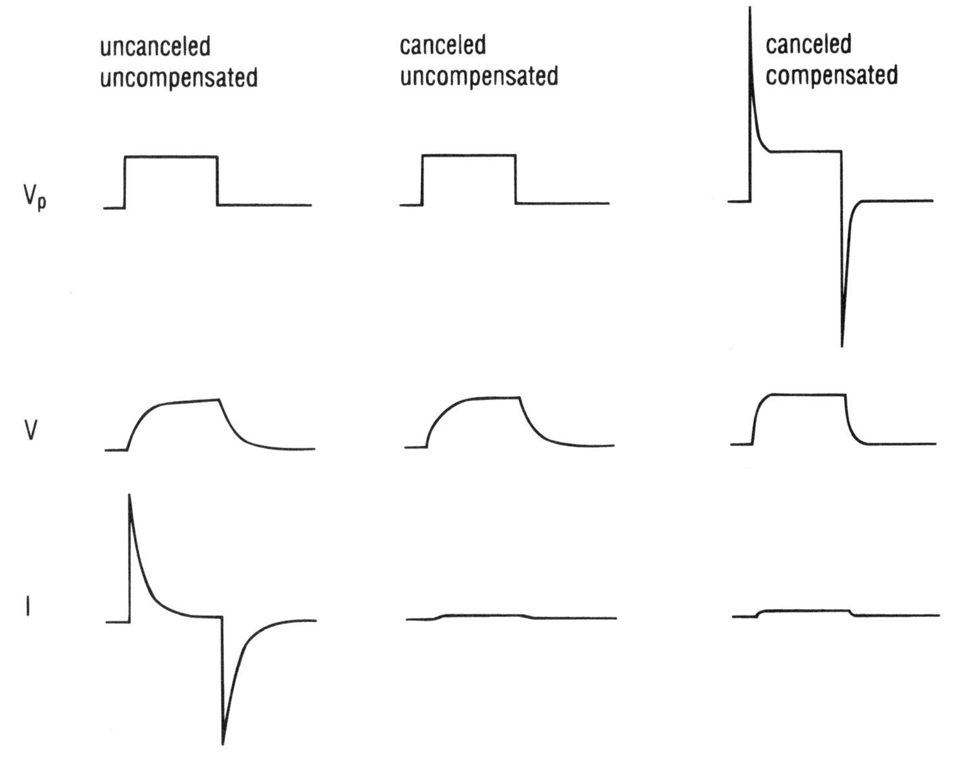

Figure 5. Pipette and cell potential during capacitance cancellation and series resistance compensation. In the uncancelled, uncompensated situation (left column), V_P is rectangular, while V and I are exponential functions of time with the same time constant τ (equation 3). Cancellation of the cell capacitive current does not change the time course of V (middle column). After series resistance compensation, however, V reaches more quickly its steady state (right column). This is achieved by adding an overshooting component to V_P.

capacitive currents will saturate quickly for increasing pulse amplitudes, as series resistance compensation results in faster and larger capacitive currents. Saturation of capacitive currents will delay effective voltage-clamp control.

3.5. Capacitance Cancellation Together with Series Resistance Compensation

When both techniques are employed, the problem arises that capacitance cancellation eliminates part of the current, which does not appear past the first amplifier. As a result, the gain in voltage-clamp speed expected from series resistance compensation during voltage pulses may be lost. Fortunately, this problem can be fixed by adding again the signal corresponding to the capacitive current to the recorded current at the end of the recording chain when shaping the correction voltage that is added to the command potential (more details on this problem can be found in Chapter 4, this volume).

3.6. Current Clamp

Current-clamp measurement can be performed with the same basic circuit with an additional feedback loop. The loop connects the current output to the command input of the pipette potential and provides negative feedback for constant-current operation, as described by Sigworth (Chapter 4, this volume).

The same procedure should be used to effectively subtract the pipette and holder capacitance as in voltage-clamp recordings. If this is not done, voltage signals following current steps will be artificially slowed down.

3.7. Noise

The dominating source of background noise in a whole-cell recording is likely to be the membrane capacitance C in series with the access resistance R_S. The power spectrum $S(f)$ of such a combination is given by

$$S(f) = 4kTR_s(2\pi fC)^2/[1 + (2\pi fR_sC)^2]$$

For typical values of R_S and C (see above), this term is much higher than the noise of the membrane conductance over most of the frequency range of interest. This is documented in Fig. 6, which shows the noise power spectra of two bovine adrenal chromaffin cells.

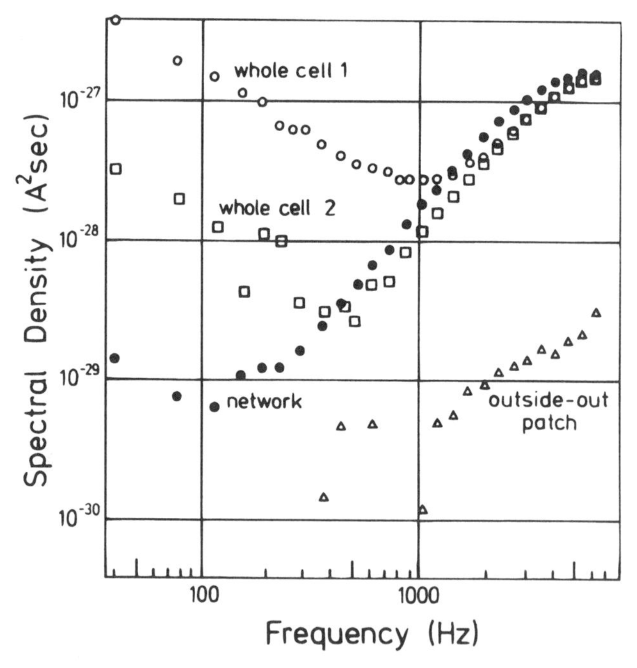

Figure 6. Average noise-power spectra of different recording situations. Power spectra of current records were taken from two adrenal chromaffin cells in the whole-cell configuration. Both cells were bathed in standard saline containing 15 μM TTX. Each had a membrane capacitance between 5.7 and 5.8 pF and a series resistance of about 5.2 MΩ. The latter value was calculated from the time constant of the capacitive current and from the cell membrane capacitance. In the recordings from cell 1, the pipette was filled with a Cs^+-containing internal solution, and the holding potential was zero. During the recording of cell 2, the pipette was filled with a K^+-containing internal solution, and the cell was held at −70 mV. For comparison, the power spectrum of a resistor–capacitor network with the following values is shown: series resistance 5.2 MΩ, capacitance 5.85 pF, parallel resistance 8.5 GΩ. When the pipette was withdrawn from cell 1, an outside-out patch formed. Its noise power spectrum is included in the figure. A background noise spectrum (including amplifier and pipette holder noise as well as noise from the tape recorder on which the data were temporarily stored) was subtracted from the whole-cell and outside-out spectra. This control was obtained from a recording with a pipette mounted onto the amplifier but not immersed in solution. The spectrum of the resistor–capacitor network was corrected for the noise of the open-input amplifier. All curves are averages of 300 individual spectra.

Both cells had a spectral density rising with the square of the frequency above 1 kHz and saturating between 5 and 10 kHz according to the above equation. The power spectrum of a resistor–capacitor network with similar parameters is shown for comparison. Cell 1 was held at 0 mV and had appreciable noise attributable to ionic channels, which dominates the noise power below 1 kHz. Still, in the range 40 Hz to 1 kHz, the noise from the series combination R_SC is ten times larger than the conductance noise (the logarithmic display distorts the relative contributions). Cell 2, which was held at −70 mV under more "physiological" conditions, was very quiet. Noncharacterized single-channel inward currents (Fenwick *et al.,* 1982b, Fig. 21) at low frequency could be observed; their noise dominated the power spectrum at frequencies below 400 Hz.

All spectral densities measured in the whole-cell configuration are one to two orders of magnitude higher than those measured with patches. For comparison, a spectrum from an outside-out patch is also shown Fig. 6. It was measured on a patch taken from cell 1 after completion of the whole-cell measurements.

The background noise of cell 2 corresponds to an rms value of 0.15 pA in the frequency range 0 to 400 Hz and to approximately 1.5 pA_{rms} for 0 to 4 kHz. Thus, single channels of relatively long duration (such as ACh-induced channels) can be well resolved in a whole-cell recording (see Fenwick *et al.,* 1982a), whereas short single-channel events (such as Na^+ channels) are lost. Still, the dynamic range of such a measurement is remarkable. This is illustrated in Fig. 7, which shows Na^+ currents in response to depolarizing voltage pulses in a chromaffin cell at 3.5-kHz bandwidth. The left column shows the response to various small depolarizations (from −70 mV holding) at a relatively sensitive current scale. Discrete fluctuations caused by individual Na^+ channels can be seen in the uppermost record, but they are not clearly resolved. As the membrane is depolarized further, these fluctuations add up to sizable, highly fluctuating currents. In the right column, traces from larger depolarizations are shown, but at a ten-times-smaller current scale. On this scale, the background fluctuations are no longer visible (see segments before pulses); maximum inward currents rise to 1400 pA, which is approximately 600 times the rms noise of background.

3.8. Equivalent Circuit for a Complex Cell

Some cells cannot be described by the simple equivalent circuit of Fig. 2. Brain neurons, for example, have complicated dendritic arborizations that are not instantly charged if a square voltage pulse is applied to the soma. Cerebellar Purkinje cells are a good example. In these neurons, the capacitive current response to a hyperpolarizing voltage pulse is the sum of two exponentials. In such a case capacitive current cancellation of the fast component can be performed as illustrated in Fig. 8. Series resistance compensation leads to an acceleration of the time constant of decay of the slow component. The situation can be modeled with a two-compartment equivalent circuit (Fig. 8). The first compartment, with potential V_1, is represented by the soma and proximal dendrites. The second compartment, with potential V_2, corresponds to the main part of the dendritic tree. At the onset of the voltage pulse, V_1 follows a double exponential with time constants τ_1 and τ_2, whereas V_2 follows a single exponential (after a short delay) with time constant τ_2. After series resistance compensation V_1 follows a step function, but V_2 remains an exponential function of time. However, the time constant of this exponential, τ'_2, is shorter than τ_2. The acceleration of V_2 can be followed by monitoring the canceled capacitive current trace. It can be shown that τ'_2 is also the time constant of filtering

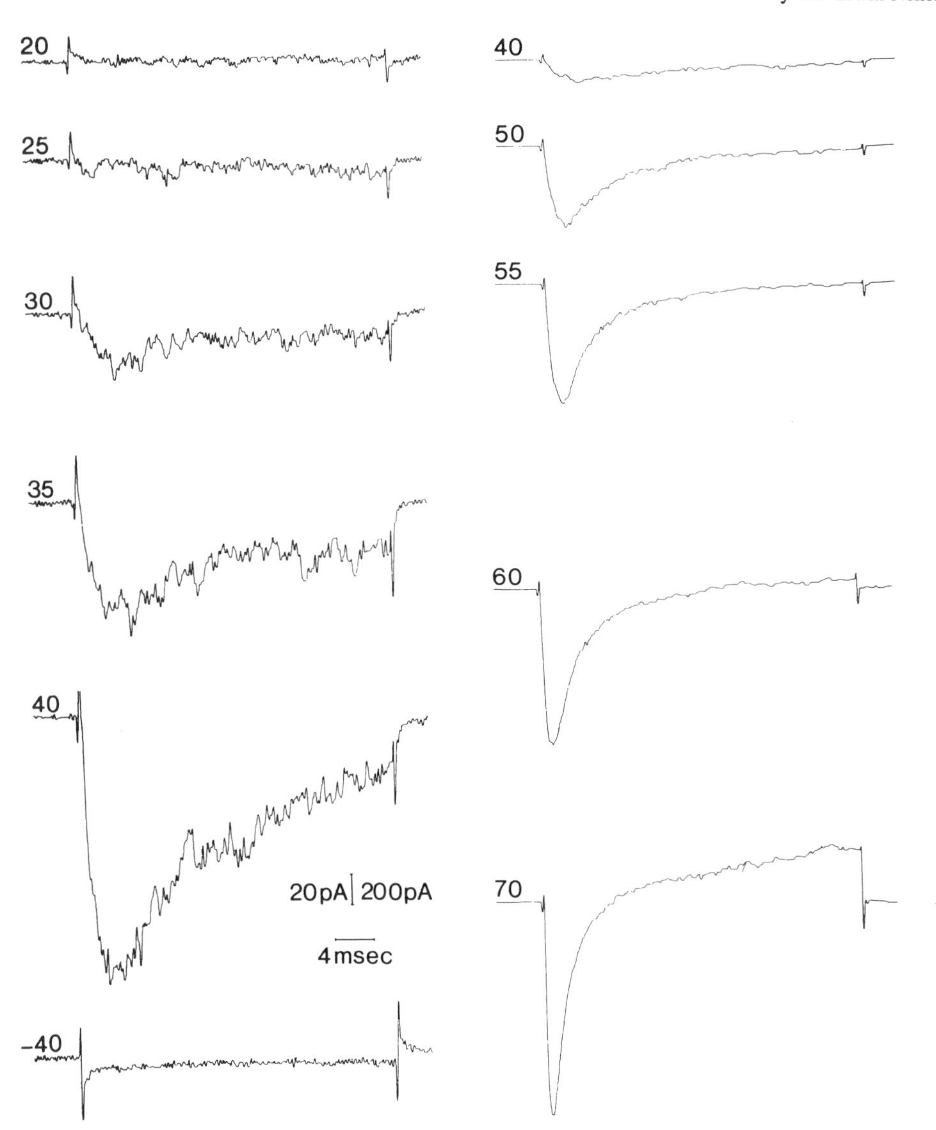

Figure 7. Dynamic range of current measurement in the whole-cell clamp. Sodium currents from a chromaffin cell with membrane capacitance of 5.8 pF (R_s = 5 MΩ) were elicited by depolarizing pulses of various amplitudes (given in millivolts at the start of each trace); holding potential was −70 mV. Current amplitudes cover the range from the single-channel level (≈2 pA) to 1400 pA; 3.5-kHz bandwidth; 21°C. The bath contained normal physiological saline; the pipette was filled with a K^+-containing internal solution. A hyperpolarizing response is shown in the left lower corner. It shows a biphasic residual capacitive artifact resulting from a slight imbalance of the τ setting in the capacitive cancellation network. Following that, a slow capacitive component is seen, which carries approximately 10 *fCb,* corresponding to 0.4 pF. A large part of this component probably arises from the pipette, because isolated patches or sealed pipettes show similar artifacts. The largest currents in this figure are subject to clamp errors of approximately 7 mV.

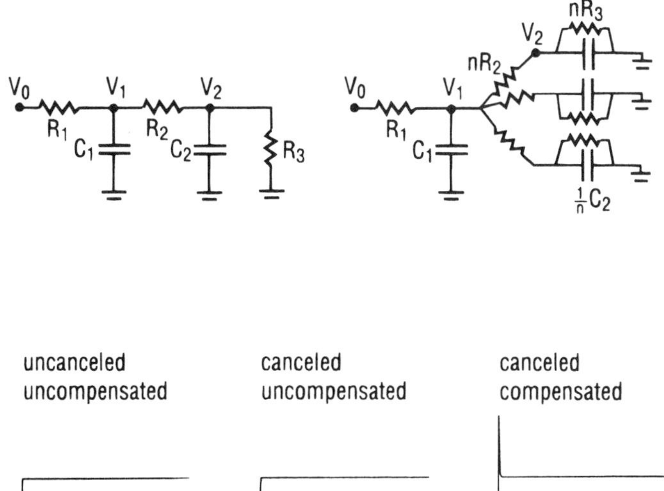

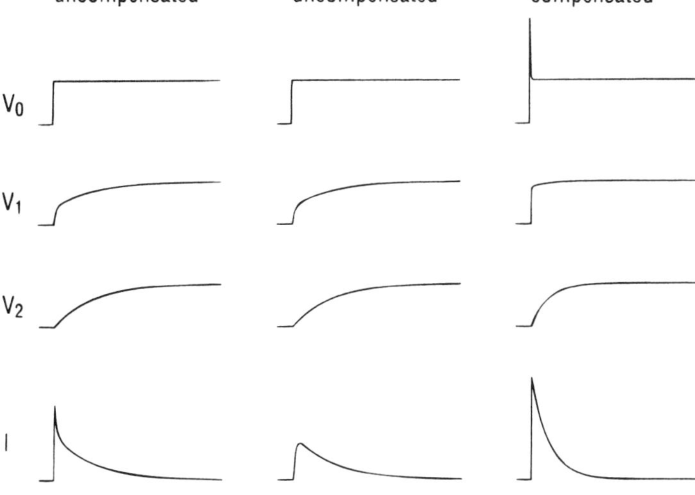

Figure 8. Capacitive currents in Purkinje cells. Upper panels: Two-compartment equivalent circuit modeling the passive properties of cerebellar Purkinje cells. V_0 is the pipette potential, and R_1 the pipette resistance. The first compartment (soma and proximal dendrites) is at a potential V_1. It has an infinite membrane resistance and a capacitance C_1. The second compartment (distal dendrites) is at a potential V_2. It has a resistance R_3 and a capacitance C_2. In the scheme on the right, the second compartment has been subdivided into *n* equivalent subcompartments representing individual branches of the dendritic tree. The two variants of the equivalent circuit are not distinguishable in somatic recordings. Lower panels: Capacitive current responses for the two-compartment equivalent circuit. The upper traces represent the potential V_0 applied to the pipette; the next traces show the voltage transients in compartments 1 and 2; the lower trace shows the capacitive current. In the uncanceled, uncompensated situation, V_1 is the sum of two exponential functions of time, with time constants τ_1 and τ_2. V_2 rises after a short delay with the slower time constant τ_2. *I* is biphasic, with time constants τ_1 and τ_2. In the recording that was used to make the calculations, τ_1 and τ_2 were 0.26 and 4.51 msec (Llano *et al.*, 1991, Fig. 2). From the time constants and amplitude ratio of the capacitive current the following values were calculated: R_1 = 8.9 MΩ, R_2 = 5.6 MΩ, C_1 = 76 pF, C_2 = 311 pF. Capacitive current cancellation was performed by removing the fast component of the capacitive current, as shown in the middle traces, without any gain in the speed of voltage control. Series resistance compensation of resistance R_1 (right) led to a substantial acceleration of V_1 and V_2. Note the overshoot of V_0. Calculations show the effect of 90% compensation. The slow time constant takes the new value τ'_2 = 2.0 msec.

of dendritic signals originating in compartment 2 (Llano *et al.*, 1991). Thus τ'_2 represents the limitation of effective voltage control introduced by dendriting filtering.

Whereas Purkinje cells can be satisfactorily modeled with a two-compartment equivalent circuit, other neurons cannot. In hippocampal pyramidal cells, for instance, three exponentials are needed to model the capacitive current decay. In such complex cases a cable analysis may be more appropriate than a multicompartment model (Jackson, 1992).

4. The Whole-Cell Recording Configuration from a Chemical Point of View

4.1. Modeling Diffusion between Pipette and Cell Compartments

Whole-cell recording establishes simultaneously an electrical and a chemical pathway to the cell interior. It is obviously desirable to determine to what extent, and at which speed, do pipette and cell contents equilibrate. Given the prevailing volume ratio (~10 μl for the pipette and ~1 pl for the cell), it is clear that eventually the pipette solution will dictate the concentration of all diffusible substances. However, the time of equilibration may vary greatly depending on the exact cell geometry and on the size of the diffusible substance. The only rigorous approach to this problem is to apply the equations of diffusion to the geometry of the pipette–cell assembly. This can be done by dividing this assembly in a number of compartments, as illustrated in Fig. 9A, and by solving the diffusion equations in each compartment (Oliva *et al.*, 1988). We consider below an alternative solution to the problem, which has the advantage of highlighting the parameters that determine the speed of diffusion.

4.2. A Solution for a Two-Compartment Model

To simplify calculations, one can assume two homogeneous compartments separated by a transition zone where concentration gradients occur. The first of these compartments

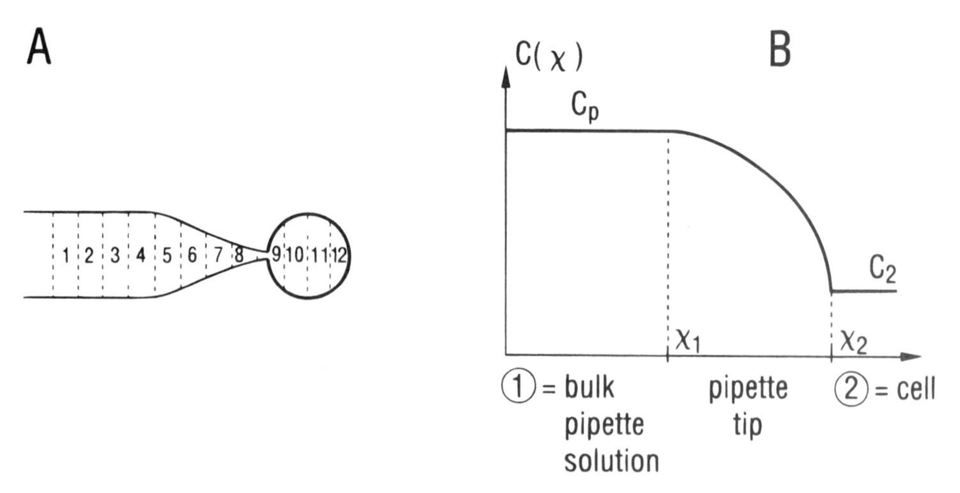

Figure 9. Pipette–cell equilibration. A: A multicompartment model of the pipette–cell assembly. B: Simplified diffusion model with two compartments separated by a transition zone representing the pipette tip.

is the bulk of the pipette solution, where the original concentration c_P of a given substance stays constant simply because the pipette volume is so much larger than the cell volume. At the other end of the diffusion path, the concentration c_2 of this substance in the cell may also be considered homogeneous. This is because concentration gradients can dissipate much faster in the cell (typical diameter $\geq$ 10 μm) than in the pipette tip (inner diameter ~1 μm). Compartments ① and ② are linked by a resistive region represented by the pipette tip (Fig. 9B). During the loading process of the cell c_2 evolves from its original value c_O to the final value c_P. To simplify the calculations, we further assume that diffusion is a steady-state process in the sense that accumulation of the diffusing substance in the pipette tip can be neglected. The following derivation can then be made, after Mathias *et al.* (1990).

Let us consider the diffusion process in the pipette tip. According to Fick's law the flux of diffusing substance at position x is

$$\Phi(x) = -D\, dc(x)/dx \tag{4}$$

where D is the diffusion constant of the substance, and $c(x)$ is the concentration at position x. By definition, the flux is the ratio of the amount transferred at position x, $J(x)$, to the cross-section area of the pipette in x, $A(x)$:

$$\Phi(x) = J(x)/A(x) \tag{5}$$

$J(x)$ is independent of x, because we assumed that the substance is transferred at the same time through all x positions of the pipette tip without accumulation anywhere but in compartment ②. Thus, $J(x) = J$.

Finally, the resistance of the pipette tip is the sum of elementary resistances represented by slabs of fluid of area $A(x)$ and length dx:

$$R_S = \int_{x_1}^{x_2} \frac{\rho}{A(x)}\, dx \tag{6}$$

where ρ is the resistivity of the pipette solution. Combining equations 4 and 5 yields

$$-D\, dc(x)/dx = J/A(x)$$

By integration from $x = x_1$ to $x = x_2$, we obtain

$$D[c(x_1) - c(x_2)] = J \int_{x_1}^{x_2} \frac{dx}{A(x)}$$

Combining with equation 6 yields

$$D[c(x_1) - c(x_2)] = (J/\rho)R_S$$

Now $c(x_1) = c_p$, and $c(x_2) = c_2$. Therefore,

$$c_P - c_2 = \frac{JR_S}{D\rho} \tag{7}$$

Since J is the flux to compartment ②,

$$J(x) = J = v \frac{dc_2}{dt}$$

where v is the cell volume. Combining with equation 7 yields

$$\frac{dc_2}{dt} = \frac{D\rho}{R_S v} (c_P - c_2). \tag{8}$$

The solution to this differential equation is an exponential function of time with a time constant

$$\tau = R_S v / D\rho \tag{9}$$

Equation 9 shows that the time constant of equilibration of a substance is proportional to the cell volume and to the pipette resistance and inversely proportional to the diffusion coefficient of this substance. Some of these predictions have been tested. In chromaffin cells, Pusch and Neher (1988) have shown that, inasmuch as R_S remained constant, the time course of equilibration of a variety of fluorescent compounds was exponential. For an R_S value of 10 MΩ, time constants of exchange varied from about 4 sec (for K^+) to 1 min for substances of molecular weight around 1000. Furthermore, in conformity to equation 9, equilibration times were inversely proportional to the diffusion coefficient D and proportional to the pipette resistance R_S. From their data a value of 240 Ω cm is calculated for the pipette resistivity by using equation 9. This is about fourfold higher than the resistivity of the pipette solution. The results suggest that the effective resistivity of the cytoplasm sucked into the pipette tip is about four times higher than that of the bath solution. This is in line with the above-mentioned result that the ratio R_S/R_{in} of the pipette access resistances measured in the bath and during WCR is at best of the order of 2 to 3.

4.3. Extension to Multicompartment Cells

The above analysis reveals an analogy between the speed of voltage changes and that of diffusion. In both cases the equilibrium is reached, for a round cell, as an exponential function of time. In both cases the time constant of the exponential is proportional to R_S (see equations 3 and 9). The analogy can be extended to more complex cases. Thus, the two-compartment model of Fig. 8 predicts not only a two-exponential capacitive current but also a two-exponential loading curve during pipette–cell equilibration. In the latter case, the time constants of loading are a function of the series resistances R_1 and R_2, of the diffusion constant of the diffusing substance, and of the volumes v_1 and v_2 of the two compartments. Approximate values for the time constant t_1 and t_2 of equilibration may be obtained provided that equilibration proceeds faster in the first compartment than in the second ($R_1 v_1 \ll R_2 v_2$).

$$t_1 = v_1/[D\rho(1/R_1 + 1/R_2)]$$

$$t_2 = v_2(R_1 + R_2)/D\rho$$

(The equation for t_2 reduces to equation 9 in the case that $R_1 = 0$.)

This is to be compared with the corresponding values of the decay time constants of the capacitive current under similar assumptions ($R_1C_1 \ll R_2C_2$) (Llano *et al.*, 1991, equations 1 and 2):

$$\tau_1 = C_1/(1/R_1 + 1/R_2)$$

$$\tau_2 = C_2/(R_1 + R_2)$$

For each compartment, the capacitance is thus replaced by the ratio $v/D\rho$.

4.4. Pipette Filling Solutions for WCR

A standard pipette solution for whole-cell recording contains a pH buffer (e.g., 10 mM HEPES, adjusted to pH 7.4), a Ca buffer (e.g., 10 mM EGTA + 1 mM Ca, giving $Ca_i \simeq 10$ nM), MgATP (~2 mM; in order to have ATPases active), free Mg^{2+} (~1 mM, as a cofactor for many cytosolic processes). GTP (~0.1 mM) is included in cases where processes involving G proteins are studied. ATP and GTP are poorly stable. Pipette solutions should therefore be kept frozen when they contain either nucleotide. During experiments, thawed solutions should be kept on ice. For long-duration (>30 min) recordings, it is advisable to use an ATP-regenerative system in order to obviate ATP degradation in the recording pipette. Because cells normally contain millimolar concentrations of glutathione, a reducing agent, one probably should (but often one neglects to do so) include ~5 mM glutathione as well. Inactivation of K^+ (Ruppersberg *et al.*, 1991) and Na^+ channels (Strupp *et al.*, 1992) depends on the redox potential of the internal solution and will be abnormal unless glutathione is duly supplemented.

The standard intracellular anion is Cl^-. In cases where it is desired to keep a negative equilibrium potential for Cl^-, a large fraction of the internal Cl^- is replaced by an impermeant ion. Glutamate, MOPS, and isethionate are often used for this purpose. F^- is also used as an intracellular anion. It is often permeant through Cl^- permeable channels. The advantage of F^- over Cl^- is that seal formation and recording stability are often better. However, F^- ions make complexes with traces of Al^{3+}, leading to the formation of AlF_4^-, a potent activator of G proteins. F^- ions also act as Ca^{2+} buffer, so that they may interfere with Ca-dependent processes. Thus, the use of F^- is inappropriate for the study of signaling pathways involving G proteins or intracellular messengers.

4.5. Perforated Patch Recording

It often occurs that an important cell function disappears during whole-cell recording as a result of the loss of unknown diffusible factors into the recording pipette ("washout"). One solution to this problem is to establish the electrical connection between cell and pipette not by suction, but by incorporation of a channel-forming substance in the cell-attached mode ("perforated patch"). There is no washout if the critical diffusing molecule is unable to cross the exogenous channel (Fig. 10A). Nystatin, a polyene antibiotic with antifungal profile, is commonly used for this purpose (Horn and Marty, 1988; Korn *et al.*, 1991). This choice was derived from the following properties of nystatin channels (Kleinberg and Finkelstein, 1984, and references herein). (1) They are 8 Å in diameter. Molecules larger than this (mol. wt. ≥ 300) do not pass through. (2) They are permeable to all small monovalent cations and anions (however the permeability is larger for cations than for anions; $P_{Na}/P_{Cl} \sim 10$). This makes it possible to control the internal K^+, Na^+, and Cl^- concentrations, which are equal to the pipette solution values after a few minutes of recording. Cs^+ can be readily substituted for K^+ if one wishes to block K^+ currents. Ca^{2+} ions are not permeant. (3) They are not voltage dependent. This ensures a good behavior of the recording system when the command (pipette) potential is changed.

Nystatin inhibits the formation of seals; it is therefore important to fill the pipette tip with a nystatin-free solution, while the rest of the pipette is filled with nystatin-containing solution (Fig. 10B). Filling the pipette tip with nystatin-free solution is achieved by dipping the pipette for ~ 2 sec into that solution. Then the shank of the pipette is

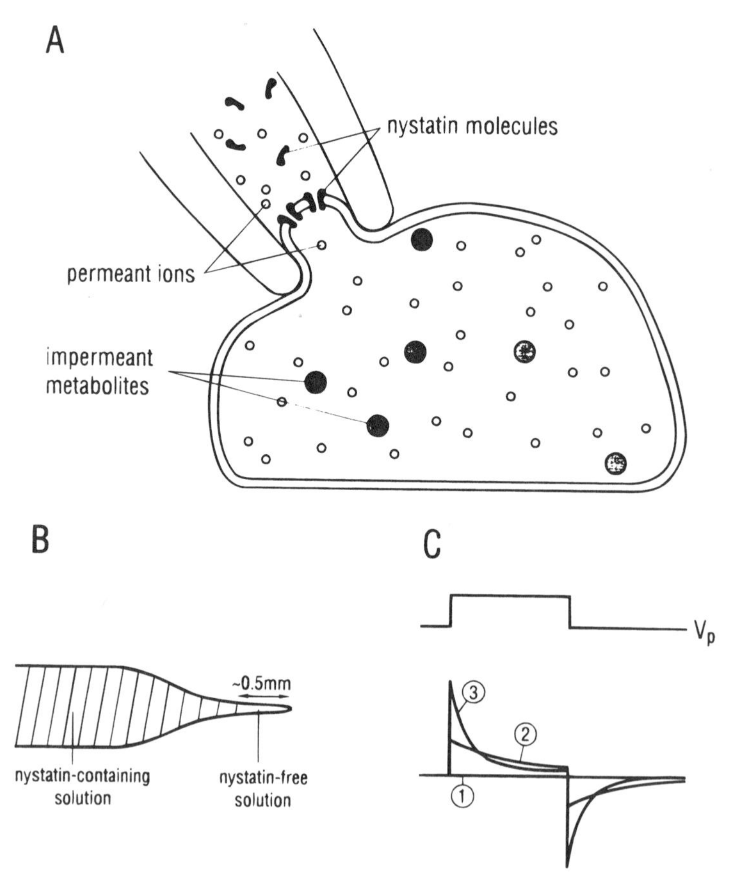

Figure 10. Perforated patch recording. A: Diagram illustrating the principle of the method. B: Pipette-filling procedure for perforated patch recording. C: Schematized current responses to voltage pulses during establishment of perforated-patch recording.

filled from the back with nystatin-containing solution. Bubbles can be removed in the usual way by gently tapping the pipette held upright. Practice has shown that this does not eliminate the nystatin-free region. Once the pipette is filled, it is secured in the pipette holder and approached to the cell in the usual manner. Two points have to be kept in mind at this stage, however. First, only very modest pressure should be applied to the pipette interior in order to avoid premature filling of the pipette tip with nystatin. Second, the approach should be fairly rapid for the same reason. Constant perfusion of the bath is helpful in avoiding accumulation of nystatin during the approach. Making the seal proceeds as usual by applying gentle suction to the pipette interior. As nystatin diffuses to the pipette tip, it gradually increases the pipette–cell conductance. This process can be conveniently monitored by observing current response to repetitive voltage steps (Fig. 10C). One should record sequentially responses such as ①, ②, and ③. The whole process may take 2–10 min. In ①, there are no nystatin channels in the patch. In ② and ③, there is an electrical communication between cell and pipette, so that the cell capacitor is loaded by the voltage jumps. The situation is the same as that described in Fig. 2, except that R_S should be replaced with the resistance R_P of the patch of membrane with its nystatin channels. (In fact, the resistance is $R_S + R_P$, because both resistances are in series. But R_P is usually 5 or 10 times larger than R_S, so that it is the predominant term.) Equation 3 can then be rewritten as $\tau = R_P C$. Thus, as R_P decreases during the course of the incorporation of nystatin channels, τ decreases (②→③). At the same time I_{in} increases, as predicted by equation 1.

Nystatin is a poorly water-soluble substance that tends to aggregate. It also has a series of double bonds, which makes it very sensitive to light. In view of these properties, the following precautions are mandatory. First, sonication is required both to make the nystatin stock solution (usually 50 mg/ml in DMSO) and the final pipette solution (final nystatin concentration 100 μg/ml). Second, the stock and pipette solutions have to be protected from light carefully. During recording, the microscope light should be turned off whenever it is not required.

Other substances can be used instead of nystatin. Amphotericin B, another polyene antibiotic, is structurally close to nystatin and can be used in a similar manner (Rae *et al.*, 1991). ATP is able to permeabilize certain cells by opening channels. It was used inside of patch pipettes in the earliest version of perforated patch recording (Lindau and Fernandez, 1986). Some specific recipes and rules for handling nystatin and amphotericin solutions are given by Zhou and Neher (1993) and in Chapter V of the *Axon Guide* (Sherman-Gold, 1993).

The ratio R_S/R_{in} of the series resistance during perforated-patch recording to the initial pipette resistance is usually 10–20, e.g., substantially larger than in standard WCR. This results in a comparatively slow voltage clamp. However, one advantage of perforated patch recordings is that the value of R_S is very stable, thus allowing a more reliable series resistance compensation than in standard WCR.

4.6. Changes in Effective Membrane Voltage during WCR

During WCR, potential changes occur at the pipette–cell junction and at the cell membrane. These changes arise from several mechanisms, which are reviewed in this section. They elicit "voltage shifts" in functionally important curves, such as activation or inactivation curves of voltage-dependent currents.

4.6.1. Liquid Junction Potentials

Liquid junction potentials, i.e., potential differences between two solutions, occur at all boundaries between solutions of differing compositions. They result from the charge separation that occurs when anions and cations of different mobilities diffuse across boundaries. With physiological solutions the magnitude of such junction potential ranges from a few up to about 12 mV.

In WCR particular attention has to be paid to the junction potential existing at the tip of the recording pipette because the pipette filling solution usually is not the same as the bath solution. Before seal formation the liquid junction at the tip typically is that of a K^+-rich solution in the pipette against physiological saline in the bath. If KCl is the predominant salt in the pipette, then the liquid junction potential is only about 3 mV, the pipette being negative with respect to the bath. If, however, a larger anion is selected, such as in a K^+-glutamate-based pipette filling solution, the anion will lag behind in the diffusion such that the pipette will be negative by about 10 mV. This is relevant for subsequent WCR measurement because it is standard practice to null the pipette current in this situation by adjusting a variable offset, so that the amplifier reads both zero current and zero voltage at the beginning of the measurement while the pipette interior actually is at -10 mV. In the WCR configuration, however, the liquid junction and the concentration gradients at the pipette tip no longer exist, particularly after diffusional equilibration between the cell interior and the pipette (for details see below). Then, the pipette potential and the cell interior will still be at -10 mV when the amplifier reads zero voltage. Therefore, all voltage readings of the amplifier have to be corrected by this amount. This correction is commonly referred to as the "liquid junction potential correction."

Liquid junction potentials can either be determined experimentally (Neher, 1992) or calculated from the generalized Henderson equation (Barry and Lynch, 1991). A computer program for calculating the corrections and illustrating their use is described by Barry (1993). The most difficult part is to convince oneself of the sign of the correction. Different strategies of handling the liquid junction potential correction and other offset problems are given in Chapter 6 (this volume). Provided that the bath solution is normal saline, the rule of thumb for WCR measurements is that the actual membrane potential is more negative than the amplifier reading if the dominant anion of the pipette filling solution is less mobile than the dominant cation, and vice versa for a less mobile cation.

4.6.2. Donnan Equilibrium Junction Potential

Junction potentials exist at any interfaces between different solutions and also at the interface between pipette solution and cytoplasm. However, the gradual exchange of ions between pipette and the cell interior may cause a gradual drift in this junction potential. Let us assume that the exchange of small ions (e.g., K^+ and Cl^-) is very fast compared to that of proteins and polyanions. After equilibration of the small ions, the pipette–cell system constitutes a Donnan equilibrium with an excess of immobile particles in the cell (the proteins and polyanions). We assume that these particles are negatively charged, so that the cell potential V_1 is more negative than the pipette potential V_0. As the particles diffuse out of the cells (a process that may last many minutes, see above), the difference between V_0 and V_1 subsides.

The pipette concentration of the cation (say K^+) at the onset of the protein exchange is given by the Boltzmann equation:

$$[K]_I = [K]_0 \exp(-V/U) \tag{10}$$

with $U = 25$ mV and $V = V_I - V_0$. Likewise, the initial concentration of the small anion (say Cl^-) is

$$[Cl]_I = [Cl]_0 \exp(V/U) \tag{11}$$

In addition, electrical neutrality requires that

$$[Cl]_0 = [K]_0 \tag{12}$$

and

$$[K]_I = [P]_I + [Cl]_I \tag{13}$$

where $[P]_I$ is the number of fixed charges (in Faradays) per liter.

Equations 10 to 13 give

$$u^2 + ([P]_I/[K]_0)u - 1 = 0 \tag{14}$$

with $u = \exp(V/U)$.

This last equation in u allows us to calculate V as a function of $[P]_I$ and $[K]_0$. In the extreme case in which all negative charges of the cell are immobile, $[P]_I = [K]_0$, and one finds $V = -12$ mV. In actual cases, however, $[P]_I$ is smaller than $[K]_0$, and V is accordingly smaller.

In summary, we expect the membrane potential to be more negative than the clamp potential at the beginning of the experiment. The difference V should be at most 12 mV and should gradually decline with a time course of minutes. Liquid junction potentials of approximately this magnitude were found to occur at the tip of conventional glass microelectrodes when filled with isotonic saline (Hironaka and Morimoto, 1979).

Figure 11 shows the results of an experiment designed to test these predictions. The pipette potential was held at -100 mV throughout the experiment. Sodium currents were measured at different times after the beginning of the WCR. Both peak-current and h_∞ curves showed a negative shift after 30 min of WCR (Fig. 10B). In Fig. 10C, the shift of the activation curve (compared to the curve obtained 1 min after WCR) is plotted as a function of time. The shift reached 9 mV in 30 min. Similar results were obtained in other experiments. Signs of a negative voltage shift were also found in the analysis of Ca^{2+} currents (see Fenwick *et al.*, 1982b, Fig. 17). In addition, we found in a previous study of Na^+ currents that a negative shift of 10–15 mV occurs rather quickly (within 1 or 2 min) just after formation of an outside-out patch (Fenwick *et al.*, 1982b). This shift is probably of the same nature as that illustrated in Fig. 11. The fact that it occurs more quickly for an outside-out patch than in a WCR is consistent with the view that it is linked to a diffusion process. In a Donnan equilibrium, permeant ions are at equilibrium, i.e., have a homogeneous electrochemical potential. Therefore, the above voltage shift does *not* apply for the reversal potentials of ion-selective pathways. Thus,

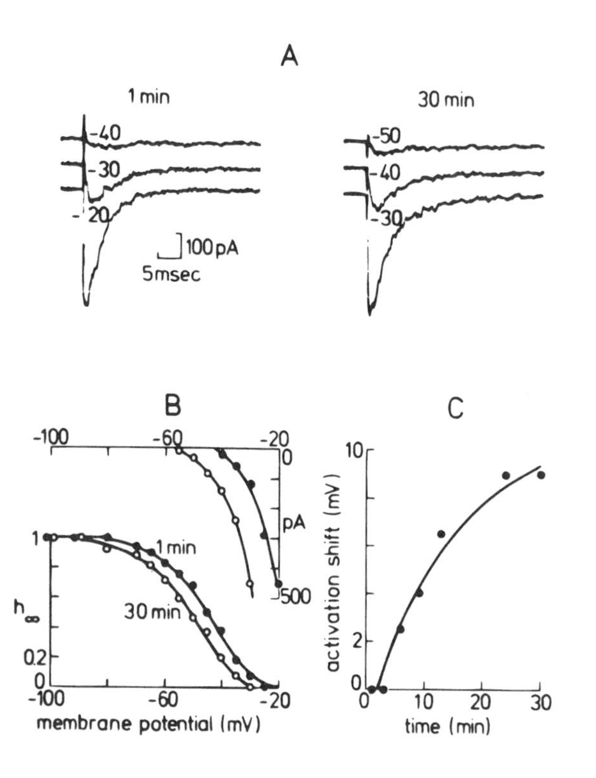

Figure 11. Slow potential shift in WCR. The pipette solution contained 140 mM KCl. The bath contained normal saline except that Ca^{2+} ions were replaced by Co^{2+} to eliminate Ca^{2+} currents. Positive pulses were given at 0.5 Hz from a holding potential of −100 mV. The amplitudes of the resulting Na^+ currents were measured at various times after the start of the WCR (taken as $t = 0$). Also h_∞ curves were obtained by measuring the current at −15 mV following an inactivating pulse 40 msec long. A: Currents at −40, −30, and −20 mV after 1 min of recording (left) and at −50, −40, and −30 mV after 30 min of recording (right). The two sets of records indicate a 10 mV negative voltage shift. B: Peak current (above) and h_∞ (below) curves obtained at 1 min (●) and 30 min (○). The curves indicate a negative shift of 9 mV and 6 mV, respectively. C: Voltage shift of the activation curve as a function of time. The shift was measured at the 300-pA level with comparison to the curve obtained at $t = 1$ min. The curve may be approximated by an exponential having a time constant of about 15 min. Cell capacitance $C = 5.4$ pF. Time constant of capacitive current $\tau = 25$ μsec.

in the case of Na^+ channels, the activation and inactivation curves are shifted, but the reversal potential is maintained.

It should be pointed out, however, that shifts along the voltage axis may occur from other effects. Thus, during the course of WCR, membrane proteins may undergo phosphorylation/dephosphorylation as a result of inevitable changes in cytosol composition. Such processes can change the apparent potential "felt" by these proteins according to the additional change of phosphate residues. Since these residues are negatively charged, the activation and inactivation curves of the phosphorylated channel are shifted toward positive potentials (Perozo and Bezanilla, 1990). In frog skeletal muscle, shifts in the activation curve of Na^+ channels were found to depend on the nature of anions, following the lyotropic series (Dani *et al.,* 1983).

5. Comparison of Whole-Cell Recording with Other Electrophysiological Methods to Record from Cells

Initially it was thought that a major limitation of WCR was that the requirement to form a tight seal necessitated a perfectly clean cell membrane. Many physiological preparations, where the cell of interest is covered by other cells or by extracellular material, seemed out of range. Recently, however, procedures have been developed that largely overcome this limitation (see Chapter 8, this volume). In view of these developments the comparison of WCR with traditional microelectrode recordings has gained some new interest.

In bovine chromaffin cells, WCR measurements give an input resistance of 1–10 GΩ and a resting potential of about −60 mV (Fenwick *et al.,* 1982a). This compares with values of up to 500 MΩ and about −60 mV with microelectrodes (Brandt *et al.,* 1976). Thus, WCR gives a higher resting resistance than conventional recordings. A similar discrepancy is often found in other preparations. It is often argued that the larger conductance seen with microelectrodes is genuine, because an artifactual leak conductance should bring the apparent resting potential close to 0 mV, whereas recorded resting potentials are similar with the two methods. Therefore, the low conductance seen in WCR is attributed to the loss of a physiologically important conductance as a result of washout. But this hypothesis is unlikely in chromaffin cells because the low input conductance is observed immediately on establishment of WCR, i.e., before any substance could have left the cell. Furthermore, the input conductance is similar in standard WCR and in perforated patch recording. As an alternative to the washout hypothesis, it can be proposed that the discrepancy is in fact caused by a leakage conductance around the site of impalement in microelectrode recordings. Along the leakage pathway, Ca ions enter the cell, where they activate Ca-dependent K channels. The combination of the additional Ca-dependent K conductance and the depolarizing leakage conductance yields an input conductance much larger than in WCR although the measured resting potentials are similar.

Recent WCR results in central neurons likewise yield input resistance values much larger than previous microelectrode measurements (e.g., hippocampal granule cells: Edwards *et al.,* 1989; Staley *et al.,* 1992; cerebellar Purkinje cells: Llano *et al.,* 1991). Such large values may have important functional implications. In a situation of low background synaptic activity, individual synaptic signals arising at distant dendritic locations will be effectively propagated to the soma by simple electrotonic spread. Such signals can therefore initiate a somatic action potential without requiring active propagation along the dendrite (Stuart and Sakmann, 1994).

References

Armstrong, C. M., and Chow, R. H., 1987, Supercharging: A method for improving patch-clamp performance, *Biophys. J.* **52:**133–136.

Barry, P. H., 1993, JP Calc, a software package for calculating liquid junction potential corrections in patch-clamp, intracellular, epithelial and bilayer measurements and for correcting junction potential measurement, *J. Neurosci. Methods* **51:**107–116.

Barry, P. H., and Lynch, J. W., 1991, Liquid junction potentials and small cell effects in patch-clamp analysis, *J. Membr. Biol.* **121:**101–117.

Brandt, B. L., Hagiwara, S., Kidokoro, Y., and Miyazaki, S., 1976, Action potential in the rat chromaffin cells and effects of acetylcholine, *J. Physiol.* **263:**417–439.

Dani, J. A., Sanchez, J. A., and Hille, B., 1983, Lyotropic anions. Na channel gating and Ca electrode response, *J. Gen. Physiol.* **81:**255–281.

Edwards, F. A., Konnerth, A., Sakmann, B., and Takahashi, T., 1989, A thin slice preparation for patch-clamp recordings from neurones of the mammalian central nervous system, *Pflügers Arch.* **414:**600–612.

Fenwick, E. M., Marty, A., and Neher, E., 1982a, A patch-clamp study of bovine chromaffin cells and of their sensitivity to acetylcholine, *J. Physiol.* **331:**577–597.

Fenwick, E. M., Marty, A., and Neher, E., 1982b, Sodium and calcium channels in bovine chromaffin cells, *J. Physiol.* **331:**599–635.

Hamill, O. P., Marty, A., Neher, E. Sakmann, B., and Sigworth, F. J. 1981, Improved patch-clamp techniques for high-resolution current recording from cells and cell free patches, *Pflügers Arch.* **391:**85–100.

Hironaka, T., and Morimoto, S., 1979, The resting membrane potential of frog sartorius muscle, *J. Physiol.* **297:**1–8.

Horn, R., and Marty, A., 1988, Muscarinic activation of ionic currents measured by a new whole-cell recording method, *J. Gen. Physiol.* **94:**145–159.

Jackson, M. B., 1992, Cable analysis with the whole-cell patch-clamp. Theory and experiment, *Biophys. J.* **61:**756–766.

Kleinberg, M. E., and Finkelstein, A., 1984, Single-length and double-length channels formed by nystatin in lipid bilayer membranes, *J. Membr. Biol.* **80:**257–269.

Korn, S. J., Marty, A., Connor, J. A., and Horn, R., 1991, Perforated patch recording, *Methods Neurosci.* **4:**264–273.

Lindau, M., and Fernandez, J. M., 1986, IgE-mediated degranulation of mast cells does not require opening of ion channels, *Nature* **319:**150–153.

Llano, I., Marty, A., Armstrong, C. M., and Konnerth, A., 1991, Synaptic- and agonist-induced excitatory currents of Purkinje cells in rat cerebellar slices, *J. Physiol.* **434:**183–213.

Mathias, R. T., Cohen, I. S., and Oliva, C., 1990, Limitations of the whole cell patch clamp technique in the control of intracellular concentrations, *Biophys. J.* **58:**759–770.

Neher, E., 1992, Correction for liquid junction potentials in patch-clamp experiments, *Methods Enzymol.* **207:**123–131.

Oliva, C., Cohen, I. S., and Mathias, R. T., 1988, Calculation of time constants for intracellular diffusion in whole cell patch clamp configuration, *Biophys. J.* **54:**791–799.

Perozo, E., and Bezanilla, F., 1990, Phosphorylation affects voltage gating of the delayed retifier K^+ channel by electrostatic interactions, *Neuron* **5:**685–690.

Pusch, M., and Neher, E., 1988, Rates of diffusional exchange between small cells and a measuring patch pipette, *Pflügers Arch.* **411:**204–211.

Rae, J., Cooper, K., Gates, G., and Watsky M., 1991, Low access resistance perforated patch recordings using amphotericin B, *J. Neurosci. Methods* **37:**15–26.

Ruppersberg, J. P., Stocker, M., Pongs, O., Heinemann, S. H., Frank, R., and Koenen, M., 1991, Regulation of fast inactivation of cloned mammalian $I_K(A)$ channels by protein phosphorylation, *Nature* **352:**711–714.

Sherman-Gold, R. (ed.), 1993, *The Axon Guide,* Axon Instruments, Inc., Foster City, CA.

Staley, K. J., Otis, T. S., and Mody, I., 1992, Membrane properties of dentate gyrus granule cells: Comparison of sharp microelectrode and whole-cell recordings, *J. Neurophysiol.* **67:**1346–1358.

Strupp, M., Quasthoff, S., Mitrovic, N., and Grafe, P., 1992, Glutathione accelerates sodium channel inactivation in excised rat axonal membrane patches, *Pflügers Arch.* **421:**283–285.

Stuart, G. J., and Sakmann, B., 1994, Active propagation of somatic action potentials into neocortical pyramidal cell dendrites, *Nature* **367:**69–72.

Zhou, A., and Neher, E., 1993, Mobile and immobile coalcium buffers in bovine adrenal chromaffin cells, *J. Physiol.* **469:**245–273.

Chapter 3

Guide to Data Acquisition and Analysis

STEFAN H. HEINEMANN

1. Introduction

This chapter should provide a first guide to the acquisition and analysis of patch-clamp data. Except for the sections on single-channel analysis, the procedures and considerations also hold for data obtained using other voltage-clamp methods. It is assumed that the reader is familiar with the standard methods and terminology of patch-clamp electrophysiology. Because many of the problems that arise during data analysis can be avoided by a proper design of the experiment, including data acquisition, we start by deriving some criteria that should be considered before actually starting to record data.

Performing electrophysiological experiments and, in particular, analyzing them are tasks made easier using a high degree of automation, which can be provided by an increasing variety of computer hardware and software. This means that, except for very simple applications, one has to decide what kind of computer system, including peripherals and software packages, should be used for the experiments. Because of the rapid turnover of hardware and software products it is impossible to provide a complete overview of the available components. Therefore a continually updated list of products, specifications, and vendors is deposited on a public-domain data base. Access to this data base is described in Chapter 1 of this volume. More information on specific products and analysis methods can be found in French and Wonderlin (1992), Dempster (1993), and in brochures and reference manuals such as *The Axon Guide* (Axon Instruments) from various vendors of patch-clamp hardware and software. The major aim of this chapter is to derive criteria for the development or purchase of acquisition and analysis software.

The next section illustrates what kinds of analysis tasks exist and how they can be accomplished with various software configurations. In particular, it is important to consider which features are essential for a successful experiment and whether these features should be supplied by dedicated acquisition and analysis programs or by general-purpose programs.

After a brief introduction to what should be considered during the acquisition of current data (Section 2), an overview of analysis procedures applied to single-channel and macroscopic current data is given. Only the basic principles, advantages, and disadvantages are discussed. For more detailed theoretical treatments and for numerical implementations the reader should refer to the cited literature and to the theoretical analysis chapters later in this volume.

STEFAN H. HEINEMANN • Max Planck Society, Research Unit, "Molecular and Cellular Biophysics," D-07747, Jena, Germany.
Single-Channel Recording, Second Edition, edited by Bert Sakmann and Erwin Neher. Plenum Press, New York, 1995.

Section 6 is a buyers' guide for computer hardware and software. The intimate connection between hardware and software, which poses considerable restrictions once decided on, is discussed. In particular, an introduction is given to the different approaches to solving software problems. These range from the purchase of ready-to-go programs for electrophysiology to the development of user software from scratch.

1.1. Levels of Analysis

The ultimate aim of the digital acquisition of electrophysiological data is to allow both quantitative and qualitative analysis of the biological system under consideration. Evaluating the change in shape of compound action potentials on alteration of experimental conditions is an example of qualitative analysis, whereas the determination of the equilibrium binding constant of a molecule to an ion channel based on the measurement of single-channel kinetics is highly quantitative. Although for the first example only data recording, timing, and display features are required, the second example demands more dedicated single-channel analysis functions.

In Fig. 1 an outline of various levels of analysis is given. This starts with the conditioning and acquisition of the data and is followed by display and preanalysis. The major part comprises analysis dedicated to the very specific problems arising in electrophysiology at the level of raw data or on parameters that were derived at earlier stages of analysis. At all levels analysis may have become so generalized that the use of multipurpose graphics or spreadsheet programs is possible. Alternatively, very specialized analysis tasks may arise

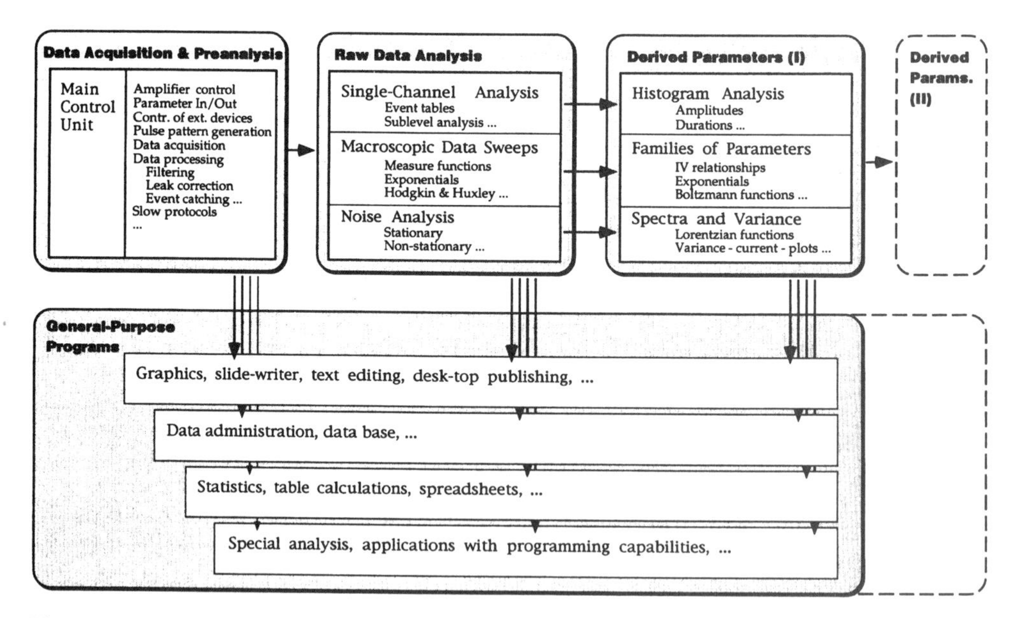

Figure 1. Schematic diagram illustrating the first levels of analysis. The arrows indicate data flow: horizontal arrows show data transfer in a highly specified format, allowing fast access and optimal information transparency as accomplished by sets of dedicated programs; vertical arrows represent the data flow in a less specified format (e.g., ASCII tables) to general-purpose programs, possibly at the expense of convenience, speed, and transparency.

whose solution cannot be implemented in standard software packages and that therefore also necessitate general-purpose programs or programming tools.

Besides those features that are absolutely essential for doing a certain kind of experiment, one should not underestimate the importance of program features that make life easier and which help to avoid errors. In general, the experimenter should have an overview of parameters and control functions that are necessary for decision making during an experiment or during the course of analysis. This includes that the experimenter should get the "right" impression of the data already recorded; i.e., one has to get a feel for the quality of the data. This can be achieved by proper display of the data traces and the structure of the recorded data file as well as by access to on-line analysis options that help one to decide whether to discard records right away or whether to continue with the experiment. Despite the usefulness of having information available at any time, it is very important to be able to "forget" about other parameters that are not currently so important but that may become so at a later stage. This is achieved by a proper configuration of the input parameters, which are stored in the background together with the data so that the experimenter does not need to worry about them during the experiment. Just consider the number of mistakes one can make if the gain setting of a patch-clamp amplifier has to be typed into the computer (or even noted on a piece of paper) after each change. Effort spent in automating such things and designing an appropriate user interface not only saves time but also reduces the error rate considerably and thus has a substantial impact on the success of the work.

Many problems in electrophysiology occur frequently and require specialized analysis features involving considerable programming. In such cases, dedicated analysis software is recommended. This has the advantage that it can be optimally adapted to the structure of the raw data, is commercially available, and is usually faster than general-purpose programs. In Fig. 1 dedicated software is grouped according to levels of complexity, starting with the analysis of raw data (e.g., current versus time). At the next level, parameters derived during primary analysis or transformed data (e.g., event histograms, power spectra) are analyzed further. More levels can be added as necessary. In the following sections some applications of raw data analysis, including analysis of single-channel events, of macroscopic relaxation experiments, and of current fluctuations, are discussed.

Another important consideration is the interface to analysis programs. One should try to use only a few programs that allow easy access to the data generated. This will not be a problem if there are dedicated analysis programs with a similar layout to the acquisition program and that support the same kind of data structure. The acquisition program would be responsible for the experiment control and the recording of pulsed or continuous data together with all parameters necessary to completely reconstruct the experimental configuration, including remarks made by the experimenter. One or several specialized analysis programs (e.g., single-channel analysis, pulse data analysis, noise analysis) would then read such data directly, thereby also providing access to the accompanying parameters. Particularly after specialized analyses, data should be exported as spreadsheet files in ASCII format or other more condensed formats to multipurpose curve-fitting or graphics programs for particular applications.

1.2. Analysis Starts before the Experiment

Signal theory provides a great variety of algorithms and methods for processing electrophysiological data and, in particular, eliminating or to compensating for distortions of the signals. As detailed in later sections, several methods have been developed over the years for the reconstruction of single-channel current events when the recorded signals are obscured by background noise or when too few data samples were recorded. However, these are always time-consuming and approximate methods. Similarly, a lot of effort is usually required—if it is possible at all—to reconstruct the exact experimental conditions at analysis time when one forgot to note the parameters, or if it is difficult to decipher the handwriting in a notebook. Yet another problem is the formating of data so that they can be recognized by several different analysis routines. Therefore, it is a good idea to plan an experiment or a group of experiments very carefully before starting to record data.

Such general experimental design should consider the kind of parameters to be measured and the precision required. The projected task will set limits to the experiment and to the analysis procedures that may not be immediately obvious. Important questions to ask are: Do the data actually contain the information needed for the derivation of the desired parameters? Are there software tools available to extract the desired information, and in which form do these software tools accept the data? The last, but not least important question is how conveniently data acquisition and analysis can be performed.

2. Data Acquisition and Preanalysis

Data acquisition and preanalysis are usually part of complex software packages that consist of main control units for regulating the program flow and thereby the execution of an experiment. In early versions of such programs, pulse pattern generation, stimulation, and the recording of current traces were the major tasks, but many more complex functions, such as versatile displays, on-line analysis, and analysis at later stages, have been added as the applications became more demanding.

2.1. Data Acquisition

Electrophysiological signals are usually transformed to analogue voltages by an amplifier unit. These voltages have to be recorded for display purposes and for later analysis. Analogue signals can be recorded using, for example, an oscilloscope, a chart recorder, or a frequency-modulated magnetic tape. However, because the data should ultimately be imported into a computer for analysis purposes, sooner or later the analogue signals have to be converted to digital numbers. In this section this analogue-to-digital conversion, the filtering of data, and the storage and retrieval of electrophysiological data are discussed.

The sampling of raw data and some signal processing, such as digital filtering and compression, constitute only a small part of acquisition programs. Issues that must be considered in this context are the maximal/minimal sampling speed, the maximum number of sample points that can be output and sampled at any one time, and whether and at what rate continuous data recording is supported.

2.1.1. Analogue-to-Digital Conversion

Because the digitization of data for internal representation on a computer is an important issue, one should consider this step very early in the planning stage of an experiment. Digitization is achieved using an *analogue-to-digital converter* (AD converter), which is connected to the computer bus. In some cases an AD converter is already built into the amplifier (EPC-9, HEKA Elektronik) or into a recording unit (pulse code modulator, PCM; or digital audio tape, DAT; see below).

2.1.1a. *Dynamic Range.* Usually AD converters accept analogue voltages in the range of ± 10 V, but, because the data will be manipulated using a digital system, AD converters are very often scaled for a maximum range of ± 10.24 V. Therefore, the signal output of the amplifier must not exceed these values. On the other hand, the signal should span as much of this voltage range as possible in order to increase voltage resolution. A measure for the efficient use of an electronic device (e.g., amplifier or AD converter) is the dynamic range, which is related to the ratio of the largest and the smallest signals that can be measured. Consider, for example, a patch-clamp amplifier that can register, at the highest gain (1000 mV/pA) and at a given bandwidth, signals as small as 10 fA. Because of the output limit of 10 V, the maximum measurable signal will be 10 pA. This gives a dynamic range of $20 \cdot \log(10 \text{ pA}/10 \text{ fA})$ dB = 60 dB (decibels). To increase the dynamic range, patch-clamp amplifiers have built-in gain functions. With a maximal signal of 200 pA (at the lower gain of 50 mV/pA), and the minimal signal as mentioned above, the amplifier with the reduced gain now has a dynamic range of 86 dB. For the measurement of even larger currents one would need to change the feedback resistor of the amplifier headstage, which in turn would reduce the resolution as a result of an increase in the current noise contributed by the feedback resistor.

Given the dynamic range of the amplifier, we now have to consider the voltage resolution of the AD converter. Usually, the converter represents a voltage level in the range of ± 10.24 V as a number composed of 8, 12, or 16 bits; 8-bit boards have now been largely replaced by 12-bit technology, leading to a resolution of 5 mV/bit. Modern converters offer 16 bits, but the two highest-resolution bits are usually obscured by instrumentation noise, such that they actually offer an effective 14-bit resolution (1.25 mV/bit). For some extreme applications, converters with even higher resolution can be used at the expense of sampling speed. An AD board with an effective 14-bit resolution provides a dynamic range of 84 dB $[20 \cdot \log(2^{14} - 1)]$, which is well matched to the dynamic range of patch clamp amplifiers.

Nevertheless, care has to be taken that the signal never saturates during data acquisition. One possible problem is the saturation of the amplifier. Although saturating low-frequency components are easily detected, saturation of very fast components is not so readily apparent. Some amplifiers, therefore, provide a clipping monitor and internal low-pass filters that can be set so as to avoid saturation. The aim then is to generate a signal at the output of the amplifier that lies within the range of ± 10.24 V and has the highest bandwidth possible without ever saturating the amplifier. This output is then filtered to accommodate the requirements of the AD conversion as specified below.

2.1.1b. *Aliasing.* Because the AD converter has only a limited sampling rate, the signal must also have a limited bandwidth. The sampling theorem (Nyquist, 1928) states that the sampling rate should be faster than twice the highest-frequency component within the signal. An ideal sine wave of 1 kHz, therefore, can only be sampled safely if the sampling frequency is greater than 2 kHz. Violation of this principle will cause a distortion of the signal, which is commonly called *aliasing,* as illustrated in Fig. 2. In the frequency domain, this is equivalent to "folding" of higher-frequency components into the frequency range accessible by the sampling device. Put more simply, such a distortion is equivalent to the appearance of low-frequency beating when one combines two high-frequency tones that are not of exactly the same frequency. The problem of aliasing can be quite serious and has to be solved by correct low-pass filtering of the data, i.e., by the elimination of high-frequency components from the signals before AD conversion.

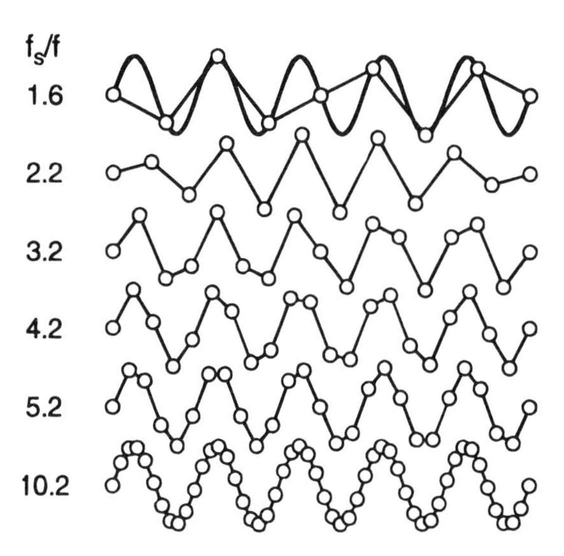

Figure 2. A sine-wave signal of frequency f (thick curve) is sampled at a frequency f_S. The sample points are indicated by circles, which are connected by straight lines. Although the sampling theory defines 2 as a minimum for the sampling factor, f_s/f, it is clearly seen that a reliable reconstruction of the signal is only possible with f_s/f greater than 4 if only a linear interpolation method is used. For small sampling factors not only is the signal reconstructed incorrectly, but inappropriate low-frequency signals also become apparent, an effect referred to as aliasing.

2.1.2. Filtering

Low-pass filtering of electrophysiological signals basically serves two purposes: the elimination of high-frequency components to avoid aliasing and the reduction of background noise in order to increase the signal-to-noise ratio.

The characteristics of low-pass filters are specified by a corner frequency, a steepness, and a type. The corner or cutoff frequency is defined as the frequency at which the power of the signal falls off by a factor of 2; i.e., the amplitude decreases by $1/\sqrt{2}$. This corresponds to an attenuation of the amplitude by -3 dB. This corner frequency may deviate from what is written on the front panel of a filter because some manufacturers use different definitions for the corner frequency. If there is uncertainty about the corner frequency of a certain filter, one should verify it with a sine-wave generator by feeding a sine wave of fixed amplitude into the filter and increasing the frequency until the amplitude of the filter output decays to $1/\sqrt{2}$ of the input signal.

The steepness of a filter is given in dB/octave, i.e., the attenuation of the signal amplitude per twofold increase in frequency (steepness given in dB/decade reflects the attenuation per tenfold increase in frequency). The steepness of a filter function is also characterized by its order, or equivalently by the number of poles. A four-pole low-pass filter has a limiting slope of 24 dB/octave or 80 dB/decade. Commonly used filters have four poles; eight-pole filters are more expensive but also more efficient. In addition, higher-order filters are better approximated by Gaussian software filters, which makes the theoretical treatment during analysis easier (see Chapter 19, this volume). Given the characteristics of a low-pass filter, one can easily see that a certain fraction of the input signal spectrum that exceeds the corner frequency will still be output by the filter. Therefore one has to use a sampling rate greater than the Nyquist minimum (twice the corner frequency). The ratio of the sampling frequency and the corner frequency of a signal is called the oversampling factor. For most applications an oversampling factor of 5 is sufficient if an eight-pole filter is used. If the exact waveform of a signal is to be reconstructed during analysis without interpolation, an oversampling factor of 10 may be more appropriate (see Fig. 2). Figure 3 illustrates how the steepness of a filter at a given corner frequency affects the response to a step input pulse in the time

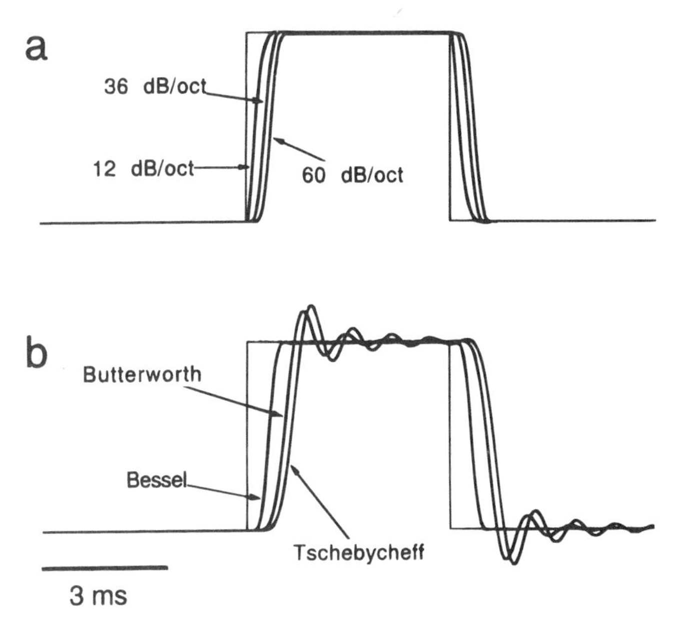

Figure 3. Analogue filtering of a square pulse of 5 msec duration. Data were sampled at a rate of 100 kHz after filtering at 1 kHz with the indicated characteristics. a: Bessel filter with a steepness of 12, 36, and 60 dB/octave. b: Bessel, Butterworth, and Tschebycheff filters with a steepness of 48 dB/octave. Bessel filters introduce the least ringing and are therefore used for single-channel analysis (see Section 3). Butterworth and Tschebycheff filters are inappropriate for single-channel analysis. However, they produce a steeper attenuation of signals outside the pass band and are therefore used for noise analysis (see Section 4).

domain. The steeper the rolloff of the filter function, the less of the high-frequency component will pass the filter; as a result, the onset after the start of the pulse is smoother, and the introduced filter delay is increased.

The most common filter types used in the processing of biological data are Bessel (for time-domain analysis) and Butterworth (for frequency-domain analysis). Bessel filters are most important in the time domain because at a given steepness they cause minimal overshoot of the output signal if the input was a step function (see Fig. 3b). In addition, Bessel filters preserve the shape of signals by delaying all frequency components equally. This is particularly important for the analysis of single-channel current events (see Section 3). For noise analysis in the frequency domain, Butterworth filters and Tschebycheff filters are used. They provide a more efficient attenuation in the frequency domain above the corner frequency but cause considerable ringing in a step response (see Fig. 3b).

So far only low-pass filters have been discussed. Similar considerations hold for high-pass filters that only attenuate low-frequency components, as required for the elimination of slow signals such as drifting baselines. They are used, for example, for noise analysis where only the fluctuating component of the signal is important and slow drifts in the mean current would cause signal saturation. These filters behave as AC coupling devices with an adjustable frequency response. In some cases a combination of both low- and high-pass filters is required. Such band-pass filters are used, for example, to set a frequency window in an amplifier in which the rms noise is measured. A filter that attenuates only a certain frequency band is called a band-reject filter. Very sharp band-reject filters centered around 50 or 60 Hz can be used to eliminate signal distortions caused by line pickup.

As shown above, digital filtering is used mostly for display and analysis purposes. However, if the computer and its peripherals are fast enough, one can always sample at the maximal rate using a fixed antialiasing filter before AD conversion and then use a digital filter for analysis at a specified bandwidth. To reduce the amount of data stored to disk, data compression can be performed after digital filtering if no important information is expected in the high-frequency range.

2.1.3. Control of Experiment Flow

An acquisition program should by today's standards not only record data but also provide a versatile toolbox that enables an experiment to be designed in advance. This is achieved in part by allowing the storage and retrieval of program configurations that meet specific requirements and by macro programming facilities, i.e., "recording and replay" of frequently repeated program events. Important features include parameter input from external devices as well as from the amplifier. In particular, the settings of the amplifier should be recorded as completely as possible so as to reconstruct the experiment at analysis time as precisely as possible.

2.1.4. Pulse Pattern Generation

The most essential part of a program for pulsed data recording is a pulse pattern generator, i.e., the program that sets up the pattern of the voltage-clamp command to be delivered to the amplifier. There should be the possibility of linking various pulses to form a family and linking individual families together to create complicated stimuli necessary, for example, to study the activation and inactivation behavior of voltage-dependent ion channels.

In addition to pulse patterns for the amplifier command voltage, other external devices such as valves or flash lamps can be controlled by pulse generators. This is achieved either by setting digital trigger pulses at specific times during the main pulse pattern or by supporting more than one analogue output channel with separate but synchronized voltage patterns.

2.1.5. Display of Traces and Relevant Information

Because of the inherent instability and variability of biological preparations, an electrophysiological experiment is a highly interactive process; decisions often have to be made during the course of a measurement. It was mentioned above how important it is to have facilities that allow the configuration of stereotypic tasks. This is no contradiction to the required flexibility, since such simplifications give the experimenter time to concentrate on the measurement, to judge incoming data, and to decide what to do next. Therefore, the presentation of data during an experiment is an important issue. Fast graphics allow the display of current traces on the computer screen almost in real time, mimicking an oscilloscope. Additional features make such display procedures superior to an oscilloscope, for example, display of traces with/without leak correction and with/without zero-line subtraction, display of leak currents and other traces that were recorded in parallel, the ability to store previously recorded traces, and comparison of incoming traces with a specified reference trace. Despite these advantages, however, a high-speed oscilloscope cannot be entirely replaced by a computer, for the traces as displayed on the screen give only a representation of the real data, depending on the coarseness of the AD conversion and the resolution of the computer display. As a result, one may not always see very fast spikes because of capacitive transients and may therefore introduce errors through signal saturation.

Besides having the current traces being displayed as realistically as possible, it is important to have access to the values of other experimental parameters, which are stored together with the raw data. Because the requirements for external parameters may vary considerably depending on the kind of experiment, it is best if only those parameters actually of interest are brought to the attention of the experimenter. The importance of data documentation cannot be underestimated. Acquisition and analysis programs should therefore provide text input for individual sweeps, families, groups of families, and the entire data file. Marks at individual data sweeps, e.g., indicating blank sweeps in nonstationary single-channel recordings, are very helpful and save a lot of time during later stages of analysis.

Increasing computer performance and decreasing data storage costs encourage the tendency to store more data than actually required. Worst cases are the storage of obviously redundant data and "bad" data that do not contain any valuable information. Although one can always say that such data can be deleted at a later stage, it should be stressed that it becomes increasingly difficult to extract a fixed quantity of information from an increasingly large data set. In other words, it takes more time to find a needle in a haystack the larger the haystack is. For this reason, acquisition and analysis programs should provide access to data recorded during the experiment to facilitate selective data deletion, compression, and averaging.

2.1.6. On-Line Analysis

The above-mentioned display features already perform a kind of data analysis. More quantitative analysis can be supported by dedicated analysis routines that process incoming data traces according to specified procedures. Examples of such on-line functions are the determination of values such as minima and maxima, time to peak, mean, variance, etc. These parameters will be determined within a specified time window conveniently set with respect to a relevant segment of the data trace, as specified in the pulse generator. The analysis results can then be displayed as numerical lists or as plots. In the latter case, several abscissa options allow the versatile generation of on-line analysis plots. For example, in a steady-state inactivation curve, one would determine the peak current within the constant test-pulse segment; the results would then be plotted as a function of the voltage of the conditioning prepulse segment, which was previously specified as a relevant abscissa segment in the pulse generator.

2.2. Data Storage and Retrieval

Long traces of single-channel recordings take up a lot of disk space and are therefore often stored in an inexpensive medium before analysis is performed section by section using a computer. Analogue signals can be stored on analogue magnetic tapes in a frequency-modulated format. Nowadays, however, signals are more often digitized before storage by a PCM and then stored on video tape using a conventional video cassette recorder (VCR)

(Bezanilla, 1985) or on digital audio tapes using modified DAT tape drives (Fig. 4a,b). These media are inexpensive and capable of storing sampled data at rates of up to 100 kHz for more than an hour. Alternatively, one can read data into the computer and transfer them directly to an inexpensive storage medium using a DAT drive hooked up to the computer. Optical WORM (write once, ready many times) disks are still rather slow but offer, like other disks, random-access capability. In most cases storage on exchangeable disks, which can be accessed like a conventional hard disk, is more practical because of shorter access delays. Hard disk cartridges or magnetooptical disks have come down in price considerably, such that they are now a feasible alternative for mass storage of electrophysiological data. For high-speed requirements these media may still be too slow, and a large hard disk may be required as an intermediate buffer.

The choice of long-term storage medium is mainly determined by the amount of data that has to be stored. If one wants to store single-channel data filtered at 10 kHz at a time resolution of 44 kHz (typical in PCM/VCR systems), one could store data for the entire running time of a video tape (e.g., 90 min). If the data are stored as 16-bit numbers (2 bytes), the sampling process amounts to a data flow of 2 bytes · 44,000/sec = 88 kbytes/sec. This

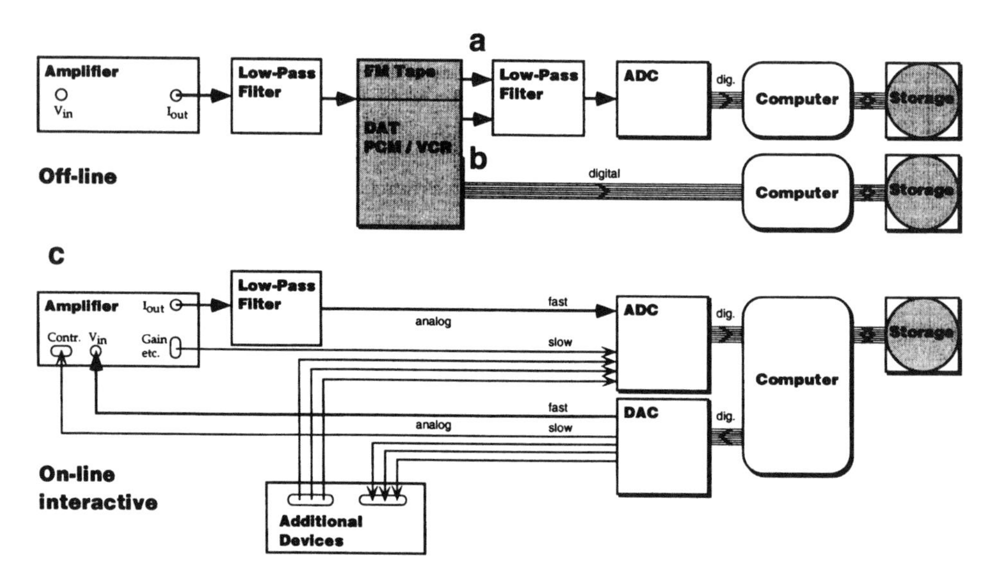

Figure 4. Possible configurations for the acquisition of patch-clamp data. In a and b the current monitor signal from the patch-clamp amplifier is passed through an antialiasing filter appropriate to the sampling frequency of a tape-recording device (FM tape, digital audio tape, or video tape). a: For off-line analysis data are replayed from the tape, passed through a filter according to the sampling rate of the ADC board, and read into a computer. b: Alternatively, using DAT drives or PCM/VCR combinations, the digitized signals can be transferred to the computer directly in order to reduce errors from a second digitization as in a. c: Configuration for an interactive, semiautomated acquisition system. The computer controls the patch-clamp amplifier and other external devices via a DA interface. The analogue signals may be grouped into fast signals (e.g., command voltage) and slow signals (e.g., setting of amplifier gain, perfusion valves). Similarly, analogue signals are acquired on a fast (e.g., current traces) and a slow (e.g., amplifier gain, filter setting, temperature) time scale. Further automation is achieved with the EPC-9 patch-clamp amplifier (HEKA Elektronik, Lambrecht, Germany) in which all functions are directly controlled by digital connection to a computer. In c, long-term data storage capabilities are mandatory, whereas in a and b the tapes can also be used as long-term storage media. Only selected sections to be studied more thoroughly at a later stage would be stored to disk or to another computer memory.

means that a recording of 13.6 sec could be stored on a floppy disk (1.2 Mbyte). An exchangeable hard disk (80 Mbytes) would be full after 15 min, a magnetooptical disk (250 Mbyte) after 47 min, and a DAT streamer tape (2 Gbyte) after 6 hr 18 min.

Low-activity single-channel recordings can be compressed considerably if only events of interest are stored. For this purpose data are sampled, passed through an event catcher, and stored selectively. The same kind of software can be adapted to the recording of spontaneous synaptic potentials or currents, for example.

Short sections of data, and especially those recorded under varying conditions, such as during a voltage-step experiment, should be stored directly on the computer's hard disk (see Fig. 4c). This is the easiest way of solving the problem of exact timing. Several software packages are available, depending on the hardware, that provide the user with many tools for keeping track of the experiment and for data analysis.

2.3. Interface to Other Programs

The kernel of an acquisition program not only controls the flow of an experiment but also defines the data structure. Because the tasks and approaches considered during the development of acquisition software vary, most programs generate their own specific data structure. Analysis programs should therefore recognize a variety of structures for data import; however, this is rarely the case. Programs that convert one data structure into another are available from several companies, but the same problems arise if the information contained in the two data structures is not compatible. It is usually possible to convert important information such as raw data traces, timing, and gain, but the conversion of accompanying information and the exact reconstruction of the pulse pattern used to evoke the stored data are often complicated or simply impossible. Therefore, it is advantageous to use dedicated analysis software (see below) from the same source as the acquisition program. Sometimes, however, this will not be possible because some specific analysis features may not be supported or may prove to be insufficient. In such cases, and for interfacing to general-purpose programs (see below), acquisition programs must allow data output in a more general format, for example, as text tables (ASCII).

3. Single-Channel Analysis

Current recordings from membrane patches with only one or a few active ion channels are suitable for the analysis of opening and closing current events. A direct transformation of the current signals, as a function of time, into amplitude histograms on a sample-to-sample basis can be used to obtain an overview of the single-channel records. Evaluation of the peaks in the histograms enables the number of single-channel current levels or the existence of sublevels to be determined.

The major task of single-channel analysis programs is to compile event lists. For that, single-channel current events have to be detected, quantified, and stored. Before the actual event analysis can be performed, the data have to be filtered so that they are suitable for this type of analysis. The output of an event-detection program is an event table that contains for each transition at least a level index, an amplitude, and a duration.

At the next level of analysis, information has to be selectively extracted from these event tables. Typical applications are the compilation of amplitude and duration histograms for certain event transitions or levels, respectively. Such histograms can be compiled and displayed in various ways and are used to fit model functions, such as Gaussian curves for amplitude histograms or multiexponential functions for lifetime histograms, to the data.

Single-channel analysis packages are commercially available for various types of computer hardware. Several other programs of this kind have been developed in various laboratories but are not commercially available. Great differences in the quality and capabilities of these programs make necessary a very careful evaluation of the program specifications or of a demo version before purchase. The following topics may help to define the requirements of the features provided by the programs.

3.1. Data Preparation

3.1.1. Digital Filtering and Data Display

Programs should take in continuous as well as pulsed single-channel data. For display and event detection, data have to be presented on the computer screen after passing through a digital low-pass filter. Usually Gaussian filters are used for this purpose (see Fig. 5) because they resemble a Bessel characteristic of high order, which is mandatory for single-channel analysis (see Fig. 3). In addition, they can be described in a mathematically compact form (see Chapter 19, this volume) and therefore can be implemented in software that executes relatively fast. Nevertheless, digital filtering demands a lot of processor time, and fast computers are therefore preferred.

Individual channel transitions have to be inspected at high time resolution, and this can result in a loss of overview when one is analyzing a long data record. Therefore, two or three simultaneous data representations at different time scales are very helpful.

At the highest time resolution, data points may appear very sparse on the screen. For a smoother appearance and for a careful analysis of single-channel records, the individual data points have to be connected using some kind of interpolation. The easiest method is to connect the values by straight lines. A better approach is to use more complicated spline functions such as cubic polynomials (see Chapter 19, this volume).

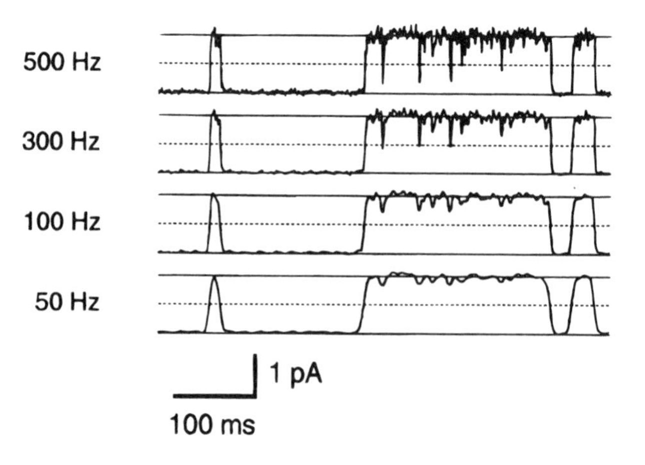

Figure 5. Single-channel current events were recorded at a rate of 2 kHz and then passed through a digital Gaussian filter with the indicated corner frequencies. With an appropriate filter setting, the brief closures during the long channel openings can be reduced in amplitude such that they do not cross the 50% threshold (dotted line).

3.1.2. Leak Correction

If single-channel events are elicited by voltage pulses, the related capacitive current transients and leak currents have to be compensated before actual single-channel analysis can be done. During the recording of nonstationary single-channel records with low activity, data sweeps without channel openings can be averaged and used as a background reference. The average of these "null" traces can then be directly subtracted from the individual data sweeps. If the noise of the averaged leak records is too large, it can be idealized by fitting theoretical functions to it. Usually polynomials or summed exponential functions are quite appropriate.

3.2. Event Detection

For single-channel analysis, an event is defined as a sudden change in current as the result of the opening or closing of an ion channel. Once detected, such an event has to be characterized and stored in an event table, which is then used for statistical analysis of channel currents (amplitudes) and channel kinetics (durations). An event has to be characterized by at least two parameters: (1) an amplitude and (2) a time (e.g., time when the amplitude reaches the 50% level). In order to facilitate analysis, one also notes a current level index, which specifies how many channels are open after (or before) the event happened (e.g., 0 = all channels closed, 1 = one channel open). Besides these "regular" events one can introduce "special" events such as sublevels relative to a normal channel opening. This would mean that a program for event table analysis could extract all sublevel durations as well as all main level durations with and without sublevel contributions. Additional information might include the current variance and an indication of how the event was detected and how the amplitude was determined (e.g., amplitude determined manually or automatically, or amplitude taken from the previous event). In order to avoid the effects of drifting baselines, two amplitudes can be stored for each event in addition to the durations used to determine them. Particularly for nonstationary single-channel event tables, the event timing must be given with respect to the start of the corresponding stimulus or to a specified pulse segment.

The main part of single-channel analysis programs consists of event-detection algorithms. Several strategies are followed in commercially available programs. Most of them rely on automatically or manually predefined baseline currents and single-channel current amplitudes; transitions are then determined as the crossing of certain critical current thresholds. Automated methods often make use of variance measurements in sliding windows in order to obtain objective criteria for threshold crossings.

3.2.1. Filtering for Single-Channel Analysis

For stationary and nonstationary recordings the hardest task is choosing the right bandwidth for analysis. This is largely determined by the signal-to-noise ratio of the records; i.e., to what extent does background noise result in the detection of false events? On the other hand, too narrow a bandwidth may cause short channel events to be missed. Because the latter aspect can only be judged if one knows what the signals look like, it may be necessary to perform a preliminary single-channel analysis at an estimated bandwidth. The distribution of the measured events will then provide an estimate of how many events were missed, and an optimal bandwidth can then be chosen (see Chapter 19, this volume).

3.2.2. Threshold-Crossing Methods

The most popular method for event detection uses a threshold that is halfway between the open and the closed current levels (see Fig. 5). These "50%-threshold" methods are easily implemented and do not require correction of event durations as long as the current reaches the full channel level, because the effect of filter delay on the time lag between the actual channel transition and the measured time at which the current signal crosses the threshold is the same for openings and closings. For events that do not fully reach the next level, a correction has to be applied (see Chapter 19, this volume). One can test the correction methods applied in single-channel analysis programs by creating artificial square-shaped single-channel events and filtering them with a filter rise time of approximately 75% of the event width. The events are then analyzed with a 50%-threshold criterion by setting the correct single-channel amplitude. An underestimation of the actual event width is an indication that the correction has not been properly implemented.

In nonstationary recordings the timing of the events has to be correct with respect to the start time of the sweep in order to yield correct first-latency intervals. All detected events therefore have to be shifted to the left by the delay introduced by the system response, including that of the digital filter used for analysis.

Threshold-crossing methods with higher than 50% levels can be used for very noisy data, although stronger filtering would be better. Levels of less than 50% can be necessary for the detection of events in very low-noise data acquired at the maximum attainable rate. Such a limitation may arise if a sampling device cannot take in data at a rate high enough for the actual time resolution of the current recording. Suppose, for example, that one wants to record single-channel events with an amplitude of 20 pA on a PCM/VCR combination with the maximum rate of 44 kHz, and the maximum corner frequency of the steep low-pass filter to be used is 10 kHz. At this bandwidth the rms background noise could be 500 fA. Thus, a detection threshold could be safely set to eight times the standard deviation, i.e., 4 pA, corresponding to a 20% threshold criterion.

3.2.3. Time-Course Fit

Threshold-crossing methods yield good results if the single-channel events are of homogeneous amplitude. When, for example, brief flickers during a burst of activity represent full channel closures, threshold-crossing events can be safely converted into event durations. Ambiguities arise if one is not certain whether the events really are of "full" amplitude, i.e., if they are too short for satisfactory amplitude determination. In such cases an idealized square-shaped event for which the amplitude and duration are allowed to vary is fitted to the data in order to enhance the resolution and reliability of the analysis (Colquhoun and Sakmann, 1985; Colquhoun, 1987; Chapter 19, this volume). For the creation of an idealized channel event, the effect of filtering according to the system transfer function must be taken into account. The time course of a step response can therefore be approximated by Gaussian-filtered square events; alternatively, measured step responses can be used. Note that this method is usually more time-consuming than simple threshold-crossing methods.

3.2.4. Automatic Data Idealization

Single-channel event detection can be automated in several ways. A kind of semiautomated method is to measure several single-channel events manually, to get an idea of the single-channel amplitude, and then to use this estimate to define a threshold-crossing criterion for an automatic search routine. Starting from a baseline segment, this would identify the first transition, measure the amplitude after the transition if the open duration is long enough, reset the amplitude of the open channel to the measured value, and continue to search for the next event (closing of the present channel or opening of another one). Sublevel events and fast flickers can cause serious problems with such algorithms. Therefore, one may use automatic methods of this kind only if (1) the data are ideally suited or (2) the experimenter observes the automatic process and interrupts it if the algorithms start to catch false events. In either case, it is very helpful if already stored data idealizations (i.e., contents of the event lists) can be superimposed on the measured data at any time. In this way the experimenter can reconfirm that the automatic algorithms yield sound results.

Several other algorithms have been developed that use the mean current and the variance in sliding windows for the detection of channel transitions. Such edge detectors can be iteratively applied to the raw data so as to optimize event detection (e.g., Sachs *et al.,* 1982; Kirlin and Moghaddamjoo, 1986; Moghaddamjoo *et al.,* 1988; Pastushenko and Schindler, 1993).

3.2.5. Maximum-Likelihood Methods

Very rigorous methods can be applied that directly maximize the likelihood of a kinetic model to describe a certain data set under consideration. Such methods use hidden-Markov algorithms and are computationally very expensive (Chung *et al.,* 1990, 1991; Fredkin and Rice, 1992; Auerbach, 1993). Since neither event detection nor compilation of dwell-time histograms is required, they promise great savings in analysis time spent by the experimenter.

3.2.6. Drifting Baseline Problems

Most of the methods mentioned above work well within the limits of their time resolution if the baseline current does not change. In many applications, however, a change in seal resistance during a long recording period cannot be avoided. The user therefore has to ensure that during single-channel analysis the actual channel-closed period is taken as the baseline. This is straightforward if the channel activity is not too high and if there are enough long-lived closed periods. Then the current levels of periods between channel openings can be used frequently to determine the new baseline (e.g., as the mean or the median of a specified number of data points). Baselines between two baseline determinations then have to be interpolated. There are also more involved methods for automatic baseline tracking (e.g., Sachs *et al.,* 1982), but the user should always verify that the algorithms did not mistake a long-lived open channel state or a substate for a new baseline.

3.3. Analysis of Histograms

After the compilation of event tables, specific information such as single-channel amplitudes and open and shut periods have to be extracted for further analysis. For display purposes, and also for comparison with theoretical predictions, the collected event information is displayed in the form of histograms. The abscissa of a histogram is divided into intervals (bins) of event amplitude or event duration for amplitude or lifetime histograms, respectively.

The number of events within a certain observation period that fall within the individual bins are counted and displayed as bars.

3.3.1. Display of Histograms and Binning Errors

There are several methods for displaying dwell-time histograms. A straightforward way is to use linear scaling for durations and the number of entries per bin. Since dwell-time distributions are usually sums of exponential functions, kinetic components are more easily appreciated if an exponential time base is used (McManus *et al.,* 1987). A very useful presentation of dwell-time histograms displays the square root of events in bins of exponentially increasing width (Sigworth and Sine, 1987; Jackson, 1992). In the case of a single-exponential distribution the *probability density function* (pdf) then peaks at the time constant, and all bins have the same theoretical scatter throughout the entire time range.

Several errors can be introduced when a histogram is constructed. Some of them are related to the discreteness of the sampled signals. If the bin boundaries are not multiples of the smallest resolvable unit (current or time, respectively), some bins may have a higher chance of being populated than others. In amplitude histograms this problem is largely eliminated if single-channel currents, based on the average of many sample points, are stored as real numbers. For dwell-time histograms this error can be serious if the exact timing of the channel-open or -shut times is based on individual samples. The problem becomes minimal if an interpolation method (e.g., a cubic spline) for the exact determination of the threshold crossing is used.

Noise in a signal that is to be displayed as histograms also causes a distortion of the results, particularly if the event distribution is highly nonlinear. Such an effect can be accounted for but is usually of minor importance (Chapter 19, this volume).

Another kind of binning error arises for large bin widths when one uses the center of the bins to compare heights directly with theoretical probability density functions (McManus *et al.,* 1987). This problem can be avoided by following a more rigorous approach using probability distribution functions (e.g., Sigworth and Sine, 1987).

3.3.2. Fit of Theoretical Functions to Histograms

There are various methods for fitting theoretical functions to histograms. They range from very simple least-squares fits to the histogram bins to the use of maximum-likelihood methods (Chapter 19, this volume; Magleby, 1992) on bins as well as on the events themselves. In the latter case, binning errors are completely eliminated. For the optimization itself various algorithms, such as simplex, steepest descent, or Levenberg–Marquardt methods, are used. For discussion of these methods and the estimation of error bounds, see e.g., Dempster (1993). An important issue for the fitting of binned distributions is the weighting of the data points. Considerable improvements over fits of linear dwell-time histograms are yielded by fitting histograms with logarithmically scaled bin widths (e.g., McManus *et al.,* 1987; Sigworth and Sine, 1987).

3.3.3. Compilation of Amplitude Histograms

The compilation of event amplitude histograms by specification of event level range and bin width is implemented in most single-channel analysis packages. An important requirement for analysis programs, however, is that they allow the selection of single-channel transitions under certain conditions in order to enhance the accuracy of analysis and to restrict analysis to subsets of the data. Here are several criteria that might be considered for the extraction of entries from event tables:

- *Event level.* Events from the baseline to the first open level can be measured most precisely because the noise level increases as more channels are open. Thus, for the compilation of an amplitude histogram, one may want to discard transitions between higher levels than the first one if there are enough events of this kind available.
- *Event duration.* Only if the duration of the closed and open time before/after the transition is long enough can a precise amplitude measurement be achieved. The precision of an amplitude histogram is therefore increased if channel events shorter than a certain duration are discarded.
- *Detection method.* In some programs both automatic event detection and amplitude measurement are supported. One might select for an amplitude histogram only those events that were manually measured or at least visually validated by the experimenter. In such cases one does not rely on events that were measured by an automatic algorithm in a possibly inappropriate way.
- *Time range.* In stationary single-channel analysis, one might want to compare events from early and late parts of the recordings in order to check for drift phenomena or effects such as the shifting of gating mode (e.g., Zhou *et al.,* 1991). In nonstationary recordings, separation of events from early and late phases of individual pulses might help to separate kinetically distinct channel components.
- *Sublevels.* If sublevels were marked as "special events" (see Section 3.2), there must be a separate way of selecting them. In order to address the question of when a sublevel occurs preferentially, it is helpful if one can extract them conditionally by specification of the previous/next event before/after the sublevel.

3.3.4. Gaussian Distributions

Usually one or more Gaussian distributions, each characterized by a mean value, I_0, a variance, σ^2, and an amplitude, a, are fitted to the histograms manually or by least-squares or maximum-likelihood methods.

$$n(I) = \sum_{i=1}^{n} \frac{a_i}{\sqrt{2\pi}\,\sigma_i} \exp\left(\frac{-(I_i - I_{0i})^2}{2\sigma_i^2}\right) \tag{1}$$

Deviations from Gaussian functions are discussed in Chapter 19 (this volume). Similarly to dwell time histograms, maximum-likelihood methods can be employed to fit functions to individual events rather than to the histogram bins.

3.3.5. Compilation and Display of Dwell-Time Histograms

As for amplitude distributions, analysis programs should provide versatile tools for the extraction of dwell times from event tables. Here are several criteria that might be considered for the selection of events:

- *Event level.* Open and closed times can be extracted from event tables. This is trivial if only one channel is active during the recording period. If multiple openings occur, however, errors are introduced for open times if they are skipped or if the superposition of two channel openings is counted as one.

- *Event amplitude.* If two clearly distinct single-channel amplitudes were measured, then the kinetics of one of the components can be characterized separately if amplitude ranges are set for the extraction of events from the event table.
- *Event duration.* For duration histograms the influence of false events resulting from background noise can be reduced if a minimum event duration is set.
- *Burst events.* If an ion channel has two closed states with clearly different dwell times, openings can appear as bursts. Approximate histograms of burst lengths and gaps between bursts can then be compiled by setting a minimum dwell time for a channel closure to be accepted. Given such a minimum gap time, event tables have to be recalculated because the remaining closed and open durations will change. After the definition of bursts, gaps and open times within bursts can then be extracted from event tables. For objective criteria of how to set such minimum gap times see Chapter 19 (this volume).
- *Time range.* During long recordings single-channel activity may change (e.g., Hess *et al.,* 1984). For an overview of stationarity, programs should provide features to display open probability, as determined in a sliding window, as function of time. Similarly, event histograms can be compiled from selected time periods to determine small changes in channel kinetics during the course of an experiment.
- *First latencies.* Nonstationary data require the analysis of first latencies, i.e., the time from a given stimulus until the first channel opening. This is an important parameter for the investigation of inactivating channels when one wants to obtain information on the activation process that is not obscured by the inactivation mechanism (e.g., Sigworth and Zhou, 1992).

More involved statistical analyses, such as the correlation of adjacent intervals (e.g., Blatz and Magleby, 1989), are usually not implemented in standard single-channel analysis programs. In such cases the event tables usually have to be reanalyzed with user-designed software.

3.3.6. Probability Density Functions

The simplest way to determine time constants from dwell-time histograms is to fit a sum of exponential functions, each characterized by a time constant, τ, and a relative amplitude, a (equation 2).

$$pdf(t) = \sum_{i=1}^{n} \frac{a_i}{\tau_i} \exp\left(\frac{-t}{\tau_i}\right), \qquad \sum_{i=1}^{n} a_i = 1 \tag{2}$$

Independently of the binning, maximum-likelihood methods can be employed to fit *pdfs* directly to the events rather than to the histograms (see Chapter 19, this volume).

3.3.7. Missed Event Correction

Because of limited time resolution there are always events missed during the detection procedure. In closed-time histograms the limited bandwidth can, as a first approximation, be compensated for by neglecting the first bins in which the nonresolved events are missing, provided that there are no very fast opening events. If the time constants under consideration are clearly longer than the sampling period, this method yields acceptable results.

Open times, however, are always badly affected by the limited time resolution, because a missed closing event will result in an overestimated open time. Several theoretical methods have been developed to overcome this problem (see Chapter 19, this volume; Roux and Sauve, 1985; Blatz and Magleby, 1986; Crouzy and Sigworth, 1990; Hawkes *et al.,* 1991).

Once a kinetic scheme has been decided on, a rigorous method can be applied, such as that of Magleby and Weiss (1990). This method uses simulated single-channel data that have been masked with background noise, filtered, and analyzed exactly as the experimental data. The resulting simulated histograms, based on simulated data, are then compared with the measured histograms. The same time-consuming process is repeated with altered kinetic parameters of the model until a satisfactory match between measured and modeled histograms is achieved.

3.4. Open-Channel Analysis

So far we have just been concerned with single-channel events that have been identified with a detection method and then characterized by an amplitude and a duration. More information can be extracted from single-channel records when single open-channel currents are recorded at varying potentials or by analyzing the current noise in individual open channels.

3.4.1. Conditional Averaging

If the ion channel open times last several milliseconds, single-channel current–voltage relationships can be acquired by the application of voltage ramps. However, channels may not stay open for the duration of an entire ramp. Therefore, single-channel programs should provide editing features that allow the extraction of data sections from traces for averaging. Such conditional averaging is illustrated in Fig. 6, which shows four single-channel responses to identical voltage ramps. Open-channel sections can be selected with a cursor or mouse-operated routine and are stored in an accumulation buffer. The number of entries per sample point is also stored and then used for the proper scaling of the averaged single-channel current–voltage relationship. Note that after this procedure the errors in the individual data points are no longer the same because of the heterogeneous averaging. Besides the open-channel sections, baseline entries can be stored and used for leak correction.

3.4.2. Open-Channel Histograms

Single-channel recordings can also be analyzed on a sample-to-sample basis by the compilation of current histograms as illustrated in Fig. 7. The peaks in such histograms indicate the main current levels, and the widths of the peaks are a measure of the current noise in the corresponding level. Deviations from the symmetry of the peaks is indicative of nonresolved events that cause a skew in the distribution (see Fig. 7b). The relative lifetime of a current level is obtained from the relative area.

Note that these methods, which use raw data traces, generally require a very stable baseline and the proper selection of data sections. Even small shifts in the baseline current

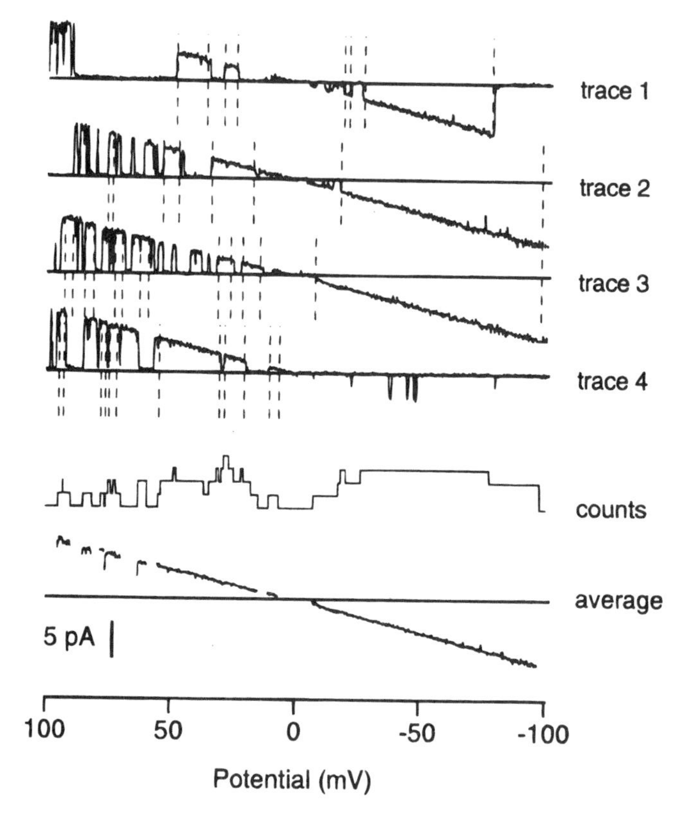

Figure 6. Conditional averaging of single-channel currents evoked by a voltage ramp from +100 to −100 mV of 200 msec duration. Open-channel sections without flickers were selected (vertical dashed lines) and accumulated. The accumulated data were then divided by the number of entries per sample point and displayed as "average," yielding a single-channel current–voltage relationship with some gaps where no open channels could be recorded.

can cause considerable broadening of the open-channel current histograms, which results in an overestimation of the noise.

More information about channel behavior can sometimes be extracted by selecting and analyzing current traces that represent only one channel state (e.g., open or closed). The fluctuating current that is measured during an open-channel period is composed of both statistical background noise and the noise arising from the flow of ions through the channel. Both of them can be described as a first approximation by Gaussian functions (equation 1). Additional processes that generate current noise will add to this Gaussian noise and can cause deviations of the histograms from the typical bell shape.

Long-lasting sublevel events, for example, can be detected directly as separate peaks in the open-channel histograms between the peaks of the main level and the baseline.

If closing events are too brief to be resolved directly they will not show up as separate peaks in open-channel histograms but will skew the histograms in the closing direction (see Fig. 7b). Assuming flicker events to cause complete channel closures, several theoretical approaches can be applied to extract information about the flicker kinetics from the skewed histograms. For heavily filtered signals, Yellen devised a method that uses the fit of β-functions to histograms to yield time constants for flicker dwell time distributions (Yellen,

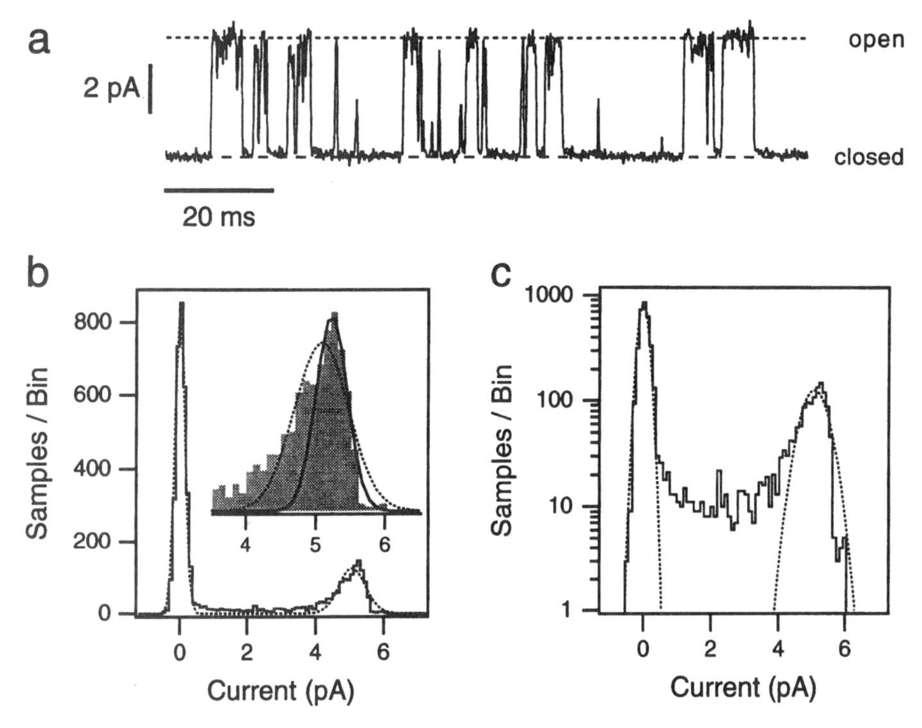

Figure 7. Analysis of single-channel recordings on a sample-to-sample basis. a: Outward currents low-pass filtered at 1 kHz and recorded at a rate of 4 kHz. b: Linear current histogram with a bin width of 0.1 pA. the dotted line represents a fit of the sum of two Gaussian functions (equation 1) to the baseline peak (center, I_{0c}: 0.01 pA, height, a_c: 858 samples, standard deviation, σ_c: 0.147 pA) and the open-channel peak (I_{0o}: 5.09 pA, a_o: 125 samples, σ_o: 0.389 pA). Because the recording is based on the activity of only one channel, the ratio of the areas under the two curves yields an open probability of $[1 + (a_c\, \sigma_c/a_o\, \sigma_o)]^{-1} = 0.28$. Although the baseline peak is reasonably well described by a Gaussian function, the open-channel peak is clearly skewed toward zero because of brief closing events. A better estimate of the peak current is obtained by fitting a Gaussian function only to the data above 5 pA, as illustrated by the solid curve in the inset (I_{0c}: 5.23 pA, a_o: 145 samples, σ_o: 0.226 pA). The current value of the peak of this histogram is indicated as a dashed line in part a. c: The same histogram as in b, but with logarithmic scaling of the ordinate.

1984). The kinetics of very fast and rare closing events far beyond the actual time resolution of the recording system can be estimated from the higher moments of the open-channel histograms (Heinemann and Sigworth, 1991).

3.4.3. Mean-Variance Methods

Particularly useful for the identification of sublevel events are the methods proposed by Patlak (1988, 1993) in which mean current and variance are calculated in a sliding window. Three-dimensional histograms of the number of entries as a function of variance and mean current allow the separation of distinct current levels as peaks whose volume and shape contain information about the kinetics of the events.

3.4.4. Open-Channel Noise

As we have seen in the previous examples, the current noise in an open channel can be much greater than the baseline noise. By analysis of the power spectra of open- and closed-channel currents, more insight can be gained into the properties of the excess current fluctuations. For nicotinic acetylcholine receptor channels, Sigworth (1985) showed that Lorentzian components in the power spectra (see equation 7) of open-channel current events (corrected for baseline spectra) were indicative of current fluctuations that are too small to be detected directly in the time domain. They could arise from slow conformational fluctuations that do not close the channel completely but rather modulate the single-channel amplitude. If the kinetic events responsible for the generation of excess noise are far faster than the time resolution of the measurement, the spectral density at low frequencies still provides information about the underlying processes. These approaches have been used for the characterization of fast, nonresolved, channel-blocking events (Heinemann and Sigworth, 1988, 1989) and even for the analysis of the shot noise generated by the statistical motion of ions as they flow through a channel (Heinemann and Sigworth, 1990).

4. Analysis of Macroscopic Currents

In this section we discuss the analysis of macroscopic currents as obtained from whole-cell recordings or from patch recordings using large pipettes. In most cases ionic currents do not occur spontaneously but must be evoked by stimuli such as a change in the membrane potential or the fast application of agonist. Because voltage-clamp experiments are most common, they are discussed in more detail. The problems of pulse pattern generation, parameter control, leak correction, and, finally, various methods of extracting information about single-channel properties from macroscopic data are addressed.

4.1. Parameter Control in Relaxation Experiments

A relaxation experiment is a type of measurement in which an experimental parameter is changed suddenly in order to perturb the equilibrium of the system under consideration. After the perturbation a new equilibrium will be reached with a time course dependent on the new experimental parameters (e.g., new potential after a voltage step). In order to characterize the kinetics of voltage-dependent ion channels, for example, the membrane potential is changed according to a pulse pattern comprised of a number of segments of variable duration and potential.

4.1.1. Voltage-Clamp Performance

For later analysis of the recorded currents, it is very important to consider how much the potential at the membrane deviates from the desired potential as specified in the pulse protocol. Typical reasons for such deviations are:

- Limitations in the voltage-clamp amplifier (e.g., from filtering of the stimulus at the input of the patch clamp amplifier)

- Poor space clamp (i.e., not all membrane areas can be held at the same potential because of unfavorable membrane topology)
- Insufficiently compensated series resistance (i.e., voltage drop across the series resistance caused by large currents could not be completely corrected by analogue methods.)

All of these problems can be rather serious when one wants to derive quantitative information on channel kinetics from the current recordings. In patch-clamp recordings the errors introduced can usually be easily estimated by considering the filtering of the stimulus and the series resistance compensation (if necessary at all). For recordings with fine-tip electrodes (two-electrode voltage clamp), deviations from the theoretical potential can be significant, but they can be measured by recording of the actual membrane potential in parallel to the membrane current. The estimated real potential profiles (patch clamp) or the measured potential profiles (two-electrode voltage clamp) can then be used as a reference when model functions are to be fitted to the recorded currents (see below).

4.1.2. Filter Delays and Rise Times

Although the actual potential profile only deviates from the theoretical one in situations with insufficient voltage clamp control, the recorded current is generally masked by distortions from low-pass filters. The delay and the rise time of a step response are usually taken as indicative of the filter characteristics. For an eight-pole Bessel filter with a cutoff frequency f_c, the (0–10%) delay and the (10–90%) rise time are each approximately $0.34/f_c$ (see Fig. 3 and Chapter 19, this volume). If the filter characteristics are not known exactly, e.g., if the signal is filtered several times between the pipette and the actual display on the computer, it might be better to use an experimentally determined step response as a reference.

If the kinetic time constants of interest are far slower than the delays and rise times introduced by filtering, then the effects of filtering can be neglected. In several cases measured time constants or start times can be corrected simply by using the filter delay. If, for example, the sigmoidal onset of current after a voltage pulse is characterized by a delay and an exponential function to the power of n, the filter delay can be subtracted from the measured delay as a first approximation. For more precise analysis, theoretical functions have to be passed through an equivalent filter before they are fitted to the recorded data.

4.2. Signal Averaging and Leak Correction

4.2.1. Signal Averaging

If evoked data are recorded repeatedly under identical conditions, the current traces can be averaged in order to increase the signal-to-noise ratio, yielding smoother data traces. The statistical noise can be reduced by a factor of $\sqrt{2}$ if the number of averaged traces is doubled. Different programs use different approaches for averaging, depending on the degree of automation. These include on-line methods that acquire and average the data sweeps and only store the average, on-line methods that show the average but store all of the individual sweeps, and off-line methods that store only the individual sweeps and leave the averaging to analysis programs at a later stage. The second method is most useful, for it gives an immediate result while allowing individual records that are impeded by extraneous noise (e.g., from a current spike caused by an electrical surge) to be discarded during off-line analysis. These off-line analysis programs should facilitate data deletion and compression.

4.2.2. Leak Correction

A change in the membrane potential is accompanied by capacitive currents, which should be canceled before data analysis. This can conveniently be done when analyzing voltage-dependent conductances.

Most programs for pulsed data acquisition, therefore, support features that allow the generation of so-called *P/n* leak correction protocols. In a voltage range where voltage-dependent channels are not active, a scaled-down version of the pulse protocol is applied n times, and the resulting current is averaged, scaled, and subtracted from that elicited by the main test pulse. This method gives good results only if the signals that have to be compensated for depend linearly on voltage. In standard applications a scaling factor of $r = 0.25$ is used, and four leak responses are added to yield the scaled *P*/4 correction record (Armstrong and Bezanilla, 1974; see Fig. 8a). Because of the subtraction of a leak correction signal from the main signal, the noise is increased by a factor of $R = \sqrt{1 + 1/nr}$, with r being the ratio of leak and test pulse amplitude. For $r = 0.25$ and $n = 4$, the noise increases by $\sqrt{2}$. Leak responses should be stored together with the raw data; this ensures that one can subsequently analyze the data with and without leak correction.

Because the leak pulses can be applied from a potential other than the "normal" holding potential, a step from the holding potential to a "leak holding potential" can also create capacitive currents. These are eliminated if one performs signal averaging in which the leak

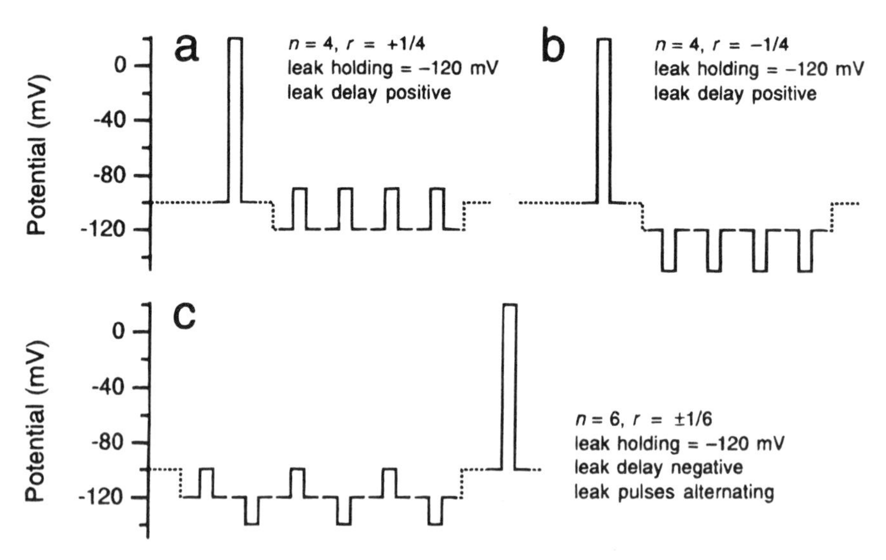

Figure 8. Pulse protocols for on-line leak correction. Panel a represents a standard *P*/4 pulse protocol with the leak pulses following the test pulse (leak delay positive). The protocol in b is similar to that in a, but the leak pulses have the opposite polarity to the test pulse. During signal averaging an alternation of the protocols shown in a and b ensures that the capacitive transients resulting from the steps from the holding potential to the leak holding potential are canceled. c: Six leak pulses, scaled by 1/6 (*P*/6 protocol), of alternating polarity precede the test pulse (leak delay negative).

pulse polarity of successive pulses alternates (Heinemann *et al.,* 1992) as illustrated in Fig. 8a,b. If no signal averaging is used, the effect of small nonlinearities around the leak holding potential can be reduced if the leak pulse polarity within a train of leak pulses is alternated (Fig. 8c). Because leak pulses might affect the currents during a test pulse or vice versa, it is best if the user has the option to record leak responses after (positive leak delay in Fig. 8) or before (negative leak delay in Fig. 8) the test pulse, respectively.

If channel activity can be abolished completely by application of a channel blocker, traces recorded after application could be used for leak correction. Analysis programs should therefore provide features to mark data sweeps to be used for leak correction of traces with channel activity.

4.3. Relaxation Experiments

For the following consideration we will assume that macroscopic currents are composed of many current events arising from the same kind of ion channel. In such cases macroscopic currents display a statistical average of many single-ion-channel events, and kinetic parameters can therefore be related to the probability functions of single-channel state transitions. The opening and closing of voltage-dependent ion channels is described theoretically in terms of kinetic Markovian schemes with channel states (e.g., open, closed, inactivated) and voltage-dependent transition rate constants (see Chapter 18, this volume). The ultimate aim of relaxation experiments is therefore to determine the state occupancies and the individual transition rates as a function of potential. Thus, pulse protocols have to be designed such that the measured macroscopic time constants can be attributed as closely as possible to microscopic time constants for state transitions.

4.3.1. Design of Pulse Patterns

A pulse pattern usually comprises a number of pulse segments of specified voltage and duration. In some cases it might be advantageous to define such segments as voltage ramps (see Fig. 6) or, for the implementation of phase-sensitive measurements, sine waves on top of a specified DC voltage (see Chapter 7, this volume). Besides a test segment, protocols usually contain at least one segment that primes the channels such that there is a defined initial condition. The kinetics of transitions among channel states are then determined.

For the investigation of time- or voltage-dependent processes either test or priming segments are varied in a systematic way, thereby creating a family of pulse patterns. From pulse to pulse individual segment durations or voltages can be altered by adding linear or exponential increments or decrements. Linear increments are widely used for segment voltages. For the investigation of steady-state inactivation properties, for example, a prepulse potential is changed, and the current is measured during a subsequent constant test pulse segment (Fig. 9c). If time constants are to be derived, e.g., the time constant for channel recovery from inactivation (Fig. 9e), exponential increments are useful, because they give rise to data points that are spaced according to their significance. Figure 9 illustrates several frequently used pulse protocols for relaxation experiments.

For analysis purposes it is quite helpful if one can specify in a pulse generator which segment is a test segment and which is a priming segment. In Fig. 9 the test segment is labeled with a "y" because in this segment a measurement has to be performed, and the result is to be displayed as an ordinate value. The label "x" denotes either a variable priming

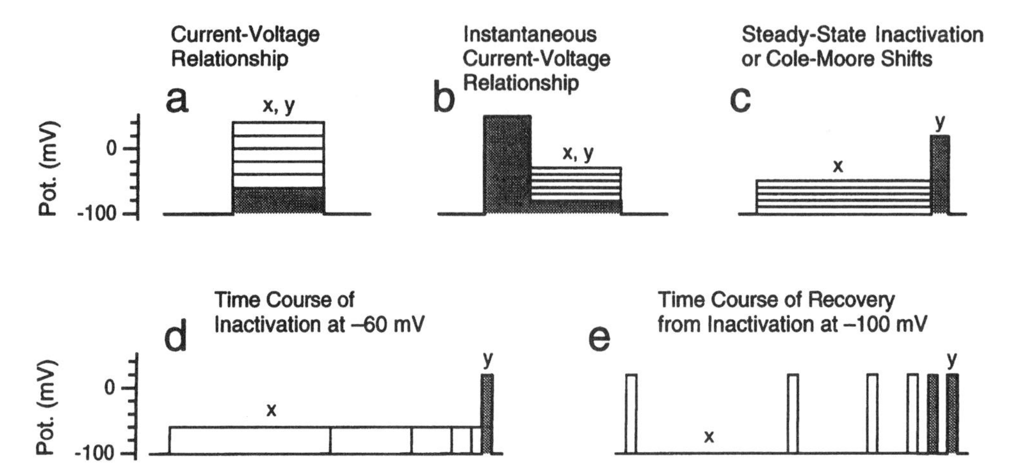

Figure 9. Pulse protocols frequently used for relaxation experiments on voltage-dependent ion channels. The first pulse is indicated by shading; successive pulses are drawn with continuous lines. The letters x and y denote which pulse segments are used as abscissa or ordinate, respectively, during secondary analysis.

segment (Fig. 9c–e) or the variable potential of the test segment (Fig. 9a,b), which are to be used as abscissa values.

4.3.2. Determination of Kinetic Parameters

Usually it is a long way from initial relaxation experiments to a kinetic scheme of channel gating. Therefore, there are several hierarchic levels for the quantitative representation of measured data and the comparison with theoretical models. Just in regard to channel activation, these levels could be as follows:

- Without any model in mind, one could describe channel activation in terms of time to peak, time to half-activation, or slope at half-activation time (see Fig. 10b). These values, determined as a function of potential, can be used as quantitative parameters to describe the channel under consideration. In further calculations these parameters can be compared with the same parameters as derived from models.

- More detailed descriptions could make use of the sums or products of several exponential functions and possibly a time delay, yielding a first estimate of how many kinetic components contribute to activation. Based on these initial guesses, analytical solutions of a devised kinetic model using idealized initial conditions can be fitted to the data. This is done during data description with Hodgkin–Huxley equations, for example (e.g., Hodgkin and Huxley, 1952), where time constants for activation are free parameters for data fit and thereby yield a more direct comparison of model and measured data (Fig. 10b).

- A more general approach is to fit all the transition rates and state occupancies of a kinetic scheme to the data directly, rather than using kinetic parameters such as relaxation constants. In this way a transition matrix (see Chapter 20, this volume) representing a particular kinetic model is fitted to entire data sweeps or even to families of sweeps without idealization of the initial conditions, yielding data descriptions that

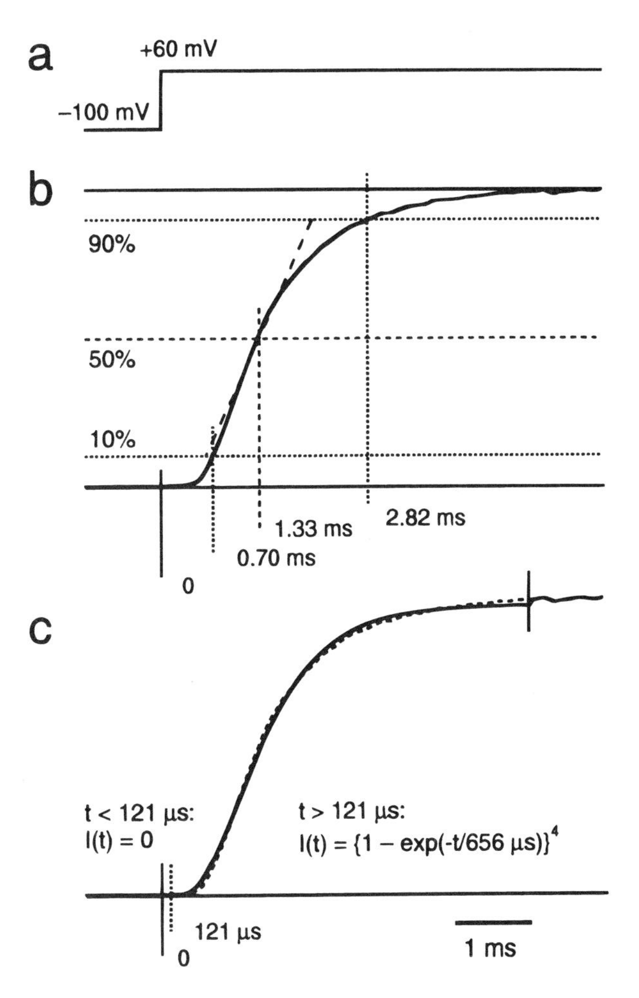

Figure 10. Examples of the description of kinetic parameters of potassium channel activation. Currents in macropatches from *Xenopus* oocytes expressing *Shaker* BΔ6-46 (Hoshi *et al.*, 1990) channels were elicited by a depolarizing step from −100 to +60 mV as shown in a. b: Activation was characterized by the time taken to reach certain current levels with respect to that taken to achieve the maximal current. Also shown is a dashed straight line through the half-maximal current. This has a slope of 55%/msec as determined by a fifth-order polynomial fit to the data in the interval between 1 and 3 msec after the beginning of the pulse. c: Same data as in b, superimposed with a fit function describing a Hodgkin–Huxley activation curve of the fourth order with an additional delay of 121 μsec. The fit was done between the two vertical solid lines. Because the data were filtered at 4 kHz with a low-pass Bessel characteristic, a filter delay of approximately 80 μsec is expected. The measured delay of 121 μsec could therefore be indicative of state transitions in addition to those described by the Hodgkin–Huxley formalism.

are not compromised by approximations. Such methods, however, are quite time consuming and are rarely implemented in standard analysis software packages.

The effects of data filtering can be considered by filtering the theoretical functions with an equivalent filter characteristic before comparison with the data. Limited clamp speed can be considered if the calculations are based on the actual potential profile rather than on the idealized voltage step (see Section 4.1.2).

Fit results derived from raw data sweeps often have to be analyzed further. Therefore analysis programs should store these results and should provide tools to display them as a function of various parameters, such as the potential of a specified pulse segment. Exponential functions, current–voltage relationships, and Boltzmann functions are often used to fit such data. For more specialized functions, tools that interpret text lines of numerical expressions (parsers) may be provided by the program; otherwise the data must be transferred to general-purpose programs (see below).

4.4. Noise Analysis

Even if a current signal recorded from a membrane patch or a whole cell is composed of many superimposed single-channel events, such that individual events cannot be detected, analysis of current fluctuations may provide information on single-channel properties such as single-channel amplitude or mean open times (e.g., Neher and Stevens, 1977). This information can be extracted from the signal either by transformation of the data into the frequency domain (Fourier analysis) or by analysis of the current variance at a given bandwidth.

In many cases valuable information can be extracted from fluctuating signals by simply considering the mean current and the current variance if the channel-open probability is small. The mean current, I, is given by the product of the single-channel current, i, the number of channels, n, and the open probability, p_o:

$$I = inp_o \tag{3}$$

Since a channel can only be open or closed, a binomial distribution applies, which has the variance:

$$\sigma^2 = i^2np_o(1 - p_o) \tag{4}$$

With equation 3, this expression can be written as:

$$\sigma^2 = iI(1 - p_o) \tag{5}$$

For very small open probabilities, equation 5 simplifies to

$$i \approx \sigma^2/I \tag{6}$$

yielding an expression for the single-channel current.

4.4.1. Power Spectra of Continuous Data

For stationary signals, i.e., signals with a DC component that does not vary with time during an experiment, spectral analysis can be performed. For this purpose the power spectra of sections of current data are calculated and averaged. The discreteness of the signals allows the use of FFT (fast Fourier transform) algorithms, which are implemented in many software packages and can be found in the procedure libraries of development systems or statistics programs (see below). Usually the power spectra are computed from 1024 data points. The minimum and maximum frequency of the spectrum are thereby set, together with the sampling interval. If the fluctuating signal exhibits large DC components, the actual fluctuation may span only a small fraction of the input range of the AD converter. In such cases AC and DC components are sampled separately with different gain settings in order to increase the dynamic range for the fluctuating signal.

The power spectra can then be compared to theoretical predictions. Given a channel with open and closed states having exponential dwell-time distributions with the mean open and mean shut times τ_0 and τ_c, respectively, the spectral density (A^2/Hz) is described by the Lorentzian function,

$$S(f) = \frac{S(0)}{1 + (f/f_c)^2} \tag{7}$$

where $S(0)$ is the low-frequency limit of the spectral density. The cutoff frequency, f_c, is related to the relaxation time constant, $\tau = (1/\tau_o + 1/\tau_c)^{-1}$, by

$$\tau = \frac{1}{2\pi f_c}. \tag{8}$$

At low channel-open probability, the single-channel current amplitude, i, is obtained from the signal variance, i.e., the integral of the power spectrum $[\sigma^2 = f_c S(0)/2]$, according to equation 6. The theoretical background and spectral functions for more complicated channel-gating schemes are discussed in detail by Neher and Stevens (1977) and DeFelice (1981). In particular, the existence of more than one Lorentzian component in the power spectrum can be considered as evidence of more than two distinct kinetic states. If the noise is determined by several exponential processes with very similar time constants (or a continuum of time constants), the power spectrum can acquire a shape that is not unambiguously described by a sum of Lorentzian functions. The use of power laws then helps to describe such $1/f^n$ noise.

4.4.2. Nonstationary Noise Analysis

Nonstationary signals can be divided into two groups. The first comprises signals that vary in amplitude as a function of time, e.g., as a result of the slow fluctuation in concentration of an activating transmitter at the membrane. These signals may be divided into smaller sections for determination of the time variance. When one considers the activity of only one class of ion channels, the variance is related to the DC current within the selected sections by the single-channel current and the channel-open probability. If the open probability is small, the single-channel current can be easily estimated according to equation 6.

If transient currents are recorded in response to identical repetitive stimulations, the ensemble variance can be calculated as a function of time. This is the variance caused by the deviation of each individual data point from the mean of many equivalent measurements. From equation 4 it is seen that the variance is zero if all the channels are either closed ($p_o = 0$) or open ($p_o = 1$). It reaches a maximum when half of the channels are open. If p_o is neither constant nor small, a plot of σ^2 versus I yields a parabola with a zero-crossing at the maximal current $I = i\,n$ and an initial slope corresponding to the single-channel current amplitude (Sigworth, 1980; see Fig. 11):

$$\sigma^2 = iI - I^2/n. \tag{9}$$

This method can be used for nonstationary records, where the variance is determined by computing the deviations of individual records from the mean, or, in order to eliminate drifts in the signal, from differences of successive records (e.g., Heinemann and Conti, 1992). Programs for nonstationary noise analysis must take into account background noise and leak. Automated identification of records with excess extraneous noise, based on objective statistical criteria (e.g., Heinemann and Conti, 1992), is desirable.

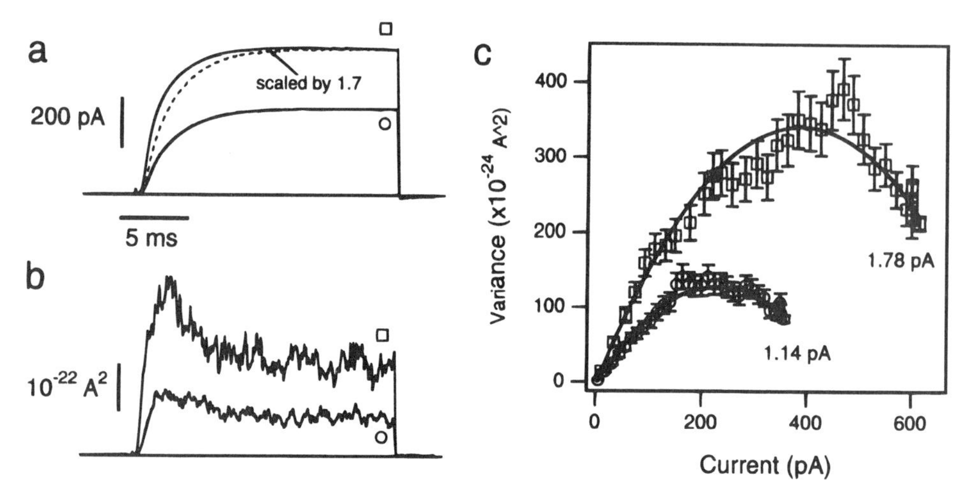

Figure 11. Nonstationary noise analysis of outward potassium currents flowing through *Shaker* BΔ6-46 (Hoshi *et al.*, 1990) channels expressed in *Xenopus* oocytes. a: Averaged currents recorded from inside-out macropatches in response to depolarizations to +80 mV. The larger current (average of 254 traces) is a control record; the smaller one (average of 356 traces) was recorded after application of sucrose to the bath, which resulted in a reduction of the current to approximately 60%. b: Averaged ensemble variance of the two experiments mentioned in a (based on 243 and 226 differences between successive current traces, respectively). c: Ensemble variance as a function of mean current. The continuous curves represent data fits according to equation 9, yielding the indicated estimates for the single-channel current and maximal channel-open probabilities of 0.78.

5. Multipurpose Programs

At many instances during execution of an experiment, during on-line analysis, or at later stages of specialized analysis, tasks arise that are readily handled by commercially available multipurpose programs. Since these programs are developed for a much larger market than patch-clamp programs, the task can be carried out with much more sophistication at a very small price. On the other hand, problems might arise during data acquisition or analysis that are so unusual that tackling them is (as yet) not implemented in such packages. In such cases one might try to extend the program that is used for acquisition and analysis (see Section 6). On many occasions, however, this is not possible, because the source code for the program might not be available, or program changes might be undesirable because even small features added to a running program may cause unwanted side effects. Alternatively, the implementation of very specialized features may simply cause an increase in complexity at the expense of user-friendliness.

In either case it is important that data and intermediate and final results can be exported from the specialized programs in such a way that they can be read by general-purpose programs for data presentation, data administration, or further analyses.

5.1. Data Presentation

Graphics programs or desktop publishing programs with graphics facilities are used to read structurally simple data files in order to generate figures for presentation. Such figures can be incorporated into text files with text-editing programs. Computer interfaces to slide writers are now widely used to generate color slides directly.

5.2. Data Administration

Data-base programs are used to generate and to manipulate complex data structures. Large data bases could, for example, be used to keep track of which channel mutants were investigated electrophysiologically, which protocols were used, and in which data files the information is stored. This will facilitate access to data files according to given experimental parameters, e.g., access to all data sweeps recorded from a certain channel mutant in a specified solution with a pulse to a specified membrane potential. For this purpose information has to be output from the acquisition or analysis program and to be interpreted by the data-base program. An alternative application is the import of information stored in a data base to the acquisition and analysis programs. The compositions of solutions used during an experiment, for example, could be imported by a dedicated analysis program to facilitate the generation of dose–response curves.

5.3. Table Calculations

Arrays of data output by acquisition or analysis programs are ideally suited to manipulation in spreadsheet programs. Such table calculation programs provide a variety of tools for data presentation and further analysis, including several statistical procedures.

5.4. Curve-Fitting Programs with Programming Capabilities

Several programs are available that support data display and the fitting of specific theoretical functions. Such specialized functions are usually composed of a set of standard functions. Alternatively, these programs provide parsers that interpret user-defined mathematical functions. Advanced programs of this kind offer their own simplified command language, which can be used to define mathematical functions or more complex analysis algorithms.

In principle, programs for data presentation, statistics, and data fitting cannot be discriminated easily because most of them offer some features for each of these purposes. However, usually these programs are particularly good for only one or two applications. One may therefore need several programs to meet all the requirements. In general, however, it should be remembered that the use of a few programs that one knows well may be better in the long run than using many. Even if these few programs do not contain all possible features, there will be fewer problems originating from data transfer, and time and money will be saved.

6. Choices for Hardware and Software

The computer market is expanding so rapidly that it is hard to keep track of all the new products and their specifications. In this section several criteria are presented that may help the reader to decide on a combination of computer, peripherals, and software that are suitable for specific experimental tasks and are compatible with the budget. The range of possible experimental tasks should be clear from the above sections and from the first chapters of this volume. The question of what should be done if the software package is insufficient in some respects will be discussed.

We will start with a description of hardware components and typical specifications that should be checked carefully before purchase. The hardware comprises the amplifier, the computer, the data storage media, and the peripherals that interface with the actual experiment, i.e., AD and DA converters. Software components such as operating systems, development software, application software, and application software with development capabilities are discussed with respect to hardware configuration. As illustrated in Fig. 12, in reality such components are not independent of each other. This means that a certain computer configuration is only compatible with some of the higher-level software available, and vice versa. During the process of evaluation of what to purchase, one will be faced with the old chicken-or-egg question. In the past, a computer and its peripherals were something precious, and software was just some imaginary quantity that could easily be copied, so the order of priority

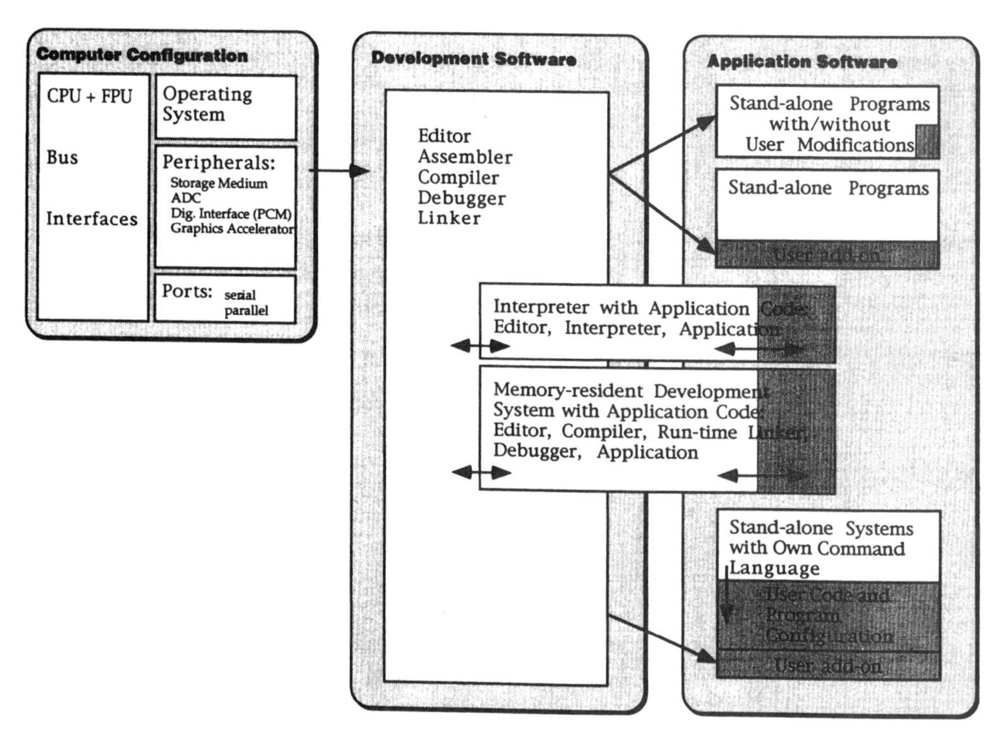

Figure 12. Organization of computer configurations and software levels. The computer hardware with all its components, including the operating system, forms a functional configuration. On top of this configuration there are roughly two more levels of software. The development software is used to generate code and to install application software, which performs such tasks as data acquisition or analysis. This separation of development and application software is only strict for compiled and linked programs that are distributed as stand-alone applications. These two kinds of software merge in so-called development systems, where development tools and application code coexist in a single running application. The arrows indicate direction of interaction. Dark shading is used to illustrate code introduced by the user. The amount of user code certainly cannot be predicted, but in general, modifications will be rare in stand-alone applications and will become mandatory in general-purpose toolbox programs that do not provide applications dedicated to electrophysiological experiments.

was hardware first and then software. However, the considerable drop in prices for computer hardware and the increased sophistication and copy protection of software tools is now clearly setting the trend for buying hardware compatible with a specified software configuration.

6.1. Criteria for the Selection of Hardware

The PDP-11 computer series (Digital Equipment Corp.) was until recently widely used for electrophysiological research applications because of its reliability and the availability of powerful real-time interfaces and programming environments. In the meantime, however, personal computers have become faster and much less expensive so that most of the applications that are now commercially available are designed to run on one of these machines. Although more sophisticated and more expensive computers, like workstations, are also used for electrophysiological experiments, most experiments are carried out using IBM-compatible and Apple Macintosh II personal computers. Therefore, only these two systems will be referred to in the following discussion.

Depending on how deeply one wants to go into the details of a computer system and its software, it can be a box with an on/off button and a stand-alone application program used for data acquisition and analysis, or it can be viewed as a highly complex modular system of hardware components and software tools that interact with each other and cannot therefore be easily considered independently of the rest. "Computer configuration" is defined here as the combination of hardware and an operating system that allows the loading of development and application software as described below.

6.1.1. Computer Configuration

The heart of a computer is its *central processing unit* (CPU), which determines speed of performance and poses the major compatibility problem among different machines. For high performance, IBM-compatible computers should be equipped with an 80486 series (Intel) processor, whereas for Macintosh computers the processor of choice is currently the 68040 CPU from Motorola. In each case the processor should be supported by a *floating point unit* (FPU), which takes over time-consuming floating point operations, thereby considerably improving performance. The type of CPU and FPU have important implications for the software that is going to run on the computer. This is the major reason for the incompatibility of applications running on IBM and Macintosh computers. A new series of PowerPCs that combine features of IBM and Macintosh computers based on a RISC (reduced instruction set computer) processor may solve some of these problems.

The CPU communicates with the environment via built-in interfaces such as serial or parallel ports. Such ports can have various specifications. Serial ports often use the RS-232 convention; a typical parallel port is an SCSI interface. Besides these standard interfaces, personal computers offer extended bus structures for the exchange of data and commands. IBM-compatible computers can have various bus structures (e.g., 16-bit AT, 32-bit EISA); for Macintosh II computers only the 32-bit NuBus is available. External devices such as AD converters, digital interfaces to a PCM, and graphics accelerators are connected to this bus. One must therefore consider the compatibility of bus and peripherals and how many bus slots are necessary.

An important determinant of computer speed and performance is the amount of disk space and the amount of *random access memory* (RAM) available. A considerable drop in the cost of both types of memory and the extended address ranges of the new generation of computers have removed some of the limitations. For a computer system that is used for real-time high-demand data acquisition, 24 Mbyte RAM and a 1 Gbyte capacity hard disk would be an appropriate configuration. Less demanding tasks can be accomplished with smaller systems.

Media for long-term data storage are hooked up either to an expansion bus or to a parallel port (e.g., SCSI). These media range from very fast removable hard disks through optical and magnetooptical devices to magnetic streamer tapes and floppy disks. The cost of all of them has come down a lot, but it is still important to consider access time and cost per megabyte before deciding which to buy. For long records of single-channel events, relatively inexpensive digital magnetic tapes, which hold several gigabytes of data, or WORM cartridges are a good choice. For data that take up less space and are not as linearly structured as single-channel records, storage media with shorter seek times and random access capabilities are desirable. For these data types removable hard disks and rewritable magnetooptical disks are appropriate.

6.1.2. Operating Systems

Operating systems are programs that manage the interactions between all components of a computer configuration, including application software, and are intimately linked to the hardware components. For personal computers there is not usually much choice of operating systems. IBM-compatible computers are equipped with MS-DOS (Microsoft disk operating system) or advanced graphics-oriented systems like *Windows* (Microsoft). The operating system of Macintosh computers is *Finder.* For both types of machine there are also other systems available (e.g., UNIX), but these are not widely used on personal computers.

6.1.3. Analogue-to-Digital Converters

Both AD and DA converters are specified by their time and voltage resolution (see Section 2.1.1). The interface between the converters and the computer bus determines the actual speed with which data can be sampled. Often AD/DA converters can sample data at a certain rate for only a short period of time, depending on the size of a buffer memory, which is then slowly emptied via the bus. In continuous recording mode the usually slower data transfer rate limits the attainable sampling rate.

A DA converter may have a problem denoted as "glitching." This results from an asynchronous switching of all relevant bits, causing the analogue output to be set briefly to an unintended value in some cases. This effect is not uncommon for DA converters and can cause serious problems if small voltage steps are distorted by short but huge voltage spikes at the transition points. Such problems are avoided by using deglitched DA converters, which, however, are more expensive. The errors caused by glitching can be reduced by (1) filtering the stimulus voltage before feeding it into the patch-clamp amplifier and (2) by the subsequent use of an optimal dynamic range for the DA conversion.

In Table I several requirements for AD/DA converters and typical specifications are summarized.

Table I. Requirements and Typical Specifications for AD/DA Converters

Requirement	Typical Specification
Number of AD and DA channels	4–8 and 2–8
Maximal sampling rate	50–200 kHz/one channel
Resolution	(8), 12, 14, 16 bits
On-board buffer	8–16k samples
DA deglitched	no/yes
Additional digital triggers (in/out)	no/yes

6.2. Criteria for the Selection of Software

Software can be grouped on several hierarchical levels. For our purposes the lowest level is the operating system (e.g., MS-DOS on IBM-compatible computers, UNIX on various workstations, or Finder on Macintosh computers). As discussed earlier, some operating systems are only available for certain computers and some application programs are only written for certain operating systems. On the next level one finds compilers/interpreters or so-called development systems. A compiler is a program that translates program code written in a specific programming language into code that can be interpreted by the computer's CPU. After this translation process the machine code is linked to operating system routines and user-defined libraries and set up such that it can be executed. High-level application programs, like those of various acquisition and analysis packages, are generated in this way. Other application software used for data analysis comprise programs for data presentation and administration, statistics, and curve fitting as outlined in Section 5.

6.2.1. Development Software

If one is not going to write one's own code for data acquisition or analysis, the steps involved in program development do not need to be considered, because commercially available acquisition and analysis programs are ready to use. If a software package that meets all requirements can be found, one should purchase it, because no special knowledge of programming is required except for the configuration of the program flow as can be determined from the program manuals. In many cases, however, such "closed" systems are not sufficient. In particular, the experienced experimenter may find it necessary to generate some extra code that complements a commercial program (e.g., by adding analysis functions or by writing extra programs in a style similar to the "master" program), or to alter the program. In some cases it may even be necessary, or at least advantageous, if one can execute such changes "on the fly," i.e., during a running experiment. In all these cases a software development system is useful because it provides an environment for writing and executing programs.

6.2.1a. ***Compiled Stand-Alone Programs.*** Figure 12 illustrates the organization of acquisition and analysis software, with potential modifications shaded. Stand-alone programs, if purchased commercially, are ready to run and cannot be modified easily. If source code, information about compilation, and linking instructions are available, the user may introduce changes, but these changes should usually be minor ones and strictly speaking should be avoided altogether because of the possibility of introducing side effects. Other stand-alone programs are provided as a set of objects (compiled code) that are linked and can be executed. These programs have built-in programming interfaces that allow user routines to be written, compiled, and linked to the system without directly interfering with the code of the main program. The user code may still cause program errors, but this method is safer than directly changing complex code that was written by someone else. For data acquisition and analysis systems, popular computer languages are C, C++, and Pascal. FORTRAN is not used much for data acquisition programs on personal computers, but, particularly for analysis purposes, it has its strengths because of the extensive subroutine libraries that are available on larger computers.

6.2.1b. ***Interpreter Systems.*** In interpreter systems program instructions are parsed, compiled, and executed line by line. Therefore, one usually has access to the source code (some parts of the programs may be supplied as object files that cannot be accessed by the user). It is quite simple then to incorporate user-specific modifications, and the program flow can be stopped and restarted at any point. A very successful product widely used in electrophysiological laboratories was the legendary Basic-23 system running on PDP-11 computers. The disadvantage of these interpreter systems is their low speed (Basic-23, therefore, made use of many subroutines written in assembly code) and the fact that the language is not very structured, leading ultimately to messy code if the program exceeds a certain size. Several interpreter systems enable the compilation and linking, once tested, of the entire code in order to generate a stand-alone program that runs faster but that can no longer be modified. This is an important option, as run time is always an issue, and a linked program is less prone to the introduction of accidental changes by inexperienced users.

6.2.1c. ***Memory-Resident Development Systems.*** On today's computers, compilation is so fast that one can afford to compile entire program modules rather than only lines. In addition, RAM has become so inexpensive that memory-resident development systems consisting of programming tools such as an editor, compiler, and debugger, which are always held in RAM, are now available. The system itself links newly compiled code to its own running application, enabling the possible immediate execution of code. Fast compilation and the essential lack of linking time makes such systems behave like a very powerful interpreter system. The advantages are clearly speed, the availability of structured programming languages suitable for large software packages, and the possibility of a high degree of user interaction. The runtime-linking, memory-resident development system *PowerMod* (HEKA Elektronik) is based on the Modula-2 programming language. Software for the control of the EPC-9 patch-clamp amplifier and the acquisition and analysis package *Pulse+-PulseFit* run in this environment.

The tool library is an important consideration if development systems are to be used for electrophysiological research. A variety of procedures operating on data arrays such as standard mathematical operations, histogram functions, or Fourier transforms are very helpful.

6.2.1d. ***Toolbox Programs.*** In contrast to development systems, which normally make use of standard programming tools such as editors, compilers, and linkers and thereby specify the use of one or several standard programming languages, some stand-alone programs provide their own comprehensive command language. This language often has a simplified structure but is supported by sets of procedures with predefined tasks. These procedures can be used as tools to set up complex programs. Efficiency is increased by the option of linking procedures written in a common programming language (e.g., Pascal or C) as additional external tools. Examples of such systems, which have been adapted to patch-clamp applications, are LabView, Igor, and ASYST.

Thus, there are a variety of options for the kind of system one should actually get. The decision as to which is most suitable will be determined by the requirements and the programming skills of the experimenter. In general, however, it should be noted that the

Data Acquisition

- **Sampling:**
  - output channels — min / max
  - input channels — min / max
  - sampling frequency — min / max
  - data flow rate (cont. mode) — max
  - event catching — y / n
- **Hardware compatibility of:**
  - AD/DA board
  - patch clamp amplifier
  - programmable filter
- **Support of external devices:**
  - input (e.g. temperature control unit)
  - output (e.g. trigger of application systems)
- **Generation of pulse patterns:**
  - # sequences in pool — max
  - # segments per pulse — max
  - segment types — constant / ramp / sine / ext. profile / ...
  - increment modes — linear / log; increase / decrease / alternate / nested / random
  - support of P/n — y / n
  - linking of sequences — y / n
- **Signal averaging:**
  - on-line averaging — y / n
  - display of cum. averages — y / n
- **Display:**
  - options for trace display — overlay / zeroline subtraction / leak subtraction on / off / compare with reference trace
  - display of data structure — y / n
  - digital filtering — y / n
- **On-line analysis:**
  - x — potential / duration / time / index / ...
  - y — min. / max. / mean / variance / ...
- **Data editing:**
  - inspection of acquired data — y / n
  - edit — delete / average / compress / scale

Single-Channel Analysis

- **Display:**
  - multiple time scales — y / n
  - digital filtering — y / n
  - spline interpolation — y / n
  - mark traces e.g. "blank" or "bad" — y / n
- **Leak correction:**
  - subtraction of averaged and smoothed blank or control traces — y / n
- **Event detection:**
  - baseline — float / auto
  - threshold — 50% / time course / use splined data
  - table entries — time / amplitude / level / pre-dur. / post-dur / sublevel / man/auto / ...
  - correction for filter delay — y / n
- **Read event tables:**
  - selection criteria — level / Δ time / class / amplitude / duration / latency / ...
  - stationarity analysis — y / n
- **Amplitude histograms:**
  - display — lin / log
  - fit — multiple Gaussians / ...
- **Dwell-time histograms:**
  - display — lin / log / log/log / sqrt/log
  - fit — mult. exponentials to bins / events
  - missed event correction — y / n
  - two-dimensional histogram — y / n
- **Open-channel analysis:**
  - open-channel histograms — y / n
  - mean-variance methods — y / n
  - conditional averaging — y / n
  - ramp analysis — y / n
- **Fits:**
  - optimization — Simplex / Levenberg-Marquardt / ...
  - standard error provided — y / n
  - fit criterion — least squares / maximum likelihood / ...
  - lim. bandwidth considered — y / n

Pulsed-Data Analysis

- **Display, data editing, and on-line analysis:**
  - see "Data Acquisition"
- **Leak correction:**
  - off-line leak subtraction — y / n
  - un-do on-line leak subtr. — y / n
- **Cursor operations:**
  - measure current & time — y / n
  - set windows for analysis — y / n
- **Fit of raw data traces:**
  - mode — manual / auto
  - fit — polyn. / expon. / Hodgkin & Huxley / ...
  - automatic fit of entire family — y / n
- **Ramp analysis:**
  - displ. as function of potential — y / n
  - determine reversal potential — y / n
  - fit of theor. functions — y / n
- **Secondary analysis:**
  - fit functions — polyn. / expon. / Boltzmann / current-voltage / dose-response / ...
- **Noise analysis:**
  - spectral analysis — y / n
  - fit functions — Lorentzian / $1/f^n$ / ...
  - non-stationary analysis — y / n
- **Fits:**
  - see "Single-Channel Analysis"
  - limited clamp speed considered — y / n

Figure 13. Checklist for essential features of software packages for data acquisition and analysis of single-channel recordings as well as pulsed data.

generation of new code is always associated with new problems and should only be considered if there is no other simpler and more economical way of solving a problem.

6.2.2. Acquisition and Analysis Systems

It is time-consuming but very important to check whether the planned experiments can actually be performed with the software under consideration. However, there is no ideal way of testing a software package other than by trying to execute an experiment with a demo version. Nevertheless, the collection of requirements and features illustrated in Fig. 13 may be used as a checklist when considering the purchase or design of acquisition or analysis software.

6.2.3. Test of the Software

A golden rule is that one should never rely blindly on software. One cannot test all the features of a complex acquisition and analysis program, but one should at least try to check critical parameters such as the voltage scaling, timing, and current gain. Errors in these parameters may indicate problems with hardware compatibility (e.g., clock rate or AD converter scaling). Analysis functions are best tested with simulated data. Alternatively, the results of the same analysis obtained using different programs can be compared.

ACKNOWLEDGMENT. I should like to thank A. Elliott and K. Gillis for their helpful comments on the manuscript and T. Schlief for his assistance in the preparation of figures.

References

Armstrong, C. M., and Bezanilla, F., 1974, Charge movement associated with the opening and closing of the activation gates of the sodium channel, *J. Gen. Physiol.* **63:**533–552.

Auerbach, A., 1993, A statistical analysis of acetylcholine receptor activation in *Xenopus* myocytes: Stepwise versus concerted models of gating, *J. Physiol.* **461:**339–378.

Bezanilla, F., 1985, A high capacity data recording device based on a digital audio processor and a video cassette recorder, *Biophys. J.* **47:**437–442.

Blatz, A. L., and Magleby, K. L., 1986, Correcting single channel data for missed events, *Biophys. J.* **49:**967–980.

Blatz, A. L., and Magleby, K. L., 1989, Adjacent interval analysis distinguishes among gating mechanisms for the fast chloride channel from rat skeletal muscle, *J. Physiol.* **410:**561–585.

Chung, S. H., Moore, J. B., Xia, L. G., Premkumar, L. S., and Gage, P. W., 1990, Characterization of single channel currents using digital signal processing techniques based on hidden Markov models, *Phil. Trans. R. Soc. Lond. B* **329:**265–285.

Chung, S. H., Krishnamurthy, V., and Moore, J. B., 1991, Adaptive processing techniques based on hidden Markov models for characterizing very small channel currents buried in noise and deterministic interferences, *Phil. Trans. R. Soc. Lond. B* **334:**357–384.

Colquhoun, D. 1987, Practical analysis of single channel records, in: *Microelectrode Techniques. The Plymouth Workshop Handbook* (N. B. Standen, P. T. A. Gray, and M. J. Whitaker, eds.), pp. 83–104, Company of Biologists Ltd., Cambridge.

Colquhoun, D., and Sakmann, B., 1985, Fast events in single-channel currents activated by acetylcholine and its analogues at the frog muscular end-plate, *J. Physiol.* **369:**501–557.

Crouzy, S. C., and Sigworth, F. J., 1990, Yet another approach to the dwell-time omission problem of single-channel analysis, *Biophys. J.* **58:**731–743.

DeFelice, L. J., 1981, *Introduction to Membrane Noise,* Plenum Press, New York.

Dempster, J., 1993, *Computer Analysis of Electrophysiological Signals,* Academic Press, London.

Fredkin, D. R., and Rice, J. A., 1992, Maximum likelihood estimation and identification directly from single-channel recordings, *Proc. R. Soc. Lond. B* **249:**125–132.

French, R. J., and Wonderlin, W. F., 1992, Software for acquisition and analysis of ion channel data: Choices, tasks, and strategies, in: *Ion Channels, Methods in Enzymology,* Vol. 207 (B. Rudy and L. E. Iverson, eds.), pp. 711–728, Academic Press, San Diego.

Hawkes, A. G., Jalali, A., and Colquhoun, D., 1991, The distribution of open and shut times in a single channel record when brief events cannot be detected, *Phil. Trans. R. Soc. Lond. A* **332:**511–538.

Heinemann, S. H., and Conti, F., 1992, Non-stationary noise analysis and its application to patch clamp recordings, in: *Ion Channels, Methods in Enzymology,* Vol. 207 (B. Rudy and L. E. Iverson, eds.), pp. 131–148, Academic Press, San Diego.

Heinemann, S. H., and Sigworth, F. J., 1988, Open channel noise. IV. Estimation of rapid kinetics of formamide block in gramicidin A channels, *Biophys. J.* **54:**757–764.

Heinemann, S. H., and Sigworth, F. J., 1989, Estimation of Na^+ dwell time in the gramicidin A channel. Na ions as blockers of H^+ currents, *Biochim. Biophys. Acta* **987:**8–14.

Heinemann, S. H., and Sigworth, F. J., 1989, Estimation of Na^+ dwell time in the gramicidin A channel. Na ions as blockers of H^+ currents, *Biochim. Biophys. Acta* **987:**8–14.
Heinemann, S. H., and Sigworth, F. J., 1990, Open channel noise. V. A fluctuating barrier to ion entry in gramicidin A channels, *Biophys. J.* **57:**499–514.
Heinemann, S. H., and Sigworth, F. J., 1991, Open channel noise. VI. Analysis of amplitude histograms to determine rapid kinetic parameters, *Biophys. J.* **60:**577–587.
Heinemann, S. H., Conti, F., and Stühmer, W., 1992, Recording of gating currents from *Xenopus* oocytes and gating noise analysis, in: *Ion Channels, Methods in Enzymology,* Vol. 207 (B. Rudy and L. E. Iverson, eds.), pp. 353–368, Academic Press, San Diego.
Hess, P., Lansman, J. B., and Tsien, R. W., 1984, Different modes of Ca channel gating behaviour favoured by dihydropyridine Ca agonists and antagonists, *Nature* **311:**538–544.
Hodgkin, A. L., and Huxley, A. H., 1952, A quantitative description of membrane current and its application to conduction and excitation in nerve, *J. Physiol.* **117:**500–544.
Hoshi, T., Zagotta, W. N., and Aldrich, R. W., 1990, Biophysical and molecular mechanisms of *Shaker* potassium channel inactivation, *Science* **250:**533–538.
Jackson, M. B., 1992, Stationary single-channel analysis, in: *Ion Channels, Methods in Enzymology,* Vol. 207 (B. Rudy and L. E. Iverson, eds.), pp. 729–746, Academic Press, San Diego.
Kirlin, R. L., and Moghaddamjoo, A., 1986, A robust running window detector and estimator for step signals in contaminated Gaussian noise, *IEEE Trans. Acoust. Speech Signal Process.* **34:**816–823.
Magleby, K. L., 1992, Preventing aritfacts and reducing errors in single-channel analysis, in: *Ion Channels, Methods in Enzymology,* Vol. 207 (B. Rudy and L. E. Iverson, eds.), pp. 763–791, Academic Press, San Diego.
Magleby, K. L., and Weiss, D. S., 1990, Estimating kinetic parameters for single channels for simulation. A general method that resolves the missing event problem and accounts for noise, *Biophys. J.* **58:**1411–1426.
McManus, O. B., Blatz, A. L., and Magleby, K. L., 1987, Sampling, log binning, fitting and plotting durations of open and shut intervals from single channels and the effect of noise, *Pflügers Arch.* **410:**530–553.
Moghaddamjoo, A., Levis, R. A., and Eisenberg, R. S., 1988, Automatic detection of channel currents, *Biophys. J.* **53:**153a.
Neher, E., and Stevens, C. F., 1977, Conductance fluctuations and ionic pores in membranes, *Ann. Rev. Biophys. Bioeng.* **6:**345–381.
Nyquist, H., 1928, Certain topics in telegraph transmission theory, *Trans AIEE* **47.**
Pastushenko, V. P., and Schindler, H., 1993, Statistical filtering of single channel ion channel records, *Acta Pharm.* **43:**7–13.
Patlak, J. B., 1988, Sodium channel subconductance levels measured with a new variance–mean analysis, *J. Gen. Physiol.* **92:**413–430.
Patlak, J. B., 1993, Measuring kinetics of complex single ion channel data using mean-variance histograms, *Biophys. J.* **65:**29–42.
Roux, B., and Sauve, R., 1985, A general solution to the time interval omission problem applied to single channel analysis, *Biophys. J.* **48:**149–158.
Sachs, F., Neil, J., and Barkakati, N., 1982, The automated analysis of data from single ionic channels, *Pflügers Arch.* **395:**331–340.
Sherman-Gold, R. (ed.), 1993, *The Axon Guide,* Axon Instruments, Inc., Foster City, CA.
Sigworth, F. J., 1980, The variance of sodium current fluctuations at the node of Ranvier, *J. Physiol.* **307:**97–129.
Sigworth, F. J., 1985, Open channel noise. I. Noise in acetylcholine receptor currents suggests conformational fluctuations, *Biophys. J.* **47:**709–720.
Sigworth, F. J., and Sine, S. M., 1987, Data transformations for improved display and fitting of single-channel dwell time histograms, *Biophys. J.* **48:**149–158.
Sigworth, F. J., and Zhou, J., 1992, Analysis of nonstationary single-channel currents, in: *Ion Channels, Methods in Enzymology,* Vol. 207 (B. Rudy and L. E. Iverson, eds.), pp. 746–762, Academic Press, San Diego.
Yellen, G., 1984, Ionic permeation and blockade in Ca^{2+}-activated K^+ channels of bovine chromaffin cells, *J. Gen. Physiol.* **84:**157–186.
Zhou, J., Potts, J. F., Trimmer, J. S., Agnew, W. S., and Sigworth, F. J., 1991, Multiple gating modes and the effect of modulating factors on the μI sodium channel, *Neuron* **7:**775–785.

Part II

METHODS

Chapter 4

Electronic Design of the Patch Clamp

F. J. SIGWORTH

1. Introduction

The patch-clamp amplifier is fundamentally a sensitive current-to-voltage converter, converting small (picoampere to nanoampere) pipette currents into voltage signals that can be observed with an oscilloscope or sampled by a computer. This chapter describes the basic principles of the current-to-voltage (I–V) converter and describes some of the technology of I–V converters having wide bandwidth and low noise. Also considered in this chapter are additions to the basic I–V converter design for capacitive-transient cancellation and series-resistance compensation, features that are particularly useful for whole-cell current recording.

The most important property of an I–V converter for single-channel recording is its noise level. Under good patch-recording conditions, the background noise introduced by the patch membrane, together with the seal conductance and noise sources in the pipette, together corresponds to the movement of a few tens of elementary charges. It is important to maintain this low noise level in the circuitry of the I–V converter because there are many types of channels whose currents are near the limit of resolution of the recording system.

In whole-cell recordings an important problem is the series access resistance between the pipette electrode and the cell interior. A high series resistance limits the speed with which voltage changes can be imposed on the cell membrane and limits the time resolution of current recordings. The access resistance is determined by the size of the pipette tip and the nature of the access to the cell interior; however, the effects of series resistance can be reduced by circuitry that is discussed later in this chapter.

2. Current-Measurement Circuitry

2.1. Current–Voltage Converter

The standard way to measure small currents is to monitor the voltage drop across a large resistor. Figure 1 shows three circuits that accomplish this. In part A, a battery with voltage V_{ref} is used to set the pipette potential, and the pipette current I_p is measured from the voltage drop I_pR across the resistor. The problem with this configuration is that the pipette

F. J. SIGWORTH • Department of Cellular and Molecular Physiology, Yale University School of Medicine, New Haven, Connecticut 06510.
Single-Channel Recording, Second Edition, edited by Bert Sakmann and Erwin Neher. Plenum Press, New York, 1995.

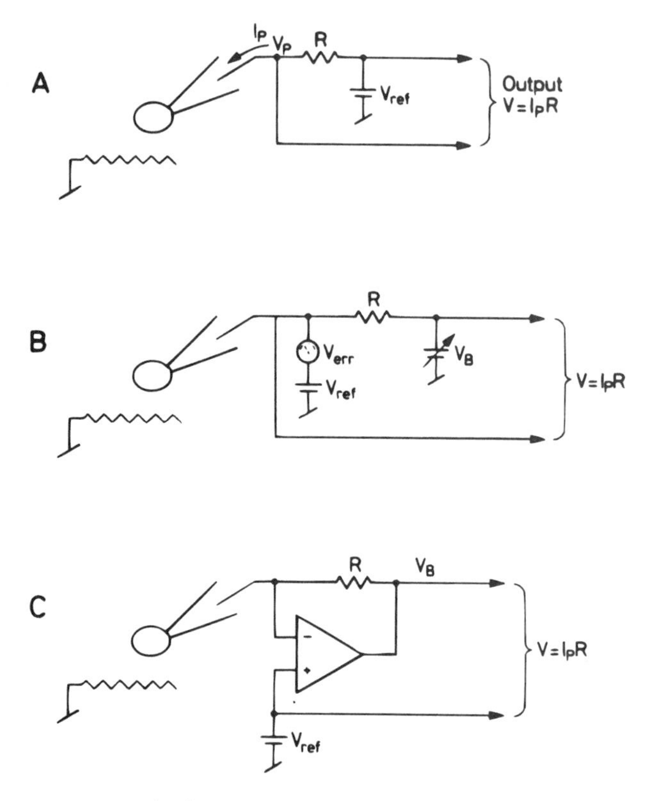

Figure 1. Current measurement circuits. A: A simple circuit in which the pipette potential is determined approximately by V_{ref}. In B, the deviation V_{err} of the pipette potential from V_{ref} is monitored, and V_B is adjusted to keep it zero. In C, the op-amp makes the adjustment automatically.

potential is not exactly equal to V_{ref} but has an error that depends on the current; when the resistor R is made large to give high sensitivity, the voltage error also becomes large.

The solution to this problem (Fig. 1B) is to measure V_p directly and to continuously adjust a voltage source V_B to bring V_p to the correct value. Provided that the adjustment is made quickly and accurately, R can be made very large for high sensitivity.

An operational amplifier (op-amp) can be used to automate the adjustment of V_B. The op-amp can be thought of as a voltage-controlled voltage source; the output voltage changes in response to differences in the voltages V_+ and V_- at the input terminals according to

$$\frac{dV_{OUT}}{dt} = \omega_A(V_+ - V_-) \tag{1}$$

The factor ω_A can be very large: for typical commercial op-amps, it is about 10^7 sec^{-1}, which means that a 1-mV difference on the input terminals causes the output voltage to slew at 10^4 V/sec. The value ω_A is the *gain–bandwidth product* of the amplifier. In this chapter, ω will be used to represent an "angular frequency" in units of radians/sec. The relationship to the more familiar frequencies f (given in Hertz) is $\omega = 2\pi f$, so that ω_A is 2π times the gain–bandwidth product f_A that is normally given in amplifier specification sheets. (Sometimes

f_A is given as the "unity-gain bandwidth" of an op-amp, which for our purposes is essentially the same thing.)

At the same time, the op-amp draws essentially no current through its input terminals. This is an important feature, since such currents would disturb the measurement. (For a more complete discussion of op-amps and feedback, see, for example, Horowitz and Hill, 1989.)

Part C of Fig. 1 shows the final current-to-voltage converter circuit. The op-amp varies its output to keep the pipette potential at V_{ref}. This action can be made very rapid and precise, so that for practical purposes V_p can be assumed to be precisely V_{ref}. This in turn allows us to measure $V_B - V_{ref}$ as shown, rather than $V_B - V_p$, to obtain the voltage drop across the resistor. The voltage differences should be the same, but the former measurement is preferable because it avoids an additional direct connection to the pipette electrode. The voltage difference is usually measured using a standard differential amplifier circuit (not shown).

2.2. Dynamics of the *I–V* Converter

For single-channel recording, suitable values for the current-measuring resistance R_f are on the order of 10–100 GΩ. Commercial resistors in this range typically have a shunt capacitance C_f of 0.1 pF (Figure 2A); the resulting time constant, $\tau_f = R_f C_f$, is the order of 1 msec and limits the time resolution of the *I–V* converter.

Assuming that the op-amp acts instantaneously ($\omega_A \rightarrow \infty$), the response characteristics of the *I–V* converter are given by a transfer function $Z_c(s)$, which can be used to give the response at V_{OUT} for any input current I_p,

$$Z_c(s) = \frac{V_{OUT}(s)}{I_p(s)} = \frac{R_f}{\tau_f s + 1} \tag{2}$$

This function can be used in two ways (see, for example, Aseltine, 1958). First, if the imaginary frequency $j\omega$ is substituted for s, the resulting magnitude and phase of the (complex-valued) Z_c give the amplitude ratio and phase shift of V_{OUT} relative to I_p. Thus, Z_c gives the "frequency response" of the circuit. For convenience, Z_c can be rewritten as

$$Z_c(s) = R_f T(s) \tag{3}$$

where T is dimensionless and has the form

$$T(s) = \frac{1}{\tau_f s + 1} \tag{4}$$

which is the transfer function of a simple low-pass filter. T is unity at low frequencies ($s \rightarrow 0$) but rolls off at high frequencies. The "corner frequency," at which the response is down by 3 dB, occurs when $\tau_f \omega = 1$.

The second use of the transfer function is to calculate the time course of V_{OUT} for an arbitrary I_p. This is done using the inverse Laplace transform. For example, the response to a step of input current is found in this way to be exponential with a time constant τ_f.

The reason for going to all the trouble of introducing the transfer function is that things become more complicated when ω_A is assumed to be finite. We define $\tau_A = 1/\omega_A$, the characteristic time constant of the op-amp and $C_t = C_f + C_{in}$, the total capacitance on the

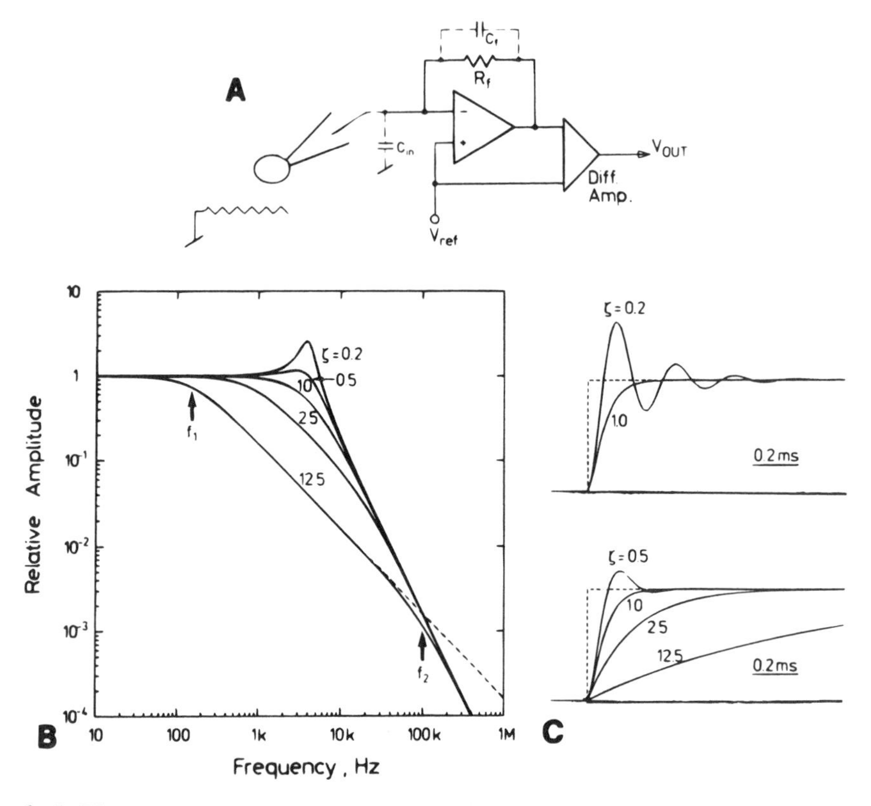

Figure 2. A: Diagram of the *I–V* converter showing the stray feedback capacitance C_f and the total input capacitance C_{in}. A unity-gain differential amplifier on the output produces the current monitor signal. B: Magnitude of the transfer function T as the damping factor ζ is varied. The curves were calculated for the circuit of part A with C_{in} = 10 pF, R_f = 10 GΩ, and f_A = 10 MHz (τ_A = 16 nsec). The ζ values of 12.5, 2.5, 1, and 0.2 correspond to C_f values of 100, 20, 8, and 1.6 fF, respectively. The dashed curve is the response from a single time constant τ_1 = 1 msec (corresponding to f_1 = 160 Hz as indicated, with C_f = 100 fF). The corresponding τ_2 value is 1.6 μsec (f_2 = 100 kHz). C: Variation of the step response time course with ζ.

input terminal. Using equation 1 to describe the op-amp's behavior, we can write the differential equation for V_{OUT} as

$$\tau_A R_f C_t \frac{d^2 V_{OUT}}{dt^2} + (\tau_A + \tau_f) \frac{dV_{OUT}}{dt} + V_{OUT} = R_f I_p(t) \tag{5}$$

The equivalent transfer-function representation of the response is

$$Z_c(s) = \frac{R_f}{\tau_A R_f C_t s^2 + (\tau_A + \tau_f)s + 1} \tag{6}$$

Once again, let us write $Z_c = R_f T$, where T can be written in the form

$$T(s) = \frac{1}{(\tau_1 s + 1)(\tau_2 s + 1)} \tag{7}$$

This is the transfer function of two simple filters in cascade: τ_1 and τ_2 are found as roots of a quadratic equation, but provided τ_A is sufficiently short, they can be approximated by

$$\tau_1 \simeq \tau_f \tag{8}$$

$$\tau_2 \simeq \frac{C_t}{C_f}\tau_A$$

The frequency response now has two corner frequencies, $\omega_1 = \tau_1^{-1}$, caused by the stray capacitance across the feedback resistor, and the higher cutoff $\omega_2 = \tau_2^{-1}$ which arises from the finite speed of the op-amp. To improve the frequency response of the *I–V* converter, one usually tries to reduce C_f. This reduces τ_1 but lengthens τ_2 at the same time, usually by about the same factor, since C_f makes only a minor contribution to C_t. If this is carried to an extreme such that τ_1 and τ_2 become comparable in size, the approximations (equation 8) are no longer valid. Instead, it is more useful to write T in the equivalent form

$$T(s) = \frac{1}{\tau_0^2 s^2 + 2\zeta\tau_0 s + 1} \tag{9}$$

which is the equation for a damped harmonic oscillator with natural frequency $\omega_0 = \tau_0^{-1}$ and damping factor ζ. These parameters are given by

$$\tau_0 = (\tau_A R_f C_t)^{1/2} \tag{10}$$

$$\zeta = \frac{1}{2}\frac{\tau_A + \tau_f}{\tau_0}$$

The surprise here is that τ_0 does not depend directly on C_f but only on the total capacitance C_f. However, since τ_A is typically very small compared to τ_f, the damping factor is proportional to C_f.

When C_f is reduced beyond a certain point, then the bandwidth of the *I–V* converter does not increase further; instead, the frequency response begins to show a peak, and the step response shows "ringing" (Fig. 2B,C). For the best transient response, ζ should be kept in the vicinity of unity: $\zeta = 1$ corresponds to a "critically damped" step response with no overshoot; $\zeta = 0.71$ gives the "maximally flat" frequency response but about 10% overshoot in the step response.

It is quite difficult to obtain a high natural frequency in a sensitive *I–V* converter. For example, if we use a 10^{10}-Ω feedback resistor, and the total input capacitance $C_t = 10$ pF to obtain a bandwidth of 10 kHz—corresponding to $\tau_0 = 16$ μsec (equation 7)—requires that $\tau_A = 2.6 \times 10^{-9}$ sec or a gain–bandwidth product of about 60 MHz for the amplifier. At the same time, the stray feedback capacitance would have to be kept to 3.2×10^{-15} F. Both of these requirements are not readily achieved.

A better strategy for wide-band recording is to design the *I–V* converter to have a nonideal but well-defined frequency response characteristic and then to correct the frequency response in a later amplifier stage. The best way to do this is to allow C_f to be large enough, and choose the op-amp to be fast enough, so that the rolloff characterized by τ_2 (equation 8) is well beyond the frequency range of interest. Within the range of interest, only the single rolloff of τ_1 is then present, and this can be compensated as is described in the next section.

Making τ_2 small has several other advantages. First, τ_2 depends on C_t, which includes contributions from the pipette and stray capacitances. Since these can vary, τ_2 depends on the experimental conditions; if τ_2 is very small, these variations can be ignored. Second, τ_2 describes the response time of the I–V converter as a voltage clamp. The transfer function relating V_p to the command voltage V_{cmd} is

$$T_{VC} = \frac{V_p(s)}{V_{cmd}(s)} = \frac{1 + \tau_f s}{\tau_A R_f C_t s^2 + (\tau_A + \tau_f)s + 1} \tag{11}$$

The denominator of equation 11 is the same as that of T. The numerator approximately cancels the factor $(1 + \tau_l s)$ in equation 7, so the clamp transfer function is approximately $T_{VC} \simeq 1/(\tau_2 s + 1)$ when $\tau_2 \ll \tau_l$.

In a practical situation, the op-amp might have a gain–bandwidth product of 10 MHz. With $C_f = 0.1$ pF and $C_t = 10$ pF, then $\tau_2 = 1.6$ μsec, giving a clamp bandwidth of 100 kHz. τ_l in this case would be 1 msec, so the first rolloff in the frequency would occur at 160 Hz. The lowest curve in Fig. 2B shows this response, which corresponds to a damping factor of 12.5.

2.3. Correcting the Frequency Response

When C_f is chosen as just described, the I–V converter will have a very nearly exponential step response with a fairly long time constant around 1 msec (Fig. 3, top trace). The role of the correction circuit is to perform an "inverse filtering" operation to recover a faster-rising response. One way of looking at the correction operation is to notice that the derivative of the exponential step response function is itself an exponential and to exploit this fact by summing the original response with a scaled copy of its derivative to recover the original form (Fig. 3, bottom trace).

Fortunately, this particular strategy works for all possible input waveforms, not just steps. The operation just described has the transfer function

$$T_{co} = \tau_c s + 1 \tag{12}$$

where τ_c is the factor scaling the derivative. When the I–V converter transfer function T (equation 7) is multiplied by T_{co}, the rolloff caused by τ_l can be canceled exactly when $\tau_c = \tau_l$.

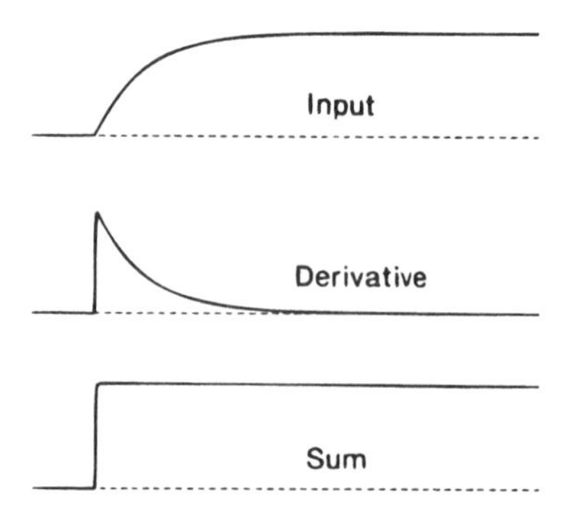

Figure 3. A strategy for correcting the response of the I–V converter. The input to the correction circuit shows a slow exponential step response. When it is summed with a scaled copy of its time derivative a nearly square step response results.

A practical compensation circuit is shown in Fig. 4A. Many implementations are possible, but this particular circuit is useful because its low-frequency gain is fixed at unity. At low frequencies, the capacitor C acts as an open circuit, and the op-amp acts just as a voltage follower. The increasing gain with frequency arises from the decreasing impedance of C; the "corner" time constant τ_c is equal to $(R_1 + R_2)C$.

The magnitude of the ideal transfer function (equation 12) increases without limit as the frequency variable s increases. This sort of behavior cannot be achieved in practice. The actual transfer function of the circuit in Fig. 4A is

$$T_{co} = \frac{\tau_c s + 1}{\tau_A \tau_c s^2 + (\tau_A + R_2 C)s + 1} \tag{13}$$

which is the desired response multiplied by a second-order transfer function that can be written in the form of equation 9 with

$$\tau_0 = \sqrt{\tau_A \tau_c} \tag{14}$$

$$\zeta = \frac{1}{2}\frac{\tau_A + R_2 C}{\tau_0}$$

The maximum useful frequency of this circuit is given immediately by $\omega_0 = \tau_0^{-1}$. If a 10-MHz op-amp were used in the circuit ($\tau_A = 0.016$ μsec) and $\tau_c = 1$ msec, τ_0 would be 4 μsec, giving a useful bandwidth of about 40 kHz. The importance of R_2 can be seen from the expression for ζ. Since τ_A is negligibly small, R_2 should be chosen to give an R_2C time constant on the order of τ_0. If R_2 is zero, there will be no damping, and the circuit will oscillate.

The correction circuit in effect "removes" the corner in the I–V converter's frequency response (Fig. 4B) and replaces it with a new, second-order rolloff at a much higher frequency.

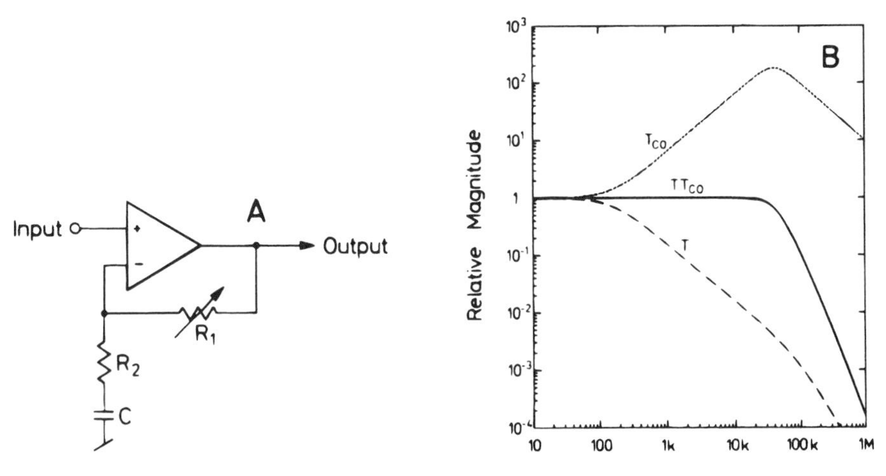

Figure 4. A: Response-correction circuit. The variable resistor R_1 allows the time constant τ_c to be varied to match the time constant τ_f of the I–V converter. B: Frequency response of the I–V converter (T), the correction circuit (T_c), and the composite (TT_c). The correction circuit response was computed for $\tau_c = 1$ msec, $\tau_A = 16$ nsec ($f_A = 10$ MHz), and $\zeta = 0.7$. The resulting overall bandwidth is 40 kHz.

The effective removal of the low-frequency corner requires that the time constant τ_c be precisely matched to the time constant τ_l of the *I–V* converter. For this reason, R_l is best made a variable resistance that is trimmed for the proper step response. The error in the step response when a mismatch is present is a relaxation of relative amplitude $(\tau_c - \tau_l)/\tau_l$ and time constant τ_l.

How far can the frequency response be extended in this way? One limitation is the second time constant τ_2 in the *I–V* converter, which can be shortened by using a sufficiently fast op-amp there. As we saw in Section 2.2, an intrinsic bandwidth of 100 kHz is not difficult to achieve. The other limitation is τ_0 of the correction circuit; op-amp speed is the limiting factor here also. To increase the gain–bandwidth product, two op-amps can be operated in cascade, with one of them provided with local feedback to make it an amplifier with fixed gain *A*. Provided the bandwidth of this amplifier stage is large compared to τ_0^{-1}, the composite can be used in the circuit as a "super op-amp" with the gain–bandwidth product increased by the factor *A*. Extension of the *I–V* converter's bandwidth by a factor of 1000 (e.g., to 100 kHz or more) can be performed, provided care is taken in the grounding and layout of the circuitry.

Up to this point, we have assumed that the *I–V* converter's response in the frequency range of interest can be represented as a single-time-constant rolloff caused by the stray capacitance of the feedback resistor. We have modeled that capacitance as a "lumped" quantity, C_f, but it is actually distributed along the length of the resistor. If the distribution is uniform (Fig. 5A), the impedance is the same as that in the parallel *R–C* model. If it is not, the impedance of the combination, and therefore the transfer function of the *I–V* converter, will show an additional "step" in its frequency dependence. Figure 5C shows a circuit for performing a two-time-constant correction for a response function of this kind. The circuit is relatively complex (requiring three op-amps) but has the advantage of being relatively easy to adjust, with minimal interaction among the controls.

Testing the frequency response requires a precise, high-impedance source of picoampere currents. Since commercial high-value resistors have substantial stray capacitance, injecting a current into the *I–V* converter with a resistor is suitable only for testing the DC gain. For dynamic testing, a capacitor is the circuit element of choice, since nearly ideal capacitors in the appropriate range of 0.01 to 1 pF are easy to make by bringing two conductors near each other. Such homemade capacitors are usually better than commercial ones, which often do not have low enough leakage conductance. For estimating small capacitances, recall that for two parallel conducting plates of area *A* having a small spacing *d*,

$$C = \epsilon_0 A/d$$

where, in appropriate units, $\epsilon_0 = 0.089$ pF/cm.

Connecting a wire to the output of a signal generator and holding it near the pipette or the input terminal of the *I–V* converter often makes an acceptable signal source. An appropriate waveform to apply in this way is a triangle wave, because the coupling capacitance carries a current that is the time derivative of the applied voltage, and the derivative of a triangle wave is a square wave. For critical use, such as in adjusting the compensation circuitry, a function generator with especially good linearity should be used. A nonlinearity of a few percent (not uncommon) results in a "drooping" of the injected current by the same amount.

An arbitrary current waveform could be injected if the test signal were electronically integrated before being differentiated by the coupling capacitance. A standard integrator circuit (Fig. 6A) will not work in practice because unavoidable DC offsets in the input signal are also integrated and quickly drive the output into saturation. A slowly acting feedback

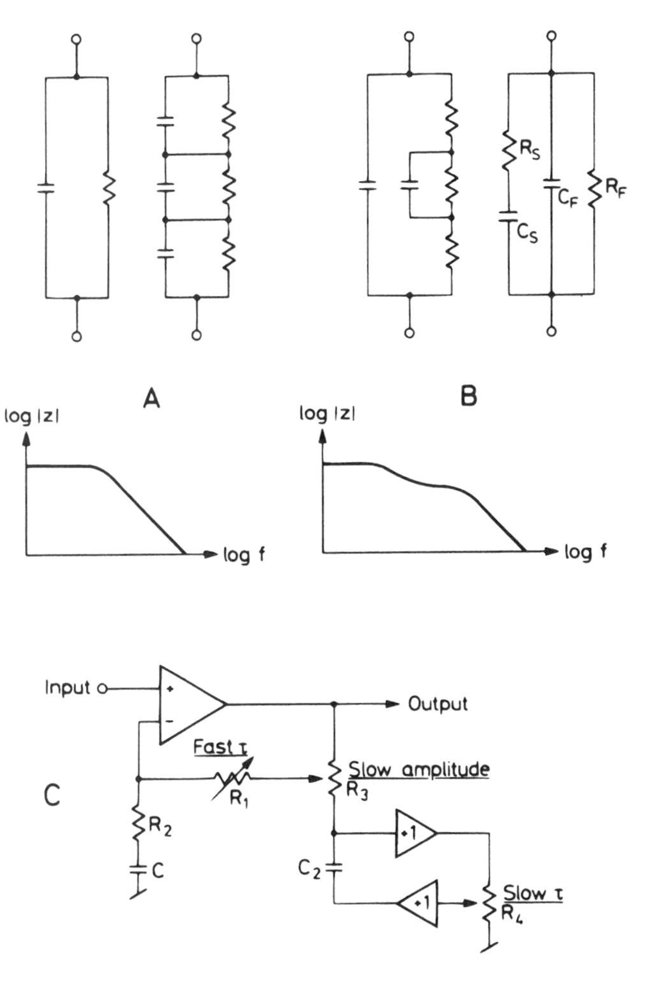

Figure 5. Distributed capacitance in the feedback resistor. A: An evenly distributed capacitance results in the same frequency dependence of Z (and thermal noise) as a lumped capacitance. B: An uneven distribution introduces an extra dispersion in Z. C: A circuit for correcting the frequency response as would result from the network in B. R_3 is chosen small compared to R_1 to minimize interaction between the controls. Because a large-valued capacitor is then needed to set the slow time constant, a large variable capacitor is synthesized using the two voltage followers and R_4 and C_1. The effective capacitance is adjustable from near zero up to that of C_2.

pathway can, however, compensate for offsets without distorting practical test signals. A circuit of this kind is shown in Fig. 6B; its feedback loop gives a lower corner frequency of 0.5 Hz and causes less than a 1% droop for pulses up to 30 msec in length. The advantage of this sort of circuit is that it allows the frequency response or the response to short pulses to be checked directly by applying sine waves or pulses to the input.

2.4. Capacitor-Feedback *I–V* Converter

Since higher-valued resistors yield increased sensitivity, why not go to the limit of infinite resistance and have only a capacitor as the feedback element in the *I–V* converter?

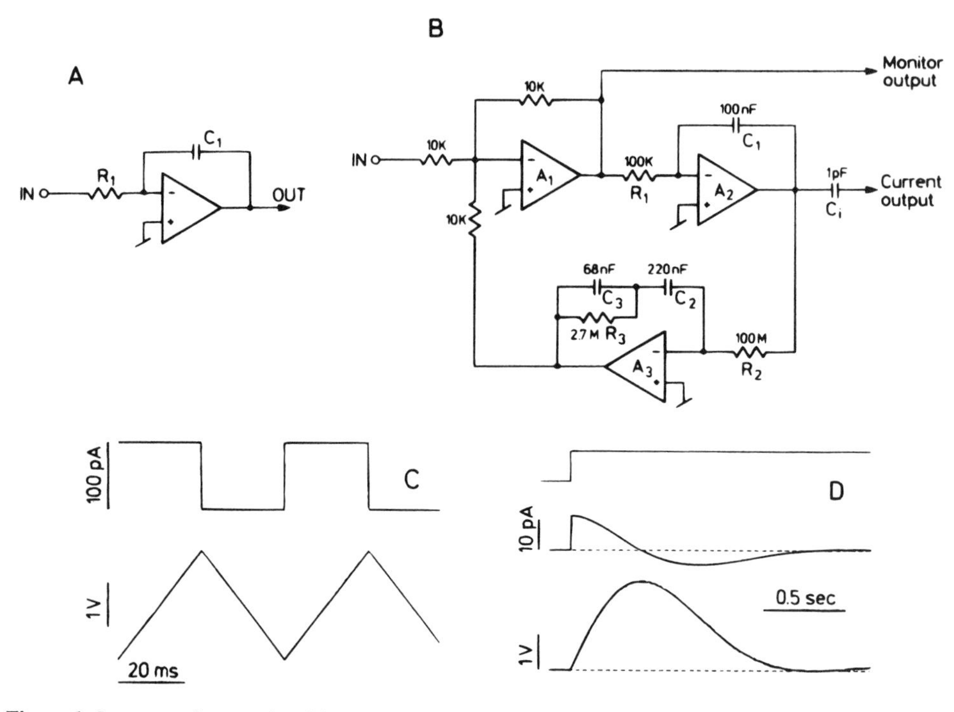

Figure 6. Integrator for test signal injection. A: Basic integrator circuit. Its output voltage is given by $I_{out} = (1/R_1C_1)\int V_{in}dt$. B: The integrator ($R_1$,$C_1$,$A_2$) is embedded in a slow feedback loop ($f_0 = 0.5$ Hz) to correct for any DC component of V_{in}. A monitor output shows the actual voltage being integrated. The current output is connected directly to the I–V converter input; its scale factor is $C_i/(R_1C_1)$, which in this case is 100 pA/V. C: Injected current (upper trace) and the voltage applied to C_i (lower trace) when a 1-V, 20-Hz square wave is applied as V_{in}. D: Response to a voltage step (top trace) on a slower time scale, showing the action of the slow feedback on the injected current (middle trace) and the integrator output (bottom trace).

This is the approach taken in capacitor-feedback or "integrating headstage" amplifier designs (Finkel, 1992). In such a design (Fig. 7) the headstage amplifier output voltage V_1 is related to the input current I_p (assuming for now that V_{cmd} is constant) by

$$\frac{dV_1}{dt} = \frac{I_p}{C_f}$$

so that the output voltage is the time integral of the pipette current. Correction of the frequency response is performed by a differentiator circuit (A_3, C_d, and R_d in the figure) so that the final output voltage V_0 of the circuit shown is given by

$$V_o = R_d \frac{C_d}{C_f} I_p. \tag{15}$$

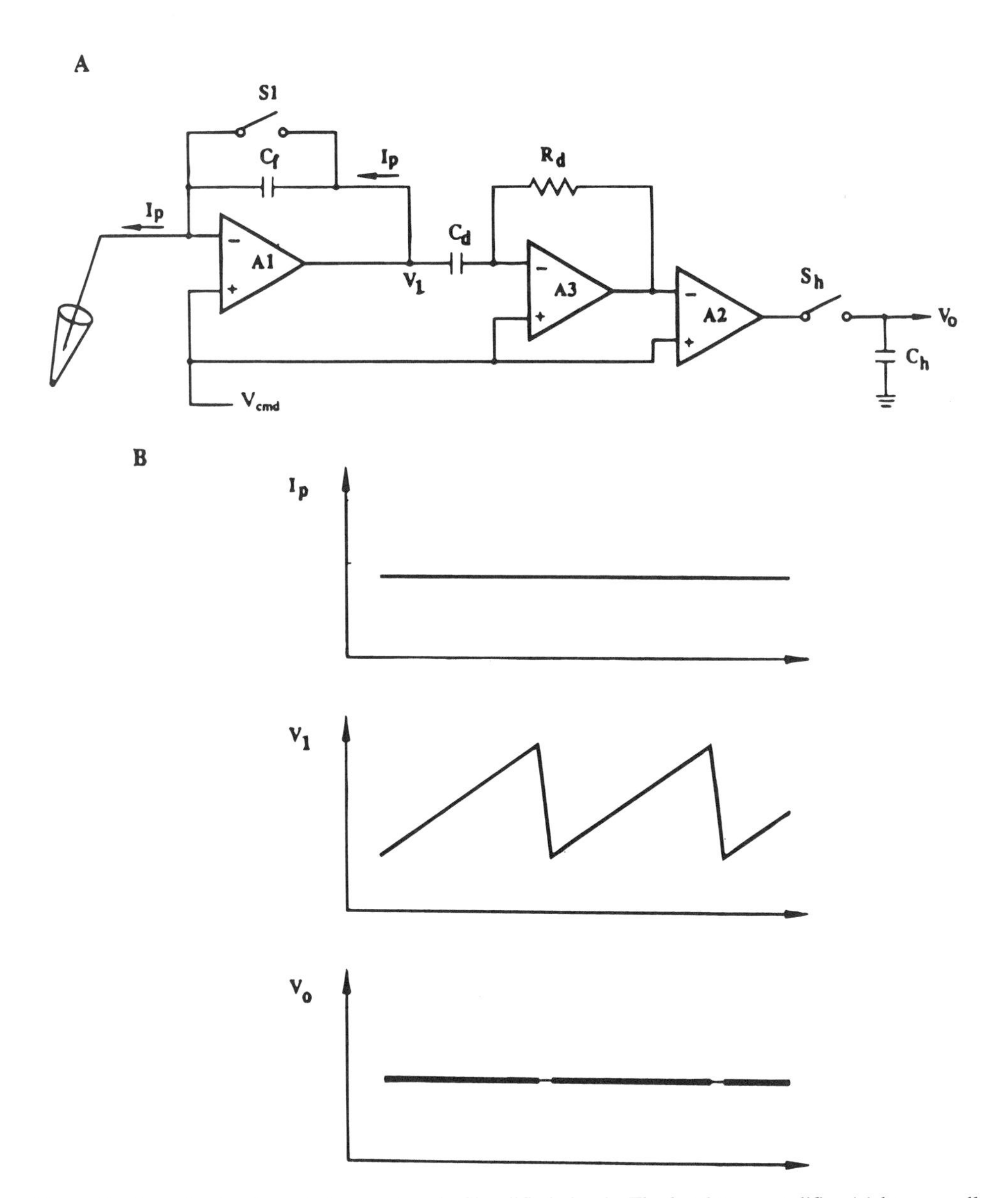

Figure 7. Capacitor-feedback *I–V* converter. A: Simplified circuit. The headstage amplifier A1 has a small capacitor C_f as its feedback element. Its output voltage V_1 is differentiated by A3. Differential amplifier A2 subtracts the command voltage from the result. Switch S1 periodically discharges C_f to prevent saturation of V_1; sampling switch S_h opens during the discharge time to mask the large transient voltages. B: Signal waveforms. A steady positive pipette current I_p results in a positive ramp at V_1 between discharges. The output voltage V_0 is held during discharges, resulting in a constant output. From Finkel (1992).

The only problem with having a capacitor as the feedback element for A_{I} is that there is no pathway for handling steady input currents. A constant current will cause the voltage V_{I} to rise linearly with time until A_{I} is driven into saturation; once this occurs, the input voltage will no longer remain equal to V_{cmd}, and the output voltage will not reflect I_{p}. The solution is to periodically discharge the feedback capacitor to bring the voltage across it back to zero. During the time that the discharge switch S_{I} is closed, the voltage at the output of A_2 does not reflect the input current. A second switch S_{h} works with a holding capacitor C_{h} to freeze the output voltage during the discharge period. The duration of the discharge period is a few tens of microseconds in commercial capacitor-feedback amplifiers.

The capacitor-feedback design has two important advantages over the conventional resistor-feedback *I–V* converter. First, the noise level can be lower because of the absence of the feedback resistor, which as we shall see in Section 3.4, can contribute substantially to the total *I–V* converter noise. Second, the dynamic range can be larger. In the resistor-feedback *I–V* converter, a large resistor value R_{f} is chosen to minimize the noise; but this large value also sets the maximum current I_{p} that can be passed without saturation of the op-amp. For example, a 50-GΩ feedback resistor limits the maximum current to 200 pA if the maximum amplifier output voltage is 10 V. In the capacitor-feedback amplifier, on the other hand, a larger input current can be handled by choosing an appropriately small value for the differentiator time constant $R_{\mathrm{d}}C_{\mathrm{d}}$, yielding a low overall gain factor in equation 15. The main problem arising with large input currents in this case is instead that the feedback capacitor must be discharged more often. A typical value of C_{f} is 2 pF; a 1-nA input current results in $dV_{\mathrm{I}}/dt = 0.5$ V/msec. If V_{I} is allowed to reach 10 V before the capacitor is discharged, the period between discharge events is 20 msec.

The capacitor-feedback *I–V* converter requires a switch to discharge the feedback capacitor; practical resistor-feedback *I–V* converters often require switches as well to change among different feedback resistors. These switch elements are connected directly to the input terminal of the *I–V* converter and therefore must be chosen carefully. Standard CMOS "analogue switch" devices have a leakage current on the order of 100 pA and capacitance on the order of 50 pF, both of which would seriously degrade the performance of an *I–V* converter for patch-clamp use. The best commercially available switch device, in terms of having low capacitance and leakage current and introducing little noise, appears to be the 2N4118 small-geometry junction field-effect transistor (FET) (Sigworth, 1994). This device has an "on" resistance of about 1 kΩ while having capacitances below 1 pF and leakage current well below 1 pA. When used in chip form, for example in a hybrid integrated circuit, these devices introduce relatively little excess noise into the amplifier.

Complications in the design of a practical capacitor-feedback *I–V* converter largely surround the issue of providing rapid capacitor discharge without causing large transient voltages at the *I–V* converter input (from temporary loss of feedback control by the op-amp) or at the output of the *I–V* converter system. Transients at the input can arise from charge injected through the capacitance of the switching FET and from switching that occurs too rapidly for the op-amp output to follow because of its slew-rate limitation. Transient voltages at the output can arise from the slow settling of either the input stage or the differentiator stage as a result of nonideal properties (e.g., dielectric absorption) in the capacitors C_{f} or C_{d}, respectively. Metal-oxide-semiconductor chip capacitors have greatly superior properties to, say, ceramic capacitors for the small (1–2 pF) feedback capacitor C_{f}. For the larger differentiator capacitor, capacitors with polystyrene or polypropylene as the dielectric appear to have the best properties. Levis and Rae (1992) give a further discussion of details in the design of capacitor-feedback headstages.

3. Background Noise in the Current–Voltage Converter

In patch-clamp recording, the background noise arises from sources in the electronic circuitry, from the pipette and holder assembly, and from the tight seal and membrane itself. In the best recording situations, the contributions from each of these noise sources are roughly equal. The noise in the electronic circuitry arises primarily from the current-measuring resistor and the amplifier in the *I–V* converter.

3.1. Noise in the Feedback Resistor

Thermal noise in the feedback resistor R_f places a lower limit on the noise level of the current-to-voltage converter. The important relationship is that any passive, two-terminal electrical network (e.g., a resistor) at equilibrium produces a noise current with the spectral density

$$S_I(f) = 4kT\,\mathrm{Re}\{Y(f)\} \tag{16}$$

when its terminals are shorted together (Nyquist, 1928). $\mathrm{Re}\{Y(f)\}$ is the real part of the admittance of the device. For a resistor, $Y = 1/R$; thus, the spectral density is inversely proportional to the resistance.

On the other hand, the voltage noise in a device of impedance Z (Z is the reciprocal of Y) is given by

$$S_V(f) = 4kT\,\mathrm{Re}\{Z(f)\} \tag{17}$$

This is the spectral density of the voltage noise present at the (open-circuited) terminals of the device. Equation 17 explains, for example, why a high-resistance microelectrode gives a noisy voltage recording.

Depending on one's point of view, either equation 16 or equation 17 can be used to characterize the noise in the *I–V* converter. For now, let us ignore C_f and assume once again that the op-amp provides stable and rapid feedback response, as we did in Fig. 1C. If we are interested in the noise at the output, equation 17 is then the relevant expression. Since by the action of the op-amp the output voltage is just the voltage drop across R_f, the noise in the output voltage is given by equation 17 and is larger when R_f is chosen to be large.

Another, more useful approach is to ask what current noise, when presented at the input of the *I–V* converter, would give rise to the observed output voltage noise. To calculate this, we make use of the following theorem (see, for example, Papoulis, 1965, p. 347). Suppose a fluctuating signal $x(t)$ is processed by a linear network (an amplifier, an impedance, etc.) having a transfer function $H(j\omega)$ to yield an output $y(t)$. Then the spectral density S_y of y is related to the spectral density S_x of x according to

$$S_y(f) = S_x(f)\,|H(j2\pi f)|^2 \tag{18}$$

In the present case, the transfer function between the input current and output voltage is just $H = R_f$, so that the relationship between the spectral densities is found to be

$$S_I(f) = \frac{S_V(f)}{R_f^2} = \frac{4kT}{R_f} \tag{19}$$

where S_I is the current noise, referred to the input. Even though S_v increases with R_f, the effective current noise decreases as R_f increases. In the general case in which the feedback element is not a pure resistance, equation 16 is the correct expression for S_I.

An expression for the effective input current noise density S_I is useful mainly because the rms background noise σ_n can be computed from it. (σ_n is, by definition, also the standard deviation of background noise fluctuations.) The noise level observed at the output of a recording system depends on its overall frequency response. If we let $T(f)$ be the transfer function of the system, including the various amplifier and filter stages, the noise variance σ_n^2 is given by

$$\sigma_n^2 = \int_0^\infty |T(f)|^2 S_I(f)df \tag{20}$$

An important special case is when $T(f)$ is unity at low frequencies but shows a sharp cutoff at some frequency f_c. If S_I is independent of frequency, then $\sigma_n^2 = f_c S_I$. In the case of a pure resistance, the rms noise current can be written, using equation 19, as

$$\sigma_n = (4kTf_c/R_f)^{1/2} \tag{21}$$

The rms noise is seen to increase as the square root of f_c and inversely as the square root of R_f. Thus, for low noise, a high value of R_f is desirable. Table I gives spectral densities and σ_n values for various values of R_f.

As was pointed out in Section 2.3, the feedback resistor in the *I–V* converter cannot be considered as a pure resistance but has appreciable stray capacitance. A parallel combination of a resistance R and a capacitance C (Fig. 5A) gives the complex admittance.

$$Y = 1/R + j\omega C \tag{22}$$

The real part of Y is therefore just $1/R$. The capacitance modifies the frequency response of the *I–V* converter, but it causes no change in the current noise as referred to the input (equation 16).

If the stray capacitance has some resistance in series with it (as in Fig. 5B), the noise situation changes dramatically. In this case, we can write

Table I. Thermal Current Noise in Resistors[a]

$R(\Omega)$	S_I (A^2/Hz)	$S_I^{1/2}$ (fA/$Hz^{1/2}$)	σ_n (pA: f_c = 1 kHz)
10 M	1.6×10^{-27}	40	1.3
100 M	1.6×10^{-28}	13	0.4
1 G	1.6×10^{-29}	4	0.13
10 G	1.6×10^{-30}	1.3	0.04
100 G	1.6×10^{-31}	0.4	0.013

[a]Noise amplitude is expressed as the spectral density S_I, its square root, and the rms current noise σ_n for a 1-kHz bandwidth.

$$Y = 1/R + Y_s \tag{23}$$

where Y_s is the admittance of the series combination of R_s and C_s,

$$Y_s = \frac{j\omega C_s}{1 + j\omega R_s C_s} \tag{24}$$

The real part of Y_s is

$$\mathrm{Re}\{Y_s\} = \frac{\omega^2 R_s C_s^2}{1 + \omega^2 R_s^2 C_s^2}$$

which approaches the value $1/R_s$ as $\omega \to \infty$. At low frequencies, the current spectral density from the entire admittance Y is that characteristic of the resistance R_f. At higher frequencies, the spectral density rises and approaches a final asymptotic value characteristic of the parallel combination of R_s and R_f. If R_s is much smaller than R_f, the increase in noise can be significant.

Practical high-value resistors usually have a distributed stray capacitance that causes excess noise like that just described. If the capacitance were distributed evenly (Fig. 5A), there would be no excess noise because the admittance is of the same form as equation 22. The unevenly distributed capacitance that is usually present gives rise to one or more frequency-dependent admittance components of the form shown in equation 24.

3.2. Noise in the Amplifier

In addition to the thermal noise in the feedback resistor, there are several important noise sources in the *I–V* converter that are associated with the operational amplifier itself. An op-amp typically consists of two or more amplifying stages in cascade. If the first stage has a relatively large gain, the noise contributions from subsequent stages, when referred to the input, can usually be neglected. The main noise sources in the op-amp used in an *I–V* converter are mainly determined by the properties of the amplifying device in the first stage, which is usually a field-effect transistor (FET).

A FET can be thought of as a conducting "channel," the effective width of which is varied by the electric field set up by a "gate" electrode (Fig. 8A). In the N-channel FET shown, the channel is a potential well in which electrons are free to carry current between the source and drain electrodes. Imposing a negative voltage on the gate restricts the width w of the channel and therefore controls the flow of current. Beyond a certain "pinch-off voltage" V_{po} (typically -1 to -5 V), the channel ceases to exist, and current flow effectively stops.

The use of the FET as an amplifier is illustrated in Fig. 8B. The battery V_s sets the operating current I_{DO} in the FET: increasing V_s makes the gate more negative, reducing the drain current. The input voltage V_{in} that is applied to the gate also modulates the drain current. An important parameter describing the FET as an amplifier is the variation of the drain current that results from small gate voltage changes. This parameter g_{fs} has units of conductance and is commonly called the "transconductance":

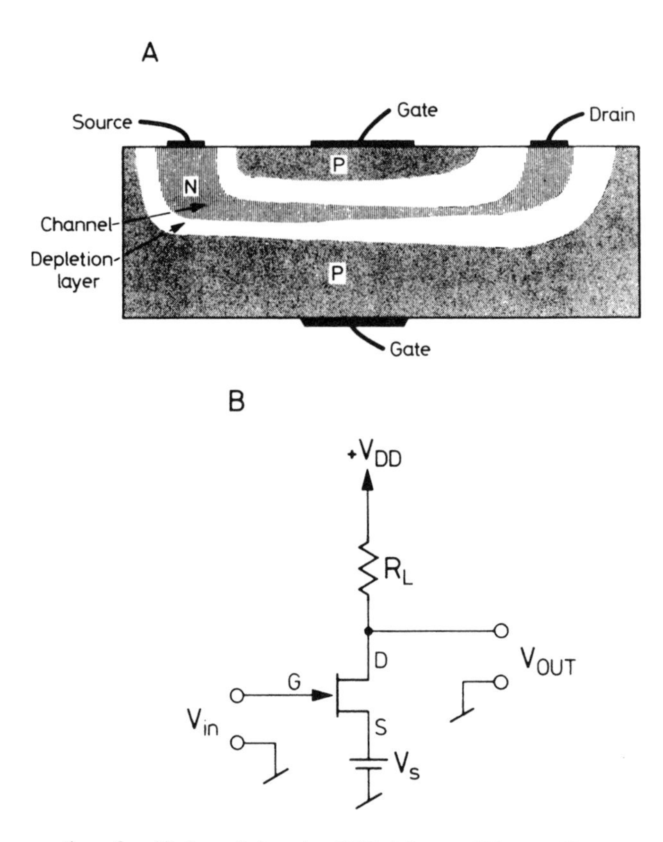

Figure 8. A: Cross section of an N-channel, junction FET. A layer of N-type silicon (having free electrons as charge carriers) serves as the channel. A negative voltage applied to the P-type gate layers broadens the depletion regions (where there are no charge carriers) at the expense of the channel width. Under normal operation, the drain voltage is more positive than the source; this results in wider depletion regions and a narrower channel near the drain, which in turn tends to make the drain current insensitive to the drain voltage. B: A simple FET amplifier. The input voltage applied to the gate controls the drain current, which in turn determines the voltage drop across R_L.

$$g_{fs} = \frac{\partial I_D}{\partial V_g} \tag{25}$$

The output voltage of the amplifier can then be written as

$$V_{out} = V_{DD} - R_L(I_{DO} + g_{fs}V_{in}) \tag{26}$$

where we have made use of the property that, for typical operating voltages, I_D is essentially independent of the drain voltage. The voltage gain of this amplifier is then

$$A_V = -R_L g_{fs} \tag{27}$$

In a typical situation with $I_{DO} = 1$ mA, $g_{fs} = 2$ mS, and $R_L = 5$ kΩ, the magnitude of the voltage gain is 10.

In normal operation no current needs to flow between the gate and the channel, because it is the electric field that controls the drain current. (Current does flow to charge and discharge stray capacitances when the voltages change, however.) In practice, the PN junction between the gate and channel has a leakage current I_g, which, in the FETs commonly used for patch-clamp amplifiers, is usually in the range of 0.01 to 1 pA. A steady current of this magnitude is not serious and can easily be compensated. However, the leakage current is temperature dependent, with a Q_{10} of about 2, so it can give rise to slow drifts in high-resolution recordings.

The gate current also is noisy because it is carried by isolated, thermally generated charge carriers. This "shot noise" has the spectral density

$$S_I = 2I_g q^* \tag{28}$$

where q^* is the effective charge of the carriers (approximately the elementary charge, 1.6×10^{-19} coulomb). Shot noise in the gate current can be a significant noise source, since the noise spectral density from a 5-pA current is the same as that from a 10-GΩ resistor. Therefore, some care is needed in choosing the particular FET type for low I_g. Also, the particular operating conditions are important: I_g increases steeply with drain voltage and also depends on the drain current.

Another noise source that becomes very important at high frequencies arises from thermal noise currents in the conducting FET channel. This noise has two effects. First, the resulting fluctuating voltages along the channel are coupled capacitively to the gate electrode, yielding an additional fluctuation in the gate current. The second effect, which is more important for our purposes, is that the drain current fluctuates when the gate voltage is held constant. This current has a spectral density like that of a resistor with value $1/g_{fs}$ (Van der Ziel, 1970)

$$S_{Id} = 4kTn_d g_{fs} \tag{29}$$

in which the constant n_d is approximately unity. The fluctuating drain current can be referred back to the input, where it is equivalent to a fluctuating gate voltage. To obtain the voltage spectral density, S_{Id} is divided by g_{fs}^2 to give

$$S_{Vg} = 4kTn_d/g_{fs} \tag{30}$$

This is the FET's voltage input noise spectral density. It is usually specified on transistor data sheets as the square root, $e_n = S_{Vg}^{1/2}$ and is usually expressed in units of nanovolts/Hertz$^{1/2}$.

This input noise voltage is negligible in itself (it is only a few nV/Hz$^{1/2}$). Its importance has to do with the charging currents that flow in the various input capacitances to cause them to follow this noise voltage. Figure 9 shows this situation in an *I–V* converter circuit. Suppose I_d shows a transient increase δ_I because of the thermal noise source i_n. To counteract this, the rest of the *I–V* converter's feedback loop will act to bring the gate voltage more negative by the amount δ_I/g_{fs}. To do this quickly, a considerable current must be forced through R_f to charge the various capacitances associated with the input.* Part of this is the

*Note that any driven shielding or other tricks to synthesize a lower apparent input capacitance will not reduce this "noise charging current" because the fluctuations arise in the FET itself. The only effective driven shielding would make use of an additional, parallel amplifier having a lower noise voltage. But it would be better to use such an amplifier as the main one in the first place.

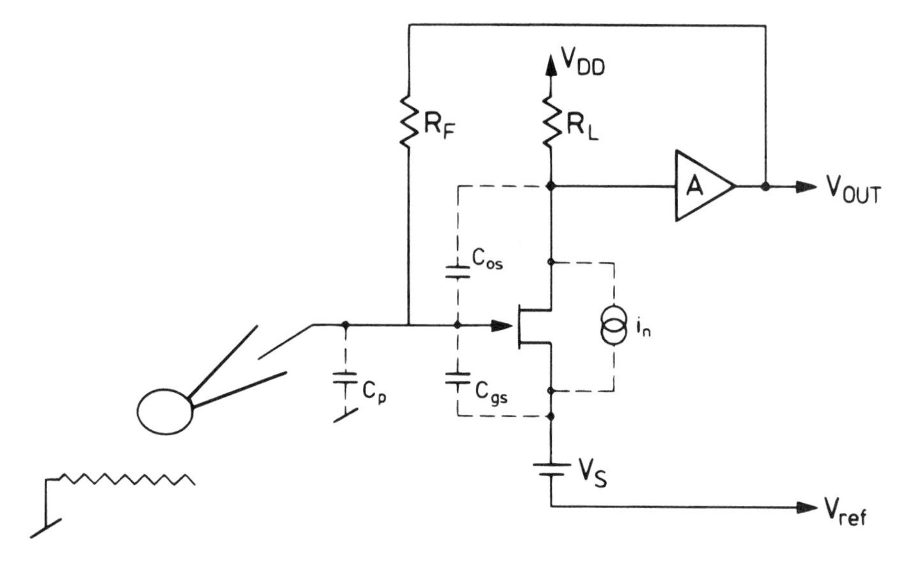

Figure 9. Model of noise effects in the *I–V* converter. Thermal noise in the FET channel is represented by the current generator i_n. C_p is the capacitance of the pipette and holder, and C_{os} and C_{gs} are the internal capacitances of the FET.

input capacitance of the FET itself, C_{is} which can be thought of as the parallel combination of the gate-source capacitance C_{gs} and the "output" (gate-drain) capacitance C_{os}. Also included is the capacitance C_p, which includes contributions from the input connector, the pipette holder, and the capacitance of the pipette itself.

The spectral density of the fluctuating current caused by the FET input noise voltage is given by

$$S_I(f) = (2\pi f C_t)^2 S_{V_g} \tag{31}$$

where C_t is the total input capacitance. This "$e_n C_t$ noise" is the dominant noise source in the *I–V* converter at higher frequencies. In a fairly good *I–V* converter having $S_V = 9 \times 10^{-18}$ V^2/Hz (i.e., $e_n = 3$ nV/Hz$^{1/2}$) and $C_t = 15$ pF, the noise from this mechanism equals that of a 10-GΩ resistor at $f = 5$ kHz.

Commercially available FETs generally show higher noise levels than would be expected from equations 28 and 30 because of noise from other sources. For example, a significant amount of noise can come from the transistor package itself. Plastic transistor packages show considerable leakage conductance and dielectric noise and are not suitable for picoampere-level input circuits. The glass-and-metal transistor packages have very low leakage conductances but can be noisy. The problem appears to be dielectric relaxation in the glass seals around the wire leads, and the noise is largest when the lead spacing is small. Figure 10A shows the spectrum of current noise from one lead of a TO-71 package, the package in which the popular NDF9406 and U401 series dual FETs are supplied. Noise from the package rises with frequency and is comparable to the total current noise from the unencapsulated U401 device alone in the range 1–10 kHz. The larger TO-78 and TO-99 packages (used for the U421 and U430 series dual FETs) have a noise spectral density that is lower by at least a factor of 2 and can be neglected for some purposes.

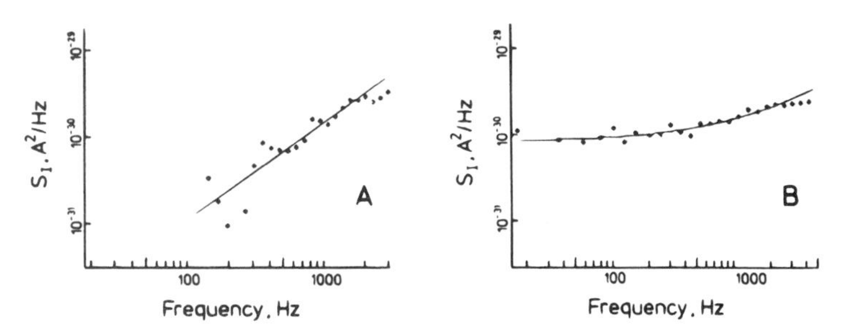

Figure 10. Current noise spectra. A: Spectral density of the noise from one lead of a TO-71 header. The top of the transistor package was sawed off, the internal bonding wires were removed, and the package was grounded. The spectrum shown is the difference with and without one lead connected to the input of a patch clamp amplifier (List EPC-5). The straight line corresponds to a spectral density proportional to f. B: Input current noise measured in an I–V converter using a U430 dual FET and a 30-GΩ feedback resistor. The total high-frequency noise in this circuit is actually less than the noise in A. The curve is drawn according to $S_f = 0.87 \times 10^{-30} (1 + f/10^3 \text{ Hz}) \text{ A}^2/\text{sec}$.

Table II summarizes the relevant properties of some FET types that are suitable for use in I–V converters. These are all dual FETs, since matched pairs of transistors are useful for making amplifiers with low offset voltages (see Section 3.3). The U421 has a very low gate current and is suitable for measuring very small, slow currents. Its considerable voltage noise makes it inferior to the other transistor types for frequencies above about 500 Hz.

The NDF9401 transistors specified by Hamill *et al.* (1981) are no longer available. The same devices are available in TO-71 packages as the NDF9406 series; however, generally better noise performance can be obtained from the U401 because it has a higher g_{fs} and lower e_n. The gate leakage current of the U401 is relatively low provided that the drain voltage is kept at about 6 V or less.

The U430 gives the best high-frequency noise performance, partly because it is in the larger TO-99 case but also because e_n is low and decreases even further with frequency above 1 kHz. Its gate current is larger, but gate currents below 0.5 pA can be obtained in selected devices operated at low drain voltages (2–4 V).

Table II. Noise-Related Parameters for Dual FET Types That Are Suitable for Patch-Clamp I–V Converters[a]

	I_z (pA)	g_{ts} (mS)	e_n (nV/Hz$^{1/2}$)	C_{ts} (pF)	Typ. I_d (mA)	Typ. I_d (V)	Case
U421	0.05	0.3	10	3	0.1	6	TO-78
NDF9406	0.3	2	4	10	1.0	5–10	TO-71
U401	0.3	3	3	10	2.0	5	TO-71
U430	1.0	8	2	13	3.0	3	TO-99

[a]Parameters for the U421 were obtained from the manufacturer's specifications; the others are based on the author's measurements and are "typical" values for one side of the dual FET. Typ. I_d and Typ. I_d are typical operating conditions for low noise. Low drain voltages and currents allow low I_g levels to be obtained for the U401 and U430, whereas use of relatively high I_d values decreases e_n in the NDF9401 and U421; c_n was measured at 1 kHz. C_{ts} (measured at 1 kHz) appeared to be higher than specified by the manufacturers. Devices that were tested were NDF9401 and U401 from National Semiconductor and U430 from Siliconix. The U421 is made by Siliconix.

3.3. Example of a Low-Noise Amplifier Design

A simple design for a complete FET-input amplifier is shown in Fig. 4 of Hamill *et al.* (1981). A similar but slightly more complicated circuit is shown in Fig. 11 and will be analyzed in detail. Basically, a U430 dual FET (Q2) is used in a differential preamplifier, the output of which is amplified further by the conventional op-amp A_1. Transistor Q1 acts as a current source for the FETs, while Q3 and Q4 set the drain voltages and isolate the load resistors R5 and R6 from the drains.

The differential amplifier configuration is a simple solution to the problem of setting up the operating conditions for an FET, such as determining the source voltage V_s (see Fig. 8B). In the differential amplifier, Q2B in effect sets V_s for Q2A (and vice versa) such that the total current flowing through Q1 is divided between the two FETs. When both input voltages V_{ref} and V_{in} are equal, equal currents should flow and develop equal voltages across the load resistors R5 and R6.

A disadvantage of the differential amplifier is that the total input voltage noise variance is the sum of the contributions from each FET. The induced e_nC_t noise has less than twice the variance, however, because only part of the input capacitance of Q2A needs to be charged by voltage noise arising in Q2B. Nevertheless, some improvement can be obtained by decoupling Q2B, for example, with large capacitors.

In the current source, a voltage divider (R2 and R3) determines a voltage to be imposed across the scaling resistor R1. For the particular values shown, the voltage is about 90% of the negative supply voltage minus one diode drop (about 0.7 V). The current flowing through

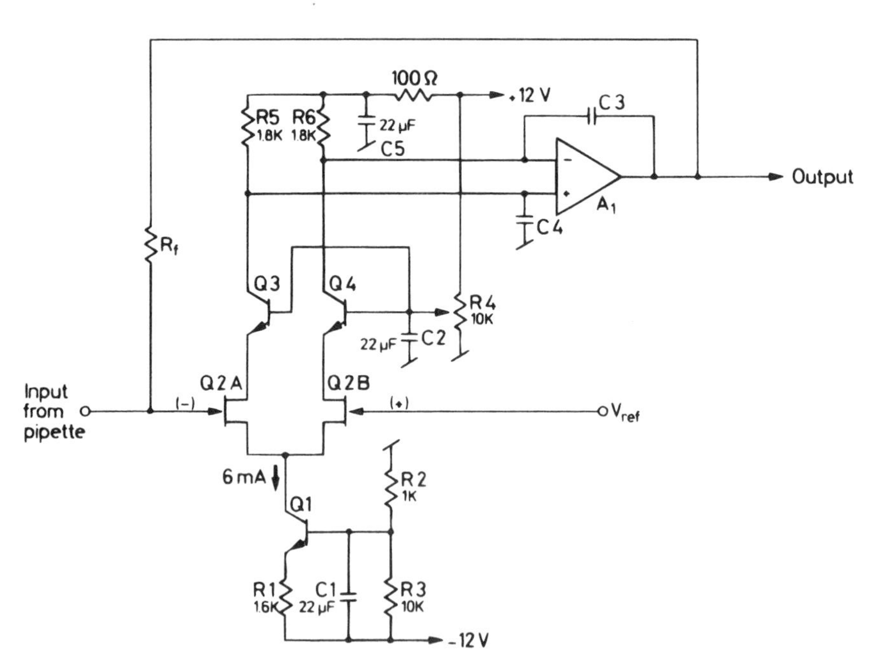

Figure 11. Schematic diagram of a composite op-amp connected as an *I–V* converter. Transistors Q1, Q3, and Q4 are 2N4401 (Motorola), and Q2 is a U430 dual FET (Siliconix). Op-amp A_1 is an LF356 (National Semiconductor). Potentiometer R4 is adjusted for the best noise level consistent with a low input current and is usually set for a drain voltage of about 3 V.

R1 to maintain this voltage, minus about 1% that is lost as base current, appears at the collector of Q1. The total current is 6 mA (i.e., 3 mA for each half of the dual FET) and is very insensitive to the voltage at the collector. The bypass capacitor C1 reduces the effect of noise in the −12 V supply.

Transistors Q3 and Q4 act as common-base amplifiers, coupling the drain currents to the load resistors R5 and R6 while keeping the drain voltages constant at the level determined by R4. This is a convenient way to set the drain voltage for the best noise performance. Also, Q3 and Q4 isolate the FET drains from voltage noise in the load resistances. R5 develops a considerable noise voltage (6 nV/Hz$^{1/2}$), which, if coupled through the gate-drain capacitance C_{os} of Q2A, would become a significant part of the input current noise at high frequencies. The particular transistor type used (2N4401, Motorola) has the much lower voltage noise of about 1 nV/Hz$^{1/2}$ at the operating current of 3 mA, making this noise source insignificant. Capacitor C2 filters the Johnson noise of R4 and noise on the +12 V supply, which otherwise would be coupled into the drains.

The presence of Q3 and Q4 is not as important when FETs with lower C_{os} are used. For the U401, which has C_{os} = 3 pF (instead of ~6 pF in the U430), the simpler circuit shown by Hamill *et al.*, gives similar performance when the current-source (8.2 kΩ) and load (10 kΩ) resistors are reduced to increase the operating current from 0.5 to 1–2 mA. The NDF9400-series transistors have an additional, internally cascode-connected FET pair that serves the same purpose as Q3 and Q4, making them unnecessary.

The common-mode rejection and amplifier input capacitance properties could be improved by driving the bases of Q3 and Q4 from a voltage referred to V_{ref} or to I_s. The derived voltage would have to have very little noise, however, to avoid losing the noise advantage of having Q3 and Q4 in the first place.

The voltage gain of the entire preamplifier is $A_V = g_{fs}R_L$, which is equal to about 12 in this circuit. Thus, the input voltage noise of ~2 nV/Hz$^{1/2}$ is magnified by this factor, and the following amplifier stages need only to have voltage noises smaller than about 24 nV/Hz$^{1/2}$. This requirement is readily met by commercial op-amps, for example, the LF356 (e_n = 12 nV/Hz$^{1/2}$ at 1 kHz) and the NE5534 (4 nV/Hz$^{1/2}$).

Because of the extra gain from the preamplifier, some thought has to be given to stabilizing the feedback loop around the composite amplifier. The bandwidth of the preamplifier extends well beyond 10 MHz, so extra phase shift is not a problem. The only requirement is to reduce the overall gain at high frequencies so that the loop-closure frequency (at which the gain of the amplifier equals the loss through the feedback network) is less than A_1's stable unity-gain bandwidth ω_A. The loop closure frequency ω_2 is just $1/\tau_2$ (see equation 8),

$$\omega_2 = \omega_A A_V \frac{C_f}{C_f + C_m} \tag{32}$$

where C_{in} is the sum of the extrinsic input capacitance (pipette, etc.) and the input capacitance of the amplifier, which is approximately equal to C_{os}.

If ω_2 is too large, the best way to reduce it is to introduce the equal-valued capacitors C3 and C4, which provide local feedback around the op-amp, reducing the gain without degrading the noise performance. With them, the op-amp becomes an integrator having a time constant τ_a = R6·C3. Provided that $\tau_a > 1/\omega_A$, i.e., that the capacitors actually slow down the amplifier, the loop-closure frequency becomes

$$\omega_2 = \frac{A_v}{\tau_a} \frac{C_f}{C_f + C_{in}} \tag{33}$$

In practice C_f is usually very small (< 0.1 pF) compared to C_{in} (5–10 pF), making $\omega_2 \ll \omega_A$, so that C3 and C4 are unnecessary. Stray capacitance, for example, between the output of A_1 and the noninverting input, sometimes causes high-frequency oscillations, however. These can usually be eliminated by including a small-valued (2–10 pF) capacitor as C3.

When an *I–V* converter like this is built, special precautions need to be taken to avoid sources of excess noise. Most importantly, everything contacting the input terminal should have low leakage and dielectric noise. The insulation of the input connector should be Teflon® or some other hydrophobic, low-loss insulator. The FET input lead and feedback resistor leads should be soldered directly to the input connector and should not touch any other surface. Other components connected to the input, such as capacitor for injecting test signals, should be checked for their noise contributions.

Consideration also should be given to noise from the power supply, which can result in an additional input voltage noise component. At low frequencies, the sensitivity to ripple on the supply lines is limited mainly by the matching of R5 and R6; for 1% resistors, the supply rejection ratio is about 600, so that 1 mV of ripple on the -12 V line would cause 1–2 μV of input voltage variation. Low-frequency voltage noise is not important except when the pipette tip is open (i.e., not sealed), in which case a few microvolts is acceptable. Above a few hundred Hertz, C1, C2, and C5 filter the supply-line noise, but it is still a good idea to have the supply voltage noise below 1 $\mu V/Hz^{1/2}$. Most integrated-circuit regulators meet this requirement.

Shielding the feedback resistor requires special attention. The most important factor is that no signals are coupled capacitively to the middle of the resistor. The capacitance from the middle of the resistor to the shield itself should also be kept small. For normal-size resistors (1–2 cm long), a suitable arrangement is to surround the resistor with shields spaced 2–5 mm away from the resistor body. The shields are best driven from V_{ref}. Several commercially available patch-clamp amplifiers use a hybrid integrated circuit for the critical components in the input stage, such as the input FET, feedback resistors, and switching FETs. The hybrid IC in the EPC-7, for example, uses a quartz substrate and is installed in a package with borosilicate glass feedthroughs. The hybrid allows the use of small chip resistors having low capacitance, simplifying the problem of resistor shielding.

3.4. Summary of Noise Sources

Figure 12 compares the spectral density of background noise in actual patch recordings with the noise of a resistor-feedback amplifier (EPC-7) and a capacitor-feedback amplifier (Axopatch 200B). In recordings using a conventional borosilicate-glass pipette (open circles) the spectral density caused by the pipette and other sources is considerably larger than the EPC-7 noise. However, when quartz pipettes are used and careful attention is given to the pipette holder, Sylgard coating, and pipette immersion, the noise can be reduced to be comparable to that of the capacitor-feedback amplifier (closed circles). The residual noise sources remaining under these conditions have been analyzed by Levis and Rae (1993). Noise sources are also discussed by Benndorf (Chapter 5, this volume).

The advantage of the capacitor-feedback amplifier is seen to be greatest at low frequencies, where the thermal noise of the resistor predominates (along with the shot noise in the gate leakage current of the input FET). The resistor noise increases with frequency because of distributed capacitance, but at frequencies above 10 kHz the disparity between resistor- and capacitor-feedback amplifiers becomes smaller as the amplifier's $e_n C_t$ noise predominates.

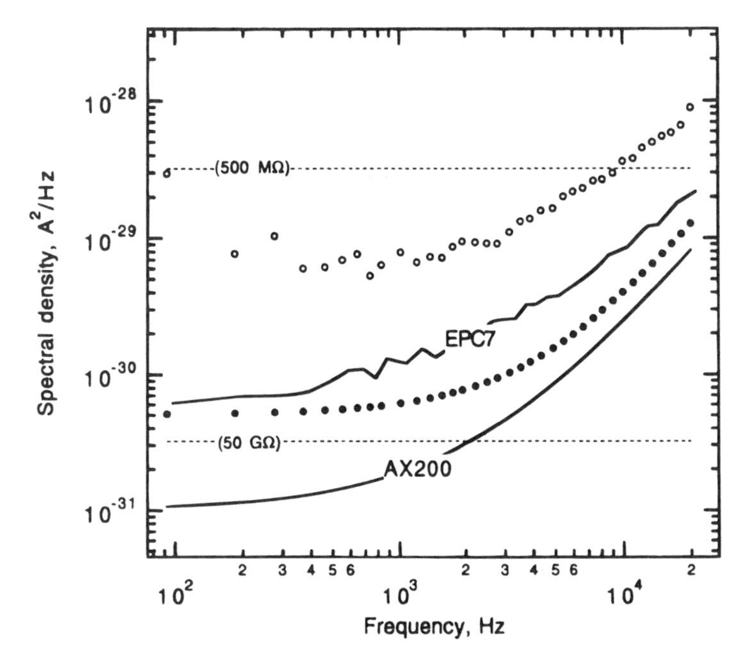

Figure 12. Total background noise in patch recordings (symbols) and the contributions of patch-clamp amplifiers (solid curves). The open circles show the spectral density in a randomly selected recording from the experiments of Sine *et al.* (1990), in which Sylgard-coated pipettes made from Corning 7040 glass were used with an EPC-7 amplifier in its 50-GΩ range. The solid circles plot the spectral density given by Levis and Rae (1993) for their "typical" recording situation using Sylgard-coated quartz pipettes and a 40-GΩ seal. Data for the Axopatch 200B amplifier spectrum are also from Levis and Rae (1993).

4. Capacitance Transient Cancellation

4.1. Overload Effects in the Patch Clamp

Step potential changes can be applied to membrane patches or (in the whole-cell recording configuration) to entire cells by applying a step command voltage as the reference voltage V_{ref} of the *I–V* converter. The converter in turn causes the pipette voltage V_p to follow V_{ref}, forcing any necessary currents through the feedback resistor to bring V_p to the correct level. The currents that flow in charging the pipette capacitance C_p (typically 2 pF or more) are typically some nanoamperes in magnitude, very large compared to single-channel currents. The large currents cause two problems. First, they generally exceed the linear range of the recording device (e.g., tape recorder or AD converter). This means that for a short interval after a voltage step, the membrane current information is lost. Second, if the current pulse is large enough to drive amplifiers within the patch clamp into saturation, serious distortions of the current monitor signal can persist even for several milliseconds after the current pulse is over.

The worst thing that can happen is that the *I–V* converter op-amp goes into saturation. Because no feedback is then possible, the pipette is no longer voltage clamped but sees the high impedance of the feedback resistor. The current monitor signal shows a large steplike response, the total duration of which equals the saturation time (trace 3 of Fig. 13).

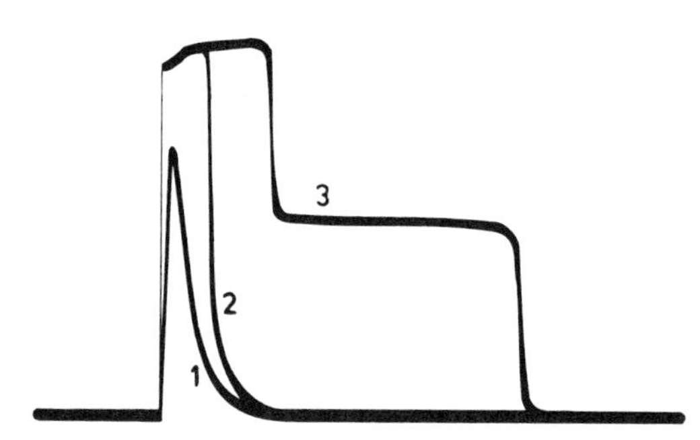

Figure 13. Response of a patch clamp to large transient currents. Current pulses with time constant 80 μsec and peak amplitudes of 1.3 nA (trace 1), 3.3 nA (trace 2), and 16 nA (trace 3) were applied to the input of a List EPC-5 patch clamp, which has R_t = 10 GΩ and a maximum V_{OUT} = 11 V. In trace 1, the current was recorded faithfully; in trace 2 the frequency-compensation circuit saturated. Trace 3 shows the initial saturation of the frequency-compensation circuit followed by a plateau during which the *I–V* converter remains in saturation.

To determine whether a given voltage step will cause saturation, we notice first that the rapid charging currents will flow primarily through the stray capacitance C_f rather than through the feedback resistor itself. The output voltage therefore shows an initial step of magnitude $V_{out} = V_{ref}\,[(C_p/C_f) + 1]$. Slower currents, such as those arising in whole-cell clamp from the cell capacitance, must also be small enough to be passed through the feedback resistor. With a 10-GΩ resistor and a 10-V maximum output voltage, the largest current that can be passed, for times longer than about τ_1, is 1 nA.

Even when the *I–V* converter does not saturate, the frequency-compensation circuit recovers slowly if it is driven into saturation by a large current transient. It remains in saturation (trace 2 of Fig. 13) until its output has the same area as the "unclipped" waveform would have had, but after recovery the signal shows no distortion.

4.2. Fast Transient Cancellation

The stray capacitance to be charged can be reduced somewhat by the use of shielding driven from V_{ref}; alternatively, V_{ref} can be kept constant, and command pulses be applied to the bath. The capacitance of the pipette wall remains, however. The best way to avoid overload effects causes by this residual capacitance is to charge C_p by a separate pathway. Figure 14A shows the basic circuit. At the same time that a voltage step is applied as V_{ref}, current is injected through C_i to bring V_p to the correct voltage. The gain A_1 of the amplifier is made adjustable to allow a range of C_p values to be compensated. If A_1 is adjusted so that

$$(A_1 - 1)C_i = C_p \tag{34}$$

then no capacitive current need be supplied by the *I–V* converter, and no capacitance transient will appear in the current monitor signal.

In practice, the temporal response of the *I–V* converter and of A_1 are not matched perfectly. This causes a biphasic current to flow in the *I–V* converter even when equation 34 is satisfied. The amplitude of this error current depends on the rate of rise of the command voltage, so it is advantageous to "round off" the command by filtering. The response times of the two pathways can be better matched by including delays in one or both of them. Figure 14B shows one way of doing this. A fixed time constant of 0.5 μsec in V_{ref} is matched by a variable time constant in the cancellation pathway. (In this circuit, the *I–V* converter loop time constant τ_2 and the time constant of the amplifier are both assumed to be smaller

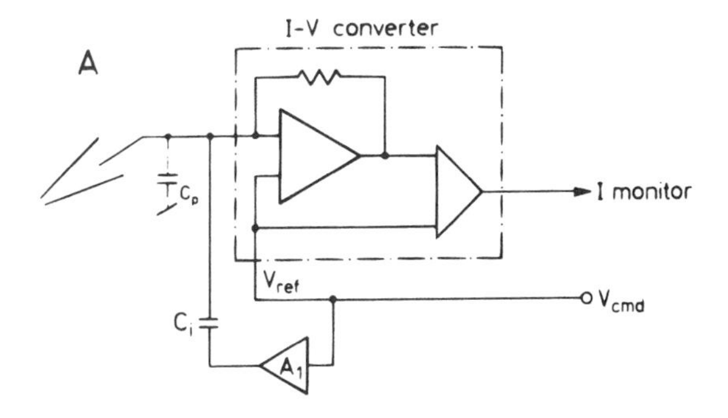

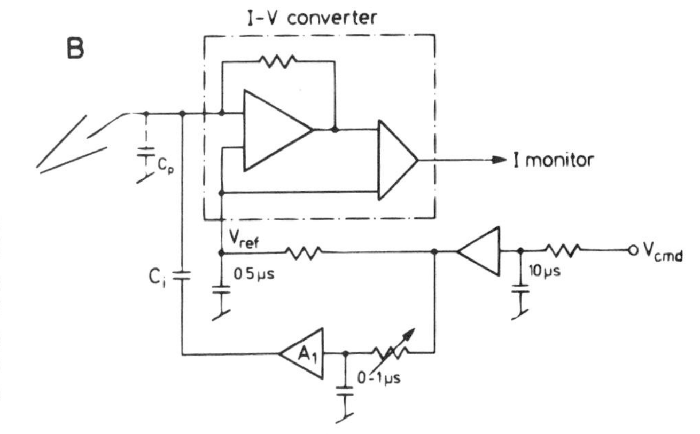

Figure 14. A: Basic fast transient cancellation circuit. The command signal, scaled by the noninverting amplifier A_I is used to inject current through C_i. B: Addition of filtering to allow the response times of the I–V converter and the cancellation circuit to be matched.

than 0.5 μsec.) The time constant of "rounding," 10 μs, is chosen to be much longer than the others. This simplifies the matching, since the effect of the short time constants on such a rounded waveform (essentially a simple time delay) is indistinguishable from the effect of the higher-order amplifier responses.

An added advantage of employing capacitance-current cancellation is that it reduces the effect of a noisy stimulus. In the same way that currents caused by voltage steps are canceled, currents caused by noise in the stimulus voltage, which would appear as C_p is clamped to V_{ref}, are also canceled by the second pathway. It should be noted that noise introduced within the second pathway, such as in the amplifier, is not canceled, however. Care should be taken to keep the amplifier output voltage noise, scaled by C_i, smaller than the $e_n C_{in}$ current noise of the I–V converter.

4.3. Slow Transient Cancellation

For whole-cell recording, it is useful to cancel the currents caused by the charging of the cell membrane capacitance C_m. Since the capacitance is charged through a substantial series resistance R_s, the charging current caused by a potential step ΔV_{ref} does not flow instantaneously. Instead, the current transient is exponential, having the amplitude $\Delta V_{ref}/R_s$ and time constant $R'_s C_m$, where R'_s is the parallel combination of R_s and the membrane resistance R_m. Since R_m is usually several orders of magnitude larger than R_s, we ignore R_m and set $R'_s = R_s$.

By analogy to the circuits of Fig. 14, the slower transient current could be canceled by introducing a resistor in series with C_i. An equivalent, but more practical, technique is to drive C_i with a low-pass filtered waveform, which can be added to the fast-compensation signal (Fig. 15). The time constant of the simple filter would be set equal to R_sC_m, and the gain A_2 would be chosen according to the ratio C_m/C_i.

To see how this works, we consider the transfer functions. The relationship between the pipette current I_p and the voltage I_p is the admittance of the R_sC_m combination (compare equation 24),

$$Y_s(s) = \frac{sC_m}{sR_sC_m + 1} \tag{35}$$

The "transfer admittance" giving the ratio of the injected current I_i to V_{cmd} is the product of the admittance of C_i (which is just sC_i), the amplifier gain A_2, and the low-pass filter's transfer function,

$$Y_t(s) = \frac{sC_iA_2}{\tau_{SC}s + 1} \tag{36}$$

where $\tau_{SC} = R_2C_2$ is the time constant of the filter. By proper choice of A_2 and τ_{SC}, the two admittances (equations 35 and 36) can be made equal, and cancellation will result. Once again, the mismatch in response times of A_2 and the I–V converter introduces an error, but this is usually insignificant because τ_{SC} is typically much longer than the response times.

5. Series Resistance Compensation

The patch recording configuration provides a nearly ideal voltage-clamp situation for single-channel currents. The access resistance to the membrane patch is typically 2–10 MΩ,

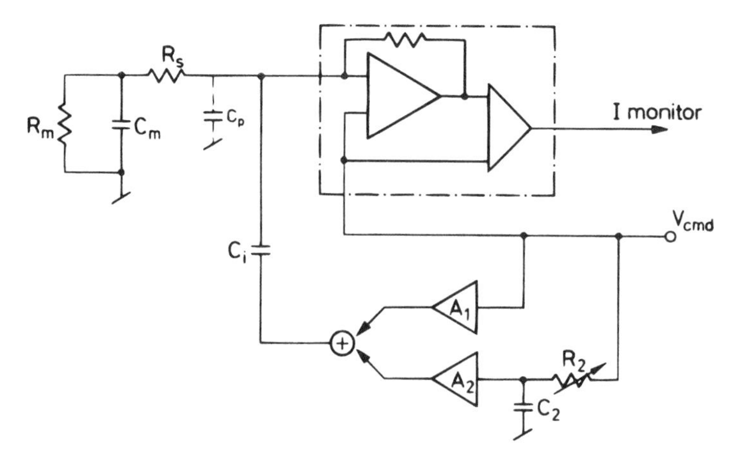

Figure 15. Slow-capacitance cancellation. The command and reference voltage V_{ref} is filtered with a time constant R_2C_2 chosen to be equal to R_sC_m. After amplification, this signal is summed with the fast-capacitance signal and injected through c_i.

and the patch capacitance is <0.1 pF, yielding a charging time constant of 1 μsec or less. This is the time constant for charging of the patch membrane when a potential step is applied to the pipette, and it is also the time constant of the pipette current arising from a step change in patch current. The voltage errors are also very small: a 10-pA channel current causes a voltage drop of at most 0.1 mV in the series resistance.

The situation is very different in the case of whole-cell recording. Instead of the patch capacitance, the entire cell's membrane capacitance contributes to the charging time constant. For a spherical cell, the approximate capacitance C_m (in picofarads) is related to the diameter d (in micrometers) by $C_m = \pi d^2/100$, so that a 20-μm cell has a capacitance of 12 pF (or more, if the membrane has substantial infoldings). The access resistance is often higher, typically 5–20 MΩ, so that time constants of several hundred microseconds are common. Voltage errors can also be considerable, since whole-cell currents larger than 1 nA are typical. These currents can cause errors of tens of millivolts in the membrane potential.

In series-resistance (R_s) compensation, the voltage error caused by the access resistance is estimated electronically, and a correction is made to the voltage applied to the pipette (Fig. 16A). This technique can reduce the apparent access resistance by a factor of 5 or 10. The maximum practical level of compensation is limited, as we will see, by the bandwidth of the current-measurement circuitry and by the precision of the fast transient cancellation.

5.1. Theory

The basic principle is the same as that commonly used in other voltage-clamp schemes (Hodgkin *et al.,* 1952). The current monitor signal is scaled by a variable factor and added to the voltage command, with the polarity corresponding to positive feedback. In many voltage-clamp configurations, the effect of the compensation on the stability of the voltage-clamp feedback loop has to be taken into account (see, for example, Sigworth, 1980). In the

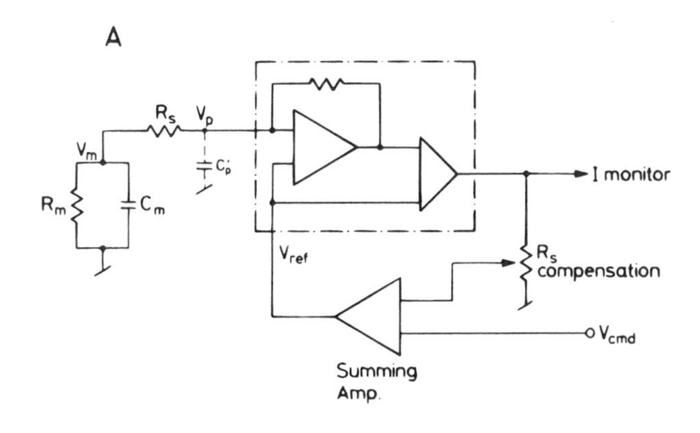

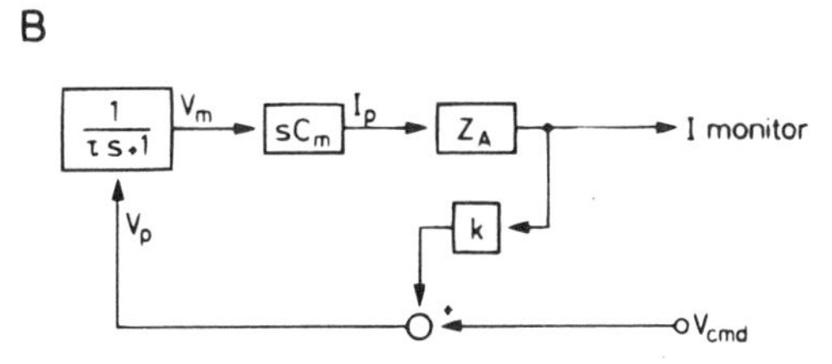

Figure 16. A: Diagram of R_s compensation. A fraction of the current monitor signal is added to the command, giving positive feedback. B: Transfer-function representation of the circuit in A.

present case, the pipette voltage is clamped by the I–V converter, which acts on a very short time scale compared to the charging time. This means that the R_s compensation feedback loop can be studied without concern about interaction.

For simplicity, we assume that the membrane resistance R_m is very large compared with R_s. For the purpose of analyzing the dynamics of the feedback loop, R_m can then be ignored. A symbolic representation of the system is shown in Fig. 16B. The command signal V_{cmd} is summed with the compensation signal to form the pipette potential V_p. This is filtered by the access time constant $\tau_a = R_sC_m$, yielding the true membrane potential V_m. The resulting pipette current is C_m times the time derivative of V_m; it is converted to a voltage according to the transfer function Z_A of the I–V converter and associated circuitry. This current monitor voltage V_{IM} is then scaled by a factor k to form the compensation voltage. The product kZ_A represents the entire frequency-dependent pathway between pipette current and compensation voltage and has units of impedance.

The transfer function of the entire feedback loop, relating V_m to V_{cmd}, is given by

$$T(s) = \frac{V_m(s)}{V_{cmd}(s)}$$

$$= \frac{1}{s[R_s - kZ_A(s)]C_m + 1} \tag{37}$$

If kZ_A is independent of s and approaches R_s from below, $T(s)$ approaches unity for all values of the frequency variable s. In practice, Z_A is frequency dependent, reflecting the limited bandwidth of the current-measuring circuitry. Taking as a simple case a single-time-constant rolloff with time constant τ_Z and letting α be the "fractional compensation," we have

$$kZ_A(s) = \frac{\alpha R_s}{\tau_Z s + 1} \tag{38}$$

and

$$T(s) = \frac{(\tau_Z s + 1)}{s^2\tau_Z\tau_a + s(1 - \alpha)\tau_a + 1} \tag{39}$$

The term $\tau_Z s$ in the numerator can usually be ignored. Putting the rest of equation 39 into the standard form for a second-order transfer function (equation 9) yields

$$\tau_0 = (\tau_Z\tau_a)^{1/2} \tag{40}$$

$$\zeta = (1 - \alpha)(\tau_a/\tau_Z)^{1/2}/2$$

Of importance here is the damping factor ζ: when it drops below unity, the response will overshoot. When α approaches unity (i.e., complete compensation), ζ approaches zero. For nonovershooting responses, therefore, α is restricted to the range

$$\alpha \leq 1 - 2(\tau_Z/\tau_a)^{1/2} \tag{41}$$

To compensate 90% of the series resistance, for example, the feedback time constant τ_Z must be 400 times shorter than τ_a. For $\tau_a = 400$ μsec, the required $\tau_Z = 1$ μsec corresponds to a bandwidth of 160 kHz in the current-monitor circuitry.

The transfer function $Z_A(s)$ shows at least a third-order rolloff in practical patch-clamp circuits, with one pole from the *I–V* converter and two more from the frequency compensation circuit (Fig. 4B). Equations 39 and 41 are nevertheless useful approximations when τ_Z is evaluated according to $\tau_Z = 1/(2\pi f_c)$ where f_c is the overall bandwidth of the current-monitor circuitry.

Somewhat higher values of α can be used at a given overall bandwidth when Z_A is specifically tailored for low phase shift in the vicinity of the frequency τ_0^{-1}. One way to do this is to insert a phase-lead network into the feedback pathway. Another is to reduce the damping of the second-order pole in the patch-clamp response. The main second-order pole is in the frequency response correction circuit, where the damping is readily controlled (see Section 2.13). This reduces the low-frequency phase shift and gives rise to the surprising situation in which an underdamped element in the feedback loop contributes to the damping of the overall loop. These tricks may be of limited utility, however, because in practice α is often limited by the accuracy of the capacitance compensation, as is discussed below.

The dynamics of R_s compensation have so far been considered from the point of view of controlling the membrane potential. The relationship between the membrane current (e.g., arising from ionic channels) and the voltage at the current monitor output V_{IM} is given by

$$V_{IM}(s)/I_m(s) = Z_A(s)T(s) \tag{42}$$

where $Z_A(s)$ is again the transfer function of the current monitor circuitry. The same transfer function T therefore also describes the effect of series resistance and its compensation on the observed current signal.

Finally, it is instructive to compare R_s compensation in the patch clamp with capacitance neutralization in standard microelectrode recording. In the equivalent circuit in Fig. 17, R_e represents the electrode resistance, and C_{in} the total input capacitance. Letting $C_t = C_{in} + C_f$, the transfer function relating the electrode potential V_e to the membrane potential V_m is (see Guld, 1962):

$$T_{CN} = \frac{V_e}{V_m} = \frac{1}{s[C_t - C_f A(s)]R_e + 1} \tag{43}$$

which is seen to have the same form as equation 37. The requirements on the amplifier transfer function $A(s)$ are therefore essentially the same as those for $Z_A(s)$. Thus, the well-known properties of capacitance neutralization circuits, such as ringing and oscillation at high neutralization settings, have their counterparts in R_s compensation.

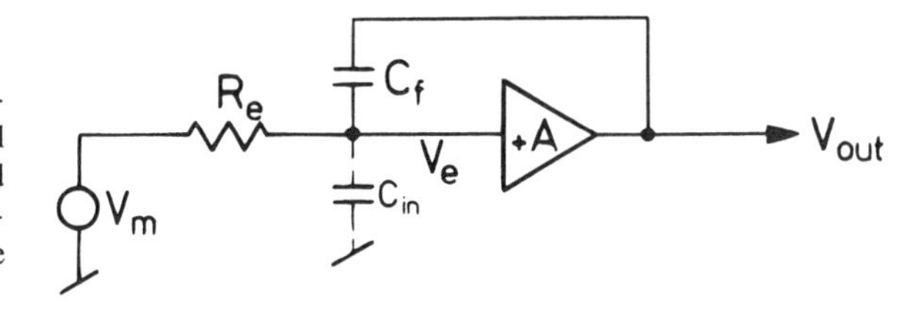

Figure 17. Model of capacitance neutralization in microelectrode potential recording. Positive feedback is applied through C_t to allow the electrode potential I_e to follow changes in the membrane potential V_m more quickly.

5.2. Effect of Fast Transient Cancellation

The analysis of the preceding section ignored the effect of the fast capacitance C_p, which arises mainly in the pipette and holder and has negligible series resistance. In practice, the charging currents of this capacitance can be canceled almost completely by the circuit described in Section 2.3, and for proper R_s compensation, this cancellation is important.

Let C'_p be the residual fast capacitance (Fig. 16); C'_p can be either positive or negative, corresponding to the cases of under- or overcompensation, respectively. When its effects are included in the feedback loop, the transfer function becomes

$$T'(s) = \frac{1}{[sC_m R_s + 1][-sC'_p + 1] - sC_m Z_C(s)} \tag{44}$$

When C'_p is negative, it adds an additional phase lag to the R_s compensation loop, slowing down and destabilizing the compensation. When C'_p is positive, it causes a phase lead, leading to increased overall damping. If C'_p becomes too large, however, oscillation results. To avoid this, C'_p is restricted to $|C'_p| \lesssim \tau_Z/R_s$. For a relatively fast R_s compensation system with τ_Z = 2 μsec and R_s = 10 MΩ, C'_p must be smaller than 0.2 pF and is best kept in the range $|C'_p| < 0.05$ pF. This is actually achievable in practice: although the capacitance C_p is typically 1–3 pF before transient cancellation, it can be made to stay constant within about 50 fF during an experiment. The most serious drifts would arise from changes in the capacitance between the pipette interior and the bath as the solution level is changed. Fortunately, however, the level of the meniscus around the pipette shank often does not change with small bath level changes, so that the capacitance is stable to within a few percent.

It is therefore important that the fast transient cancellation be set correctly, especially when a fast R_s compensation circuit is used. Slight misadjustment of the cancellation can improve the damping of the R_s compensation, but this has the disadvantage of possibly causing a distortion of the clamped membrane potential.

5.3. Incorporating Slow Transient Cancellation

Cancellation of the membrane capacitance transients is especially desirable when R_s compensation is used, because the compensation makes the transients shorter but larger in amplitude. These transients could be subtracted from the current monitor output signal, provided none of the amplifier stages are overdriven. Transient cancellation at the input as described in Section 4.3 is, however, preferable when the dynamic range is limited.

The problem with combining transient cancellation and R_s compensation is that the latter uses the current monitor signal as an estimate of the total pipette current in order to correct the pipette voltage. The transient cancellation circuit, however, injects a current into the pipette that does not appear in the current monitor signal. This current, I_{SC}, can be estimated, for example, by differentiating the voltage V_{SC} that is applied to the injection capacitor. Since the admittance of the capacitor C_i is sC_1, I_{SC} is given by

$$I_{SC} = sC_i V_{SC} \tag{45}$$

The way this signal is used is shown in Fig. 18. In this block diagram, the transfer function T_{SC} represents the filter network R_2C_2 and the amplifier A_2 shown in Fig. 15, so that $T_{SC} = A_2/(\tau_{SC}s + 1)$, where $\tau_{SC} = R_2C_2$. The output of this network is applied to the injection capacitor C_i and at the same time to a differentiator whose scale factor is C_i times the amount of series resistance to be compensated, αR_S. The result is added to the original command V_{cmd} to form V'_{cmd}. If V_{cmd} is a voltage step, V'_{cmd} has a large overshoot whose effect is to charge the membrane capacitance quickly. Because of the positive feedback through T_{SC}, the overshoot has an initial value $(1 - \alpha)^{-1}$ times the original step amplitude.

The signal V'_{cmd} then serves as the command for a conventional R_S-compensation system. If A_2 and τ_{SC} are adjusted correctly, and if k is adjusted simultaneously with α to provide the same degree of compensation, as in equation 38, the current to charge C_m will be exactly provided through the injection capacitor, and no current will flow into the I–V converter. If there is an error, or if there is an ionic current in the membrane, it will be clamped with compensated series resistance through the action of the feedback through V_2.

Other, simpler configurations, in which, for example, the network T_{SC} is driven from V_p instead of V'_{cmd}, do not provide proper R_s compensation. The somewhat complicated configuration shown in Fig. 18 does, however, have an additional benefit. If τ_{SC} is chosen to be equal to the charging time constant R_sC_m, it does not need to be changed when the fractional R_s compensation, α, is varied. As α increases toward unity, the actual charging transient is accelerated, but at the same time, the additional positive feedback involving V_1 causes the injected current transient to be accelerated the same amount. The knobs that set the parameters for the slow transient cancellation therefore do not need to be changed when R_s compensation is put into operation.

A practical implementation of the slow-capacitance cancellation and R_s compensation system is shown in Fig. 19. The three adjustments provided to the user are C-slow (allows the membrance capacitance C_m to be matched), G-series (to match the series conductance), and the fraction α of R_s compensation. The G-series and α adjustments are dual potentiometers that adjust both the R_s compensation (correction) pathway and the slow-transient cancellation circuitry (prediction pathway). The latter results in an overshooting command voltage that rapidly charges the cell membrane capacitance. The result is like that of the "supercharging"

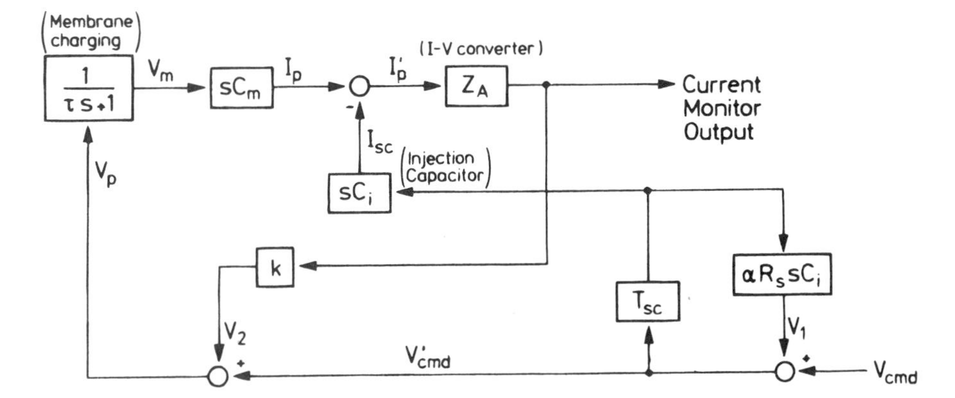

Figure 18. Combination of slow-capacitance cancellation and R_s compensation. The slow-capacitance filter network T_{sc} drives both the current-injection capacitor and a differentiating network ($\alpha R_s s C_1$), which effectively computes the voltage drop across a resistor R_s caused by the injected current. The modified command voltage V'_{cmd} is applied to the remainder of the system (compare Fig. 16B).

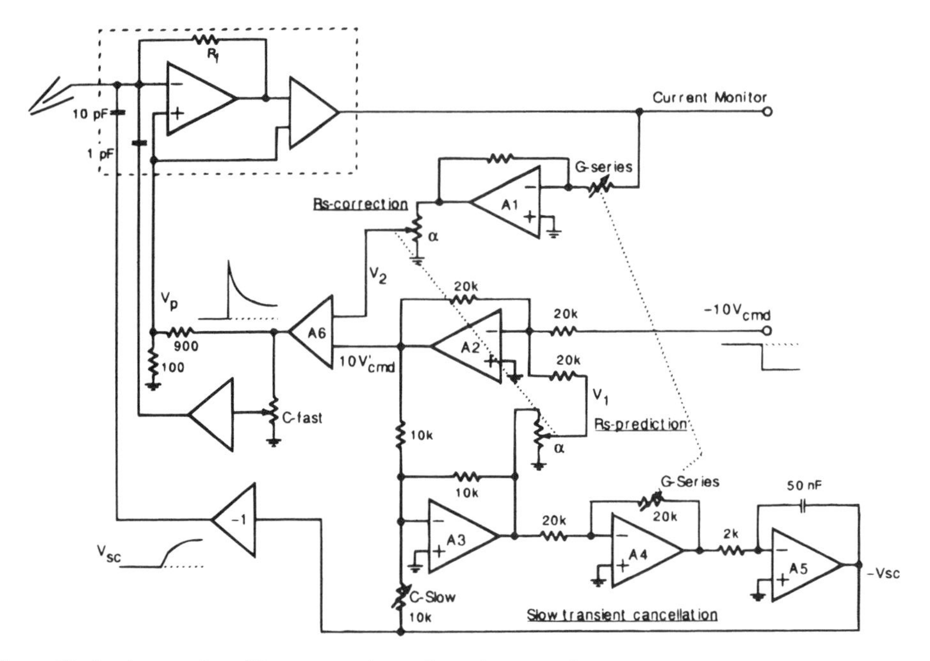

Figure 19. Implementation of R_s compensation and transient cancellation in the EPC-7 patch-clamp amplifier. The commanded pipette potential V_p is the sum of the series-resistance correction signal V_2 and the modified command voltage V'_{cmd}. The voltage for slow transient cancellation is obtained from V'_{cmd} through the state-variable loop encompassing A3, A4, and A5. The voltage V_1 that predicts the R_s error caused by the injected current is obtained from the state-variable loop as well and is summed with the command in A2. Positive feedback through A2 and the state-variable loop results in an overshooting V'_{cmd} in response to a step V'_{cmd}, resulting in "supercharging." The fast and slow transient cancellation circuits are shown using separate injection capacitors. With the values shown, the full-scale C-slow setting is 100 pF, and the maximum G-series setting is 1 μS.

technique of Armstrong and Chow (1987). The current required to charge the membrane capacitance is delivered through a 10-pF capacitor into the input terminal, so that the charging current will not be sensed by the current-monitor circuitry.

The EPC-9 computer-controlled patch-clamp amplifier uses a circuit similar to that in Fig. 19 to perform capacitance cancellation and R_s compensation except that the variable elements in the circuit are multiplying digital-to-analogue converters, which are under computer control. A detailed description of the EPC-9 circuitry is given in Sigworth (1994); computer algorithms for automatically determining the correct settings of the capacitance cancellation and R_s controls are described by Sigworth *et al.* (1994).

References

Armstrong, C. M., and Chow, R. H., 1987, Supercharging: A method for improving patch clamp performance, *Biophys. J.* **52:**133–136.

Aseltine, J. A., 1958, *Transform Method in Linear System Analysis,* McGraw-Hill, New York.

Finkel, A. S., 1991, Progress in instrumentation technology for recording from single channels and small cells, in: *Cellular and Molecular Neurobiology: A Practical Approach.* (V. Chad and H. Wheal, eds), pp. 3–25. Oxford University Press, New York.

Guld, C., 1962, Cathode follower and negative capacitance as high input impedance circuits. *Proc. IRE* **50:** 1912–1927.
Hamill, O. P., Marty, A., Neher, E., Sakmann, B., and Sigworth, F. J., 1981, Improved patch-clamp techniques for high-resolution current recording from cells and cell-free membrane patches. *Pflügers Arch.* **39:**85–100.
Hodgkin, A. L., Huxley, A. F., and Katz, B., 1952, Measurement of current–voltage relations in the membrane of the giant axon of *Loligo, J. Physiol.* **116:**424–448.
Horowitz, P., and Hill, W., 1989, *The Art of Electronics,* Cambridge University Press, Cambridge.
Levis, R. A., and Rae, J. L., 1992, Constructing a patch clamp setup, *Methods Enzymol.* **207:**18–66.
Levis, R. A., and Rae, J. L., 1993, The use of quartz patch pipettes for low noise single channel recording, *Biophys. J.* **65:**1666–1677.
Nyquist, H., 1928, Thermal agitation of electric charge in conductance, *Phys. Rev.* **32:**110–113.
Papoulis, A., 1965, *Probability, Random Variables, and Stochastic Processes,* McGraw-Hill. New York.
Sigworth, F. J., 1980, The variance of sodium current fluctuations at the node of Ranvier, *J. Physiol.* **307:**97–129.
Sigworth, F. J., 1994, Design of the EPC-9, a computer controlled patch clamp amplifier. 1. Hardware. *J. Neurosci. Methods* **56** (in press).
Sigworth, F. J., Affolter, H., and Neher, E., 1994, Design of the EPC-9, a computer controlled patch clamp amplifier. 2. Software. *J. Neurosci. Methods* **56** (in press).
Sine, S. M., Claudio, T., and Sigworth, F. J., 1990, Activation of *Torpedo* acetylcholine receptors expressed in mouse fibroblasts. Single channel current kinetics reveal distinct agonist binding affinities. *J. Gen. Physiol.* **96:**395–437.
Van der Ziel, A., 1970, Noise in solid state devices and lasers, *Proc. IEEE* **58:**1178–1206.

Chapter 5

Low-Noise Recording

KLAUS BENNDORF

1. Introduction

Any analysis of single-channel current events needs sufficient resolution of the signals from the background noise, which is accomplished by reasonable low-pass filtering. An increase of the recording bandwidth is therefore possible only if the background noise in the recording system is reduced. The background noise is determined in a complex manner by a number of individual noise sources. If these noise sources are uncorrelated, the standard deviation of the total noise is given by the square root of the summed squared standard deviations of the individual noise contributions, i.e., reduction of the largest noise contribution would cause the greatest reduction of the total noise. Hence, improved noise performance of patch-clamp amplifiers with integrated headstage technology may only be utilized if all additional noise contributions are reduced to a similar degree or, better, even more.

In this chapter noise arising in the patch, the pipette, and the pipette holder is analyzed, and a strategy is derived to minimize the individual noise contributions. The patch pipettes employed were pulled from borosilicate glass, which is at variance to the approach of Levis and Rae (1992, 1993), who use quartz. The major factors that promoted low noise and stable recording were the use of thick-walled pipettes, exceptionally small pore diameters, and a low immersion depth. Though the exceptionally small pores were obtained only if the cone angle was very small and the pipette resistance correspondingly high, the *RC* noise generated by this resistance in conjunction with the distributed pipette capacitance remained reasonably low. The technique allowed recording at background noise levels of 95–120 and 630–780 fA rms in a 5- and 20-kHz bandwidth (eight-pole Bessel filter), respectively. Based on noise measurements, the actual resolution limits of the single-channel open time are discussed as a function of the bandwidth and the unitary current amplitude.

2. Types of Noise

Any statistical fluctuation of current or voltage may be quantified by the variance σ^2, its square root, the standard deviation σ (rms noise), or the power spectral density (power spectrum), which also considers the frequency (f) dependence. The variance is calculated as the integral of the power spectrum over a bandwidth f_B. Table I summarizes relevant types of current noise in the patch clamp and also provides equations 1 to 9 for their description.

KLAUS BENNDORF • Department of Physiology, University of Cologne, D-50931 Cologne, Germany.
Single-Channel Recording, Second Edition, edited by Bert Sakmann and Erwin Neher. Plenum Press, New York, 1995.

Table I. Noise sources[a]

Frequency dependence	Type of noise	Origin	Equation	
$S_I \sim f^2$	Voltage noise plus capacitance	Input voltage noise of FET in conjunction with all capacitance	$S_I(f) = 4\,\pi^2C^2f^2S_V$	(1)
		Distributed *RC*-noise of pipette	$\sigma^2(f) = \frac{4}{3}\,\pi^2C^2f_B{}^3S_V$	(2)
		RC noise of pipette resistance and patch capacitance		
		Thin-film *RC* noise on the outside or inside of the pipette		
$S_I \sim f$	Dielectric noise	Pipette holder	$S_I(f) = 8\pi kTDCf$	(3)
		Pipette glass	$\sigma^2(f) = 4\pi kTDCf_B{}^2$	(4)
S_I = const	Johnson noise (Nyquist noise)	Seal and patch	$S_I(f) = 4kT/R$	(5)
			$\sigma^2(f) = 4kTf_B/R$	(6)
	Shot noise	FET gate current	$S_I(f) = 2Iq$	(7)
		Seal and patch	$\sigma^2(f) = 2Iqf_B$	(8)
$S_I \sim 1/f^b$	$1/f$ noise (flicker noise)	Holder	$S_I(f) = a/f^b$	(9)
		Pipette in bath		

[a]Relevant current noise arising in the patch, the pipette, the holder, and the headstage of the amplifier. The individual noise contributions are listed according to the frequency dependence of their spectral density $S_I(f)$. C is a capacitance, S_V is the spectral density of voltage noise, $\sigma^2(f)$ is the noise variance, which is the integral of S_I over the bandwidth f_B, D is the dielectric loss factor, R is a resistance, q is the elementary charge, I is a current, k is the Boltzmann constant, T is the absolute temperature, and a and b are the amplitude and the exponent of the $1/f$ noise, respectively.

The types of noise considered in more detail below are (1) all current noise generated by thermal voltage noise, either in conjunction with a capacitance, inherent in a loss conductance (dielectric noise), or in a resistance (Johnson noise); (2) shot noise; and (3) $1/f$ noise. In the table provided these types of noise are grouped according to their frequency dependence, as this is the characteristic information obtained from fitting wide-band power spectra. Several physically well-defined types of noise increase theoretically either with f^2 (f^2 noise) or f (f noise) or are independent of f (flat noise); $1/f$ noise, rising proportional to $1/f^b$, with b being in the range of 1, is physically less well defined.

2.1. Current Noise Generated by Thermal Voltage Noise

If a linear network has the complex admittance Y, the general relationship between the power spectrum of current noise S_I and the power spectrum of voltage noise S_V is

$$S_I = |Y|^2 S_V \tag{10}$$

Between the patch and the headstage, two different sources of thermal voltage noise exist. One arises in the field-effect transistor (FET) of the headstage and reacts with the network from the outside. The other source is generated in any resistance within the network.

The main component of the FET voltage noise is independent of f. In conjunction with any serial RC combination with the complex admittance

$$Y = \frac{j\omega C}{1 + j\omega RC} \tag{11}$$

$|Y|$ is approximately ωC ($\omega \ll 1/RC$; $\omega = 2\pi f$), and equation 1 in Table I follows directly from equation 10. This voltage noise generates current noise whose power spectrum rises approximately in proportion to f^2. The capacitance C in equation 1 is the sum of all capacitance associated with the FET input. In the case of a capacitive feedback headstage, C sums from the gate-to-source and the gate-to-drain capacitance within the FET (10 to 15 pF), all stray capacitance (1–2 pF), the capacitance of the injection capacitor (1 pF), and the capacitance of the feedback capacitor (1–2 pF). Further capacitance from the holder and the pipette simply adds to C and reacts, together with all other capacitance, with the FET input voltage noise. Thus, noise introduced by the capacitance of the holder and the pipette is perfectly correlated with that of the other capacitance. The magnitude of voltage noise $e_n = S_V^{1/2}$ of a good FET is on the order of 2–3 nV/Hz$^{1/2}$. These values correspond to a capacitive feedback headstage as used in the Axopatch 200A amplifier (Axon Instruments, Foster City, CA). For a more detailed discussion of noise in the amplifier, the reader is directed to Sigworth (Chapter 4, this volume).

Each open-circuited complex admittance Y generates voltage noise. The power spectrum of this voltage noise, the second source of voltage noise to be considered, is

$$S_V = 4kT\,\mathrm{Re}\{1/Y\} \qquad (12)$$

$\mathrm{Re}\{1/Y\}$ is the real part of $1/Y$. This voltage noise generates current noise according to equation 10. Herein, Y may be (1) the admittance of an RC combination or network, (2) the conductance generated by lossy dielectric properties, and (3) a reciprocal ohmic resistance.

The simplest case is a series combination of R and C, as it is approximately valid for the pipette resistance and the patch capacitance. The resulting noise can be quantified directly with equation 1. RC noise also exists in the immersed pipette tip. Here, however, the situation is more complicated because R is distributed in the pipette in a cable-like fashion, and this distribution must be known to calculate the admittance. In the limit $\omega \ll 1/RC$, the cable-like distribution of RC leaves the frequency dependence of this noise unaffected ($S{\sim}f^2$). Further RC noise results from thin solution films, which may creep along the surface of the pipette wall either at the outside from the bath fluid or at the inside from the pipette fluid. The distributed resistance in the film generates noise in conjunction with the wall capacitance of the pipette. At the outside, coating with hydrophobic material as the elastomer Sylgard 184® (Dow Corning Corp., Midland, MI) may drastically reduce this noise by preventing these films. Also, any thin conducting films may generate additional noise between the pipette and the holder or within the clefts of the holder.

Thermal voltage noise is also generated by the loss conductance of a real dielectric, which is given by $Y = \omega CD$ with D being the loss factor (also called dissipation factor). This loss conductance is frequency dependent. Inserting the loss conductance in equations 12 and 10 directly yields equation 3. The power spectrum of the dielectric noise rises in proportion to f. It may be reduced either by choosing appropriate materials for the pipettes and the holder or by decreasing the capacitance C.

In the simple case where $1/Y$ is a resistance R, equations 12 and 10 yield equation 5 describing the lower-limit estimate of the thermal current noise in a resistor (Johnson or Nyquist noise). This noise is independent of f (flat or white noise). It decreases with increasing R, and this is the reason for the dramatic noise reduction that occurs during gigaseal formation. Depending on the contribution of other noise sources, a resistance of the patch–seal combination significantly higher than, e.g., 100 GΩ might become desirable in order to keep this noise source at negligible levels.

2.2. Shot Noise

Shot noise is generated when charges flow across a potential barrier. This noise increases in proportion to the mean current I and the charge q of the particles. Equations 7 and 8 may be used as a rough description, thereby ignoring any specificity of the charge translocation processes. The elementary charge has been inserted for simplicity. Two shot-noise sources are present in the patch clamp. One is associated with the gate current in the FET and cannot be manipulated by the experimentalist. The other depends on the DC current through the patch–seal combination and may be minimized by a high seal resistance and the absence of membrane inherent charge-translocating mechanisms.

2.3. 1/*f* Noise

This type of noise (also called "flicker noise") decreases with increasing frequency. In contrast to the previously described noise sources, its physical origin seems to be as heterogeneous as the physical systems are where it was found. It is therefore likely that this noise does not have a unique origin (Dutta and Horn, 1981). Equation 9 is only a phenomenological description of its frequency dependence. The power b may adopt values either larger or smaller than 1. $1/f$ noise has been included in Table I because it generally dominates the noise spectra in the patch clamp at bandwidths below several hundred Hertz to 1 kHz. The holder- and the pipette-induced noise, in particular, may make large $1/f$-noise contributions.

3. A Strategy for Reducing Noise

Several of the parameters determining noise listed in Table I are open to optimization: the total capacitance of the holder and of the nonimmersed part of the pipette, the capacitance of the immersed pipette tip, the pipette resistance, and the dielectric loss factor of the pipette glass and the holder material should be minimzed. At the same time, the seal resistance should be as large as possible, and all types of thin films at the pipette and the holder should be avoided.

Let us specify now what can be done practically with respect to each of the points listed to achieve low-noise recordings over a wide bandwidth. For comparison with the measured noise, the theoretical data, calculated with equations (1–8), must be corrected because they are valid only for an ideal filter in which the transfer function is 1 at $f \leq f_B$ and 0 at $f > f_B$. The transfer function in real filters, however, is smooth, and the bandwidth f_B is regularly related to the -3 dB value (cut-off frequency f_C). The nonideal transfer function makes the total noise exceed that of the theoretical predictions. All theoretical noise data are therefore corrected by factors specific for an eight-pole Bessel-filter (flat noise 1.04, f noise 1.3, f^2 noise 1.9).

3.1. Holder and Pipette Capacitance

Commercial unshielded polycarbonate holders have a capacitance of 1–1.6 pF; the holder used here was both shorter and thinner (Benndorf, 1993). Its capacitance was 0.52 pF. The pipettes were as short as 8 mm and had a capacitance with the tip above the bath of 0.16 pF compared to 0.57–0.60 pF measured in pipettes with the usual length of 42 mm. Minimization of the capacitance by using a smaller holder and short pipettes approximates the total capacitance reacting with the FET input to the inherent capacitance of the headstage and, according to equation 1, thus minimizes the f^2 noise.

3.2. Capacitance of the Immersed Pipette Tip

The capacitance of the immersed pipette tip is treated separately from the previous section because, under the conditions described here (<0.1 pF), its noise contribution in conjunction with the FET input voltage noise is negligible. On the other hand, this capacitance is very important for the magnitude of the distributed *RC* noise and the dielectric noise in the tip (equations 1–4). The total capacitance of the immersed tip is

$$C_p = \epsilon_o \epsilon \frac{2\pi l}{\ln(d_o/d_i)} \tag{13}$$

where ϵ_o is the vacuum permittivity (8.854×10^{-12} C/Vm), ϵ the dielectric constant, depending on the glass type, l the immersion depth, and d_o and d_i are the outer and inner diameter of the pipette, respectively. It is noteworthy that C_p does not depend on the particular shape of the pipette tip but only on the ratio d_o/d_i, which is kept fairly constant to the tip. Scanning electron micrographs (Fig. 1A) in six pipettes showed a tip d_o/d_i ratio of 3.85 ± 0.02 (mean ± SD) in comparison to 4.00 in the raw material. In order to obtain a small C_p, ϵ and l should be as small and d_o/d_i as large as possible. The dielectric constant ϵ of available glasses ranges from 3.8 for quartz to some less than 10 for high-lead glasses. The glass used here was the borosilicate glass Duran®, with $\epsilon = 4.6$. The use of uncoated quartz pipettes of equal dimensions would reduce C_p by only 17%. In reasonably well Sylgard®-coated pipettes, the difference is even less. A much stronger reduction of C_p may be achieved by decreasing the immersion depth l because it is directly proportional to C_p and may be reduced under many experimental conditions. In appropriate experimental chambers, immersion depths of 200 μm may be reached easily.

3.3. Pipette Resistance

A low pipette resistance *R* reduces the distributed *RC* noise. The pipettes designed here, however, had an unusually high resistance (50–90 MΩ when filled with a 200% Tyrode solution, specific resistance $\rho = 26$ Ωcm) to obtain exceptionally small pore diameters. Hence, the strategy to keep the distributed *RC* noise low was to reduce the capacitance. Knowing the distribution of the pipette resistance in a "prototype pipette" (Fig. 1B), the distributed *RC* noise was calculated by modeling the immersed pipette tip as a network of parallel *RC* circuits. After calculating the complex admittance *Y*, the distributed *RC* noise was obtained with equations 12 and 10. For the calculations, the d_o/d_i ratio was assumed to be constant; i.e., the slight decrease to the tip was not considered. In an uncoated pipette, the *RC* noise considerably decreases at higher d_o/d_i ratios (Fig. 2). Most unfavorable is a d_o/d_i ratio <2. Figure 3 shows the *RC* noise as a function of the immersion depth under different conditions. A reduction of the noise variance by more than one order of magnitude may be achieved simply by increasing the d_o/d_i ratio from 2 to 4, decreasing the specific

A

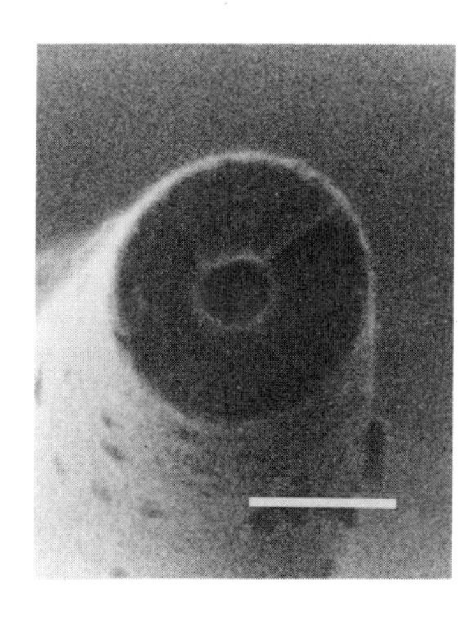

B

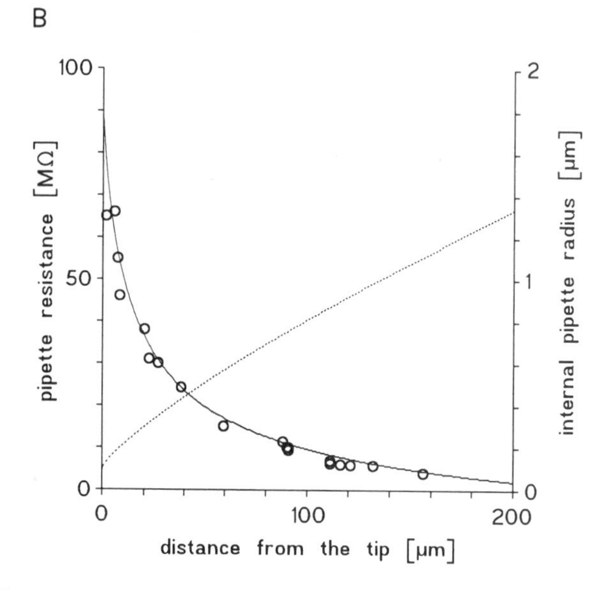

Figure 1. Geometry of thick-walled, small-pore patch pipettes pulled from borosilicate glass. A: Scanning electron micrograph of a pipette tip tilted by 5°. The resistance of the other pipette of the pair was 84 MΩ when filled with 200% Tyrode solution (ρ=26 Ωcm). A gold layer of 2.5 nm thickness was generated by evaporation. The pore is not perfectly round. The d_o/d_i ratio is 4 in the horizontal and 3.6 in the vertical direction; scale bar, 500 nm. B: Resistance (left ordinate) of the remaining pipette after breaking of the tip as function of the distance from the tip. Five pipettes were repeatedly broken (tall thick-walled tips do not splinter), and both the length of the broken segment and the resistance of the remaining pipette were measured (the data were kindly provided by Mr. T. Böhle). On the basis of a mean pore radius r_t of 100 nm, the resistance R of the remaining pipette tip was described as a function of the pipette length l (μm), related to a point 200 μm from the original tip (l = 0 μm), with

$$R(l) = \frac{\rho}{\pi} \int_0^l \frac{dx}{[\alpha(200\text{-}x)^\beta + r_t]^2} + R_{200}$$

R_{200} is the pipette resistance after breaking 200 μm. $[\alpha(200 - x)^\beta + r_t]$ is an empirical expression relating the pipette radius to the length l. The integral was solved numerically. Fitting the data yielded R_{200} = 2 MΩ, α = 0.0178, β = 0.8. The resulting internal pipette radius (dotted line, right ordinate) shows that the cone angle increases slightly to the tip.

resistivity ρ from 51 to 26 Ωcm, and coating with Sylgard®, starting 50 μm from the tip. Coating to the very tip further reduces noise. However, in pipettes with a slim cone as used here, the thickness of the Sylgard® coat at the tip is usually low because the elastomer tends to move away from the tip, and this underscores the benefit of the thick-walled glass tubing. In small-pore pipettes, coating to the very tip was also excluded because the pores were obstructed when the coating came closer to the tip than ~50 μm. This suggests that the slowly flowing Sylgard® had reached the pore before polymerization was performed. Figure 3 also shows that for the pipettes used here an immersion depth less than 200 μm has a large effect on the *RC* noise, whereas it does not become much worse when the pipette goes deeper than 200 μm.

The large pipette resistance raises the question of the time constant of the distributed *RC* circuit in response to a voltage step. In an uncoated immersed pipette, the capacitance

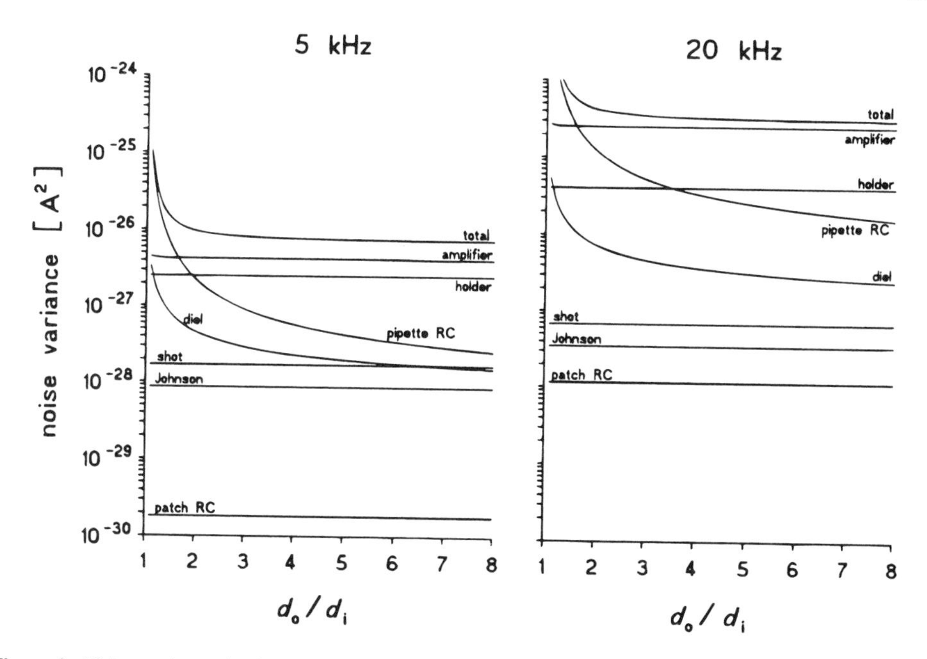

Figure 2. Noise variance in the patch clamp as a function of the ratio of the outer (d_o) to the inner (d_i) diameter of the glass at 5- and 20-kHz bandwidth. Immersion depth 200 μm. The individual noise contributions were calculated as described in the text. diel, dielectric noise of the pipette tip; shot, seal shot noise assuming a current I = 100 fA; Johnson, seal Johnson noise at a seal resistance of 1 TΩ; patch *RC*, noise associated with the patch capacitance and the pipette resistance. The amplifier noise and the holder noise are measured data and were included for comparison. The holder noise and the pipette *RC* noise are the dominating noise sources apart from the amplifier. A doubling of the d_o/d_i ratio from 2 to 4 approximately halves the *RC* noise.

is 37 fF (d_o/d_i = 4, immersion depth 200 μm). Assuming as the worst case that the large pipette resistance of 200 MΩ occurs in series with this capacitance, the time constant is 7 μsec. Taking into account both the distribution of *RC* and the reduced capacitance because of the Sylgard® coat, the true time constant will be considerably smaller and may therefore be neglected at all recording bandwidths discussed here. The additional capacitance introduced by the tiny patches is less than 1 fF and without any relevance.

3.4. Dielectric Loss

Apart from the capacitance, the dielectric noise of both the holder and the pipette is determined by the loss factor *D*. The holder used here was fabricated from polycarbonate. The loss factor *D* of this material is in the range of 10^{-3}, but considerable variability has been reported to exist between the manufacturers and the production charges (Levis and Rae, 1993). In fact, spectra revealed a relatively large *f*-noise component attributable to the holder, which suggests that the holder material used here had a *D* larger than 10^{-3}. Because of the complex geometry that holders usually have, this noise was not evaluated theoretically. In a 5-kHz bandwidth, the holder alone generated 34 fA rms noise, approximately of the *f* type, and in conjunction with the short pipette (with the tip above the bath) 52 fA.

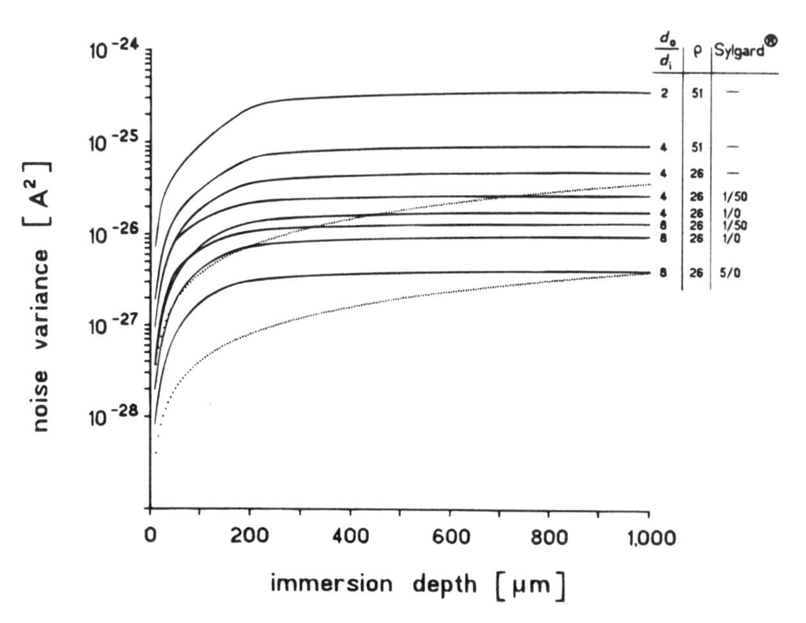

Figure 3. Theoretical pipette *RC* noise as a function of the immersion depth at increasing wall thickness and degrees of Sylgard® coating. At the right, the conditions for each curve are indicated (ρ in Ωcm). The data for Sylgard® indicate the coat thickness, in multiples of the glass wall thickness, and the starting point behind the tip (in micrometers). The tip capacitance is reduced by Sylgard® to that of the series combination of the glass and the coat. The three top curves illustrate the noise reduction for the increase of the d_o/d_i ratio from 2 to 4 and the subsequent reduction of ρ. The fourth curve shows the further effect of Sylgard® coating on the reduction of the capacitance, thereby diminishing the *RC* noise. At the immersion depth of 50 μm, the curve begins to deviate from that of the uncoated pipette. Further reduction of noise appears if coating is started from the very tip of the pipette (. . ./0). The increase of the d_o/d_i ratio from 4 to 8 and corresponding coating further reduce noise. Increasing the coat thickness to five times the glass wall thickness (5/0) reduces the noise at 1000-μm immersion depth to only 63 fA rms. The dotted lines show the corresponding dielectric noise of the pipette at the conditions of the upper and the lower *RC*-noise curve.

Duran® borosilicate glass, of which the patch pipettes were prepared, has the loss factor 3.7×10^{-3} and thus belongs to those glasses with lowest loss apart from quartz, which has a loss factor of $\sim 10^{-4}$. A technology for using quartz pipettes has been pioneered by Levis and Rae (1992, 1993). The dielectric noise variance of uncoated, equally shaped quartz pipettes would decrease by two orders of magnitude because of the small *D* (equation 4), thereby practically canceling the dielectric noise. Unfortunately, the quartz technology is very expensive, and its advantages can only be used if the dielectric noise belongs to the dominating noise sources. In the tipped pipettes used here, however, it is the *RC* noise that exceeds the dielectric noise by far at low immersion depths and 5- and 20-kHz bandwidth (Fig. 4). Only at 2 kHz are both noise types of similar magnitude. In other words, at the higher frequencies it is more effective to decrease *C,* either by using thick-walled glass material or thick Sylgard® coats, than to decrease *D* by using quartz. If pipettes with larger cone angle, and a corresponding smaller pipette resistance, are used, the *RC* noise decreases and thereby increases relatively the contribution of the dielectric noise.

When the dielectric noise levels in equally shaped quartz and borosilicate glass pipettes that have been coated with Sylgard® onto the tip (Levis and Rae, 1993) are compared, the elastomer is found to increase the noise in the quartz pipettes and decrease the noise in

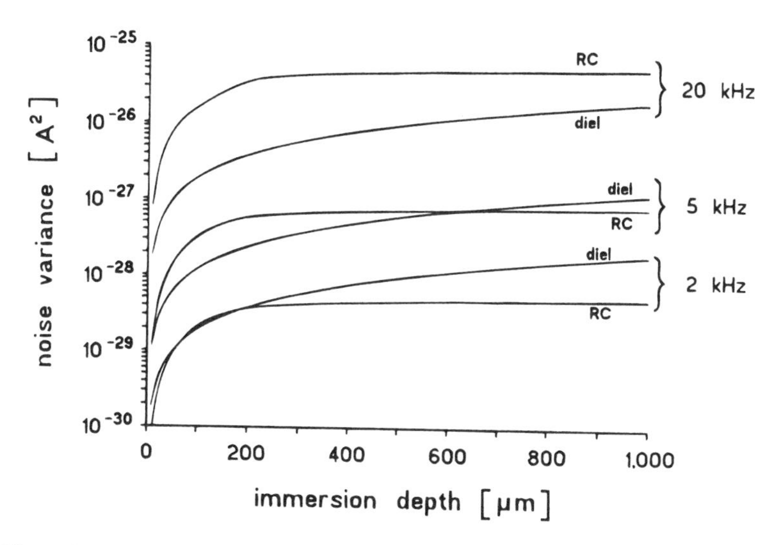

Figure 4. Theoretical noise variance of the dielectric (diel) and *RC* noise (*RC*) in a pipette tip as function of the immersion depth at three bandwidths. The pipettes were assumed not to be coated; $d_o/d_i = 4$; $\rho = 26\ \Omega$cm.

borosilicate glass pipettes. If the layer is thick, both values approximate each other. This is shown in an example where the total dielectric noise in an immersed coated pipette was evaluated as series of two (glass/quartz and Sylgard®) parallel combinations of the respective capacitance C and loss conductance $2\pi fCD$. If the capacitance of the Sylgard® ($D = 2 \times 10^{-3}$) coat is one-third that of the immersed borosilicate glass pipette, the total dielectric rms noise in the glass pipette would exceed that in the coated quartz pipette only 1.3-fold. It is concluded that thick Sylgard® coats make the dielectric noise depend more on the properties of the elastomer than on the pipette material, especially if they are established onto the tip.

The dotted lines in Fig. 3 show that the dielectric noise in the pipettes used here concomitantly decreases with the *RC* noise to negligible levels when increasing the wall thickness and using Sylgard®. These curves do not depend on the pipette resistance and therefore are also valid for patch pipettes of much lower resistance.

3.5. Pore Diameter

Reducing the pore diameter should minimize the patch membrane area in a squared fashion and the seal conductance in a linear fashion. With the small-pore pipettes used here, seals in the range of up to 4 TΩ were obtained. The Johnson noise is reduced at 5-kHz bandwidth from 28 to 9 fA rms if the seal resistance is increased from 100 GΩ to 1 TΩ (curve labeled "Johnson" in Fig. 2). Because the thermal noise in a real patch certainly exceeds the lower limit estimate, a seal resistance considerably exceeding 100 GΩ should help to reduce this noise.

The main benefit of the tiny patches used here is certainly that the probability of the appearance of any other charge-translocating processes is reduced in proportion to the patch area. Such processes may be naturally inherent in the membrane (pumps, exchangers, channels with unresolved unitary currents), or they may be generated by artifacts (membrane perturba-

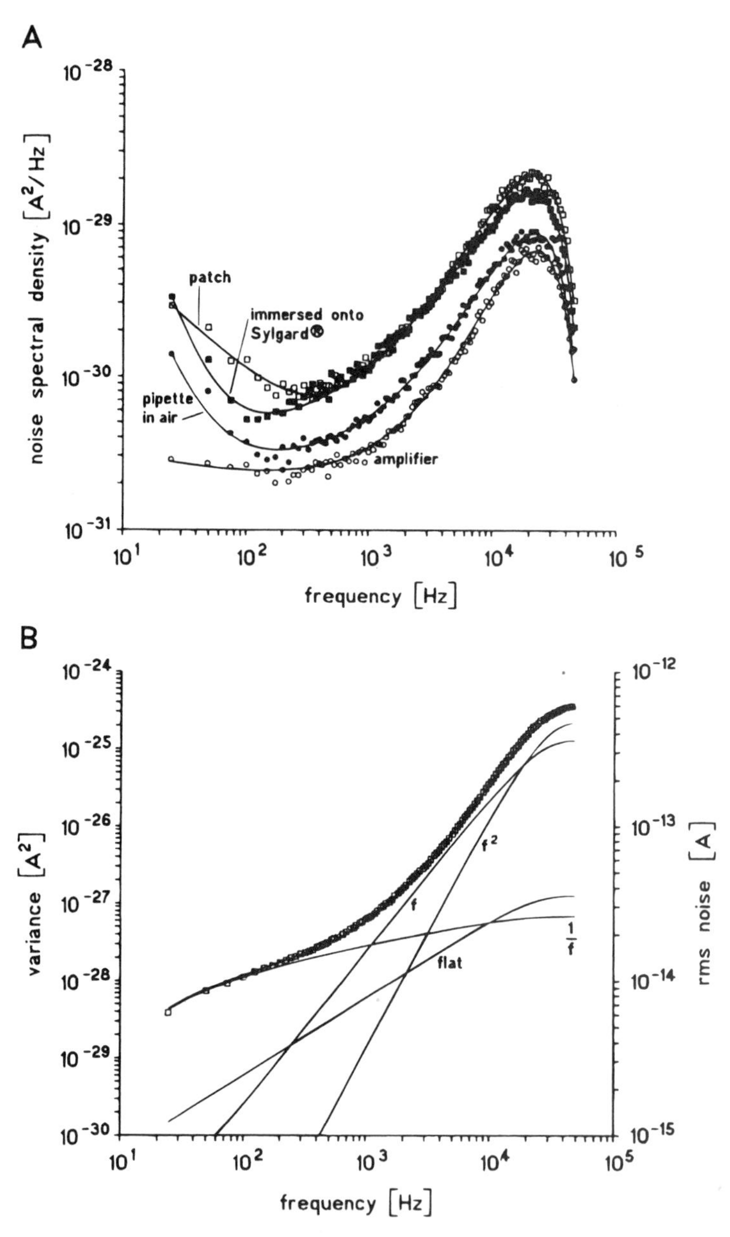

Figure 5. Measured noise. A: Power spectra of noise. The noise was measured with an Axopatch 200A amplifier in a 20-kHz bandwidth of the amplifier alone (open circles; rms noise 416 fA), with the pipette holder and a short pipette above the bath (filled circles; rms noise 489 fA), with the pipette (coated with Sylgard®) immersed in the bath (1000 μm) and closed with Sylgard® (filled squares; rms noise 759 fA, d_o/d_i = 4; pipette resistance 72 MΩ, pipette solution 200% Tyrode, seal resistance >4 TΩ), and in a cell-attached patch of a myocardial mouse cell containing no channel (open squares; rms noise 658 fA; d_o/d_i = 4; pipette resistance 80 MΩ, pipette solution 200% Tyrode, seal resistance 1.2 TΩ, coated with Sylgard®). The curves are best fit with $S_I = [a/f^b + c + df + ef^2]\, H(f)$, where *a*–*e* are parameters to quantify the noise types according to Table I and *H*(*f*) is the transfer function of the filter composed of the four-pole and eight-pole Bessel filter (Tietze and Schenk, 1985). The parameters *a* [$A^2/Hz^{(1-b)}$], *b* (dimensionless), *c* (A^2/Hz), *d*

tions, currents through the seal). In fact, the stability of TΩ seal patches could be enormously high (recording times up to 2.5 hr) but also depended on the cell type.

The *RC* noise generated by the patch capacitance and the pipette resistance is negligible in the case of the small pores despite the large resistance. Assuming a pipette resistance of 140 MΩ and a specific membrane capacitance of 1 μF/cm^2, a patch modeled as a hemisphere with 200-nm diameter would cause an rms noise of only 18 fA at 20-kHz bandwidth.

3.6. Fluid Films

Besides decreasing the pipette capacitance, coating with Sylgard® also reduces noise by avoiding the formation of thin fluid films creeping from the bath upward on the wall of uncoated pipettes (Hamill *et al.,* 1981). This effect is very important for the noise performance. Other fluid films may arise inside the pipette above the solution meniscus and also within the narrow clefts between the pipette, the holder body, and the screw cap. Such films may be avoided by carefully drying the walls before inserting the pipettes into the holder and excluding any possibility of fluid contamination within the holder via the back end of the pipette. It proved very helpful to replace the solution at the back end by paraffin or silicon oil. This oil layer strips off any solution from the wire when a pipette is removed and does not affect the electrical properties of the Ag/AgCl/solution interface.

4. Measured Noise

For measuring noise, an eight-pole Bessel filter was connected to the output of the Axopatch 200A amplifier whose internal four-pole Bessel filter was set to 50 kHz (cutoff frequency, −3 dB). In order to obtain the final cutoff frequency f_c the eight-pole Bessel filter was set to a larger cutoff frequency ($f_{c,8\text{-pole}}$) according to

$$\frac{1}{f_c^2} = \frac{1}{(50\ \text{kHz})^2} + \frac{1}{f_{c,8\text{-pole}}^2} \tag{14}$$

In the considered bandwidths $f_B \leq 20$ kHz, the characteristics of this filter combination are very close to those of an eight-pole Bessel filter alone set to f_c. All noise data given in the text correspond to this Bessel-filter characteristic.

Figure 5A illustrates power spectra calculated from noise recorded in a patch clamp with the Axopatch 200A amplifier with f_B = 20 kHz. The downward deflection at high frequencies corresponds to the filter setting. The most important results are: (1) If the holder

(A^2/Hz^2), e (A^2/Hz^3) are, for the amplifier, 3.6×10^{-31}, 0.44, 1.8×10^{-31}, 1.1×10^{-34}, 2.6×10^{-38}; for the pipette in air 5.0×10^{-28}, 1.9, 2.5×10^{-31}, 2.6×10^{-34}, 2.9×10^{-38}; for the immersed pipette onto Sylgard® 3.1×10^{-29}, 0.78, 1.1×10^{-31}, 8.0×10^{-34}, 6.6×10^{-38}; and for the patch 3.3×10^{-27}, 2.2, 4.1×10^{-31}, 7.5×10^{-34}, 4.2×10^{-38}. B: Noise variance as a function of the frequency for the noise attributable to the immersion of the pipette. The variance was calculated by integrating the difference of the spectra immersed onto Sylgard® and pipette in air in A. The variance of the individual noise contributions sums to the curve fitted to the data points. The right ordinate illustrates the corresponding rms noise levels. The respective parameters of the fitted curve are 2.1×10^{-29}, 0.79, 8.0×10^{-32}, 4.8×10^{-34}, 4.0×10^{-38}.

is attached to the headstage, $1/f$ noise dominates the spectra from the lowest frequency included (25 Hz) to several hundred Hertz. Hence, resolution at very low frequencies depends preferentially on $1/f$ noise. Both parameters of the $1/f$ noise, a and b, were found to be variable to a high degree. The difference among five pipettes sealed with Sylgard® and six membrane patches was not significant. (2) Attachment of the holder and the pipette to the headstage predominantly increased f noise, which suggests that the holder material was not optimal. (3) At intermediate frequencies, f noise also dominates the spectra of both the Sylgard®-closed pipette and the membrane patch as well as the difference spectrum corresponding to the immersion of the pipette tip (Fig. 5B). The relative contribution of f noise was found to be larger than predicted by the theory (cf. Fig. 4). (4) The flat noise in the patch could not be distinguished from that in the Sylgard®-closed pipette, which indicates that the noise arising in the seal and in the patch was very small.

With the technique described (short patch pipettes, $d_o/d_i = 4$, Sylgard® coat starting 50 μm behind the tip), rms noise levels in actual patch-clamp recordings at 5- and 20-kHz bandwidth of 95–120 and 630–780 fA, respectively, were regularly achieved.

5. Perspectives

Further improvement of the noise performance could be expected by increasing the wall thickness of the pipette and minimizing the size of the holder. Figure 6A illustrates a scanning electron micrograph of a pipette tip pulled from Duran® glass with a d_o/d_i ratio of 8 ($d_o = 2.0$ mm, $d_i = 0.25$ mm). The d_o/d_i ratio at the tip is approximately 7; i.e., the favorable wall thickness for lower dielectric and RC noise is largely preserved (cf. Fig. 3). With this glass tubing, pipettes with a resistance between 40 and 600 MΩ ($\rho = 51$ Ωcm) could be pulled, and seals were obtained easily, also without fire-polishing.

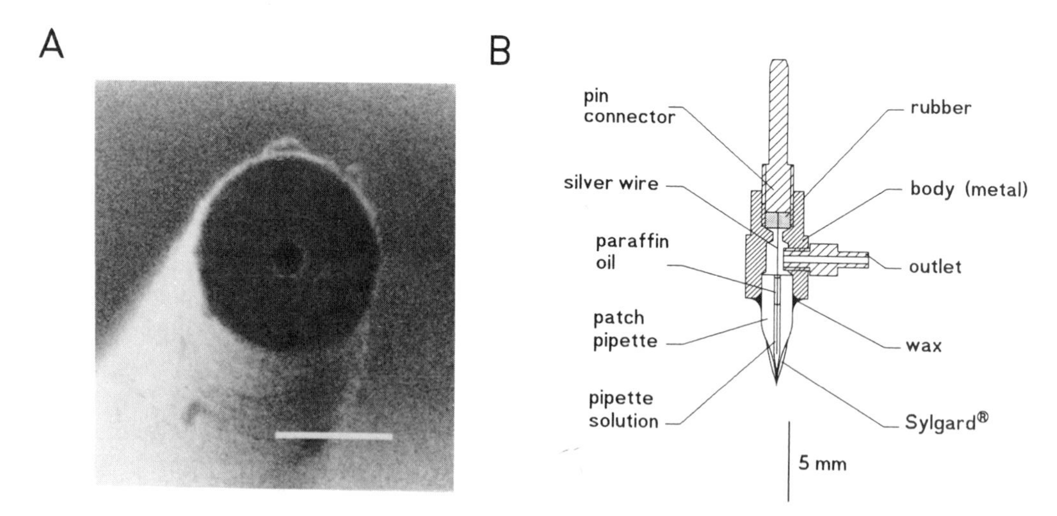

Figure 6. Further minimization strategy. A: Scanning electron micrograph of a pipette tip pulled from glass tubing with $d_o/d_i = 8$. At the tip this value is only slightly reduced to ~7, pipette resistance is 570 MΩ ($\rho = 26$ Ωcm), pore diameter is 120 nm, scale bar is 500 nm. Same preparation for electron microscopy as in Fig. 1A. B: Schematic drawing of a minimum pipette holder fabricated from silver. The short pipettes are fixed with wax.

Figure 6B illustrates a minimum-sized pipette holder that has been successfully used. The holder, fabricated from silver (other metals can also be used), has been reduced practically to the metallic pin connector of a conventional pipette holder provided with a small outlet, a silver wire, and a part taking up 1 mm of the pipette. The small pipette was fixed with wax. This was performed by first positioning a small wax ring (outer diameter 4 mm, inner diameter 2 mm) on the holder opening, then inserting the pipette, and finally fixing it by briefly touching the wax with a small soldering iron. A simple candle paraffin was used, but waxes with the lowest loss factor are certainly more appropriate. The holder with the pipette was then attached to the headstage. When the angle of the holder axis to the chamber bottom was about 45°, the patch clamp could be performed as usual without altering the size of the headstage case. With this pipette–holder combination, the rms noise level, with the pipette tip above the bath, was at the borderline to be distinguishable from the headstage noise. With a thick Sylgard® coat to the tip, the pore closed with Sylgard®, and the pipette tip immersed, noise levels as low as 81 fA (5-kHz bandwidth) were reached. Conclusively, it is possible to achieve a total noise close to that of the headstage alone with this minimization strategy.

6. Noise-Dependent Resolution Limits of Channel-Open Time

6.1. The Half-Amplitude Threshold Technique

Two principal limits exist for the resolution of the channel-open time. (1) If openings become shorter than the filter rise time $t_r = 0.3321/f_c$, they progressively decrease in amplitude, and the open time measured at 50% of the fully open level also decreases, reaching zero when the amplitude is 50%. (2) If the probability that noise deflections reach the detection threshold Φ is not negligible, these artificial noise events either split or shorten an opening. This probability rises with the duration of the opening and the noise amplitude. Assuming the noise at the open level to equal the background noise (variance, σ^2) and an exponential distribution of the open times t_o, with τ_o being the estimate of the mean, error of τ_o is less than 10% if for the upper limit the false-event rate $\lambda_f = 0.05/\tau_o$ is assumed. An upper-limit estimate for the measurability of τ_o is then (cf. Chapter 19, this volume)

$$\tau_o = \frac{0.05}{kf_c} \exp\left(\frac{\Phi^2}{2\sigma^2}\right) \tag{15}$$

where k is a constant depending on the type of noise (0.849 flat noise; 1.25 f^2 noise). A plot of this relationship for the noise of the patch in Fig. 5A (filled squares) is illustrated in Fig. 7. An example may explain the plot (dotted lines): Openings with the amplitude $A_o = 2$ pA are measured at the 50% threshold $\Phi = 1$ pA at $f_c = 10$ kHz. Mean open times may be reliably measured (brace at the ordinate) between somewhere less than 2000 μsec and the lower limit of 33 μsec, here set to the rise time t_r of the filter.

6.2 The Baseline Method

Channel-open times may also be determined from the distribution of the opening-induced gaps in the middle of the baseline noise even if this noise is substantially large, as

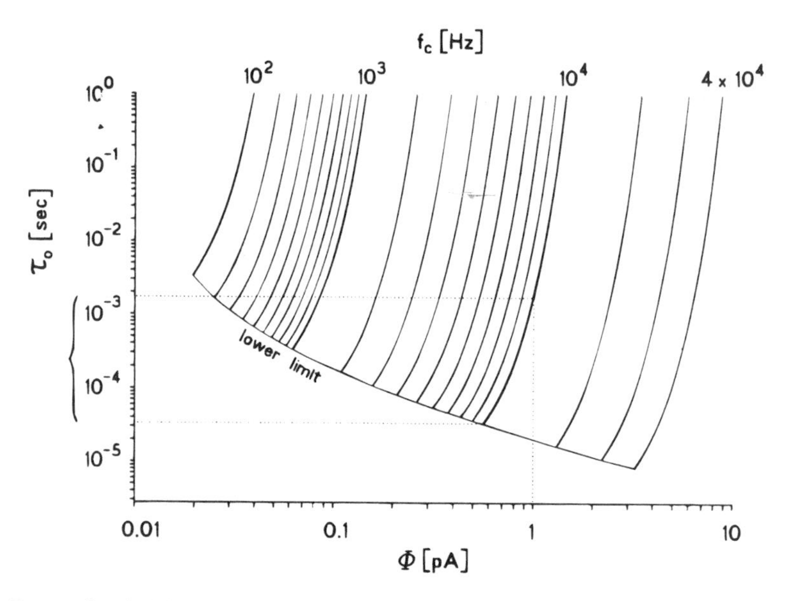

Figure 7. Range of evaluable open time τ_o as a function of the threshold Φ (Φ is $A_o/2$ for the half-amplitude method and A_o for the baseline method) and the cutoff frequency f_c, calculated with equation 15 for the noise in the patch of Fig. 5A. The noise variance σ^2 is related to the parameter f_c by the integral of the fitted spectrum over the full bandwidth with the parameters of the patch given in the legend to Fig. 5A. The upper limit for measuring τ_o is obtained by the intersection of the vertical line with the curve of the respective f_o; the lower limit is given by the lower end of the respective f_c curve.

shown in Fig. 8A for the activity of Na channels at 35°C and 20-kHz recording bandwidth. In contrast to usual open-time histograms, the distribution of dwell times measured at the baseline contains both channel events (t_o) and noise events (t_n). t_n is distributed in a complicated fashion (Rice, 1954); its mean is approximately $1/(2kf_c)$ (Papoulis, 1991). In regard to the openings, it is assumed for simplicity that the filter generates "trapezium-like" events instead of the more complex time course provided by the transfer function of real filters. The rise time of one transition is assumed to be t_r. With respect to the original open time t_o, measurement of the open time in the middle of the baseline leads to three types of alterations. (1) Independent of noise, all original openings exceeding t_r are prolonged constantly by t_r. (2) If the trapezium-like opening starts or ends just at the time when the noise is deflected in the direction of the channel opening, the opening is artificially prolonged. Since noise events with large t_n are met statistically more often, the mean prolongation $\bar{t}_{pr}$ must be some larger than $\bar{t}_n/2$, which is therefore used in the following as lower limit estimate for $\bar{t}_{pr}$. At $f_c = 20$ kHz ($k = 1.25$), this lower limit estimate for $\bar{t}_{pr}$ is 10 μsec. (3) If the trapezium-like opening starts or ends just at the time when the noise is deflected in the direction where the channels do not open, the opening is artificially shortened. The mean shortening $\bar{t}_{sh}$ per transition is

$$\bar{t}_{sh} = \frac{\sigma_n t_r}{A_o} \tag{16}$$

At $f_c = 20$ kHz with $t_r = 17$ μsec and typical values for σ_n and A_o of 0.7 and 2.8 pA, respectively, the mean shortening per transition is only 4 μsec; i.e., the mean noise-induced

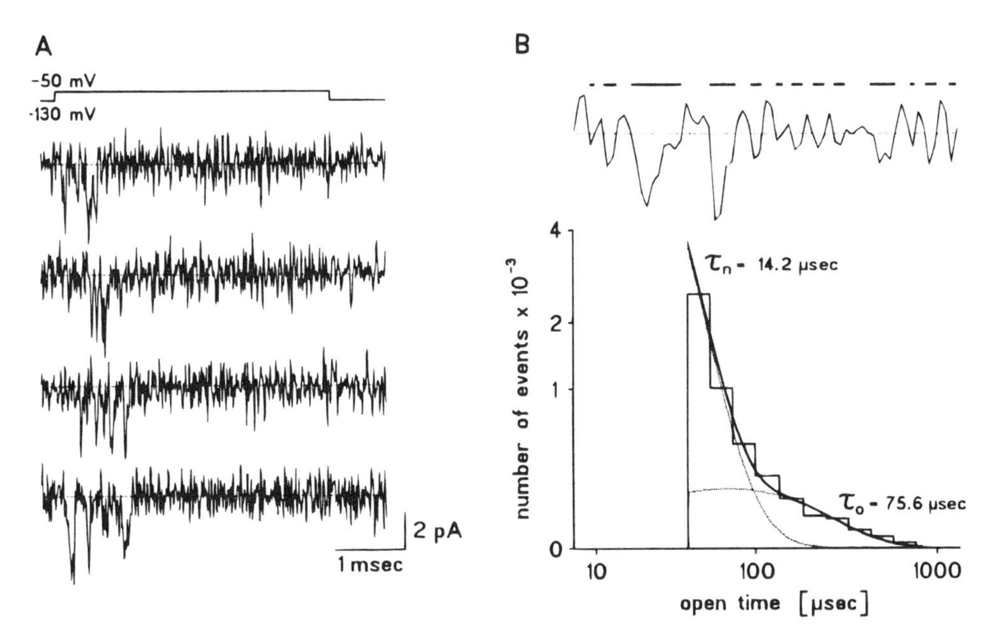

Figure 8. Determination of the mean channel-open time with the baseline method. A: Patch recordings in a TΩ seal of voltage-dependent Na channels at 35°C in a cell-attached patch of a myocardial mouse cell; bandwidth 20 kHz. Correction for leakage and capacitive currents was performed by subtracting averaged blanks selected from the neighborhood of the actual record. B: Time histogram built from all intervals in the middle of the baseline in which the trace is deflected in the direction of the channel openings (bars above the trace in the inset); square-root-transformed ordinate, logarithmic abscissa. Simultaneous fit with two exponentials yielded the indicated time constants for the mean open time (τ_o) and for the exponential decay (τ_n) of the noise events. Data were recorded and analyzed with the ISO2 software.

prolongation of the measured τ_o exceeds the noise-induced shortening. In summary, if measured at the baseline, each original opening of sufficient amplitude and a duration $t_o > t_r$ is considerably prolonged by $t_r + \bar{t}_{pr}$ and insignificantly shortened by $\bar{t}_{sh}$. Since all longer openings are prolonged on average by the same amount, and if the fit starts only at the time $2t_r + \bar{t}_{pr} - \bar{t}_{sh}$, the time constant of the distribution of the original openings and of the openings evaluated at the baseline is equal. For the values in the example at $f_c = 20$ kHz, all original open times t_o are prolonged on average by 23 μsec; i.e., the shortest opening reaching approximately full amplitude with $t_o = t_r = 17$ μsec is represented in the histogram by an event with the expected value of 40 μsec.

For the purpose of separation of the t_o from the t_n distribution, an exponential (time constant τ_n) can be fitted with high accuracy to the t_n distribution if starting at the time $2t_r + \bar{t}_{pr} - \bar{t}_{sh}$. τ_n was found to be very consistent with $\bar{t}_n/2 < \tau_n < t_r$. In the case of a much larger noise event number than opening event number, a separate fit of the t_n distribution may be performed, ignoring the opening events. Alternatively, the t_n distribution may be fitted to the oppositely directed noise events. If τ_o substantially exceeds τ_n, the corresponding t_n bins can be clipped for fitting the t_o distribution. If τ_o does not exceed τ_n substantially, as in the example of Fig. 8B, the t_n and t_o distributions should be fitted simultaneously. In the case when the contribution of the noise events substantially exceeds that of the opening events, reasonable separability was obtained if $\tau_o > 2\tau_n$. Improved resolution is possible if the noise event number is relatively smaller.

Two essential conditions for utilizing the baseline method are the following. (1) The seal should be of high quality. Problems arising from slow drifts of leakage (and capacitance in pulsed recordings) can be overcome by subtraction of a sliding blank, which is the average of a specified number of blanks recorded as close to the actual trace as possible (ISO2 patch clamp software, MFK-Computer, Frankfurt/M, Germany). (2) The number of openings should be sufficient to recognize the distribution in the histogram because the method does not identify the individual openings to be evaluated but only the distribution in addition to that of the noise events. If enough events are available, the baseline method provides two advantages:

(1) At a given unitary current amplitude A_o, the noise may be twice as large as with the half-amplitude threshold technique; i.e., the recording bandwidth may be considerably increased. If the shortest resolvable opening event number is in the order of the noise event number because of a large open probability, these shortest events are included in the procedure to determine τ_o. With Φ being replaced by A_o, the range for reliable determination of τ_o in Fig. 7 is increased. With respect to the indicated example in the figure, the baseline method would allow either opening the filter to $f_c \sim 17$ kHz, having the same upper limit at an improved lower limit, or measuring at 10 kHz with an upper limit out of scale and the same lower limit. At intermediate frequencies both limits are improved to intermediate degrees.

(2) The advantage of the baseline method is even more important if the open levels are heterogeneous. If the filter frequency is chosen such that the lowest levels to be included have acceptable false-event rates according to Fig. 7, larger levels may be evaluated at the same time with a practically negligible error. For the above example (f_c = 20 kHz, σ_n = 0.7 pA), reduction of the amplitude of equally long openings from 2.8 pA to 1.4 pA would reduce the measured τ_o by only 4 μsec. As a consequence, in cardiac Na channels the mean open time was very consistent among the patches, even if evaluated at 35 °C (Benndorf, 1994), and at room temperature, shorter open times were found than those measured with a conventional threshold technique (Benndorf and Koopmann, 1993).

Multiexponential distributions of the channel-open time can be quantified with the baseline method in a similar way if only the false-event rate for the longer events is low enough (cf. Fig. 7). The method should also be appropriate for the analysis of burst kinetics. Here, the number of noise events may be kept low by excluding closed sojourns, which would allow evaluation of openings as short as t_r at a maximally open filter according to Fig. 7.

7. Appendix

7.1. Fabrication of Small-Pore Patch Pipettes

The tipped patch pipettes used here were pulled on a computer-controlled puller (DMZ Zeitz Universal, Augsburg, Germany) in a three-stage pull from thick-walled (d_o 2.00 mm, d_i either 0.50 or 0.25 mm) borosilicate glass tubing without filament (Hilgenberg GmbH, Malsfeld, Germany). The tips were not fire-polished because they were too small for optical control. Filling of the pipettes, 40 mm long at this stage, was carried out from the rear end by a thin plastic tube moved to the tip as close as possible. The pipettes were then connected with a short piece of silicon tubing to a 20-ml syringe. Repeated application and relief of a negative pressure in combination with vigorous tapping with a finger promoted an upward migration of bubbles of various size. After removal of the last bubble, the pipettes were shortened to the final length of 8 mm by first sawing them in a clockmaker's turning lathe and subsequently breaking them while the front part (~10 mm) was immersed in pipette solution to avoid destruction of the tip. Pipettes were shortened for two reasons: to decrease the capacitance and to facilitate the placement of a thin wire (diameter 100 μm) as close to the tip as possible to reduce the series resistance. Finally, the pipettes were dried on the outside, and the solution at the rear end was replaced by paraffin or silicon oil.

7.2. Interfering Signals

Low-noise measurements require improved screening of interfering signals. The line pickup of 50 Hz was excluded most successfully by completely closing the cage around the setup with sheet metal. Grids were sometimes insufficient. The amplifier and the filter were separately shielded. When working at high bandwidths, computer screen refresh frequencies of ~16 and ~32 kHz were often prominent in the spectra. These frequencies, not always detectable in the traces by eye, may increase the total noise. It is therefore helpful to calculate power spectra of the noise from time to time. These high-frequency signals seem to interfere directly with the amplifier or with attached cables. It sometimes helps to move the screen away from the amplifier by 1 m. In the $1/f$-noise range, disturbing interference could arise from discontinuity of the bath solution flow. If a flowing bath solution is desired, the fluid level should be kept as constant as possible.

ACKNOWLEDGMENTS. I am grateful to Mr. R. Koopmann for his advice and support through all the stages of this work, to Mr. A. Draguhn, Mr. P. Scherer, Mr. T. Böhle, and Mr. D. W. Hilgemann for comments on the manuscript, to Mr. W. Mackowiak for the expert help with the scanning electron microscopy, and to Ms. D. Metzler and Ms. R. Kemkes for technical assistance and preparing the illustrations. This work was supported by the Deutsche Forschungsgemeinschaft, Be 1250/1−4. The author is a Heisenberg-fellow of the Deutsche Forschungsgemeinschaft.

References

Benndorf, K., 1993, Multiple levels of native cardiac Na^+ channels at elevated temperature measured with high-bandwidth/low noise patch clamp, *Pflügers Arch.* **422:**506–515.

Benndorf, K. 1994, Properties of single cardiac Na channels at 35° C. *J. Gen. Physiol.*, (in press).

Benndorf, K., and Koopmann, R., 1993, Thermodynamic entropy of two conformational transitions of single Na^+ channel molecules, *Biophys. J.* **65:**1585–1589.

Dutta, P., and Horn, P. M., 1981, Low frequency fluctuations in solids: $1/f$ noise, *Rev. Mod. Phys.* **53:**497–516.

Hamill, O. P., Marty, A., Neher, E., Sakmann, B., and Sigworth, F. J., 1981, Improved patch-clamp techniques for high-resolution current recording from cells and cell-free membrane patches, *Pflügers Arch.* **391:**85–100.

Levis, R. A., and Rae, J. L., 1992, Constructing a patch clamp setup, *Methods Enzymol.* **207:**18–66.

Levis, R. A., and Rae, J. L., 1993, The use of quartz patch pipettes for low noise single channel recording, *Biophys. J.* **65:**1666–1677.

Papoulis, A., 1991, *Probability, Random Variables and Stochastic Processes,* 3rd ed., McGraw-Hill, New York.

Rice, O. S., 1954, Mathematical analysis of random noise, in: *Selected Papers on Random and Stochastic Processes,* Dover, New York.

Tietze, U., and Schenk, C., 1985, *Halbleiter Schaltungstechnik,* 7th ed., Springer-Verlag, Heidelberg.

Chapter 6

Voltage Offsets in Patch-Clamp Experiments

ERWIN NEHER

1. Introduction

Offset voltages of various origins have to be considered in patch-clamp experiments. Some of the offsets are constant during a typical experiment, such as amplifier input offsets; some are variable, such as liquid junction potentials, depending on ionic conditions. Some arise in the external circuit (i.e., in the patch pipette, in the experimental chamber, or at the silver chloride electrodes), and some arise in the patch-clamp amplifier. Typical magnitudes are ± 30 mV for amplifier offsets, up to 100 mV (depending on Cl^- concentrations) for electrode offsets, and up to ± 15 mV for liquid junction potentials at interfaces between different solutions. It is standard practice to compensate amplifier and electrode offsets by performing a reference measurement before the pipette is sealed to a cell. This is done by adjusting a variable offset (V_O), which, in the amplifier, is added to the command voltage (V_C), such that there is zero current flow at $V_C = 0$. This protocol is correct, provided that none of the offsets mentioned above changes during the experiment. If, however, the pipette solution is different in its composition from the bath solution (as is usually the case for whole-cell measurements), a liquid junction potential will be present at the pipette tip during the reference measurement. This will no longer be the case during the test measurement, when the liquid junction is replaced by the membrane under study.

Likewise, other liquid junction potentials that may appear or disappear during an experiment when solution changes are performed (see Neher, 1992) have to be taken into account. Also, a membrane resting potential may be in series with the membrane under study (in cell-attached configuration).

2. Analysis of the Offset Problem

For analyzing this situation Fig. 1A displays an equivalent circuit that contains five voltage sources, two in the external current pathway and three in the internal pathway (internal to the amplifier): V_F, a fixed offset representing electrode potentials; V_V, a variable offset representing liquid junction potentials; V_A, the fixed amplifier offset, V_O an adjustable offset of the patch-clamp amplifier; and V_C, the command potential. Part A represents the situation during the reference measurement. This is why V_V, V_O, and V_C carry a superscript R to indicate

ERWIN NEHER • Department of Membrane Biophysics, Max-Planck Institute for Biophysical Chemistry, Am Fassberg, D-37077 Göttingen, Germany.
Single-Channel Recording, Second Edition, edited by Bert Sakmann and Erwin Neher. Plenum Press, New York, 1995.

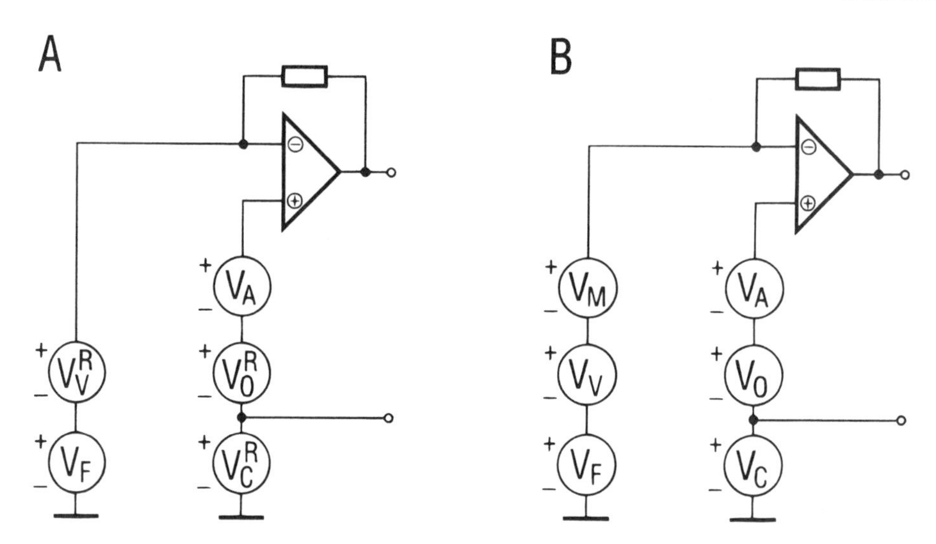

Figure 1. Equivalent circuit of a patch-clamp measurement. Part A represents the situation during a reference measurement, and part B that during a test measurement. In each branch of the circuit a variable offset and a fixed offset are considered in addition to the quantities of primary interest, which are membrane potential V_M and the command potential V_C. In the external branch the fixed offset V_F comprises mainly the electrode potentials, i.e., the voltage drops at the silver-chloride surface, which are assumed to be protected from solution changes during an experiment. The variable offset V_V is the sum of liquid junction potentials and, if applicable, the potential of a membrane in series with the patch (in cell-attached configuration). In the branch internal to the amplifier (right side), the fixed offset is the amplifier offset V_A, and the variable one is an adjustable offset voltage V_O, which is considered to be in series with the command potential V_C. In part A the quantities V_V, V_O, and V_C carry a superscript R to indicate values at the time of the reference measurement. Note that the sign convention for V_M agrees with the physiological one for whole-cell and outside-out recordings but not for cell-attached and inside-out patches.

values in effect at this time. Part B represents the situation during the test measurement. Now, the circuit includes the membrane potential V_M. Here, V_V, V_O, and V_C are considered to change during the course of an experiment. The latter is available as an output signal for data storage and analysis. V_V may change whenever solutions or configurations are changed. In the following it is shown that V_C will always be identical to V_M if V_O is changed in parallel with V_V during an experiment.

During the reference measurement, V_O is varied until current flow in the external circuit is zero. Then,

$$V_V^R + V_F = V_A + V_O^R + V_C^R \tag{1}$$

and

$$V_O^R = -V_A - V_C^R + V_V^R + V_F \tag{2}$$

During a test measurement (Fig. 1B), we have

$$V_M = V_A + V_O + V_C - V_V - V_F \tag{3}$$

For the purpose of the following discussion, we write both V_O and V_V in terms of their values at the time of the reference measurement and an increment to that:

$$V_O = V_O^R + \Delta V_O \qquad V_V = V_V^R + \Delta V_V \tag{4}$$

Then,

$$V_M = V_C - V_C^R + \Delta V_O - \Delta V_V \tag{5}$$

It is seen that the reference measurement eliminates all fixed offsets such that we have to consider only variable ones and V_C^R, the command potential at the time of the reference measurement.

3. Three Strategies for Handling the Offset Problem

3.1. *A Posteriori* Correction

Most commonly researchers do not consider corrections until the time of publication of results. In that case the reference measurement is performed while the command potential is set to zero ($V_C^R = 0$), and the offset voltage is unchanged during the rest of the experiment ($\Delta V_o = 0$). Then, equation 5 reduces to

$$V_M = V_C - \Delta V_V \tag{6}$$

This means that the potential read from the amplifier has to be corrected by subtracting all the changes in offsets (ΔV_V) that occurred between the reference measurement and the test measurement.

During standard whole-cell measurements the only change in offset potential is the disappearance of the liquid junction potential V_L at the pipette tip. Liquid junction potentials for a given solution pair can be determined experimentally or calculated from the Henderson equation (Barry and Lynch, 1991; Neher, 1992; Ng and Barry, 1994; see also program JPCalc by Barry, 1994). V_L is defined as the potential of a reference solution (usually Ringer) with respect to the test solution (such as the pipette filling solution). This is opposite to the sign convention of Fig. 1. Since V_L disappears, however, another sign inversion has to be applied, with the result that numbers for V_L found in tables of the above references are equal to ΔV_V of equation 6 and therefore have to be subtracted from V_C. This holds for whole-cell and outside-out recording configurations in which the sign convention of Fig. 1 agrees with the physiological one.

A further complication arises with cell-attached and inside-out configuration because here the reading of the patch-clamp amplifier by convention is taken as the negative of the membrane potential, considering the orientation of the membrane. It would be straightforward to correct the reading of the amplifier and then invert it, since the orientation of the membrane is irrelevant to the offset problem. However, if correction is done only at publication time, the numbers have been inverted already, such that the correction too has to be inverted. Configuration-specific recipes are given by Neher (1992).

In the more general case, the quantity ΔV of equation 6 should be considered as the sum of all offset changes occurring between the reference and the test measurement. It will therefore also be called the correction sum below. It may include additional liquid junction potentials, if bath solutions are being changed (Barry and Lynch, 1991; Neher, 1992), and the resting potential (in cell-attached configuration). When the correction sum is formed, all voltages should be added with the polarities given in Fig. 1. As explained above, numbers for the standard liquid junction potential corrections should be added to the correction sum with the sign found in tables of Barry and Lynch (1991) and Neher (1992) because of a double sign inversion. The correction sum, then, should be subtracted from the amplifier reading in whole-cell and outside-out configurations or added to the inverted amplifier reading in cell-attached and inside-out configurations.

3.2. On-Line Correction

Equation 5 suggests a simple way to do the correction on line in cases where no other offset changes occur except for the standard liquid junction potential corrections. In that case, the command potential can be set to the negative of the expected liquid junction potential correction during the reference meassurement ($V_C^R = \Delta V_V$) such that equation 5 simplifies to

$$V_M = V_C + \Delta V_O \tag{7}$$

which means that $V_M = V_C$, if the offset potential setting is left untouched during the measurement. Thus, the amplifier displays a corrected voltage that can be interpreted with a positive or negative sign depending on the configuration.

3.2.1. An Example: Whole-Cell Recording with a K-Glutamate-Based Pipette Filling Solution

The liquid juction potential of such a solution is given by Neher (1992) as +10 mV, such that the command potential should be set to −10 mV during reference measurement. If the so-called "search Mode" is used for the reference measurement, the offset potential should be adjusted such that V-command reads −10 mV while current is zero.

3.2.2. An Example: Cell-Attached Recording on a Cell That Is Assumed to Have −60 mV Resting Potential

In this case, the pipette is usually filled with bath solution such that no liquid junction potential correction has to be applied. To correct for the resting potential, ΔV_V has to be taken as −60 mV (the resting potential, which is in series with the patch, has the same polarity as assumed in Fig. 1). Thus, the command potential should be at +60 mV during the reference measurement. The amplifier readings during the measurement have to be inverted in order to obtain membrane potentials according to the physiological convention.

3.3. On-Line Correction Using a Computer-Controlled Patch-Clamp Amplifier

3.3.1. General

Equation 6 also implies that any changes in offset potential (ΔV_V) that may occur during the course of a measurement can be compensated by changing V_O in parallel. This does require a calibrated offset voltage, however. In the following, a procedure is given that has been implemented on a computer-controlled patch-clamp amplifier (EPC-9, HEKA Elektronik, Lambrecht, Germany), which allows one to compensate for offsets throughout an experiment.

At the time of the reference measurement the command potential is first set to the negative value of a software variable (termed *LJ* for liquid junction), which is specified by the user. Then, an automatic search (the so-called ZERO procedure) finds the setting of an internal calibrated offset voltage that yields zero pipette current. If the user has actually entered the liquid junction potential, then this procedure is analogous to the on-line correction mentioned above.

Furthermore, the EPC-9 controlling software forces V_o to change in parallel with *LJ* whenever the user changes *LJ* implying $\Delta LJ = \Delta V_o$, such that, together with the initial condition, $LJ(0) = -V_C^R$, we obtain

$$LJ(t) = LJ(0) + \Delta LJ = -V_C^R + \Delta V_O \quad (8)$$

or

$$\Delta V_O = V_C^R + LJ(t) \quad (9)$$

Inserting equation 9 into equation 5, we see that V_C^R cancels out, so that

$$V_M = V_C + LJ(t) - \Delta V_V \quad (10)$$

Thus, the user can provide for $V_M = V_C$ and thereby achieve a valid correction at any time during the measurement by setting the variable *LJ* to ΔV_V irrespective of what had been set at the time of the reference measurement. ΔV_V has to be calculated by the sum of all changes in offset potentials that occur between the reference measurement and the test measurement, as explained above. Some further examples for calculating ΔV_V will be given below. The procedure described can readily be simulated using any patch-clamp amplifier if the command potential is generated by a computer-controlled digital-to-analogue converter. In that case, the stimulus input to the amplifier should be subdivided by software into an offset part and a command part.

The above considerations did not specify a particular patch-clamp configuration, since voltage offsets, as viewed from the measuring amplifier, are independent of the orientation of the membrane. As outlined under Section 3.1, the offset-corrected potentials should be interpreted either directly (in whole-cell and outside-out configuration) or after a sign inversion (in cell-attached or inside-out configuration). The EPC-9 allows the user to select modes, and this sign inversion is performed together with a sign inversion of the resulting current such that both membrane current and membrane voltage are displayed and output according to the physiological convention. This is done automatically, if cell-attached mode has been selected.

3.3.2. Examples of Offset Handling

Below, two standard examples of experiments are described, one representing a whole-cell recording employing two different bath solutions, the other representing cell-attached recording. In both cases the term "correction sum" refers to the quantity (ΔV_V), to which the variable *LJ* should be set in order to achieve a correction according to equation 10. As explained above, the correction sum is the sum of the liquid junction potential according to the definition of Barry and Lynch (1991) and of other offsets using polarities given in Fig. 1. In both examples the aim is to perform the experiment in a way that the amplifier voltage output (V_C) gives the correct membrane voltage at all times, both in magnitude and in polarity.

3.3.2a. Whole-Cell Recording. It is assumed that the measurement is started in Ringer solution with a patch pipette containing mainly K-glutamate. For this combination of solutions a value for the liquid junction potential of +10 mV can be found in the literature (Neher, 1992). Thus, *LJ* should be set to +10, and a zero-current search operation be performed before seal formation. This will provide for correct voltage readings in subsequent whole-cell recordings as long as ionic conditions remain constant. As explained above, *LJ* can actually be set to any value during the ZERO; however, it should be changed to 10 mV before the start of the whole-cell recording.

Later on, during the same experiment, the bath solution is assumed to be changed to a sulfate-rich Ringer by local perfusion. Thus, a liquid junction potential will develop at the interface between the two solutions in the bath. In Table 1 of Neher (1992), we find a value of +6 mV for a solution pair standard Ringer versus sulfate Ringer. This polarity is opposite to what is required for the calculation of the correction, since we need the potential of the solution close to the pipette (sulfate-rich) with respect to the more remote solution. Thus, we calculate the correction sum to be 4 mV and correspondingly set *LJ* to 4 mV during the episode in sulfate Ringer. When we switch back to Ringer solutions, we set *LJ* back to 10 mV. It should be noted that a similar correction is also required if the whole bath is perfused, since a liquid junction potential arises at the reference electrode unless a "bleeding" KCl reference is used (see Neher, 1992).

3.3.2b. Cell-Attached Recording. We assume the same solutions as above but want to perform a cell-attached measurement on a cell that we know to have −50 mV resting potential. Therefore, we form the sum of the pipette liquid junction potential correction (10 mV) and the resting potential (−50 mV; we can use the "physiological" polarity since the intracellular compartment is closer to the pipette than the extracellular one) and set *LJ* to −40 mV during the cell-attached measurement. Then the amplifier will display the correct magnitude of the membrane potential. Its polarity has to be inverted in order to conform with the physiological convention. This is done automatically if "cell-attached mode" has been selected.

4. Conclusions

Unfortunately, offset corrections, particularly those for junction potentials occurring during solution changes in the course of an experiment, are paid little attention in the current patch clamp literature. The problems have been pointed out before (Barry and Lynch, 1991; Neher, 1992). This chapter is intended to demonstrate that procedures are at hand to conveniently handle them.

Particularly for ion selectivity studies and for comparison of properties of voltage-activated channels, an accuracy of voltage readings better than a few millivolts is mandatory. This cannot be obtained unless care is taken to perform valid offset corrections.

References

Barry, P. H., 1994, JPCalc, a software package for calculating liquid junction potential corrections in patch-clamp; intracellular, epithelial and bilayer measurements and for correcting junction potential measurements, *J. Neurosci. Methods* **51:**107–116.

Barry, P. H., and Lynch, J. W., 1991, Liquid junction potentials and small cell effects in patch-clamp analysis, *J. Membr. Biol.* **121:**101–117.

Neher, E., 1992, Correction for liquid junction potentials in patch clamp experiments, *Methods Enzymol.* **207:**123–131.

Ng, B., and Barry, P. H., 1994, Measurement of additional ionic mobilities for use in junction potential corrections in patch-clamping and other electrophysiological measurements, *Proc. Aust. Neurosci. Soc.* **5:**141.

Chapter 7

Techniques for Membrane Capacitance Measurements

KEVIN D. GILLIS

1. Introduction

Understanding the process whereby cells transduce an external signal to a secretory response ("stimulus–secretion coupling": Douglas, 1968) has been an important topic of research for many years. The understanding of early events in the cascade in excitable cells, whereby an external signal evokes an electrical response mediated by ion channels, has certainly been revolutionized by the development of the patch-clamp technique. Extensions of the technique, however, have also provided surprising flexibility in reporting events late in the cascade whereby intracellular Ca^{2+} and other second messengers lead to exocytosis. In 1982, Neher and Marty reported that the patch-clamp technique together with basic impedance analysis could be used to monitor membrane (electrical) capacitance as a single-cell assay of exocytosis and endocytosis. Since exocytosis involves the fusion of secretory granule membrane with the plasma membrane and a corresponding increase in surface area, an increase in membrane capacitance is observed. The excess membrane is reclaimed in the process of endocytosis, which leads to a corresponding decrease in capacitance. Present techniques can detect changes in capacitance on the order of a femtofarad, allowing the fusion of single secretory granules with diameters greater that about 200 nm to be resolved. The temporal resolution possible is on the order of milliseconds (e.g., Breckenridge and Almers, 1987); therefore, capacitance-recording techniques can almost achieve the resolution of synaptic preparations, where the electrical response of a postsynaptic cell serves as a reporter of secretion. Resolution limits of membrane capacitance estimation techniques are discussed in greater detail in Section 5.

The greatest disadvantage of membrane capacitance measurements is that they only report the net change in membrane surface area, the difference between the dynamic processes of endocytosis and exocytosis. If the rate of endocytosis is slow, then in principle exocytosis can be clearly discerned. However, experiments that rapidly elevate intracellular Ca^{2+} concentration through release from caged compounds suggest that endocytosis can also have a rapid (<1 sec), Ca^{2+}-dependent component (Neher and Zucker, 1993; Thomas *et al.*, 1994). Another note of caution is that changes in capacitance can result from factors other than an increase in membrane surface area. For example, the mobilization of charges during gating of voltage-dependent channels produces a change in specific capacitance (Fernandez *et al.*, 1982;

KEVIN D. GILLIS • Department of Membrane Biophysics, Max-Planck Institute for Biophysical Chemistry, Am Fassberg, D-37077 Göttingen, Germany.
Single-Channel Recording, Second Edition, edited by Bert Sakmann and Erwin Neher. Plenum Press, New York, 1995.

Horrigan and Bookman, 1993). Whereas these phenomena provide opportunities for studying gating behavior and endocytosis, they highlight that care must be taken in quantitatively relating evoked increases in capacitance to secretion. Practical aspects of membrane capacitance measurements, including possible causes of errors and artifacts, are discussed in Section 6.

Amperometric techniques avoid these complications by measuring an actual secreted product (see Chapter 11, this volume). However, amperometry can not detect many secreted substances of interest and can measure release at high temporal resolution only from a fraction of the cell surface (Chow *et al.,* 1992). Therefore, we view the two techniques as complementary in providing single-cell assays of secretion.

1.1. Sample Recordings of Membrane Capacitance

Figure 1 depicts sample records of membrane capacitance changes in a nonexcitable cell (A) and an excitable cell (B).

Figure 1A was recorded from a mast cell that was stimulated to secrete by including GTP-γ-S in the pipette solution. The discrete steps in capacitance correspond to the fusion of individual granules with the plasma membrane. The steps are clearly resolvable because mast cells have very large granules (approximately 0.7 μm in diameter).

Figure 1B was recorded from a rat pancreatic B cell. Here the capacitance change was evoked by a 50-msec depolarizing pulse. The stimulus voltage consisted of a sine wave superimposed on a hyperpolarized holding potential (−70 mV). The resulting current sinusoid was analyzed using a "phase-sensitive detector" to estimate changes in capacitance (C trace). The sinusoid was interrupted in order to apply a pulse to +10 mV, which resulted in the indicated Ca^{2+} current (i_{pc} trace). The change in capacitance during the depolarization cannot be measured because of the activation of voltage-dependent ionic channels; rather, the difference in capacitance before and after the depolarization is noted. The capacitance increase can be related to the number of Ca^{2+} ions that entered the cell during the depolarization by integrating the Ca^{2+} current. The G_{ac} trace indicates changes in membrane or pipette resistance and is used as a control to ensure that changes in resistive parameters are not mistaken for changes in capacitance. Insulin-containing granules have a diameter of about 200 nm; therefore, a 20-fF increase in capacitance is thought to correspond to the fusion of about 16 granules.

1.2. The Equivalent Circuit of a Cell in the Patch-Clamp Recording Configuration

Techniques for estimating membrane capacitance presented in this chapter rely on the accuracy of the equivalent circuit of a cell in the whole-cell or perforated-patch recording configurations depicted in Fig. 2. It is important to note that such a simple model only applies to cells that are approximately spherical without significant "neurite-like" membrane projections. More complex models are treated in Chapter 2 (this volume).

C_p represents the sum of pipette and other sources of capacitance at the input of the patch-clamp amplifier and is usually on the order of several picofarads. Since the contribution to the current by C_p can usually be effectively canceled using pipette capacitance compensation within patch-clamp amplifiers, it is usually neglected in circuit analysis.

R_a is the series or "access" resistance, which results from the geometry of the pipette

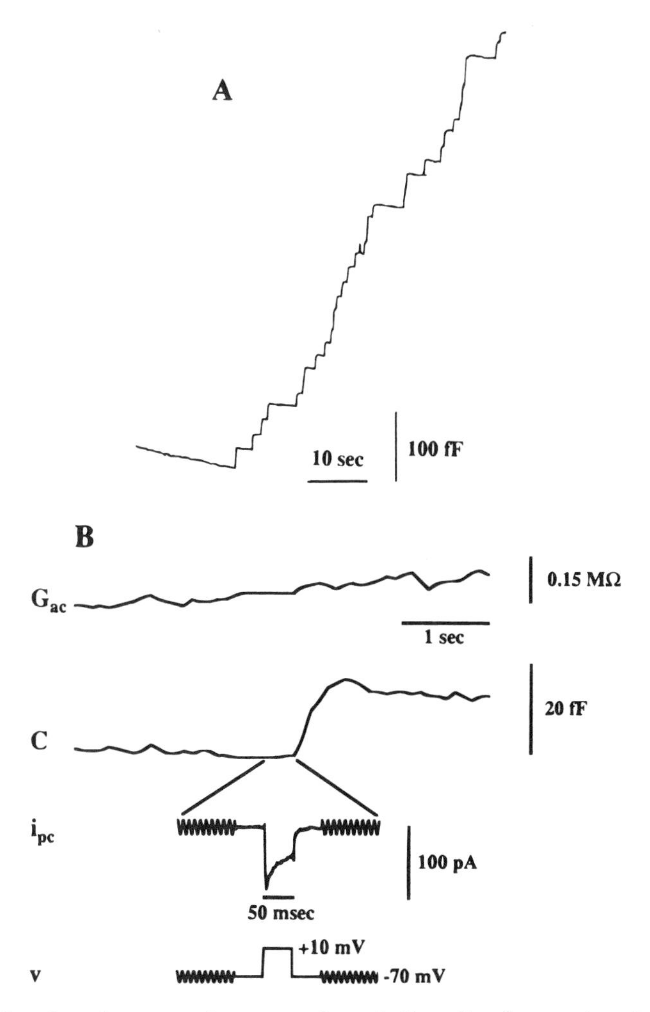

Figure 1. Examples of membrane capacitance recordings. A: Recording from a rat mast cell stimulated to secrete by the inclusion of GTP-γ-S in the patch pipette. B: Recording from a rat pancreatic B cell stimulated by membrane depolarization. Note that the C and G_{ac} traces are displayed with a different time scale than the i_{pc} and v traces. The lock-in amplifier output was filtered with a time constant of 200 msec, which accounts for the rise time of the C trace. Changes in the G_{ac} trace indicate changes in either R_a or R_m (see Section 3.3). The calibration bar indicates the expected displacement of the G_{ac} trace for a change in R_a of 0.15 MΩ. The sinusoids are represented schematically. Both records were obtained using the piecewise linear technique (Section 3.3).

plus any obstruction at the tip by adherent membrane (in the case of conventional whole-cell recording) or the resistance of the perforated membrane underneath the pipette (in the case of perforated-patch recording). The value of R_a is typically on the order of 10 MΩ. The reciprocal of R_a will be denoted G_a and will be used when it makes an expression more readable.

R_m, the membrane resistance, is principally determined by the properties of ionic channels within the cell membrane. Since R_m is highly nonlinear at potentials that activate voltage-dependent channels in excitable cells, it is important that the voltage stimulus not activate

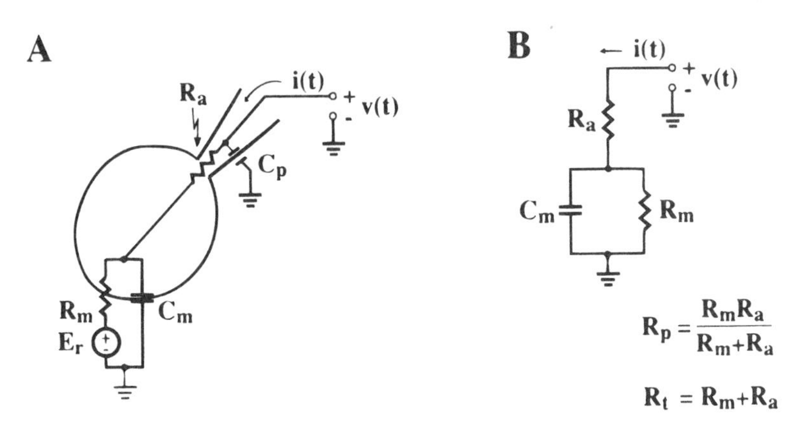

Figure 2. A: The equivalent circuit of a cell in the whole-cell or perforated-patch recording configuration. The physical origin of each circuit element is discussed in the text. B: Simplified three-component equivalent circuit. R_p (the parallel combination of R_a and R_m) and R_t (the series combination of R_a and R_m) are used in some equations for convenience of notation.

these conductances. R_m is typically on the order of 1 GΩ. The reciprocal of R_m will be denoted G_m.

E_r is the voltage responsible for any DC current present at the holding potential of the cell. Since R_m is often nonlinear, it is important to make the distinction that (for the purposes of this model) E_r is the zero-current potential *extrapolated* from the value of R_m found at the holding potential, and not necessarily the *actual* zero-current potential. E_r can be ignored in an AC analysis of the equivalent circuit but is important if the DC current is used in the estimation process.

C_m is the membrane capacitance, which is the parameter of greatest interest in monitoring exocytosis. Under normal conditions, biological membranes have a uniform thickness on the order of 5 nm and a relative dielectric constant of 2–3. Modeling the biological membrane as a parallel-plate capacitor, the specific membrane capacitance can be calculated (and experimentally measured) to be roughly 1 μF/cm^2 (Cole, 1968). Therefore, a cell with a diameter of 13 μm has a capacitance of about 5 pF.

1.3. Parameter Estimation Using Capacity Transient Neutralization Circuitry of Patch-Clamp Amplifiers

A scheme for electrical parameter estimation using patch-clamp techniques is presented in Fig. 3. A voltage stimulus is applied to the pipette and cell, and the current response is reported using a patch-clamp amplifier. In parallel, the same stimulus is applied to a model of the passive electrical parameters of the cell, and the consequent predicted current is subtracted from the actual current. The resulting "error current," $i_{pc}(t)$, can then be used to adjust the parameters of the model until the error is minimized according to some chosen criterion. The overall process is similar to the balancing of a Wheatstone bridge.

An example of this process is the adjustment of "capacity transient neutralization" circuitry during whole-cell recording. Common patch-clamp amplifiers contain many of the components of Fig. 3, including a basic model of the series resistance and membrane

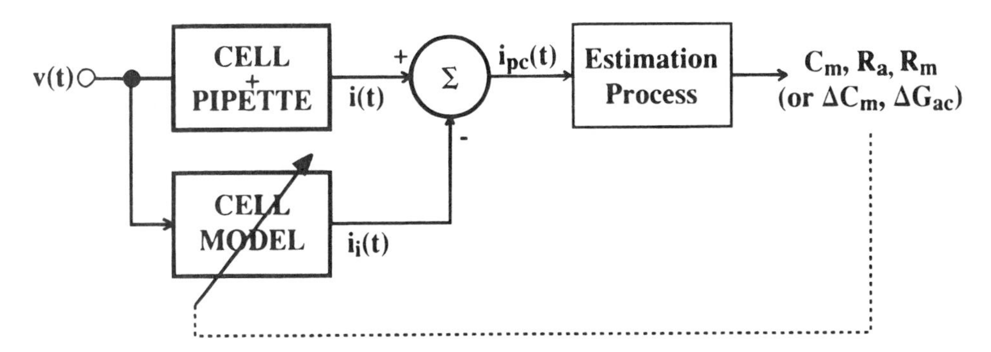

Figure 3. Electrical parameter estimation using patch-clamp techniques. A voltage-clamp stimulus is applied to both a cell and an electrical model of the cell. The current response of the cell model [$i_i(t)$] is compared to the actual response of the cell [$i(t)$]. The difference [$i_{pc}(t)$, the "error current"] is used to adjust parameter estimates of the cell model. For most capacitance estimation techniques, $i_{pc}(t)$ is the signal output from the patch-clamp amplifier. For the case of sine-wave stimulation, a phase-sensitive detector (e.g., lock-in amplifier) is an important component of the "estimation process" block.

capacitance of a cell during whole-cell recording. The primary purpose of this circuitry is to electronically subtract the transient currents that charge C_m during steps of voltage. Canceling these transient currents (which are usually much larger than the ionic currents) allows the dynamic range of the recording apparatus to be expended on the more "interesting" ionic currents.

The adjustment of capacitance compensation circuitry proceeds as follows. Usually a square-wave voltage stimulus is applied that alternates between hyperpolarized voltage levels that do not activate nonlinear membrane conductances. The experimenter manually adjusts two potentiometers until the transient currents that charge the membrane capacitance are zeroed. The two potentiometers are calibrated so that, after proper adjustment, the value of series conductance and membrane capacitance can be read directly from the dials. The remaining current evoked by voltage-clamp pulses to depolarized potentials is the nonlinear ionic current of interest that is recorded during the course of an experiment.

Of course, high-resolution techniques for monitoring changes in membrane capacitance do not rely on manual estimation. For example, the EPC-9 "cap. track" feature applies voltage step stimuli and applies an algorithm to the resulting $i_{pc}(t)$ signal. The settings of capacitance compensation circuitry are then automatically and continuously updated by loading appropriate values into digital-to-analogue converters (Sigworth *et al.,* 1995).

However, it is not always convenient (nor necessary) to adjust capacitance compensation circuitry continuously to follow changes in C_m and R_a during the course of an experiment. In this case $i_{pc}(t)$ is processed to generate parameter estimates while the settings of the capacitance compensation circuitry (reflecting "baseline" values of R_a and C_m) remain constant. Since $i_{pc}(t)$ is radically altered by the use of capacitance compensation, the estimation process must also be given information about the settings of this circuitry (called "C_{slow}" and "G_{series}" in the EPC-7/9). If the circuit parameters change significantly during the course of the experiment, then the capacitance compensation settings can occasionally be updated.

1.4. Types of Voltage-Clamp Stimuli Used to Measure Membrane Capacitance

The aforementioned case of square-wave stimulation is discussed in more detail in Section 2. Another type of stimulus that has been used to measure membrane capacitance is a "pseudorandom binary sequence" (PRBS) of voltages. Similar to square-wave stimulation, the voltage is alternated between two levels. However, the length of time that is spent at each level is ideally an "unpredictable" random variable. The resulting stimulus spectrum approximates white noise, i.e., with equal-amplitude frequency components over a wide frequency band. The spectrum of the resulting current signal is then directly related to the admittance spectrum of the equivalent circuit, which can be fitted to a model to yield parameter estimates. This method is discussed in Section 4.

Perhaps the most popular technique currently in use for high-resolution measurement of capacitance changes employs a sine-wave stimulus about a hyperpolarized holding potential. Typically, the amplitude of the sinusoid is about 25 mV, and the frequency is on the order of 1 kHz. The amplitude and phase of the resulting sinusoidal current can be resolved using a "phase-sensitive detector," which can be implemented in software or in hardware using a lock-in amplifier (Neher and Marty, 1982; Joshi and Fernandez, 1988). Basic circuit analysis can then be used to calculate parameter estimates (the "Lindau–Neher" technique, Section 3.2). Alternatively, changes in capacitance can be estimated by making the approximation that small changes in equivalent circuit parameters result in linear changes in the sinusoidal current (the "piecewise-linear" technique, Section 3.3).

2. Square-Wave Stimulation: The "Time Domain" Technique

Consider the case in which the pipette potential is stepped from zero to some voltage V_Δ. The current through the equivalent circuit of Fig. 2B will consist of an initial transient (I_o) that relaxes to a steady-state value (I_{ss}) with an exponential time course; i.e.,

$$i(t) = (I_o - I_{SS})\exp(-t/\tau) + I_{SS} \qquad (1)$$

where t is the time after the step in voltage. Immediately after the voltage step, the impedance of C_m approaches zero, and V_m is dropped entirely across R_a, resulting in an initial current given by:

$$I_o = V_\Delta/R_a \qquad (2)$$

At steady state, C_m is fully charged to its new voltage and draws zero current; therefore, the overall impedance is just the series combination of R_a and R_m, giving a steady-state current of:

$$I_{SS} = V_\Delta/(R_a + R_m) \qquad (3)$$

The time constant of charging of C_m is given by:

$$\tau = R_p C_m \qquad (4)$$

where R_p is the parallel combination of R_a and R_m; i.e.,

$$R_p = \frac{R_a R_m}{(R_a + R_m)} \qquad (5)$$

Therefore, if capacitance compensation circuitry is disabled, equation 1 can be fitted to the current response to obtain I_o, I_{ss}, and τ. Equations 2–5 can be solved to yield estimates of the circuit parameters:

$$R_a = V_\Delta / I_o$$
$$R_m = (V_\Delta - R_a I_{SS}) / I_{SS}$$
$$C_m = \tau(1/R_a + 1/R_m) \tag{6}$$

Complications to the above scheme include that filtering in the stimulus pathway prevents the voltage change from being an ideal step and filtering of the patch-clamp amplifier output causes $i(t)$ to deviate from the ideal case of equation 1. In addition, uncanceled pipette capacitance can contribute to the peak value of the current transient. Therefore, it is important to minimize the stimulus filtering, for example, the 2-μsec time-constant filter of the EPC-7/9 is used. The output filtering of the patch-clamp amplifier is set to a value on the order of 30 kHz, and a sampling interval of 5–10 μs is used. It is also important to coat pipettes with a hydrophobic compound (such as Sylgard®) and to attempt to compensate the pipette capacitance electronically as completely as possible. Further details can be found in Lindau and Neher (1988).

The interval between voltage steps must be at least on the order of 5τ in order to allow time for the complete charging/discharging of C_m. Therefore, the maximum theoretical time resolution is on the order of 500 μsec per C_m estimate. However, the estimates are often actually generated intermittently at a rate of a few Hertz to allow the computer time to implement the estimation algorithm and to perform other tasks. The noise performance of square-wave estimation techniques will be discussed further in Section 5.3.

2.1. Electronic Transient Neutralization Approximates a Three-Element Network with Two Elements

The calculation of parameter estimates while using capacitance compensation circuitry can be done in a number of ways within the general scheme of Fig. 3. In any case, the algorithm must be supplied with the G_{series} and C_{slow} values used for compensation.

One alternative would be for the algorithm to "add back" the current that was subtracted [$i_i(t)$] and proceed with the algorithm already discussed. Here $i_i(t)$ is given by:

$$i_i(t) = V_\Delta G_{series} \exp(-t/\acute{\tau})$$
$$\acute{\tau} = C_{slow}/G_{series} \tag{7}$$

Another possibility is for the algorithm to operate directly on the residual current to estimate the changes in G_{series} and C_{slow} necessary to renull the current transient. The problem is that capacitance compensation approximates a three-element network with two elements by neglecting G_m. Since $i_i(t)$ is injected through a capacitor, it is impossible to supply a DC current (I_{ss}) indefinitely, and only the transient component of the current is subtracted (compare equation 7 with equation 1). For the case where R_m is much greater than R_a, I_{ss} is relatively unimportant, and equation 7 is a close approximation to equation 1. When I_{ss} is not negligible, some choice has to be made about the desired residual current.

Adjusting the circuitry to yield a rectangular residual current will result in values of

C_{slow} and G_{series} that underestimate their corresponding equivalent circuit values according to Lindau and Neher (1988)*:

$$G_{series} \simeq G_a(1 - R_a/R_m)$$
$$C_{slow} \simeq C_m(1 - 2R_a/R_m) \quad (8)$$

The approach used in the EPC-9 is to set C_{slow} equal to C_m and G_{series} equal to G_a. In this case the residual current is given by:

$$i_{pc}(t) \simeq I_{SS}[1 - (1 + t/\tau)e^{-t/\tau}] \quad (9)$$

Current that deviates from this response is processed to update C_{slow} and G_{series} values (Sigworth *et al.*, 1995).

3. Sinusoidal Excitation

At the present time, the most popular techniques for high-resolution measurement of C_m use a sinusoidal voltage stimulus about a hyperpolarized DC potential. The magnitude and phase shift of the resulting current sinusoid are then analyzed using a phase-sensitive detector to produce estimates of C_m or ΔC_m. It is important to note, however, that a single sinusoid provides only *two* independent pieces of information (magnitude and phase or, equivalently, real and imaginary current components). Therefore, another piece of information is necessary to model the *three*-component network of Fig. 2B. The Lindau–Neher technique (Section 3.2) obtains this information from the DC current. Alternatively, the "piecewise linear" technique (Section 3.3) does not attempt to determine all three parameters. Rather, it is assumed that, at an appropriate frequency, small changes in R_m or R_a have little effect on the estimation of changes in C_m.

Because AC analysis of current is not a common tool in the repertoire of electrophysiologists, an introduction to complex impedance analysis is presented in the next section.

3.1. Introduction to Complex Impedance Analysis

Consider a sinusoidal voltage applied across a circuit element. If the element is an ideal resistor, the resulting current is simply given by Ohm's law:

$$v(t) = V_o \cos\omega t$$
$$i_R(t) = \frac{v(t)}{R} = \frac{1}{R} V_o \cos\omega t \quad (10)$$

As is the case with all linear circuit elements, the current is a sinusoid with the same frequency as the voltage. In the case of a resistor, the voltage and current are in phase (they reach

*One reason for the underestimation of the parameters is that C_m is only charged to $R_m/(R_a + R_m)$ of the applied voltage step (V_Δ).

maximum values at the same time), and the ratio of peak voltage to current is R. Now consider the case of a sinusoidal voltage applied across a capacitor:

$$i_C(t) = C\frac{dv(t)}{dt} = -\omega C V_o \sin\omega t \tag{11}$$

Since an ideal capacitor is also a linear circuit element, the current has the same frequency as the voltage; however, there is a 90° phase shift between the voltage and current. In order to simplify the algebra involved in dealing with trigonometric functions, complex notation is often used. Central to this notation is the use of Euler's formula:

$$e^{j\omega t} = \cos\omega t + j\sin\omega t \tag{12}$$

where $j = \sqrt{-1}$. The sinusoidal voltage and resulting current for a capacitor (for example) can then be rewritten as:

$$\begin{aligned} v(t) &= \mathrm{Re}[V_o e^{j\omega t}] \\ i_C(t) &= \mathrm{Re}[(j\omega C)V_o e^{j\omega t}] \end{aligned} \tag{13}$$

We can define the expressions in brackets as the complex voltage and complex current, respectively. The ratio of complex voltage to complex current is defined as the *impedance* of a circuit element. It turns out that impedance can be used to calculate complex current from complex voltage exactly the same way as Ohm's law relates current to voltage. In fact, the units for impedance are the same as the units for resistance (ohms). The impedance of a capacitor is given by:

$$Z_C = 1/j\omega C = -j/\omega C \tag{14}$$

where $-j$ accounts for the 90° phase lead of the current relative to the voltage and $1/\omega C$ is the ratio of peak voltage to peak current. The impedance of a resistor is simply given by:

$$Z_R = R \tag{15}$$

where the absence of an imaginary component indicates that there is no phase shift between current and voltage. The impedance of an inductor is given by:

$$Z_L = j\omega L \tag{16}$$

where j indicates that the current lags the voltage by 90°.

The inverse of impedance is referred to as *admittance* and is usually denoted Y.

Individual circuit elements can be combined to determine the overall impedance of a network by applying the same rules that govern resistor networks; i.e., impedances in series add, whereas impedances in parallel add in an inverse manner:

$$\frac{1}{Z_{\text{total}}} = \frac{1}{Z_1} + \frac{1}{Z_2} + \cdots$$

or

$$Y_{\text{total}} = Y_1 + Y_2 + \cdots \tag{17}$$

The rules of complex algebra are then applied to determine the current–voltage relationship. For example, by applying Euler's equation, a complex number such as admittance can be written in polar form:

$$Y = |Y| e^{j\angle Y}$$

where

$$|Y| = \sqrt{\mathrm{Re}^2[Y] + \mathrm{Im}^2[Y]}$$

and

$$\angle Y = \tan^{-1}\left(\frac{\mathrm{Im}[Y]}{\mathrm{Re}[Y]}\right) \tag{18}$$

Multiplying the admittance by the complex voltage and taking the real part gives the current as a function of time:

$$i(t) = \mathrm{Re}[YV_o e^{j\omega t}] = \mathrm{Re}[|Y| V_o e^{j(\omega t + \angle Y)}] = |Y| V_o \cos(\omega t + \angle Y) \tag{19}$$

Thus, in the polar representation of the complex admittance, the magnitude gives the ratio of peak current to peak voltage, and the phase angle gives the phase shift between the voltage and current sinusoids.

Consider now the case of a resistor R_m and capacitor C_m in parallel. The total admittance is the sum of the individual admittances:

$$Y = Y_{C_m} + Y_{R_m} = j\omega C_m + \frac{1}{R_m} \tag{20}$$

Using equation 19, the total current is readily given by:

$$\begin{aligned} i(t) &= I_o \cos(\omega t + \alpha) \\ I_o &= V_o\sqrt{1/R_m^2 + \omega^2 C_m^2} \\ \alpha &= \tan^{-1}(\omega R_m C_m) \end{aligned} \tag{21}$$

If a resistor (R_a) is placed in series with the parallel combination of R_m and C_m, then the equivalent circuit of Fig. 2B is obtained. The total admittance is given by:

$$Y(\omega) = \frac{\dfrac{1}{R_a}\left(j\omega C_m + \dfrac{1}{R_m}\right)}{\dfrac{1}{R_a} + \dfrac{1}{R_m} + j\omega C_m} \tag{22}$$

or more compactly:

$$Y(\omega) = \frac{(1 + j\omega R_m C_m)}{R_t(1 + j\omega R_p C_m)} \tag{23}$$

where:

$$R_t = R_m + R_a$$
$$R_p = \frac{R_m R_a}{R_m + R_a} \tag{24}$$

In many situations, the value of R_m is a hundredfold or more greater than the value of R_a, in which case $R_t \approx R_m$ and $R_p \approx R_a$.

At low frequencies, C_m draws very little current, and the total admittance approaches the series combination of R_m and R_a ($1/R_t$). At high frequencies, the admittance of C_m approaches infinity, effectively shorting R_m to result in an overall admittance of $1/R_a$. At intermediate frequencies, the admittance of C_m is much greater than $1/R_m$ and much less than $1/R_a$; therefore, the overall magnitude of the admittance is dominated by C_m. The magnitude of the admittance versus frequency is depicted in Fig. 4 for a typical set of parameter values. Note in this example that for frequencies between 100 and 1000 Hz, the magnitude of the admittance of the equivalent circuit is nearly equivalent to the magnitude of the admittance of C_m alone; therefore, one might expect that sinusoidal excitation in this range of frequencies

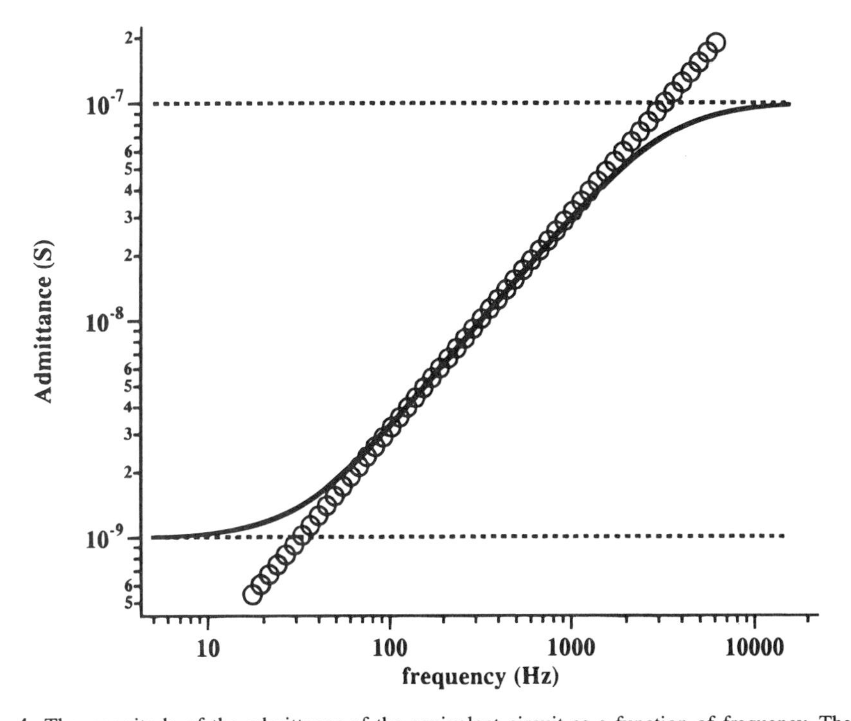

Figure 4. The magnitude of the admittance of the equivalent circuit as a function of frequency. The solid line indicates the magnitude of the admittance calculated from equation 23 with R_a = 10 MΩ, R_m = 1 GΩ, and C_m = 5 pF. The circles indicate the admittance of a 5-pF capacitor alone. The top dotted line indicates the admittance of a 10-MΩ resistor, and the bottom dotted line indicates the admittance of a 1-GΩ resistor.

would be optimal for the estimation of C_m. It is important to note, however, that the *phase* of the admittance can differ quite a bit from that of a pure capacitor (90°), even at these intermediate frequencies.

3.2. Parameter Estimation from Basic Circuit Analysis: The Lindau–Neher Technique

The measured current sinusoid can be separated into a component in phase with the stimulus voltage (real part) and a component 90° out of phase with the voltage (imaginary part) using a phase-sensitive detector (lock-in amplifier) implemented in hardware or software. If capacitance compensation circuitry is disabled, then these current components divided by the stimulus voltage amplitude give the real and imaginary admittance values. One can rewrite equation 23 to give the real and imaginary components of admittance for the three-element model of Fig. 2B*:

$$Y(\omega) = A + jB \tag{25}$$

with

$$A = \frac{1 + \omega^2 R_m R_p C_m^2}{R_t(1 + \omega^2 R_p^2 C_m^2)}; \qquad B = \frac{\omega R_m^2 C_m}{R_t^2(1 + \omega^2 R_p^2 C_m^2)} \tag{26}$$

The DC current is given by:

$$I_{dc} = \frac{V_{dc} - E_r}{R_t} \tag{27}$$

where V_{dc} is the DC holding potential, which is summed with the sinusoidal voltage. In the original formulation of the L–N technique, E_r is assumed to have a constant value for a given cell type and ionic composition of the pipette solution (e.g., 0 mV). I_{dc} can be measured by low-pass filtering or digital averaging of the current and is used to calculate R_t using equation 27. A and B are the 0° and 90° outputs of the phase-sensitive detector normalized to the amplitude of the voltage sinusoid. From these three quantities, the three circuit parameters can be calculated:

$$R_a = \frac{A - G_t}{A^2 + B^2 - AG_t}$$

$$R_m = \frac{1}{G_t}\frac{(A - G_t)^2 + B^2}{A^2 + B^2 - AG_t}$$

$$C_m = \frac{1}{\omega_c B}\frac{(A^2 + B^2 - AG_t)^2}{(A - G_t)^2 + B^2} \tag{28}$$

*The real and imaginary components can be found by multiplying the numerator and denominator of equation 23 by the complex conjugate of the denominator.

Here G_t is defined as the inverse of R_t for ease of notation, and $\omega_c = 2\pi f_c$, where f_c is the frequency of the stimulus sinusoid.

In certain situations, the assumption of a constant, known value for E_r may not be appropriate. For example, maneuvers that activate ionic channels at the holding potential may result in a shift in E_r. In order to address this possibility, Okada *et al.* (1992) stepped the holding potential by 14 mV every 500 msec. The difference in the DC current was related to R_t by:

$$R_t = \frac{\Delta V_{dc}}{\Delta I_{dc}} \tag{29}$$

In principle, this modification of the L–N technique should ensure that C_m estimates are insensitive to changes in E_r that are slow compared to the period of the steps in holding potential (here 1 sec). In this type of analysis, the frequency of the steps in holding potential is limited to several Hertz because AC effects at higher frequencies invalidate the simple expression in equation 29.

Another approach that has been suggested to determine R_m independently of E_r is the use of two sinusoidal frequencies (Rohlicek and Rohlicek, 1993; Donnelly, 1994). Techniques using two sinusoids are discussed in Sections 4.1 and 5.1.2.

It is important to note, however, that errors in E_r are important only in cases where R_m becomes quite small. In other words, an error in E_r produces an error in the estimate of G_t, but G_t has only a small impact on C_m estimates unless it approaches within an order of magnitude of G_a. For example, consider a 1600-Hz sinusoid applied about a DC holding potential of -70 mV stimulating a cell under conditions where $R_m = 1$ GΩ, $R_a = 10$ MΩ, and $C_m = 5$ pF. The assumed value of E_r is -30 mV, but the activation of ionic channels results in an actual E_r value of -50 mV. This error in E_r will produce an error in C_m of about 1.5 fF, certainly a negligible value. If $R_m = 0.1$ GΩ, the same error in E_r leads to an error in C_m of 150 fF (about 3%). Thus, if the activation of membrane conductances is expected during the course of an experiment, E_r should be set to the zero-current potential of the expected conductance.

3.2.1 Extension of the L–N Algorithm to Allow the Use of Capacity Transient Neutralization Circuitry

As mentioned previously, the real and imaginary components of the current are proportional to the admittance values only when capacitance compensation circuitry is disabled. However, the use of capacitance compensation is very desirable in cases where changes in C_m are evoked by depolarizing voltage-clamp pulses. If supplied with the values of G_{series} and C_{slow} used, the algorithm can "add back" the real and imaginary current components that were subtracted. (Measured and compensated currents are additive; see Fig. 3.) Compensated real and imaginary admittance (A_{comp} and B_{comp}, respectively) are given by:

$$A_{comp} = \frac{\omega_c^2 C_{slow}^2 / G_{series}}{1 + (\omega_c C_{slow}/G_{series})^2}$$

$$B_{comp} = \frac{\omega_c C_{slow}}{1 + (\omega_c C_{slow}/G_{series})^2} \tag{30}$$

After each adjustment, values of C_{slow} and G_{series} can be manually entered through the user interface of the software that implements the L–N algorithm. With the EPC-9, another alternative is to have the software automatically read the values.

3.2.2. Compensating for Phase Delays

The desired phase setting of the phase-sensitive detector (PSD) is 0° in order for the in-phase output to indicate the real component of current and the orthogonal output to give the imaginary current component. However, low-pass filtering in the stimulus input and current output stages introduces phase shifts (delays) that must be compensated for. Patch-clamp amplifiers typically include a basic low-pass filter in the stimulus pathway, and the output signal is always filtered. In addition, a sinusoid generated by filtering a signal generated by a digital-to-analogue converter will be shifted in phase relative to the original waveform.

Determining the overall phase shift can be done empirically. For this purpose, one end of a resistor is inserted into the input of the patch-clamp amplifier headstage, and a stimulus voltage sinusoid is applied. Any current sinusoid caused by stray capacitance is zeroed using C_{fast} (pipette capacitance) compensation before the other end of the resistor is clipped to the signal ground. After grounding the end of the resistor, the current sinusoid should be entirely in phase with the stimulus voltage; i.e., it should have only a real component (see equation 15). The phase offset is found by adjusting the phase setting of the lock-in amplifier or digital PSD until the imaginary output is zero and the real output is positive.

The value of the resistance used should be small enough that any parasitic capacitance present will introduce a negligible phase shift (say ≤ 1 MΩ); therefore, the patch clamp amplifier has to be set to a low gain. An alternative technique is to use a small-value capacitor (say 5 pF) rather than a resistor and adjust the phase offset until the real component is zero and the imaginary component is positive. Adjustment using a capacitor has the advantage that a capacitor more closely approximates an ideal circuit element.

It is important to note that the phase offset changes when any of a number of alterations are made to the recording conditions. The frequency of the sinusoid, the stimulus filter setting, the output low-pass filter setting, and the number of points per sine-wave cycle (if the sine wave is generated by an digital-to-analogue converter) all affect the phase offset. A change in any of these parameters requires that the phase offset be redetermined.

Alternatively, the phase shift can be calculated analytically based on the electronics of the particular patch-clamp amplifier and low-pass filters used. However, this approach is practical only when the software is provided with all relevant instrument settings.

3.2.3. Compensating for Parasitic Elements Present after Seal Formation

A number of parasitic elements not included in the basic equivalent circuit of Fig. 2B may be present during whole-cell recording. These include the seal resistance, an imperfectly canceled pipette capacitance, and dielectric losses in the pipette and pipette holder. In order to account for these effects, the real and imaginary components of current are zeroed after seal formation but before patch rupture or perforation (Lindau and Neher, 1988). This process does not account for any change in seal resistance that may accompany patch rupture, and the capacitance of the patch is included in the compensation even though it is not present during whole-cell recording or has little effect during perforated-patch recording. In general, however, these effects appear to be small for tight seals. One indication of the correct handling of the problems mentioned is that changes in R_a (such as when applying suction to clear the pipette of obstructing membrane) occur with little corresponding changes in estimated C_m values.

In the case of perforated-patch recording, the assumption that the patch itself has a very high resistance when the offsetting is performed is troublesome because the patch begins to perforate very shortly after seal formation. Therefore, one can dip the pipette tip in antibiotic-free solution for many seconds in order to produce a sufficient delay in the onset of patch permeabilization to allow the zeroing of the real and imaginary current components immediately after seal formation. Alternatively, one can skip the offsetting process and "hope for the best," with the ultimate test being the lack of correlation between reported changes in R_a and C_m.

3.3. Piecewise-Linear Techniques for Estimating Changes in Membrane Capacitance

This method was actually the first technique that was used for high-resolution measurement of changes in C_m related to exocytosis (Neher and Marty, 1982). Although it initially appears to be somewhat *ad hoc,* it is related to well-established estimation algorithms (Gillis, 1993). Perhaps the most attractive feature of the P-L technique is that changes in C_m are directly proportional to the signal output from a lock-in amplifier set to an appropriate phase.

The piecewise-linear technique is based on the approximation that small changes in C_m produce a linear change in the sinusoidal current. This linearization is performed by approximating $\Delta I/\Delta C_m$ by the partial derivative: $\partial I/\partial C_m = U\partial Y/\partial C_m$, where U is the amplitude the sinusoidal voltage. One must also consider the effect of changes in R_a and R_m on the admittance; therefore, the full set of partial derivatives is given by:

$$\begin{aligned}
\partial Y/\partial C_m &= j\omega_c T^2(\omega_c) \\
\partial Y/\partial R_a &= -(1/R_m + j\omega_c C_m)^2 T^2(\omega_c) \\
\partial Y/\partial R_m &= -T^2(\omega_c)/R_m^2
\end{aligned}$$

where

$$T(\omega) = R_m/R_t(1 + j\omega R_p C_m)^{-1} \simeq (1 + j\omega R_a C_m)^{-1} \tag{31}$$

where the approximate expression for $T(\omega)$ assumes $R_m >> R_a$; therefore, $R_t \approx R_m$, and $R_p \approx R_a$.

The linear approximation states that small changes in each of the three parameters induces changes in the sinusoidal current given by:

$$\Delta I \simeq \left[\frac{\partial Y}{\partial C_m}\Delta C_m + \frac{\partial Y}{\partial R_a}\Delta R_a + \frac{\partial Y}{\partial R_m}\Delta R_m\right]U \tag{32}$$

Large changes in any of the parameters, however, change the values of the partial derivatives (see equation 31). Therefore, equation 32 is valid only for small changes in the parameters; if cumulative parameter changes become significant, then the partial derivatives can be redetermined, and equation 32 can be reapplied for this new set of "baseline" parameter values. This type of approximation is therefore called "piecewise-linear."

What is needed now is a way to invert equation 32 so that a change in the current sinusoid can be related back to the change in C_m that produced it. Also, the method used will have to distinguish changes in sinusoidal current produced by changes in C_m from those originating from deviations in R_a or R_m. Insight into how this is possible will be provided by applying equations 31 and 32 to an example.

Consider a cell with baseline values of C_m = 5 pF, R_a = 15 MΩ, and R_m = 2 GΩ. The stimulus sinusoid has an amplitude of 15 mV and a frequency of 1 kHz. Application of equation 23 reveals that the baseline sinusoidal current (I_m) has an amplitude of 423 pA and is shifted 64° relative to the stimulus (assuming capacitance compensation is turned off). Equations 31 and 32 reveal that a 0.1-pF increase in C_m would cause a change in I_m equivalent to the addition of a sinusoid with an amplitude of 7.7 pA and a phase 39.6° relative to the stimulus. A 1-MΩ increase in R_a would contribute a sinusoid of 12.1 pA $\angle -52.2°$, and a 0.5-GΩ increase in R_m would add a sinusoid of 1.5 pA $\angle 129.6°$. Figure 5 presents these changes in current in vector form. The length of the vector indicates the amplitude of the added sine wave, and the angle of the vector to the positive x axis gives the phase of the added sinusoid relative to the stimulus voltage. Since the equivalent circuit is linear, the frequency of each sinusoid is the same as that of the stimulus (f_c).

A striking feature of Fig. 5 is that the phase of $\partial Y/\partial C_m$ is orthogonal (at right angle

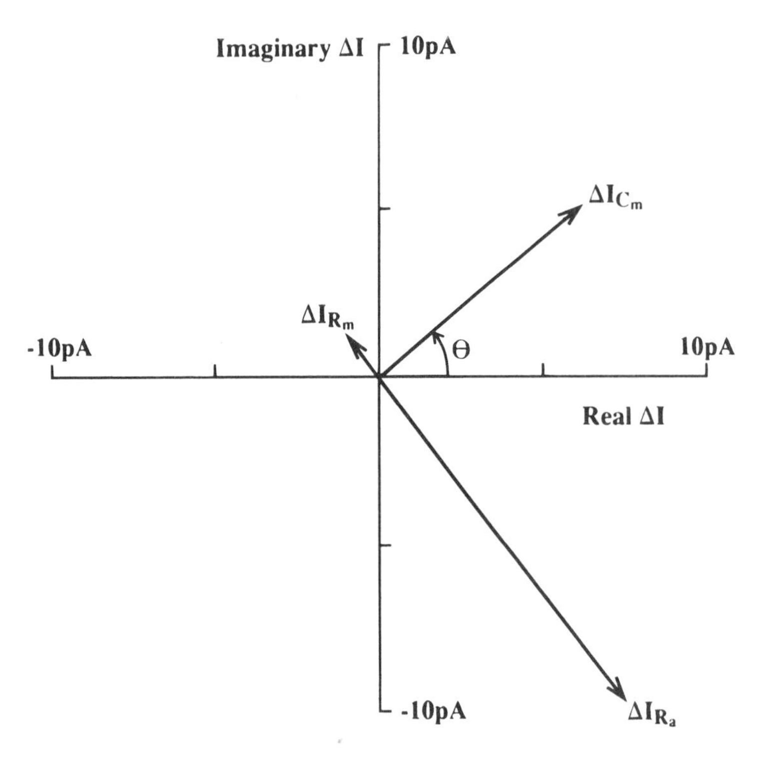

Figure 5. Vector changes in the sinusoidal current introduced by small changes in each of the three circuit elements from Fig. 2B. See Section 3.3 for conditions.

to) $\angle\partial Y/\partial R_m$. For these typical parameter values and frequency, $\angle\partial Y/\partial C_m$ is also nearly orthogonal (within 2°) to $\angle\partial Y\partial R_a$. In general, the phases of the partial derivatives are given by:

$$\begin{aligned}
\theta &= \angle\partial Y/\partial C_m = 90° - 2\tan^{-1}(\omega_c R_p C_m) \\
\eta &= \angle\partial Y/\partial R_a = 180° + 2\tan^{-1}(\omega_c R_m C_m) - 2\tan^{-1}(\omega_c R_p C_m) \\
&\simeq -2\tan^{-1}(\omega_c R_a C_m) = \theta - 90° \\
\xi &= \angle\partial Y/\partial R_m = 180° - 2\tan^{-1}(\omega_c R_p C_m) = \theta + 90°
\end{aligned} \tag{33}$$

The fact that θ is orthogonal to ξ and nearly orthogonal to η suggests that a phase-sensitive detector (PSD) can be used to measure changes in C_m independently of changes in R_a or R_m. The operation of PSDs is described in detail in Appendix A.

The output of a PSD is proportional to $N\cos(\alpha - \beta)$, where α is the phase setting of the PSD, and N is the amplitude, and β is the phase of the input signal. Therefore, a maximum output (N) results when $\beta = \alpha$, whereas the output is zero if β is orthogonal to α. If α is set to θ, then the output of the PSD is maximally sensitive to changes in C_m, insensitive to changes in R_m, and nearly insensitive to changes in R_a. Usually a "two-phase" PSD is used, where one channel is set to phase α and the other is set to $\alpha + 90°$. If $\alpha = \theta - 90°$, then one output (C) is proportional to changes in C_m and the other output (G_{ac}) is proportional to a combination of changes in R_a and R_m according to:

$$\begin{aligned}
C &= U|T(\omega_c)|^2\omega_c\Delta C_m \\
G_{ac} &\simeq U|T(\omega_c)|^2[\omega_c^2 C_m^2\Delta R_a - \Delta R_m/R_m^2]
\end{aligned} \tag{34}$$

where the expression for G_{ac} assumes that $\omega_c >> 1/R_m C_m$.

One final observation from the example presented in Fig. 5 is that, for frequencies near 1 kHz (and for R_m in the gigohm range), even large changes in R_m have a relatively small impact on the sinusoidal current or on the partial derivatives $\partial Y/\partial C_m$ and $\partial Y/\partial R_a$.

Two pieces of information are required in order to use a PSD (lock-in amplifier) to implement equation 34 to measure changes in C_m. First, some method is needed in order to find θ, the desired phase setting of the PSD. Second, some sort of calibration is required to scale the C output of the PSD (i.e., the $\omega_c|T(\omega_c)|^2$ scaling factor needs to be determined).

3.3.1. Setting the Phase of the PSD

From consideration of Fig. 5, it is evident that the correct phase setting of the PSD is important to prevent changes in R_a or R_m from appearing as a change in C_m. A general rule of thumb is that a 6° error in phase results in about 10% of changes in the G_{ac} trace to show up as changes in the C trace. Thus, monitoring the G_{ac} trace serves as an important control, since correlated changes in both the C and G_{ac} traces are indicative either of phase problems or of a cell that is not well modeled by the three-component network of Fig. 2B. More precisely, an error in phase α_{err} will result in an artifactual change in C_m given by:

$$\frac{\Delta C_{art}}{C_m} \simeq \sin(\alpha_{err})\cdot\left(\frac{1}{\omega_c R_m C_m}\frac{\Delta R_m}{R_m} - \omega_c R_a C_m\frac{\Delta R_a}{R_a}\right) \tag{35}$$

Considering the example illustrated in Fig. 5, a 6° error in phase would result in a (negligible) 2-fF increase in the C trace as a result of a ΔR_m of 0.5 GΩ and a 16-fF decrease in the C trace due to a ΔR_a of 1 MΩ.

Another important question is: how much can parameter values change before a new phase should be determined? Consider the baseline values given in the previous example. From equation 33, the desired phase setting (θ) will change by 6° if C_m changes by 0.7 pF or R_a changes by 2 MΩ. R_m would have to drop to a value of about 100 MΩ in order to produce a similar error.

The Neher–Marty Technique

θ can be determined by a clever use of capacitance compensation circuitry (Neher and Marty, 1982). Recall that, on proper nulling of transient currents, the capacitance compensation potentiometers are calibrated to indicate baseline C_m and R_a values. A small displacement of the C_{slow} potentiometer can be thought of as simulating a change of capacitance by unbalancing the compensation network. The phase of the PSD, α, can be adjusted until output G_{ac} is insensitive to these simulated changes in C_m. Calibration can be performed by noting the change in output C on a predetermined offset of the C_{slow} potentiometer (e.g., 0.1 pF).

The patch-clamp amplifier can be modified to allow the quick and accurate offsetting of the C_{slow} setting. For example, in an EPC-7, the C_{slow} potentiometer (used as a variable resistor) has a resistance of 10 kΩ, corresponding to a full-scale reading that is selected to be either 10 or 100 pF. A 100-Ω resistor is placed in series with this potentiometer, and a switch is placed in parallel with the resistor. The resistor is shorted by closing the switch; this 100-Ω change in the resistance value corresponds to a 0.1-pF change in the C_{slow} setting if the selected full-scale reading is 10 pF.

Another alternative is to calculate θ directly from the C_{slow} and G_{series} settings.

The main disadvantage of the Neher–Marty technique is that the accuracy of the phase determination relies on the accuracy of the C_{slow} and G_{series} settings, which must be readjusted whenever there is a significant change in either R_a or C_m. Each determination of phase is time consuming. First, the capacitance compensation potentiometers are readjusted (by nulling capacitive currents in response to a train of voltage steps); then the phase is redetermined by dithering the C_{slow} switch while adjusting α. This motivated the development of automated methods for determining the phase setting.

Series Resistance Dithering: The "Phase-Tracking" Technique

An actual induced change in series resistance (rather than a simulated change in C_m) can be used to find the appropriate phase setting for the PSD. Usually the bathing solution of the cells is grounded in electrophysiological measurements. In the phase-tracking technique (Fidler and Fernandez, 1989), a resistor (R_Δ, about 1 MΩ) is placed between the bath and ground. R_Δ is normally shorted with a digitally controlled switch (often a relay). Opening the switch induces a change in series resistance (see Fig. 6A). If the phase of the PSD (α) equals zero, then the induced change in the outputs is given by (neglecting C_p and C_s of Fig. 6A):

$$\Delta C \simeq U \left| \frac{\partial Y}{\partial R_a} \right| \Delta R_a \sin\eta$$

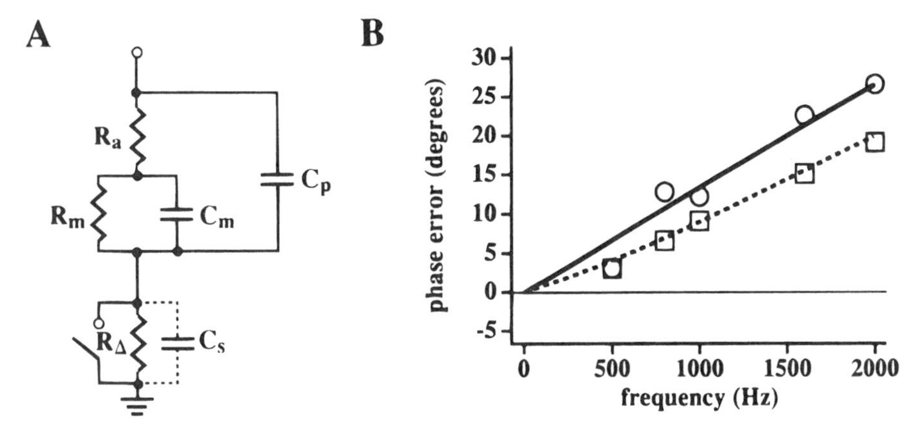

Figure 6. Phase errors introduced by series resistance dithering. A: The equivalent circuit of the cell including series resistance dithering. B: Plot of the phase error versus frequency. Symbols were measured from a model circuit with R_a = 11.6 MΩ, C_m = 10 pF, and C_p = 2 pF. The circles were obtained with R_a = 0.1 MΩ, whereas the squares were measured with R_Δ = 1 MΩ. The solid line was calculated from equation 38, with R_Δ = 0.1 MΩ and C_s = 5 pF. The dashed line was calculated for R_Δ = 1 MΩ and C_s = 5 pF. Phase errors were estimated as the deviations from the "ideal phase," which was assumed to be twice the measured phase of the admittance minus 90° (see equation 41).

$$\Delta G_{ac} \simeq U \left| \frac{\partial Y}{\partial R_a} \right| \Delta R_a \cos\eta \tag{36}$$

where $\Delta R_a = R_\Delta$. The desired phase setting of the PSD can then be set according to:

$$\alpha = \eta = \tan^{-1}(\Delta C/\Delta G_{ac}) \tag{37}$$

It is important to note, however, that pipette capacitance (C_p) can cause a serious phase error. See the following section.

Although η does not exactly equal $\theta - 90°$ (equation 33), it can be argued that η is actually a preferable phase setting under common experimental conditions (Joshi and Fernandez, 1988). The choice of phase setting $\theta - 90°$ (where the C output is insensitive to changes in R_m) or η (where C is insensitive to changes in R_a) depends on whether changes in R_a are expected to cause more intereference than changes in R_m. In many cases η is preferable because changes in R_m have a small impact on the sinusoidal current unless R_m is quite small (within an order of magnitude of R_a, see the example presented in Fig. 5).

C traces are also calibrated by noting the change induced by offsetting the C_{slow} potentiometer at the beginning of the experiment. If large changes in parameter values occur during the course of an experiment, the C trace can be recalibrated automatically by noting the change in the G_{ac} trace on each induced change in series resistance (Fidler and Fernandez, 1989).

3.3.1a. Phase Errors Related to Pipette Capacitance and the Noninfinitesimal Magnitude of Series Resistance Change. A basic assumption of the phase-tracking technique is that the computer-controlled switch induces a change in impedance that is purely resistive and that this change in resistance is sufficiently small so that $\Delta Y/\Delta R_a \approx \partial Y/\partial R_a$. Deviations from these assumptions can be expected to introduce a small phase error. A much more serious error, however, results from neglecting the effect of pipette capacitance (C_p).

As discussed in the introduction, C_p is usually neglected in circuit analysis because the current drawn by this element is electronically subtracted using "pipette capacitance neutralization" circuitry. When a resistor is inserted between the bath and ground, however, the voltage across C_p no longer equals the stimulus voltage, and pipette capacitance neutralization circuitry no longer balances the current through C_p (see Fig. 6A). The total resulting phase error is given by (as derived in Appendix B):

$$\alpha_{err} = 2\tan^{-1}\left(\frac{\omega_c R_a C_p}{1 + C_p/C_m}\right) - \tan^{-1}[\omega_c R_\Delta(C_m|T(\omega)|^2 + C_p + C_S)] \quad (38)$$

where C_s is the stray capacitance contributed by the relay or other sources that is in parallel with R_Δ.

The first error term results from the "unbalancing" of the current through C_p, whereas the second term includes errors caused by C_s and the noninfinitesimal magnitude of R_Δ. The error in phase resulting from C_s alone is approximately $-\omega_c R_\Delta C_s$; thus [for $T(\omega_c) \approx 1$], the error from the nonzero magnitude of R_Δ is mathematically equivalent to an additional stray capacitance with a magnitude equal to the membrane capacitance.

Since the total phase error is the difference between two error terms, it is conceivable that certain combinations of experimental parameters will result in an overall negligible error in phase. For example, consider a case with a low series resistance and a pipette well coated with Sylgard®: R_a = 5 MΩ, C_p = 1.5 pF, C_m = 10 pF, R_Δ = 0.5 MΩ, C_s = 5 pF, and f_c = 1 kHz; then the total phase error is about 1.9°. Under most circumstances this magnitude of error would be considered acceptable. Now consider a case with a higher series resistance and slightly higher pipette capacitance (R_a = 10 MΩ, C_p = 2.5 pF, other values identical to the previous example); the total phase error is then about 11.7°, a value that is clearly unacceptable. The sensitivity of the phase error to R_a suggests that perforated patch recording (with typical R_a values of 15–25 MΩ) should be particularly prone to phase errors when using series resistance dithering to set the phase.

Figure 6B presents phase error data obtained using a model circuit and the DR-1 resistance dithering unit (Axon Instruments). The model circuit included a 1.5-pF capacitor to represent the pipette capacitance. The circles were measured using a R_Δ value of 0.1 MΩ, and the squares were obtained with R_Δ = 1 MΩ. Note that a larger value of R_Δ actually results in smaller errors. The lines are the expected errors calculated from equation 38, with C_s estimated to be 5 pF.

3.3.1b. Variance in Phase Estimates ("Phase Jitter"). Another limitation of the phase-tracking technique is that the phase value may vary slightly each time it is determined because of measurement noise, even if the cell parameters remain constant. Consider an equation for the update of the phase (α) with each induced change in series resistance (ΔR_a):

$$\alpha_n = \alpha_{n-1} + \tan^{-1}\left(\frac{\Delta C}{\Delta G_{ac}}\right) \quad (39)$$

where ΔC and ΔG_{ac} are the changes in the values output by the PSD upon ΔR_a.

If α is set to the correct phase, then ΔC should be zero. Thermal noise in R_a, however, introduces a minimum rms "phase jitter" (σ_α) given by:

$$\sigma_\alpha \approx \sigma_{c,t}\left|\frac{d\alpha}{dC_m}\right| \approx \frac{\sigma_{c,t}}{\omega_c C_m^2 R_\Delta} \approx \frac{(4kTR_a B_N)^{1/2}/(|T(\omega_c)|U)}{\omega_c C_m R_\Delta} \quad (40)$$

$\sigma_{c,t}$ is the minimum thermal noise of C_m estimates, and B_N is the "noise bandwidth" of the PSD. Both of these terms are described in more detail in Section 5.1. As an example, consider a case where $R_a = 10$ MΩ, $C_m = 5$ pF, and $R_\Delta = 0.5$ MΩ. A 1-kHz, 10-mV amplitude sinusoidal stimulus is applied, and 100 cycles are averaged before and after the change in resistance; therefore, $B_N = 10$ Hz (see Section 5.1.3). Under these conditions, the minimum value of phase jitter is about 0.5°. In actively secreting cells, σ_α can be expected to be severalfold larger.

3.3.2. Setting the Desired Phase from the Phase of the Admittance

Combination of equations 23 and 33 reveals that the phase sensitive to changes in capacitance ($\eta + 90°$) is given by:

$$\theta \simeq \eta + 90° = 2\beta - 90° \tag{41}$$

where β is the phase of the admittance of the cell $[\angle Y(\omega_c)]$, which can be determined experimentally from:

$$\beta = \tan^{-1}(B/A) \tag{42}$$

where A and B are the real and imaginary components of admittance measured with a PSD in the same manner as for the Lindau–Neher technique (Section 3.2). Thus, capacitance compensation can occasionally be turned off, and equations 41 and 42 can be used to determine a new phase value (Zierler, 1992). Capacitance compensation can then be resumed until another phase determination is needed.

It is important to note, however, that phase delays in the patch-clamp amplifier and low-pass filters must be accounted for, or large errors in phase can occur (see Section 3.2.2).

As with other piecewise-linear methods, calibration can be performed by offsetting the C_{slow} setting. Alternatively, the gain can be calculated from the phase according to:

$$\frac{\Delta C}{\Delta I} \simeq \frac{1}{|dY/dC|U} = \frac{1}{\omega_c |T(\omega_c)|^2 U} \simeq \frac{1 + \tan^2(\pi/2 - \beta)}{\omega_c U} \tag{43}$$

The use of equation 43 allows the software to calibrate the C trace on line without knowing the C_{slow} and G_{series} settings of the patch-clamp amplifier.

3.3.3. Discontinuities in Capacitance Traces Accompany Phase Changes

Usually in the piecewise-linear technique, the bulk of the sinusoidal current is nulled out using capacitance compensation circuitry before the experiment begins. However, the current can become substantially uncompensated during the course of an experiment if major changes in R_a or C_m occur. In this case, each change in phase can lead to discontinuities in the C and G_{ac} traces as the unnulled current is projected onto the rotated axes. Elementary trigonometry yields (for small $\Delta\alpha$):

$$\Delta C \simeq -G_{ac,b}\Delta\alpha \qquad \Delta G_{ac} \simeq C_b\Delta\alpha \tag{44}$$

where C_b and $G_{ac,b}$ indicate their respective values before $\Delta\alpha$. If gain changes accompany $\Delta\alpha$ (Fidler and Fernandez, 1989), then a slightly more complicated relationship holds:

$$\Delta C \simeq (R - 1)C_b - RG_{ac,b}\Delta\alpha$$
$$\Delta G_{ac} \simeq (R - 1)G_{ac,b} + RC_b\Delta\alpha \tag{45}$$

where R is the ratio of the gain after the phase change to the gain before the phase change. Thus, displacements in the C and G_{ac} traces can be corrected using equation 45; remaining discontinuities suggest that parameter changes occurred during the resistor-dithering period.

4. Stimulation with Signals Containing Multispectral Components

As previously mentioned, the main problem with using a stimulus that contains only one sinusoid frequency is that, in principle, no more than two parameters can be uniquely determined. The Lindau–Neher algorithm gets an additional piece of information from the DC current, but in this case the zero-current potential (E_r) must be considered (Okada *et al.*, 1992). A more spectrally rich stimulus can be used to determine all three parameters of Fig. 2B. In addition, the additional information available can be used to evaluate the "goodness of fit" of the equivalent-circuit model to the actual response of the cell. Square-wave stimulation presented in Section 2 is an example of a stimulus with multispectral components that is analyzed in the "time domain." The present section concentrates on "frequency-domain" analysis.

It is important to point out that the use of more frequency components does not necessarily make C_m estimates more accurate or lower in noise. In fact, the opposite can be true if the estimation algorithm is not carefully designed. In general, the contribution of each frequency component to the C_m estimate should be weighted according to its signal-to-noise ratio. The optimal frequency range for generating C_m estimates will be presented in Section 5.1.1. In addition, the use of spectrally rich stimuli provides more pieces of information than are required to determine the three parameters; therefore, some sort of fitting of measured data to the model is required. This complicates the estimation algorithm and leaves open the possibility of convergence to false parameter values if poor initial estimates are given.

4.1. Stimulation with Two Sinusoids

The use of two sinusoids has been proposed for cases where large changes in R_m might contaminate C_m estimates (Rohlicek and Rohlicek, 1993; Donnelly, 1994). With two sinusoids, one has four pieces of information (the real and imaginary component at each frequency) in order to solve for three parameters (R_a, R_m, and C_m). One can "fit" the four measurements to the data using some criteria such as minimizing the squared error. Alternatively, a nonunique algebraic solution for the parameters can be arbitrarily chosen (Rohlicek and Rohlicek, 1993; Donnelly, 1994; Rohlicek and Schmid, 1994).

Principles of phase-sensitive detection (see Appendix A) can easily be extended to determine the real and imaginary components of the two sinusoids as long as the ratio of the two frequencies is an integer. In this case the maximum parameter estimation rate is equal to the lower frequency.

The frequencies and amplitudes of the sinusoids can be chosen to try and minimize the noise of C_m estimates; therefore, the discussion of these choices is delayed until Section 5.1.2.

4.2. Stimulation with a Pseudorandom Binary Sequence

It is possible to create a signal containing a very broad range of frequencies simply by alternating the voltage between two levels (say $V_{dc} + M$ and $V_{dc} - M$) at random intervals. Such a signal can be generated with hardware using digital shift registers with feedback (Clausen and Fernandez, 1981) or by outputing values stored in computer memory with a digital-to-analogue converter.

The smallest possible time interval between level changes (δ) determines the highest frequency component of the signal. If the probability of changing levels at each time increment (integer multiple of δ) is 0.5, then the power spectrum of the resulting signal is given by:

$$W(f) = \delta M^2 \left[\frac{\sin(\pi f \delta)}{\pi f \delta} \right]^2 \tag{46}$$

The spectrum of a PRBS is quite flat for frequencies up to about $0.3/\delta$; therefore, the signal is a good approximation of "white noise." The advantage that this type of signal has over gaussian-distributed white noise is that the amplitude of the stimulus is absolutely restricted for PRBS. It was noted earlier that the stimulus voltage must never exceed the threshold for significant activation of voltage-dependent conductances if the assumption of linearity is to remain tenable in excitable cells. Gaussian-distributed white noise always has a finite (although small) probability of reaching values many standard deviations from the mean value.

A PRBS is also optimized for delivering the maximum power for a given voltage excursion. Recall that the power of a signal is proportional to the square of its rms (root mean square) value. The rms value for PRBS is identical to the amplitude of the voltage excursion (U), whereas the rms value of a sinusoid is only $U/2^{1/2}$. It will be demonstrated in Section 5 that the minimum rms noise of C_m estimates is inversely proportional to the amplitude of the applied stimulus. However, this potential advantage is not realized because much of the power of a conventional PRBS signal is contained at frequencies outside the range that is optimal for estimating C_m (see Section 5.1.1).

Frequency-domain methods for estimating parameter values use a fast Fourier transform to obtain the complex current in response to stimulation with a PRBS. The complex admittance is then found by dividing the complex current by the previously stored Fourier transform of the PRBS signal. Parameter values can be obtained by fitting the admittance to the theoretical relationship given in equation 23 (Fernandez *et al.*, 1984).

5. Resolution Limits of Membrane Capacitance Measurements

In order to maximally resolve changes in C_m, careful attention to noise sources is required. The following analysis emphasizes stimulation with a single sine wave, but concepts explored here are applicable to more spectrally rich excitations also.

Sources of noise that limit the resolution of C_m measurements certainly include those of biological origin such as constitutive exocytosis/endocytosis and the fluctuation of ionic channels in the cell membrane. Many times, ionic channels are largely blocked by the contents of the pipette solution and by the maintenance of a hyperpolarized potential during C_m measurements. Even if these biological sources of noise are minimal, however, there are fluctuations in C_m estimates caused by thermal noise in the equivalent circuit of Fig. 2B, which are discussed in Section 5.1. Sources of "excess" noise that the experimenter has some control over are discussed in Section 5.2.

5.1. Thermal (Johnson) Noise Limits the Resolution of C_m Measurements

Any energy-dissipating physical system exhibits fluctuations of thermal origin. Whereas C_m, as an energy storage component, has no thermal noise associated with it by itself, the resistors R_m and particularly R_a contribute the Johnson noise that provide the lower bound in the variance of C_m estimates.

In general, the power spectral density of the current noise (S_I) is 4 kT times the real part of the admittance of a network. For the case of the equivalent circuit of Fig. 2B, $S_I(f)$ is given by:

$$S_I(f) = 4kT\,\mathrm{Re}\{Y(\omega)\} = 4kT\frac{1 + \omega^2 R_m R_p C_m^2}{R_t(1 + \omega^2 R_p^2 C_m^2)} \tag{47}$$

Note that at low frequencies, the noise density is dominated by R_t ($= R_m + R_a \approx R_m$). At high frequencies, the noise density is determined by R_a. At intermediate frequencies (appropriate for C_m measurements), the noise density increases with frequency.

The current noise variance is related to S_I according to:

$$\sigma_I^2 = \int_0^\infty |Hf)|^2 S_I(f)df \tag{48}$$

Where $H(f)$ accounts for the relative weight of frequencies used in forming parameter estimates. In our case, $H(f)$ is dominated by the phase-sensitive detector, which acts as a band-pass filter about f_c, the stimulus frequency. For a narrow-band filter, $H(f)$ approaches zero for frequencies other than f_c.* In this case:

$$\sigma_I^2 = S_I(f_c)B_N$$

where

$$B_N = \int_0^\infty |H(f)|^2 df \tag{49}$$

B_N is defined as the "noise bandwidth" of the estimator.

The current noise can be related to the noise of C_m estimates by using the same

*Equation 49 can be derived by assuming $H(f)$ equals $B_N\delta\ (f - f_c)$, i.e., zero everywhere except for $f = f_c$. However, numerical integration of equation 48 indicates that equation 49 is a close approximation under a wide variety of common recording conditions.

"piecewise-linear" approximation as in Section 3.3 (see Gillis, 1993, for a more rigorous derivation of σ_c neglecting R_m):

$$\frac{\sigma_I}{\sigma_{c,t}} \simeq \left|\frac{\partial I(\omega_c)}{\partial C_m}\right|$$

$$\therefore \sigma_{c,t} \simeq \frac{\sigma_I}{\left|\frac{\partial I(\omega_c)}{\partial C_m}\right|} = \frac{\sigma_I}{\left|\frac{\partial Y(\omega_c)}{\partial C_m}\right| U} \tag{50}$$

Thus, the use of stimulus frequencies at which the admittance is sensitive to changes in C_m ($\partial Y/\partial C_m$ large) will also result in C_m estimates that are less sensitive to current noise. From equation 31, recall that $\partial Y/\partial C_m$ increases with frequency until $1/2\pi R_p C_m$, after which it decreases with increasing frequency.
Combining equations 31, 47, 49, and 50 shows that the rms noise of C_m estimates from thermal noise ($\sigma_{c,t}$) is given by:

$$\sigma_{c,t} = \frac{(4kTB_N)^{1/2}(1 + \omega_c^2 R_m R_p C_m^2)^{1/2}(1 + \omega_c^2 R_p^2 C_m^2)^{1/2} R_t^{3/2}}{\omega_c U R_m^2} \tag{51}$$

Figure 7 plots $\sigma_{c,t}$ versus stimulus frequency and actual measured noise levels recorded from a model circuit.

5.1.1. The Optimal Frequency Range of a Stimulus Used to Estimate C_m

Note in Fig. 7 that at intermediate frequencies, $\sigma_{c,t}$ assumes a constant, minimal value because both σ_I and $\partial Y/\partial C_m$ increase linearly with frequency. This optimal frequency range is between frequencies f_m and f_p, which are defined as:

$$f_m = \frac{1}{2\pi\sqrt{R_m R_p}C_m}$$

$$f_p = \frac{1}{2\pi R_p C_m} \tag{52}$$

Thus, for values of the parameters such as those indicated in the legend to Fig. 7, the optimal frequency range is between several hundred Hertz and several kilohertz (Lindau and Neher, 1988). Measurements with model circuits (Fig. 7), however, suggest frequencies closer to f_p are better in practice. Note that other frequency ranges may be optimal for preparations with different "typical" parameter values.

For cases where the value of R_m is quite large, the frequency used may be much greater than f_m; in addition, $R_t \approx R_m$ and $R_p \approx R_a$. This results in a thermal capacitance noise ($\sigma_{c,t}$) given by:

$$\frac{\sigma_{c,t}}{C_m} \simeq \frac{(4kTR_a B_N)^{1/2}}{U\,|T(\omega_c)|} \simeq \frac{\text{thermal voltage noise of } R_a}{\text{amplitude of applied voltage sinusoid}} \tag{53}$$

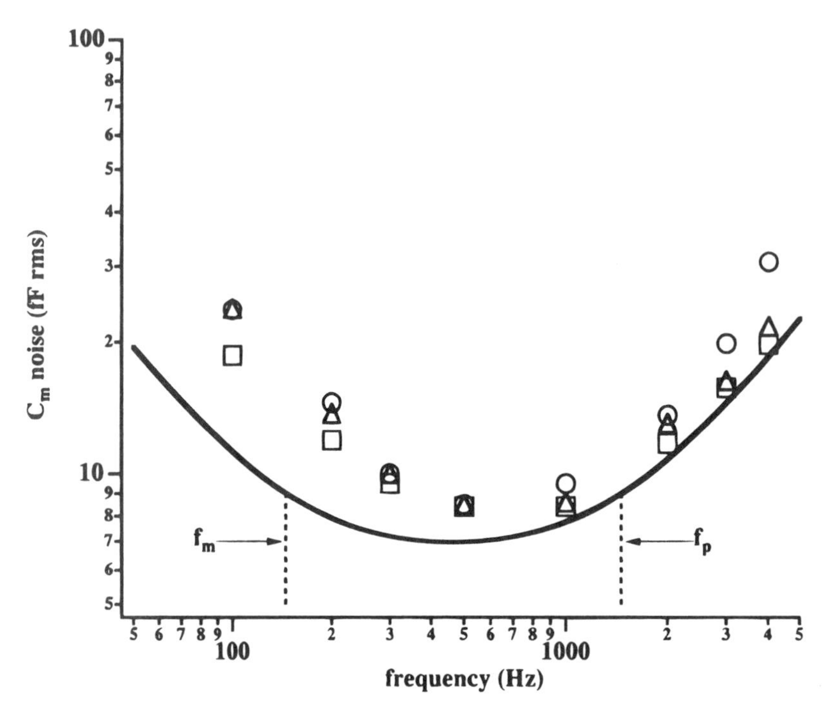

Figure 7. C_m noise versus frequency measured using a model circuit. The curve was calculated from equation 51 using parameter values of $R_a = 5\ \text{M}\Omega$, $R_m = 0.5\ \text{G}\Omega$, $C_m = 22$ pF, $B_N = 100$ Hz, $U = 10$ mV. The circles were measured using no capacitance compensation and a 0.5-GΩ feedback resistor (low gain of patch-clamp amplifier). The triangles were measured with capacitance compensation and a 0.5-GΩ feedback resistor. The squares were obtained with capacitance compensation and a 50-GΩ feedback resistor (high-gain range of patch-clamp amplifier). The optimum frequency range for sinusoidal stimulation is between f_m and f_p, which were calculated from equation 52. See Section 5.3.2 for discussion.

Here the expression has a satisfying intuitive interpretation: the signal-to-noise ratio (SNR) of the capacitance estimate equals the SNR of the applied voltage sinusoid. Equation 53 differs from equation 26 given by Lindau and Neher (1988) in that "U" represents the *amplitude* rather than the rms value of the applied voltage sinusoid. In addition, the $T(\omega_c)$ term can be thought of as assigning a noise penalty caused by the (usually small) fraction of the sinusoidal voltage that drops across R_a instead of across C_m.

5.1.2. The Choice of Amplitudes and Frequencies for Techniques Using Two Sine Waves

Two algorithms have recently been proposed for estimating the equivalent-circuit parameters from the real and imaginary current components determined at two frequencies (Rohlicek and Rohlicek, 1993; Rohlicek and Schmid, 1994; Donnelly, 1994). In Donnelly's algorithm (1994), the frequency ratio is fixed at 2. The Rohlicek algorithm (1993) allows the frequency ratio to be any integer value, although a value of 2 was also used in practice. Both groups used stimulus sinusoids of equal amplitude, although, in principle, amplitude ratios other than 1 can be used. Monte-Carlo simulations (data not shown) indicate an amplitude ratio of 1 may be appropriate for these algorithms because each sinusoid's contribution to the C_m estimate is roughly equal.

As might be expected from the analysis presented in the previous section, lowest C_m noise is obtained if both frequencies are between f_m and f_p. The main justification for using two sinusoids is cases where R_m becomes quite small, in which case f_m might change quite a bit during the course of an experiment. In this case the greater of the two frequencies should probably be set to something like half the expected f_p value, and the lower frequency should be set according to the desired frequency ratio.

A frequency ratio greater than 2 provides lower noise estimates of R_m because lower frequencies are optimal for estimation of this parameter. On the other hand, C_m estimates become noisier as the frequency ratio increases if the lower frequency drops below f_m. In addition, the maximum estimation rate is equal to the lower frequency.

Because there are four measured quantities and three unknown parameters, both of the published algorithms represent nonunique algebraic solutions that do not weigh the contribution of each sinusoid to the C_m estimate according to its signal-to-noise ratio. Monte-Carlo simulation suggests that both algorithms have at least twice the rms capacitance noise as the Lindau–Neher algorithm applied under comparable conditions (data not shown). Better noise performance can be expected as algorithms are developed that optimally fit the four measured quantities to the three-parameter model.

5.1.3. The Noise Bandwidth Under Common Recording Conditions

The noise bandwidth (B_N, defined in equation 49)* depends on how the phase-sensitive detector (PSD) is implemented. If the PSD integrates m sine waves (common in software PSDs, e.g., Herrington and Bookman, 1993):

$$|H(f)| = \left| \frac{\sin(\pi m(f - f_c)/f_c)}{\pi m(f - f_c)/f_c} \right| \tag{54}$$

$$\therefore B_N \approx f_c/m$$

If a single-time-constant (τ) low-pass filter is used to filter the output of the PSD (common in analogue lock-in amplifiers):

$$|H(f)| = \{1 + [2\pi\tau(f - f_c)]^2\}^{-1/2} \tag{55}$$

$$\therefore B_N = 1/2\tau$$

5.2. Excess Noise Sources

Excess noise sources can be reduced by careful experimental design. A "figure of merit" is the ratio of the rms magnitude of an excess noise source to the minimal thermal noise

*The "noise bandwidth" should not be confused with the "3dB bandwidth" (f_{3dB}) of fluctuations in C_m. f_{3dB} for a digital PSD $\approx 0.45f_c/m$, whereas f_{3dB} for an analog LIA = $1/(2\pi\tau)$.

($\sigma_{c,t}$). Assuming independent noise sources, the total measurement noise is given by:

$$\sigma_{c,total} = \sigma_{c,t}\left[1 + \left(\frac{\sigma_{c,1}}{\sigma_{c,t}}\right)^2 + \left(\frac{\sigma_{c,2}}{\sigma_{c,t}}\right)^2 + \cdots\right]^{1/2} \tag{56}$$

5.2.1. Noise in the Patch-Clamp Amplifier Headstage

Thermal current fluctuations in the feedback resistor of a patch-clamp amplifier headstage contributes an excess noise ($\sigma_{c,Rf}$) given by*:

$$\frac{\sigma_{c,R_f}}{\sigma_{c,t}} = \frac{(R_t/R_f)^{1/2}(1 + \omega_c^2 R_p^2 C_m^2)^{1/2}}{(1 + \omega_c^2 R_m R_p C_m^2)^{1/2}} \tag{57}$$

For the low-gain range of a patch-clamp amplifier such as an EPC-7, $R_f = 500$ MΩ. "Typical" parameter values of $R_a = 10$ MΩ, $C_m = 5$ pF, $R_m = 1$ GΩ, and $f_c = 1$ kHz yield $\sigma_{c,Rf}/\sigma_{c,t} = 0.45$. Substitution into equation 57 suggests that the excess noise from R_f is about 10%. Thus, at "typical" frequencies, capacitance measurements are unlikely to be greatly affected by thermal noise of the feedback resistor (Lindau, 1991). Low-frequency measurements, however, can exhibit significant excess noise when a low R_f value is used. For example, with the same parameter values given above, a stimulus frequency of 100 Hz results in a 67% increase in noise over the thermal minimum (also see Fig. 7). Noise contributed by a resistive feedback patch-clamp amplifier operating in a high-gain range (e.g., $R_f = 50$ GΩ) or by a capacitive feedback amplifier (e.g., Axopatch 200) should be completely negligible.

5.2.2. Quantization Noise

This source of excess noise is the limited resolution of analogue-to-digital converters (ADCs).

5.2.2a. In the Lindau–Neher Technique Without Capacitance Compensation. Quantization errors can be significant when the L–N algorithm is implemented by sampling the two outputs of an analogue lock-in amplifier with 12-bit ADCs (Lindau and Neher, 1988). The quantization noise ($\sigma_{c,q}$) is approximately given by (Gillis, 1993):

$$\sigma_{c,q} \approx \frac{C_{m,fs}/2}{2^n} \tag{58}$$

where $C_{m,fs}$ is the capacitance value that results in saturation of the ADCs and n is the number of bits of resolution of the ADC.† For example, for a full-scale capacitance ($C_{m,fs}$) of 8 pF and a 12-bit ADC, $\sigma_{c,q} \approx 1$ fF. Under common conditions quantization noise can be several femtofarads (see Fig. 9).

*The power spectral density of the current in the feedback resistor ($S_{I,Rf}$) equals $4kT/R_f$. By analogy with the derivation of equation 51: $\sigma_{c,Rf}/\sigma_{c,t} = (S_{I,Rf}/S_I)^{1/2}$.

†Note, when using the Lindau–Neher technique with an analogue lock-in amplifier, the range of the ADCs should be set to 0 to full-scale voltage, because the outputs of the LIA are always positive. If the range is set to ± full-scale voltage, then the effective value of n is reduced by 1.

From consideration of equation 58, it is apparent that the use of 16-bit ADCs should not result in significant quantization noise (Lindau, 1991). It should be noted that equation 58 does not apply to cases in which the phase-sensitive detection is implemented in software. Such estimates should be less sensitive to quantization errors because a number of sampled current points are processed to estimate A and B (see Appendix A). This processing leads to a certain amount of "averaging out" of quantization errors (see the following section).

5.2.2b. In Digital Phase-Sensitive Detectors. With Joshi and Fernandez's (1988) algorithm for implementing a digital PSD (four points per sine-wave cycle) the excess quantization noise is approximately given by (Gillis, 1993):

$$\frac{\sigma_{c,q}}{\sigma_{c,t}} \simeq \frac{\delta I/6^{1/2}}{(4kTf_c/R_a)^{1/2}} \simeq \frac{\text{quantization current noise}}{\text{thermal current noise of } R_a} \tag{59}$$

where δI is the current resolution of the ADC. For example, for a patch-clamp amplifier gain of 10 mV/pA, +10 V full scale voltage, and a 12-bit ADC, $\delta I = 2$ nA/4096. If $R_a = 10\ \text{M}\Omega$, then $\sigma_{c,q}/\sigma_{c,t} \approx 0.16$. Even better performance should be expected for digital PSDs with more points per sine wave. Thus, no dramatic improvement can be expected on replacement of 12-bit by 16-bit ADCs.

5.2.3. "Aliasing Noise": Choosing the Number of Points per Sine-Wave Cycle

In implementing a phase-sensitive detector in software, an important design parameter is the number of points per sine-wave cycle (p_c) that is used. The sampling rate ($f_s = p_c f_c$) and the number of calculations required to generate each C_m estimate increase linearly with p_c. Thus, hardware limitations will provide an upper limit to the parameter. Commonly, the same number of points per cycle are output from the DAC in constructing the sine wave as are sampled by the ADC to be used in the phase-sensitive detection algorithm. Thus, there are two issues involved in choosing a minimum value of p_c: the generation of a smooth sine wave and the maintenance of a sampling rate fast enough to avoid "aliasing."

The sine wave is usually created by low-pass filtering of the signal output from a DAC. The raw output of the DAC is a series of voltage levels that remain constant between sampling intervals (giving the signal a "staircase" appearance). In order to generate a smooth sine wave, the low-pass filter that follows the DAC output should have a cutoff frequency greater than f_c and a sharp enough roll-off that frequencies above $(p_c - 1) f_c$ are attenuated to the background noise level. Certainly this criterion is easier to fulfill for a large p_c value; it follows intuitively that more points make a better sine wave. It is worth mentioning, however, that a pure sine-wave input is not essential to the operation of the PSD as long as high-frequency components do not saturate the patch-clamp amplifier. In addition, filtering in the stimulus pathway will also lengthen the rise time of a depolarizing step.

The most important reason for choosing a p_c value greater than some minimum value is to avoid aliasing. Frequency components greater than $f_s/2$ alias as lower-frequency signals, which can lead to an increase in measurement noise. The noise of C_m estimates is particularly sensitive to aliasing because current noise is greatest at high frequencies (see equation 47). The low-pass filter that precedes the ADC certainly needs to pass frequencies at least as high as the stimulus sine wave frequency (f_c). In fact, the cutoff frequency of the filter should be set to at least $2f_c$ if the desired time resolution of C_m is on the order of the period of the sine wave ($1/f_c$). In this case a p_c value of 4 results in a sampling rate ($f_s = 4f_c$) that exactly satisfies the minimum Nyquist sampling rate (twice the cutoff frequency of the low-pass filter). Because real-world low-pass filters do not absolutely eliminate frequencies above their cutoff values, some increase in noise from aliasing will occur.

Figure 8 presents some measurements using a model cell network that show the effect of p_c on the C_m noise level. The use of four points per sine wave results in excess noise except when the low-pass filter was set to only 1 kHz (1.25 f_c), a value not acceptable if maximum time resolution is desired. It is apparent from Fig. 8 that a minimal value of p_c of about 10 is desirable in practical situations (also see Herrington and Bookman, 1993).

A similar problem can occur in analogue lock-in amplifiers that use square-wave demodulation for phase-sensitive detection (see Appendix A). In this case, the signal input to the lock-in amplifier (i_{pc}, the patch-clamp amplifier signal) should be filtered to eliminate frequencies above $3f_c$.

5.2.4. Degradation of Resolution as a Result of Noncontinuous Data Acquisition

Often the stimulus waveform is applied intermittently at a rate of a few Hertz to allow the computer time to implement the estimation algorithm and to perform other tasks. In this case, the estimates are significantly noisier than those generated by techniques that average the signal over the entire period between estimates. For example, a 50-msec-duration waveform applied five times per second (25% "duty cycle") results in an rms C_m noise twice as

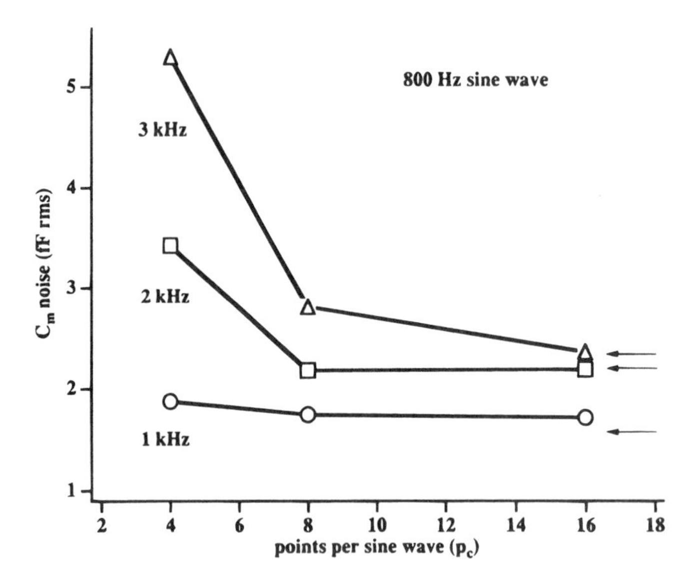

Figure 8. Noise of C_m estimates as a function of the number of points per sine-wave cycle (p_c). The frequency of the stimulus sinusoid was 800 Hz, and the cutoff (−3 dB) frequency of the low-pass filter that precedes the phase-sensitive detector was set to either 1 kHz (circles), 2 kHz (squares), or 3 kHz (triangles). The Nyquist sampling frequency corresponds to a p_c value of 2.5 (circles), 5.0 (squares), or 7.5 (triangles). (p_c values must actually be integers.) The excess noise at small p_c values results from aliasing. The arrows indicate the minimum thermal noise level calculated from equation 53 with B_N determined by a numerical integration of the $H(f)$ function given in equation 54. (The approximation $B_N \approx f_c/m$ is not valid when the cutoff frequency of the low-pass filter is less than $f_c + f_c/m$.) Measurements were made from a model circuit: $C_m = 4.5$ pF, $R_a = 5.3$ MΩ, $U = 25$ mV, with C_m estimates generated from data collected from individual sine-wave periods ($m = 1$).

large as that obtained with continuous acquisition. Because continuous acquisition over long periods of time is not always practical, a reasonable compromise is to use continuous acquisition during periods of greatest interest (such as immediately before and after a depolarizing pulse) and then use intermittent acquisition during periods where resolution is not as critical.

5.3. Results of Noise Measurements

The rms noise of C_m estimates was measured in both actual cells and cell models in order to illustrate low-noise techniques and to compare different methods of estimating C_m or ΔC_m.

5.3.1. Actual Cells Can Have C_m Noise Levels near the Theoretical Minimum

Figure 9 presents measurement noise data obtained from actual cells using an analogue lock-in amplifier. The data were obtained from portions of records selected for low noise. The

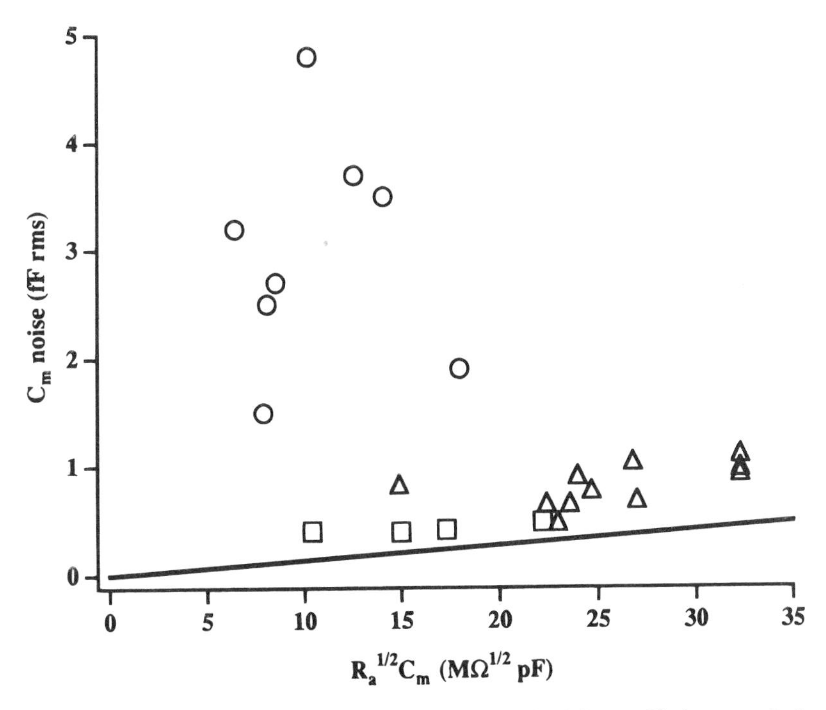

Figure 9. Actual cells can have C_m noise levels near the theoretical minimum. Circles were obtained using the Lindau–Neher algorithm with an analogue lock-in amplifier sampled with 12-bit ADCs. The theoretical quantization noise under conditions of these measurements (from equation 58) was about 2.7 fF. The squares were obtained using the L–N algorithm with a digital PSD (16-bit ADCs). The triangles were recorded using the piecewise-linear technique with an analogue lock-in amplifier. The line indicates the approximate minimum noise level calculated from equation 53 with $T(\omega) \approx 1$, $B_N = 2.5$ Hz, and $U = 15$ mV. Circles and triangles were obtained from rat pancreatic B cells. Squares were obtained from bovine adrenal chromaffin cells.

triangles were obtained using the piecewise-linear technique (with capacitance compensation). They demonstrate that actual cells can be recorded with C_m noise levels near the thermal minimum. The circles were obtained using the Lindau–Neher technique with 12-bit ADCs and an analogue lock-in amplifier. Here, quantization noise dominates. The squares indicate values obtained using the L–N algorithm with a software PSD and 16-bit ADCs, demonstrating that this technique can also achieve near-ideal noise levels. It should be pointed out, however, that actively secreting cells often have noise levels severalfold higher than the minimum thermal noise value.

5.3.2. Results with Model Circuits

Figure 7 presents noise measurements made using the Lindau–Neher algorithm with a software phase-sensitive detector (PULSE software) using the MC-9 model circuit (HEKA). The solid line indicates the theoretical minimum noise level (from equation 51) using parameter values indicated in the legend. The squares were obtained using a high gain (50-GΩ feedback resistor) and capacitance compensation (thus, equations 30 were applied). Very nearly ideal noise levels at high frequencies ($\geq$1 kHz) were obtained under these conditions. Lower frequencies exhibited "excess noise," perhaps due to low frequency noise sources not modeled by equation 47. The triangles were obtained under the same conditions, except a smaller value feedback resistor (0.5 GΩ) was used. As expected from equation 57, the noise is significantly higher only at low frequencies ($<$300 Hz). The open circles were obtained with the smaller feedback resistor and no capacitance compensation. The excess noise at low frequencies (compared to the squares) is caused by the feedback resistor, but there is also excess noise above 2 kHz when compared to measurements made with capacitance compensation. One possible explanation for this observation is the presence of a small amount of high-frequency noise in the stimulus pathway of the EPC-9, which is canceled when capacitance compensation is used.

Noise measurements were also made using other C_m estimation techniques on the same model circuit. An analogue lock-in amplifier that integrates the signal over one sine-wave cycle gave a noise level of about 7.5 fF at a frequency of 800 Hz* compared with 8.4 fF measured at 1 kHz with a digital PSD. Thus, analogue lock-in amplifiers may be slightly less noisy than (properly configured) software PSDs, but the difference (10–15%) is not very significant. The piecewise-linear method gave nearly identical noise performance as the Lindau–Neher technique.

Noise in C_m estimates generated with square-wave stimulation were also measured using the *cap. track* software of the EPC-9. Under comparable conditions as in Fig. 7 ($\pm$10-mV voltage steps over a 10-msec interval), this technique gave an rms noise level of about 6.9 fF. By comparison, the minimum noise value obtained with a 10-mV-amplitude sine wave using the Lindau–Neher technique (software PSD) was 8.4 fF. Thus, square-wave stimulation can achieve noise levels comparable to sinusoidal techniques. The slightly lower noise level of the time-domain technique (15–20%) may be because a square wave has a ($\approx$40%) greater rms value for a given voltage excursion than a sine wave. For example, taking only the first two terms of the Fourier series, a 1-kHz, 10-mV-amplitude square wave contains a 12.7 mV amplitude sinusoid at 1 kHz and a 4.2 mV sine wave at 3 kHz.

*This rms noise value was obtained by dividing the actual measured value by $8^{1/2}$ in order to allow direct comparison with results obtained in Fig. 7, where B_N = 100 Hz; see equation 54.

6. Practical Application, Hints, and Warnings

6.1. Choosing the Technique

The first consideration in undertaking capacitance measurements is the choice of the technique. Certainly the decision will hinge on the type of experiments that are planned. However, because there are a variety of ways of performing the same measurement, a large part of the decision will depend on how well each available technique fits in with the lab's existing hardware, software, and individual preferences.

6.1.1. Square-Wave versus Sinusoidal Stimulation

If changes in C_m are to be monitored during the course of an experiment without the need for high time resolution, then estimation from square-wave stimulation may be appropriate. This is particularly convenient if one already uses software that includes this feature. The limited time resolution of some software that uses square-wave stimulation (on the order of 5 Hz) results from the "down time" for calculations between sweeps of stimuli. Even large changes in R_m and R_a should have little effect on C_m estimates as long as they are slow compared to the estimation rate and the equivalent circuit of Fig. 2B is a valid model of the cell. Perhaps the greatest advantage, however, is that the "goodness of fit" of the estimates can be readily evaluated by seeing how well the current transient is canceled by the capacitance compensation circuitry. In fact, it is a good idea to test the quality of transient cancelation using capacitance compensation regardless of the C_m method used in order to verify the appropriateness of the model of Fig. 2b. Perhaps the greatest disadvantage is that some algorithms that use square-wave stimulation can converge to clearly false values if poor initial "guesses" are given. Thus, this technique can be less "robust" in response to large, sudden parameter changes than the Lindau–Neher technique. Another disadvantage is that a square wave contains high-frequency components that are filtered; therefore, the current response of equation 1 is only an approximation.

6.1.2. One Sinusoid versus Two

The two principal techniques that use a single sine wave are the piecewise-linear technique (Section 3.3) and the Lindau–Neher technique (Section 3.2). The P-L technique is not well suited to the case where R_m becomes quite small (within an order of magnitude of R_m) because it becomes impossible to find a phase where changes in C_m can be monitored independently of changes in both R_a and R_m ($\eta \neq \theta + 90°$, see equation 33). As mentioned in Section 3.2, the Lindau–Neher technique uses the DC current to estimate R_t based on the assumption of a constant reversal potential (E_r). This assumption may lead to errors if both (1) E_r shifts during the course of an experiment and (2) R_m is quite small, approaching within an order of magnitude of R_a. The additional piece of information needed to estimate E_r can be obtained by stepping the holding potential (Section 3.2, Okada *et al.*, 1992). Other types of stimuli that would be appropriate under these conditions are a square wave, PRBS, or techniques that use two sinusoids.

Currently published algorithms for estimating C_m using two sinusoids (Rohlicek and Rohlicek, 1993; Donnelly, 1994; Rohlicek and Schmid, 1994) can be expected to be about twice as noisy as the Lindau–Neher algorithm (or square-wave stimulation) under comparable conditions (Section 5.1.2). Therefore, these algorithms may not be optimal for general use. Nevertheless, techniques that use two sinusoids may be useful for cells that undergo dramatic changes in R_m, since attaining a low C_m noise level is much less important than accurately estimating C_m in the face of large changes in R_m.

6.1.3. Implementing the PSD in Hardware versus Software

A hardware lock-in amplifier can be used to perform a good part of the signal processing required and thereby reduce the computational burden on the computer hardware (Lindau and Neher, 1988). When used with the piecewise-linear technique, it also provides an analogue signal that is directly proportional to changes in C_m, that can be displayed directly on a chart recorder and recorded with virtually any type of data acquisition system. A hardware lock-in amplifier may also be slightly (~10%) lower in noise than a software PSD (see Section 5.3.2).

Phase-sensitive detection in software, however, certainly has the advantage of greater flexibility. As computer hardware continues to climb the performance/price curve, commonly available and affordable computers are able to perform the required calculations in real time. Implementation in software allows the elimination of a several-thousand-dollar instrument and the automation of gain and phase adjustment. In addition, a software "virtual instrument" allows all the "front panel controls" (frequency, gain, phase) to be stored along with the data or to be used for further processing, such as the application of the Lindau–Neher algorithm.

In order for the software PSD to be useful, it must be tightly integrated with software used for data acquisition, voltage-clamp pulsing (if desired), and data analysis. Writing such software "from scratch" is a daunting challenge. Commercial and public-domain software is now beginning to fill the void but is always designed around a particular computer and data-acquisition interface.

6.1.4. The Lindau–Neher Algorithm versus Piecewise-Linear Techniques

There are several advantages of the Lindau–Neher algorithm. Most importantly, it provides C_m estimates that are independent of changes in either R_a or R_m. Large variations in any of the parameters can be followed without changing the phase setting of the PSD. Actual values of all three parameters are generated rather than just relative changes. In addition, external resistor dithering and calibration using capacitance compensation circuitry are not required.

Nevertheless, P-L techniques can be more convenient under certain circumstances. As mentioned previously, an analogue signal proportional to changes in C_m can be generated directly as the output of an analogue lock-in amplifier. Calibration by offsetting capacitance calibration can actually be quite convenient because it does not require taking into account the gains of the lock-in and patch-clamp amplifiers and the frequency and amplitude of the stimulus sinusoid. Also, software does not have to take into account the settings of the capacitance compensation (C_{slow} and G_{series}) in order to estimate changes in C_m (Section 3.2.1). In addition, since the phase is found empirically, the actual phase value (which takes into account delays generated by various low-pass filters, Section 3.2.2) does not have to be determined. These conveniences should become less important as software is developed that allows computer control of all pertinent experimental settings (a "virtual instrument," e.g., EPC-9) or direct reading of the relevant parameters (via instrument "telegraphing").

6.2. The Recording Configuration

6.2.1. Square-wave Stimulation and PRBS

Square-wave stimulation and PRBS use high-frequency information in forming C_m estimates. Therefore, filtering in the stimulus pathway should be minimal (e.g., a 2-μsec time constant), and the output filtering should also be set to a high value (about 30 kHz). The sampling rate should also be very fast (say 100 ksamples/sec or higher). If possible, a 16-bit ADC should be used. Thus, a reasonably high-performance (and expensive) data acquisition system is needed. For generating estimates at a fast rate (say greater than 10/sec), a high-performance computer will also be required.

In many cases the measurement is performed with capacitance compensation disabled (an exception is square-wave stimulation in the EPC-9); therefore, the patch clamp amplifier is usually set to a low gain in order to prevent the amplifier from saturating.

6.2.2. Sine-Wave Stimulation

The frequency of the sine wave should be selected to be between frequencies f_m and f_p calculated from equation 52 using estimates of R_a, R_m, and C_m that are "typical" for the cell type used. In general, we recommend a value closer to f_p (e.g., $0.3 f_p$) because of the possibility of "excess noise" sources of lower frequencies.

6.2.2a. Using an Analogue Lock-in Amplifier. A dual-phase lock-in amplifier is required. Custom designed lock-in amplifiers can generate the stimulus sine wave as well as perform the PSD operation (Lindau and Neher, 1988; Gillis, 1993). If a commercially available model is used, then a function generator will also be required to provide the sine wave, and some sort of summing junction will be needed if voltage-clamp pulses are to be added to the sinusoid.

The phase can be set either manually or digitally if the instrument has a GPIB (general purpose interface bus) capability. For phase adjustment using a piecewise-linear technique, the phase setting of the instrument can be left at some arbitrary value, and a basic software "rotation of axes" can be performed to adjust the effective phase to any desired value (Lindau, 1991).

The signal input to the lock-in amplifier (the signal output from the patch-clamp amplifier, i_{pc}) should be filtered to eliminate frequencies greater than $3f_c$ (see Appendix A). A cutoff frequency of $2f_c$ is generally recommended.

The outputs of a lock-in amplifier are often filtered with a basic single-pole low-pass filter. The time constant can be set as the usual tradeoff between noise and speed of response. For calculating the minimum thermal noise of C_m estimates (equation 51 or 53), the noise bandwidth is $1/2\tau$ (equation 55). The 3-dB bandwidth of the lock-in outputs is $1/2\pi\tau$; therefore, if the outputs are sampled at 10 Hz, a τ of about 50 msec is appropriate.

If maximum time resolution is required (on the order of $1/f_c$), then a single-time-constant filter may not be appropriate. This is because the phase-sensitive detection involves a "demodulation" operation that results in a signal spectrum that has a prominent component at $2f_c$, which is not sufficiently attenuated with a single-pole filter. This harmonic component can be eliminated by analogue averaging of the signal over an integral number of sine-wave periods (Breckenridge and Almers, 1987; also see Appendix A).

6.2.2b. Using a Software PSD. Two design choices unique to a software PSD are the number of points per sine wave (p_c, Section 5.2.3) and the number of sine waves averaged to produce an estimate (m). Because p_c is selected to avoid aliasing (see Section 5.2.3), a good choice would be to set the cutoff frequency of the low-pass filter that follows the current output of the patch-clamp amplifier to about $2f_c$ and use a p_c value of 10 or greater. The value of m is chosen as a tradeoff between measurement noise and time resolution. The frequency that estimates are generated is f_c/m, which is identical to the "noise bandwidth" that can be used to estimate the minimum thermal noise limit from equation 51 (see Section 5.1.3).

Usually a low-pass filter follows the DAC output in order to produce a smooth sine wave (Section 5.2.3). A cutoff frequency of about $2f_c$ yields a clean sine wave if the p_c value is 10 or greater. Keep in mind that the low-pass filter will slow the rise time of a voltage step. (The 0–90% rise time is about $0.7/f_{-3dB}$ for an eight-pole Bessel filter.) If a fast rise time is essential, use only the stimulus filter built into the patch-clamp amplifier, but make sure the output filter has a sharp cutoff (e.g., at least four poles) and use a large p_c value to avoid aliasing. Make sure the patch-clamp amplifier does not saturate from the high-frequency components of the stimulus.

A 12-bit ADC may be sufficient for implementing a software PSD (Section 5.2.2b), and the maximum sampling rate needed is only about 50 kHz. However, a 16-bit data acquisition system, if available, is preferable.

6.2.2c. Implementing the Lindau–Neher Technique. If the L–N algorithm is implemented with an analog lock-in amplifier, and capacitance compensation circuitry is not used, a 16-bit data acquisition system is recommended to avoid "quantization noise" (Section 5.2.2a). If only a 12-bit ADC is available, set the gain of the lock-in amplifier (and/or patch-clamp amplifier) as high as possible without saturating. If capacitance compensation circuitry is used, equations 30 can be used to account for the effect of this circuitry on the sinusoidal current. If any significant membrane conductances are expected to be activated during the course of an experiment, then E_r should be set to the zero-current potential of the expected conductance.

6.2.2d. Implementing the Piecewise-Linear Technique. Series resistance dithering (Section 3.3.1) is a convenient way of automating phase determinations. However, a serious error in phase can result if R_a and C_p (pipette capacitance) are not very small (see Section 3.3.1). In some cases, a large value series resistor (R_Δ) reduces both the mean error (equation 38) and the statistical variance of the phase ("phase jitter," equation 40). A value of about 0.5–2.0 MΩ is appropriate. The number of sine waves averaged before and after the resistor is switched should be on the order of 100.

Discontinuities in C traces can accompany phase changes (Section 3.3.5). Equation 45 can be used to correct these displacements if they are of concern.

6.3. Suggestions for Low-Noise Recording

Consideration of equations 51 and 53 suggests the most effective way to minimize noise is to increase the amplitude of the stimulus sinusoid (U). The limit to this approach is that the most positive excursion of the stimulus should not exceed the potential at which voltage-dependent channels are activated in excitable cells (Neher and Marty, 1982). Thus, a more hyperpolarized holding potential (DC offset) can allow a larger U value to be used. In nonexcitable cells, U is limited by the voltage that the membrane can support without dielectric breakdown. Smaller values of R_a result in lower noise also, but the impact is limited because the noise goes with the square root of R_a; nevertheless, high values (not uncommon in perforated-patch recording) are detrimental to noise performance. $\sigma_{c,t}$ increases linearly with the baseline capacitance (C_m); therefore, smaller cells can be selected if noise is critical.

For detecting unitary fusion events, recording from an "on-cell" patch may be useful (Neher and Marty, 1982). Since fusion events are rare in "on-cell" recording, however, one is losing the signal as well as the noise.

The use of hydrophobic ions to increase the specific capacitance of the membrane has been proposed for high-resolution studies of membrane capacitance (Oberhauser and Fernandez, 1993). The charged groups of these ions increase the dielectric constant of the membrane as the hydrophobic portions of the ions become incorporated into the membrane core. Incubation of mast cells with dipicrylamine increases the baseline capacitance of the cell by 2.5-fold, whereas step increases in capacitance from the fusion of individual granules is enhanced sevenfold (Oberhauser and Fernandez, 1993). The better incorporation of the ions into granule membranes than in the plasma membrane is important, since equal loading would not result in any improvement in the signal-to-noise ratio (see equation 53).

The noise bandwidth (B_N) is under the control of the experimenter; reducing B_N provides the classic tradeoff of lower measurement noise for a slower responding system. Equations 54 and 55 give the value of B_N under common recording conditions.

Quantization errors are a significant source of excess noise when implementing the Lindau–Neher technique with an analogue lock-in amplifier sampled with 12-bit ADCs (Lindau and Neher, 1988). When one is implementing the piecewise-linear technique using an analogue lock-in amplifier and 12-bit ADCs, a general rule of thumb is that a 100-fF change in capacitance (e.g., during a calibration) should produce a change in the C output of at least 1 V for quantization noise to be negligible. Reconstruction of small, single-granule fusion events requires severalfold higher gain.

"Aliasing noise" can be eliminated by using a large number of points per sine wave cycle (p_c) and an appropriate cutoff frequency in the low-pass filter that follows the patch-clamp amplifier (Section 5.2.3). In most cases four samples per period are not sufficient; we recommend 10 or more.

Noncontinuous data acquisition can also degrade resolution (see Section 5.2.4).

6.4. Some Sources of Errors and Artifacts

6.4.1. Monitor G_{ac} or R_a

The most important thing that can be done to guard against errors or artifacts in C_m measurements is to monitor changes in resistive parameters along with C_m changes. For the piecewise-linear technique, the G_{ac} signal should be recorded, and for the Lindau–Neher technique, R_a (and, if it undergoes large changes, R_m) estimates serve as the control. Parallel (or antiparallel) changes in C_m and the resistive parameters are indicative of either (1) an incorrect phase setting or (2) a cell that is not well modeled by the simple equivalent circuit of Fig. 2B.*

It is quite simple to apply this test over short intervals, such as before and after a brief depolarizing pulse, because the basic assumption is that changes in C_m should occur without accompanying changes in resistance (see Fig. 1B). Over the course of minutes, however, actual changes in resistive parameters can be expected to occur. In this case, the main criterion

*An exception to this general rule is the case where the insertion of ionic channels into the plasma membrane accompanies vesicle fusion. If this is believed to be the case, extreme caution must be applied to rule out the possibility of artifact (Zorec and Tester, 1993).

is that changes in capacitance and resistance should not be well correlated, i.e., increase and decrease in parallel or antiparallel.

Often, even under good conditions, there is a small amount of correlation between estimated changes in resistance and capacitance, so what is considered an "acceptable" level? When using the piecewise-linear technique, correlated changes in C and G_{ac} can be quantitated as a phase error (α_{err}). A 5% correlation between uncalibrated* C and G_{ac} traces corresponds to roughly a 3° phase error, which perhaps should be considered as near the upper limit of acceptability. With the Lindau–Neher technique, the interpretation is not as straightforward. However, an α_{err} value can also be calculated from equation 35 for abrupt changes in resistive parameters that appear as small changes in C_m. For the case of an abrupt change in C_m (such as following a depolarizing pulse) that also appears as a small change in R_a, equation 35 can be reconfigured as:

$$\frac{\Delta R_{a,art}}{R_a} \simeq \tan(\alpha_{err}) \cdot \left(\frac{1}{\omega_c R_a C_m} \frac{\Delta C_m}{C_m} \right) \tag{60}$$

It should be emphasized that α_{err} is only a "figure of merit" and does not necessarily correspond to a genuine phase error, since the actual problem may be that the cell is not well represented by the three-component equivalent circuit. If α_{err} actually does represent phase problems, then the error in the 0° phase setting of the Lindau–Neher technique (Section 3.2.2) is $\alpha_{err}/2$ because of the relationship expressed in equation 41.

6.4.2. Watch the Fluid Level of the Bath

The pipette capacitance (C_p) of Fig. 2A can be minimized by coating the pipette with a hydrophobic substance and by maintaining a reasonably low bath level. Since the contribution to the current by C_p can be largely eliminated by C_{fast} compensation, it is usually neglected in circuit analysis. Major changes in fluid level, however, can change the C_p value and cause havoc in the parameter estimates. Thus, flushes of the bathing solution can introduce artifacts. This is particularly troublesome if the fluid level becomes high enough to reach the uncoated region of the pipette.

If series resistor dithering is used to determine the phase in the piecewise-linear technique, it is important to minimize C_p by keeping the solution level low (see equation 38).

6.4.3. Separating Exocytosis from Endocytosis

As mentioned in the introduction, the "Achilles heel" of C_m measurements is that they report only the net rate of exocytosis minus endocytosis. Following a depolarizing voltage pulse, C_m often increases abruptly and then, in a quite variable manner, undergoes a much slower decrease. This is usually interpreted as exocytosis followed by a much slower phase of endocytosis. If this slow rate of endocytosis can be extrapolated back to the time of depolarization (i.e., is constant), then exocytosis should be clearly discernable from endocytosis. However, some experiments suggest that a fast *decrease* in C_m can occur shortly after

*Here, "uncalibrated" means C and G_{ac} are taken directly from the outputs of the PSD with the same gain applied to each channel (identical units). That is the case for C and G_{ac} defined by equation 34.

an increase in internal Ca^{2+} to high levels (Neher and Zucker, 1993; Thomas *et al.*, 1994). This presumed "fast" phase of endocytosis makes quantitative interpretation of increases in C_m as exocytosis problematic. Quantitative interpretation is even more suspect for slow increases in C_m (such as those that accompany dialysis of stimulatory substances through the patch pipette). Amperometric techniques (see Chapter 11, this volume) should help address this issue by measuring an actual secreted substance.

6.4.4. Gating Charge Artifacts

As mentioned previously, the most positive excursion of the voltage stimulus should not exceed the threshold for the significant activation of voltage-dependent channels because R_m becomes highly nonlinear. However, C_m can also have a voltage dependence as a result of the mobilization of charges in polarizable membrane proteins. For example, the mobilization of "gating charges" of voltage-dependent channels at depolarized potentials leads to an increase in C_m (e.g., Fernandez *et al.*, 1982).

This contribution to C_m can be expected to have a bell-shaped dependence on the DC value of the stimulus, with the maximum obtained at a potential where the gating charges are most likely to move within the electric field (near −40 mV for the case of Na^+ channels: Fernandez *et al.*, 1982). Because capacitance measurements used to monitor exocytosis are chiefly concerned with changes in C_m measured at a hyperpolarized DC potential, a small *constant* contribution by mobile charges is not of much concern. However, problems can occur in interpreting C_m changes immediately after a depolarizing pulse because the voltage dependence of gating charge movement can shift upon channel inactivation (Fernandez *et al.*, 1982; Horrigan and Bookman, 1993).

For example, Na^+ channel gating charges are more mobile at a hyperpolarized potential following a depolarizing pulse. In this case, C_m may have an added component following repolarization that decays as the gating charges become immobilized during recovery from inactivation. Since this transient component does not reflect changes in membrane surface area, care must be taken in interpreting records immediately following a depolarizing pulse. Horrigan and Bookman (1993) found that the transient component decays with a time constant of about 70 msec at −80 mV. Tetrodotoxin does not block the transient component but lengthens the time constant of decay to about 0.6 sec. Dibucaine, which blocks the mobilization of Na^+ channel gating charges, also eliminates the C_m transient. The amplitude of the transient is about 8.3 fF for every nA of peak Na^+ current. This phenomenon is chiefly of concern in cells with a high density of voltage-dependent channels or under conditions where small secretory responses are measured with high time resolution.

6.4.5. Can Exocytosis Be Measured during Membrane Depolarization?

It certainly would be desirable to obtain information about C_m changes *during* a depolarization, since this is when the rate of exocytosis should be maximal in excitable cells. Because of the activation of voltage-dependent ionic channels and the accompanying nonlinearities, however, parameter estimates generated during a depolarization are usually ignored. Rather, the *difference* in C_m before and after a depolarizing pulse is used to infer exocytosis during the pulse (e.g., Fig. 1B).

Recently, Lindau *et al.* (1992) examined C_m estimates made during a depolarization to infer the rate of exocytosis. They argue that C_m estimates generated during later stages of a long depolarizing pulse (seconds in duration) are valid because there is little change in estimated C_m (or dC_m/dt) on repolarization to the holding potential. This can be expected to be the case if voltage-dependent conductances are inactivated after a long period of depolarization; if R_m is large, a small amount of nonlinearity becomes unimportant. The contribution of gating charge movement to C_m can be expected to become unimportant after a several-second depolarization to a potential where the voltage-dependent conductances are maximally active (since the gating charges are immobilized at these potentials). However, it seems unlikely that much useful information can be obtained from records obtained during brief depolarizations (or under any conditions where voltage-dependent conductances are large) using a single sinusoid stimulus.

Appendix A: Phase-Sensitive Detection

Consider a sinusoidal signal of amplitude N and phase β:

$$i(t) = N\cos(\omega_c t + \beta) \tag{61}$$

A phase-sensitive detector resolves this sinusoid into components in phase (X_1) and 90° out of phase (X_2) with a predetermined setting α, i.e.,

$$i(t) = N\cos(\omega_c t + \beta) = X_1\cos(\omega_c t + \alpha) + X_2\sin(\omega_c t + \alpha) \tag{62}$$

Elementary trigonometry yields values for X_1 and X_2 given by*.

$$\begin{aligned} X_1 &= N\cos(\alpha - \beta) \\ X_2 &= N\sin(\alpha - \beta) \end{aligned} \tag{63}$$

We will concentrate on determining X_1, since X_2 follows in a similar fashion. If the input signal is multiplied by a cosine wave at the desired phase angle, then the result (X_m) consists of a DC value proportional to X_1 plus a component at twice the original frequency:

$$\begin{aligned} X_m &= i(t)\cos(\omega_c t + \alpha) \\ &= N\cos(\omega_c t + \beta)\cos(\omega_c t + \alpha) \\ &= \frac{N}{2}\cos(\alpha - \beta) + \frac{N}{2}\cos(2\omega_c t + \alpha + \beta) \end{aligned} \tag{64}$$

The second harmonic component can be removed by low-pass filtering of X_m. Another alternative is to average X_m over one sinusoid period (T_c):

$$\frac{1}{T_c}\int_0^{T_c} i(t)\cos(\omega_c t + \alpha)dt = \frac{N}{2}\cos(\alpha - \beta) \tag{65}$$

*If the PSD is to be calibrated to output the rms value of the input sinusoid, X_1 and X_2 need to be divided by $\sqrt{2}$.

Thus, X_1 can be determined by correlating the input signal with a sinusoid at the desired phase setting:

$$X_1 = \frac{2}{T_c}\int_0^{T_c} i(t)\cos(\omega_c t + \alpha)dt \tag{66}$$

Equation 66 can be implemented in hardware or software (Neher and Marty, 1982; Joshi and Fernandez, 1988; Herrington and Bookman, 1993). For a software PSD that samples p_c values per cycle (overall sampling rate is p_c/T_c), the integral is approximated as a sum according to:

$$X_1 \simeq \frac{2}{p_c}\sum_{l=0}^{p_c-1} i(lT_c/p_c)\cos(2\pi l/p_c + \alpha) \tag{67}$$

The cosine function can be calculated in advance and stored as a look-up table with the number of entries determined by the desired phase resolution. For example, 3600 values are stored for a phase resolution of 0.1°. α determines which entries (p_c in number) are selected from the table in implementing equation 67. Sampled data values [i (l T_c/p_c)] are then simply multiplied by the corresponding table value, summed, and multiplied by the scaling factor $2/p_c$. See Herrington and Bookman (1993) for tips on efficient implementation of equation 67.

For implementing a PSD in hardware, it may be more convenient to correlate $i(t)$ with a square wave with a period of T_c, since this involves simply alternating a gain of +1 or −1 (Lindau and Neher, 1988). Consider the Fourier series representation of a square wave shifted in time by $\alpha T_c/2\pi$:

$$X_s(t) = \frac{4}{\pi}\sum_{l=0}^{\infty}\frac{\cos[(2l + 1)\omega_c t + \alpha]}{2l + 1} \tag{68}$$

If this signal is multiplied by $i(t)$, we get:

$$X_g(t) = \frac{2N}{\pi}\cos(\alpha - \beta) + \frac{2N}{\pi}\cos(2\omega_c t + \alpha + \beta) + \cdots \tag{69}$$

Thus, a value proportional to X_1 can be obtained by low-pass filtering of $X_g(t)$. However, the single-pole low-pass filters commonly found in analogue lock-in amplifiers may not sufficiently attenuate the sine-wave harmonics. If $X_g(t)$ is electronically integrated over the sine-wave cycle (Breckenridge and Almers, 1987), all harmonic contaminants are eliminated. In this case X_1 is given by:

$$X_1 = \frac{\pi}{2T_c}\int_0^{T_c} X_g(t)dt \tag{70}$$

Note that if the spectrum of $i(t)$ has components (such as noise) at frequencies that are odd harmonics of f_c, they will also contribute to the DC value determined by X_1. Therefore, it is best to low-pass-filter $i(t)$ to eliminate frequencies greater than or equal to $3f_c$ when using an analogue lock-in amplifier.

See Lindau and Neher (1988), Gillis (1993), and Zorec *et al.* (1991) for further details on the construction of analogue phase-sensitive detectors.

Appendix B: Derivation of Equation 38

Consider the equivalent circuit of Fig. 6A for the case where the resistor R_Δ is not shorted. The admittance of the three-component equivalent circuit of Fig. 2B (Y_m) is in parallel with the pipette capacitance (admittance Y_p). This network, in turn, is in series with the parallel combination of R_Δ and C_s (admittance Y_Δ). The total admittance is then given by:

$$Y_t = \frac{(Y_m + Y_p)Y_\Delta}{Y_m + Y_p + Y_\Delta} \tag{71}$$

where:

$$\begin{aligned} Y_\Delta &= 1/R_\Delta + j\omega_c C_S \\ Y_p &= j\omega_c C_p \\ Y_m &\simeq \frac{j\omega_c C_m}{1 + j\omega_c R_a C_m} \end{aligned} \tag{72}$$

Here R_m is neglected from the expression for Y_m because in many cases $R_m \gg R_m$. The use of pipette capacitance compensation circuitry subtracts the current through C_p *if the voltage across C_p equals the stimulus voltage.* Thus, the effect of this circuitry can be accounted for by subtracting Y_p from Y_t. The change in total admittance (ΔY) that is induced by closing the switch is found by simply subtracting Y_m (the admittance of the network with Y_Δ shorted) from the total admittance. Thus:

$$\Delta Y = \frac{(Y_m + Y_p)Y_\Delta}{Y_m + Y_p + Y_\Delta} - Y_p - Y_m \tag{73}$$

Algebraic manipulation yields:

$$\Delta Y = \frac{-Y_m^2(1 + Y_p/Y_m)^2}{Y_m + Y_p + Y_\Delta} \tag{74}$$

Ideally, the phase of ΔY should be given by (equation 33):

$$\eta = \angle dY/dR_a = \angle -Y_m^2 \tag{75}$$

Therefore, the phase error (α_{err}) is given by:

$$\alpha_{err} = 2\angle(1 + Y_p/Y_m) - \angle(Y_m + Y_p + Y_\Delta) \tag{76}$$

Substitution and simplification yields:

$$\alpha_{err} = 2 \tan^{-1}\left(\frac{\omega_c R_a C_p}{1 + C_p/C_m}\right) - \tan^{-1}[\omega_c R_\Delta(C_m |T(\omega)|^2 + C_p + C_S)] \quad (77)$$

ACKNOWLEDGMENTS. I am indebted to Erwin Neher, Julio Fernandez, Robert Chow, David Barnett, and Richard Bookman for useful discussions. Robert Chow participated in some of the noise measurements presented in Figs. 8 and 9. Part of this work was supported by NIH grant DK37380 to Stanley Misler during my graduate period in his laboratory. This chapter was written while Kevin Gillis was an Alexander von Humboldt research fellow.

References

Breckenridge, L. J., and Almers, W., 1987, Currents through the fusion pore that forms during exocytosis of a secretory vesicle, *Nature* **328:**814–817.

Chow, R. H., Rüden, L. v., and Neher, E., 1992, Delay in vesicle fusion revealed by electrochemical monitoring of single secretory events in adrenal chromaffin cells, *Nature* **356:**60–63.

Clausen, C., and Fernandez, J. M., 1981, A low-cost method for rapid transfer function measurements with direct application to biological impedance analysis, *Pflügers Arch.* **390:**290–295.

Cole, K. S., 1968, *Membranes, Ions and Impulses,* University of California Press, Berkeley.

Donnelly, D. F., 1994, A novel method for rapid measurement of membrane resistance, capacitance, and access resistance, *Biophys. J.* **66:**873–877.

Douglas, W. W., 1968, Stimulus–secretion coupling: The concept and clues from chromaffin and other cells, *Br. J. Pharmacol.* **34:**451–474.

Fernandez, J. M., Bezanilla, F., and Taylor, R. E., 1982, Distribution and kinetics of membrane dielectric polarization, *J. Gen. Physiol.* **79:**41–67.

Fernandez, J. M., Neher, E., and Gomperts, B. D., 1984, Capacitance measurements reveal stepwise fusion events in degranulating mast cells, *Nature* **312:**453–455.

Fidler, N., and Fernandez, J. M., 1989, Phase tracking: An improved phase detection technique for cell membrane capacitance measurements, *Biophys. J.* **56:**1153–1162.

Gillis, K. D., 1993, Single cell assay of secretion using membrane capacitance measurements: The modulation of calcium-triggered insulin granule exocytosis by cyclic 3′,5′-adenosine monophosphate (cAMP), D.Sc. Dissertation, Washington University, University Microfilms, Ann Arbor, MI.

Herrington, J., and Bookman, R. J., 1993, *Pulse Control v3.0: Igor XOPs for Patch Clamp Data Acquisition,* University of Miami, Miami.

Horrigan, F. T., and Bookman, R. J., 1993, Na channel gating charge movement is responsible for the transient capacitance increase evoked by depolarization in rat adrenal chromaffin cells. *Biophys. J.* **64:** A101.

Joshi, C., and Fernandez, J. M., 1988, Capacitance measurements: An analysis of the phase detector technique used to study exocytosis and endocytosis, *Biophys. J.* **53:**885–892.

Lindau, M., 1991, Time-resolved capacitance measurements: Monitoring exocytosis in single cells, *Q. Rev. Biophys.* **24:**75–101.

Lindau, M., and Neher, E., 1988, Patch-clamp techniques for time-resolved capacitance measurements in single cells, *Pflügers Arch.* **411:**137–146.

Lindau, M., Stuenkel, E. L., and Nordmann, J. J., 1992, Depolarization, intracellular calcium and exocytosis in single vertebrate nerve endings, *Biophys. J.* **61:**19–30.

Neher, E., and Marty, A., 1982, Discrete changes of cell membrane capacitance observed under conditions of enhanced secretion in bovine adrenal chromaffin cells, *Proc. Natl. Acad. Sci. USA* **79:**6712–6716.

Neher, E., and Zucker, R. S., 1993, Multiple calcium-dependent processes related to secretion in bovine chromaffin cells, *Neuron* **10:**21–30.

Oberhauser, A. F., and Fernandez, J. M., 1993, Hydrophobic ions amplify the capacitative currents used to measure exocytosis with the patch clamp technique, *Biophys. J.* **64:**A234.

Okada, Y., Hazama, A., Hashimoto, A., Maruyama, Y., and Kubo, M., 1992, Exocytosis upon osmotic swelling in human epithelial cells, *Biochim. Biophys. Acta* **1107:**201–205.

Rohlicek, V., and Rohlicek, J., 1993, Measurement of membrane capacitance and resistance of single cells two frequencies, *Physiol. Res.* **42:**423–428.

Rohlicek, V., and Schmid, A., 1994, Dual-Frequency method for synchronous measurement of cell capacitance, membrane conductance and access resistance on single cells, *Pflügers Arch.* **428:**30–38.

Sigworth, F. J., Neher, E., and Affolter, H., 1995, Design of a computer-controlled patch clamp amplifier. 2. Internal software, *J. Neurosci. Methods,* in press.

Thomas, P., Lee, A. K., Wong, J. G., and Almers, W., 1994, A triggered mechanism retrieves membrane in seconds after Ca^{2+}-stimulated exocytosis in single pituitary cells, *J. Cell Biol.* **124:**667–675.

Zierler, K., 1992, Simplified method for setting the phase angle for use in capacitance measurements in studies of exocytosis, *Biophys. J.* **63:**854–856.

Zorec, R., and Tester, M., 1993, Rapid pressure driven exocytosis–endocytosis cycle in a single plant cell, *FEBS Lett.* **333:**283–286.

Zorec, R., Henigman, F., Mason, W. T., and Kordas, M., 1991, Electrophysiological study of hormone secretion by single adenohypophyseal cells, *Methods Neurosci.* **4:**194–210.

Chapter 8

Patch-Pipette Recordings from the Soma, Dendrites, and Axon of Neurons in Brain Slices

BERT SAKMANN and GREG STUART

1. Introduction

The brain-slice technique (Yamamoto and McIlwain, 1966; Andersen *et al.*, 1972; Alger *et al.*, 1984) has greatly facilitated the investigation of the electrical properties of neurons and the analysis of synaptic transmission between neurons in the central nervous system (CNS). This is because in brain slices neurons remain healthy, and their connections are preserved to a certain extent while at the same time technical problems encountered in *in vivo* experiments, such as mechanical instability and difficulties in modifying the extracellular environment, are overcome. To combine the brain-slice technique with the power of the patch-clamp technique therefore offers many advantages.

At present there are three main methods available for making patch-pipette recordings from neurons (or glial cells) in brain slices. The "cleaning" method (Edwards *et al.*, 1989) and the "blind" method (Blanton *et al.*, 1989) and their advantages and disadvantages are described in the original publications. More recently a third technique, which is a hybrid of these two previous methods, has been developed (Stuart *et al.*, 1993). We refer to this method as the "blow and seal" technique. The surface membrane of the target cell to be recorded from is cleaned of surrounding neuropile by the application of positive pressure to the recording pipette, similar to the "blind" technique, but the recording pipette is positioned under visual control, as with the "cleaning" technique. This chapter focuses on the "blow and seal" method, which, when combined with infrared differential interference contrast (IR-DIC) video microscopy (Dodt and Zieglgänsberger, 1990; Stuart *et al.*, 1993), permits the recording of membrane potential and currents not only from the relatively large cell body of neurons in the CNS but also from small processes such as dendrites and axons. Furthermore, recordings can be made with two pipettes from different parts of the same neuron, providing the possibility of observing the spread of electrical signals within a single neuron.

2. Preparation of Brain Slices

The general procedures for removing brain tissue for slicing and the storing of brain slices are very similar to those described previously (Alger *et al.*, 1984; Edwards *et al.*,

BERT SAKMANN and GREG STUART • Department of Cell Physiology, Max-Planck Institute for Medical Research, D-69120 Heidelberg, Germany.
Single-Channel Recording, Second Edition, edited by Bert Sakmann and Erwin Neher. Plenum Press, New York, 1995.

1989) and are given here in short form. We recommend the use of a vibrating tissue slicer to obtain brain slices with healthy cells near the surface. Slices up to 500-μm thickness can be used, but for the best visibility of neuronal processes we recommend 200 to 300-μm-thick slices.

2.1. Preparation of Tissue for Slicing

The animal is decapitated, and the appropriate part of the brain or spinal cord is removed. This procedure should not take more than 1 to 1.5 min. Following removal, the tissue is immediately submerged in ice-cold oxygenated saline of the following composition (in mM): 125 NaCl, 25 $NaHCO_3$, 25 glucose, 2.5 mM KCl, 1.25 NaH_2PO_4, 2 $CaCl_2$, 1 $MgCl_2$. Cooling of the tissue is particularly important, presumably as this minimizes damage from anoxia and improves the texture of the tissue for slicing. If, as with large animals (e.g., more than 2-month-old rats), more time is required for removal of the brain, ice-cold physiological saline can be poured over the brain as soon as the skull is open.

2.2. Slicing

A vibrating tissue slicer (e.g., Vibracut, FTB, Weinheim, Germany; Vibroslicer, Campden Instruments, Sileby, England) is used to cut slices. Mechanical stability of the tissue during slicing is essential. For this purpose a larger block of tissue containing the region of interest is cut by hand. A surface of this block, trimmed parallel to the desired orientation of the slices, is then glued to the stage of the slicer (e.g., for parasagittal slices the brain can simply be glued onto the midline). Firm, instant attachment can be achieved by using a thin film of cyanoacrylate glue. The slicing chamber is then immediately filled with ice-cold physiological saline. The slicing chamber is routinely precooled in a freezer to −20°C. Tissues that are too small to be glued directly to the stage of the slicer (e.g., newborn rat spinal cord) can first be embedded in agar (Takahashi, 1978). The agar (2% dissolved in physiological saline) is cooled to below 40°C before use. Careful application of ice-cold physiological saline then facilitates cooling and solidifying of the agar. A block of agar, cut to contain the tissue at the correct orientation, can then be glued to the stage of the slicer and immersed in ice-cold physiological saline as described above.

After the brain is sliced down roughly to the required level, at least one slice must be discarded before slices of a uniform thickness are obtained. The slicing procedure is monitored continuously, and the forward speed of slicing is adjusted so that the tissue is never pushed by the blade. Care must be taken to reduce the damage to neurons and neuronal processes close to the surface of the slice during the slicing procedure, as the possibility of seeing fine details of neurons decreases with increasing distance from the slice surface and is limited to the first 40 to 50 μm from the surface of the slice. Less damage appears to occur to neurons and neuronal processes close to the surface of slices from young animals (2 to 3 weeks old), possibly because of the relative lack of connective tissue and myelination in the brains of these animals. "Damaged" cells, when present, can be identified easily by their rough, crinkled appearance and high-contrast membrane not easily dimpled by the approaching patch pipette tip. "Healthy" cells, on the other hand, look "smooth" and have a soft membrane that can easily be dimpled by the patch pipette tip.

2.3. Incubation of Slices

After sectioning, each slice is immediately placed in a holding chamber containing oxygenated physiological saline (see above), which is placed in a water bath at a temperature of 37°C for 30 to 60 min and are then stored at room temperature until required. In order to ensure efficient oxygenation and continuous movement of the solution around the slices, a submerged holding chamber has been designed in which the slices are bubbled from below. The holding chamber, which is made from simple disposable parts (120-ml glass beaker, plastic petri dish with the bottom removed and replaced by a gauze net, a 5-ml plastic syringe with the end cut off, and two plastic Pasteur pipettes) that are easily cleaned or replaced and can hold up to about 12 slices, is illustrated in Fig. 1A. Slices are transferred into and out

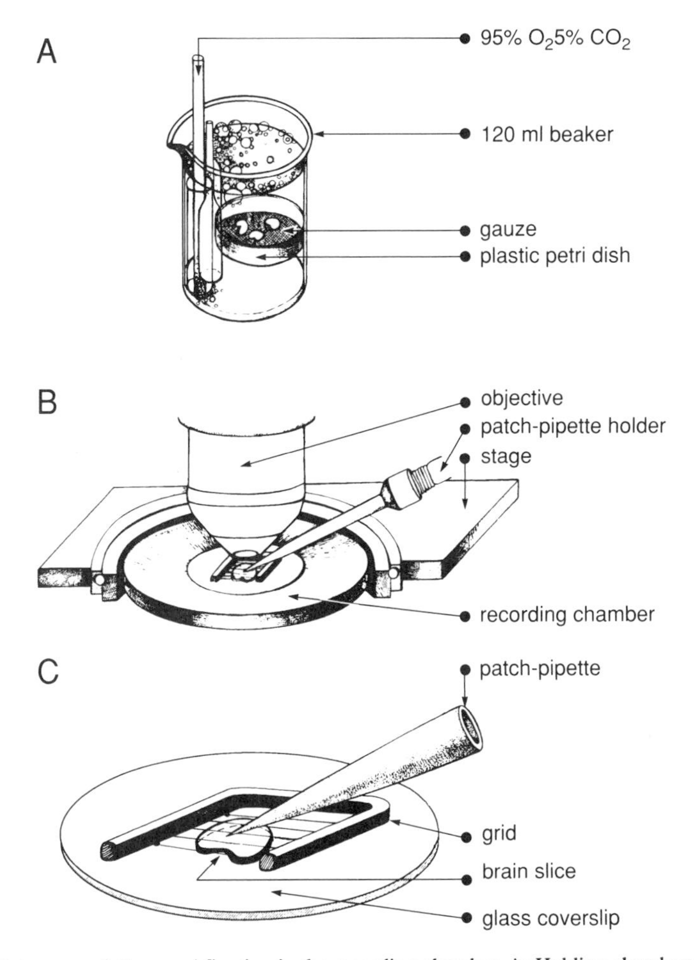

Figure 1. Maintenance of slices and fixation in the recording chamber. A: Holding chamber used for storing brain slices during incubation in a water bath at 37°C and for storage prior to transfer to the recording chamber. This chamber is made from a glass beaker filled with physiological saline and bubbled with 95% O_2/5% CO_2. The slices rest on a gauze net covering the bottom of a plastic pertri dish from which the bottom has been removed. B: Schematic diagram of the circular, rotatable recording chamber with brain slice, grid, and patch pipette as seen during recording. C: Fixation of a brain slice in recording chamber by a grid of nylon threads glued to a U-shaped platinum frame.

of the holding chamber using the large end of a Pasteur pipette. The condition of the tissue is optimal over the first 3 or 4 hr, however, stable recordings can still be obtained 10 to 12 hr after slicing.

2.4. Mechanical Fixation of Brain Slices for Recording

One slice is placed into a circular, glass bottomed recording chamber made from an acrylic plastic ring glued to a circular piece of microscope slide glass or a large coverslip (Fig. 1B). The chamber holds a volume of about 1 ml and during recording is perfused with oxygenated saline (see Section 2.1) at a flow rate of 1 to 2 ml/min. The slice is held in place with a grid of parallel nylon threads. The U-shaped frame of the grid is made from 0.5-mm platinum wire flattened with a vice (Konnerth *et al.,* 1987). Fine nylon stockings provide a convenient material for making the threads. A hole is made in the stocking so that a "ladder" forms. This results in a parallel array of compound threads that can then be tightly stretched over a ring (about 2 cm diameter) and clamped in place. Under a dissecting microscope one of these compound threads is then separated into individual fine fibers, which are arranged on the ring to form parallel threads about 0.4 mm apart. The platinum frame, having been coated with a very thin film of cyanoacrylate glue, is then balanced across these parallel threads, where it is left until the glue dries. Such a grid, placed over the slice, holds it firmly in position on the bottom of the recording chamber (Fig. 1C). The dimensions of the frame and the distance between the fibres can be varied according to the size of the preparation and recording chamber. A frame of 7 × 9 mm is suitable for hippocampal, neocortical, and cerebellar slices.

3. Visualization of Nerve Cells Using IR-DIC

3.1. Optical Setup

Brain slices are viewed with an upright compound microscope using a high-numerical-aperture (NA) water immersion objective, corresponding differential interference contrast (DIC) optics, and a high-NA condenser. We use a Zeiss Axioskop fitted with either a 40× 0.75-NA or 63× 0.9-NA water immersion objectives (Achroplan, Zeiss, Germany) with corresponding DIC optics and a 0.9-NA condenser (Zeiss). Unless manipulators for recording or stimulating pipettes are mounted directly onto the microscope stage, a fixed-stage microscope (e.g., Zeiss Axioskop FS) is best, as the relative positions of the slice and recording or stimulating pipettes do not change when the microscope focus is changed. In addition, this microscope can be mounted on a sliding table so that it can be moved away from the preparation, facilitating the changing of pipettes and access to the preparation (see Stuart *et al.,* 1993).

The optical pathway used for IR-DIC video microscopy is shown schematically in Fig. 2. An infrared (IR) filter (λ_{max} = 780 nm, RG9, Schott, Germany) is inserted in the light path prior to the DIC polarizer (the infrared filter can be moved in and out of the light path as required). Illumination is by a tungsten light source in conjunction with a 100-W, 12-V power supply. Infrared illumination necessitates the use of an IR-sensitive video camera and observation of pipette movements and neurons on a video monitor. The video camera that offers the best resolution is a Newvicon C2400 07-C (Hamamatsu, Japan). This camera

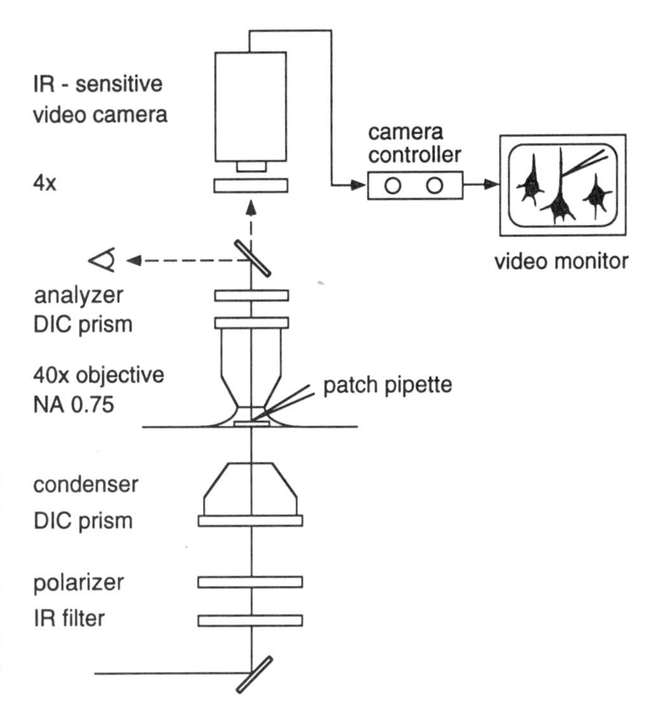

Figure 2. Optical setup for IR-DIC video microscopy. Schematic drawing of the optical setup used for electrophysiological recording from brain slices using infrared differential interference contrast video microscopy. Slices are illuminated with infrared (IR) light by placement of an IR filter (λ_{max} = 780 nm) in the light path before the polarizer. Neurons are viewed with a Zeiss 40× or 63× water immersion lens (numerical aperture of 0.75 and 0.9, respectively) and corresponding differential interference contrast (DIC) optics (two DIC prisms, one polarizer and one analyzer). The image is further magnified 4× and detected by an infrared-sensitive video camera (Newvicon C2400-07-C, Hamamatsu, Japan) and displayed on a standard black-and-white video monitor via the camera controller.

comes with a camera controller that provides analogue contrast enhancement and shading correction. Less expensive video cameras also sensitive in the IR range can also be used. For patch-pipette recording from small processes, the image is further magnified either 2.5 or 4 times prior to detection by the IR-sensitive video camera. To switch between IR-DIC video microscopy and, for example, fluorescence microscopy, we use a binocular phototube fitted with two C-mount adapters for TV cameras (e.g., binocular phototube 25 with two outputs, Zeiss). This allows observation with either IR light or detection of epifluorescence by, for example, a CCD camera (e.g., CH 250 A, Photometrics, Tucson, AZ).

3.2. Mechanical Setup

When a fixed-stage microscope is used, micromanipulators for holding recording and stimulating pipettes are either attached to columns (e.g., X95 Profile, Spindler und Hoyer, Göttingen, Germany) that are themselves mounted to an antivibration table (Physik Instrumente, Waldbronn, Germany) or can be mounted directly to the antivibration table. The circular recording chamber is placed into a circular recess on a fixed stage that is independent of the microscope (see Stuart *et al.*, 1993). The microscope rests on an X–Y table that slides on ball-bearing railings (SKF, Schweinfurt, Germany), allowing the field of view to be changed without changing the position of the slice or the recording or stimulating pipettes. This X–Y table can be moved mechanically or can be driven by programmable stepping motors (Physik Instrumente). If a fixed-stage microscope cannot be used (e.g., when recordings are made in combination with a confocal laser scanning microscope), micromanipulators are then attached directly to the microscope stage. When recordings are made from dendrites or axons, it is advantageous if the recording chamber is free to rotate so that dendrites or axons can be aligned at right angles to the advancing patch pipette. Experiments are usually done at room temperature but can be performed at higher or lower temperatures by simply heating or cooling the saline solution. This can be achieved by encasing 30 or 40 cm of the perfusion tubing in a heated or cooled water jacket just prior to its entry into the recording chamber.

3.3. Viewing with IR-DIC

The procedure used for viewing slices with IR-DIC video microscopy is as follows. Initially, the slice is illuminated with visible light and viewed through the oculars. Most critical for high resolution is exact Köhler illumination of the object of interest. The polarizer should be adjusted to a position where the image appears darkest and the DIC prism above the objective fully offset. Once the area of interest has been selected and Köhler illumination achieved, the IR filter is moved into position, the light path switched to the IR-sensitive video camera, and the image viewed on the video monitor. Best resolution is obtained with the condenser diaphragm fully open. Images can be printed immediately using a thermal video printer (P68E, Mitsubishi, Japan) or can be stored on video tape or on computer using a frame grabber.

4. The "Blow and Seal" Technique

4.1. Patch-Pipette Techniques

The procedure for the formation of high-resistance (>5 GΩ) seals onto the soma or processes of neurons in brain slices is illustrated in Fig. 3. Positive pressure is continually applied to the patch pipette as it is advanced through the slice under visual control at an angle that depends on the objective used. This angle is approximately 20° to the horizontal for the 40× objective (Zeiss; working distance 1.9 mm) and 15° for the 63× objective (Zeiss; working distance 1.3 mm). A wave of pressure can be seen as the pipette tip is advanced slowly through the slice (Fig. 3A). The amount of positive pressure applied depends on the size of the pipette used and is smaller for larger patch pipettes. Typically, pressures between 20 to 150 mbar are used for pipettes of between 2 and 10 MΩ resistance. Pressure is applied by mouth and can be monitored by a pressure sensor (104 PC, Honeywell, Seligenstatt, Germany). The desired amount of pressure can be maintained simply by closing a valve. Once the pipette tip touches the membrane of the structure or process to be patched (as seen by the dimpling of the membrane or the small displacement of the structure or process by the tip of the advancing patch pipette), the pressure is immediately released, and slight negative pressure (20 to 50 mbar) is applied by mouth until a high-resistance seal is established (Fig. 3B). The release of positive pressure should be associated with an approximate 1.5 to 2-fold increase in tip resistance. Hyperpolarization of the patch pipette by 60 mV is routinely used to aid seal formation. The success rate for formation of high-GΩ seals with this method can be as high as 100% when recordings are made from large structures such as the soma but is lower for recordings from smaller structures. It is possible, however, to use this technique to record regularly from neuronal structures in brain slices as small as 1 to 2 μm in diameter.

Rupture of the patch membrane to obtain a whole-cell recording configuration is made

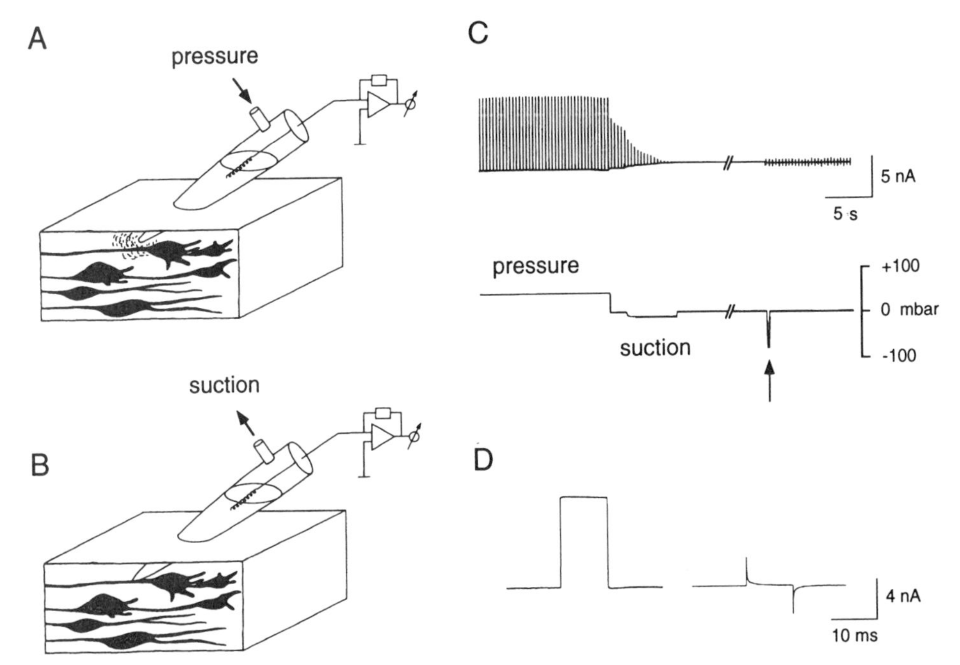

Figure 3. Schematic drawing of "blow and seal" method of recording from neurons in brain slices together with the pipette current and the applied pressure and suction during gigohm seal formation. A: Pipette tip entering the slice at shallow angle during the application of pressure to the pipette. This generates a pressure wave of pipette solution in front of the pipette tip, which can be seen to clean the surface of the neuronal membrane. B: Pipette tip sealing to a neuronal cell dendrite during the application of suction to the recording pipette. C: Recording of pipette current (top) in response to 10-mV test pulses and pressure (bottom) applied to the recording pipette during the approach of the patch pipette to the neuron and following release of positive pressure and during application of suction to form a high-gigohm seal with the somatic membrane of a layer V pyramidal cell. The pipette capacitance was then compensated at high current gain (not shown), and the membrane patch was ruptured by application of a short (100-msec) pulse of suction (arrow). The fast capacitive transients observed in response to the test pulse following rupture of the membrane patch are attenuated. D: Pipette current in response to a 10-mV test pulse before seal formation (left) and after rupture of the membrane patch to obtain the whole-cell recording configuration (right). The open pipette resistance was 1.1 MΩ, and the whole-cell access resistance was estimated to be 3.5 MΩ. C and D are from the same experiment.

by the application of brief pulses of suction to the patch pipette, either by mouth or generated by a miniature vacuum pump (G/028, Brey, Memmingen, Germany). The time course of pressure application and changes in the patch pipette response during the formation of a somatic whole-cell recording are illustrated in Fig. 3C. The whole-cell recording configuration was obtained by rupture of the patch membrane by a brief (100-msec) pulse of suction (arrow). The pipette current in response to a 10-mV test pulse before and after formation of a whole-cell recording is shown in Fig. 3D. The open pipette tip resistance and estimated whole-cell access resistance in this example were 1.1 and 3.5 MΩ, respectively.

To obtain patch-pipette recordings from small neuronal processes, the procedure is identical; however, it requires the use of patch pipettes with smaller tip openings (~10 MΩ open tip resistance) and higher positive pressures as the pipette is advanced through the slice (up to 150 mbar). The use of patch pipettes with higher tip resistances leads consequently to higher access resistances during whole-cell recording. Access resistances during dendritic whole-cell recording are typically around 40 to 50 MΩ (range 20 to more than 100 MΩ, depending on the initial patch pipette resistance).

Examples of recordings from the soma or dendrites of pyramidal cells in the neocortex are shown in Fig. 4. In Fig. 4A,C, the pressure wave from the recording pipette can be seen to "clean" the somatic or dendritic neuronal membrane just prior to the formation of a high GΩ seal to the soma or dendrite (Fig. 4B,D).

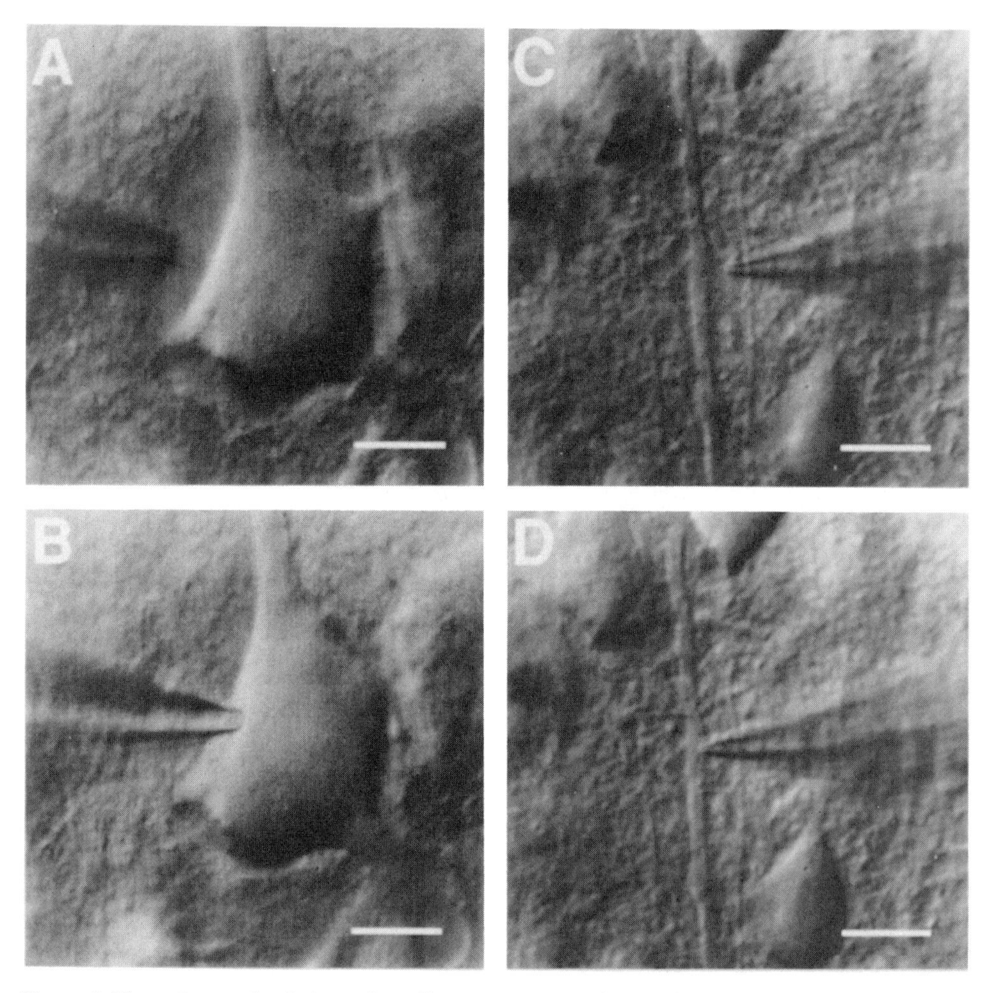

Figure 4. Photomicrographs of pipette tip sealing onto soma or apical dendrite of layer V pyramidal neurons in rat neocortex brain slices. Scale bar 10 μm in all pictures. A: Positive pressure applied to the patch pipette (approximately 50 mbar) can be seen to "clean" the surface of the somatic membrane of this neuron. The cell body of this neuron is approximately 20 μm from the surface of the slice. B: Release of positive pressure and application of slight negative pressure results in gigohm seal formation between the tip of pipette and the soma membrane. C: Positive pressure applied to the patch pipette is seen to "clean" the surface of the dendrite from neuropile near the pipette tip (290 μm from the soma). D: Release of positive pressure and application of slight suction leads to sealing of the pipette tip to the dendrite membrane.

4.2. Verification of Dendritic or Axonal Recordings

Verification that patch-pipette recordings are in fact made from either the dendritic or axonal membrane of a particular neuron can be confirmed by the rapid diffusion of the fluorescent dye Lucifer yellow (Sigma) from the patch pipette into the neuron from which the recording is made following rupture of the patch membrane by brief pulses of suction (Fig. 5).

4.3. Problems Associated with Recording from Small Neuronal Structures

Technical problems associated with patch-pipette recordings from small neuronal structures include the need for mechanically very stable recording conditions and the ability to

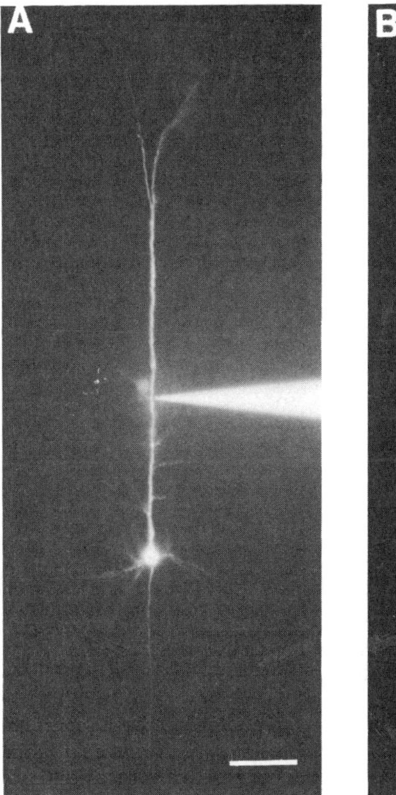

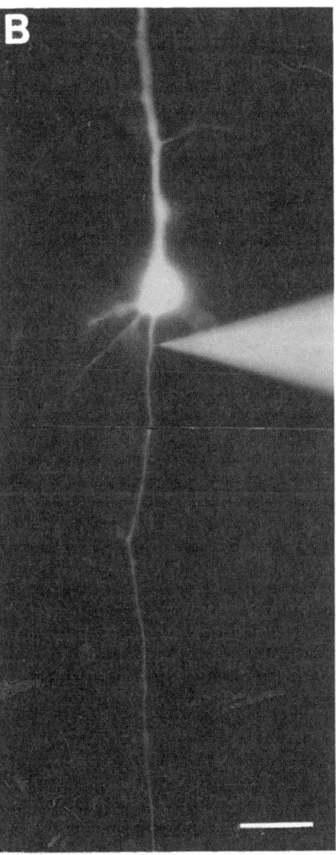

Figure 5. Verification of dendritic and axonal recordings by filling of layer V pyramidal neurons with the fluorescent dye Lucifer yellow from the recording pipette. A: Fluorescence photomicrograph of a layer V pyramidal cell following intracellular filling with the fluorescent dye Lucifer yellow via a dendritic pipette located 230 μm from the soma of this neuron. Scale bar 100 μm B: Fluorescence microphotograph of a layer V pyramidal cell following intracellular filling with Lucifer yellow via an axonal pipette located approximately 15 μm from the edge of the soma of this neuron. Scale bar, 40 μm. [For color figure see insert following the table of contents.]

make very fine movements of the patch pipette. For this we recommend a high-quality micromanipulator (Märzhäuser, Wetzlar, Germany; Luigs & Neumann, Ratingen, Germany; Newport, Darmstadt, Germany). In addition, although low-access-resistance whole-cell recordings can readily be obtained from the soma of neurons using large patch pipettes (1–2 MΩ open tip resistances), whole-cell recording from small structures requires the use of patch pipettes with small tip diameters, which leads consequently to higher access resistances during whole-cell recording. Access resistances during dendritic whole-cell recording are typically around 40 to 50 MΩ (range 20 to more than 100 MΩ). While this should not cause problems for excised patch-clamp recordings, it may slow the rate with which the membrane capacitance is charged and hence the speed of the voltage clamp during whole-cell recordings (see Chapter 2, this volume). In current clamp, the combination of the access resistance and the patch pipette capacitance acts as a low-pass filter. With high-access-resistance (i.e., $>\sim 20$ MΩ) whole-cell recordings this can cause significant filtering of the voltage signal. Under these conditions it is recommended that an amplifier with electrode capacitance neutralization in the current-clamp mode is used (e.g., Axoclamp 2A, Axon Instruments, Foster City, CA).

5. Examples of Recordings from Dendrites

Some examples are given of dendritic whole-cell recordings and of recordings from cell-attached and excised dendritic membrane patches from the apical dendrites of layer V pyramidal neurons in rat neocortical brain slices.

5.1. Cell-Attached Patch Recordings

Cell-attached recordings from the apical dendritic membrane of layer V pyramidal cells in the rat neocortex revealed that the dendritic membrane of these neurons contains both inward and outward voltage-activated currents. These currents were evoked following step depolarizations from a holding potential negative to the resting membrane potential (Fig. 6A).

5.2. Whole-Cell Recordings

Following obtaining a cell-attached recording, access into the apical dendrite of a layer V pyramidal neuron could be obtained by rupture of the dendritic membrane patch by brief pulses of suction. In current-clamp, action potentials could be evoked by the application of positive current pulses (Fig. 6B). These action potentials have recently been shown to originate near the soma and then to actively "back-propagate" into the dendrites (Stuart and Sakmann, 1994).

5.3. Outside-Out Patch Recordings

Dendritic outside-out patches can easily be obtained by withdrawing the patch pipette from the slice after establishing a dendritic whole-cell recording. In this configuration, single-channel currents could be activated by the application of glutamate (1 μM in the presence of 10 μM CNQX) to the dendritic membrane patch (Fig. 6C). These currents had a main

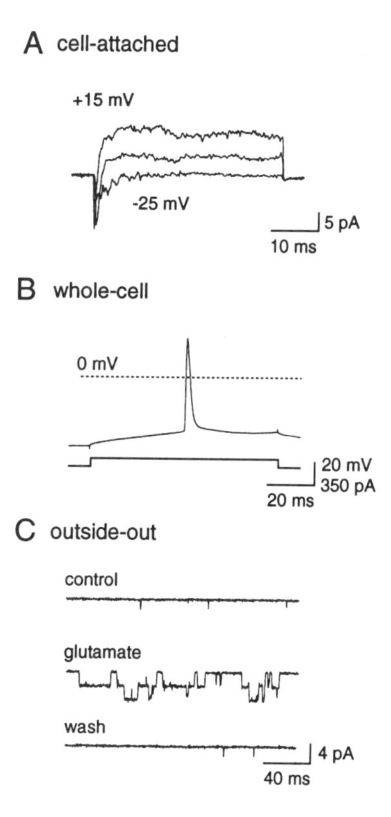

Figure 6. Recordings of electrical activity from apical dendrites of rat layer V neocortical pyramidal cells. A: Cell-attached patch recording. Activation of inward Na^+ and outward K^+ currents during the application of depolarizing voltage steps to a dendritic cell-attached patch 113 μm from the soma. Depolarizing voltage pulses to membrane potentials of approximately −25 mV, −5 mV, and +15 mV (assuming a resting membrane potential of −65 mV) were applied from a holding potential of approximately −115 mV. Leak and capacitive currents were subtracted on line, and each trace represents the average of 5 to 10 sweeps. B: Whole-cell recording. Dendritic action potential elicited by a 130-pA current step applied during a dendritic whole-cell recording 148 μm from the soma. Resting membrane potential −62 mV, voltage was recorded with a microelectrode amplifier (Axoclamp 2A, Axon Instruments, Foster City, CA) using electrode capacitance neutralization and bridge balance. C: Outside-out patch recording. Activation of single-channel currents by the application of glutamate (1 μM) to a dendritic outside-out patch isolated 112 μm from the soma. The patch pipette was filled with a K-gluconate-based intracellular solution and held at −60 mV. The three traces were taken before (control), during (glutamate), and after (wash) application of glutamate-containing extracellular solution. All solutions contained 10 μM CNQX plus 1 μM glycine and were nominally Mg^{2+} free. Downward deflections represent channel openings.

open-state conductance of approximately 50 pS and presumably represent the activation of dendritic NMDA receptor channels. Dendritic outside-out patches can also be exposed to agonists using fast application techniques (see Chapter 10, this volume; Spruston *et al.*, 1995), where agonists are applied in a fashion that is intended to mimic the rapid release of transmitter that occurs during synaptic transmission.

5.4. Two-Pipette Recordings

As both the soma and the dendrite, or axon, of the same neuron can be visually identified, it is possible to make simultaneous recordings from different parts of the same neuron. Figure 7A illustrates one such recording where simultaneous recordings were made from the soma and the dendrite of the same cell. Filling of the same pyramidal neuron with two different fluorescent dyes from the soma and the dendrite was used to verify that the recordings were made from the same cell. Following extracellular stimulation of distal afferent inputs, EPSPs can be recorded with both pipettes (Fig. 7B). The subthreshold EPSP recorded by the dendritic pipette occurs earlier and is faster and larger than the EPSP recorded by the somatic pipette, as expected if the activated synapses are located close to the dendritic recording pipette. When the stimulus is suprathreshold, the EPSP recorded in the soma, although smaller in amplitude than that recorded in the dendrite, elicits an action potential that is observed first at the soma and recorded by the dendritic pipette with a clear delay (Fig. 7C). These

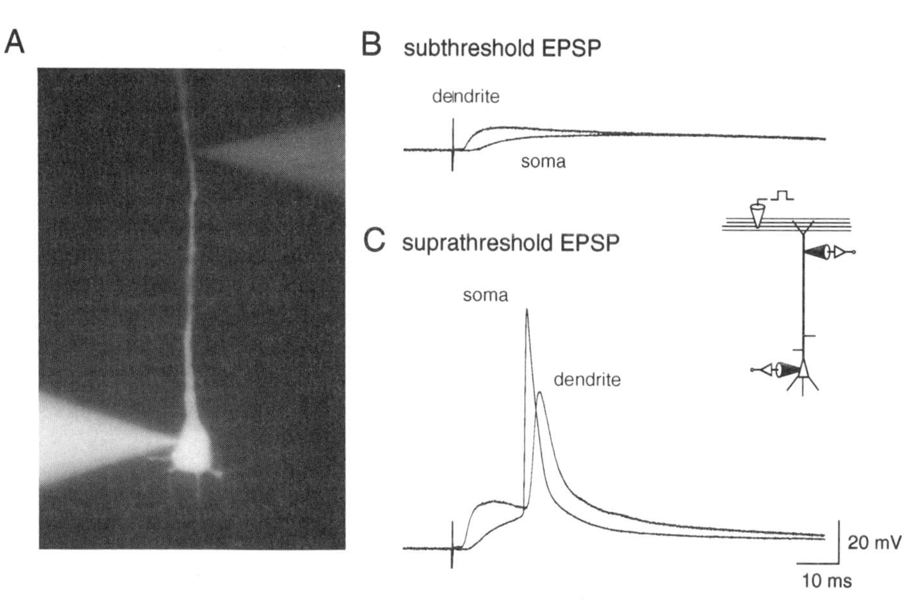

Figure 7. Simultaneous recording with two pipettes from the soma and dendrite of the same layer V pyramidal neuron in a rat neocortical slice. A: Simultaneous filling of the same layer V pyramidal neuron from the dendrite and the soma with different colored fluorescent dyes, Cascade blue at the soma and Lucifer yellow in the dendrite. The dendritic recording was made 190 μm from the soma. Scale bar is 40 μm. B: Subthreshold EPSPs recorded simultaneously from the dendrite and the soma of the same layer V pyramidal neuron following extracellular electrical stimulation in layer I (stimulus artifact precedes EPSPs). Dendritic recording 525 μm from the soma. C: Suprathreshold stimulation in layer I evokes an EPSP followed by action potential. Same experiment as in B. The action potential initiated by this EPSP occurs first at the soma. Simultaneous whole-cell voltage recordings were made using two microelectrode amplifiers (Axoclamp 2A, Axon Instruments, Foster City, CA). Inset represents schematic diagram of experimental arrangement. Calibration is the same for B and C. [For color figure see insert following the table of contents.]

experiments have been used to identify the site of action potential initiation in layer V pyramidal neurons (Stuart and Sakmann, 1994).

6. Discussion and Conclusions

The techniques described in this chapter for making patch-pipette recordings from neurons and their processes in brain slices offer several advantages over previously described methods. The method is easy to implement and causes little damage to cells and their processes (or connections) during seal formation. When combined with IR-DIC video microscopy, the improved resolution offers the possibility of recording from small cellular structures and processes such as dendrites or axons of neurons in the mammalian central nervous system *in vitro.* This increase in resolution also enables the possibility to record from more than one location in the same cell.

Through the application of IR-DIC video microscopy and patch clamp techniques to record from different regions of the neuronal membrane, the electrophysiological properties of different channel subtypes in these different regions can be directly assessed. In addition, the ability to identify and record from less common cell types is greatly facilitated by the increased resolution offered by IR-DIC video microscopy. This increased resolution has also proved useful for single-cell PCR experiments from identified neuronal types in brain slices (see Chapter 16, this volume), which requires careful visual monitoring of the harvesting of cytoplasm.

In summary, this chapter describes techniques for making electrical measurements using patch pipettes from visually identified neurons or neuronal processes in brain slices. Compared with other techniques, when combined with IR-DIC video microscopy, the "blow and seal" method offers many advantages, such as the ease with which patch pipette recordings can be made, the possibility of identifying the neuronal cell type prior to recording, and, finally, the ability to visualize and record electrical activity from different neuronal compartments or from more than one site in the same neuron.

References

Alger, B. E., Dhanjal, S. S., Dingledine, R., Garthwaite, J., Herderson, G., King, G. L., Lipton, P., North, A., Schwartzkroin, P. A., Sears, T. A., Segal, M., Whittingham, T. S., and Williams, J., 1984, Brain slice methods, in: *Brain Slices* (R. Dingledine, ed.), pp. 381–437, Plenum Press, New York.

Andersen, P., Bland, B., Skrede, K., Sveen, O., and Westgaard, R., 1972, Single Unit dischange in brain slices maintained *in vitro, Acta Physiol. Scand.,* **84:**1–2a.

Blanton, M. G., Lo Turco, J. J., and Kriegstein, A. R., 1989, Whole cell recording from neurons in slices of reptilian and mammalian cerebral cortex, *J. Neurosci. Methods* **30:**203–210.

Dodt, H.-U., and Zieglgänsberger, W., 1990, Visualizing unstained neurons in living brain slices by infrared DIC-videomicroscopy, *Brain Res.* **537:**333–336

Edwards, F. A., Konnerth, A., Sakmann, B., and Takahashi, T., 1989, A thin slice preparation for patch clamp recordings from neurons of the mammalian central nervous system. *Pflügers Arch.* **414:**600–612.

Konnerth, A., Obaid, A. L., and Salzberg, B. M., 1987, Optical recording of electrical activity from parallel fibres and other cell types in skate cerebellar slices *in vitro, J. Physiol.* **393:**681–702.

Spruston, N., Jonas, P., and Sakmann, B., 1995, Dendritic glutamate receptor channels in rat hippocampal CA3 and CA1 pyramidal neurons. *J. Physiol.* **482:**325–352.

Stuart, G. J., and Sakmann, B., 1994, Active propagation of somatic action potentials into neocortical pyramidal cell dendrites, *Nature* **367:**69–72.

Stuart, G. J., Dodt, H.-U., and Sakmann, B., 1993, Patch clamp recordings from the soma and dendrites of neurons in brain slices using infrared video microscopy, *Pflügers Arch.* **423:**511–518.

Takahashi, T., 1978, Intracellular recording from visually identified motoneurons in rat spinal cord slices, *Proc R Soc Lond* [*Biol*] **202:**417–421.

Yamamoto, C., and McIlwain, H., 1966, Electrical activities in thin sections from the mammalian brain maintained in chemically-defined media *in vitro, J. Neurochem.* **13:**1333–1343.

Chapter 9

Patch Clamp and Calcium Imaging in Brain Slices

JENS EILERS, RALF SCHNEGGENBURGER, and
ARTHUR KONNERTH

1. Introduction

The aim of this chapter is to describe the requirements for combining the patch-clamp technique with the fluorometric monitoring of changes in ion concentration in single cells in brain slice preparations (Llano *et al.*, 1991; Kano *et al.*, 1992; Konnerth *et al.*, 1992; Schneggenburger *et al.*, 1993a,b). Additional details of the patch-clamp technique in brain slices and of the different procedures for fluorometric ion measurements are given in other chapters (Chapters 8 and 16) of this volume and elsewhere (Smith *et al.*, 1983; Tsien and Poenie, 1986; Neher, 1989).

The use of these techniques for neurons in brain slices is generally somewhat more difficult than their use for cultured or acutely isolated dispersed cells for the following reasons. (1) In slices the cell body and the dendrites of a neuron are usually not located in the same plane of focus. Therefore, out-of-focus light will often cause a blurring of the images and will thus complicate the quantitative estimation of the changes in Ca^{2+} concentration. (2) The usual patch-clamp procedure in brain slices of approaching cells by ejecting a strong jet of pipette solution (to avoid clogging of the patch pipette) bears the risk of an excessive release of indicator dye into the tissue surrounding the cell. (3) There is in general a high level of autofluorescence resulting mostly from damaged cells at the surface of the slice. Moreover, this autofluorescence is not constant throughout an experiment (often lasting more than 60 min) but, like the fluorometric indicator dye itself, undergoes a gradual "bleaching" if the ultraviolet (UV) excitation light is too intensive.

2. Setup

The original method for patch-clamp recordings from identified neurons in brain slices involved the use of upright microscopes equipped with long-working-distance water immersion objectives (Edwards *et al.*, 1989). With this approach it is possible to visualize directly neurons within the top layers of the slice and, if necessary, to "clean" the neurons that will be patched (see also Chapter 8, this volume). In addition, this approach makes it quite easy to perform fluorometric measurements such as Ca^{2+} imaging simultaneously with patch-

JENS EILERS, RALF SCHNEGGENBURGER, and ARTHUR KONNERTH • I. Physiological Institute, University of Saarland, D-66421 Homburg/Saar, Germany. *Current address for R.S.:* Neurobiology Laboratory, Teacher's Training College, 46 Rue d'Ulm, F-75005 Paris, France.
Single-Channel Recording, Second Edition, edited by Bert Sakmann and Erwin Neher. Plenum Press, New York, 1995.

clamp recordings. Here we describe two versions of experimental setups involving upright microscopes, one with a fixed stage and a movable objective (Fig. 1) and a second system with the conventional fixed objective and movable stage (Fig. 2).

In order to be able to manipulate one or more recording and stimulation pipettes in these setups it is convenient to use objectives with a working distance longer than about 0.8 mm (see Fig. 1 of Konnerth, 1990). Microscope objectives fulfilling this requirement are, of course, limited in some of their optical properties, including the numerical aperture (NA) and the capacity to pass UV light. Despite these limitations, there are several commercially available water immersion objectives (e.g., from Zeiss, Olympus, Nikon) that give good results. One explanation for the high rate of successful recordings obtained with this approach is that neurons close to the top layer of the slice preparation can be directly visualized, identified, and carefully selected for recording according to the orientation of their dendrites within the plane of focus of interest. Approaches involving the "blind" patch approach (Blanton *et al.,* 1989) have in principle the advantage of the use of inverted microscopes with a large selection of high-quality objectives, but even for these recordings the number of suitable objectives is often limited by the need for a working distance of more than 0.3–0.4 mm.

2.1. Single-Photomultiplier Detector System

This is the simplest system for fluorometric monitoring of changes in ion concentration in combination with patch-clamp recordings. We describe here a Ca^{2+}-detection system based on the most widely used fluorometric indicator dye, fura-2 (Grynkiewicz *et al.,* 1985). At least two modifications of the standard setup for patch-clamp recordings in slices (Edwards *et al.,* 1989; Konnerth, 1990) are necessary for this approach (Fig. 1). The first requirement is a photomultiplier tube (PMT) that can easily be attached to the standard binocular tubes of most types of commercially available microscopes. A second, perhaps not always necessary but very useful, modification consists in separating the illumination unit for the UV excitation light from the body of the microscope to avoid the vibrations generated by the rotating filter wheel of the dual-wavelength excitation system. Guiding the UV light through appropriate flexible optical fibers (available, for example, from T.I.L.L. Photonics, Munich, Germany) provides sufficiently intensive UV light for most applications.

The filter-wheel-based illumination unit was designed to provide alternating excitation light at two wavelengths (Neher, 1989) in order to perform ratiometric Ca^{2+} measurements with the indicator dye fura-2 (Grynkiewicz *et al.,* 1985). In our system the excitation light is provided by a xenon lamp (Osram XBO 75W/2) and a rotating filter wheel (Luigs & Neumann, Ratingen, Germany) that can be equipped with two filter sets selecting two different wavelengths for excitation (e.g., about 360 nm and 380 nm when using the indicator dye fura-2; see Grynkiewicz *et al.,* 1985). A flexible optical fiber feeds the excitation light into the epifluorescence pathway of the microscope (Fig. 1). A diaphragm mounted in the small optical bench connecting the fiber optics to the microscope allows a precise illumination of selected regions of the recording field.

An additional option that can be provided in such an experimental setup is a UV flash system used for the release of "caged" compounds either intracellularly, such as "caged calcium" (Kao *et al.,* 1989), "caged ATP," or "caged IP_3" (Kaplan and Somlyo, 1989), or extracellularly, such as "caged glutamate" (Callaway and Katz, 1993). As a UV flash system we have successfully used either a high-energy xenon arc flash lamp (Gert Rapp Optoelektronik, Hamburg, Germany) or an illumination system consisting of a conventional mercury

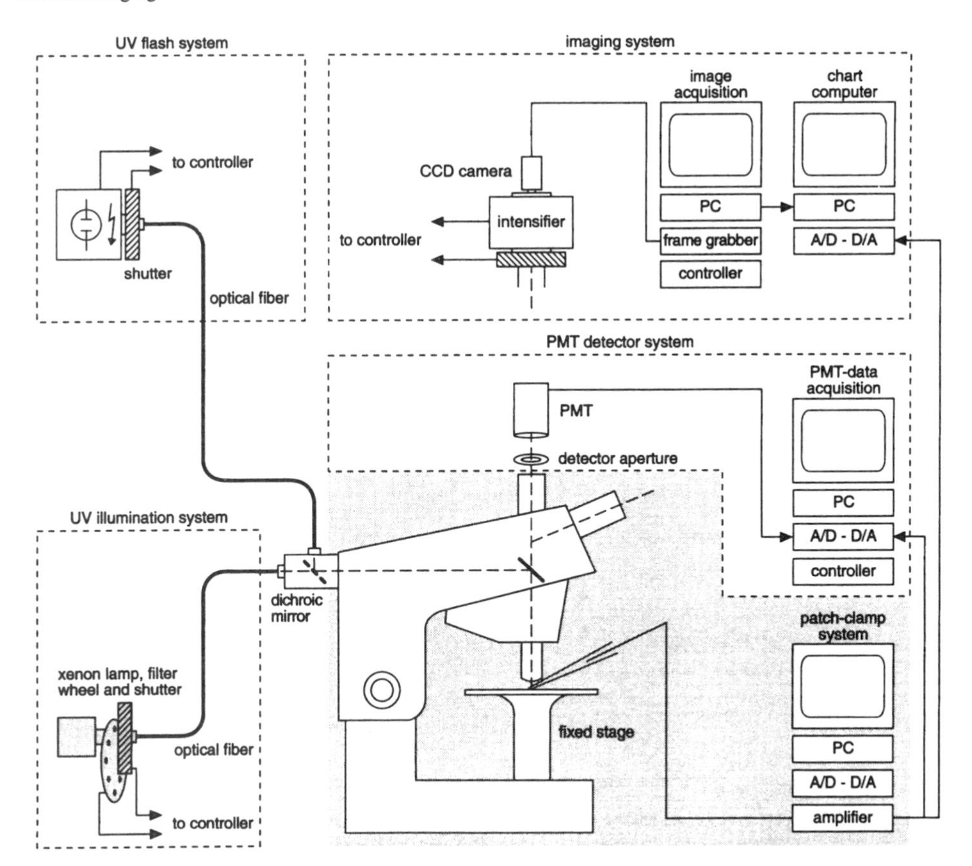

Figure 1. Schematic diagram of the experimental setup used for simultaneous patch-clamp recordings and fluorometric Ca^{2+} measurements in neurons from brain-slice preparations. The diagram within the shaded region represents a standard setup for patch clamp in brain slices based on an upright microscope with a fixed stage (Edwards *et al.*, 1989) and a computer-controlled system for the acquisition and on-line display of data. For fluorometric monitoring of $[Ca^{2+}]_i$, either a photomultiplier tube (PMT)-based system ("PMT detector system") or, alternatively, an imaging system may be used. Each of these systems consists of a fluorescence detector (PMT or intensified CCD camera, respectively), a computer-based device for display, acquisition, and storage of data, and a "controller" unit for the UV excitation. Note that the imaging system includes a "chart computer" that is used for on-line display and storage of patch-clamp and fluorometric recordings (see Fig. 5). The UV illumination unit is equipped with a xenon lamp, a filter changer wheel, and a shutter. The (optional) UV flash system contains a mercury lamp and an electronically controlled fast shutter. Both illumination systems are coupled to the microscope via suitable UV optical fibers.

lamp (HBO 100, Zeiss, Göttingen, Germany) equipped with a fast shutter (Uniblitz, Vincent Associates, Rochester, NY) that was coupled to the microscope via an optical fiber and an appropriate dichroic mirror (see Dreessen, 1992). The use of caged compounds in neurons in slice preparations (Malenka *et al.*, 1992) is in principle similar to their use in dispersed cells.

2.2. Imaging of Changes in Intracellular Calcium Concentration

Spatially resolved information on the changes in intracellular free Ca^{2+} concentration ($[Ca^{2+}]_i$) can be obtained by using an imaging device instead of the PMT. We describe here a setup with a digital video imaging system used in our laboratory (see also Lewis and Cahalan, 1989). This imaging device uses as a detector a microchannel plate intensifier (Hamamatsu Photonics, Japan) combined with a conventional CCD (charge-coupled device) video camera (Fig. 1), but, in principle, any type of camera, for example, variable-scan cooled CCD cameras (Connor, 1986; Lasser-Ross *et al.,* 1991), can be used in combination with the patch-clamp technique.

The general problems in these experiments are the display and the storage of data arising simultaneously from the whole-cell patch-clamp recording and the imaging system. In order to control the experiment adequately, it is useful to be able to extract all relevant information on line. One possibility that was used initially was to construct on-line background-corrected ratio images (Lewis and Cahalan, 1989) to have a rough estimate of the localization and the extent of the changes in Ca^{2+} concentration. With this approach the quantitative estimation of the Ca^{2+} signals and their direct correlation to the corresponding whole-cell currents or potentials was done off line (Llano *et al.,* 1991; Kano *et al.,* 1992).

A more convenient approach is the on-line display of the whole-cell current responses combined with fluorescence intensity signals integrated over one or more regions of interest (Figs. 4 and 5). For this purpose we use a computer-based "chart recorder" system in our setup (Fig. 1). This personal computer (Macintosh) receives the fluorescence data via a serial port line, while the patch-clamp data are digitized with an AD converter (ITC 16, Instrutech, New York). All data are displayed and processed on line (Fig. 5) with an appropriate computer program (X-Chart, HEKA Electronics, Lambrecht, Germany).

2.3. Patch-Clamp and Confocal Microscopy in Brain Slices

Confocal laser scanning microscopes (CLSMs) offer the advantage of improved lateral and axial resolution as well as the ability to obtain fine optical sections through the specimen (Fine *et al.,* 1988). As a consequence, the contribution of out-of-focus fluorescence arising from regions just above or below the optical plane of interest is greatly reduced. This is particularly advantageous for imaging in slices because most neurons extend their dendrites within a large volume of tissue, and out-of-focus dendrites may often generate large noninter-pretable fluorescence signals.

Various CLSMs with different lasers producing excitation light either in the UV range or in the visible light range are available today. However, designing suitable UV optics for CLSMs is complicated, and UV lasers are still quite expensive, so that excitation with visible light is much more common. A limitation of CLSM systems using visible-light excitation is given by the fact that the preferable indicator dyes available for these systems today, such as fluo-3 and calcium green-1 (Tsien and Waggoner, 1990), do not exhibit the emission (or excitation) shifts necessary for the ratiometric estimation of the Ca^{2+} concentration.

Figure 2 illustrates an example of a setup combining the patch clamp and a CLSM. This setup is quite similar to that used for conventional imaging (Fig. 1) except for the microscope, which is equipped with the standard movable stage and fixed objective. This requirement is imposed by the rigid connection between the CLSM (Odyssey, Noran, USA) and the upright microscope (Axioskop, Zeiss, Germany). In this system several manipulators for positioning the patch and stimulation pipettes need to be attached to the movable stage. Although this setup is a little less stable than that shown in Fig. 1, recordings lasting for about 60–90 min can routinely be obtained.

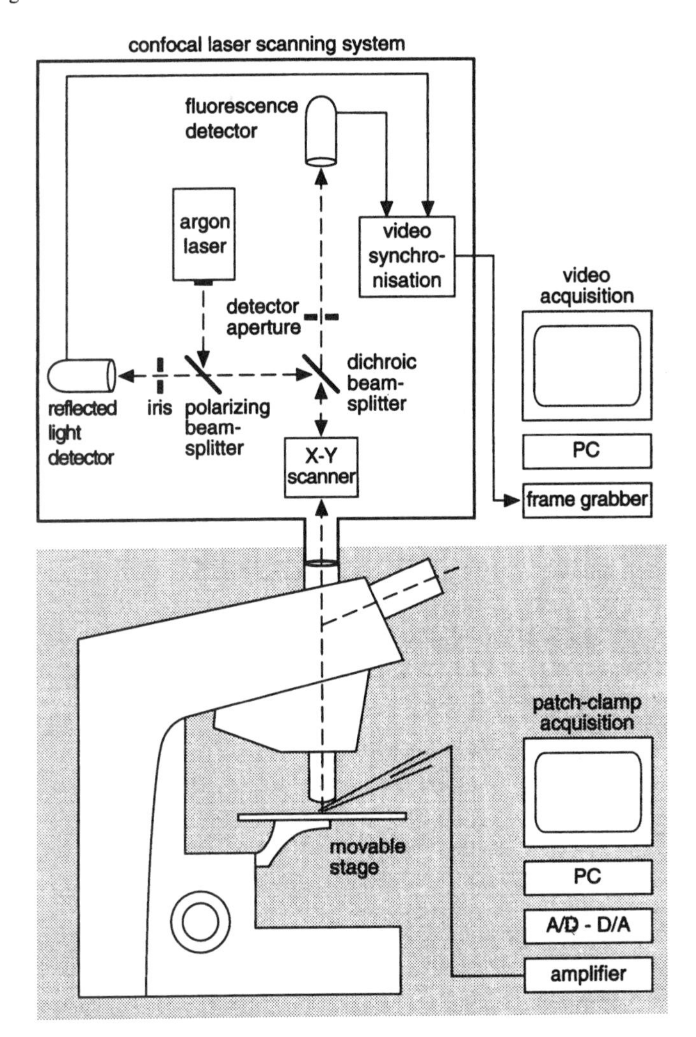

Figure 2. Schematic diagram of a setup for simultaneous confocal laser scanning microscopy and patch-clamp recordings in brain slices. The diagram within the shaded region represents a setup for patch clamp in brain slices based on a standard upright microscope with a "movable stage." The confocal laser scanning system is rigidly connected to the microscope. An argon laser provides the excitation light at the wavelength of 488 nm used for the indicator dyes fluo-3 and calcium green-1 (see Fig. 7). The *x–y* scanner directs the laser beam over the specimen, and the resulting epifluorescence light is transmitted to the fluorescence detector. The "reflected light detector" collects the light reflected by the slice following laser illumination and is useful for the fine positioning of the pipettes. The output video signal is digitized by a frame-grabber board connected to a personal computer ("video acquisition").

3. Procedures and Techniques

3.1. Pipette Solutions and Indicator Dyes

In most applications of whole-cell patch clamp and simultaneous fluorometric Ca^{2+} measurements, the indicator dye is loaded into the cells via the recording patch pipette (see also Section 3.2). For this purpose, the free dyes (and not the membrane-permeable esters) are simply added to the pipette solution. Depending on the design of the experiment, pipette solutions with various ionic compositions (see also Section 3.2) and various concentrations of the indicator dyes can be used. For example, for recording voltage-gated Ca^{2+} currents and the corresponding changes of the intracellular Ca^{2+} concentration (see Fig. 5), the following pipette solution may be used (in mM): 130 CsCl, 20 TEA-Cl, 10 HEPES, 2 Na_2ATP, 2 $MgCl_2$, 0.2 Na_3GTP, 0.2 fura-2; pH 7.3. The pH and the ionic strength largely influence the Ca^{2+}-binding characteristics of most indicator dyes (Grynkiewicz *et al.,* 1985) and should therefore be carefully checked before each experiment.

For Ca^{2+} monitoring in combination with patch-clamp recordings in brain slices we have used fura-2 for the ratiometric PMT- and video-imaging-based measurements and calcium green-1 and fluo-3 for the CLSM-based recordings. All dyes were purchased from Molecular Probes (Eugene, OR). Despite their different fluorometric properties (excitation shift in the case of fura-2 versus increase in emission intensity for fluo-3 and calcium green-1), these three dyes have rather similar dissociation constants (K_D) for the binding of Ca^{2+} (K_D is 189 nM for calcium green-1, 224 nM for fura-2, and 316 nM for fluo-3; see Grynkiewicz *et al.,* 1985; Haugland, 1992). These relatively low dissociation constants of around 200–300 nM allow the measurement of changes in intracellular free Ca^{2+}-concentrations ($[Ca^{2+}]_i$) near the usual resting $[Ca^{2+}]_i$ (around 10^{-7} M). However, it should be emphasized that these estimations of the dissociation constants critically depend on several parameters including the ionic composition, the ionic strength, and the pH of the solution. It should also be noted that because of their high affinity for Ca^{2+}, these dyes act as Ca^{2+} buffers, even when they are present at low intracellular concentrations, and may thereby influence the kinetics and amplitude of $[Ca^{2+}]_i$ changes (Neher, 1988; Neher and Augustine, 1992).

Therefore, the dye concentration used for a given experiment needs to be carefully chosen. Faithful recordings of the dynamics of $[Ca^{2+}]_i$ changes require low concentrations (50–100 μM), but the estimation of the total Ca^{2+} influx requires large concentrations (1 mM or more) of the indicator dye. According to Neher and Augustine (1992), with increasing intracellular concentrations, the dye will compete more successfully with the endogenous Ca^{2+} buffers and eventually override their binding capacity for Ca^{2+}. At high dye concentrations, the kinetics of changes in $[Ca^{2+}]_i$ will be prolonged, and their amplitude largely reduced; however, the Ca^{2+}-sensitive fluorescence signals will give a direct measure of the total Ca^{2+} influx (Neher and Augustine, 1992). The minimal concentration at which the dye will bind virtually all of the entering Ca^{2+} depends on the following factors: (1) the dissociation constant (K_D) of Ca^{2+} binding to the indicator, (2) the basal $[Ca^{2+}]_i$ and the changes in $[Ca^{2+}]_i$, and (3) the endogenous $^{2+}$ buffer capacity of the cell type studied. For example, in medial septal neurons in forebrain slices it was shown that this condition can be easily achieved at the concentration of 1 mM fura-2 for a $[Ca^{2+}]_i$ level of less than 200 nM (for further details see Schneggenburger *et al.,* 1993a; see also Section 4.2).

For the faithful monitoring of the physiological changes in $[Ca^{2+}]_i$, as low an intracellular dye concentration (50–100 μM) as possible needs to be used (Neher and Augustine, 1992). However, for monitoring $[Ca^{2+}]_i$ changes simultaneously in the soma and in the dendrites of a neuron, it is often necessary to use higher indicator concentrations (of 200 μM or more) in the pipette solution (see Figs. 3, 4, and 7). This will make it possible to obtain sufficiently high dye concentrations in neuronal processes that are remote from the site of whole-cell recording. For example, when the dye is delivered to the soma, the most common configuration, the dye concentration in remote dendritic branches will only equilibrate with the pipette solution after a considerable time delay (see Fig. 3). Usually, even at the time of measurement (30–60 min after obtaining the somatic whole-cell configuration), the dye concentration at distant dendritic sites will not have completely equilibrated with the dye concentration of the pipette solution. In addition, because of the small diameter of the fine dendrites and the often significant background autofluorescence levels of the surrounding tissue (see Section 3.3), larger dye concentrations are needed in order to obtain spatially resolved fluorescence changes with a reasonable signal-to-noise ratio.

3.2. Dye-Loading Procedures

Loading the indicator dye into the somata (Konnerth *et al.*, 1992) or dendrites (Stuart *et al.*, 1993) of neurons via the patch pipette is similar to the loading procedure described in detail by Pusch and Neher (1988). In the largely intact neurons with long and fine dendritic processes in slice preparations, the diffusion of the dye is slow, and the loading procedure may last several tens of minutes. Moreover, in slices there is often a gradual increase of the series resistance as a result of "resealing" problems (Edwards *et al.*, 1989), which delays the loading of remote dendrites even more.

Figure 3 shows a sequence of images of a Purkinje neuron in a cerebellar slice at different times during the loading procedure. The pipette solution contained 200 μM fura-2. The fluorescence images were taken at 360-nm excitation light, near the isosbestic point of fura-2 (see Tsien and Poenie, 1986). The brightness of the dendrites increased gradually with time, and the remote dendrites could only be resolved reasonably well after about 30 min of dye loading (Fig. 3E). More specifically, in this experiment the times of half-maximal loading with fura-2 were determined for three dendritic regions (Fig. 3F) and were found to be 3 min, 16 min, and 32 min for regions 1, 2, and 3, respectively. Rexhausen (1992) analyzed the time course of accumulation of fura-2 in the dendrites of Purkinje neurons and found that it can be well explained by a passive diffusion process, assuming a two-compartment model consisting of a somatic and a lumped dendritic component.

The loading time constant is obviously dependent on the access resistance between the patch pipette and the cell interior (Pusch and Neher, 1988). For rapid loading and good voltage control, a low access resistance is desirable; however, "long-term" whole-cell recordings (lasting longer than 60 min) are very difficult to obtain when the resistances of the patch pipettes are lower than about 1 MΩ. Therefore, for recordings from cerebellar Purkinje or hippocampal pyramidal neurons, we routinely use patch pipettes with resistances of 1 to 3 MΩ with resulting access resistances of around 4–10 MΩ that allow fluorometric recordings from dendrites after about 30–40 min (Fig. 3).

3.3. Background Subtraction

The accurate subtraction of background fluorescence is particularly critical for Ca^{2+} imaging in slice preparations because of the significant levels of background fluorescence from the surrounding tissue. The background fluorescence results partially from the release of dye-containing pipette solution just prior to the formation of the seal and, more importantly, from endogenous fluorophores in the surrounding tissue excited by light in the UV range (such as pyridine nucleotides, e.g., NADH, NADPH) or in the visible range (flavins; see Tsien and Waggoner, 1990).

The release of indicator dye from the patch pipette can be minimized by taking rather simple precautions such as filling the tip of the patch pipette with a solution with no dye or

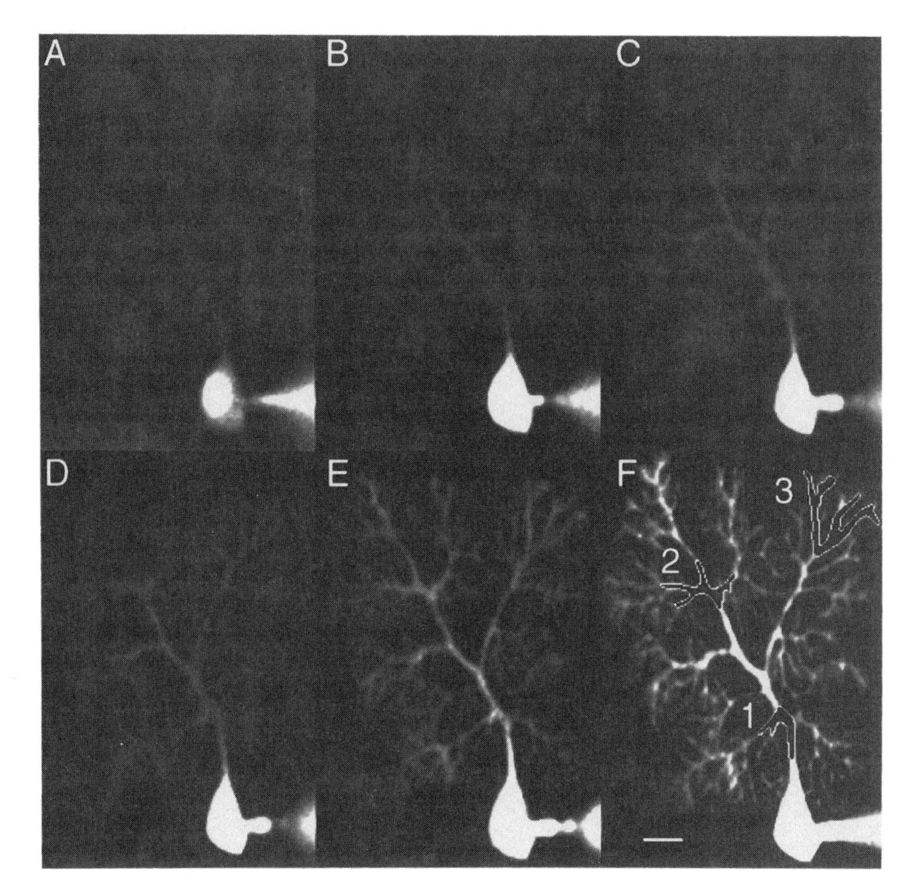

Figure 3. Time course of fura-2 loading of a whole-cell-clamped Purkinje neuron in a rat cerebellar slice. Fluorescence images were recorded with constant camera and intensifier gain settings throughout the experiment at an excitation wavelength of 360 nm. The images shown in A–F correspond to the following time points after achievement of the whole-cell configuration: A, 10 sec; B, 1 min; C, 2 min; D, 4 min; E, 30 min; and F, 120 min. Images in A–C are averages of 32, and in D–F are averages of 128 images, respectively. The access resistance was 5.5 MΩ at the beginning of whole-cell recording, increased to 6.7 MΩ after 4 min, to 13 MΩ after 20 min, and remained stable for the rest of the experiment. In panel F, 1, 2, and 3 indicate the regions for which the half-maximal fura-2 loading time was estimated. The scale bar corresponds to 15 μm. (Modified, with kind permission, from Rexhausen, 1992).

applying minimal levels of pressure to the patch pipette when approaching the cell. Whenever possible, we also prefer to "clean" the neurons (Edwards *et al.*, 1989) before patching in order to accelerate the process of gigaseal formation. With these precautions, the background fluorescence usually becomes constant shortly after the seal formation.

Background fluorescence from autofluorescence is known to be tissue specific. For example, it was found that background fluorescence is different in the various layers of hippocampal slices (Regehr and Tank, 1992). The autofluorescence intensity can also vary during the time course of the experiment, probably as a result of bleaching caused by the UV excitation light. Therefore, we find it absolutely necessary to perform background measurements throughout the experiment. With the imaging, this can be accomplished by approximating the background fluorescence from defined pixel regions near the cell under study (see, for example, Fig. 4). For recordings with the PMT detector system, the problem of background fluorescence bleaching seems to be less critical, probably because lower excitation intensities are used.

3.4. Calibration Procedures for Ratiometric Calcium Measurements

According to Grynkiewicz *et al.* (1985), the concentration of intracellular free Ca^{2+} ($[Ca^{2+}]_i$) can be calculated using the relationship:

$$[Ca^{2+}]_i = K_{eff} \cdot (R - R_{min})/(R_{max} - R) \quad (1)$$

in which K_{eff} is an "effective binding constant," R is the ratio of fluorescence at 360 nm to that at 380 nm excitation wavelength, and R_{min} and R_{max} are the limiting fluorescence ratios at zero Ca^{2+} and at high Ca^{2+} concentrations, respectively. The calibration constants K_{eff}, R_{min}, and R_{max} need to be determined directly for each experimental setup, since they depend on the specific optical characteristics of each system (see also Neher, 1989).

For most applications, we prefer an *in vivo* calibration procedure in which neurons in slices are loaded via the recording patch pipette with calibration solutions containing defined, buffered Ca^{2+} concentrations. Calibrations are done in every experimental setup for each type of neuron and are frequently repeated. The general procedure is similar to that described in detail for isolated cells (Almers and Neher, 1985; Neher, 1989). A serious limitation of the *in vivo* calibration procedure in neurons with an extensive dendritic arborization is the reduced ability to control $[Ca^{2+}]_i$ levels in distant dendrites (Fig. 3). There is not only a long delay before the dye concentration in the dendrites equilibrates with that of the pipette solution, but the large dendritic surface-to-volume ratio may cause an increased transmembrane transport rate for Ca^{2+}. Especially when calibration solutions with high Ca^{2+} concentrations are used (e.g., 10 mM Ca^{2+} to determine R_{max}), the active extrusion of Ca^{2+} (see also Mathias *et al.,* 1990) in dendrites will result in an underestimation of R_{max}.

Therefore, we restrict our calibration experiments to the somata and proximal dendrites, that is, to sites that can be expected to be under more immediate diffusional control from the whole-cell patch pipette. Nevertheless, for experiments using solutions with high $[Ca^{2+}]_i$ for measuring R_{max}, the access resistance needs to be as low as possible. A frequent difficulty of these experiments results from a sudden increase in the access resistance ("resealing") that is accompanied by an instant decrease in the fluorescence ratio. This indicates the presence of strong Ca^{2+} extrusion mechanisms even in the cell bodies of some neurons (particularly in cerebellar Purkinje neurons). In recordings with a stable and low (3–6 MΩ) access resistance, the somatic fluorescence ratios at intermediate or high Ca^{2+} concentrations were found to equilibrate in neurons with large dendrites within 10–15 min of whole-cell recording. This contrasts with the faster equilibration time course (around 1 min) found in mast cells (Neher, 1989). It should be noted that for the determination of R_{min}, the access resistance is less critical.

3.5. Dye Bleaching and Phototoxic Damage

In combined whole-cell and fluorometric recordings, bleaching is a major problem neither for fura-2 nor for fluo-3 or calcium green-1. This is illustrated for fura-2 in the example shown in Fig. 5, in which, for a continuous recording lasting for more than 1 min, there was no decline in the isosbestic fluorescence signal obtained by exciting with 360 nm UV light, which is a direct measure for the fura-2 concentration at that particular site. In most recordings there was no significant bleaching-dependent decline in fluorescence intensity even for continuous recordings lasting for 20–30 min. This suggests that even if bleaching occurred, its rate was smaller than that of the dye-loading process via the patch pipette.

There seems to be no significant phototoxic damage when using fura-2. However, the use of fluo-3 and calcium green-1 with the high excitation intensities produced by the laser of the CLSM is much more critical. Under our recording conditions, continuous illumination at the high intensities that are required for detecting changes in fine dendrites and spines (Fig. 7) may already produce an irreversible characteristic damage after 10–15 sec. This phototoxic damage is proportional to the illumination intensity and duration and to the concentration of the indicator dye. Typically, the first sign of this damage is the rapid (time constant of about 1 sec) development of an irreversible outward current (of about 1 nA amplitude in cerebellar Purkinje neurons), which is often followed by a persistent increase in Ca^{2+} concentration. The phototoxic damage can be partially prevented by switching from continuous to intermittent illumination and by reducing the area of the cell that is illuminated. It is, in addition, useful to avoid the illumination of the cell body (which contains a large amount of dye) whenever possible (Fig. 7).

4. Examples and Applications

4.1 Changes in $[Ca^{2+}]_i$ in Soma and Dendrites of Hippocampal Pyramidal Neurons

Figures 4 and 5 illustrate the use of an imaging system (as described in Section 2.1) that allows the spatial resolution of fura-2 fluorescence signals in combination with whole-cell patch-clamp recordings. A tight-seal whole-cell recording from a CA1 pyramidal neuron in a rat hippocampal slice was made using a pipette solution containing 200 μM fura-2. The slice was illuminated alternately with 360-nm and 380-nm wavelength excitation light, and the fluorescence signals were integrated on-line for several regions of interest chosen just prior to the recording: a somatic region (1), a region covering the proximal part of the apical dendrite (2), and a background region near the neuron (3) (Fig. 4; corresponding changes in fluorescence in Fig. 5). At the time indicated by the arrow, a depolarizing voltage step to 0 mV (800 msec duration, from a holding potential of −60 mV) was applied, which evoked an inward Ca^{2+} current (see Fig. 5A). The resulting influx of Ca^{2+} is reflected by the decrease of the fura-2 fluorescence at the Ca^{2+}-sensitive excitation wavelength at 380 nm (abbreviated F_{380} in Fig. 5B; closed symbols). The fura-2 fluorescence at the excitation wavelength of 360 nm (near the isosbestic point of fura-2) remained, as expected, constant during the Ca^{2+} influx (see Fig. 5B, open symbols).

Figure 5C,D displays the somatic and dendritic changes in $[Ca^{2+}]_i$ that were calculated using equation 1 from the background-subtracted changes in fluorescence. All traces shown in Fig. 5 were displayed on line on the monitor of the chart computer shown schematically in Fig. 1. In addition, there is the possibility of recording and storing a sequence of images (similar to that shown in Fig. 4). These images, if necessary, can be analyzed in detail off line.

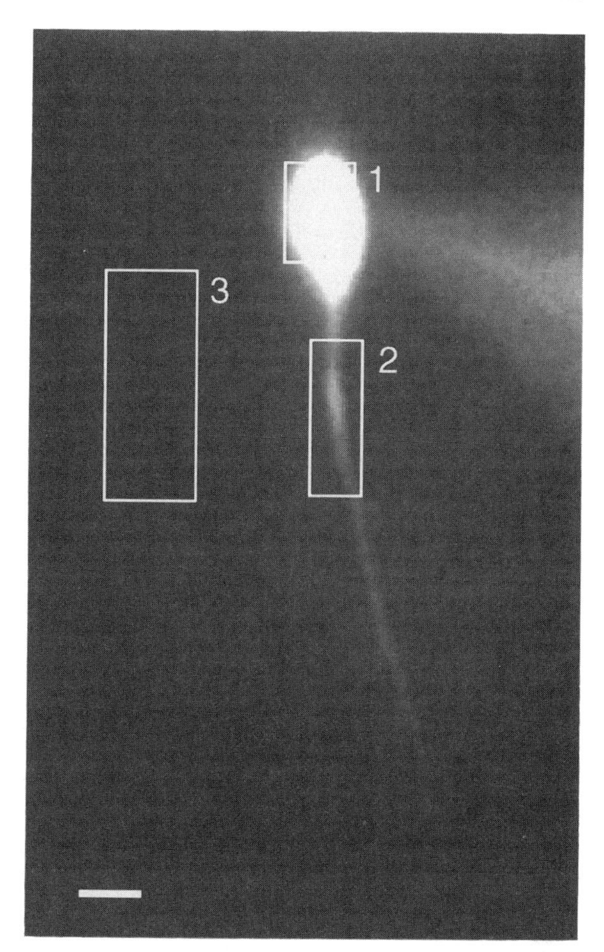

Figure 4. Fluorescence image of a CA1 pyramidal neuron from a rat hippocampal slice. The neuron was loaded with a 200 μM fura-2-containing pipette solution for 20 min. The image (average of $n = 128$) was recorded at an excitation wavelength of 380 nm. The three regions selected for the on-line measurement of the fura-2 fluorescence changes (see Fig. 5) represent a somatic region (1), a region of the proximal apical dendrite (2), and a region for background fluorescence (3). The scale bar corresponds to 10 μm.

4.2. Calcium Flux Measurements

A useful application of combined patch-clamp and fluorometric Ca^{2+} recordings is the direct measurement of transmembrane Ca^{2+}-fluxes. This is achieved by loading the cells with high concentrations (1 mM or more) of the fluorometric dye fura-2 via the patch pipette. Under these conditions, most of the incoming Ca^{2+} is captured by the indicator dye, and the total Ca^{2+}-flux can be estimated directly from the changes in Ca^{2+}-sensitive fluorescence (Neher and Augustine, 1992). Such Ca^{2+} flux measurements have been used to determine the Ca^{2+} component of nonselective cation currents, which was named the *fractional Ca^{2+} current* (Schneggenburger *et al.*, 1993a; Zhou and Neher, 1993). In brain slices, this approach was first used in rat medial septal neurons for studying the fractional Ca^{2+} current through glutamate receptor channels. Since medial septal neurons lack an extensive dendritic tree, imaging was not necessary, and therefore these experiments were performed with a single-detector PMT system (Schneggenburger *et al.*, 1993a,b).

Figure 6A shows the time course of loading of a medial septal neuron with 1 mM fura-2. The time course of the increase in intracellular fura-2 concentration was calculated from the Ca^{2+}-insensitive fluorescence signal at 360 nm excitation wavelength (Neher and

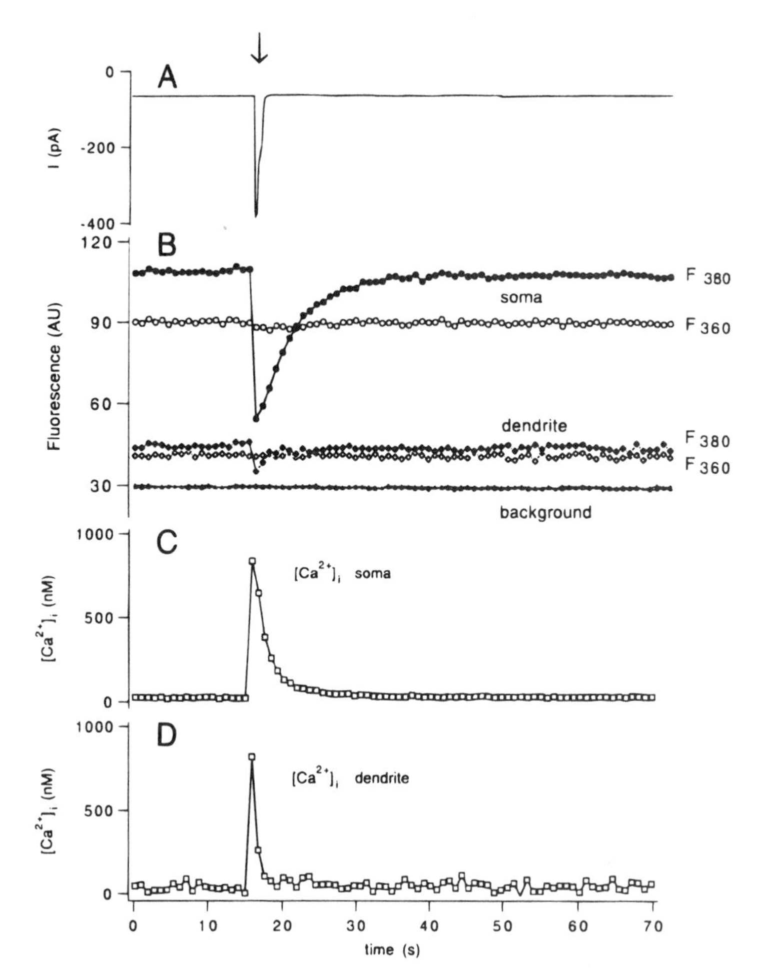

Figure 5. Simultaneous recording of the whole-cell current, the spatially resolved fura-2 fluorescence signals, and the corresponding changes in $[Ca^{2+}]_i$. Recording from the CA1 hippocampal pyramidal neuron shown in Fig. 4. A: At the time indicated by the arrow a depolarizing pulse (to 0 mV for 800 msec, holding potential −60 mV) evoked an inward current. The whole-cell current was sampled at 5 Hz. B: The fura-2 fluorescence at 360 nm (F_{360}) and 380 nm (F_{380}) excitation wavelengths (sampled at 1.2 Hz) from the three regions indicated in Fig. 4, which correspond to a somatic region (upper two traces), a dendritic region (middle two traces), and a background region (lower, superimposed two traces). Normalized fluorescence intensities (total fluorescence of a region divided by the number of corresponding pixels) were given in arbitrary units (AU). C,D: The somatic and dendritic change in $[Ca^{2+}]_i$ caused by the depolarizing pulse delivered through the patch pipette. $[Ca^{2+}]_i$ was calculated on line according to equation 1, after subtraction of the background fluorescence.

Augustine, 1992). Figure 6B shows two Ca^{2+} currents evoked at a low (135 μM) and at a high (830 μM) intracellular fura-2 concentration (at the time points marked with 1 and 2 in Fig. 6A, respectively). Despite the similar Ca^{2+} charge transported during these two Ca^{2+} currents (indicated by the shaded areas in Fig. 6B), the corresponding changes in Ca^{2+}-sensitive fluorescence (Fig. 6C, top) and $[Ca^{2+}]_i$ (Fig. 6C, bottom) were markedly different for the two fura-2 concentration levels. This was a result of the concentration-dependent Ca^{2+}-buffering capacity of fura-2. When the intracellular concentration of fura-2 and, thus,

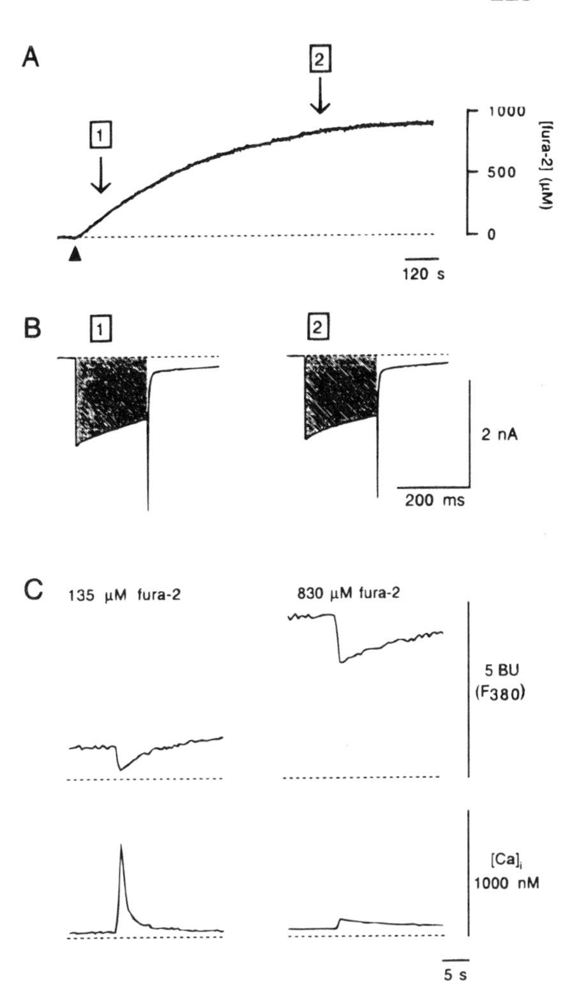

Figure 6. Concentration dependence of the Ca^{2+}-buffering capacity of fura-2. A: Time course of loading a medial septal neuron in a rat forebrain slice with a 1 mM fura-2-containing pipette solution. The intracellular fura-2 concentration was calculated from the Ca^{2+}-independent fluorescence (at 360 nm). The beginning of whole-cell recording is indicated by the arrowhead. B: Voltage-gated Ca^{2+} currents evoked at two time points during the loading phase (see arrows in A), at 135 μM and at 830 μM fura-2 respectively. The integral of the Ca^{2+} currents is indicated by the shaded areas. The currents were evoked by 200-msec voltage pulses to −10 mV from a holding potential of −80 mV. C: Ca^{2+}-sensitive fluorescence changes at 380 nm excitation wavelength (top traces, in bead units, BU; see Schneggenburger *et al.*, 1993a) and the corresponding $[Ca^{2+}]_i$ changes at 135 μM and at 830 μM fura-2 concentration. The composition of the pipette solution was as described in Section 3.1 except that 1 mM fura-2 was used. (Modified, with permission, from Schneggenburger *et al.*, 1993a).

the Ca^{2+}-buffering capacity became sufficiently large (Fig. 6C, right panel), virtually all of the incoming Ca^{2+} was bound by the dye (see also Neher and Augustine, 1992). The concentration of fura-2 at which Ca^{2+} fluxes can be directly determined from the changes in Ca^{2+}-sensitive fluorescence was determined to be less than 1 mM fura-2 in medial septal neurons (Schneggenburger *et al.*, 1993a).

4.3. Localized Dendritic Calcium Signals

In many central neurons, excitatory synaptic signals are transmitted via dendritic spines. It is also known that many forms of synaptic plasticity critically depend on the intracellular Ca^{2+} concentration (Kano *et al.*, 1992; Konnerth *et al.*, 1992; Bliss and Collingridge, 1993). Therefore, it is of special interest to evaluate the Ca^{2+} homeostasis and dynamics at the level of fine dendritic branches and ultimately at the level of spines (Gamble and Koch, 1987; Müller and Connor, 1991; Eilers *et al.*, 1994; Murphy *et al.*, 1994). It is, in general, difficult

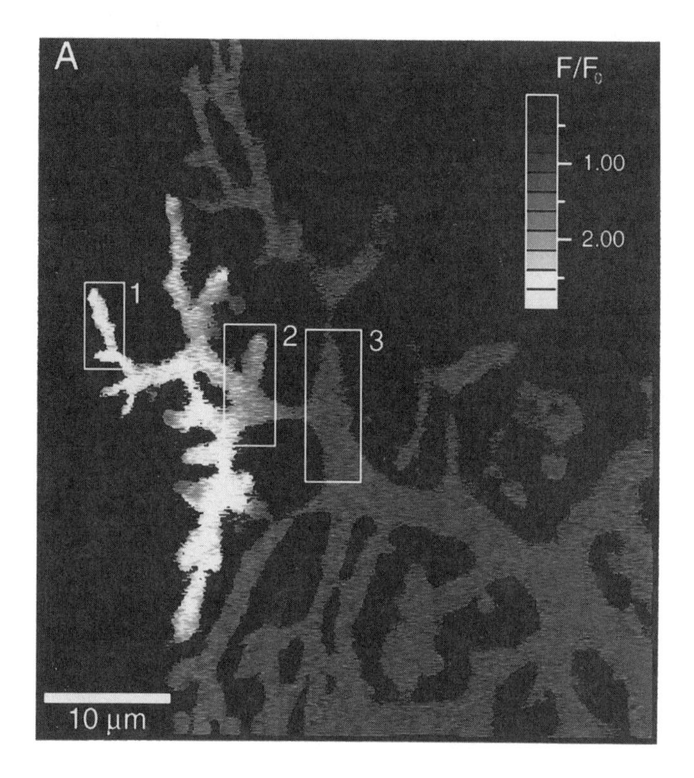

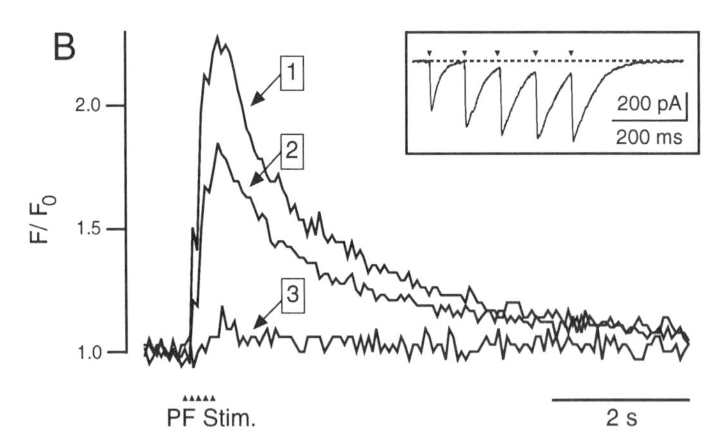

Figure 7. Local dendritic Ca^{2+} signaling in a Purkinje neuron of a cerebellar slice. Whole-cell recording after 60 min of loading with 500 μM calcium green-1 in combination with confocal imaging. A: Grayscale fluorescence ratio image F/F_0 (fluorescence F divided by the basal fluorescence F_0 before stimulation) corresponding to the peak signal in B. B: Repetitive (five stimuli at 10 Hz) parallel fiber stimulation evoked characteristic excitatory synaptic currents (see inset) associated with a transient change in fluorescence. Traces 1, 2, and 3 correspond to the similarly marked regions in A. Arrowheads indicate the time points of synaptic activation.

to resolve Ca^{2+} signals in fine dendrites and spines with conventional imaging (Kano *et al.*, 1992; Konnerth *et al.*, 1992; Miyakawa *et al.*, 1992; but see Müller and Connor, 1991). A much better spatial resolution can be obtained with CLSMs (Alford *et al.*, 1993; Eilers *et al.*, 1994).

Figure 7 illustrates an experiment in which whole-cell patch clamp of a Purkinje neuron in a cerebellar slice was combined with topical synaptic stimulation and CLSM imaging. A train of excitatory postsynaptic currents was evoked by the repetitive stimulation of a beam of parallel fibers (see inset of Fig. 7B). Despite the lack of any "regenerative component" (Miyakawa *et al.*, 1992), the parallel-fiber-mediated excitatory postsynaptic signals were associated with large, highly localized changes in fluorescence (Fig. 7). Although these changes in fluorescence cannot be readily translated into the corresponding changes in $[Ca^{2+}]_i$, their amplitudes are comparable to the peak amplitudes we observe in large portions of the dendritic tree during climbing fiber stimulation (not shown). It has been shown using fura-2 imaging that climbing-fiber-mediated $[Ca^{2+}]_i$ transients have amplitudes of several hundreds of nanomolar (Knöpfel *et al.*, 1990; Kano *et al.*, 1992; Konnerth *et al.*, 1992). The mechanisms underlying the localized parallel-fiber-mediated Ca^{2+} changes are not fully understood, but the voltage dependence of these signals indicates that dendritic voltage-gated Ca^{2+} channels are critical for their generation (Eilers *et al.*, 1994).

ACKNOWLEDGMENTS. We thank Dr. O. Garaschuk for providing unpublished data for Figs. 4 and 5 and Dr. T. Plant for critically reading the manuscript. We also acknowledge the contribution of J. Biedermann, Dr. J. Dreessen, and U. Rexhausen during the development of the approaches for monitoring calcium in brain slices. This work was supported by grants from the Bundesministerium für Forschung und Technologie, the Deutsche Forschungsgemeinschaft, the Human Frontiers Science Program Organization and the European Community to A.K.

References

Alford, S., Frenguelli, B. G., Schofield, J. G., and Collingridge, G. L., 1993, Characterization of Ca^{2+}-signals induced in hippocampal CA1 neurones by the synaptic activation of NMDA receptors, *J. Physiol.* **469:**693–716.

Almers, W., and Neher, E., 1985, The Ca signal from fura-2 loaded mast cells depends strongly on the method of dye-loading, *FEBS Lett.* **192:**13–18.

Blanton, M. G., LoTurco, J., and Kriegstein, A. R., 1989, Whole cell recording from neurons in slices of reptilian and mammalian cerebral cortex, *J. Neurosci. Methods* **30:**203–210.

Bliss, T. V., and Collingridge, G. L., 1993, A synaptic model of memory: Long-term potentiation in the hippocampus, *Nature* **361:**31–39.

Callaway, E. M., and Katz, L. C., 1993, Photostimulation using caged glutamate reveals functional circuitry in living brain slices, *Proc. Natl. Acad. Sci. USA* **90:**7661–7665.

Connor, J. A., 1986, Digital imaging of free calcium changes and of spatial gradients in growing processes in single mammalian central nervous system cells, *Proc. Natl. Acad. Sci. USA* **83:**6179–6183.

Dreessen, J., 1992, Quantitative Messung der intrazellulären Ca^{++}-Konzentration mit einem Bildverarbeitungssystem bei der Untersuchung der synaptischen Plastizität in den Purkinje-Zellen des Kleinhirns, Ph.D. Thesis, University of Hannover.

Edwards, F. A., Konnerth, A., Sakmann, B., and Takahashi, T., 1989, A thin slice preparation for patch clamp recordings from neurones of the mammalian central nervous system, *Pflügers Arch.* **414:**600–612.

Eilers, J., Augustine, G. J., and Konnerth, A., 1994, Calcium signals in fine dendrites and spines of cerebellar Purkinje neurons, *Pfügers Arch.* **426:**R27.

Fine, A., Amos, W. B., Durbin, R. M., and McNaughton, P. A., 1988, Confocal microscopy: Applications in neurobiology, *Trends Neurosci.* **11:**346–351.

Gamble, E., and Koch, C., 1987, The dynamics of free calcium in dendritic spines in response to repetitive synaptic input, *Science* **236:**1311–1315.

Grynkiewicz, G., Poenie, M., and Tsien, R., 1985, A new generation of Ca^{2+} indicators with greatly improved fluorescence properties, *J. Biol. Chem.* **260:**3440–3450.

Haugland, R. P., 1992, *Handbook of Fluorescent Probes and Research Chemicals* (K. Larison, ed.), Molecular Probes, Inc., Eugene, OR.

Kano, M., Rexhausen, U., Dreessen, J., and Konnerth, A., 1992, Synaptic excitation produces a long-lasting rebound potentiation of inhibitory synaptic signals in cerebellar Purkinje cells, *Nature* **356:**601–604.

Kao, J. P. Y., Harootunian, A. T., and Tsien, R. Y., 1989, Photochemically generated cytosolic calcium pulses and their detection by fluo-3, *J. Biol. Chem.* **264:**8179–8184.

Kaplan, J. H., and Somlyo, A. P., 1989, Flash photolysis of caged compounds: New tools for cellular physiology, *Trends Neurosci.* **12:**54–59.

Knöpfel, T., Vramesic, C., Staub, V. C., and Gähwiler, B., 1990, Climbing fiber responses in olivo-cerebellar slice cultures. II. Dynamics of cytosolic calcium in Purkinje cells, *Eur. J. Neurosci.* **3:**343–348.

Konnerth, A., 1990, Patch-clamping in slices of mammalian CNS, *Trends Neurosci.* **13:**321–323.

Konnerth, A., Dreessen, J., and Augustine, G. J., 1992, Brief dendritic calcium signals initiate long-lasting synaptic depression in cerebellar Purkinje cells, *Proc. Natl. Acad. Sci. USA* **89:**7051–7055.

Lasser-Ross, N., Miyakawa, H., Lev-Ram, V., Young, S. R., and Ross, W. N., 1991, High time resolution fluorescence imaging with a CCD camera, *J. Neurosci. Methods* **36:**253–261.

Lewis, R. S., and Cahalan, M. D., 1989, Mitogen-induced oscillations of cytosolic Ca^{2+} and transmembrane Ca^{2+} current in human leukemic T cells, *Cell Regul.* **1:**99–112.

Llano, I., Dreessen, J., Kano, M., and Konnerth, A., 1991, Intradendritic release of calcium induced by glutamate in cerebellar Purkinje cells, *Neuron* **7:**577–583.

Malenka, R. C., Lancaster, B., and Zucker, R. S., 1992, Temporal limits on the rise in postsynaptic calcium required for the induction of long-term potentiation, *Neuron* **9:**121–128.

Mathias, R. J., Cohen, I. S., and Oliva, C., 1990, Limitations of the whole cell patch clamp technique in the control of intracellular solutions, *Biophys. J.* **58:**759–770.

Miyakawa, H., Lev-Ram, V., Lasser-Ross, N., and Ross, W. N., 1992, Calcium transients evoked by climbing fiber and parallel fiber synaptic inputs in guinea pig cerebellar Purkinje neurons, *J. Neurophysiol.* **68:**1178–1189.

Müller, W., and Connor, J. A., 1991, Dendritic spines as individual neuronal compartements for synaptic Ca^{2+} responses, *Nature* **354:**73–76.

Murphy, T. H., Baraban, J. M., Wier, W. G., and Blatter, L. A., 1994, Visualization of quantal synaptic transmission by dendritic calcium imaging, *Science* **263:**529–532.

Neher, E., 1988, The influence of intracellular calcium concentration on degranulation of dialysed mast cells from rat peritoneum, *J. Physiol.* **395:**193–214.

Neher, E., 1989, Combined fura-2 and patch clamp measurements in rat peritoneal mast cells, in: *Neuromuscular Junction* (L. Sellin, R. Libelius, and S. Thesleff, eds.), pp. 65–76, Elsevier, Amsterdam.

Neher, E., and Augustine, G. J., 1992, Calcium gradients and buffers in bovine chromaffin cells, *J. Physiol.* **450:**273–301.

Pusch, M., and Neher, E., 1988, Rates of diffusional exchange between small cells and a measuring patch pipette, *Pflügers Arch.* **411:**204–211.

Regehr, W. G., and Tank, D. W., 1992, Calcium concentration dynamics produced by synaptic activation of CA1 hippocampal pyramidal cells, *J. Neurosci.* **12:**4202–4223.

Rexhausen, U., 1992, Bestimmung der Diffusionseigenschaften von Fluoreszenzfarbstoffen in verzweigten Nervenzellen unter Verwendung eines rechnergesteuerten Bildverarbeitungssystems, Diploma Thesis, University of Göttingen.

Schneggenburger, R., Zhou, Z., Konnerth, A., and Neher, E., 1993a, Fractional contribution of calcium to the cation current through glutamate receptor channels, *Neuron* **11:**133–143.

Schneggenburger, R., Tempia, P., and Konnerth, A., 1993b, Glutamate- and AMPA-mediated calcium influx through glutamate receptor channels in medial septal neurons, *Neuropharmacology* **32:**1221–1228.

Smith, S. J., MacDermott, A. B., and Weight, F. F., 1983, Detection of intracellular Ca^{2+} transients in sympathetic neurones using arsenazo III, *Nature* **304:**350–352.

Stuart, G. J., Dodt, H. U., and Sakmann, B., 1993, Patch-clamp recordings from the soma and dendrites of neurons in brain slices using infrared video microscopy, *Pflügers Arch.* **423:**511–518.

Tsien, R. Y., and Poenie, M., 1986, Fluorescence ratio imaging: A new window into intracellular ionic signalling, *Trends Biochem. Sci.* **11:**450–455.

Tsien, R. Y., and Waggoner, A., 1990, Fluophores for confocal microscopy: Photophysics and photochemistry, in: *Handbook of Biological Confocal Microscopy* (J. B. Pawley, ed.), pp. 169–178, Plenum Press, New York.

Zhou, Z., and Neher, E., 1993, Calcium permeability of nicotinic acetylcholine receptor channels in bovine adrenal chromaffin cells, *Pflügers Arch.* **425:**511–517.

Chapter 10

Fast Application of Agonists to Isolated Membrane Patches

PETER JONAS

1. Introduction and History

At a synapse, the transmitter is stored in synaptic vesicles and is released into the synaptic cleft almost instantaneously upon fusion of these vesicles with the presynaptic membrane. Subsequently, the transmitter diffuses to ligand-gated ion channels in the postsynaptic density, binds to them, and thereby causes channel activation. Unfortunately, we have estimates neither of the exact amount of transmitter in the synaptic vesicle nor of the concentration in the synaptic cleft reaching the postsynaptic receptors, and in some cases even the identity of the transmitter is unknown. These questions may be addressed by modeling of release and diffusion. Such a theoretical approach, however, is based on several assumptions, some of which lack experimental evidence.

An alternative approach is to mimic synaptic release using fast application techniques: by comparing the shape of the current evoked by pulses of different length of the putative transmitter with the synaptic current, estimates of the time course of transmitter in the synaptic cleft can be obtained. This presents a technical challenge because it is necessary to apply the putative transmitter almost as rapidly as at an intact synapse. After 1980, a few techniques were reported that allowed fairly rapid solution exchange on whole-cell somata. One system was built up from a long cylindrical tubing with a small lateral hole through which the cell was inserted. The tubing was filled with control solution, and the lower end of the tubing was immersed in test solution. The solution was changed by applying suction to the upper end using a solenoid-driven valve (Krishtal *et al.*, 1983; Akaike *et al.*, 1986). Other systems were based on U-tubes (Krishtal and Pidoplichko, 1980; Fenwick *et al.*, 1982) or double-barreled application pipettes (Johnson and Ascher, 1987). The exchange times that proved feasible were in the range from 100 msec down to a few milliseconds.

It soon became evident that very fast exchange could be achieved only on excised membrane patches. By use of Y-tube methods, exchange times of 1 msec were obtained with outside-out patches (Brett *et al.*, 1986). A breakthrough was the development of "liquid filament" switches based on an extremely sharp interface between two laminarly flowing solutions, which was rapidly moved across the patch (Franke *et al.*, 1987; Maconochie and Knight, 1989). For the first time, solution could be exchanged within about 100 μsec. In

PETER JONAS • Department of Cell Physiology, Max-Planck-Institute for Medical Research, D-69120 Heidelberg, Germany.
Single-Channel Recording, Second Edition, edited by Bert Sakmann and Erwin Neher. Plenum Press, New York, 1995.

A

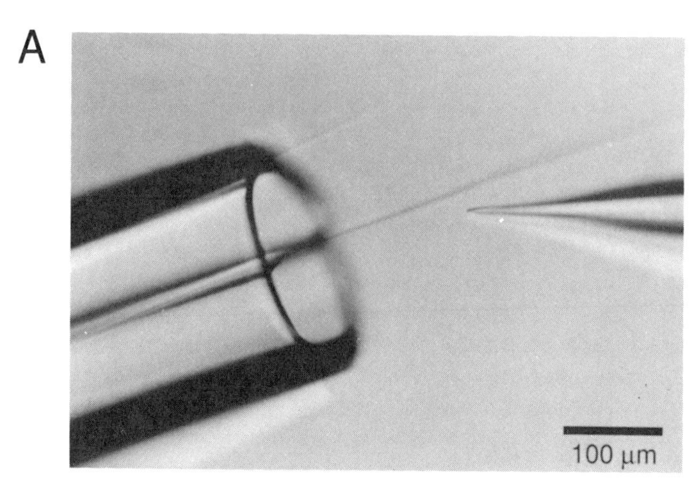

B

C

Figure 1. Experimental setup. (A) Application pipette and recording pipette under experimental conditions; light-microscopic view from top. Application pipette barrels were perfused with normal rat Ringer solution (NRR) and 10% NRR respectively. Note the sharp interface between the two solutions. (B) Application pipette attached to the piezoelectric element. Tubing connected to the pipette can be seen on the right.

the method of Franke *et al.* (1987), test solution flowed out of a single-barreled application pipette into a continuously perfused bath, and the application pipette was moved by a piezo translator. In the method of Maconochie and Knight (1989), two jets of solution from separate pipes were directed toward the membrane patch, and the liquid filament was moved by increasing the flow through one while decreasing the flow through the other pipe, using solenoid-driven valves.

When we (Colquhoun *et al.,* 1992) and others attempted to mimic excitatory synaptic transmission in the CNS mediated by α-amino-3-hydroxy-5-methyl-4-isoxazolepropionate (AMPA)-type glutamate receptor channels, several requirements had to be satisfied. (1) Extremely fast solution exchange was essential, because AMPA-type glutamate receptor channels are gated more rapidly than any other ligand-activated ion channel. (2) Application of very brief synapse-like agonist pulses was necessary. In the original system of Franke *et al.* (1987), where the removal of the test solution occurred by bath perfusion, it turned out to be difficult to obtain a rapid and complete washout of test solution; responses to brief pulses ($\leq$ 1 msec) with this system were only occasionally reported (Dudel *et al.,* 1990). Trussell and Fischbach (1989) tried to overcome these difficulties by using an additional suction pipette, which, however, complicated the experimental situation. (3) The application system had to fit under a water immersion objective with small working distance (about 1.5 mm) because we intended to combine fast application and patch-clamp measurements in brain slices requiring upright microscope optics (Chapter 8, this volume). Techniques using multiple application pipettes were thus inconvenient.

The fast application system we use is based on a double-barreled application pipette made from theta glass tubing, both channels of which are continuously perfused with control and test solution (Colquhoun *et al.,* 1992; Fig. 1A). The interface between the two solutions can be moved across the patch by a piezoelectric element to which the application pipette is attached. The design fulfills all requirements mentioned above and particularly enables us routinely to apply very brief pulses of agonist.

2. Application Pipettes

2.1. Theta Glass Tubing

Dual-channel theta glass tubing (so called because the cross section resembles the Greek letter theta) is provided by several different suppliers. We use borosilicate theta glass from Hilgenberg [Malsfeld, Germany, mostly outer diameter (O.D.) 2.0 mm, inner diameter (I.D.) 1.4 mm, and septum thickness 0.1167 mm, provided on special request]. Glass tubing from other sources (Sutter, Novato, CA; Clark Electromedical Instruments, Pangbourne, England; WPI Instruments, Sarasota, FL) may be equally satisfactory. The 2.0-mm O.D. glass has the advantage that polyethylene tubing can be fed easily into the back side of the pipette (see

Note the black neoprene tubing covering the piezo translator. View from front. (C) Mechanical setup for holding the piezo translator. Note the two translation stages onto which the brass cylinder carrying the piezo element is mounted. The lever for moving the application pipette in the *Z* dimension is visible on the right and bottom; the whole system is shown in the low position. Also note the digital micrometer for monitoring the position of the application pipette in the *X* dimension. The neoprene tubing was removed from the piezo translator for the purpose of illustration. View from side.

below). Outer diameters of 1.6 mm can be also used; as the pulling procedure just scales down the original cross-section geometry (Brown and Flaming, 1986), identical tips can be produced. The use of thick-walled glass tubing is advantageous because the application pipette is mechanically more stable, minimizing oscillations in the solution exchange. Thin septa are preferable because the interface between the solutions flowing down the two barrels of the application pipette is sharper.

2.2. Pulling and Breaking

The theta glass tube is pulled to an O.D. of about 300 μm in one step. Electrode pullers providing symmetrical narrowing and short shanks are preferable. The raw piece is then cut into two parts by the use of a diamond pencil. This step is critical, because tip irregularities will cause turbulence of flow. The best results are obtained when the breaking is performed under visual control using a 10 × binocular microscope. Both ends of the glass are fixed by magnets on a metal table to slightly bend the central part of the tubing. The raw piece is then carefully scratched with the diamond pencil in the region where the diameter is smallest. Sometimes, it is just necessary to touch the glass, in other cases the tubing has to be rotated and scratched again until the two parts separate. Up to 80% of all raw pieces give acceptable tips (see Fig. 1A). A typical double-barreled application pipette shows a tip O.D. of 300 μm and a shank length of about 5 mm (Fig. 1A,B).

2.3. Tubing Connectors

To connect the two barrels of the application pipette with the solution reservoirs, we use polyethylene (PE) microtubing (Portex, England; O.D. 0.61 mm, I.D. 0.28 mm). The tubings are inserted into the back ends of the application pipette barrels and shifted as far as possible toward the tip. Subsequently, the space between the pipette wall and the PE tubing is filled with glue. Two-component epoxy resins (like UHU-Plus Endfest 300) appear to be most durable. The glue is filled into a 1-ml syringe and is squeezed into the space between PE tubing and glass wall via a 0.4-mm O.D. hypodermic needle. High pressure is required to eject the resin because of its viscosity; we use a custom-made screw-driven ejection device for this purpose. The spaces between PE tubing and glass wall should be filled almost entirely (1) to minimize the dead volume close to the application pipette tip (which is critical when the solution fed into the pipette is changed), (2) because it is difficult to replace gas in this volume by aqueous solution—unfilled space thus presents a potential source of gas bubbles, and (3) because solution may leak out if a connection to the back ends of the pipette remains. Ejection has to be stopped when the epoxy resin is about 4 mm away from the ends of the PE tubings because it tends to expand toward the tip while hardening.

2.4. Mounting the Application Pipette to the Piezo Translator

To fix the application pipette at the piezo translator, we use the following design. A small brass or perspex block with square cross section and a central hole (part I) serves to hold the application pipette (pipette, PE tubings, and brass or perspex block form the so-called application tool). A slightly bigger connector made from brass or perspex (part II), into which the small block exactly fits, is screwed onto the piezo element; part I can be tightly fixed within part II by a small screw (Fig. 1B). The application tool can be mounted and removed quickly. A precise vertical orientation of the septum of the application pipette is advantageous for judging the sharpness of the interface and for mapping out the optimal position of the recording pipette. Once the septum has been adjusted vertically, the application pipette is glued into part I. In case the pipette breaks, part I can be reused; remaining glass and glue can be removed after heating. Ideally, the mass of part I and II should be low, because extra mass reduces the resonance frequency of the system and thereby slows the movement (see below). A compromise between sufficient mechanical stability and low mass has to be reached (5–10 g total is appropriate).

2.5. Mounting the Piezo Translator to the Setup

The piezo translator is held horizontally in a brass cylinder. This cylinder allows rotation around a vertical axis, making it possible to swing the tool away from the recording chamber (Fig. 1C). The brass cylinder is mounted onto translation stages by which the position can be changed in *X* and *Y* dimensions. The whole system is attached to a lever by which it can be moved up and down (in the *Z* dimension). This makes it easy to remove the tool quickly from the bath and to put it back into exactly the same position, minimizing the risk of breaking. A third translation stage allows us precisely to adjust the height of the tool. During the day of the experiment, the application tool resides in a safe parking position within the recording chamber from which it is moved after a good outside-out patch is isolated; to facilitate bringing the tool into the field of view, a digital micrometer that monitors its position can be used. Different angles between the application pipette and the frontal plane (in which the recording pipette is located) were tried; the solution exchange appears to be fastest when this angle is about 150° (Fig. 1A). The angle between recording and patch pipettes and the horizontal plane is dictated by the small working distance of the water immersion objective; usually it is set to about 25°.

2.6. Perfusion of Application Pipettes

There are two possibilities to establish continuous perfusion of both barrels of the application pipette. The first method uses gas pressure. The solution reservoirs are connected to N_2 via a pressure regulator and a pressure meter (Bosch, Stuttgart, Germany) by which the flow can be adjusted very precisely. The pressure is typically set to 70 mbar. The second possibility is to use a commercially available infusion pump (e.g., Infors, Bottmingen, Switzerland). Up to six 30-ml syringes can be loaded after slight modification of the original design. A flow velocity of 100–150 μm/msec at the application pipette tip is what patches typically tolerate. Thus, for an application pipette with about 120-μm barrel width, a perfusion rate of about 10 ml/hr (per barrel) is appropriate. Establishment of additional bath perfusion (about 5 ml/min) is very important to remove the agonist from the biological preparation (e.g., the brain slice). Bath perfusion should be roughly parallel to the direction of solution flow out of the application pipette to avoid turbulence.

Only filtered solutions (0.2-μm pore size) should be used for pipette perfusion. If the filters are attached directly to the outflow of the reservoirs, the risk of destroying the patch by small particles is minimized. Unless specific precautions are taken, gas bubbles may present a continuous source of trouble; they will immediately destroy the patch when they come off the tip of the application pipette. To avoid bubble formation, it is recommended that the solutions be held at room temperature for several hours before use. Degassing for a few minutes using a vacuum pump also works. In addition, the application pipette should be perfused for 30 min or more before the beginning of the experiment to remove all gas bubbles from the system.

To exchange the solutions perfusing the control and test barrel of the application pipette, we use two rotary dial selectors, each consisting of six inflow and six outflow tubings. One of the outflow tubings of each of the selectors is connected to the respective barrel of the application pipette; the other outflow tubings are placed into a waste beaker. By setting the selectors in different positions, it is possible to perfuse the barrels of the application pipette with any possible combination of solutions. The pieces of the rotary dial selector should be covered with grease to avoid leak from one channel to another. The time necessary for complete exchange between two perfusing solutions is about 20 sec for the perfusion rates typically used.

2.7. Cleaning of Application Pipettes

Dirt attached to the application pipette tip is unfavorable because it leads to turbulence of the laminar flow. To avoid contamination with dirt, application pipettes have to be rinsed with distilled water after every day of experiment. If, despite of these precautions, the application pipette gets contaminated with dirt, it has to be cleaned chemically. Chromosulfuric acid appears to be the most effective substance for cleaning application pipettes. It is sucked into the application pipette and ejected again after a few seconds; this procedure should be repeated until the pipette is clean. Chromosulfuric acid is a rather hazardous substance and should not be left in the pipette too long because it destroys the epoxy glue; it should be washed out completely using distilled water.

2.8. Sharpness of the Interface

Interface thicknesses mapped with an open recording pipette (see below) typically are a few micrometers. Close to the application pipette tip, turbulence disturbs the interface. We therefore position the recording pipette about 100 μm away from the application pipette tip. Even under ideal conditions with perfectly laminar flow, diffusion of molecules across the interface will reduce its sharpness. The thickness of an ideal interface can be estimated theoretically by Einstein's random walk equation ($t = s^2/2D$, t representing time, s the mean distance the molecules diffuse during t, and D the diffusion coefficient). When the flow velocity is 100 μm/msec, and the diffusion coefficient is $D = 6\ 10^{-10}\ \mathrm{m^2/sec}$ (for acetylcholine or glutamate), the thickness of the ideal interface at 100-μm distance from the application pipette tip can be estimated as 1.1 μm.

3. Piezoelectric Elements and Power Supplies

3.1. Piezoelectric Elements

Fast piezo translators and power supplies are essential to generate brief pulses of test solution. If only application of longer pulses is intended, slower equipment might be sufficient. Three different designs of piezo translators are used to achieve large excursions (the excursion that can be achieved with a single piezo crystal is extremely small): (1) stacked designs, comprised of several piezo crystals built in series, (2) hybrid designs, which amplify the movement using a lever, and (3) bimorph designs, which are based on bending of a bimaterial strip. In our experience, stacks are superior to hybrids or bimorphs because translation is fast and reproducible. We use piezo elements from PI (Physik Instrumente, Waldbronn, Germany or Costa Mesa, CA); another supplier is Burleigh Instruments (Fishers, NY). There are several types of stacks, which differ in their resonance frequency f_0 and in the maximum possible excursion. For applying brief pulses of test solution, a piezoelectric element with a high resonance frequency and a relatively small excursion is preferable (like the P-245.20, excursion 20 μm, f_0 15 kHz; the P-245.30, excursion 40 μm, f_0 11 kHz; or the P-245.50, excursion 80 μm, f_0 8 kHz) to be able to move the application pipette back and forth rapidly. If the piezo translator were charged with infinite current, the time for expansion would be roughly $1/3\ f_0^{-1}$. Values for f_0 are tabulated in the manuals for the piezo translators. The effective resonance frequency, however, will be slightly smaller when the application tool is attached to the piezo element. For example, the mass of a P-245.20 piezo crystal is $m = 77$ g; if the mass of the application tool is $M = 5$ g, then f_0 will be reduced by a factor $\sqrt{[m/(m + 2M)]}$, i.e., from 15 to 14.1 kHz. This would give 24 μsec expansion time.

3.2. Power Supplies

Fast movements of the piezo translator require a high-voltage power supply, which allows rapid charging and discharging. We use the P-272 piezo power switch from PI (maximal output current 10 A). It has a TTL input, and the output voltage (and thereby the excursion) can be adjusted between 100 and 1000 V with a potentiometer. The excursion of the piezo translator is usually optimal at about 20 to 30 μm. Higher excursions may cause oscillations in the solution exchange. Different charge/discharge resistors can be selected; the switches are set to either position 2 (470 Ω) or 3 (2.2 kΩ), corresponding to the slower possible modes. The piezo translator is charged with a time constant $\tau = RC$, where C represents its capacitance and R is the charging resistance. For example, C of a stack translator P-245.20 is 100 nF; if R is 470 Ω (piezo power switch, position 2), then $\tau = 47$ μsec. More rapid charging usually leads to vibrations of the tool and oscillations in the solution exchange. Care should be taken to switch off the power supply of the piezo translator before the apparatus connected to its trigger input, because otherwise the piezo crystal may be damaged.

4. Tests of the Reliability of Solution Exchange

4.1. Open-Tip Response

Application pipettes can be tested by perfusing the two barrels with normal rat Ringer (NRR) and 10% NRR solution, respectively. Because of the different refractory indices of the two solutions, the interface between them is clearly visible in the light-microscopic view (Fig. 1A). The sharpness of the liquid filament gives a first indication of the quality of the application pipette. Subsequently, the speed of solution exchange is determined from the open-tip response during activation of the piezo element, using an open patch pipette filled with potassium-rich intracellular solution. As the interface moves over the open tip of the recording pipette, a change in liquid junction potential occurs that leads to a change of the pipette current in the voltage-clamp mode (Fig. 2A). Low positive pressure (about 10 mbar) has to be applied to the recording pipette interior; otherwise diffusion of external solution into its tip gives rise to slow components

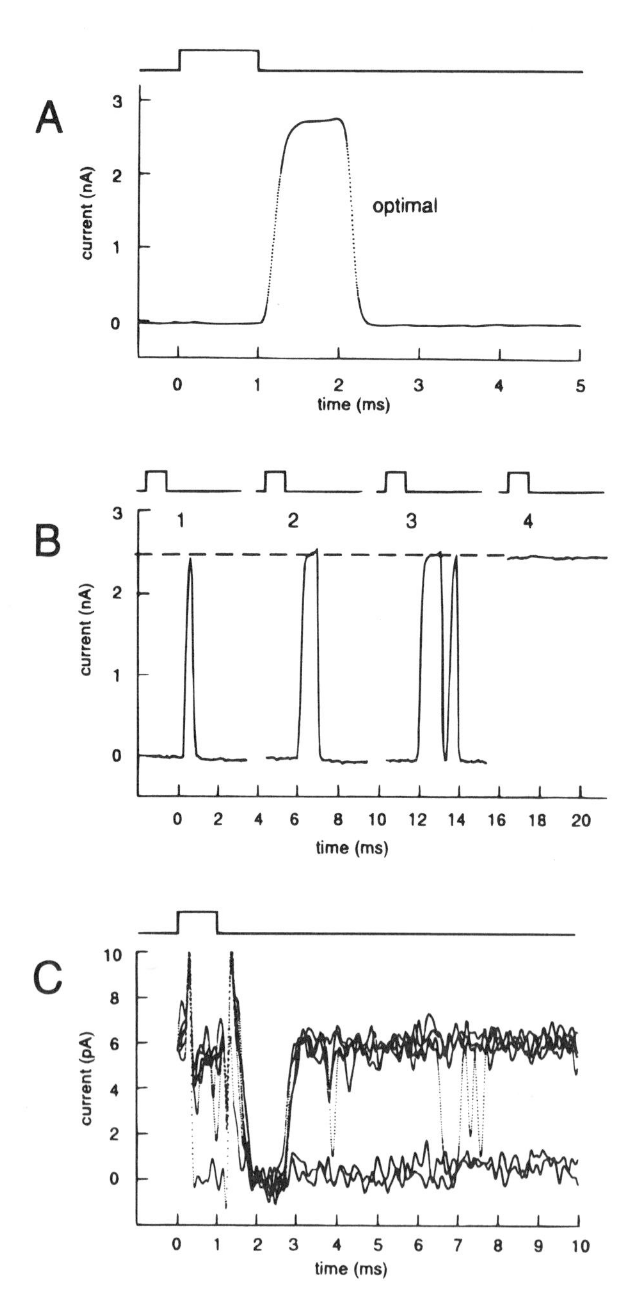

Figure 2. Test of solution exchange. (A) Open-tip response, solution exchange between NRR and 10% NRR (1-msec pulse). Optimal position of the recording pipette. Electrical pulse applied to the piezo element is shown on top. The latency between electrical pulse and solution exchange is mostly from the time it takes the solution to flow from the application pipette opening to the recording pipette tip. (B) Open-tip response, 1-msec pulse, for different distances of the recording pipette tip from the interface. Excursion of the piezo translator set to 30 μm. The recording pipette was moved in about 7-μm steps toward the interface. The optimal position of the application pipette would be position 2. The dashed horizontal line indicates the current level when the recording pipette tip is completely immersed in 10% NRR. (C) Solution exchange at an intact membrane patch. The opening of a high-conductance potassium channel triggers a solution exchange between NRR and a potassium-rich external solution (1-msec pulse). Note that the current change caused by the switch between solutions is only slightly slower than the spontaneous transitions between open and closed states of the channel. Membrane potential 0 mV. Adapted from Colquhoun *et al.*, 1992.

of solution exchange. The properties of individual application pipettes are somewhat variable; about one-third of all application pipettes tested are usable. In the best case, 20–80% exchange times of 100 μsec or less can be achieved for 5-MΩ recording pipettes (the exchange being somewhat slower when the recording pipette is larger).

4.2. Mapping out the Optimal Position

The optimal position of the recording with respect to the application pipette has to be found under visual control using a 40× water immersion objective and an eyepiece micrometer. To begin with, the recording pipette can be located in a region about 100 μm away from the application pipette tip and 20 μm away from the interface. In the *Z* dimension, the position of the recording pipette has to be slightly below the center of the application pipette tip to account for the angle between the tool and the horizontal plane (about 20°, see above).

The optimal distance between recording pipette and interface has to be mapped out carefully. When the recording pipette is moved toward the interface while 1-msec electrical pulses are repeatedly applied to the piezo translator, the open-tip responses change in a characteristic manner (Fig. 2B). (1) The first measurable response is much briefer than the electrical pulse fed into the piezoelectric element and does not reach the steady-state level. In this position, AMPA-receptor-mediated currents elicited by glutamate pulses in intact patches would not be fully activated. (2) The optimal position is a few micrometers further toward the interface. The length of the solution pulse is then identical to the length of the electrical pulse, and rising and decaying phase are monotonic and free from oscillation artifacts. In the optimal position, the open tip responses to pulses of different length (1, 10, 100 msec) should look almost square, and the maximum amplitudes obtained with brief and long pulses should be identical. This optimal position is typically about 10 μm away from the interface, depending on the excursion of the piezo translator. (3) When the recording pipette is moved further toward the interface, oscillation artifacts appear in the decaying phase, indicating that removal of the test solution is delayed and incomplete. In this position, the decay of AMPA-receptor-mediated currents activated by brief glutamate pulses in intact patches would become slower and wobbly. (4) Finally, when the recording pipette touches the interface, the baseline shifts upward. In this position, AMPA-receptor-mediated currents activated by glutamate pulses in intact patches would disappear because AMPA receptors become desensitized at equilibrium (Colquhoun *et al.*, 1992).

Once the optimal position of the recording with respect to application pipette has been found, it should be documented in order to be able to reproducibly find it at a later stage when a good outside-out patch is available. A characteristic reference point at the application pipette is one of the edges between the septum and the circular glass wall. With the eyepiece micrometer, the coordinates of the recording pipette tip with respect to this reference point can be given precisely. It is strongly recommended to record the open-tip response before every day of experiment and at the end of every successful experiment after the patch is blown off. In addition to verifying that the agonist application was fast enough in the particular experiment, this also makes it possible to identify the exact time when the solution exchange occurred.

4.3. Test at an Intact Membrane Patch

Solution exchange at an intact patch will inevitably be slower than at a bare recording pipette tip. To perform a more stringent test of the performance of the application system, we decided to apply pulses of potassium-rich external solution to an outside-out patch while a large-conductance (calcium-activated) potassium channel is open (Fig. 2C; Colquhoun *et al.*, 1992). The most elegant way to perform the experiment is to trigger the movement of the application pipette by the opening of the channel using an event detector (e.g., Al 2020A, Axon Instruments, Foster City, CA). In our experience the solution exchange at the intact membrane in most cases will be only slightly slower than at the open recording pipette tip. However, it is recommended to verify this at least a few times with membrane patches isolated under conditions identical to those in the experiments with agonist application. The rise time of the AMPA-receptor-mediated current activated by glutamate also represents a reliable indicator of the speed of solution exchange in intact membrane patches; 20–80% rise times of currents activated by pulses of 1 mM glutamate are usually very brief, between 200 and 600 μsec. In about 10% of all patches obtained from cells in brain slices, however, the rise time is much longer. This appears to present a problem with the patch rather than with the application pipette, as can be verified by checking the open-tip response at the end of the experiment. In some cases, dirt or membrane fragments attached to the patch can be identified visually. Patches with such slowly rising responses should be rejected.

4.4. Influence of Patch Size and Geometry on Solution Exchange Time

The larger the patch, the slower the solution exchange, for two reasons. (1) In large patches, the receptors in different regions are not reached by the agonist at the same time. This prolongs the exchange time. (2) Because of the viscosity of the solution, the flow velocity close to the membrane is reduced, creating an unstirred layer. In large patches, the unstirred layer is thicker. Assuming that the patch has a spherical geometry with radius r, the velocity profile can be approximated as $v(h) = v_\infty \{1 - [r/(r + h)]^3\}$ with v representing flow velocity (Maconochie and Knight, 1989). The solution exchange can be envisioned to proceed in two steps. At some distance from the membrane it occurs by convection; from there to the surface where the flow velocity is 0, it occurs by diffusion. The time necessary for solution exchange at a distance h from the patch surface is $t_1(h) = s/v(h)$, where s is interface thickness. The time required for diffusion from there to the patch surface is $t_2(h) = h^2/2D$ (see above). The minimum of $t_1(h) + t_2(h)$ represents an approximation of the exchange time (Maconochie and Knight, 1989). It is evident from these considerations that the larger r, the slower the solution exchange. Because of the Ω shape of inside-out patches, application is necessarily slower than for outside-out patches. If not much suction is applied to form the seal, however, fairly rapid solution exchange can be also achieved on inside-out patches (see Markwardt and Isenberg, 1992).

4.5. Vibrations and Charging Artifacts

Occasionally vibrations and charging artifacts superimpose on the agonist-activated currents, even when the solution exchange is totally adequate. These two types of artifacts

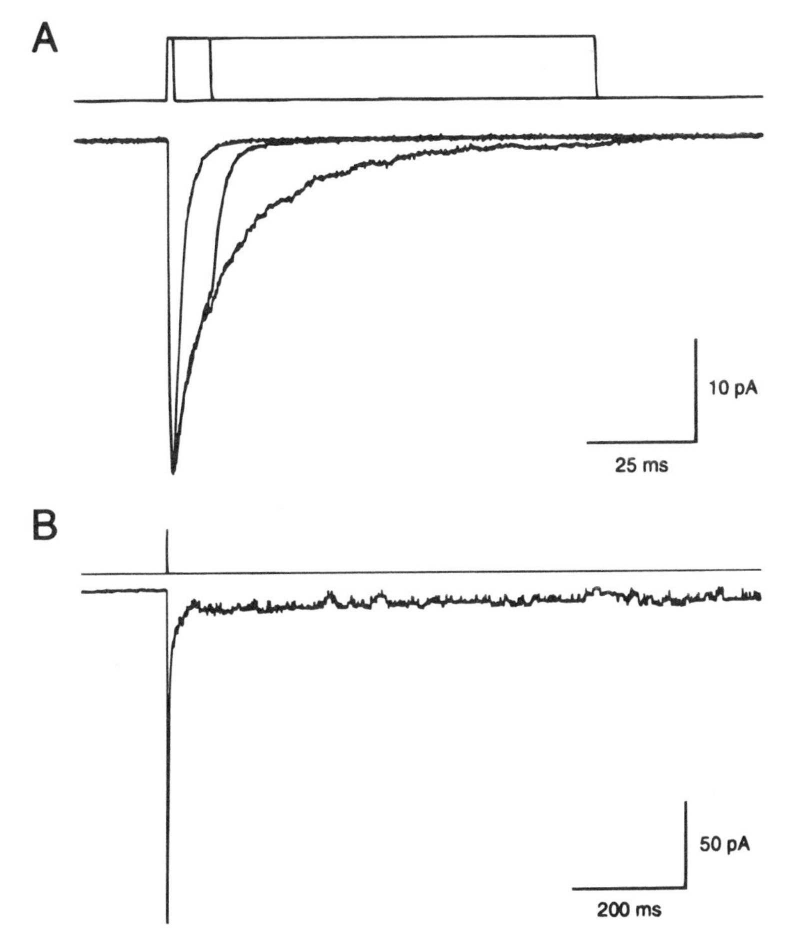

Figure 3 Examples. (A) AMPA-receptor-mediated current activated by glutamate pulses (1 mM) of different lengths. External solution NRR; 8–14 single records averaged. Membrane potential −50 mV. (B) Dual-component current. Note the openings and closings of single NMDA receptor channels in the late phase of the recording. External solution NRR without magnesium ions; 10 μM glycine added. Single trace. Membrane potential −80 mV. CA3 dendritic membrane patches, open tip responses recorded at the end of the experiment are shown on top. Corner frequency of eight-pole low-pass filter set to 3 kHz. Adapted from Spruston *et al.*, 1995.

are troublesome when the agonist-activated currents are small (e.g., because of low channel density) or when resolution at the single-channel level is required. High-frequency vibrations appear shortly after the piezo translator movement. There seem to be multiple components of these vibrations, including microphonia, conduction of the vibration over the metal parts in between application and recording pipette (including the basement), and conduction of the vibration over the surface of the bath solution. Microphonia can be reduced by covering the piezo translator with thick neoprene tubing, which significantly decreases the click noise that occurs as the piezo translator moves. The conduction of vibration over the basement can be minimized by inserting a piece of silicone between the piezo element and the brass cylinder or, alternatively, below the brass cylinder. The conduction of the vibration over the surface of the bath solution can possibly be reduced by changing the fluid level or the height of the water immersion objective. Charging artifacts appear as spikes when a voltage step is applied to the piezo element; they can be minimized by appropriate shielding and grounding or by setting the switches for the charge/discharge time constant of the power switch to slower values (see above).

5. Examples

Figure 3 shows a few applications of the technique under optimal conditions. AMPA-receptor-mediated currents evoked by pulses of 1 mM glutamate of different length are illustrated in Fig. 3A. The 20–80% rise time of the current is about 300 μsec. The peak currents evoked by 1-, 10-, and 100-msec pulses superimpose, but the decay time constants are clearly different (deactivation time constant about 2.5 msec; desensitization time constant about 10 msec; Colquhoun *et al.,* 1992; Spruston *et al.,* 1995). In Fig. 3B, a dual-component current activated by a brief glutamate pulse is shown. Openings of single NMDA receptor channels constituting the slow component can be clearly resolved (Fig. 3C). The open-tip response recorded at the end of the experiment is shown above each of the traces. Various types of measurements can be performed, including double-pulse experiments, prepulse experiments, dose–response curves, etc. (Colquhoun *et al.,* 1992; Spruston *et al.,* 1995).

ACKNOWLEDGMENTS. I thank Prof. B. Sakmann for continuous support and Drs. A. Villarroel, D.-S. Koh, and N. Spruston for critically reading the manuscript. I also thank D. Müller, H. P. Maier, and K. Schmidt for help with the construction of the fast application devices.

References

Akaike, N., Inoue, M., and Krishtal, O. A., 1986, “Concentration-clamp” study of γ-aminobutyric-acid-induced chloride current kinetics in frog sensory neurones, *J. Physiol.* **379:**171–185.

Brett, R. S., Dilger, J. P., Adams, P. R., and Lancaster, B., 1986, A method for the rapid exchange of solutions bathing excised membrane patches, *Biophys. J.* **50:**987–992.

Brown, K. T., and Flaming, D. G., 1986, *Advanced Micropipette Techniques for Cell Physiology,* John Wiley & Sons, New York.

Colquhoun, D., Jonas, P., and Sakmann, B., 1992, Action of brief pulses of glutamate on AMPA/kainate receptors in patches from different neurones of rat hippocampal slices, *J. Physiol.* **458:**261–287.

Dudel, J., Franke, C., and Hatt, H., 1990, Rapid activation, desensitization, and resensitization of synaptic channels of crayfish muscle after glutamate pulses, *Biophys. J.* **57:**533–545.

Fenwick, E. M., Marty, A., and Neher, E., 1982, A patch-clamp study of bovine chromaffin cells and of their sensitivity to acetylcholine, *J. Physiol.* **331:**577–597.

Franke, C., Hatt, H., and Dudel, J., 1987, Liquid filament switch for ultra-fast exchanges of solutions at excised patches of synaptic membrane of crayfish muscle, *Neurosci. Lett.* **77:**199–204.

Johnson, J. W., and Ascher, P., 1987, Glycine potentiates the NMDA response in cultured mouse brain neurons, *Nature* 325:529–531.

Krishtal, O. A., and Pidoplichko, V. I., 1980, A receptor for protons in the nerve cell membrane, *Neuroscience* **5:**2325–2327.

Krishtal, O. A., Marchenko, S. M., and Pidoplichko, V. I., 1983, Receptor for ATP in the membrane of mammalian sensory neurones, *Neurosci. Lett.* **35:**41–45.

Maconochie, D. J., and Knight, D. E., 1989, A method for making solution changes in the sub-millisecond range at the tip of a patch pipette, *Pflügers Arch.* **414**:589–596.
Markwardt, F., and Isenberg, G., 1992, Gating of maxi K^+ channels studied by Ca^{2+} concentration jumps in excised inside-out multi-channel patches (myocytes from guinea pig urinary bladder), *J. Gen. Physiol.* **99**:841–862.
Spruston, N., Jonas, P., and Sakmann, B., 1995, Dendritic glutamate receptor channels in rat hippocampal CA3 and CA1 pyramidal neurons, *J. Physiol.* (in press).
Trussell, L. O., and Fischbach, G. D., 1989, Glutamate receptor desensitization and its role in synaptic transmission, *Neuron* **3**:209–218.

Chapter 11

Electrochemical Detection of Secretion from Single Cells

ROBERT H. CHOW and LUDOLF VON RÜDEN

1. Introduction

Electrochemical methods based on the oxidation or reduction of specific transmitters enable exquisitely sensitive measurements of secretion from single cells. We start with a brief review of the history of their application to single cells. Then we focus on basic principles. The remainder of the chapter concerns how to make measurements and how to analyze the results.

Although electrochemical methods have been in use for over 20 years to study *in vivo* changes of neurotransmitter concentrations in the brain (readers interested in the early history should refer to the book by Justice, 1987), they have been applied only recently to single cells (Leszczyszyn *et al.,* 1990; Wightman *et al.,* 1991; Tatham *et al.,* 1991; Chow *et al.,* 1992; Jankowski *et al.,* 1992; Alvarez de Toledo *et al.,* 1993). In the case of the adrenal chromaffin cells, the electrochemical approach is more sensitive than capacitance measurements (see Chapter 7, this volume) for detecting individual secretory quanta (Wightman *et al.,* 1991; Chow *et al.,* 1992). Other advantages compared to the capacitance method are that (1) exocytosis can be monitored without interference from overlapping endocytosis (von Rüden and Neher, 1993), (2) the released product is directly monitored (not a model-dependent parameter such as electrical capacitance), (3) voltage-clamp control is not necessary, and, therefore, there are fewer restrictions on cell shape, and (4) the methods are "noninvasive" in the sense that the cell is not subject to the dialysis of cytosolic components that occurs with (whole-cell) patch-clamp measurements.

Many secreted products are readily oxidizable. Among these, the most intensively studied are norepinephrine (see Fig. 1), epinephrine, dopamine, and serotonin. Derivatives of these species, as well as nitric oxide (Malinski and Taha, 1992; Iravani *et al.,* 1993), ascorbic acid, and uric acid, are also oxidizable. Peptides and proteins containing the amino acids tyrosine, tryptophan, and cysteine can, at least in theory, be oxidized (e.g., enkephalin, Armstrong-James *et al.,* 1981; somatostatin, Crespi, 1991), but the amount released is usually small, and diffusion of these compounds is slow, thus severely limiting their detection. In

ROBERT H. CHOW and LUDOLF VON RÜDEN • Department of Membrane Biophysics, Max-Planck-Institute for Biophysical Chemistry, Am Faßberg, D-37077 Göttingen-Nikolausberg, Germany. Present address of Dr. Chow: Department for Molecular Biology of Neuronal Signals, Max-Planck-Institute for Experimental Medicine, D-37075 Göttingen, Germany. Present address of Dr. von Rüden: Howard Hughes Medical Institute, Department of Molecular and Cellular Physiology Beckman Center, B153 Stanford University Stanford, CA 94305-5428 USA.
Single-Channel Recording, Second Edition, edited by Bert Sakmann and Erwin Neher. Plenum Press, New York, 1995.

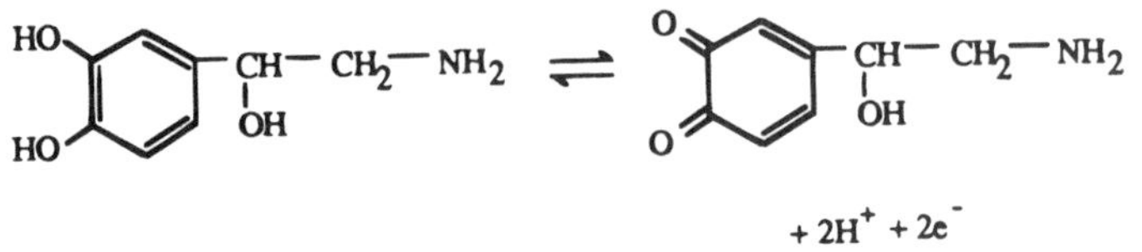

HO, HO — CH — CH_2 — NH_2, OH ⇌ O, O — CH — CH_2 — NH_2, OH

$+ 2H^+ + 2e^-$

Figure 1. Oxidation of norepinephrine (left) to its quinone product (right) leads to the loss of two protons and two electrons.

the case of insulin, the oxidation of cysteine disulfide bonds can be accelerated with the help of a catalyst (Kennedy *et al.*, 1993). Furthermore, some secreted products, although not readily oxidizable, can be chemically or enzymatically converted to compounds that are. Examples include glucose (Kawagoe *et al.*, 1991b; Wang and Angnes, 1992), acetylcholine (Kawagoe *et al.*, 1991b), and glutamate (Pantano and Kuhr, 1993). It is not yet clear whether the rate of enzyme conversion in these cases is fast enough to enable detection of secretion from single cells.

The first electrochemical electrodes used in neuroscience were made from carbon paste packed into fine (50-μm to 1.6-mm diameter) Teflon® tubes (Kissinger *et al.*, 1973). These electrodes were used to monitor monoamine and ascorbic acid concentrations in brain tissue. Although several other materials have also been tried, carbon fibers have become the material of choice for making electroactive electrodes in neurophysiological studies, particularly for single-cell studies. The first description of fabricating carbon-fiber electrodes was reported in 1978 by Gonon *et al.* (1978), and this was soon followed by other publications (Ponchon *et al.*, 1979; Armstrong-James and Millar, 1979). Since then there have been further improvements and modifications (Gonon *et al.*, 1984; Kelly and Wightman, 1986; Kawagoe *et al.*, 1993).

Electrochemical measurements of secretion from single cells were first reported in 1990. Millar's laboratory showed that stimulation of chromaffin cells, which secrete catecholamines, resulted in prominent oxidation current signals that were clearly related to a rise in intracellular calcium (Duchen *et al.*, 1990). Soon afterward, Wightman's laboratory demonstrated simultaneous recording of secretion by two carbon-fiber electrodes placed on opposite sides of a single chromaffin cell (Leszczyszyn *et al.*, 1990). Stimulating the cell with acetylcholine led to a shower of spiking oxidative transients recorded at both electrodes. The transients recorded at one electrode were not synchronous with the signals at the other electrode, raising the tantalizing possibility that each electrode was monitoring highly local events and that each spike represented the release of catecholamines from a single vesicle.

Subsequently, investigators have confirmed that carbon-fiber microelectrodes detect quantal secretory events, not only from chromaffin cells (Wightman *et al.*, 1991; Chow *et al.*, 1992) but also from mast cells (Duchen, 1993; Alvarez de Toledo *et al.*, 1993), pancreatic β cells (Kennedy *et al.*, 1993), cartoid body glomus cells (Urena *et al.*, 1994), and pineal cells (Marin and Tabares, 1993). Figure 2 illustrates examples of the individual current spikes. They are often preceded by a slower pedestal, called the "foot" signal (Chow *et al.*, 1992; Neher, 1993; Alvarez de Toledo *et al.*, 1993). In chromaffin cells, the foot signal was attributed to the slow escape of catecholamine molecules through the early fusion pore (the aqueous canal that initiates the connection between the secretory vesicle interior and the cell exterior; Almers, 1990).

Alvarez de Toledo's laboratory (1993) has confirmed that secretion does indeed occur through the early fusion pore in beige mouse mast cells. These cells have vesicles large enough (mean diameters of 2.5 μm) that the conductance of the fusion pore can be inferred

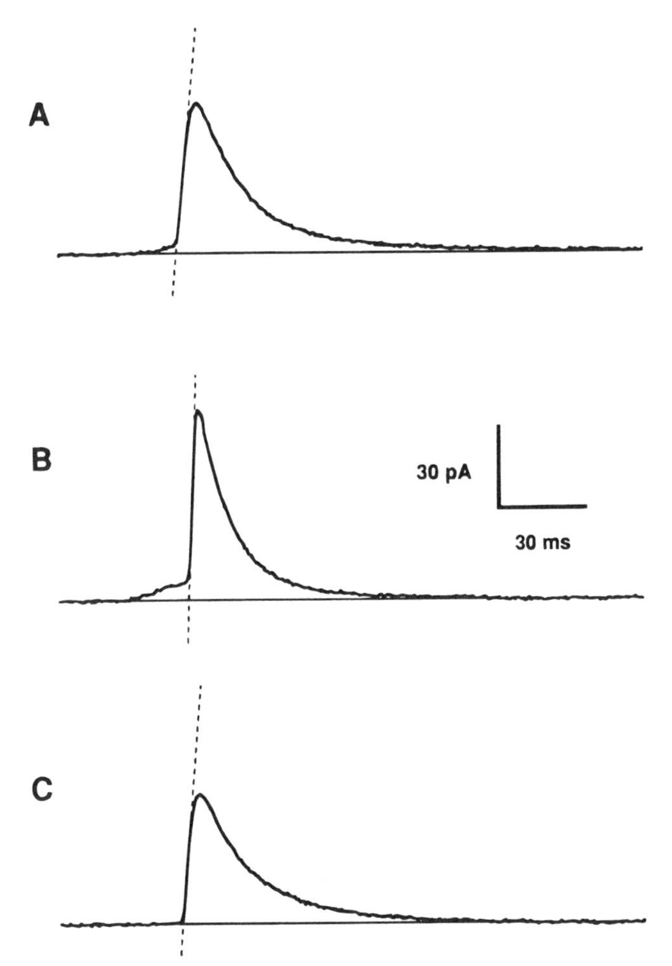

Figure 2. Examples of individual amperometric current transients with different foot durations. C has no discernible foot signal. The lines drawn through the rising phase are extrapolated back to the baseline in order to determine the end of a foot signal (see Section 4.1.2. and Fig. 10 for a detailed description of foot duration analysis and Fig. 9 for further examples of foot signals). Scale bars in panel B also apply to A and C.

from capacitance measurements before it dilates irreversibly (Almers, 1990). Shortly after a secretory vesicle fuses to the plasma membrane, the conductance of the fusion pore sometimes "flickers" open and closed several times or hesitates before it explodes open (Fernandez *et al.,* 1984; Breckenridge and Almers, 1987). Only when such flickering or hesitation was noted in the capacitance record did Alvarez de Toledo and colleagues record an electrochemical foot signal. Thus, the foot signal occurs because the initially narrow fusion pore limits the rate of escape of transmitter (serotonin, which is cosecreted with histamine from mast cell granules).

The smaller chromaffin cell vesicles have a shorter associated foot signal than the mast cell vesicles, yet the fraction of transmitter released during the foot phase compared to the remainder of the quantal signal is larger. Extrapolating this observation to the tiny (about 40-nm diameter) synaptic vesicles, Alvarez de Toledo and colleagues (1993) suggest that much of a synaptic vesicle's contents (about 8%) could be released during the foot signal. If the transmitter is freely diffusible (which is not the case for mast cell and chromaffin cell granules), then over 50% would be released during transient fusion events. Almers *et al.* (1989) have performed related calculations in which they found that synaptic vesicles should empty their (freely diffusible) transmitter through an undilated fusion pore (conductance of 300 pS, similar to the mean initial conductance of fusion pores in mast cells) with a time constant of only 250 μsec. Thus, completed fusion may not be necessary for synaptic transmission, allowing vesicle membrane to be recycled without ever undergoing full fusion ("kiss-and-run" exocytosis; see Ceccarelli and Hurlbut, 1980; Torri-Tarelli *et al.*, 1985; Neher, 1993).

Other details of the time course of release are surprising. Chow *et al.* (1992) found that the spike phase, which presumably occurs after the fusion pore has dilated, is still much slower than expected for freely diffusing catecholamine released instantaneously from a point source. In order to fit the time course with a diffusion model (see Section 2.4, below, and the Appendix), it was necessary to use a diffusion constant 20 times slower than that of freely diffusing catecholamine or to assume that the distance from release site to detector was greater than estimated experimentally. Subsequently, Wightman's laboratory has shown that the time course of catecholamine release from chromaffin cells can be influenced by pH and external calcium concentration (Jankowski *et al.*, 1993). Their findings lend support to the idea that catecholamine release requires dissolution of a storage matrix that binds catecholamine and that the rate of dissolution can be altered. Thus, the release of catecholamine is not instantaneous. At least two mechanisms contribute to the retardation of release: dilation of the fusion pore and dissolution of a storage matrix. Interestingly, in mast cell granules the rate of fusion pore dilation may be accelerated by expansion of the dissolving storage matrix (Monck *et al.*, 1991).

Combining the patch-clamp technique and electrochemical detection has provided additional information. In excitable cells, voltage-clamp depolarizations can be used to activate voltage-dependent calcium currents and thereby stimulate secretion. The duration of the stimulus is precisely controlled, allowing one to adjust the amount of calcium injected by each depolarization such that, on average, only one quantal event (or only a few events) is elicited per stimulus. Under these conditions, one can record the latency of individual quantal events (the time between the beginning of the stimulus and the beginning of an event) and evaluate the frequency of events. As is true for synapses, quantal secretion in chromaffin cells obeys Poisson statistics (Chow *et al.*, 1992). In comparison with synapses, however, the events occur with a much longer latency—in many instances up to 100 msec after the depolarization has ended.

The long secretory latency is surprising if secretory vesicles and calcium channels are colocalized, as they appear to be in neuromuscular junctions and squid giant synapses (Adler *et al.*, 1991; Augustine *et al.*, 1991; Simon and Llinas, 1985). With colocalization, the calcium concentration at the vesicle fusion machinery should rise and collapse within at most a few tens of microseconds, and the concentration should reach as high as a hundreds of micromolar or even millimolar (Chad and Eckert, 1984; Fogelson and Zucker, 1985; Simon and Llinas, 1985). The rate of secretion, which is thought to have a high-order dependence on calcium concentration (Dodge and Rahamimoff, 1967; Augustine and Charlton, 1986; Dudel, 1989; Zucker *et al.*, 1991; Heinemann *et al.*, 1993), should therefore also rise and fall nearly as rapidly.

The latter observations thus raise the question of whether secretory granules and calcium channels might be less strictly colocalized in neuroendocrine cells than in synaptic active zones. In this context, it is interesting to note that in bovine chromaffin cells, EGTA at millimolar concentrations can block secretion (Neher and Marty, 1982), suggesting that the distance between calcium channels and vesicles is great enough that exogenous buffer can compete with the fusion apparatus for calcium. In contrast, EGTA concentrations up to 80 mM are not able to block secretion in squid giant synapse (Adler *et al.*, 1991), supporting strict colocalization there.

On the other hand, there are data suggesting the existence of "hot spots" of exocytosis on the chromaffin cell surface, perhaps indicative of structural specializations. Using carbon-fiber electrodes with 2-μm-diameter tips, Schroeder *et al.* (1993) have found highly localized hot spots where exocytosis occurs. Independent data, also from chromaffin cells (Monck *et al.*, 1994), point to "hot spots" of calcium entry—focal domains of submembrane calcium elevation—which suggests that calcium channels may be clustered. Finally, Artalejo *et al.* (1994) have presented data suggestive that the fusion machinery is colocalized with one of three types of calcium channel found in bovine chromaffin cells. It will be interesting to see how the diverse observations will be reconciled.

2. Principles

2.1. Electrochemical Detection

Detection of oxidizable secreted products requires an appropriate detector. Electrochemists call an electrode having the desired properties a "polarizable electrode" (see the discussion in Chapter 7 of Crow, 1988). When such an electrode is immersed in physiological saline, electrical current is unable to flow readily across the solid–liquid interface. In the electrode the carriers of current are electrons, whereas in solution the carriers are ions dissolved in solution, for example, Na^+, K^+, Cl^-. Unless the electrode surface can undergo a rapid reaction with one of the dissolved species to "convert" electrons to ions and vice versa (as occurs in the case of silver/silver chloride electrodes immersed in chloride-containing saline solution), charge will not readily traverse the interface. Thus, when a voltage is applied to a polarizable electrode, excess charge accumulates at the surface facing the solution, rendering the surface electroactive. Of course, for electroneutrality the excess charge must be balanced by an equal charge of opposite sign on the solution side. Mobile counterions in solution are attracted electrostatically to the interface and align themselves there in a structured "double layer." Electroneutrality is achieved over a finite distance, with the electrical field falling away with a characteristic space constant, called the Debye length (about 9 Å in physiological saline). At equilibrium, little or no net current flows. On the other hand, when oxidizable (or reducible) molecules diffuse to the surface of the electrode, electrons are transferred, leading to current flow.

At any fixed voltage, a dynamic equilibrium is established between the electrode surface and the reduced and the oxidized forms of the reactive species in solution. The ratio of these species depends on the electromotive "force" or tendency of the electrode to transfer electrons, which could lead to favoring one species over the other. The *redox potential* is the potential, referred to a standard hydrogen electrode, at which half of the molecules are in the oxidized form and half in the reduced form at equilibrium. To favor conversion of all molecules to the oxidized form, one must apply a voltage that exceeds the redox potential. The speed of electron transfer depends on the amount by which the applied voltage exceeds the redox potential and also on the intrinsic "ease" with which the chemical gives up electrons. To accelerate an intrinsically slow transfer, one increases the applied voltage (an *overpotential*).

Typical materials used for the electrodes include carbon, platinum, and gold. Carbon has become the choice of neurophysiologists because, compared to the metals, it has more stable electrochemical properties. Furthermore, carbon fibers are readily available with diameters of 5 to 35 μm (and meters long)—ideal for measurements with single cells. At such dimensions the fibers are highly rigid and easily threaded into small capillaries without the threat of irreparable kinking.

The surface chemistry of graphite carbon fibers is highly complex (McCreery, 1991; Kawagoe *et al.*, 1993). The bulk carbon consists of densely packed, concentric cylinders of aromatic carbon rings interlinked in "chicken-wire" configuration, a cut surface has many functional groups (phenolic hydroxyl, carbonyl, quinone, and carboxylic acid groups). Some or all of these functional moieties may be involved in electron transfer reactions.

2.2. Quantitative Electrochemistry

One useful feature of electrochemical measurements is that the number of molecules that have been oxidized or reduced can be quantified if only one electroactive species is involved and if the number of electrons transferred per molecule is known. The relationship between the total charge transferred and the number of molecules reacted is known as Faraday's law:

$$Q = \int Idt = \frac{zFM}{N_A} = zeM \tag{1}$$

Here, Q represents the total charge involved in the redox reaction, which is obtained by integrating the current (I) transient; M represents the number of molecules reacted; z is the number of moles of electrons transferred per mole of compound reacted; F is Faraday's constant, 96,485 coul/mol; N_A is Avogadro's number, 6.023×10^{23}; and e is the elementary charge, 1.6×10^{-19} coul.

Of course, if mixtures of reactants are involved, the situation becomes complicated. Fortunately, for many cells there is only one reactive species released, or one reactive species predominates (an example is the chromaffin cell; Leszczyszyn *et al.*, 1991).

The quantitative nature of electrochemical signals enabled estimates to be made of the number of catecholamine molecules released from single chromaffin cell granules (Wightman *et al.*, 1991; Chow *et al.*, 1992). The mean value—approximately 3 million molecules—was in surprisingly close agreement with estimates obtained previously by combining electron micrographic counts of the number of vesicles per cell and biochemical measurements of the total catecholamine per cell (Phillips, 1982).

2.3. Voltammetric Techniques

In order to oxidize (or reduce) compounds in solutions, a voltage must be applied to the electrochemical detector. The measurement of the current response to an applied voltage is called "voltammetry"—to be contrasted with "potentiometry," the measurement of equilibrium potentials at electrode interfaces for specified electrochemical conditions (for example, with pH electrodes). Voltammetry is analogous to voltage-clamp recording in electrophysiology.

Depending upon the purpose, either a DC voltage or some periodic voltage waveform is chosen. We will discuss only two of the possible approaches, amperometry and fast cyclic voltammetry.

2.3.1. Amperometry

In the simplest case, constant-voltage amperometry (not to be confused with chronoamperometry, another technique, which is beyond the scope of this chapter), one applies a DC voltage. Typically, the chosen voltage exceeds the redox potential of the compound of interest by at least 200 mV—an "overpotential" in order to speed the rate of oxidation. At sufficiently large potentials, the oxidation reaction becomes limited by the rate of mass transport to and from the electrode surface; that is, the time course of the signal is determined not by the rate of electron transfer (fast) but by the time course of diffusion of reactant in the vicinity of the electrode surface (slow).

Of the different voltammetric techniques, amperometry gives the highest-time-resolution measurements of secretion. It was the approach used to obtain measurements of the foot signals in chromaffin and mast cells (Chow *et al.,* 1992; Alvarez de Toledo *et al.,* 1993). On the other hand, amperometry gives little information about the molecule or molecules being oxidized. Information about the chemical species being studied can be obtained using fast cyclic voltammetry.

2.3.2. Fast Cyclic Voltammetry

In fast cyclic voltammetry, a periodic voltage pattern (triangle wave) is applied to the electrode, first in the absence of reactant for a "background trace" and then in the presence of reactant. The background trace, which represents the currents charging the double-layer capacitance (and oxidation or reduction of carbon surface moieties), is subtracted from the subsequent traces to reveal the currents that result from oxidation or reduction of the reactant, the so-called "Faradaic current." This current is plotted against the applied potential.

Figure 3 illustrates the traces from a cyclic voltammogram, recorded on pressure-pipette application of norepinephrine (Fig. 3B–D) or serotonin (Fig. 3E–F) to a carbon fiber. The applied potential appears in Fig. 3A. The ramp rate for the segment from −700 mV to +900 mV was 400 V/sec (the rates in the published literature have ranged from about 100 to 500 V/sec; Stamford, 1990). The background current (solid line) and, superimposed, the current record in the presence of norepinephrine (dotted line) are illustrated in Fig. 3B; the difference trace (trace recorded in the presence of norepinephrine minus the background trace) is shown in Fig. 3C, and the corresponding current–voltage trace, also called a *voltammogram* or *cyclic voltammogram,* appears in Fig. 3D. The corresponding traces for serotonin appear in Fig. 3E (difference trace) and 3F (voltammogram).

Readers should be aware that according to standard electrochemical convention (which is not strictly followed in neuroscience applications of cyclic voltammetry), the axes in voltammograms are inverted compared with those in modern electrophysiology (but not compared to the classical works by, for example, Hodgkin and Huxley, 1952). More specifically, the definition of a positive voltage is the same as in modern electrophysiology (when a positive potential is applied to a "working" electrode, positive ions in solution will flow from the electrode to the bath or reference electrode), but the left side of the horizontal axis in voltammograms is considered positive. On the other hand, current is defined as positive when cations move from the bath into the working electrode (exactly the opposite of the convention used in whole-cell recordings), and positive current (current measured when a reactant is reduced at the working electrode) is above the horizontal axis in voltammograms. To make matters somewhat more confusing, in amperometry, oxidation current (negative current flow according to electrochemical convention) is displayed as an upward deflection

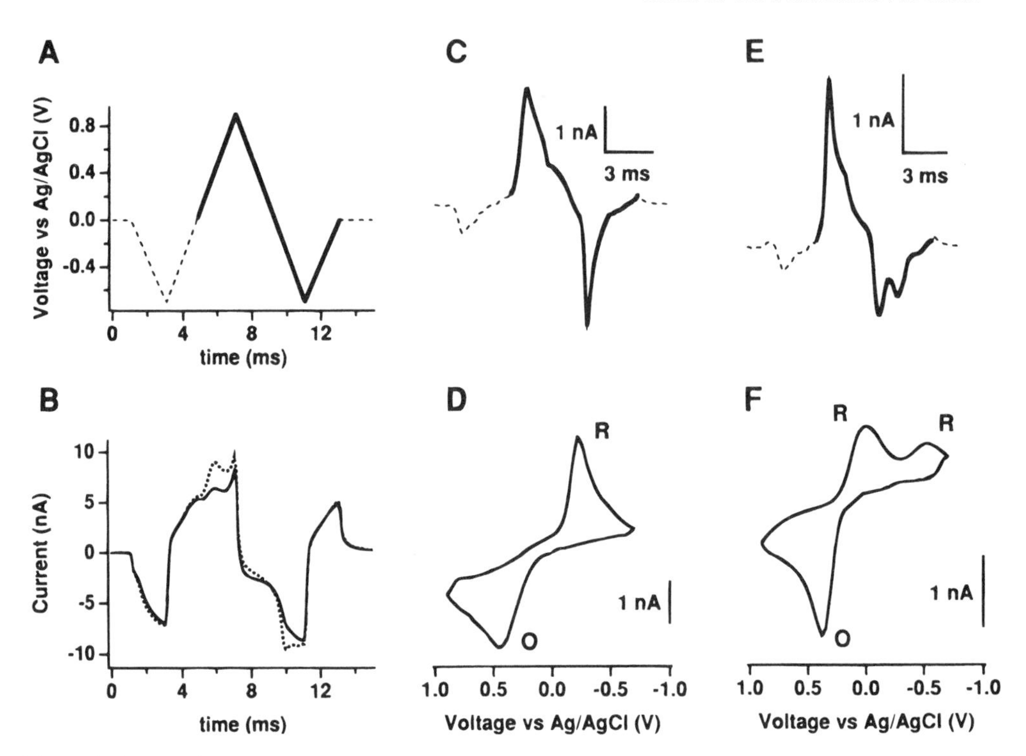

Figure 3. Fast cyclic voltammograms of norepinephrine and serotonin. (A) Applied voltage waveform. The potential varies between −700 and +900 mV with respect to a Ag/AgCl reference electrode. The rate of voltage change in the ramps is 400 V/sec. (B) Current measured in response to the voltage shown in A, in the presence (dotted curve) or absence (solid curve, "background" trace, recorded in Ringer solution) of norepinephrine (1 mM norepinephrine was applied via a pressure-application pipette from a distance of about 50 μm). (C) Difference trace obtained by subtracting the solid trace in B from the dotted trace. (D) Plot of difference current (from C) versus the applied voltage (from A) to obtain a voltammogram. O, oxidation peaks. R, reduction peaks. (E and F) The difference trace and the cyclic voltammogram, respectively, for serotonin. There are two reduction peaks for serotonin, compared to only one for norepinephrine. Further details in text.

versus time. We have chosen to display the current–voltage diagrams (voltammogram) according to standard electrochemical convention. Readers who are confused by the convention can turn the page upside down to examine Fig. 3D and F. This restores the conventions of electrophysiology.

To better understand the shape of the current–voltage plot for norepinephrine (Fig. 3D), consider the thick solid segment of the difference trace in Fig. 3C. (The applied voltage corresponding to the current trace is shown in Fig. 3A also as a thick solid segment.) At the start of the segment, the voltage is 0 mV. As the potential sweeps in the positive direction, the electromotive force progressively favors oxidation, leading ultimately to oxidation of norepinephrine. The oxidation current grows larger as the applied voltage increases beyond the redox potential, until the rate of consumption of reactant at the electrode exceeds the rate of resupply by diffusion. Local depletion of the reactant extends progressively further away from the electrode surface, reducing the concentration gradient that drives the diffusion of reactant to the electrode surface. Thus, the current actually gets smaller. The voltage then ramps in the negative direction. If the speed of the ramp is fast enough that the oxidized product has not diffused away by the time the voltage passes the threshold for reduction, one observes a reduction current.

Since specific compounds have specific redox potentials and rates of electron transfer, cyclic voltammetry of compounds can lead to distinct "signature" voltammograms. The identifying characteristics are the location and number of the peaks (both oxidizing and reducing). The exact location of a given peak is a complicated function of parameters such as the nature of the reactant, sensor surface area and geometry, the speed of the ramp, and the state of the electrode surface as well as the type of reference electrode (Justice, 1987). The ability to distinguish between two compounds is illustrated by comparing Fig. 3D and Fig. 3F, which show the current–voltage traces obtained with norepinephrine and serotonin, respectively. Both compounds have a single oxidation peak (designated O). Note, however, that the cyclic voltammogram for serotonin shows two reduction peaks (designated R), whereas that for norepinephrine shows only one reduction peak. On the other hand, a voltammogram for ascorbic acid (not shown) has no reduction peaks at all (see, for example, Stamford, 1990).

Although cyclic voltammetry readily differentiates between certain agents, it is limited in its ability to distinguish among the various monoamine transmitters. For instance, some catecholamines such as norepinephrine and dopamine are nearly indistinguishable by cyclic voltammetry.

Although cyclic voltammetry was used in the first recordings from single cells (Duchen *et al.,* 1990; Leszczyszyn *et al.,* 1990; Tatham *et al.,* 1991), most subsequent studies have used amperometry. This is because, with identified cells in culture, the nature of the secreted transmitter is usually not as interesting as the timing and amount of secretion. For studies of the kinetics of secretion, amperometry is superior (see above).

More details about the theory and practical details of fast cyclic voltammetry can be found in Justice (1987), Millar and Barnett (1988), and Stamford (1990). Numerous other types of periodic waveforms have been applied, each having its own advantages and disadvantages. Readers interested in learning more about such approaches should start with the introductory chapter in the book by Justice (1987). Two approaches that have appeared recently, which are not reviewed in this book, are fast differential ramp voltammetry (Millar and Williams, 1990) and continuous-scan cyclic voltammetry (Millar *et al.,* 1992).

2.4. A Little Diffusion Theory

In amperometry, one chooses (ideally) a command potential such that the rate of oxidation of individual molecules (secretory product) at the electrode surface is much faster than the rate of arrival of the molecules (i.e., the diffusion-limited rate). Thus, amperometric current signals for single vesicles reflect the time course of transmitter release and diffusion from the release site to the detector.

Diffusion-based signals get slower as the path over which molecules must diffuse increases. The slowing is not a linear function of distance: from the Einstein–Smolochowski equation for one-dimensional diffusion

$$\langle x^2 \rangle = 2D\tau \quad \text{or} \quad \tau = \frac{\langle x^2 \rangle}{2D} \tag{2}$$

(where $\langle x^2 \rangle$ is the mean squared displacement, D is the diffusion constant, and τ is time), we see that the time for a diffusing particle to travel some distance, x, goes with the square of x (see, for example, Atkins, 1990).

Consider for a moment the idealized case of instantaneous (δ distribution) release of transmitter from an infinite "release" plane and diffusion to a parallel infinite "detector" plane. In this case, let us say that the release surface "reflects" molecules that strike it, whereas the detector surface consumes ("absorbs") molecules that hit it. These conditions resemble those for vesicle fusion near the center of the disk surface of the carbon-fiber electrode. The signals will have a time course given by the infinite series (see Appendix, and Chow *et al.*, 1992)

$$\vec{J}_a = \lim_{n\to\infty} \sum_{i=-n}^{n} (-1)^i \frac{Mx(1-2i)}{2t\sqrt{\pi D t}} e^{-x^2(1-2i)^2/4Dt} \tag{3}$$

Here, J is the flux (molecules per second) into the detecting plane; M is the number of molecules released at the release plane at time $t = 0$; x is the distance separating the two planes; and D is the diffusion constant. This sum converges rapidly, and in practice, summing over $i = -7$ to $i = 8$ is sufficient for conditions designed to mimic our experimental conditions (but longer distances and times require more terms). The flux can be converted to current by multiplying by z, the number of charges per molecule, and e, the elementary charge. Figure 4 shows the predicted time course of signals for several separation distances. As expected, diffusional "smearing" has a steep dependence on distance.

Further attributes of the signals can be seen by normalizing the time axis with the dimensionless variable

$$T = \frac{t}{(x^2/2D)} = \frac{t}{\tau}$$

where τ is defined in equation 2. Rewriting equation 3, we obtain

$$\vec{J}_a = \left[\frac{MD\sqrt{2}}{\sqrt{\pi}x^2}\right]\left[\lim_{n\to\infty} \sum_{i=-n}^{n} (-1)^i \frac{(1-2i)}{T\sqrt{T}} e^{-(1-2i)^2/2T}\right] \tag{4}$$

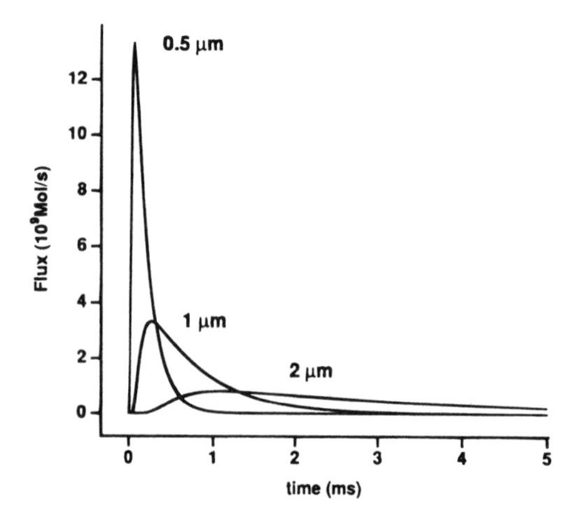

Figure 4. Predicted time course of quantal signals. Equation 3 in the text was used to simulate the flux of molecules at a detecting plane following instantaneous release of molecules from a parallel release plane. Three different separation distances were used: 0.5, 1, 2 μm. $M = 3$ million molecules. $D = 6 \times 10^{-6}$ cm²/sec, close to the experimentally determined value for catecholamines freely diffusing in solution (Gerhardt and Adams, 1982).

The term in the first pair of brackets can be thought of as the magnitude of the event, and the term in the second pair of brackets is the normalized time course. By rewriting the equation in this form, we see that it should be possible, with appropriate scaling of the time course and the amplitude, to superimpose the entire time course of signals originating from sites at different distances from the detector (see, also, Schroeder *et al.*, 1992). Deviations from the idealized shape can be used as evidence that release of transmitter is not instantaneous. In addition, we can extract some simple rules of thumb regarding the signal shape. On the nondimensional time axis, the time to peak is about 0.33; the 50%-to-90% risetime is 0.089; the time for half of the molecules to strike the detector is 0.75; and the half-width of an event is about 0.85. The halfwidth/risetime (50–90%) ratio should be constant, about 9.6. We can convert normalized time back to real time by multiplying by τ.

As discussed in Section 1, release of transmitter is probably not instantaneous. What are the limits on our being able to recognize details of noninstantaneous release? For any given distance separating the cell and detector, there is a certain amount of diffusional smearing in the observed signal, and the degree of smearing determines whether or not we can discern details of the release time course. In a typical experiment, the electrode is positioned from about 0.5 to 5 μm from the cell surface. What is the minimum duration of release that would be discerned as noninstantaneous if release occurred in a rectangular pulse (the more complicated foot-spike pattern of release is discussed below) A simple rule for the limits of detection at a given separation distance is that the release duration must be approximately equal to or exceed the predicted halfwidth for instantaneous release for the given distance. Figure 5 shows an example of signals at a detector located at 0.5 μm (part 5A) and 5 μm (part 5B) for release of 1-msec duration. The signal detected at 0.5 μm is recognizable as a diffusionally "filtered" rectangular pulse. The pulse duration of 1 msec exceeds the predicted half-width at this distance by a factor of about 5.7. The signal detected at 5 μm, on the other hand, is barely distinguishable from one caused by instantaneous release (the two signals are superimposed in Fig. 5B). In this case, the release duration of 1 msec is much less than the predicted half-width of 18 msec. Note that, as suggested by equation 4, signals resulting from instantaneous release originating from 0.5 and 5 μm have identical shapes after scaling for peak amplitude and time to peak.

For the case of release with an initial "foot" and then a spike, the situation is more complicated. As the distance separating the release site and detector increases, the rising edge of foot signal becomes progressively slower and eventually drops below the noise level, while the spike remains clearly discernible. To illustrate this, we performed simulations in which 10% of the total molecules were released at a flat rate over 2 msec and the remainder were released as an instantaneous packet, immediately following the "foot". The signal at the detector was obtained by convolving the release time course with the time course given by equation 3. Figure 5C shows the simulated oxidation "current" at the detector plane located at 0.5 μm and 1 μm away from the release site. Both signals have the foot-spike pattern, and, in the absence of noise, it appears that the foot is about 2 msec long in both cases.

An expanded view of the first 2.5 msec of the signals is shown in figure 5D. Two other signals are appended, for release originating at 0.25 and 0.5 μm from the detector, as well as a "noise level" (10 pA, in this case). The time course of release is identical in all cases. The important message is that the foot signal progressively drops below the noise level, while the spike is still readily discernible. This would lead to underestimation of the foot durations. At 5 μm, the foot is not discernible at all (not shown). Clearly, if one plans to study foot signals, one must place the electrode tip as close to the cell as possible, and, in the analysis, it is useful to select events with the fastest rise times.

In the actual experimental situation, the geometry is much more complex than that

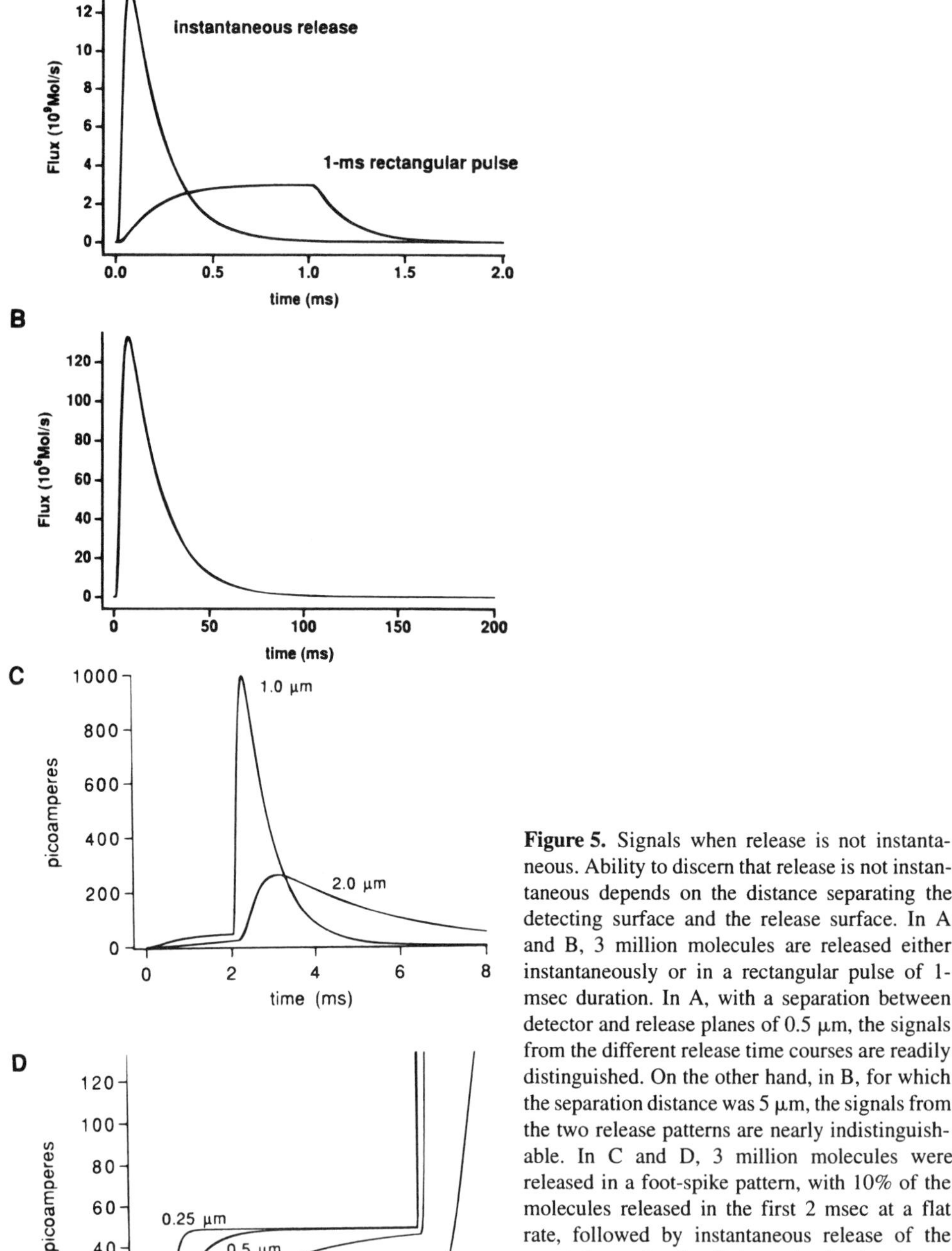

Figure 5. Signals when release is not instantaneous. Ability to discern that release is not instantaneous depends on the distance separating the detecting surface and the release surface. In A and B, 3 million molecules are released either instantaneously or in a rectangular pulse of 1-msec duration. In A, with a separation between detector and release planes of 0.5 μm, the signals from the different release time courses are readily distinguished. On the other hand, in B, for which the separation distance was 5 μm, the signals from the two release patterns are nearly indistinguishable. In C and D, 3 million molecules were released in a foot-spike pattern, with 10% of the molecules released in the first 2 msec at a flat rate, followed by instantaneous release of the remaining molecules. C shows the simulated amperometric signals recorded at 1.0 and 2.0 μm. An expanded view of the foot signals is shown in D. A threshold "noise" is appended (dashed line), as well as two other signals, corresponding to release at 0.25 and 0.5 μm.

assumed in the diffusion model. The situation would be better described as a spherical cell with the disk end (the detector surface) of a cylinder placed near the cell and with the center of the cell and the cylinder axis being coaxial. Because it is not possible to place the carbon-fiber electrode tip next to the cell so as to have a uniform distance between the detecting surface and the cell surface (even if the electrode surface were planar, cell curvature and irregularities in the cell surface would still lead to variable separation distances), one expects that the signals will show considerable variations in shape, at the very least from the differing separation distances. From the geometric considerations, one also recognizes that the signals will become attenuated more rapidly than simple distance dictates, because, as separation distance increases, more molecules simply escape detection at the finite detector surface (and diffuse away into the effectively infinite sink of the surrounding medium; Schroeder *et al.*, 1992). However, when the electrode tip is placed less than 1 μm of the cell surface, at closest approach, the diffusion model is a useful construct for thinking about the signals recorded amperometrically, especially the signals with a fast risetime (those that presumably arise nearest the detector). At such distances, foot signals become resolvable (Chow *et al.*, 1992). On the other hand, for separation distances of 5 μm or greater, Schroeder *et al.* (1992) have shown that signal shapes are indistinguishable from those predicted for instantaneous release.

3. Setup

Making electrochemical measurements from single cells requires many of the same components used for patch-clamp recording (see Chapter 1, this volume), including a microscope with a stage mount for a recording chamber, adequate shielding from electrical interference, and vibration isolation. This is convenient, as we generally combine patch-clamp and electrochemical measurements. The major additions are the carbon-fiber electrode, an appropriate amplifier with an initial high-gain headstage onto which to mount the electrode, and a data acquisition system. We will focus only on the equipment additional to the standard patch-clamp setup. Although many different combinations of equipment parts are possible, we focus on the equipment with which we have experience. This does not imply endorsement of any particular name brands. Alternative recommendations for the equipment can also be found in the electrochemical literature (see, for example, Kawagoe *et al.*, 1993).

3.1 Electrodes

3.1.1. Design Criteria

Several features are important for carbon-fiber electrodes that can be used for single-cell measurements. A single carbon-fiber strand must be held firmly in the electrode with an electrically insulating material so that one end is exposed and can be placed near a cell while the other end is easily connected electrically to an amplifier system. A detecting surface of 10 to 100 μm^2 is desirable and is readily obtained by transecting the carbon fiber, to expose a flat disk surface. If insulation extends as far as the tip, it must not be so bulky as to hinder placing the disk end near the cell surface.

Several alternative approaches have been developed meeting the design criteria. Most of these employ glass pipettes that are cannulated with a carbon-fiber strand and then pulled on standard electrode pullers (Gonon *et al.*, 1984; Kawagoe *et al.*, 1993; Millar, 1991). In some constructions, glue is used to ensure a firm seal between the glass and the carbon fiber at the electrode tip (Gonon *et al.*, 1984; Kawagoe *et al.*, 1993). The carbon fiber at the tip is cut to the desired length with fine scissors or a razor blade or it is etched either electrochemically (e.g., Kawagoe *et al.*, 1991a) or with an electric spark (Millar, 1991). With electrochemical etching, in combination with electrodeposition of insulating polymers, it has been possible to obtain micrometer or submicrometer tips starting with carbon fibers of greater diameter (Kawagoe *et al.*, 1991a). Some investigators bevel the carbon fiber tips to obtain a near-planar surface (Kawagoe *et al.*, 1993).

We have developed a method for making the carbon-fiber electrodes that involves polyethylene insulation. These electrodes are easily and quickly made, have very low electrical noise, and can be reused multiple times by recutting the tip. They can be beveled as well. Chemical treatments, however, may damage the polyethylene insulation.

3.1.2 Electrode Fabrication

3.1.1a. Materials.

- Polyethylene tubing, 0.28 mm ID and 0.60 mm OD (Portex, England)
- Carbon fibers (8-μm diameter, Amoco Perfomance Products, Inc., Greenville, SC, or Courtald, Ltd., England)
- Glass capillaries (Drummond microcaps, 50-μl size, Drummond Scientific Co., Broomall, PA, U.S.A.)
- Two pairs of fine forceps (watchmaker's size 3 to 5)
- Soldering iron with fine tip and adjustable temperature control (Weller, model WMCPEC, The Cooper Group, Besigheim, Germany)
- Dissecting microscope
- Stand for mounting soldering iron
- Methanol
- Surgical scalpel blades
- Epoxy glue

3.1.2b. Procedure. Single carbon-fiber strands are selected and cut to lengths of about 6 to 8 cm. The carbon fibers are very rigid and have high tensile strength; however, they are very brittle. Handling them is made easier by covering the tips of the forceps with soft plastic or rubber tubing such as that obtained by desheathing the insulation from electrical wires (Millar, 1991).

Polyethylene (PE) tubing of 0.28 mm ID and 0.60 mm OD is cut into 3-cm-long segments. Each PE segment is dipped into methanol or ethanol and allowed to fill by capillary action. The solvent reduces static attraction between the plastic and the carbon fiber during the next (cannulation) step. Care must be taken to fill the tubing completely. With a dissecting microscope and bright illumination for visualization, the tip of a single carbon fiber is inserted into one end of a PE segment and pushed through, so that the carbon fiber extends out equally from each end. Then the solvent is removed by wicking it away by touching a piece of tissue paper carefully to the end of the tubing or by gently tapping the tubing with another pair of forceps. (Be careful not to break the carbon fiber!)

The soldering iron is switched on, heated to 200° to 250°C, and mounted such that the soldering iron tip is visible at low power under the dissecting microscope. A PE tube previously cannulated with a carbon fiber is grasped firmly at each end with two pairs of forceps. Resting one's hands on the microscope stage helps stabilize them during the next steps. The middle section of the tubing is held about 1 to 2 mm above the soldering iron tip (do not touch the tip) until it melts. As it melts, the plastic becomes transparent. The region directly above the heat narrows, and a bead-like expansion forms to each side. When the plastic has reached this stage, tension is applied with the forceps to stretch the plastic about 1 to 2 mm. It should narrow further to about 0.5-mm diameter. Then the plastic is moved quickly away from the heat and allowed to cool for about 30 sec. The electrode at this stage is illustrated in Fig. 6A.

Next, the narrowed region of the plastic is again put near the hot soldering iron tip until it is molten, and then it is touched to the hot metal for less than a second. This results in the rapid melting away and vaporization of plastic at the point of contact. Plastic further away will retract and bead up. The carbon fiber will appear to be exposed at the point of contact (figure 6B), although it is actually coated with a submicrometer layer of plastic (as demonstrated in scanning EM photographs, von Rüden, 1993). Again, the plastic tubing is moved away and allowed to cool. The carbon fiber exposed in the center region is transsected with a scalpel blade. This results in two electrode tip assemblies, each with a carbon fiber extending out the back (Fig. 6C,D).

Each tip assembly must be glued into a glass capillary (for mounting on a patch-clamp headstage). We use Drummond Microcap tubes, which are narrower than standard patch-pipette glass and therefore accommodate the PE tube more snugly. A single glass capillary is dipped into methanol or acetone and allowed to fill about 2 to 3 cm. The carbon fiber

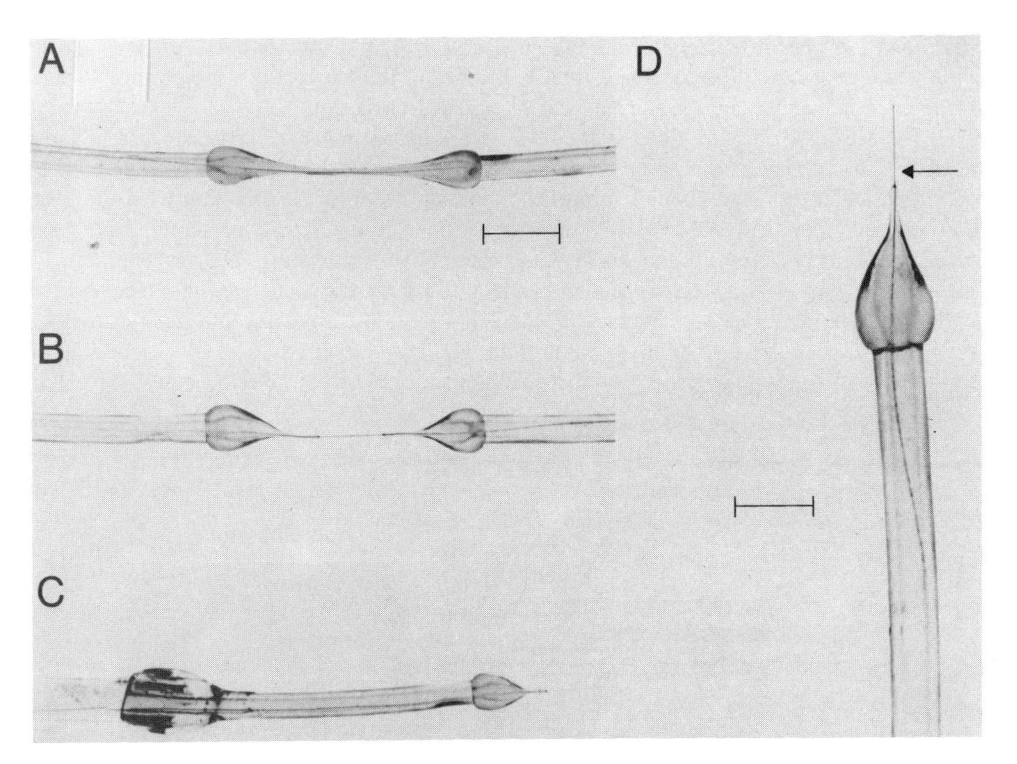

Figure 6. Preparation of polyethylene-insulated carbon-fiber electrodes. (A) After the first stage of heating. The scale bar, 3 mm, applies to A and B. (B) After the second stage of heating. (C) Completed carbon-fiber electrode, with the tip assembly glued into a glass capillary. (D) Tip of an electrode at higher magnification. The arrow indicates approximately where an electrode would be cut prior to an experiment. Scale bar, 2 mm.

extending out the back of a tip assembly is carefully cannulated into the glass tube and inserted until the PE plastic has also entered. Be careful not to break the carbon fiber during this cannulation procedure. Most of the solvent is wicked away by tissue paper, and the remainder is allowed to evaporate over about 15 min. The plastic tip assembly is pulled outward until the plastic tubing just barely remains inside the glass capillary, and then two-component glue (such as 5-min epoxy) is applied to cement the junction. A completed electrode is shown in Fig. 6C, and a schematic of all components is shown in Fig. 7. The electrodes can be used within about 30 min.

3.1.3. Preparing and Mounting the Electrode

Immediately prior to an experiment, the tip of an electrode is cut again about 100 μm from the point where the plastic insulation visibly thickens (see arrow in Fig. 6D). This can be performed under a dissecting microscope with a fine pair of iris scissors or with a scalpel blade. When using iris scissors, both the scissors and the electrode can be mounted on manipulators for optimal control. When a scalpel blade is used; fix the electrode to a glass slide with a piece of modeling clay so that the electrode is angled gently downward and the carbon-fiber tip is resting lightly on the table surface. Avoid bending the carbon-fiber tip, and avoid scraping off the insulation. For cutting, steady the scalpel blade against the edge of a glass microscope slide and "roll" the blade smoothly and continuously along its cutting edge over the carbon fiber. After the cut, inspect the cut end for irregularities. Most of the time, the end is nearly planar, but if not, recut the tip until it is more satisfactory. A new blade or nearly new blade should be used to ensure a clean cut.

The electrode is back-filled with 3 M KCl solution and mounted on a patch-clamp headstage in the same fashion as a conventional patch pipette. The silver wire should be advanced no further than about 1 mm of the junction between the PE tubing and the glass capillary in order to avoid breaking the carbon fiber. The silver wire should have been carefully chlorided to ensure low noise. As an alternative, for making electrical contact with the amplifier, use a hypodermic needle to inject liquid mercury, conducting silver paint or colloidal graphite (Bio-Rad, Cambridge, MA) into the glass capillary and push in a silver or copper wire. In the case of silver paint or collidal graphite, it is necessary to solder a pin to the back of the wire to establish electrical contact between the electrode and the patch-clamp headstage.

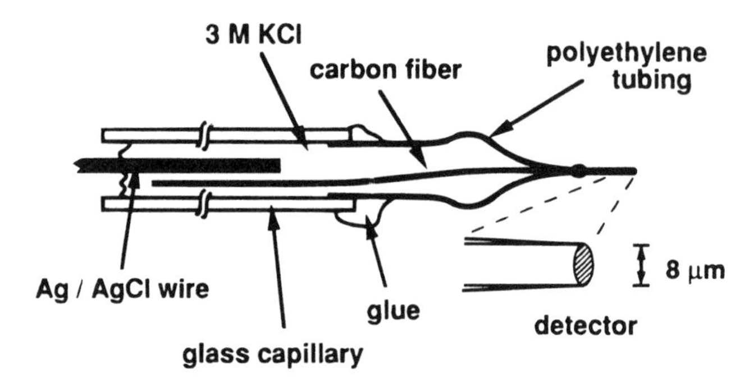

Figure 7. Schematic of a carbon-fiber electrode.

3.1.4. Electrode Properties

When 800 mV DC is applied to the carbon-fiber electrode immersed in Ringer solution with an Ag/AgCl pellet for a bath reference electrode, there is a slowly decaying current transient, lasting hundreds of seconds. Although the exact nature of the reactions occurring is unknown, presumably the current results first from the charging of the double-layer capacitance, and then from oxidation of the carbon surface exposed to Ringer solution. After about 5 min, the DC current is 10 pA or less (usually about 4 to 5 pA), at which time we begin recordings. If there is no initial current transient or if there is a maintained DC current greater than 100 pA, there may be a bad electrical connection or the bath electrode may not be connected. Otherwise, the electrode may be bad, or it may require recutting. Typical current noise levels are about 400 to 1000 fA rms with an output filter (eight-pole bessel) cutoff frequency of 3 kHz (using an EPC-7 at gain 100 mV/pA).

As has been found for other carbon-fiber electrodes (Fox *et al.*, 1980), the polyethylene-insulated electrode behaves essentially as a resistor in series with a resistor and capacitor that are in parallel. The series resistance is about 100–500 kΩ and originates at the junction between the carbon fiber and the KCl solution (the resistance of the carbon fiber itself is negligible in comparison with that arising from the carbon/KCl junction, amounting to only about 3–4 kΩ/cm), and the capacitance ranges from about 3 pF to about 40 pF. The parallel resistance is in the tens to hundreds of gigohms range, depending on the transection of the tip.

The DC limiting current in the presence of catecholamine should be proportional to the concentration of catecholamine and to the radius of the detecting surface (Dayton *et al.*, 1980). As noted previously by others using cut-end carbon-fiber electrodes, the DC limiting current is, in fact, usually greater by up to 80% than that predicted for a disk surface having diameter equal to that of the carbon fiber. This is probably because a higher surface area is exposed when the cut is irregular (Dayton *et al.*, 1980).

3.2. Instrumentation

3.2.1. Amplifier

In the amperometric mode, the signals from single vesicles have peak amplitudes in the range of one picoampere up to several nanoamperes and have risetimes from tens of microseconds to tens of milliseconds (depending on the distance separating the release site from the electrode). Thus, the amplifier used for recordings must have significant gain and wide bandwidth (in the range of 3 to 5 kHz) to capture the events with reasonable fidelity. Patch-clamp amplifiers are well suited for the purpose and do not require modification for this application. For amperometric measurements, we have been using an EPC-7 or EPC-9 amplifier at gain 5 to 50 mV/pA and with an output filter of 3 kHz. A DC voltage of 8 V, supplied by a DC power supply, is applied as an external stimulus. (The amplitude is scaled down by a factor of 10 by the patch-clamp amplifier.)

Although high amplification is possible with amperometry, a lower gain is necessary (determined empirically) when fast periodic wave forms (such as in cyclic voltammetry) are used in order to avoid amplifier saturation. The electrode has significant double-layer capacitance, so voltage ramps and other signals with rapid dV/dt lead to significant capacitive currents. In principle, it should be possible to neutralize part of the capacitive currents (but not all, as the capacitance behaves somewhat nonlinearly for extreme voltage steps) and thus increase the gain. This procedure, although performed routinely in patch-clamp experiments, has not been employed in the electrochemical field, and we have not used it consistently.

3.2.2. Grounding and Bath Electrodes

The bath reference electrode is a thoroughly chlorided silver wire or a silver/silver chloride pellet. A pellet is preferable, as it ensures a more stable reference potential in the face of currents that would tend to deplete the bath electrode of silver chloride. An alternative reference electrode is a saturated calomel electrode (SCE) or sodium-saturated calomel electrode (SSCE). The bath electrode must be connected to the headstage by a low-resistance wire. If patch-clamp measurements are being made simultaneously with another amplifier, the two headstages must be connected together by the ground pin, and there should be only one bath electrode.

3.2.3. Pulse Generator/Data Acquisition

As indicated above, for amperometry the DC voltage command can be from a DC power supply. Alternatively, the voltage can be supplied by one of the many commercial pulse generators available. For fast cyclic voltammetry, one needs either a gated wavefunction generator or some other pulse generator system that can supply a triangle wave form. Typical cyclic voltammograms use a sweep rate of 100 to 400 V/sec, and one cycle is completed in about 10 to 50 msec. The parameters of the waveform that we use were given in Fig. 3A. It is advisable to wait at least six cycle lengths, preferably ten, between successive cyclic voltammograms to replenish the transmitter that has been depleted at the electrode surface during the preceding scan (Kawagoe *et al.,* 1993). We have been using an Instrutech ITC-16 pulse/acquisition system, controlled by a MacIntosh Quadra 800 computer, running IGOR (Wavemetrics, Lake Oswego, OR) and the Pulse Control XOPs (J. Herrington and R. J. Bookman, University of Miami).

Data acquisition can be performed most conveniently with an acquisition system combined with the pulse generator. With such a combination, one can acquire sweeps of data with precisely timed stimuli. However, it is also possible to acquire continuous data with a video cassette recorder (VCR) or a digital audio tape (DAT) recorder. In the case of continuous recording, it is necessary to record the electrochemical current and the applied voltage, or else a trigger mark to indicate timing of triggered waveform voltages, for off-line analysis.

3.3. Recording Configuration

In our experiments, we generally combine whole-cell patch clamp and amperometry. The headstages of the patch clamp and of the electrochemical detector are mounted on manipulators on opposite sides of the microscope stage, allowing the respective electrodes to be maneuvered on opposite sides of the cell. Figure 8 shows a view of the two electrodes positioned at a single bovine chromaffin cell. The patch electrode is generally filled with a solution to isolate and preserve calcium currents. Secretion is stimulated by applying step depolarizations to activate calcium channels. At the same time, the patch-clamp amplifier applies a sinusoidal wave, which is used to measure the cell membrane capacitance (Chapter 7, this volume). The carbon-fiber electrode provides an independent measure of secretion.

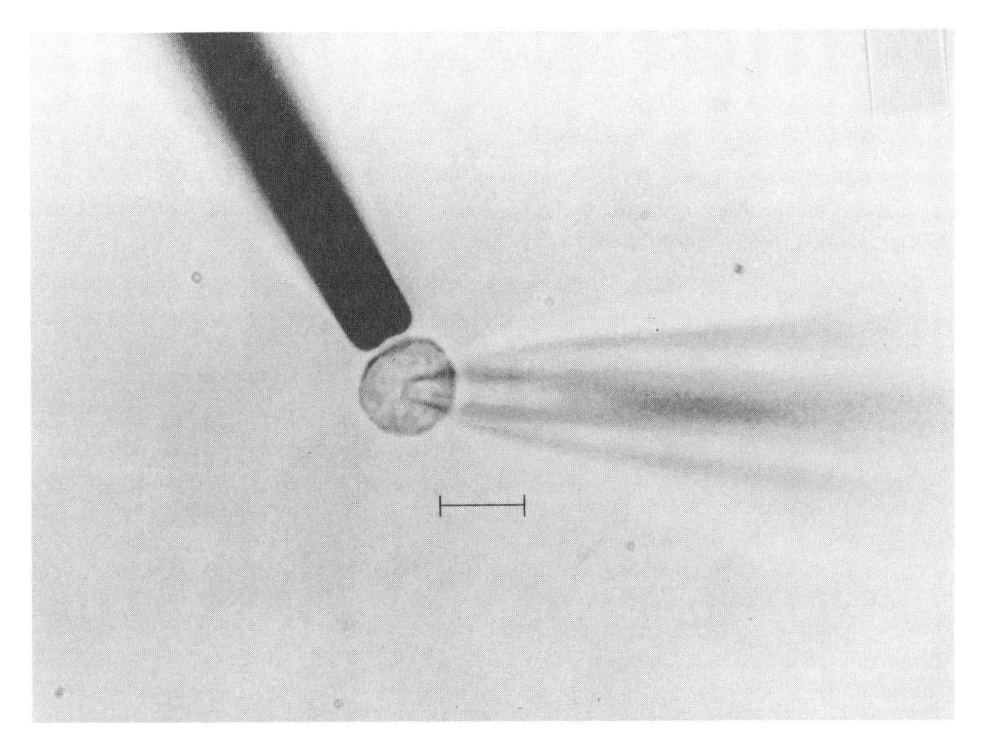

Figure 8. Typical experimental configuration. Chromaffin cell with a patch pipette sealed to it in the whole-cell configuration and a carbon fiber electrode (black) approaching the cell from the upper left. Scale bar, 10 μm.

Because of the steep dependence of the signal amplitude on distance, it is advisable to locate the carbon-fiber tip 5 μm or less from the cell surface. It is even possible to press the electrode tip gently against the cell membrane, without damaging the cell, and to make recordings from this location. In this case, however, the electrode tip tends to adhere to the cell membrane, and at the end of the measurements, it will be difficult to assure that the tip is clean enough to use in further measurements. Adherent materials will act as a diffusion barrier and consequently lead to reduced sensitivity. We have found that recutting the tip of the electrode permits further measurements.

One can perform electrochemical measurements, of course, independently of patch-clamp measurements. In this case, stimulation of secretion can be done by local application of agonists or KCl (to depolarize the cell and activate calcium currents) by means of a delivery pipette located a few micrometers from the cell. Either pressure from a pneumatic system or iontophoretic current pulses can be used to deliver agents. Alternatively, one can exchange the entire bath for an agonist-containing solution. In all cases, the applied agent must itself be screened for electrochemical activity to avoid mistaking oxidation of the applied agent for secretion. In cyclic voltammetry, changes in bath level will induce capacitance changes, which will interfere with voltammograms. Nonuniform rates of bath flow can influence amperometric recording by convective removal of reactant from the electrode, which appears as a slow baseline drift. Therefore, it is advisable to maintain the bath level as constant as possible, with uniform flow rate, or to avoid solution exchange during recordings.

4. Experimental Considerations and Analysis

4.1 Amperometry

The following comments are based on our experience with bovine chromaffin cells. The applicability of specific approaches should be carefully reassessed for each new cell type.

4.1.1. Rapid Screen for Secretion

If the goal is to show simply whether or not secretion has occurred in response to a stimulus, one need only record the output of the amperometric headstage and, simultaneously, a signal that indicates the timing of the stimulus. Since amperometric monitoring, in contrast to capacitance measurements, does not require patching the cell, experiments can be performed relatively quickly, allowing one to screen rapidly the effects of different agents on secretion. Other noninvasive techniques such as fura-2 Ca^{2+} measurements can be combined at the same time (von Rüden *et al.,* 1993).

In comparing secretion under different conditions, certain precautions are necessary. Cell-to-cell variability in secretory responses is quite significant compared to the variability in one cell. Thus, in studying how a drug or experimental maneuver affects secretion, it is simpler to study how the response changes in the same cell in the "control" and "test" conditions. Keep in mind that, with repeated stimuli in the same cell, depletion of readily releasable vesicles can occur (von Rüden and Neher, 1993), necessitating a recovery period between successive stimuli.

The possible existence of "hot spots" of secretion (Schroeder *et al.,* 1993) mandates that the electrode tip be maintained in the same position for a given cell in all the conditions tested. With the electrode maintained in the same position, secretory responses exhibit stationarity (uniform probability of release) for up to 10 min of repeated short depolarizations (stimuli designed to be nondepleting, as in Chow *et al.,* 1992), even in the whole-cell configuration.

The magnitude of the secretory response can be determined in one of several ways. For low-intensity stimuli one can count the number of quantal events elicited per stimulus. This is possible only if there is no overlap of individual events or if overlap is not too extensive. If secretion is continuous at a low level, one can monitor the frequency of quanta by counting the number of events occurring in discrete time bins. When stimuli elicit large responses, it is possible to integrate the amperometric current response, which in such a case consists of a complex summation of "near" and "far" events.

Integration of the amperometric current also makes possible direct comparison of amperometry and capacitance measurements (von Rüden and Neher, 1993). Both, then, are cumulative records of secretion. But the time courses are generally not identical, reflecting distinctions between the two methods. Whereas capacitance measurement monitors the membrane surface area and its increment on vesicle membrane addition, amperometry is a measure of the actual release of transmitter. In chromaffin cells there is a delay of about 5 msec between single-vesicle fusion and transmitter release, as revealed by cross-correlation studies (Klingauf *et al.,* 1994). Furthermore, the capacitance technique follows the sum of changes from both exo- and endocytosis, whereas amperometry detects exclusively exocytosis. Therefore, a comparison of the time courses can yield information on the rate of endocytosis.

4.1.2. Single Events

Evaluating the features of the individual secretory events from chromaffin cells requires using stimuli at sufficiently low intensity that single events do not overlap significantly in time. Most parameters evaluated for individual events (peak amplitude, integral, half-width) are similar to those for synaptic potentials or currents. However, several features of the quantal signals in chromaffin cells make it inappropriate to use the standard programs for analyzing synaptic currents/potentials. First, the long and variable latencies imply that the quanta do not ordinarily summate. Thus, the conventional amplitude histogram analysis used for synapses cannot be applied. Furthermore, the presence of the foot signals complicates detecting the start of an event and makes the integration of the events more difficult. The span of the risetime should be chosen to avoid including the foot signals. We use the 35%-to-90% risetime or the 50%-to-90% risetime.

For constructing latency histograms (Chow *et al.,* 1992), a multichannel acquisition system is desirable (although, if the stimuli can be reproducibly triggered at the same time in each amperometric sweep, one-channel acquisition is sufficient). One channel is dedicated to recording the stimulus intensity, duration, and timing, while another is used to record the amperometric current response. (Another channel can be used to record the calcium current, if one is using step depolarizations to elicit secretion.) As discussed above, the signal intensity should be titrated down to levels such that the individual events do not overlap. It is advisable to record a short amperometric "baseline" period prior to the stimulus in order to assess the baseline noise. The latency is measured on or off line as the time between the start of the stimulus and the time at which the amperometric current signal first exceeds the baseline noise by 2 standard deviations.

In order to study foot signals one should analyze events with a fast risetime, because these events, presumably originating near the detector, will more faithfully indicate the time course of release. Use fast sampling, at least 4 kHz, in order to get enough points on the rising phase of the signals, and avoid excessive filtering (i.e., use a cutoff frequency above 2 kHz).

Foot signals are most confidently identified in amperometric events that occur when the probability of overlap of individual events is very low. For example, dialysis of chromaffin cells with solutions in which free calcium is buffered to levels of only a few micromolar leads to a slow rate of secretion (Augustine and Neher, 1992). On the other hand, during depolarizations, when calcium levels near the membrane are expected to be elevated, the probability of superimposed signals is very high and distinguishing "true" foot signals from overlap is very difficult.

A useful criterion for excluding superposition of independent events is that the declining phase of a unitary event that is preceded by a "true" foot should be smooth (no inflections) and should return completely to the baseline. Examples of two cases where this criterion is fulfilled appear in Fig. 9A,B. Note the smooth decline to baseline of each event despite the prominent current plateau preceding the rapid upstroke. The probability that two independent events would end simultaneously in the observed manner is very low.

Conceptually, one might expect that a vesicle could fuse reversibly, with flickering, and that the fusion pore could close again, leading to a "stand-alone" foot signal. In mast cells, amperometric signals corresponding to such reversible fusions have been confirmed with simultaneous capacitance measurements (Alvarez de Toledo *et al.,* 1993). In chromaffin cells, the signal-to-noise ratio does not permit simultaneous measurements of amperometric events and single-vesicle capacitance steps (in whole-cell mode). However, one does rarely find amperometric signals that have a pattern that is expected of stand-alone foot signals. An

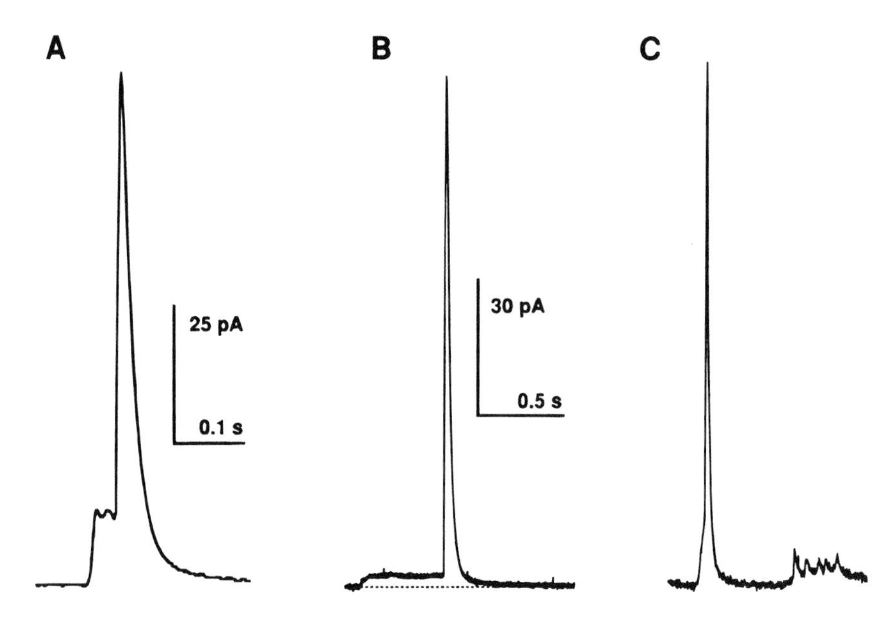

Figure 9. Amperometric events with unusual foot signals. (A) Two short "flickers" in the foot preceding an amperometric event. The decay of current after the event shows no inflections that would, if present, suggest overlap of two independent events. (B) An event with an extraordinarily long foot (more than 500-msec duration). Scale bars in B also apply to C. (C) After a more conventionally shaped event, multiple small current transients occur, possibly representing fusion pore flickering and incompleted fusion of a vesicle.

example is shown in Fig 9C. Following the more conventionally shaped event in the beginning of this trace, there is a jagged signal consisting of multiple small peaks, each only a few picoamperes in amplitude. The risetime of these individual peaks is comparable to that of the preceding large event. The probability that each of the little peaks is a separate vesicle fusing is very low, as the probability that five vesicles would fuse so close together and have nearly identical size is extraordinarily low.

What is a convenient definition of foot duration? Figure 10 shows an amperometric event with a line drawn through points in the rising phase located at 35% and 60% of the peak and another line through the baseline. The line through the baseline (dotted line) is fitted to the record in a section preceding the event, at a location where there is no evidence of a secretion signal. As indicated in the figure, the intersection of these two lines serves as a convenient mark for the end of the foot signal. The start of the foot signal is defined as the time at which the amperometric signal exceeds the baseline noise by 2 standard deviations (the threshold is indicated by the thin solid line above the baseline).

Even for idealized instantaneous release, the fast upstroke of an event is preceded by a very small "native" foot signal. The duration of this slow phase, however, is always a fixed fraction of the risetime. In general, for a given risetime, we can calculate how long the foot duration would be for instantaneous release (assuming a typical signal-to-noise ratio):

$$\text{native foot duration} = 0.33 \times 50\text{–}90\%\ \text{risetime} \tag{5}$$

Experimental foot signals lasting longer than predicted by equation 5 indicate that transmitter release has been retarded, for example, by leak through the initially narrow fusion pore.

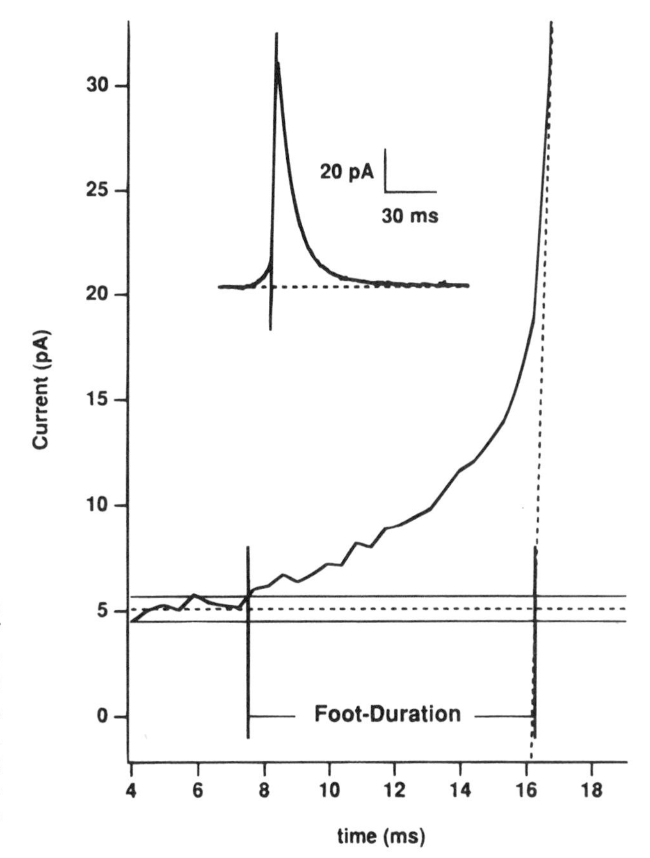

Figure 10. Determination of the foot duration. High-resolution image of the foot signal. The lines used to define the end of the foot signal and the beginning of the foot signal are described in the text. The inset shows the complete amperometric event with a line (solid) on the rising phase extrapolated back to the baseline (dotted).

Following these guidelines, the duration of foot signals is approximately exponentially distributed (Chow *et al.,* 1992).

4.2. Cyclic Voltammetry

For cyclic voltammetry, one should ideally have a channel to record the periodic voltage and one to record the current response. Records of the "baseline" current, recorded prior to secretion, must be subtracted either on line or off line from subsequent records. Finally, the current must be plotted as a function of the applied potential. Because of the lower time resolution and sensitivity of fast cyclic voltammetry as compared to amperometry, we suggest using this method only if the chemical nature of the released compound is of special interest.

The peak amplitude of cyclic voltammograms is proportional to the concentration of reactant down to about 1 μM for catecholamines and indoleamines (Ponchon *et al.,* 1979). This has been used as a means to estimate concentrations of specific transmitters in the brain. However, although the peak amplitude of cyclic voltammograms is meaningful in the context of uniform bulk concentration, as in the case of brain tissue, it should be used with caution in studies of release from single cells. Calibration is performed with solutions of uniform bulk concentration. On the other hand, with single-vesicle release near the carbon fiber, concentration is highly nonuniform in the gap between the cell surface and the detector. Thus, the calibration would not be applicable, and the result would be an underestimate of true concentration.

5. Conclusions and Prospects

We have briefly surveyed the application of electrochemistry to detecting secretion from single cells. The voltammetric methods of amperometry and cyclic voltammetry are exquisitely sensitive for measuring the release of readily oxidizable transmitters such as monoamines and indoleamines and their derivatives. Whereas amperometry offers high-time-resolution measurements of secretion, cyclic voltammetry offers a means (although limited) to distinguish among some classes of transmitters.

Compared to capacitance measurements, voltammetry has the advantages of measuring signals related only to exocytosis. It is also less invasive than capacitance measurements with whole-cell patch clamp (although the perforated-patch approach makes less invasive capacitance measurements possible), and it does not require that the cell shape be amenable to good voltage clamp. On the other hand, the electrochemical approach does have the limitation that only a few transmitters are readily oxidizable. The electrodes detect transmitter in a highly localized domain (within about 5 μm of the tip). This can be an advantage for studying "hot spots" of secretion. However, if secretion is highly nonuniform over the cell surface, then it can be a disadvantage, as secretion monitored at any given location may not be representative of the secretion elsewhere in the same cell.

A number of recent innovations extend the range of substances that can be detected or increase the selectivity of measurements. Unfortunately, most of these approaches lead to slower electrode kinetics. Immobilization of enzymes on the surface of the carbon fibers has enabled monitoring levels of glucose and acetylcholine (Kawagoe *et al.,* 1991b; Pantano *et al.,* 1991) and glutamate (Pantano and Kuhr, 1993). The response time is limited by many factors including the limited number of enzyme molecules that can be fixed to the surface, slow enzyme turnover, and the finite rate at which the substrate binds the enzyme and the product then diffuses to the carbon surface. Nevertheless, some of the electrodes that have been constructed have response times of hundreds of milliseconds. Ruthenium, a platinum-like catalyst, can be coated on carbon-fiber electrodes to accelerate electron transfer reactions. This has made possible detection of insulin release from single pancreatic β cells (Kennedy *et al.*, 1993). Porphyrin-coated electrodes have been used to detect NO (Malinski and Taha, 1992), although Iravani *et al.*, (1993) have shown that untreated carbon fibers can also detect NO. Electrochemical pretreatment of carbon-fiber electrodes enhances sensitivity to numerous agents, and, by shifting the redox potentials of certain compounds, it sometimes improves the separation of voltammetric signals of different compounds in complex mixtures of reactive species. Nafion is a negatively charged polymer that can be coated onto carbon to prevent negatively charged molecules (such as ascorbate and urate) from reaching the electrode surface. Thus, as demonstrated by Gerhardt *et al.* (1984), the electrode becomes selective for neutral or positively charged molecules. This is a significant advantage for measurements in the complex milieu of brain extracellular fluid.

Although we have focused on applications of electrochemical detection to secretion from single nonneuronal cells, clearly the methods can be applied to neurons that secrete oxidizable transmitters. Several papers have already been published on studies of secretion in brain slices (O'Connor and Kruk, 1991; Palij and Stamford, 1992) and in rat tail artery sympathetic nerve terminals (Gonon *et al.*, 1993). Neurons secrete at synapses, which are typically only 1–2 μm in size. These delicate anatomic specializations do not lend themselves readily to being probed by carbon-fiber electrodes of conventional size (5–10 μm). Thus, as for *in vivo* brain studies, electrochemical investigations of single or clustered synapses depend on overflow of transmitter from the synaptic cleft. Consequently, time resolution of these studies is on a time scale of seconds, as opposed to milliseconds. If a neuronal synapse is available, recording of postsynaptic signals may be more straightforward than attempting electrochemical measurements.

The application of voltammetric techniques to studies of single-cell secretion has already yielded important insights into mechanisms of secretion. The promise is great for continued advances, particularly as voltammetry is combined with patch-clamp techniques. Although the first single-cell studies focused on chromaffin and mast cells, voltammetry will soon be applied to many more cell types, as evidenced by recent publications on pancreatic β cells (Kennedy *et al.,* 1993) carotid body glomus cells (Urena, *et al.*, 1994), and pineal cells (Marin and Tabares, 1993), particularly as progress in modifying electrode sensitivity and specificity is moving at a high pace.

Appendix

The goal of this appendix is to outline the derivation of equation 3 in the text (see also Chow *et al.,* 1992). We use an approach that is analogous to the method of images, a standard approach in electrostatics (Feynman *et al.,* 1964; Jackson, 1975). In voltammetric recording from single cells, a carbon-fiber electrode tip is placed near a cell. If the distance between the electrode tip and the cell surface is small compared to the radius of the electrode and of the cell, and if we consider for the moment release and detection only in the central region of the detector, we can approximate the geometry of the release and detecting surfaces as two parallel infinite planes. In amperometry, the voltage of the carbon fiber is set sufficiently large that transmitter is oxidized as fast as it diffuses to the electrode surface. That is, the concentration of transmitter at the electrode surface is zero at all times. In the discussion to follow, we call such a surface "absorbing." It is to be contrasted with a "reflecting" surface, such as the cell surface, which is impermeant and nonsticky for molecules that strike it.

For this derivation, we assume that particles (transmitter) are released instantaneously (δ distribution) at the release plane and that they diffuse freely to the absorbing plane, where they are oxidized. This is a one-dimensional diffusion problem, since we consider only the component of the movement perpendicular to the release plane. The solution of the diffusion equation for instantaneous release in one dimension is (Crank, 1975)

$$C = \frac{M}{2\sqrt{\pi D t}}\, e^{-x^2/4Dt} \tag{6}$$

C represents the concentration at defined time, t, and distance, x, from the plane of release (at $x = 0$). M is the number of molecules released, and D is the diffusion constant.

If there is an impermeable barrier at $x = 0$ such that the particles can diffuse only in the direction of positive x, the concentration distribution is given by doubling the result of equation 6

$$C = \frac{M}{\sqrt{\pi D t}}\, e^{-x^2/4Dt} \tag{7}$$

This is possible because equation 6 is symmetrical about $x = 0$, and the boundary conditions of a reflecting (impermeable) surface

$$\partial C/\partial x = 0, \qquad x = 0 \tag{8}$$

are fulfilled.

Let us suppose that the detecting (absorbing) plane is located at $x = d$. The boundary condition for an absorbing plane is

$$C = 0, \qquad x = d \tag{9}$$

How can we construct a mathematical expression such that the concentration at $x = d$ is maintained at zero at all times? Imagine that at the same instant that particles are released at $x = 0$, an identical number of "antiparticles" are released at $x = 2d$, i.e., on the opposite side and equidistant from the absorbing plane, relative to the "normal" particles. These antiparticles have the same diffusion constant as their normal counterparts and therefore diffuse with identical mean speed; however, if a particle and an antiparticle collide, they "annihilate" one another. We can write a mathematical statement that will give the same result by assigning a "negative" concentration to particles released at $x = 2d$:

$$C = \frac{M}{\sqrt{\pi Dt}} e^{-x^2/4Dt} - \frac{M}{\sqrt{\pi Dt}} e^{-(x-2d)^2/4Dt} \tag{10}$$

In Fig. 11A we see that the concentration profiles initially start out as concentration spikes about the "planes" of release, one with "positive" concentration and one with "negative" concentration. Successive snapshots of the concentration profiles (increasing numbers indicate progressively later times, and a number and its prime represent identical time for positive and negative concentrations, respectively) reveal the expected diffusional broadening. However, at the point halfway between the two release sites, where the diffusion of particles and antiparticles overlaps and annihilation occurs, the net concentration always remains zero (thick line in Fig. 11B). Thus, we have found a way to meet the absorbing "plane" boundary conditions (equations 9).

Unfortunately, although the present expression fulfills the conditions for an absorbing plane at $x = d$, the boundary conditions of reflection at $x = 0$ (equations 8) will be violated at late times. This can be seen in Fig. 11B. As the wave of antiparticles (thin line, designated 4′) sweep in the negative direction across the plane at $x = 0$, they annihilate the normal particles (thin line, designated 4), leading to a nonzero concentration gradient at $x = 0$.

We can again meet boundary conditions of reflection at $x = 0$ by releasing antiparticles at $x = -2d$. These antiparticles are released at the same time as the particles at $x = 0$. The symmetry of antiparticle release (equidistant, on either side of $x = 0$) about the plane $x = 0$ ensures that the concentration gradient at $x = 0$ remains zero. But, at late times, this would lead to nonzero concentration at $x = d$ (violation of equations 9). To ensure that the boundary conditions for the absorption plane at $x = d$ are fulfilled, we add another virtual plane at $x = 4d$ but this time release normal particles. The antisymmetry about $x = d$ ensures zero concentration at all times but disturbs the boundary conditions for reflection at $x = 0$. Ultimately, to satisfy the boundary condition of reflection and absorption at the required planes, the expression must be written as an infinite sum of terms of positive and negative concentration (an alternating series), each representing release of particles or antiparticles at locations $x = n2d$, where n is an integer.

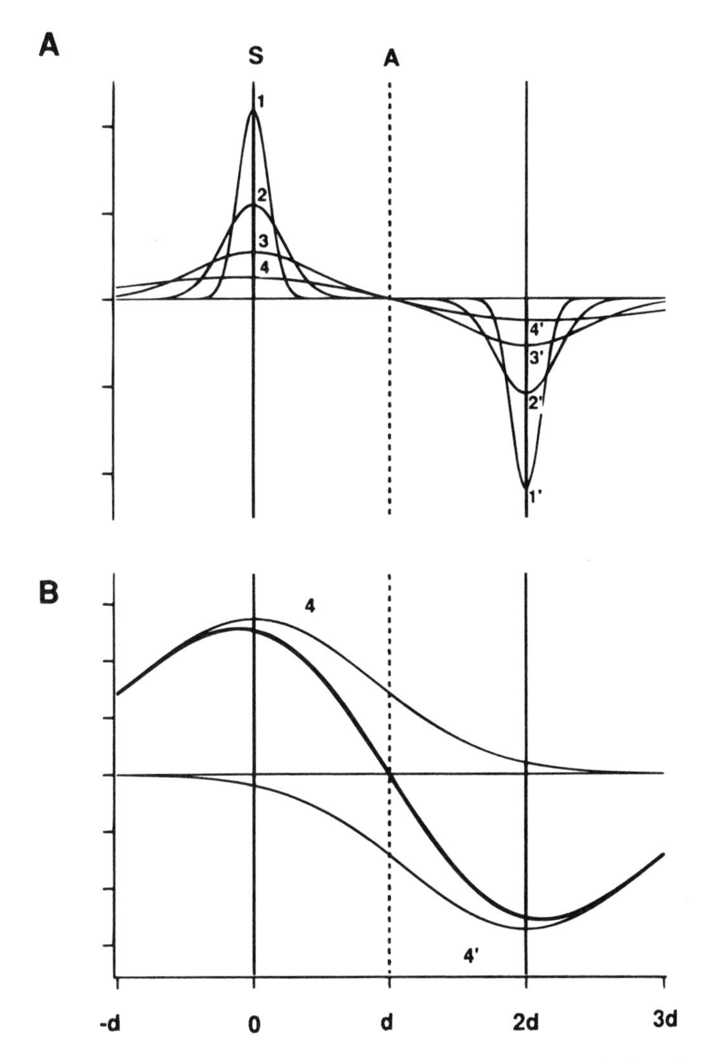

Figure 11. Diffusional broadening of concentration profiles. (A) Diffusional broadening of particles released from plane S, and antiparticles released at the same time from the plane located at $2d$, at four successive times (numbered chronologically 1–4 and 1′–4′, respectively). The particles and antiparticles collide and annihilate one another in plane A, resulting in zero concentration. Displayed is the sum of their concentrations at each time. (B) Enlarged view of the net concentration profile of particles at time 4 (fat line) and, in addition, the concentration profiles of particles and antiparticles (thin lines). The slope of concentration at plane S of the net concentration profile deviates from zero, thus disturbing the boundary conditions for the reflection plane at S. See Appendix for details.

The flux of particles at the absorbing plane is determined by Fick's law:

$$\vec{J}_a = -D \frac{\partial C}{\partial x}\bigg|_{x=d} \tag{11}$$

Thus, computation of the final flux requires summing the derivatives with respect to x of each concentration term in the infinite sum and multiplying the sum by $-D$. The final form of the solution is given by equation 3. The absolute value of the terms decreases steadily to zero; thus, this equation satisfies a basic test of convergence for alternating series (see, for example, Boas, 1983).

ACKNOWLEDGMENTS. We gratefully acknowledge François Gonon, Zhuan Zhou, Stan Misler, and the members of our laboratory for critically reading the manuscript. R.H.C. is a Howard Hughes Medical Institute Physician Postdoctoral Fellow. The work described here was begun when R.H.C. was an Alexander von Humboldt Research Fellow. L.vR. is an EMBO Fellow and was a recipient of a fellowship from the Graduiertenkolleg "Signalvermittelter Transport von Proteinen und Vesikeln."

References

Adler, E. M., Augustine, G. J., Duffy, M. P., and Charlton, M. P., 1991, Alien intracellular calcium chelators attenuate neurotransmitter release at the squid giant synapse, *J. Neurosci.* **11:**1496–1507.

Almers, W., 1990, Exocytosis, *Annu. Rev. Physiol.* **52:**607–624.

Almers, W., Breckenridge, L. J., and Spruce, A. E., 1989, The mechanism of exocytosis during secretion in mast cells, in: *Secretion and Its Control* (G. S. Oxford and C. M. Armstrong, eds.), Rockefeller University Press, New York, pp. 269–282.

Alvarez de Toledo, G., Fernández-Chacóon, R., and Fernández, J. M., 1993, Release of secretory products during transient vesicle fusion, *Nature* **363:**554–558.

Armstrong-James, M., and Millar, J., 1979, Carbon fiber microelectrodes, *J. Neurosci. Methods* **1:**279–287.

Armstrong-James, M., Fox, K., Kruk, Z. L., and Millar, J., 1981, The electrochemical detection of enkephalins in bulk solution and following iontophoresis, *J. Physiol.* **313:**38P.

Artalejo, C. R., Adams, M. E., and Fox, A. P., 1994, Three types of Ca^{2+} channel trigger secretion with different efficacies in chromaffin cells, *Nature* **367:**72–76.

Atkins, P. W., 1990, *Physical Chemistry,* 4th ed., pp. 766–773, Oxford University Press, Oxford.

Augustine, G. J., and Charlton, M. P., 1986, Calcium dependence of presynaptic calcium current and post-synaptic response at the squid giant synapse, *J. Physiol.* **381:**619–640.

Augustine, G. J., and Neher, E., 1992, Calcium requirements for secretion in bovine chromaffin cells, *J. Physiol.* **450:**247–271.

Augustine, G. J., Adler, E. M., and Charlton, M. P., 1991, The calcium signal for transmitter secretion from presynaptic nerve terminals, *Ann. N.Y. Acad. Sci.* **635:**365–381.

Boas, M. L., *Mathematical Methods in the Physical Sciences,* 2nd ed., pp. 15–16, John Wiley & Sons, New York.

Breckenridge, L. J., and Almers, W, 1987. Final steps in exocytosis observed in a cell with giant secretory granules, *Proc. Natl. Acad. Sci. U.S.A.* **84:**1945–1949.

Ceccarelli, B., and Hurlbut, W. P., 1980, Vesicle hypothesis of the release of quanta of acetylcholine, *Physiol. Rev.* **60:**396–441.

Chad, J. E., and Eckert, R., 1984, Calcium domains associated with individual channels can account for anomalous voltage relations of Ca-dependent responses, *Biophys. J.* **45:**993–999.

Chow, R. H., Rüden, L. von, and Neher, E., 1992, Delay in vesicle fusion revealed by electrochemical monitoring of single secretory events in adrenal chromaffin cells, *Nature* **356:**60–63.

Crank, J., 1975, *The Mathematics of Diffusion,* 2nd ed., Oxford University Press, Oxford.

Crespi, F., 1991, *In vivo* voltammetric detection of neuropeptides with micro carbon fibre biosensors: Possible selective detection of somatostatin, *Anal. Biochem.* **194:**69–76.

Crow, D. R., 1988, *Principles and Applications of Electrochemistry,* 3rd ed., Chapman and Hall, London.

Dayton, M. A., Brown, J. C., Stutts, K. J., and Wightman, R. M., 1980, Faradaic electrochemistry at microvoltammetric electrodes, *Anal. Chem.* **52:**946–950.

Dodge, F. A., Jr., and Rahamimoff, R., 1967, Cooperative action of calcium ions in transmitter release at the neuromuscular junction, *J. Physiol.* **193:**419–432.

Duchen, M. R., 1993, Voltammetric detection of quantal secretory events from isolated rat peritoneal mast cells, *J. Physiol.* **467**:2P.

Duchen, M. R., Millar, J., and Biscoe, T. J., 1990, Voltammetric measurement of catecholamine release from isolated rat chromaffin cells, *J. Physiol.* **426**:5P

Dudel, J., 1989, Calcium dependence of quantal release triggered by graded depolarization pulses to nerve terminals on crayfish and frog muscle, *Pflügers Arch.* **415**:289–298.

Fernandez, J. M., Neher, E., and Gomperts, B. D., 1984, Capacitance measurements reveal stepwise fusion events in degranulating mast cells, *Nature* **312**:453–455.

Feynman, R. P., Leighton, R. B., and Sands, M., 1964, *Lectures on Physics,* Volume 2, pp. 6–8 to 6–9, Addison-Wesley, Reading, MA.

Fogelson, A. L., and Zucker, R. S., 1985, Presynaptic calcium diffusion from various arrays of single channels: Implications for transmitter release and synaptic facilitation, *Biophys. J.* **48**:1003–1017.

Fox, K., Armstrong-James, M., and Millar, J., 1980, The electrical characteristics of carbon fibre microelectrodes, *J. Neurosci. Methods* **3**:37–48.

Gerhardt, G. A., and Adams, R. N., 1982, Determination of diffusion coefficients by flow injection analysis, *Anal. Chem.* **54**:2618–2620.

Gerhardt, G. A., Oke, A. F., Nagy, G., Moghaddam, B., and Adams, R. N., 1984, Nafion-coated electrodes with high selectivity for CNS electrochemistry, *Brain Res.* **290**:390–395.

Gonon, F., Cespuglio, R., Ponchon, J. L., Buda, M., Jouvet, M., Adams, R. N., and Pujol, J. F., 1978, Measure électrochimique continue de la libération de dopamine réalisée in vivo dans le néostriatum du rat. *C. R. Acad. Sci.* **286**:902–904.

Gonon, F., Buda, M., and Pujol, J. F., 1984, Treated carbon fiber electrodes for measuring catechols and ascorbic acid, in: *Measurement of Neurotransmitter Release in Vivo* (C. A. Marsden, ed.), pp. 153–171, John Wiley & Sons, New York.

Gonon, F., Msghina, M., and Stjärne, L., 1993, Kinetics of noradrenaline released by sympathetic nerves, *Neuroscience* **56**:535–538.

Heinemann, C., Rüden, L. von, Chow, R. H., and Neher, E., 1993, A two-step model of secretion control in neuroendocrine cells, *Pflügers. Arch.* **424**:105–112.

Hodgkin, A. L., and Huxley, A. F., 1952, Currents carried by sodium and potassium ions through the membrane of the giant axon of *Loligo, J. Physiol.* **116**:449–472.

Iravani, M. M., Kruk, Z. L., and Millar, J., 1993, Electrochemical detection of nitric oxide using fast cyclic voltammetry, *J. Physiol.* **467**:48P.

Jackson, J. D., 1975, *Classical Electrodynamics,* 2nd ed., John Wiley & Sons, New York.

Jankowski, J. A., Schroeder, T. J., Holz, R. W., and Wightman, R. M., 1992, Quantal secretion of catecholamines measured from individual bovine adrenal medullary cells permeabilized with digitonin, *J. Biol. Chem.* **26**:18329–18335.

Jankowski, J. A., Schroeder, T. J., Ciolkowski, E. L., and Wightman, R. M., 1993, Temporal characteristics of quantal secretion of catecholamines from adrenal medullary cells, *J. Biol. Chem.* **268**:14694–14700.

Justice, J. B., Jr., ed., 1987, *Voltammetry in the Neurosciences: Principles, Methods, and Applications.* Humana Press, Clifton, NJ.

Kawagoe, J. L., Jankowski, J. A., and Wightman, R. M., 1991a, Etched carbon-fiber electrodes as amperometric detectors of catecholamine secretion from isolated biological cells, *Anal. Chem.* **63**:1589–1594.

Kawagoe, J. L., Niehaus, D. E., and Wightman, R. M., 1991b, Enzyme-modified organic conducting salt microelectrode, *Anal. Chem.* **63**:2961–2965.

Kawagoe, K. T., Zimmerman, J. B., and Wightman, R. M., 1993, Principles of voltammetry and microelectrode surface states, *J. Neurosci. Methods* **48**:225–240.

Kelly, R. S., and Wightman, R. M., 1986, Bevelled carbon-fibre ultramicroelectrodes, *Anal. Chim. Acta* **187**:79–87.

Kennedy, R. T., Huang, L., Atkinson, M. A., and Dush, P., 1993, Amperometric monitoring of chemical secretions from individual pancreatic β-cells, *Anal. Chem.* **65**:1882–1887.

Kissinger, P. T., Hart, J. B., and Adams, R. N., 1973, Voltammetry in brain tissue—a new neurophysiological measurement. *Brain Res.* **55**:209–213.

Klingauf, J., Chow, R. H., Heinemann, C., Zucker, R. S., and Neher, E., 1994, Sources of secretory delay, *Biophys. J.* **66**(2):A55.

Leszczyszyn, D. J., Jankowski, J. A., Viveros, O. H., Diliberto, E. J., Jr., Near, J. A., and Wightman, R. M., 1990, Nicotinic receptor-mediated catecholamine secretion from individual chromaffin cells, *J. Biol. Chem.* **265:**14736–14737.

Leszczyszyn, D. J., Jankowski, J. A., Viveros, O. H., Diliberto, E. J., Jr., Near, J. A., and Wightman, R. M., 1991, Secretion of catecholamines from individual adrenal medullary chromaffin cells, *J. Neurochem.* **56:**1855–1863.

Malinski, T., and Taha, Z., 1992, Nitric oxide release from a single cell measured *in situ* by a porphyrinic-based microsensor, *Nature* **358:**676–678.

Marin, A., and Tabares, L., 1993, Single secretory events recorded by electrochemical methods from rat pineal cells, *Biophys. J.* **64:**A195.

McCreery, R. L., 1991. Carbon electrodes: Structural effects on electron transfer kinetics, in: *Electroanalytical Chemistry,* Vol. 17 (A. J. Bard, ed.), pp. 221–374, Marcel Dekker, New York.

Millar, J., 1991, Simultaneous *in vivo* voltammetric and electrophysiological recording with carbon fiber microelectrodes, *Methods Neurosci.* **4:**143–154.

Millar, J., and Barnett, T. G., 1988, Basic instrumentation for fast cyclic voltammetry, *J. Neurosci. Methods* **25:**91–95.

Millar, J., and Williams, G. V., 1990, Fast differential ramp voltammetry: A new voltammetric technique designed specifically for use in neuronal tissue, *J. Electroanal. Chem.* **282:**33–49.

Millar, J., O'Connor, J. J., Trout, S. J., and Kruk, Z. L., 1992, Continuous scan cyclic voltammetry (CSCV): A new high-speed electrochemical method for monitoring neuronal dopamine release, *J. Neurosci. Methods* **43:**109–118.

Monck, J. R., Oberhauser, A. F., Alvarez de Toledo, G., and Fernandez, J. M., 1991, Is swelling of the secretory granule matrix the force that dilates the exocytotic fusion pore? *Biophys. J.* **59:**39–47.

Monck, J. Escobar, A., Robinson, I., Vergara, J., and Fernandez, J. M., 1994, Pulsed-laser image of rapid Ca^{2+} signaling in excitable cells, *Biophys. J.* **66:**A351.

Neher, E., 1993, Secretion without full fusion, *Nature* **363:**497–498.

Neher, E., and Marty, A., 1982, Discrete changes of cell membrane capacitance observed under conditions of enhanced secretion in bovine adrenal chromaffin cells, *Proc. Natl. Acad. Sci. U.S.A.* **79:**6712–6716.

O'Connor, J. J., and Kruk, Z. L., 1991, Fast cyclic voltammetry can be used to measure stimulated endogenous 5-hydroxytryptamine release in untreated rat brain slices, *J. Neurosci Methods* **38:**25–33.

Palij, P., and Stamford, J. A., 1992, Real time monitoring of endogenous noradrenaline release in rat brain slices using fast cyclic voltammetry: 1. Characterization of evoked noradrenaline efflux and uptake from nerve terminals in the bed nucleus of stria terminalis, pars ventralis, *Brain. Res.* **587:**137–146.

Pantano, P., and Kuhr, W. G., 1993, Dehydrogenase-modified carbon-fiber microelectrodes for the measurement of neurotransmitter dynamics. 2. Covalent modification utilizing avidin–biotin technology, *Anal. Chem.* **65**(5)**:**623–630.

Pantano, P., Morton, T. H., and Kuhr, W. G., 1991, Enzyme-modified carbon fiber microelectrodes with millisecond response times, *J. Am. Chem. Soc.* **113:**1832–1833.

Phillips, J. H., 1982, Dynamic aspects of chromaffin granule structure, *Neuroscience* **7:**1595–1609.

Ponchon, J., Cespuglio, R., Gonon, F., Jouvet, M., and Pujol, J., 1979, Normal pulse polarography with carbon fiber electrodes for *in vitro* and *in vivo determination of catecholamines, Anal. Chem.* **51:**1483–1486.

Schroeder, T. J., Jankowski, J. A., Kawagoe, K. T., Wightman, R. M., Lefrou, C., and Amatore, C., 1992, Analysis of diffusional broadening of vesicular packets of catecholamines released from biological cells during exocytosis, *Anal. Chem.* **64:**(24)**:**3077–3083.

Schroeder, T. J., Jankowski, J. A., and Wightman, R. M., 1993, in: *7th International Symposium on Chromaffin Cell Biology and Pharmacology,* p. 134, University of Ottawa, Ottawa.

Simon, S. M., and Llinas, R., 1985, Compartmentalization of the submembrane calcium activity during calcium influx and its significance in transmitter release, *Biophys. J.* **48:**485–498.

Stamford, J. A., 1990, Fast cyclic voltammetry: Measuring transmitter release in "real time," *J. Neurosci. Methods* **34:**67–72.

Tatham, P. E. R., Duchen, M. R., and Millar, J., 1991, Monitoring exocytosis from single mast cells by fast voltammetry, *Pflügers Arch.* **419:**409–414.

Torri-Tarelli, F., Grohovaz, F., Fesce, R., and Ceccarelli, B., 1985, Temporal coincidence between synaptic vesicle fusion and quantal secretion of acetylcholine, *J. Cell. Biol.* **101:**1386–1399.

von Rüden, L., 1993, Untersuchungen zu verschiedenen Phasen der Sekretion in chromaffinen Zellen, Ph.D. Thesis, University of Göttingen, Göttingen, Germany.

von Rüden, L., and Neher, E., 1993, A Ca-dependent early step in the release of catecholamines from adrenal chromaffin cells, *Science* **262:**1061–1065.

von Rüden, L., García, A. G., and López, M. G., 1993, The mechanism of Ba^{2+}-induced exocytosis from single chromaffin cells, *FEBS Lett.* **336**(1)**:**48–52.

Wang, J., and Angnes, L., 1992, Miniaturized glucose sensors based on electrochemical codeposition of rhodium and glucose oxidase onto carbon-fiber electrodes, *Anal. Chem.* **64:**456–459.

Wightman, R. M., Jankowski, J. A., Kennedy, R. T., Kawagoe, K. T., Schroeder, T. J., Leszczyszyn, D. J., Near, J. A., Diliberto, E. J., Jr., and Viveros, O. H., 1991, Temporally resolved catecholamine spikes correspond to single vesicle release from individual chromaffin cells, *Proc. Natl. Acad. Sci. U.S.A.* **88:**10754–10758.

Urena, J., Fernandez-Chacon, R., Benot, A. R., Alvarez de Toledo, G., and Lopez-Barneo, J., 1994, Hypoxia induces voltage-dependent Ca^{2+} entry and quantal dopamine secretion in carotid body glomus cells, *Proc. Natl. Acad. Sci. U.S.A.,* in press.

Zucker, R. S., Delaney, K. R., Mulkey, R., and Tank, D. W., 1991, Presynaptic calcium in transmitter release and posttetanic potentiation, *Ann. NY Acad. Sci.* **635:**191–207.

Chapter 12

Technical Approaches to Studying Specific Properties of Ion Channels in Plants

RAINER HEDRICH

1. Introduction: Why Do Plants Need Ion Channels?

Plants have no bones and muscles; consequently, they do not possess, sodium and ACh-receptor channels. They are, however, equipped with unique voltage- and ligand-activated K^+ channels, anion channels, and nonselective channel types that constitute the basis for a wide spectrum of plant-specific functions essential to circumventing the inability to move (for review see Hedrich and Schroeder, 1989; Hedrich *et al.*, 1994; Schroeder *et al.*, 1994). These functions can be grouped into three categories with respect to the duration of a physiological response such as signaling, volume and turgor regulation, as well as growth, development, and reproduction. Biophysical and biochemical processes in the second to minute range such as leaf and stomatal movements represent the fastest responses to environmental signals that have been recognized in plants (Hill and Findlay, 1981; Schroeder and Hedrich, 1989; Assmann, 1993; Jones, 1994). They are, however, still slow compared to their animal counterparts.

1.1. Solute Transport

1.1.1. Signaling

Fast, transient ion fluxes or excitability seem to be restricted to some species or specialized cell types. Transient potential changes result from mechanical, electrical, or chemical stimulation or alterations in temperature or light (Buff, 1854; Burdon Sanderson, 1873; Blinks, 1937a,b; Curtis and Cole, 1937; Lüttge and Pallagky, 1969). Stimulus-dependent action potentials (APs), both single, often propagating, and repetitive potential changes, have been found (Thiel *et al.*, 1992).

In this category voltage- and time-dependent channels are elementary for transient changes in the membrane potential. Primary targets in plant signal transduction involve receptor(s)–ion channels and ion pumps located in the plasma membrane (Schroeder *et al.*, 1984, 1987; for review, see Hedrich *et al.*, 1994). From the voltage-dependent activation and inactivation cycles identified for a guard cell anion channel (GCAC1), excitability was

RAINER HEDRICH • Institute for Biophysics, University of Hannover, D-30419 Hannover, Germany.
Single-Channel Recording, Second Edition, edited by Bert Sakmann and Erwin Neher. Plenum Press, New York, 1995.

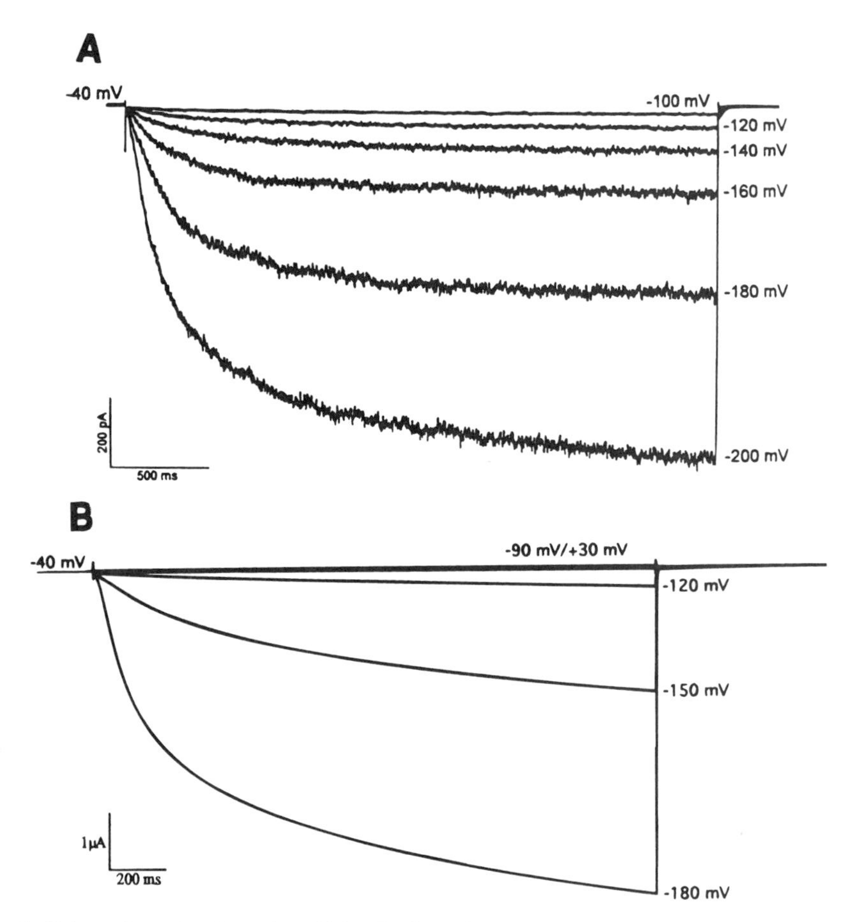

Figure 1. Voltage-dependent, inward-rectifying K^+ channels in guard cell protoplasts (*in vivo*) and in different heterologous expression systems (*in vitro*). (A) K^+ channels in *Solanum tuberosum* guard cells *in vivo*. Inward rectifying K^+ currents elicited by 3-sec voltage pulses to hyperpolarized potentials. (B) Guard cell K^+ channels *in vitro*. Functional expression of KST1, a guard cell inward rectifying K^+ channel, in *Xenopus* oocytes. (C) Current–voltage relationship of the steady-state current of KST1 *in vivo* and *in vitro*. (D) Single voltage-dependent 5-pS K^+ channels (a) in oocyte plasma membrane patches underlie the macroscopic hyperpolarization-activated inward currents (b) following KAT1 cRNA injection (cf. B). (E) Functional expression of KAT1 in Sf9 cells. Inward K^+ currents were elicited by 810-msec pulses to hyperpolarized potentials. In noninfected/transformed cells, 750-msec voltage pulses activated background currents only (I. Marten, F. Gaymard, J. B. Thibaud, H. Sentenac, and R. Hedrich, unpublished data). (F) Functional expression of KAT1 in spheroplasts of the yeast *trk1/trk2* double mutant (VEC) Inward K^+ currents were elicited by a 1-sec voltage pulse from −60 mV to −200 mV in 20-mV increments (A. Bertl, unpublished data).

predicted for this cell type before APs were recorded (Hedrich *et al.,* 1990; Thiel *et al.,* 1992; Gradmann *et al.,* 1993).

In contrast to the biology of animal cells, the sodium ion does not play a vital role in land plants.* Furthermore, inactivating K^+ or Ca^{2+} channels have not yet been discovered (Schroeder *et al.,* 1987; Cosgrove and Hedrich, 1991). Thus, Ca^{2+}- and voltage-dependent,

*For stress imposed through the increase in sodium salt concentration, see literature on "salt stress," e.g., Maathius and Prins. (1990 Pantoja *et al.*, Schachtman *et al.*, 1991) and references therein.

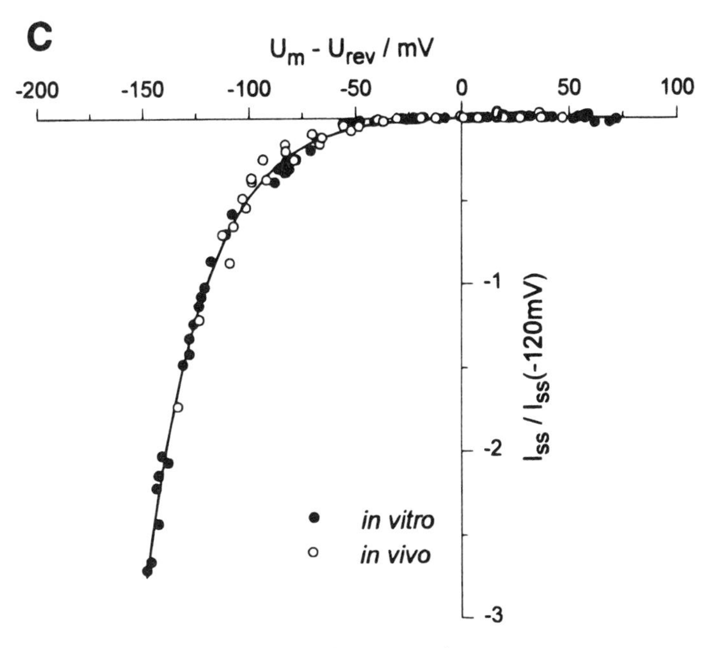

Figure 1. *Continued.*

inactivating anion channels may represent the major depolarizing activity in higher plants and giant algae (Beilby, 1989; Kolb *et al.*, 1995). In addition to the inactivation of the anion channel, outward-rectifying K^+ channels may also contribute to the shape of the repolarization phase (Schroeder, 1989; Stoeckel and Takeda, 1989a; van Duijn *et al.*, 1993). This garantees charge movement either without or accompanied by only minor salt fluxes.

1.1.2. Volume and Turgor Regulation

Fluid regulation is based on short-term ion fluxes. Although excitability is still an exception restricted to specialized cell types, short-term ion fluxes accompanying diurnal cycles in cell volume and turgor, storage and release of metabolites, or changes in pH represent a general feature of plant cells.

1.1.2a. K^+ Uptake Channels. Volume and turgor increase is mediated through the accumulation of K^+ salts, which in turn drive water uptake. K^+ influx at the level of the plasma membrane is catalyzed by voltage-dependent (inward-rectifying) K^+ channels (Sentenac *et al.*, 1992; Anderson *et al.*, 1992; for review, see Schroeder *et al.*, 1994). In some cell types (e.g., of the root) additional high-affinity uptake systems are postulated (Maathius and Sanders, 1993; Kochian and Lucas, 1993; Gassmann *et al.*, 1993). The isolation of two genes, KAT1 and AKT1, encoding K^+ channel function was possible because of their ability to restore K^+ uptake in K^+-transport-deficient yeast mutants (Sentenac *et al.*, 1992; Anderson *et al.*, 1992). Both clones, which share a high degree of homology with *shaker*-type K^+ channels from animals, were isolated from a cDNA library of *Arabidopsis thaliana* seedlings. When KAT1 and AKT1 were expressed in *Xenopus* oocytes, Sf9 cells, and yeast, two-electrode voltage-clamp and/or patch-clamp studies revealed voltage-dependent inward rectifying K^+ channels (Fig. 1; Schachtman *et al.*, 1992; I. Marten, personal communication; Bertl *et al.*, 1994;

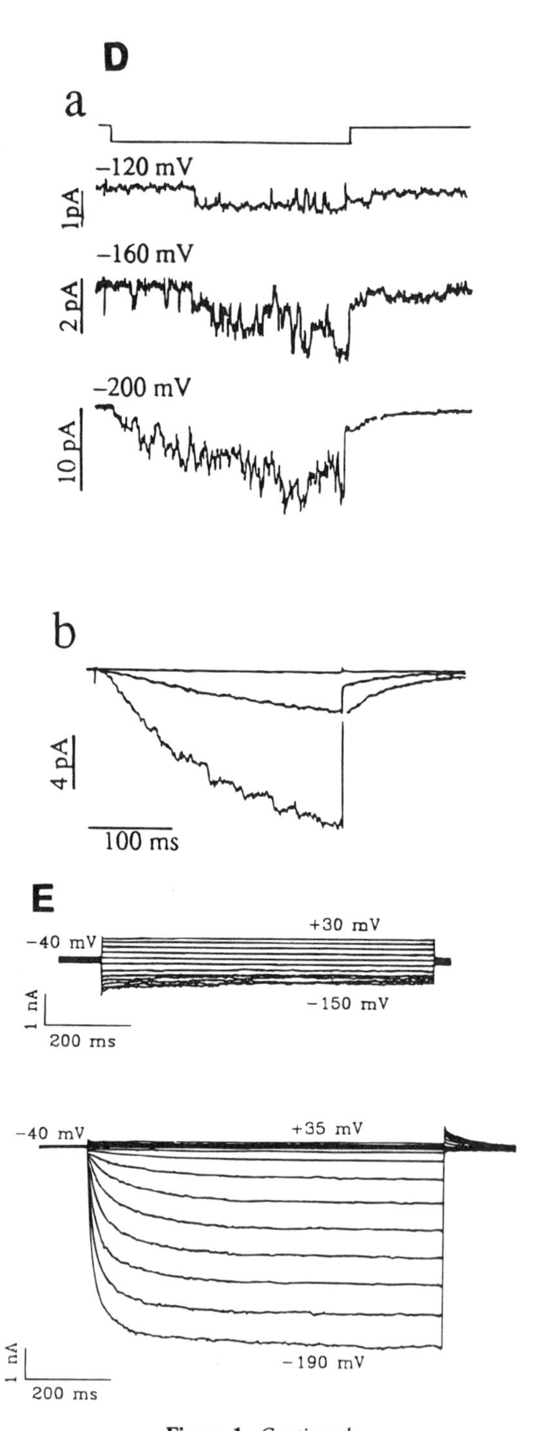

Figure 1. *Continued.*

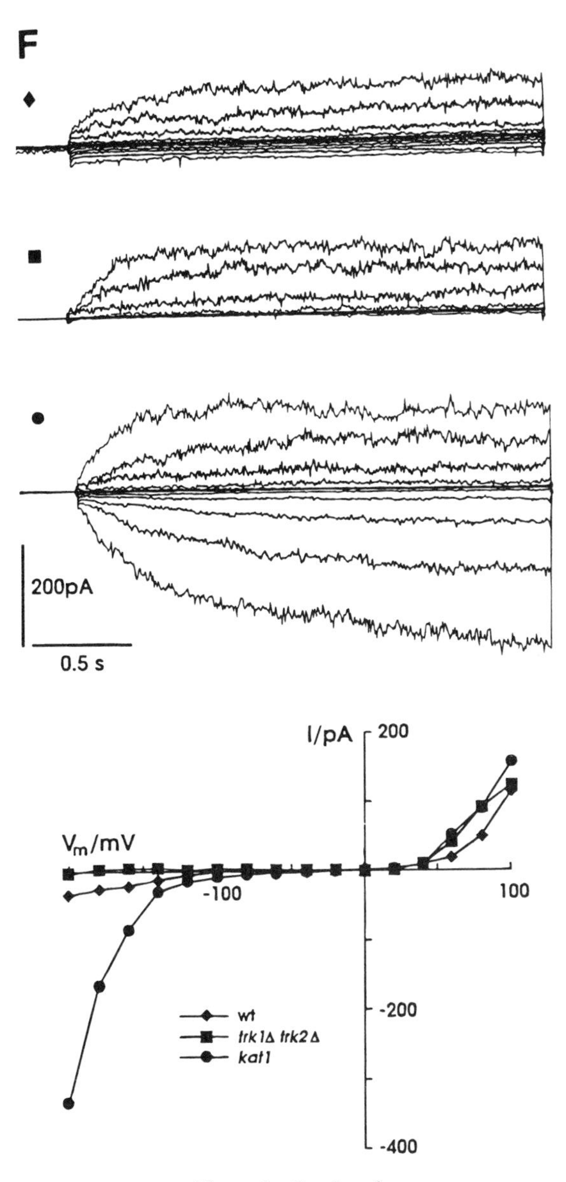

Figure 1. *Continued.*

Hedrich *et al.,* 1994). Voltage dependence, kinetics, selectivity, and block of the *Arabidopsis* channels and the recently identified homologous clones from guard cells of *Solanum tuberosum* resembled those of K^+ channels *in vivo* (Hedrich *et al.,* 1994; Müller-Röber *et al.,* 1995). All these K^+ channels have in common that cytoplasmic Mg^{2+} is not required for inward rectification but contain an intrinsic voltage sensor like the outward-rectifying *shaker*-type channels (Hedrich *et al.,* 1994). On the other hand, they share the current direction and susceptibility to voltage-dependent block by Cs^+ and Ba^{2+} with their animal counterparts (Ho *et al.,* 1993; Kubo *et al.,* 1993; Hedrich *et al.,* 1994). In contrast to the animal inward rectifier, activation is slow ($t_{1/2}$ = 100–200 msec; see Fig. 1), and the threshold potential (around −100 mV) is almost insensitive to changes in the extracellular K^+ concentration but sensitive to the external pH (Schroeder and Fang, 1991; Blatt, 1992; Hedrich *et al.*, 1994).

Exchanging of domains between the plant inward rectifier and *shaker*-type channels as well as labeling with epitope-specific antibodies may help to understand the current paradox of K^+ channels with *shaker*-like structure but inward rectification.

1.1.2b. The H^+-ATPase Creates the Driving Force. Because of the activity of the plasma membrane H^+-ATPase, the membrane potential is sufficiently negative (−100 to −250 mV; Blatt, 1991; Lohse and Hedrich, 1992) to drive channel-mediated K^+ uptake. In the presence of MgATP, this electroenzyme provides a constitutive activity for the establishment and maintenance of an electrical and pH gradient (Lohse and Hedrich, 1992; cf. Assmann *et al.*, 1985; Serrano *et al.*, 1988). On stimulation by blue light, red light, the fungal toxin fusicoccin, or the growth hormone auxin, proton currents through the pump are increased. The H^+ pump represents a hyperpolarizing membrane conductance and is itself voltage dependent. Increasing the membrane potential positive to the reversal potential of the pump results in a steady rise in proton current, which reaches a plateau around −100 mV (Lohse and Hedrich, 1992). Therefore, the H^+-ATPase creates the driving force for voltage-dependent and H^+-coupled uptake of nutrients (see Sauer and Tanner, 1993 Schroeder *et al.*, 1994, for review) and provides a major repolarizing conductance of the plasma membrane within an action potential (Gradmann *et al.*, 1993).

1.1.2c. K^+ and Anion Release. Signal transduction in response to light, phytohormones, and pathogens often involves solute fluxes and volume changes. During volume and turgor decrease, salt efflux through voltage-dependent K^+ and anion channels is induced by membrane depolarization (Keller *et al.*, 1989; Schroeder and Hagiwara, 1990b). Depolarization results from activation of Ca^{2+}-permeable channels on one hand (Cosgrove and Hedrich, 1991; Lohse and Hedrich, 1992; Thuleau *et al.*, 1994) and hormone- or CO_2/malate-induced activation of anion channels on the other (Marten *et al.*, 1991; Hedrich and Marten, 1993; Hedrich *et al.*, 1994). Voltage-dependent anion channels in, e.g., guard cells called GCAC1 are closed at hyperpolarized potentials but activate positive to −100 mV with a half-activation potential $V_{1/2}$ of about −50 mV (Fig. 2; slope factor 20 mV; gating charge 2; Kolb *et al.*, 1995). In the presence of growth hormones or CO_2/malate, however, the voltage dependence of GCAC1 is shifted toward the resting potential of the cell (Fig. 2B; see also Marten *et al.*, 1991; Hedrich and Marten, 1993). In other cell types various different anion channels might catalyze anion efflux during prolonged stimulation (see Hedrich, 1994, for review).

Outward rectifying K^+ channels, in contrast to the inward rectifier, sense the extracellular K^+ concentration (Blatt, 1992; Bertl *et al.*, 1994). During a rise in K^+ concentration the threshold potential shifts positive ($V_{1/2}$ = 7 mV, z = 2 in 11 mM K^+; Schroeder, 1989). Thus, strong depolarizations will cause net salt release.

1.1.3. Long-Term Salt Uptake

Long-term salt uptake is essential for turgor formation and maintenance and thus creates the driving force for growth, development, and reproduction. Furthermore, turgor and volume regulation are prerequisites for salinity adaptation and control of the water status.

Similar to short-term (reversible) salt transport, long-term transport requires noninactivating K^+ transporters, a property already described for inward-rectifying K^+ channels, and K^+/H^+ symporters (Schroeder *et al.*, 1987; Schroeder *et al.*, 1994). K^+ uptake is accompanied by anion influx in cotransport with protons. Although the Cl^- uptake system, is not yet

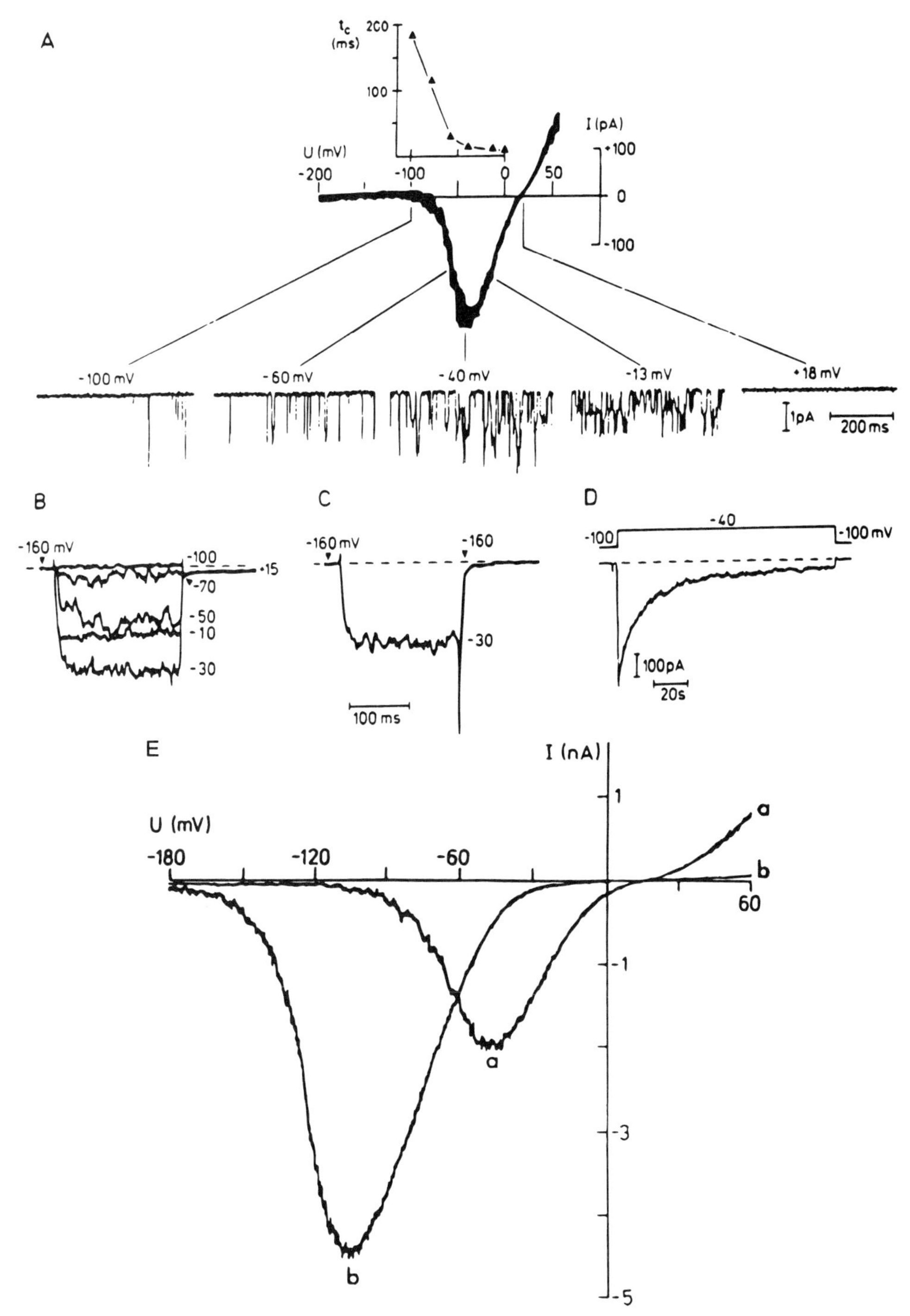

Figure 2. Whole-cell and single-channel anion currents mediated by *Vica faba* GCAC1, the guard cell anion channel 1, during voltage ramps (A and E) and steps (B–D) toward depolarized potentials. (Reproduced from Hedrich *et al.,* 1990.) (B) Activation kinetics. (C) Deactivation kinetics. (D) Inactivation kinetics. (E) Shift in the voltage sensor of GCAC1 through CO_2/malate action (b, a = malate-free) on the plasma membrane of guard cells of *Vicia faba.* (Reproduced from Hedrich and Marten, 1993.)

understood at the electrical or molecular level, one would expect similarities with a recently identified six-transmembrane-domain NO_3^- transporter (Tsay *et al.*, 1993). During cell growth, K^+ uptake, anion uptake, and/or organic acid synthesis and water influx allow volume increase and turgor formation, whereas tension-sensitive channels (see below) might fine-tune the location of the steady state (Falke *et al.*, 1988; Cosgrove and Hedrich, 1991; Maurel *et al.*, 1993; Kammerloher *et al.*, 1994; Kaldenhof *et al.*, 1993).

1.1.4. Vacuolar Ion Transport

Because of the lack of hydrostatic pressure and osmotic gradient across the lysosomal–vacuolar membrane (Wendler and Zimmermann, 1985a,b), prolonged salt fluxes through the plasma membrane require coordination of ion transport processes across both diffusion barriers. Functionally analogous to the plasma membrane H^+ pump, the vacuolar enzyme and the PP_i-driven H^+ pump generate an electrical and pH gradient across the endomembrane.* In contrast to the small-sized vacuoles in animal cells, the central vacuole of plants can be accessed directly by the patch clamp technique (Hedrich *et al.*, 1986a; Hedrich and Becker, 1994). The major conductance in this membrane results from a Ca^{2+}-dependent slowly activating outward rectifier (called SV-type channel for slow vacuolar; Hedrich and Neher, 1987). Depending on the plant species, cell type, and developmental state, the single-channel conductance varies from 60–80 pS in *Beta vulgaris* taproot vacuoles to 280–300 pS in *Vicia faba* guard cell vacuoles (Hedrich *et al.*, 1988; Schultz-Lessdorf and Hedrich, 1994). Likewise, differences in the relative permeability were described, ranging from SV-type channels highly selective for K^+ in *Acer pseudoplatus* (Colombo *et al.*, 1988) to homologous channels in *Vicia faba* guard cells which are almost equally permeable for K^+ and Cl^- (Schulz-Lessdorf and Hedrich, 1994). Whereas SV-type channels are activated by a rise in the cytoplasmic Ca^{2+} concentration, a fast-activating weakly cation-selective channel (called FV-type for fast vacuolar) is already active at resting Ca^{2+} levels.

In coexistence with Ca^{2+}-sensitive SV- and FV-type channels, hyperpolarization-activated Ca^{2+} channels have been found (Johannes *et al.*, 1992; Allen and Sanders, 1994). During salt transport plasma and vacuolar membrane Ca^{2+} conductances may control cytoplasmic Ca^{2+} concentration and thus the magnitude and direction of ion and water fluxes across both membranes.

1.2. Nonchannel Solute Transporters

Besides fluxes of the macronutrients K^+, Cl^-, and Ca^{2+}, which have been shown to be channel-mediated, plants require transmembrane and long-distance transport mechanisms for sources of reduced carbon (sugars), nitrogen, sulfur, phosphorus, and the various micronutrients (Marschner, 1986; Clarkson, 1989).

Recently, a number of carriers specific for sugars (Sauer and Tanner, 1993; Riesmeier *et al.*, 1992), amino acids (Frommer *et al.*, 1994), PO_4/phosphoglycerate (Flügge *et al.*, 1989), NO_3^- (Tsay *et al.*, 1993), and NH_4^+ (Ninnemann *et al.*, 1994), have been cloned and functionally expressed in heterologous systems such as yeast mutants or *Xenopus* oocytes

*Please note, that since 1992, the polarity convention for this membrane has been inverted; it is now given relative to the extracytoplasmic space (Bertl *et al.*, 1992).

or reconstituted into liposomes (Sauer and Tanner, 1993; Schwarz *et al.,* 1994). Voltage-dependent currents carried by protons in cotransport with hexoses have been analyzed (Boorer *et al.,* 1992; Aoshima *et al.,* 1993). Because these cotransporters are only weakly electrogenic, *in vivo* studies on the electrical properties are, however, restricted to cells containing high transporter densities or to transgenic plants (G. Lohse, N. Sauer, and R. Hedrich, unpublished data).

1.3. Water Channels: Transporting the Solvent

Under certain circumstances such as situations that require large and rapid volume changes, the membrane permeability to water can be limiting. Recently, members of a gene family that encodes channels permeable to water only have been isolated from animal and plant cells (Preston *et al.,* 1992; Maurel *et al.,* 1993; Kammerloher *et al.,* 1994; R. Kaldenhof personal communication). In plants, gene products with defined localization in the vacuolar or plasma membrane were identified. Initially this channel type has been isolated from red blood cells, with which the plant channels share homologies. The physiological role of water transport was concluded from the amplitude of osmotic swelling of *Xenopus* oocytes following expression of the aquaporin (cf. Zhang and Verkman, 1990).

Blue light and other signals that induce stomatal opening have been shown to increase the transcriptional activity for a member of the water channel family in *Arabidopsis* guard cells. The cellular localization, expression level, and gene product were identified through promotor analysis, antisense *in situ* hybridization, and immunogold labeling (Kaldenhoff *et al.,* 1994).

2. Experimental Procedures

Plant-specific tasks require ion channels and ion pumps with plant-specific abundance or features in the various membrane systems. I will thus concentrate on how to access ion channels and pumps in different plant membranes, both in cell and organelle preparations and in the individual patch-clamp conditions.

The hard- and software requirements for a plant patch-clamp setup are practically identical to those described for studies on animal cells. One difference is the more negative command voltage generally required for voltage activation of plant ion channels, since the resting potential of the majority of plant cells and the activation threshold of, e.g., inward K^+ channels are negative to -100 mV (Hedrich and Becker, 1994). Furthermore, the kinetics of the majority of voltage-dependent plant channels are at least a magnitude slower than those of excitable cells in animals.

In recent years the patch-clamp technique has been extended from excitable to nonexcitable cells as well as to nonanimal systems including higher plants, lower plants, algae, fungi, and bacteria. Development of appropriate procedures and patch-clamp conditions turned out not to be straightforward. Thus, attempts to patch-clamp the membrane surface of protoplasts may initially fail (Fairley and Walker, 1989; Fairley *et al.,* 1991), since skills on accessing the membrane in question and high-resistance seal formation have neither been developed nor adapted from related systems. In addition cytoplasmic and osmotic requirements for keeping an individual cell or channel alive must be defined.

To facilitate the start into the field of plant-specific solute transport, different strategies to patch-clamp the plasma membranes and intracellular organelles are outlined in this chapter.

2.1. The Plasma Membrane

2.1.1. Isolation of Protoplasts: Cell-Wall-Free Organisms

Irrespective of the method used to isolate protoplasts, cell-specific physiological properties such as photosynthesis/light sensitivity (mesophyll cells), hormone sensitivity, secretion (aleuron cells), cytoplasmic streaming, and volume regulation should be monitored to prove viability and physiological competence (Zeiger and Hepler, 1977; Schnabl, 1981; Bush *et al.,* 1988; Spalding *et al.,* 1992).

Because protoplasts are osmotically very active and burst rather easily, the correct choice and concentration of the plasmolyticum are important factors throughout protoplast isolation. However, obtaining nice-looking protoplasts in osmotically balanced media, does not guarantee gigaseal formation and functional integrity. Often media of low osmotic pressure, just sufficient to prevent disintegration of swollen protoplasts, are better suited.

2.1.1a. Naturally Occurring Protoplasts. Natural protoplasts can be directly isolated from developing fruits of several *Amaryllidaceae* such as *Clivia* and *Haemanthus* without the use of exogenous lytic enzymes (Stoeckel and Takeda, 1989b, and references therein). Within a certain stage of development endocellulases release viable and patchable endosperm protoplasts from ovules that are slit open with a razor blade.

2.1.1b. Protoplasts Released by Enzymatic Treatment. Incubation of cells or tissue with osmotically adjusted enzyme cocktails releases protoplasts from most tissues. Variation of osmotic pressure, enzyme cocktail, and incubation time with respect to the composition and thickness of the wall surrounding the various cell types in different developmental states is required (Raschke and Hedrich, 1990; L. Van Volkenburgh personal communication).

2.1.1c. Preparation of Specialized Cells or Homogeneous Tissues. Correlation of transport activity to cell- or tissue-specific properties requires (1) cellular markers to identify the individual cell type after tissue dissociation, (2) tissues formed by cells homogeneous in size, shape, and function, or (3) preenrichment of objects in question. Therefore, in an initial step the cell type of interest should be enriched by techniques taking advantage of the size (filtration), density (e.g., starch in chloroplasts or amyloplasts, by centrifugation), composition and thickness of the cell wall (composition of enzyme cocktail and incubation time), osmotic pressure (degree of plasmolysis), or mechanical stability (cell disintegrators, e.g., Waring Blendor; see below and Raschke and Hedrich, 1990; Bush *et al.,* 1988; L. Van Volkenburgh, personal communication). Contaminations by other cell types do not present problems as long as cells of interest can be distinguished by individual markers (e.g., presence and number of chloroplasts, color, or autofluorescence).

2.1.1d. Identification of Cell-Specific Protoplasts. This is not a problem for cells/protoplasts isolated from uniform tissues, such as guard cells, epidermal cells, mesophyll cells, hair cells, coleoptile cells, or aleuron cells. However, in case of protoplast release from heterogeneous tissues such as roots and shoots, identification even within physiologically defined zones is not yet established (for progress in combined use of mechanical and enzymatic steps, see Elzenga and Keller, 1991; Wegner and Raschke, 1994). Here an experimental approach analogous to the brain slice technique (see Chapter 8, this volume) or cell- and side-specific antibody staining of peripheral proteins or lectins should provide further advances in plant cell research.

2.1.1e. Procedure. As an example, of a cell type that occupies 0.1–1% of the leaf's volume, and even less compared to the fresh weight of the plant, the isolation and purification of guard cell protoplasts is described:

1. Enrichment of viable guard cells starts with the isolation of epidermal pieces. Epidermal tissue can be obtained either by peeling it from leaves with detachable epidermis or by fractioning leaf tissue after it has been minced in a blender. There are only a few species that possess an easily detachable epidermis with guard cells (Raschke and Hedrich, 1990). Using the blender method it is possible to prepare guard cell protoplasts from leaves of various species with even less or nondetachable epidermis. Guard cell suspensions obtained by the blender method (Raschke and Hedrich, 1990) contain xylem elements; thus, enzyme concentrations and the incubation time need to be increased with respect to isolated epidermal peels as starting material.

2. Remove major veins from expanding leaves with razor blades. Rinse the remaining laminae with deionized water and collect them in ice-cold water or buffer in a Waring Blendor. Mince batches of leaf tissue in buffer or distilled water with crunshed ice for 15–30 sec each, pour onto a mesh screen (200 μm), and rinse with cold buffer. Repeat blending two to four times (dependent on detachability of epidermis) with cold washes in between to isolate an epidermal fraction with functional guard cells but with the majority of common epidermal and mesophyll cells ruptured (Raschke and Hedrich, 1990; Becker *et al.*, 1993).

3. During the long-term exposure of the sample to isosmotic enzyme cocktails, cell walls are completely degraded, and protoplasts are released into the medium. After enzyme-free washes, protoplasts can be transferred into the recording chamber. The degree of agitation and the incubation time of the cells/preprotoplasts depend on the enzyme resistance of the walls, which is, however, limited by the level of irritation tolerated by the individual cell type. Incubation temperatures should be equivalent to the growing conditions.

4. The quick enzyme-digestion method was developed to produce protoplasts from plant tissues with a high rate of gigaseal formation and channel activity with minimal exposure to digestive enzymes (see Elzenga and Keller, 1991; Vogelzang and Prins, 1992). Cells are exposed to digestive enzymes only long enough to weaken the walls but not to completely remove them. Tissue sections are placed onto a plasmolyzing solution containing the enzyme cocktail. Incubation is carried out at constant temperature for several minutes, followed by two 5-min washes in enzyme-free solution. Protoplasts are released following transfer of the tissue into a solution osmotically adjusted to permit swelling of the protoplasts. As a result of partial recovery of turgor and volume, the weakened cell walls are ruptured. Protoplasts float freely in the solution, settle and stick to the glass bottom of the recording chamber.

This quick isolation method has been used successfully on different cell types from various tissues, producing protoplasts that seal within minutes and display channel activity.* When the quick procedure is applied to any new tissue, the most critical parameter to adjust is the osmolarity of the releasing bath solution. It is also necessary to adjust the duration of the digestion and wash steps. These will depend both on enzyme activity and on the quality of the cell wall material, which varies not only among species but also with growing conditions. Proper adjustment of bath osmolarity and digestion time should result in immediate release (within seconds) of protoplasts with very little accompanying debris.

5. Efforts to replace enzymatic treatments with mechanical- or laser-based isolation procedures were induced by the following considerations:

Stimulation of the cell during enzyme-induced cell wall lysis might occur through the release of elicitors (e.g., cell wall fragments) similar to those present during pathogen invasion

*Short-term incubation with the enzyme cocktail, gigaseal formation, and channel activity are, however, not generally equivalent to vitality or guarantees of physiological relevance of the responses recorded. For aleuron protoplasts, a 3-day differential enzyme treatment maintained hormone-controlled α-amylase secretion and K^+ channel activity (Bush *et al.*, 1988).

and breakdown of the cellulose matrix. Since the enzyme cocktail, on the other hand, presents a mixture of more than 20 different cellulases, hemicellulases, and pectinases isolated from cell wall lysates of various fungi, contaminations (oligopeptides, glycoproteins, or oligoglycorunides) could have elicitor activity as well (for review, see Ebel and Cosio, 1994). Thus, further purification of the enzymes might at least remove exogenous fungal elicitors.

Proteases present in the enzyme mixture might cut extracellular domains of the membrane proteins. However, differences between K^+-channel properties studied through patch-clamp measurements on guard cell protoplasts (Schroeder *et al.,* 1987; Hedrich and Schroeder, 1989; Hedrich and Becker, 1994 for review) on one hand and microelectrode recordings on guard cells surrounded by a wall (Blatt, 1991, for review) on the other have not been observed so far.

Protoplasts are spherical. Consequently, information on the cell type the protoplast originated from and the former cell polarity (e.g., at the growing tip; Taylor and Brownlee, 1992) is lost. Furthermore, number, size, and cytoplasmic orientation of organelles may have changed.

Turgor, the main driving force for growth, development, reproduction, and movement, is lost. Again, turgorless protoplasts and microelectrode-impaled small cells, maintaining residual differences in hydrostatic pressure exhibit comparable channel properties (cf. Schroeder *et al.,* 1987; Blatt, 1991).

2.1.2 Enzyme-Free Isolation of Protoplasts and Vacuoles

2.1.2a. Macrosurgery. An alternative method to gain access to plasma and vacuolar membranes is based on the removal of the cell wall using macroscopic tools. This technique was successfully applied to higher plants by Klercker in 1892. More recently it was used for the fast isolation of vacuoles directly from intact tissues (Coyaud *et al.,* 1987), of protoplasts from *Elodea* leaves, barley roots, and maize mesocotyl (Miedema, 1992; A. H. DeBoer, personal communication). It was applied to the giant internodal cells from *Chara* (Laver, 1991; Thiel *et al.,* 1993) and *Eremosphaera* (G. Schönknecht, personal communication). Because of the size of the dissecting tools and the degree of protoplast withdrawl from the cell wall following plasmolysis, this technique is not suitable for much smaller cells or cells of low abundance.

Macrosurgery will be illustrated by the giant unicellular alga *Eremosphaera viridis,* since its cell wall is highly resistant to degradation by common enzyme cocktails (Linz and Köhler, 1993). Preinserting a capillary and osmotic steps, as described in the following, permit access to the plasma membrane with patch pipettes: (1) Select and fix a single spherical alga (about 150 μm) with suction pipettes (Fig. 3). (2) Reduce turgor pressure by perfusion with hypertonic buffer. (3) Insert an empty glass micropipette a few minutes after starting perfusion of the hypertonic sorbitol-containing solution, when the alga still has enough turgor to allow impalement. Either use broken micropipettes with tip diameters of several micrometers or push pipettes relatively deep into the cell. (4) After 10 to 15 min, remove the pipette again and slowly decrease the osmotic strength of the bath solution by gradually adding sorbitol-free bath solution to the sorbitol buffer. (5) Application of slight negative pressure through the suction pipette squeezes a subcellular bleb up to 70 μm in diameter out of the cell (see micrograph in Fig. 3). These subprotoplasts, which are still in cytoplasmic continuum with the mother cell, frequently contain chloroplasts and maintain cytoplasmic streaming.

In *Chara,* which is also enzyme-resistant, a knife cannula, fine forceps, or normal patch

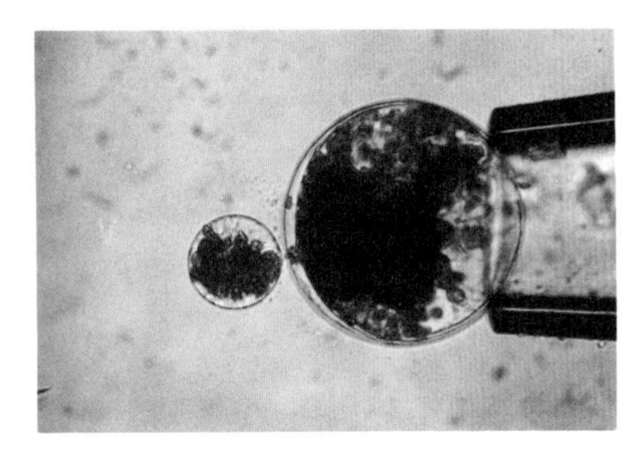

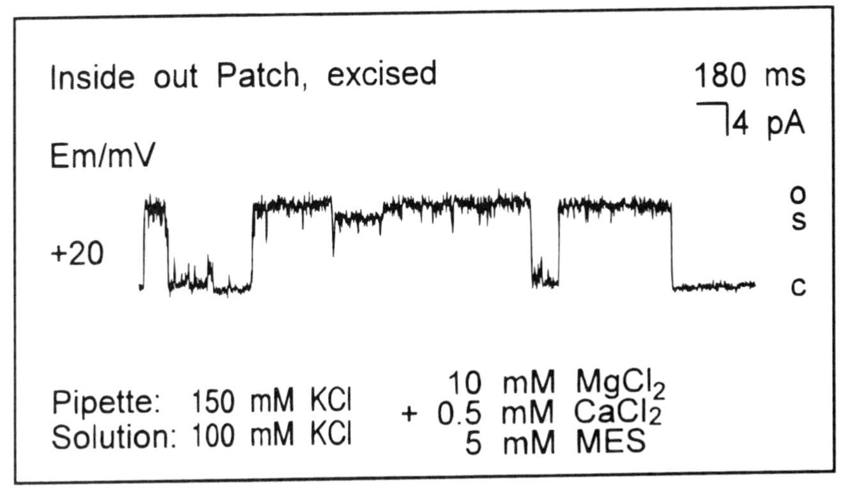

Figure 3. Macrosurgery releases cytoplasmic blisters from *Eremosphaera viridis.* (photograph by N. Sauer and G. Schönknecht, for details see text.) [For color figure see insert following the table of contents.] set in main text only, not color insert

pipettes can be advanced onto the cell wall in an area where the protoplast had withdrawn from the wall during plasmolysis. Continuous "rubbing" with the pipette tip along the axis of the internodal cell produce cuts that are large enough to advance a fresh pipette through the wall onto the plasma membrane. (Coleman, 1986, modified by Laver, 1991, and Thiel *et al.,* 1993).

2.1.2b. Laser Microsurgery. More localized perforation of cell walls sorrounding higher plant cells and algae can be obtained using UV lasers. Localized deletions result from movements of the organisms relative to the laser focus (for a detailed description of the laser microbeam system, see Greulich and Weber, 1992; Weber and Greulich, 1992). The successful removal of cell wall material with a surgical laser depends on the absorption of the laser light at the wavelength used. Lignified cell walls (DeBoer *et al.,* 1994) and the wall at the tip of root hairs (Kurkdijan *et al.,* 1993) readily absorb the UV light and thus are easily cut. The absorption efficiency can be improved through dyes that bind specifically to cell wall components (e.g., CFW 345 nm; see DeBoer *et al.,* 1994; Kurkdjian *et al.,* 1993). It should, however, be guaranteed that the dye binding is restricted to the cell wall and does not interfere with ion transport. In the case of lack of channel activity after dye-facilitated protoplast release (DeBoer *et al.*, 1994), another preparation technique and reporter channels from well-known cell types are required as a control.

For laser microsurgery the following steps (Fig. 4) require less variation than for macrosurgery on algal cells. (1) Expose cells or the whole tissue within the recording chamber to hypotonic medium in gentle, progressive plasmolysis steps. This way the cells remain viable, the plasma membrane stays intact, cellular polarity is basically unchanged (Fig. 4a), and cell–cell interaction through plasmodesmata can be maintained. Thus cell pairs within suspension cultures could lend themselves to double-whole-cell patch-clamp studies on symplastic pathways for the passage of nutrients, hormones, and electrical stimuli across the "apoplastic cleft."

After several minutes, depending on the cell size and osmotic gradient, the protoplast shrinks, withdrawing the plasma membrane from the cell wall (2–3 μm is sufficient; DeBoer *et al.*, 1994). (2) In the second step focused laser pulses should perforate the wall without affecting the integrity of the protoplast (Fig. 4a, arrow).

(3) Extrusion of the protoplast or parts thereof can be controlled by a regulated, gradual

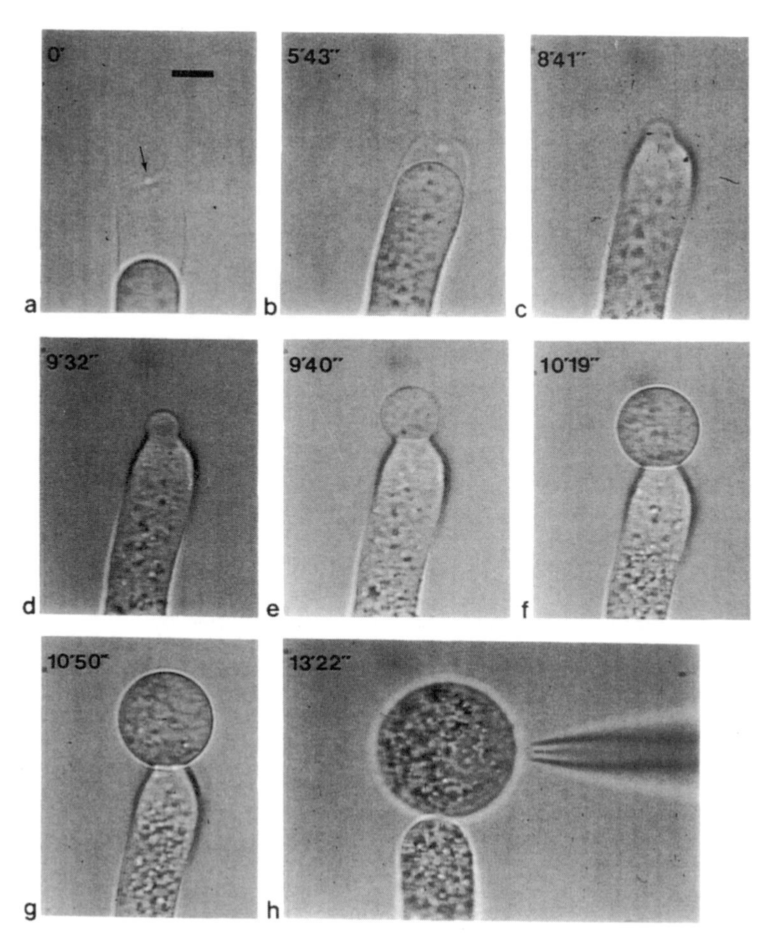

Figure 4. Release of lily pollen tip protoplasts following laser microsurgery. (Reproduced from DeBoer *et al.*, 1994; for details see text.) [For color figure see insert following the table of contents.]

decrease of the osmolarity and by variation of the calcium concentration in the extracellular solution (Fig. 4b–h; see also Taylor and Brownlee, 1992; Kurkdjian *et al.,* 1993). Adjustment of the osmotic pressure of the medium allows control over protoplast protrusion, ranging from just exposing a patch of membrane to complete release.

Following release of protoplasts or cell wall-free membrane patches, difficulties in the formation of high-resistance seals were still present in the majority of the cell types tested (Laver, 1991; DeBoer *et al.,* 1994). This raises the question of whether or not a coat of, perhaps, protein matrix (wall–membrane linkers, cf. Pont-Lezica *et al.,* 1993) that would otherwise have been removed by lytic enzymes is still present.

2.1.2c. Patch-Clamp Configuration and Ionic and Metabolic Conditions for the Study of Ion Channels and Ion Pumps in the Plasma Membrane. Two standard solutions containing salts of either permeant or impmereant anions have been used most often when ionic currents of new cell types have been explored: (in mM) 100–150 KCl or K-gluconate, 2–5 $MgCl_2$, 2 ATP, 0.1–1.0 EGTA, 5–20 HEPES, pH 7.2 in the pipette and 10 KCl or K-gluconate, 2 $MgCl_2$, 1 $CaCl_2$, 10 MES, pH 5.6 in the bath. For individual channel types the following modifications of the standard solutions given above, are suggested:

1. K^+ channels. For the measurement of K^+-selective single channels in excised patches and whole cells, replace chloride by gluconate or glutamate. Reduce the concentration of free Ca^{2+} to the micromolar level by addition of EGTA to prevent resealing of the membrane across the pipette tip. Ba^{2+} blocks hyperpolarization- and depolarization-activated K^+ channels (Schroeder *et al.,* 1987).
2. Cation channels. See K^+ channels. In order to study the exclusion limit of nonselective cation channels, replace K^+ by other monovalent and divalent cations. For Ca^{2+}-activated cation channels, increase cytoplasmic free Ca^{2+} above 1 μM (Stoeckel and Takeda, 1989b).
3. Ca^{2+} channels. In protoplasts from carrot suspension cultures, voltage-dependent Ca^{2+} channels are activated in the presence of 50 mM $CaCl_2$ in the bath and 100 mM KCl in the pipette (Thuleau *et al.*, 1994). Since voltage-dependent anion channels coexist in the same membrane (Barbara *et al.,* 1994), salts of impermeant anions should be used.
4. Inward-rectifying ICRAC-like Ca^{2+} currents. These currents can be studied in the whole-cell configuration in the absence of ATP in the pipette and high Ca^{2+} concentrations in the bath (e.g., 20 mM Ca-gluconate; Cosgrove and Hedrich, 1991; Lohse and Hedrich, 1992).
5. Anion channels. The internal solution should contain 100–150 mM K^+ or TEA salts such as TEA-Cl or K_2-malate, 2 mM MgATP, and pH and Ca^{2+} buffer as given above. The external solution should be composed of K^+-free anion salts such as $CaCl_2$, Tris-Cl or TEA-Cl, together with Ca^{2+} or Mg^{2+} and pH buffer.
6. Proton pumps. For electrogenic H^+ pumps in the plasmalemma of whole cells, use bath solutions containing 50 mM N-methylglucamine-glutamate, 5 mM $MgCl_2$, 10 mM HEPES, pH 7.0. Fusicoccin at 0.1–1 μM and auxins such as 1-NAA at 1–10 μM will stimulate H^+ extrusion. The internal solution should be identical to the bath solution plus 1–10 mM MgATP.

2.2. Endomembranes

There is no general strategy to prepare and expose different organelles or endomembranes for patch clamping. Instead, the specific structure of each organelle has to be considered.

2.2.1. Release of the Central Lysosomal–Vacuolar Compartment

Until recently, the intracellular location of this organelle has complicated the study of the electrical properties of the vacuolar membrane from higher plant cells. Improvement of methods for the isolation of stable organelles has made them accessible to patch-clamp techniques.

2.2.1a. Mechanical Isolation from Intact Tissue. A macrosurgery method for the isolation of small numbers of intact vacuoles directly from intact tissue is the fastest procedure described (Coyaud *et al.,* 1987). The surface of a freshly cut tissue slice is rinsed with buffered and osmotically adjusted solutions to wash the liberated vacuoles directly into the recording chamber. Stable vacuoles will be obtained in media (identical to the later bath solution) about 100 mOsmol higher than the osmotic pressure of the tissue. Thus, fresh vacuoles can be isolated for each experiment, and vacuole isolation and seal formation can be performed within 2–5 min.

2.2.1b. Microsurgery. Another application of microsurgery lasers is the immediate release of vacuoles from particular protoplasts by focusing the laser microbeam toward a region of the protoplast where cytoplasm or smaller organelles provide a gap between the vacuolar and the plasma membrane (see above).

2.2.1c. Selective Osmotic Shock of Isolated Protoplasts. As an example for vacuole release from preselected, enzymatically isolated protoplasts (see above) within the recording chamber, the isolation of single guard-cell vacuoles will be described. (1) Apply an aliquot of the protoplast suspension to the isotonic bathing solution within the recording chamber. (2) After protoplasts sediment and adhere to the glass bottom of the chamber, release vacuoles from single preselected protoplasts by controlled osmotic shock with hypotonic solution (Fig. 5). (3) Add hypotonic solution through an application pipette with a tip diameter of 4–6 μm while steadily perfusing the recording chamber with bathing solution (Fig. 5B; cf. Gambale *et al.,* 1994). To reduce capillary forces coat pipettes with Sigmacote (Sigma Chemicals Corp.). Position the pipette in the close neighborhood of the selected protoplasts. Continuous efflux of hypotonic solution from the pipette causes protoplast swelling and, after 3–5 min, disintegration of the plasma membrane (Fig. 5B,C). Stabilize the vacuole osmotically for the patch-clamp measurement.

2.2.1d. Patch-Clamp Configuration and Ionic–Metabolic Conditions to Examine Ion Channels and Ion Pumps in the Vacuolar Membrane. A patch-clamp survey of the electrical properties of the vacuolar membrane from a large variety of plant materials has demonstrated that the presence of voltage-dependent ion channels and electrogenic pumps are general features of ion transport in higher plant vacuoles. Essential experimental conditions to study dominant ion transporters in this endomembrane are listed:

1. Ca^{2+} channels. So far hyperpolarization-activated Ca^{2+} channels could be measured only in the inside-out configuration with 50 mM $CaCl_2$ in the bath and 100 mM KCl in the pipette. These channels are efficiently blocked by Gd^{3+}.
2. Ca^{2+}-activated channels. Slowly-activating, Ca^{2+}-dependent vacuolar channels following depolarization (named SV-type channels; Hedrich and Neher, 1987) should be recorded in solutions containing 100–200 mM KCl or KNO_3 on both membrane sides. The bathing media should include 5 mM $MgCl_2$, 0.1–10 mM $CaCl_2$, and 5 mM Tris-MES (4-morpholinoethanesulfonic acid) or citrate-KOH buffered to pH 7.5. The vacuole should equilibrate with 5 mM $MgCl_2$, 1 mM $CaCl_2$, and 5 mM MES-Tris, pH 5.5, or citrate-KOH, pH 3.5 and 4.5 (cf. Gambale *et al.,* 1994). In order to study the Ca^{2+} permeability of this channel, physiological K^+ and Ca^{2+} solutions should be replaced by (Ward and Schroeder, 1994; Schulz-Lessdorf and

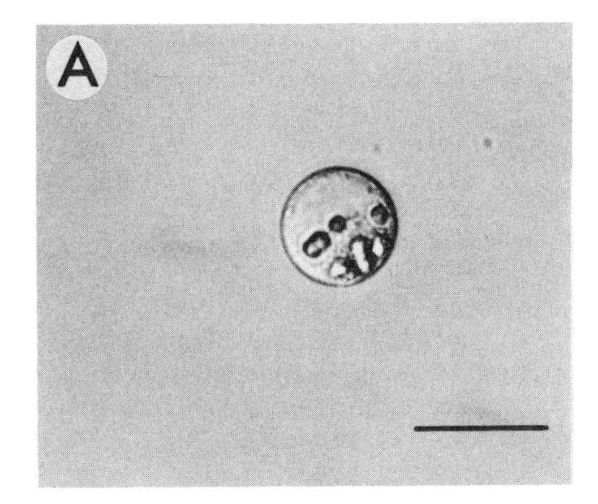

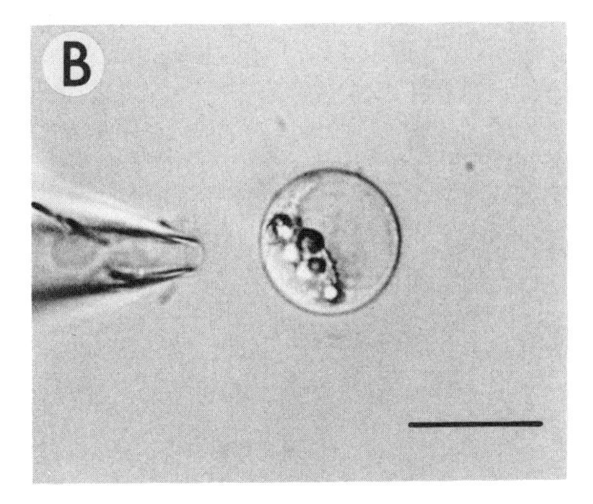

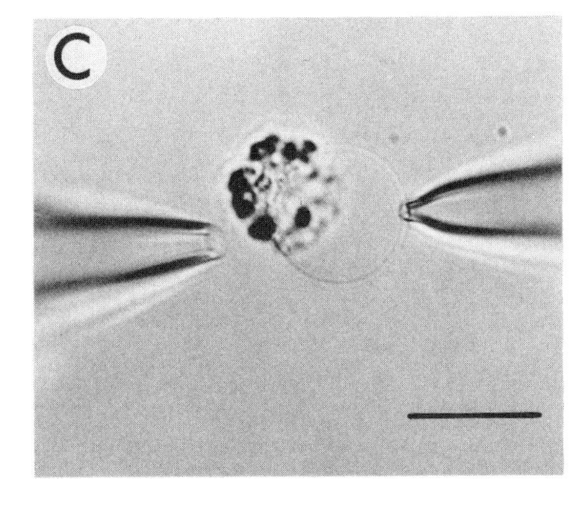

Figure 5. Isolation of guard cell vacuoles. (A) Protoplast of *Vicia faba.* (B) Osmotic swelling of the protoplast by application of hypotonic solution (pipette on the left). (C) After the osmotic shock of the protoplast, the vacuolar membrane is accessible for patch-clamp studies (patch pipette on the right). (Bar, 20 μm in A–C). (Photograph by B. Schulz-Lessdorf and R. Hedrich; for details see text.)

Hedrich, 1994) 10–50 mM $CaCl_2$ on the vacuolar side. SV-type channels are efficiently blocked by Zn^{2+} at micromolar concentrations (Hedrich and Kurkdjian, 1987; Gambale *et al.*, 1994).

3. Anion channels. So far no anion-selective ion channels have been discovered in vacuoles of higher plant cells. A 21-pS channel has, however, been observed in *Chara* (Tyerman and Findlay, 1989).
4. Electrogenic H^+ pumps (ATPase and pyrophosphatase). Currents through electrogenic H^+ ATPase and PP_iase will be resolved by increasing the cytoplasmic (extravacuolar) Ca^{2+} level to 1 mM. Under these conditions, SV-type channels are activated and dominate the membrane conductance. Since SV channels are active at voltages positive to +10 mV, in the voltage range negative to this treshold the activity of ion pumps can be studied with high resolution (Hedrich *et al.*, 1986a,b, 1989; Coyaud *et al.*, 1987; Gambale *et al.*, 1994). Following block of SV channels by, e.g., calmodulin antagonists (Weiser and Bentrup, 1993; Bethke and Jones, 1994; Schulz-Lessdorf and Hedrich, 1994), the voltage range of investigation might be expanded. Because both electroenzymes require inorganic phosphate for their individual pump cycle, specific V-type ATPase inhibitors such as bafilomycin a_1 (Youshimori *et al.*, 1991) may be used in experiments where the current direction through the PP_i-driven pump is inverted (cf. Gambale *et al.*, 1994).

2.2.2. Photosynthetic Membrane Systems

2.2.2a. Osmotic Swelling. A suitable method for highly folded membranes is the osmotic swelling technique. This technique has been successfully applied to thylakoids. To patch-clamp chloroplast membranes, two problems need to be solved: (1) giant chloroplasts should be used, or normal-sized chloroplasts have to be swollen to reach a suitable size, and (2) the outer and inner envelopes have to be separated. The latter can be solved by osmotic swelling of chloroplasts (Schönknecht *et al.*, 1988).

Isolate giant chloroplasts from protoplasts of *Peperomia metallica.* Incubate leaf slices for about an hour in an enzyme cocktail containing 270 mM sorbitol and 1 mM $CaCl_2$. Wash released protoplasts with 350 mM sorbitol and 1 mM $CaCl_2$ and separate them from debris by filtration through 200- and 25-μm nylon nets followed by centrifugation (7 min at 100 *g*). Resuspend the protoplasts in wash medium and store on ice. Mix a protoplast aliquot rapidly with volumes of bath solution in the recording chamber. After 15 min of osmotic swelling in the dark, plasma membranes, vacuoles, and thylakoid envelopes rupture, and thylakoids form large blebs. Perfuse the chamber with bath solution, and use blebs adhering to the bottom of the chamber for patch-clamp studies.

2.2.2b. The Hydration Technique: Fusion of Membrane Samples to Giant Liposomes. Here membrane fragments are reconstituted by addition of exogenous lipids. The dehydration/rehydration cycle, which is the essential step in the hydration technique, provides a general approach to investigate cellular membranes inaccessable for patch pipettes. So far, quite a large number of ion channel types have been investigated with this technique (for summary, see Keller and Hedrich, 1992). Depending on the dilution by exogenous lipids, channel analysis may range from the study of ion channels in a purely synthetic environment to channels in an environment that partially resembles the physiological situation.

Sonicate azolectin to clarity in the presence of 5 mM Tris-HCl, pH 7.2. Freeze and thaw aliquots in dry acetone twice. This process yields multilamellar liposomes. Mix aliquots of endomembrane vesicles with freeze-thawed azolectin at a desired protein-to-lipid ratio and pellet the mixture.

To form giant liposomes, resuspend the pelleted membranes into a buffer containing 10 mM MOPS and 5% (w/v) ethylene glycol, pH 7.2. Place aliquots of the suspension onto a clean glass slide and dehydrate in a desiccator at 10°C for 4 hr. Rehydrate the dehydrated lipid film on the glass slide at 4°C overnight with an electrolyte solution. If multilamellar liposomes are not directly suitable for gigohm seal formation, induce unilamellar blisters by adding a few microliters of the rehydrated suspension to a 20 mM $MgCl_2$ buffer in the patch-clamp chamber. The presence of $MgCl_2$ causes the liposomes to collapse and unilamellar blisters to emerge.

2.2.2c. Fusion of Membrane Protein to Giant Liposomes. Prepare small liposomes in a first step (see above). Freeze-thaw liposomes once and mix with purified membrane proteins. Freeze thylakoid/liposome mixtures at −80°C and thaw at 4°C. Spread a membrane sample on a glass slide and dehydrate at 4°C for 45–60 min in an exsiccator. Rehydrate in a petridish with the bottom covered with water-saturated paper. Add the electrolyte solution to be used in the patch-clamp measurements to the partially dried sample on the slide. After 1 hr giant liposomes with incorporated thylakoid membrane proteins will be observed. Giant vesicles are suitable for patch-clamp measurements after a modified hydration procedure (Criado *et al.*, 1983; Keller *et al.*, 1988; Enz *et al.*, 1993; Schwarz *et al.*, 1994).

2.3. Patch-Clamp Configurations Applicable to Different Plant Membranes

In principle all patch-clamp configurations can be applied to protoplasts and vacuoles from the majority of the different cell types (compare Hamill *et al.*, 1981 to Raschke and Hedrich, 1990). However, on free-floating or nonadhesive objects, isolation of cell-free patches requires more rigorous procedures (e.g., withdrawal of protoplast or vacuole from the membrane patch underlying the pipette by a strong solute current or air bubbles). The size of whole-cell pipettes should be adapted to the cell size, since a reasonable portion of the pipette tip is occupied by a relatively large membrane bleb before gigaseals are formed (Fig. 6A, B). With respect to the sort of glass suited for patch-clamp studies on plant cells,

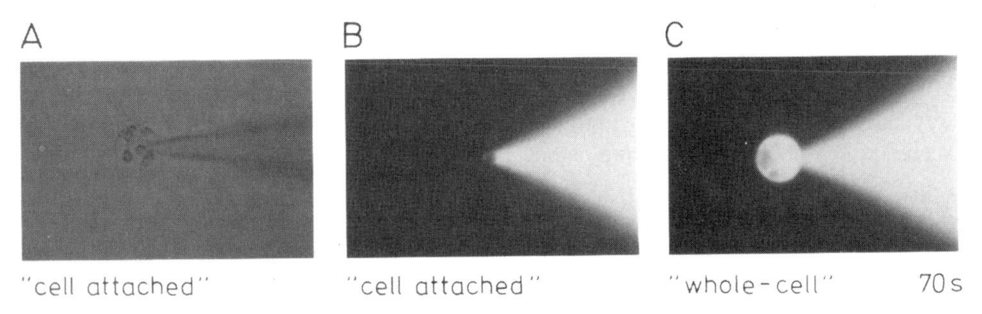

Figure 6. Equilibration of the fluorochrome Lucifer yellow with the cytoplasm of guard cells after establishment of the whole-cell configuration. (A) Cell-attached configuration transmission micrograph and (B) fluorescence micrograph with 1 μM Lucifer yellow included in the pipette solution. Note formation of an omega-shaped membrane patch in the pipette tip during the sealing process. (C) Equilibration of the pipette solution with the cytoplasm is indicated by a steady fluorescence of the cell 10 min after the establishment of the whole-cell configuration (reproduced from Marten *et al.*, 1992). [For color figure see insert following the table of contents.] set in text only.

no systematic comparison has been performed. Since 1983 the author has used borosilicate glass (Kimax, Kimble Products Ltd.) for single-channel and whole-cell recordings on protoplasts, vacuoles (Hedrich *et al.*, 1988), yeast, and various animal cells as well as for macropatches on *Xenopus laevis* oocytes (Hedrich *et al.*, 1994).

2.3.1. Protoplasts

In order to gain low-resistance access on one side and high current resolution on the other, use 2–5 MΩ pipettes for whole-cell measurements and 6–8 MΩ pipettes for single-channel recordings. Seals form more easily after tip-filling with high-Ca^{2+} solutions (0.1–1 mM). In some cell types osmotic gradients of about 10%, pipette hypertonic, facilitate seal formation and stability of whole cells. After the resistance between the pipette and the membrane exceeds 10 MΩ, hold the pipette potential negative at about −10 to −40 mV to increase seal formation. Establish the whole-cell configuration through short suction pulses and/or breakdown voltage pulses ("zap"). Follow the establishment of steady-state conditions with respect to the ionic composition by monitoring the change in resting/reversal potential through double-pulse sequences or fast voltage ramps (Hedrich *et al.*, 1988; Marten *et al.*, 1992). Equilibration of cytoplasmic effectors/blockers or buffers applied through patch pipettes should, however, be monitored by autofluorescence or marker fluorochromes (Fig. 6; Marten *et al.*, 1992; Hedrich *et al.*, 1988). Form excised patches on withdrawal of the pipette from the protoplast. In case the protoplasts do not stick to the glass bottom, withdraw the cell by a very strong pulse of bath solution or even directing of some air bubbles toward the protoplast surface.

The variability in successful seal formation on different protoplast preparations, or sometimes even preparations of the same cell type, as a result of contamination, vesiculation, or wound response might be reduced by the following modifications of the standard procedure. If the membrane surface was contaminated by proteins and slime from the enzyme cocktail (not yet removed through centrifugation) or broken protoplasts, reduce the number of protoplasts throughout the isolation procedure to improve the situation. If vesicles form in the pipette tip upon suction, lower the bath temperature (16–18°C), decrease the osmolarity of the bath solution, and/or use smaller pipettes. If within the first 1–2 min after application of suction the resistance does not exceed 10 MΩ, increase the incubation time with the enzyme cocktail or add glucose throughout the preparation. Fill the very tip of the pipette with 1 mM Ca^{2+} in addition to the internal solution before backfilling.

Use tissue sections from a single plant only once, since "wounding" in some plants may induce systemic defense reactions such as modification of the cell wall or secretion of protein or slime that resists enzyme degradation of the cell wall.

2.3.2. Vacuoles

Apply slight suction through the patch pipette to induce spontaneous seal formation on the vacuolar membrane. Establish the whole-vacuole configuration by breaking the underlying membrane through alternate ±0.6 to 1-V pulses for 1–3 msec (Hedrich *et al.*, 1986a; Hedrich and Neher, 1987). After the patch pipette gains access to the lumen of the vacuole, the pipette solution equilibrates with the vacuolar sap. When solutions are used with symmetrical ion compositions on both sides of the membrane, steady-state conditions are indicated by a resting potential of 0 mV (which will be reached in 150 mM KCl within 1–5 min for a 20-

pF vacuole). Monitor equilibration of vacuolar substances of higher molecular weights with the patch pipette through changes in autofluorescence (Hedrich *et al.,* 1988) or through introduction of exogenous fluorochromes of size and chemical structure similar to compounds of physiological or biophysical interest (cf. Marten *et al.,* 1992). Inside-out and outside-out patches can be excised without changing the bath composition (e.g., Ca^{2+} concentration) before pulling back the pipette.

2.3.3. Giant Liposomes

Giant liposomes, like many vacuole preparations that lack a cytoskeleton, neither require nor allow large negative pressures during seal formation but often seal spontaneously following attachment of patch pipettes. So far, ion channels in liposomes can only be studied in the attached and inside-out configuration. In order to study ion pumps or other low-turnover transporters, current recordings from larger membrane surface areas need to be performed (Hedrich *et al.,* 1986, 1989; Hedrich and Schroeder, 1989).

2.4. Heterologous Expression

2.4.1. Heterologous Expression of Plant Ion Transporters: Functional Expression in Yeast

Various yeast strains such as *Saccharomyces cerivisiae* and *Shizosaccharomyces pombe* are becoming suitable systems for stable expression and manipulation of both animal and plant genes (Sentenac *et al.*, 1992; Anderson *et al.*, 1992; Riesmeier *et al.*, 1992; Sauer and Tanner, 1993). Characterization of the electrical properties of heterologously expressed ion channels in the plasma membrane of yeast spheroplasts requires either a low density or lack of endogenous channels (for high background resistance use *trkl/trk2* double mutant, see Anderson *et al.*, 1993; Bertl *et al.*, 1994) in the voltage range of interest or a well-established differential pharmacology.*

In order to study the electrical properties of the protoplast membrane, the membrane surface area has to be increased, and the adhesiveness for patch electrode attachment has to be improved. Protoplasts can be liberated from most strains of *Saccharomyces* by partial digestion of the cell walls from late log-phase cells with a zymolase/glucuronidase cocktail in osmotically adjusted media (see Bertl and Slayman, 1990). After incubation for 45 min at 30°C, released protoplasts can be harvested by centrifugation, washed, and resuspended in 200 mM KC1, 10 mM $CaC1_2$, 5 mM $MgC1_2$, 10 mM glucose, and 5 mM MES/Tris, pH 7.2, at 25°C. During protoplast incubation the cell diameters and vacuolar size should increase three- to fivefold within days. For experiments on *vacuolar membranes*, use 2- to 3-day-old protoplasts, which release vacuoles suitable for patch recordings following selective osmotic shock (Bertl and Slayman, 1992). Experiments on the *plasma membranes* yield the highest success rates of gigaseal formation with protoplasts incubated for only 1–2 hr. Apply small aliquots of this suspension to the recording chamber containing an electrolyte of interest

*For endogenous plasma and vacuolar membrane channels and low-conductance yeast mutants as well as for expression of inward-rectifying plant K^+ channels such as KAT1- and AKT1-homologs in K^+ transport-deficient mutants [see Fig. 1D; Bertl and Slayman (1992, 1993) and Bertl *et al.* (1994)].

plus 2–10 mM $CaCl_2$. Various recording configurations (attached and excised patches and whole-cell configuration) can be established. Lift nonadhesive protoplasts by moving them off the bottom by slight positive pressure through the recording pipette. Gigaohm seals will form with time while suction is maintained or slowly increased, analogous to the procedure described for plant protoplasts.

2.4.2. Functional Expression in Animal Cells

Expression of plant solute and water transporters in *Xenopus* oocytes, Sf9, or COS cells should be performed in analogy to strategies developed for animal transporters (see Stühmer, 1993; Cao *et al.*, 1992; Schachtman *et al.*, 1992, Hedrich *et al.*, 1994; Kammerloher *et al.*, 1994).

2.5. Heterologous Expression of Animal Transporters in Plants

2.5.1. Functional Expression in *Chara*

To date a number of systems are utilized for functional expression of ion channels. Cytoplasmic droplets of giant green algae, described by Lühring (1986), could provide a new tool for heterologous expression not only of plant but possibly also of animal channels when they are pharmacologically distinct from endogenous channel types or are operating in different voltage windows. Cultures of *Chara australis* are inexpensive and easy to handle [for culturing conditions see Lühring (1986) and references therein]. The thalli grown for about 6 weeks after planting have internodal cells $\geq$10 cm in length and 1–1.5 cm in diameter. After reduction of turgor, mRNA injection can be performed as described for the *Xenopus* oocytes. Translation and functional insertion of foreign channel proteins into the lysovacuolar membrane requires 12–24 hr. (H. Lühring). The appearance of plasma membrane ion channels in the endomembrane (Fig. 7A) seems to be a result of the absence of targeting or even mistargeting with non-*Chara* proteins (cf. mistargeting of nonoocyte proteins, Maurel *et al.*, 1993).

Patch-clamping cytoplasmic droplets might provide access to the functional analysis of cloned ion channels. It should be noted that these subcellular structures, and vacuoles in general, became the system of choice in green laboratories when beginners first try the patch-clamp technique because of their extremely clean membrane surface and low protein density. This guarantees immediate, often spontaneous, formation of high-resistance seals even with non-fire-polished pipettes and without suction (cf. Krawczyk, 1978).

These artificial systems contain metabolically active cytoplasm (e.g., light-induced particle/chloroplast motion) surrounded by a lysovacuolar membrane with the former cytoplasmic side facing the bath solution. The intrinsic channel activity is comprised of a 150-pS K^+ channel and 21-pS Cl^- channel (Lühring, 1986; Bertl, 1989; Tyerman and Findlay, 1989; Klieber and Gradmann, 1993; for plasma membrane channels see Laver, 1991; Thiel *et al.*, 1993). Endogenous channel activity can be suppressed through application of Na^+ or Cs^+ to the media. Depending on the type of channel to be expressed in the endomembrane, individual stimulation protocols and/or differential pharmacological tools have to be developed. The nature of the newly appearing channels should be determined on the basis of their characteristic pharamacology, voltage dependence, and unit conductance (e.g., see the muscle nAChR channel in Fig. 7A). The characteristic features of the nAChR channels compare

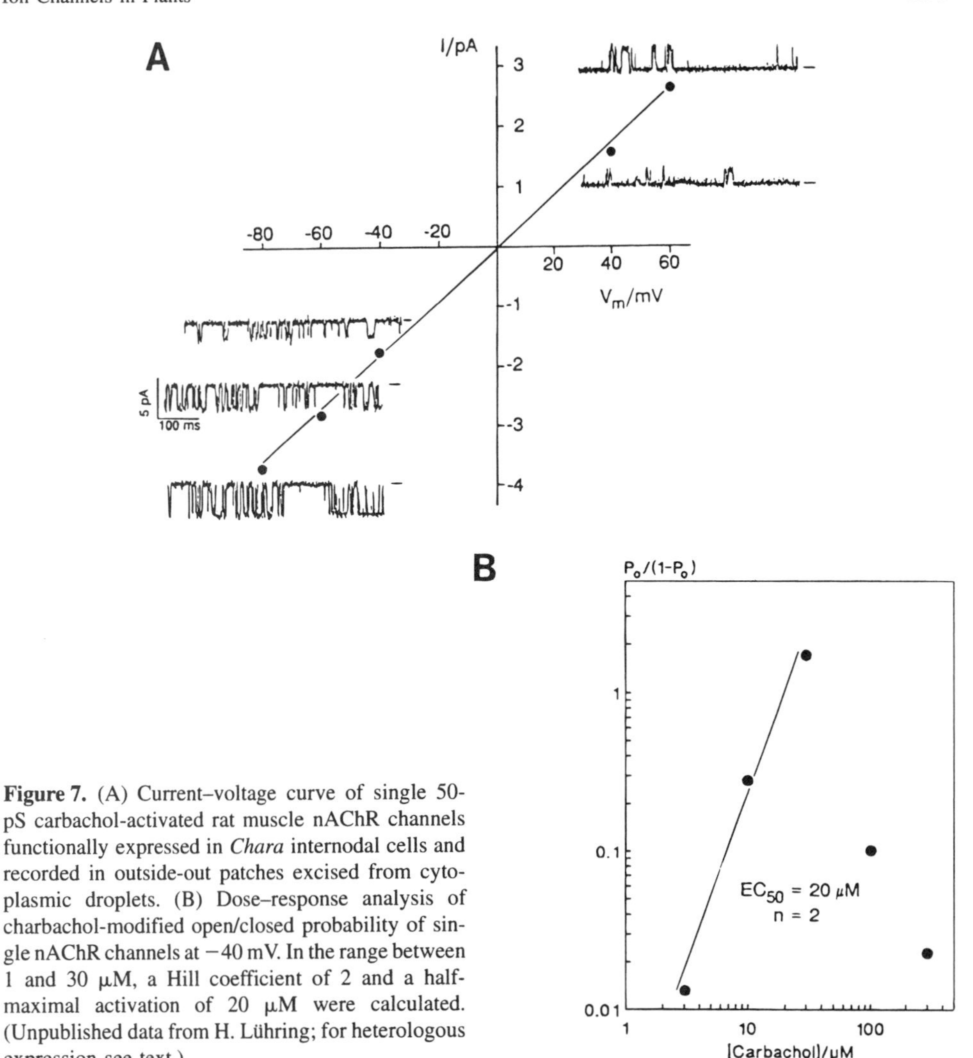

Figure 7. (A) Current–voltage curve of single 50-pS carbachol-activated rat muscle nAChR channels functionally expressed in *Chara* internodal cells and recorded in outside-out patches excised from cytoplasmic droplets. (B) Dose–response analysis of charbachol-modified open/closed probability of single nAChR channels at −40 mV. In the range between 1 and 30 μM, a Hill coefficient of 2 and a half-maximal activation of 20 μM were calculated. (Unpublished data from H. Lühring; for heterologous expression see text.)

favorably when expressed in *Chara, Xenopus,* or rat muscle *in vivo* (H. Lühring, personal communication).

2.5.2. RNA Injection

Remove incrustations if present on the cell wall by overnight incubation of *Chara* in pond water of pH 6 as a first step. Separate internodal cells under water in a second step. To overcome turgor (about 10 bar) for pressure injection of mRNA, in a third step, increase the osmotic pressure of the bath medium. Loss of turgor will be recognized by shrinkage or flattening of the cylindrical cell body. In the final step, inject cRNA as described for *Xenopus* oocytes (Stühmer, 1993). Several minutes after injection transfer cells into the sorbitol-free expression medium to recreate turgor.

2.5.3. Isolation of Cytoplasmic Droplets

After 12–24 hr of expression, layer cells on paper tissue, wipe dry, and fix with a small droplet of vaseline on a microscopic slide such that part of the cell will protrude over the edge of the support. After turgor is lost again, cut the node of the protruding end and tilt the support so that the open end is submersed into the bath solution, which is adjusted to balance the osmotic pressure of the released cytoplasmic droplets.

3. Outlook

Starting from cell-specific cDNA libraries, ion channels will be isolated on the basis of their molecular structure or biological function. Approaches based on the structure may involve low-stringency screening with, e.g., S4, S6, H5, or ankyrin-binding motives. Promoter analysis, transgenic plants, and *in situ* hybridization following screens of genomic libraries derived from whole seedlings will allow correlations between channel genes and defined cell types or developmental states. In turn, direct patch-clamp recordings on the identified cells *in vivo* will help to characterize cell- or state-dependent properties (cf. Müller-Röber *et al.,* 1994).

Approaches based on channel function will take advantage of complementation analysis on the numerous transport (uptake)-deficient yeast mutants (cf. Anderson *et al.,* 1992; Sentenac *et al.,* 1992). Comparison of phenotypes of, e.g., transgenic *Arabidopsis thaliana,* with those of well-described mutants in this species could provide insights into the physiological role of individual channel types.

ACKNOWLEDGMENTS. I would like to thank E. Neher for discussion and criticism and A. Bertl (yeast channels, Göttingen), A. H. DeBoer (laser microsurgery, Amsterdam), H. Lühring (expression in *Chara,* Jülich), G. Schönknecht (macrosurgery, Würzburg), G. Thiel (macrosurgery, Göttingen), and L. van Volkenburgh (quick enzyme-digestion, Seattle) for a supply of unpublished data and technical details. The work was supported by DFG grants to the author.

References

Allen, G. J., and Sanders, D., 1994, Two voltage-gated calcium release channels coreside in the vacuolar membrane of broad bean guard cells, *The Plant Cell* **6:**685–694,

Anderson, J. A., Huplikar, S. S., Kochian, L. V., Lucas, W. J., and Gaber, R. F., 1992, Functional expression of a probable *Arabidopsis thaliana* potassium channel in *Saccharomyces cerevisiae, Proc. Natl. Acad. Sci. U.S.A.* **89:**3736–3740.

Aoshima, H., Yamada, M., Sauer, N., Komor, E., and Schobert, C., 1993, Heterologous expression of the H^+/hexose cotransporter from *Chlorella* in *Xenopus* oocytes and its characterization with respect to sugar specificity, pH and membrane potential, *J. Plant Physiol.* **141:**293–297.

Assmann, S. M., 1993, Signal transduction in guard cells, *Annu. Rev. Cell. Biol.* **9:**345–375.

Assmann, S. M., Simoncini, L., and Schroeder, J. I., 1985, Blue light activates electrogenic ion pumping in guard cell protoplasts of *Vicia faba, Nature* **318:**285–287.

Barbara, J-G., Stoeckel, H., and Takeda, K., 1994, Hyperpolarization-activated inward chloride current in protoplasts from suspension-cultured carrot cells, *Protoplasma* **180:**136–144.

Becker, D., Zeilinger, C., Lohse, G., Depta, H., and Hedrich, R., 1993, Identification and biochemical characterization of the plasma-membrane H^+-ATPase in guard cells of *Vicia faba* L., *Planta* **190:**44–50.

Beilby, M. J., 1989, Electrophysiology of giant algal Cells, *Methods Enzymol.* **174:**403–443.

Bertl, A., 1989, Current–voltage relationship of a sodium-sensitive potassium selective channel in the tonoplast of *Chara corallina, J. Membr. Biol.* **132:**183–199.

Bertl, A., and Slayman, C. L., 1990, Cation-selective channels in the vacuolar membrane of *Saccharomyces:* Dependence on calcium, redox state, and voltage, *Proc. Natl. Acad. Sci. U.S.A.* **87:**7824–7828.

Bertl, A., and Slayman, C. L., 1992, Complex modulation of cation channels in the tonoplast and plasma membrane of *Saccharomyces cerevisiae:* Single-channel studies, *J. Exp. Biol.* **172:**271–287.

Bertl, A., and Slayman, C. L., 1993, Gating and conductance in an outward-rectifying K^+ channel from the plasma membrane of *Saccharomyces cerevisiae, J. Membr. Biol.* **132:**183–199.

Bertl, A., Blumwald, E., Coronado, R., Eisenberg, R., Findlay, G., and Gradmann, D., 1992, Electrical measurements on endomembranes, *Science* **258:**873–874.

Bertl, A., Anderson, J. A., Slayman, C. L., and Gaber, F. R., 1994, Inward and outward rectifying potassium currents in *Saccharomyces cerevisiae* mediated by endogenous and heterologously expressed ion channels, *Proc. Natl. Acad. Sci. U.S.A.* (in press).

Bethke, P. C., and Jones, R. L., 1994, Ca^{2+}-Calmodulin modulates ion channel activity in storage protein vacuoles of barley aleurone cells, *The Plant Cell* **6:**277–285.

Blatt, M. R., 1991, Ion channel gating in plants: Physiological implications and integration for stomatal function, *J. Membr. Biol.* **124:**95–112.

Blatt, M. R., 1992, K^+ channels of stomatal guard cells. Characteristics of the inward rectifier and its control by pH, *J. Gen. Physiol.* **99:**615–644.

Blinks, L. R., 1937a, The direct current resistance of *Nitella, J. Gen. Physiol.* **13:**495–508.

Blinks, L. R., 1937b, The effect of current flow on bioelectric potential of *Nitella, J. Gen. Physiol.* 20:229–265.

Boorer, K. J., Forde, B. G., Leigh, R. A., and Miller, A. J., 1992, Functional expression of a plant plasma membrane transporter in *Xenopus* oocytes, *FEBS Lett* **302:**166–168.

Buff, H., 1854, Ueber die Elektrizitaetserregung durch lebende Pflanzen, *Ann. Chem. Pharmacie.* **89:**76–89.

Burdon Sanderson, J., 1873, Note on the electrical phenomena which accompany irritation of the leaf of *Dionaea muscipula, Proc. R. Soc. Lond.* **21:**495–496.

Bush, D. S., Hedrich, R., Schroeder, J. I., and Jones, R. L., 1988, Channel-mediated K^+ flux in barley aleurone protoplasts, *Planta* **176:**368–377.

Cao, Y., Anderova, M., Crawford, N. M., and Schroeder, J., 1992, Expression of an outward-recifiying potassium channel from maize mRNA and complementary RNA in *Xenopus* oocytes, *Plant Cell* **4:**961–969.

Clarkson, D. T., 1989, Movements of ions across roots, in: *Solute Transport in Plant Cells and Tissues* (D. Y. Baker and J. L. Hall, eds.), Longman Scientific and Technical, Hallow Essex, UK.

Coleman, H. A., 1986, Chloride currents in *Chara*—a patch clamp study. *J. Membr. Biol.* **93:**55–61.

Colombo, R., Cerana, R., Lado, P., and Peres, A., 1988, Voltage-dependent channels permeable to K^+ and Na^+ in membranes of *Acer pseudoplatanus, J. Membr. Biol.* **103:**227–236.

Cosgrove, D. J., and Hedrich, R., 1991, Stretch-activated chloride, potassium, and calcium channels coexisting in the plasma membranes of guard cells of *Vicia faba L., Planta* **186:**143–153.

Coyaud, L., Kurkdjian, A., Kado, R., and Hedrich, R., 1987, Ion channels and ATP-driven pumps involved in ion transport across the tonoplast of sugar beet vacuoles, *Biochim. Biophys. Acta* **902:**263–268.

Criado, M., and Keller, B. U., 1987, A membrane fusion strategy for single-channel recordings of membranes usually non-accessible to patch-clamp pipette electrodes, *FEBS Lett.* **224:**172–176.

Curtis, H. J., and Cole, K. S., 1937, Transverse electric impedance of *Nitella, J. Gen. Physiol.* **21:**189–201.

DeBoer, A. H., Van Duijn, B., Giesberg, P., Wegner, L., Obermeyer, G., Köhler, K., and Linz, K., 1994, Laser microsurgery: A versatile tool in plant (electro)physiology, *Protoplasma* (in press).

Ebel, J., and Cosio, E., 1994, Elicitors of plant defense responses, *Int. Rev. Cytol.* **148:**1–36.

Elzenga, J. T. M., and Keller, C. P., 1991, Patch clamping protoplasts from vascular plants, *Plant Physiol.* **97:**1573–1575.

Enz, C., Steinkamp, T., and Wagner, R., 1993, Ion channels in the thylakoid membrane (a patch-clamp study), *Biophys. Acta Biochim.* **1143:**67–76.

Fairley, K. A., and Walker, N. A., 1989, Patch clamping corn protoplasts—gigaseal frequency is not improved by Congo red inhibition of cell wall regeneration, *Protoplasma* **153:**111–116.

Fairley, K., Laver, D., and Walker, N. A., 1991, Whole-cell and single-channel currents across the plasmalemma of corn shoot suspension cells, *J. Membr. Biol.* **121:**11–22.

Falke, L., Edwards, K. L., Pickard, B. G., and Misler, S. A., 1988, A stretch-activated anion channel in tobacco protoplasts, *FEBS Lett.* **237:**141–144.
Flügge, U. I., Fischer, K., Groß, A., Lottspeich, F., Eckerskorn, C., and Sebald, W., 1989, The triose phosphate-3-phosphoglycerate-phosphate translocator from spinach chloroplasts: Nucleic acid sequence of a full length cDNA clone, *EMBO J.* **8:**29–46.
Gambale, F., Kolb, H. A., Cantu, A. M., and Hedrich, R., 1994, The voltage-dependent H^+-ATPase of the sugar beet vacuole is reversible, *Eur. Biophys. J.* **22:**399–403.
Garrill, A., and Lew, R. R., 1992, Stretch-activated Ca^{3+} and Ca^{2+}-activated K^+ channels in the hyphal tip plasma membrane of the oomydete *Saprolegnia ferax, J. Cell. Sci.* **101:**721–730.
Gassmann, W., Ward, J. M., and Schroeder, J. I., 1993, Physiological roles of inward-rectifying K^+ channels, *Plant Cell* **5:**1491–1493.
Giesberg, P., and DeBoer, A. H., 1994, Protoplasts from xylem parenchyma cells of maize mesocotyl: Isolation method from patch-clamp studies, *Protoplasma* (in press).
Gradmann, D., Thiel, G., and Blatt, M., 1993, Electrocoupling of ion transporters in plants, *J. Membr. Biol.* **136:**327–332.
Greulich, K. O., and Weber, G., 1992, The light microscope on its way from an analytical to a preparative tool, *J. Microsc.* **167:**127–151.
Hamill, O. P., Marty, A., Neher, E., Sakmann, B., and Sigworth, F. J., 1981, Improved patch-clamp techniques for high-resolution current recording from cells and cell-free membrane patches, *Pflügers Arch.* **391:**85–100.
Hedrich., R., 1994, Voltage-dependent chloride channels in plant cells: identification, characterization, and regulation of a guard cell anion channel, in: *Chloride Channels,* Current Topics in Membranes, vol. 42 (W. B. Guggino, ed.), Academic Press, New York, pp. 1–33.
Hedrich, R., and Marten, I., 1993, Malate-induced feedback regulation of plasma membrane anion channels could provide a CO_2 sensor to guard cells, *EMBO J.* **12:**897–901.
Hedrich. R., and Becker, D., 1994, "Green circuits": the potential of ion channels in plants, *Annual Rev. Mol. Biol,* in press.
Hedrich, R., and Neher, E., 1987, Cytoplasmic calcium regulates voltage-dependent ion channels in plant vacuoles, *Nature* **329:**833–836.
Hedrich, R., and Schroeder, J. I., 1989, The physiology of ion channels and electrogenic pumps in higher plants, *Annu. Rev. Plant Physiol.* **40:**539–569.
Hedrich, R., Flügge, U. I., and Fernandez, J. M., 1986a, Patch-clamp studies of ion transport in isolated vacuoles, *FEBS Lett.* **204:**228–232.
Hedrich, R., Schroeder, J. I., and Fernandez, J. M., 1986b, Patch-clamp studies on higher plants: A perspective, *Trends Biochem. Sci.* **12:**49–52.
Hedrich, R., Barbier-Brygoo, H., and Felle, H., 1988, General mechanisms for solute transport across the tonoplast of plant vacuoles: A patch-clamp survey of ion channels and proton pumps, *Bot. Acta* **101:**7–13.
Hedrich, R., Kurkdjian, A., Guern, J., and Flügge, U. I., 1989, Comparative studies on the electrical properties of the H^+ translocating ATPase and pyrophosphatase of the vacuolar–lysosomal compartment, *EMBO J.* **8:**2835–2841.
Hedrich, R., Busch, H., and Raschke, K., 1990, Ca^{2+} and nucleotide dependent regulation of voltage dependent anion channels in the plasma membrane of guard cells, *EMBO J.* **9:**3889–3892.
Hedrich, R., Moran, O., and Conti, F., 1994, Voltage-dependence and high-affinity Cs^+ block of a cloned plant K^+ channel, *EMBO J.* (in press).
Hedrich, R., Moran, O., Conti, F., Busch, H., Becker, D., Gambale, F., Dreyer, I., Kuch, A., Neuwinger, K., and Palme, K., 1995, Inward rectifier potassium channels in plants differ from their animal counterparts in response to voltage and channel modulators, *Eur. Biophys. J.,* in press.
Hill, B. S., and Findlay, G. P., 1981, Power of movements in plants: The role of osmotic machines, *Q. Rev. Biophys.* **14:**173–222.
Hille, B., 1992, *Ionic Channels of Excitable Membranes,* Sinauer Associates Inc., Sunderland, Massachusetts.
Ho, K., Nichols, C. G., Lederer, W. J., Lytton, J., Vassilev, P. M., Kanazirska, M. V., and Hebert, S. C., 1993, Cloning and expression of an inwardly rectifying ATP-regulated potassium channel, *Nature* **362:**31–38.
Iijima, T., and Hagiwara, S., 1987, Voltage-dependent K channels in protoplasts of trap-lobe cells of *Dionaea muscipula, J. Membr. Biol.* **100:**73–81.
Jan, L. Y., Jan, Y. N., 1992, Structural elements involved in specific K^+ channel functions, *Annu. Rev. Physiol.* **54:**537–565.

Johannes, E., Brosnan, J. M., and Sanders, D., 1992, Parallel pathways for intracellular Ca^{2+} release from the vacuole of higher plants, *Plant J.* **2**(1):97–102.

Jones, A. M., 1994, Surprising signals in plant cells, *Science* **263:**183–184.

Kaldenhoff, R., Koelling, A., and Richter, G., 1993, A novel blue light- and abscisic acid-inducible gene of *Arabidopsis thaliana* encoding an intrinsic membrane protein, *Plant Mol. Biol.* **23:**1187–1198.

Kaldenhoff, R., Kölling, A., Meyers, J., Karmann, U., Ruppel, and Richter, G., The *blue light*-responsive *AthH2* gene or *Arabidopsis thaliana* is primarily expressed in expanding as well as in differentiating cells and encodes a putative channel protein of the plasmalemma, *The Plant Journal* **7:**1–9.

Kammerloher, W., Fischer, U., Plechottka, G. P., and Schäffner, A. R., 1994, Water channels in the plant plasma membrane cloned by immunoselection from a mammalian system, *Plant J.* (in press).

Keller, B. U., Hedrich, R., Voz, W., and Urvado, M., 1988, Single channel recordings of reconstituted ion channel proteins: An improved technique, *Pflügers Arch.* **411:**94–100.

Keller, B. U., and Hedrich, R., 1992, Patch clamp techniques to study ion channels from organelles, *Methods Enzymol.* **207:**673–681.

Keller, B. U., Hedrich, R., and Raschke, K., 1989, Voltage-dependent anion channels in the plasma membrane of guard cells, *Nature* **341:**450–453.

Klercker, J., 1892, Eine Methode zur Isolierung lebender Protoplasten, *Ofverigt Kongl. Vetenskaps-Akad. Forhandl. Stockh.* **9:**463–475.

Klieber, H. G., and Gradmann, D., 1993, Enzyme kinetics of the prime K^+ channel in the tonoplast of *Chara:* Selectivity and inhibition, *J. Membr. Biol.* **132:**253–265.

Kochian, L. V., and Lucas, W. J., 1993, Can K^+ channels do it all? *Plant Cell* **5:**720–721.

Kolb, H. A., Marten, I., and Hedrich, R., 1995, GCACl a guard cell anion channel with gating propeties like the HH sodium channel, *J. Membr. Biol.* (in press).

Krawczyk, S., 1978, Ionic channel formation in a living cell membrane, *Nature* **273:**56–57.

Kubo, Y., Baldwin, T. J., Jan, Y. N., and Jan, L. Y., 1993, Primary structure and functional expression of a mouse inward rectifier potassium channel, *Nature* **362:**127–133.

Kurkdjian, A., Leitz, G., Manigault, P., Harim, A., and Greulich, K. O., 1993, Non-enzymatic access to the plasma membrane of *Medicago* root hairs by laser microsurgery, *Cell Sci.* **105:**263–268.

Laver, D. R., 1991, A surgical method for accessing the plasma membrane of *Chara australis, Protoplasma* **161:**79–84.

Levina, N. N., Lew, R. R., Heath, I. B., 1994, Cytoskeletal regulation of ion channel distribution in the tip-growing organism *Saprolegnia ferax, J. Cell Science* **107:**127–13.

Linz, K. W., and Köhler, K., 1993, Isolation of protoplasts from the coccal green alga *Eremosphaera viridis* De Bary for patch-clamp measurements, *Bot. Acta* **106:**469–472.

Lohse, G., and Hedrich, R., 1992, Characterization of the plasma membrane H^+, ATPase from *Vicia faba* guard cells. Modulation by extracellular factors and seasonal changes, *Planta* **188:**206–214.

Lühring, H., 1986, Recording of single K^+ channels in the membrane of cytoplasmic drop of *Chara australis. Protoplasma* **133:**19–28.

Lüttge, U., and Pallagky, C. K. 1969, Light-triggered transient changes of membrane potentials in green cells in relation to photosynthetic electron transport, *Z. Pflanzenphysiol.* **61:**58–67.

Maathius, F. J. M., and Prins, H. B. A., 1990a, Patch clamp studies on root cell vacuoles of a salt tolerant and salt sensitive *Plantargo* species, *Plant Physiol.* **92:**23–28.

Maathius, F. J. M., and Prins, H. B. A., 1990b, Electrophysiological membrane characteristics of the salt tolerant *Plantargo maritima* and the salt sensitive *Plantargo media, Plant Soil* **123:**233–238.

Maathius, F., and Sanders, D., 1993, Energization of potassium uptake in *Arabidopsis thaliana, Planta* **191:**302–307.

Marschner, H., 1986, *Mineral Nutrition in Higher Plants,* Academic Press, London.

Marten, I., Lohse, G., and Hedrich, R., 1991, Plant growth hormones control voltage-dependent activity of anion channels in plasma membrane of guard cells, *Nature* **353:**758–762.

Marten, I, Zeilinger, C., Redhead, C., Landry, D. W., Al-Awqati, Q., and Hedrich, R., 1992, Identification and modulation of a voltage-dependent anion channel in the plasma membrane of guard cells by high-affinity ligands, *EMBO J.* **11:**3569–3575.

Maurel, C., Reizer, J., Schroeder, J. I., and Chrispeels, M. J., 1993, The vacuolar membrane protein y-TIP creates water specific chanels in *Xenopus* oocytes, *EMBO J.* **12**(6):2241–2247.

Miedema, H., 1992, *In Search of the Proton Channel. An Electrophysiological Study on the Polar Leaves of* Elodea densa *and* Potamogeton lucens. Thesis, University of Groningen.

Müller-Röber, B., Ellenberg, J., Provart, N., Becker, D., Busch, H., Dietrich, P., and Hedrich, R., 1995 Voltage-dependent properties of guard cell K^+ channels *in vivo* and *in vitro, EMBO J.* (in press).

Ninnemann, O., Journoux, T.-C., and Frommer, W. B., 1994, Identification of a high affinity NHCl transporter from plants, *EMBO J.* **13:**3464–3471.

Pantoja, O., Dainty, J., and Blumwald, E., 1989, Ion channels in vacuoles from halophytes and glycophytes, *FEBS Lett.* **255:**92–96.

Pont-Lezica, R. F., McNally, J. G., and Pickard, B. G., 1993, Wall-to-membrane linkers in onion epidermis: Some hypotheses, *Plant Cell. Environ.* **16:**111–123.

Preston, G. M., Carroll, T. P., Guggino, W. B., and Agre, P., 1992, Appearance of water channels in *Xenopus* oocytes expressing red cell CHIP28 protein, *Science* **256:**385–387.

Raschke, K., and Hedrich, R., 1990, Patch-clamp measurements on isolated guard-cell protoplasts and vacuoles, *Methods Enzymol.* **174:**312–330.

Riesmeier, J. W., Willmitzer, L., and Frommer, W. B., 1992, Isolation and characterization of a sucrose carrier cDNA from spinach by functional expression in yeast, *EMBO J.* **11:**4705–4713.

Rincon, M., and Boss, W. F., 1987, Myo-inositol trisphosphate mobilizes calcium from fusogenic carrot (*Daucus carota L.*), protoplasts, *Plant Physiol.* **83:**395–398.

Sauer, N., and Tanner, W., 1993, Molecular biology of sugar transporters in plants, *Bot. Acta* **106:**277–286.

Schachtman, D. P., and Schroeder, J. I., 1994, Structure and transport mechanism of a high affinity potassium uptake transporter from higher plants, *Nature* **370:**655–658.

Schachtman, D. P., Tyerman, S. D., and Terry, B. R., 1991, The K^+/Na^+ selectivity of a cation channel in the plasma membrane of root cells does not differ in salt-tolerant and salt-sensitive wheat species, *Plant Physiol.* **97:**598–605.

Schachtman, D. P., Schroeder, J. I., Lucas, W. J., Anderson, J. A., and Gaber, R. F., 1992, Expression of an inward-rectifying potassium channel by the *Arabidopsis* KATl cDNA, *Science* **258:**1654–1658.

Schnabl, H., 1981, The compartmentation of carboxylating and decarboxylating enzymes in guard cells, *Planta* **152:**307–313.

Schönknecht, G., Hedrich, R., Junge, W., and Raschke, K., 1988, A voltage-dependent chloride channel in the photosynthetic membrane of a higher plant, *Nature* **336:**589–592.

Schroeder, J. I., 1989, Quantitative analysis of outward rectifying K^+ channels in guard cell protoplasts from *Vicia faba, J. Membr. Biol.* **107:**229–235.

Schroeder, J. I., 1994, Physiology and molecular structure of K^+ transporters in plants, *Annu. Rev. Plant Physiol.* (in press).

Schroeder, J. I., and Fang, H. H., 1991, Inward-rectifier K^+ channels in guard cells provide a mechanism for low-affinity K^+ uptake, *Proc. Natl. Acad. Sci. U.S.A.* **88:**11583–11587.

Schroeder, J. I., and Hagiwara, S., 1990a, Repetitive increases in cytosolic Ca^{2+} of guard cells by abscisic acid activation of nonselective Ca^{2+} permeable channels, *Proc. Natl.-Acad. Sci. U.S.A.* **87:**9305–9309.

Schroeder, J. I., and Hagiwara, S., 1990b, Voltage-dependent activation of Ca^{2+}-regulated anion channels and K^+ uptake channels in *Vicia faba* guard cells, in: *Calcium in Plant Growth and Development,* Vol. 4 (R. T. Leonard and P. K. Hepler, eds.), pp. 144–150, American Society of Plant Physiologists Symposium Series, Rockville

Schroeder, J. I., and Hedrich, R., 1989, Involvement of ion channels and active transport in osmoregultion and signaling of higher plant cells, *Trends Biochem. Sci.* **14:**187–192.

Schroeder, J. I., Hedrich, R., and Ferandez, J. M., 1984, Potassium-selective single channels in guard cell protoplasts of *Vicia faba, Nature* **312:**361–362.

Schroeder, J. I., Raschke, K., and Neher, E., 1987, Voltage-dependence of K^+ channels in guard-cell protoplasts, *Proc. Natl. Acad. Sci. U.S.A.* **84:**4108–4112.

Schroeder, J. I., Ward, J. M., Gassmann, W., 1994, Perspectives on the physiology and structure of inward rectifying K^+ channels in higher plants: biophysical implications for K^+ uptake, *Annu. Rev. Biophys. Biomol. Struct.* **23:**441–471.

Schulz-Lessdorf, B., and Hedrich, R., 1994, Protons and calcium modulate SV-type channels in the vacuolar-lysosomal compartment—interaction with calmodulin antagonists, *J. Gen. Physiol.* (in press).

Schwarz, M., Gross, A., Steinkamp, T., Flügge, U. I., and Wagner, R., 1994, Ion channel properties of the reconstituted chloroplast triose phosphate/phosphate translocator. *J. Biol. Chem.* (in press).

Sentenac, H., Bonneaud, N., Minet, M., Lacroute, F., Salmon, J-M., Gaymard, F., and Grignon, C., 1992, Cloning and expression in yeast of a plant potassium ion transport system, *Science* **256:**663–665.

Serrano, E. E., Zeiger, E., and Hagiwara, S, 1988, Red light stimulates an electrogenic proton pump in *Vicia* guard cell protoplasts, *Proc. Natl. Acad. Sci. U.S.A.* **85:**436–440.

Spalding, E. P., Slayman, C. L., Goldsmith, M. H. M., Gradmann, D., and Bertl, A., 1992, Ion channels in *Arabidopsis* plasma membrane. Transport characteristics and involvement in light-induced voltage changes, *Plant Physiol.* **99:**96–102.

Stoeckel, H., and Takeda, K., 1989a, Voltage-activated, delayed rectifier K^+ current from pulvinar protoplasts of *Mimosa pudica, Pflügers Arch.* **414:**5150–5151.

Stoeckel, H., and Takeda, K., 1989b, Calcium-activated, voltage-dependent, nonselective cation currents in endosperm plasma membrane from higher plants, *Proc. R. Soc. Lond. B* **2137:**213–231.

Stühmer, W., 1993, Electrophysiological recording from *Xenopus* oocytes, *Methods Enzymol.* **207:**319–339.

Sze, H., 1985, H+-translocating ATPases: advances using membrane vesicles. *Ann. Rev. Plant Physiol.* **36:**175–208.

Taylor, R. R., and Brownlee, C., 1992, Localized patch clamping of plasma membrane of a polarized plant cell, *Plant Physiol.* **99:**1686–1688.

Tazawa, M., Shimmen, T., and Mimura, T., 1987, Membrane control in the *Characeae, Anny. Rev. Plant Physiol.* **38:**95–117.

Thiel, G., MacRobbie, E. A. C., and Blatt, M. R., 1992, Membrane transport in stomatal guard cells: The importance of voltage control, *J. Membr. Biol.* **126:**1–18.

Thiel, G., Homann, U., and Gradmann, D., 1993, Microscopic elements of electrical excitation in *Chara:* Transient activity of Cl-channels in the plasma membrane. *Membr. Biol.* **134:**33–66.

Thuleau, P., Ward, J. M., Ranjeva, R., and Schroeder, J. I., 1994, Voltage-dependent calcium-permeable channels in the plasma membrane of a higher plant cell, *EMBO J.*, **13:**2970–2975.

Tsay, Y. F., Schroeder, J. I., Feldmann, K. A., and Crawford, N. M., 1993, The herbicide sensitivity gene CHL1 of *Arabidopsis* encodes a nitrate-inducible nitrate transporter, *Cell* **72:**705–713.

Tyerman, S., and Findlay, G. P., 1989, Current–voltage curves of single Cl^- channels which coexist with two types of K^+ channel in the tonoplast of *Chara corallina, J. Exp. Bot.* **40:**105–117.

Van Duijn, B., Ypey, D. L., and Libbenga, K. R., 1993, Whole-cell K^+ currents across the plasma membrane of tobacco protoplasts from cell-suspension cultures, *Plant Physiol.* **101:**81–88.

Vogelzang, S. A., and Prins, H. B. A., 1992, Plasmalemma patch clamp experiments in plant root cells: Procedure for fast isolation of protoplasts with minimal exposure to cell wall degrading enzymes, *Protoplasma* **171:**104–109.

Ward, J. M., and Schroeder, J. I., 1994 Calcium-activated K^+ channels and calcium-induced calcium release by slow vacuolar ion channels in guard cell vacuoles implicated in the control of stomatal closure, *The Plant Cell*, **6:**669–683.

Weber, G., and Greulich, K. O., 1992, Manipulation of cells, organells, and genomes by laser microbeam and optical trap, *Int. Rev. Cytol.* **133:**1–41.

Wegner, L., and Raschke, K., 1994, Ion channels in the xylem–parenchyma cells of barley roots: A procedure to isolate protoplasts from this tissue and a patch-clamp exploration of salt passage ways into xylem vessels, *Plant Physiol* **105:**799–813.

Weiser, T., and Bentrup, F-W., 1993, Pharmacology of the SV-channel in the vacuolar membrane of *Chenopodium rubrum suspension cells, J. Membr. Biol.* **136:**43–54.

Wendler, S., and Zimmermann, U., 1985a, Compartment analysis of plant cells by means of turgor pressure relaxation: I. Theoretical considerations, *J. Membr. Biol.* **85:**121–131.

Wendler, S., and Zimmermann, U., 1985b, II. Experimental results on *Chara corallina., J. Membr. Biol.* **85:**133–142.

Yoshimori, T., Yamamoto, A., Yoshinori, M., Futai, M., and Tashiro, Y., 1991, Bafilomycin A_1, a specific inhibitor of vacuolar-type H^+-ATPase, inhibits acidification and protein degradation in lysosomes of cultured cells, *J. Biol. Chem.*, **866:**17707–17713.

Zeiger, E., and Hepler, P. K., 1977, Light and stomatal function: blue light stimulates swelling of guard cell protoplasts, *Science* **196:**887–889.

Zhang, R., and Verkman, A. S., 1991, Water and urea permeability properties of *Xenopus* oocytes: Expression of mRNA from toad urinary bladder, *Am. J. Physiol.* **260:**26–34.

Chapter 13

The Giant Membrane Patch

DONALD W. HILGEMANN

1. Introduction

This chapter describes methods to form and excise membrane patches with diameters of 12 to 40 μm, capacitances of 2 to 15 pF, and seal resistances of 1 to 10 GΩ. The formation of such "giant" membrane patches (Hilgemann, 1989) has been successful with most cells that allow gigohm seals to be formed with conventional patch-clamp techniques. Cell types employed to date for inside-out patches include *Xenopus* oocytes, cardiac myocytes, skeletal myocytes, tracheal myoctes, *Lymnaea* snail neurons, SF9 cells, chromaffin cells, pancreatic acinar cells, and some blood cells. Up to now, a successful procedure to form giant outside-out patches routinely has been developed only for *Xenopus* oocytes (Rettinger *et al.*, 1994).

The giant-patch techniques were developed initially for studies of electrogenic membrane transporters (Hilgemann, 1990), offering the new advantages of fast voltage clamp of a large membrane area with gigohm seals, free access to the cytoplasmic side, and rigorous control of the solution composition on both membrane sides. The techniques have now been used successfully in a wide range of applications, which include the recording of transporter currents (e.g., Matsuoka and Hilgemann, 1992; Doering and Lederer, 1993; Rettinger *et al.*, 1994) and channel currents (e.g., Hilgemann, 1989; P. Ruppersberg, personal communication), transporter and channel charge movements (Hilgemann *et al.*, 1991; Hilgemann, 1994; E. Stephani, personal communication), and single-channel recording of low-density channels (Nagel *et al.*, 1992; Hwang *et al.*, 1994).

2. Giant-Patch Methods

The methodological details given in this chapter are in part modified from previous descriptions (Collins *et al.*, 1992; Hilgemann, 1992; Matsuoka *et al.*, 1993). These changes reflect much further experience of this and several other laboratories with the giant-patch techniques.

2.1. Pipette Preparation

The preparation of pipettes must be tailored to the cell type and the specific application. Standard patch pipette pullers, including the Brown/Flaming type, can be used to prepare

DONALD W. HILGEMANN • Department of Physiology, University of Texas Southwestern Medical Center at Dallas, Dallas, Texas 75235-9040.
Single-Channel Recording, Second Edition, edited by Bert Sakmann and Erwin Neher. Plenum Press, New York, 1995.

pipettes with large-diameter tip openings (10–20 μm), at least with some types of glass (e.g., thin-walled borosilicate). However, the success rate in obtaining tips without fragmentation is often low, and tip diameters usually cannot be accurately controlled. Better success rates can be achieved with computer-controlled pullers. However, a low-budget solution employing a modified microforge is our method of choice.

Standard patch pipettes are first prepared, and the tip is then cut to the appropriate size with the technique to be described. Usually, we use borosilicate glass, but glasses with more advantageous electrical properties can also be used (e.g., Corning 7052 and aluminasilicate glass capillaries). We prefer a tubing with a relatively large inner diameter (OD 2.0 mm; ID 1.5 mm). Loss of solution with prolonged application of positive pressure remains negligible, and the large inner diameter leaves space for insertion of polyethylene tubes, as desired, for a KCl bridge and pipette perfusion devices. Figure 1A shows a schematic diagram of the microforge employed.

A relatively thick platinum wire (0.3–0.6 mm) is used, and a foot-switch control of the heating element is essential. A bead of soft glass is first melted onto the wire, and this is renewed, as necessary, after 20 to 60 pipette cuts. One such glass is Corning 8161, which has a high lead content; the still softer "solder glasses," which adhere well to most surfaces, have also been reported to be advantageous (Rettinger *et al.,* 1994).

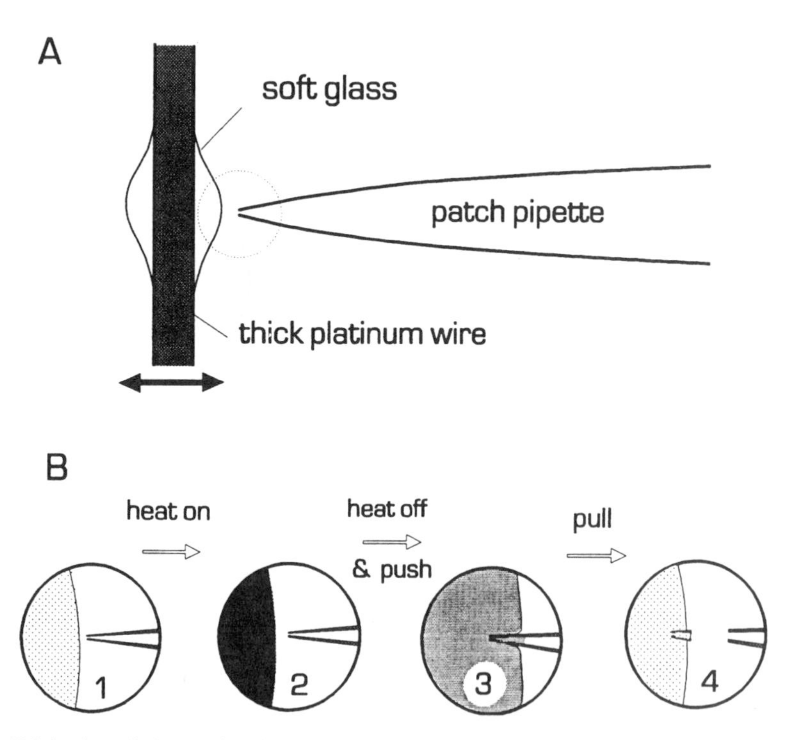

Figure 1. Fabrication of pipettes for giant patches. *A:* Microforge. The heating element is made from 0.3- to 0.6-mm platinum (or equivalent) wire with foot-switch control. A bead of soft glass is melted onto the wire. *B:* Pipette-cutting technique. The pipette tip is viewed with a long-working-distance 40× lens, and the heat level is adjusted so that an electrode tip can be melted at a close distance from the wire. *Field 1:* The pipette tip is placed close to the focused edge of the soft glass bead. *Field 2:* The heating element is activated until the pipette tip begins to recede. *Field 3:* The heater is turned off, and after about 0.5 sec, the pipette tip is pushed into the melted glass bead. *Field 4:* After cooling, the pipette tip is retracted.

We find that the properties of Corning 8161 can be improved by addition of a small amount of solder glass but that solder glass (Corning 7567) alone is not usable with the pipettes we employ. Other glasses that are suitable are the LaF and NaF glass types (Schott Glaswerke, Mainz; A. Jeromin, personal communication) and some "soda" glasses.

Easy micromanipulation of the pipette tip is advantageous, and the forge wire must be movable in the horizontal plane. The heater power supply is set just high enough that a pipette tip in close proximity to the soft glass can be melted upon application of full heat. Figure 1B describes the method of cutting the pipette tip. A patch pipette tip is positioned close to the glass bead, such that the edge of the glass bead and the pipette tip are in sharp focus in the same focal plane (Field 1). The foot switch is activated, and heat is applied until the tip begins to recede (Field 2). The heat is then switched off, and, after an approximately 0.5-sec waiting period, the softened glass bead is pushed into the pipette tip (Field 3). In the next seconds, the forge and tip are held in place while the glass bead hardens, and the forge wire is then retracted to cut the pipette tip (Field 4). Often, spontaneous retraction of the forge wire during cooling is adequate to cut the tip. With practice, clean cuts are obtained routinely at any desired tip diameter. For diameters larger than ~15 μm, or when fragmentation occurs, the process can usually be repeated until a clean cut is obtained at the desired diameter. By repeating the cutting and fire-polishing sequence, the tip can be sculptured to various shapes. Of most interest, the walls can be thickened, and the angle of descent of the tip can be steepened, to form pipettes with extremely low access resistance (~50 kΩ). This is highly desirable for fast voltage clamp and noise reduction.

Figure 2 shows some of the usual pipette shapes employed, along with the membrane configurations obtained. Membrane position can generally be visualized clearly when the pipettes are held in a plane nearly perpendicular to the light path. The membrane configuration depends on the pipette shape, the technique of seal formation, the pipette coating employed, and the nature of the cells employed. Figure 2A shows the tip shape used for routine purposes with cardiac membrane and small cells (12–20 μm inner tip diameters). The membrane usually remains close to the pipette tip; when it rises higher into the pipette, gigohm seals are usually not obtained. Especially in the beginning stages with cardiac myocytes, it is advantageous to limit fire polishing to just visible effects. Figure 2B shows a typical pipette tip after the walls are thickened and the descent is steepened by multiple cutting cycles. In this case, membrane tends to enter further into the pipette and rise up in an omega shape. For cardiac patches, inner tip diameters of 25–30 μm are employed, giving membrane capacitances of 8–14 pF. Gigohm seal formation is more demanding than with the straighter pipette shape, but this configuration is advantageous when a very large membrane area is required. Although the configuration brings the danger of generating a rim region with slow components in the voltage clamp, this danger proved to be less serious than anticipated for cardiac patches of this shape.

Figure 2C shows the usual configuration used with oocyte membrane for giant inside-out patches. Pipettes are cut with 40- to 50-μm inner diameters, and they are melted to a final diameter of 25–40 μm. When seals are made with only very light negative pressure, membrane rises into the pipette tip by less than 20 μm. Figure 2D shows the pipette tip shape found by Rettinger and Schwarz to be advantageous for outside-out oocyte patches. A relatively thick-walled capillary glass is pulled with relatively high heat such that descent at the tip is rather shallow. The tip is cut to an inner diameter of 30–35 μm and fire-polished until changes of the pipette tip are just discernible.

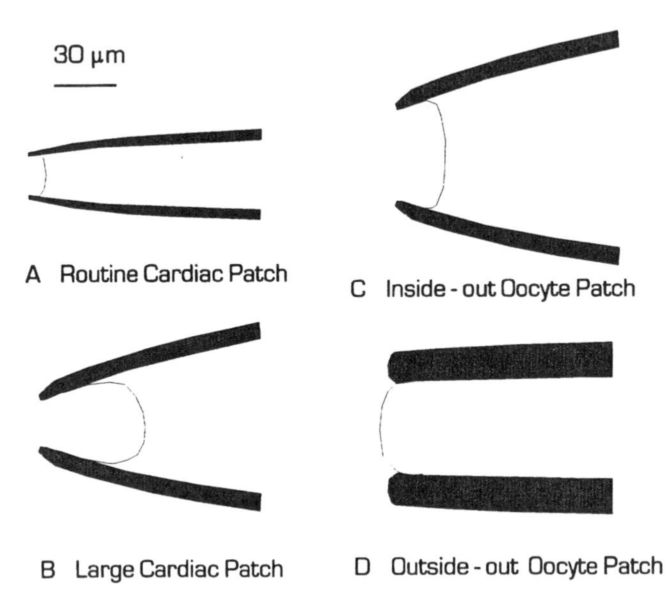

Figure 2. Pipette tip shapes used for giant patches. *A:* The most common tip shape. The pipette tip is thickened during the initial cutting procedure (Fields 1 to 2 of Fig. 1B), and after cutting the pipette tip is polished lightly. *B:* Thickened pipette tip. The tip is cut and melted several times to thicken the pipette walls and to steepen the pipette tip descent. This minimizes pipette resistance and pipette capacitance, and it favors a greater rise of membrane into the tip. *C:* Pipette tip for giant inside-out oocyte patch. The tip is cut at about 50-μm inner diameter and melted to about 35-μm diameter. Pipette resistances are usually <100 kΩ. *D:* Pipette tip for outside-out oocyte patch. A relatively thick- walled glass capillary is employed, and the pipette tip descent is very shallow. After cutting, the tip is polished lightly, with barely visible effects. Tips of this shape can be used without coating for inside-out patches; pipette resistances are relatively large.

2.2. Pipette Coating

In our experience, the application of a hydrocarbon coating to the pipette tip can enhance seal formation and greatly stabilize patches after excision. For muscle cells and for most of the small cells employed, the hydrocarbon coating is usually mandatory. With cardiac membrane, stable patches are sometimes obtained without pipette coating, but the patch membrane usually rises undesirably high into the pipette tip. When oocytes are used for inside-out patches, the coating may not be important. With bullet-like pipette tip shapes, as in Fig. 2C, a coating is sometimes mandatory. For the outside-out configuration, as implemented by Rettinger *et al.* (1994), the coating is not needed and is reported not to be useful. Also, for inside-out oocyte patches using the nearly straight pipette shape of Fig. 1D, the coating is not beneficial.

Our standard coating material consists of a mixture of light and heavy mineral oils and Parafilm® (American Can Corporation). The light mineral oil primarily facilitates seal formation, and the heavier oil and Parafilm® primarily stabilize the patches after excision. The mixture is prepared by heating equal parts of light and heavy mineral oil, mixing with it pieces of shredded Parafilm®, and stirring over heat until a uniform mixture with the consistency of a heavy syrup is obtained (about 15 min). After cooling and hardening, more (light) mineral oil is added until a viscosity comparable to that of corn syrup is obtained.

The dry pipette tip is dipped in the mixture and is back-filled with the filtered pipette solution. When reduction of the pipette tip capacitance is desired, the tip is wrapped with a highly viscous mineral oil/Parafilm® mixture to a short distance from the pipette tip. This is adequate to negate capacitance changes with changes of solution level along the tip. As desired, the tip can be washed extensively before filling with the pipette solution.

When gigohm seals cannot be obtained regularly, it can be useful to vary the hydrocarbon mixtures. Addition of α-tocopherol acetate or small amounts of decane (or hexane) can be useful. A liquid dipping solution of 95% hexane, 3% α-tocopherol acetate, and 2% (w/w) cholesterol is often helpful. Silicon oils, placticizers, and phospholipids are not useful. It is noteworthy that a dependence of seal formation on the type of glass employed is observed even when the hydrocarbon coat is used. Borosilicate glass usually seals more readily, for example, than Corning 7052.

2.3. Membrane Blebbing

Particularly with muscle cells, the induction of cell surface "blebbing" has proven to be a useful means to generate a large membrane surface that forms gigohm seals readily. With cardiac cells, giant patches with diameters much greater than the cell diameter can be formed in this way. Tissue is first digested with collagenase (or other enzymes), and single cells or tissue segments are then placed in blebbing solution: 70–130 mM KCl, 1–10 mM EGTA, 0.5–5 mM $MgCl_2$, 20 mM dextrose, and 15 mM HEPES (pH 7). This protocol is similar to that used by Standen *et al.* (1984) with skeletal muscle. Large (20–50 μm) membrane blebs usually form within 4 to 8 hr on cardiac cells stored at 4°C. In skeletal and tracheal muscle, massive bleb formation can take place within minutes. It is noted that bleb formation is not useful with some cell types. For example, gigohm seals have never been formed with blebs on *Xenopus* oocytes under similar conditions.

Of course, the use of "bleb" membrane introduces some question as to the "physiological" relevance of the membrane. For cardiac membrane, calcium current and delayed potassium current have not been observed in bleb membrane, even in the on-cell configuration. However, the densities of many currents (Na/Ca exchange, Na/K pump, Na channel, and ATP-dependent K channel) are largely as expected from whole-cell measurements.

2.4. Seal Formation: Small Cells and Blebbed Cells

The general principles of seal formation for conventional patches apply equally well to giant patches. All pipette solutions are filtered. Divalent cations facilitate seal formation, although they are not an absolute prerequisite. We find that the presence of 5 to 20 mM chloride in either the pipette solution or the bathing solution (preferably both) is essential. In preparation for seal formation, the pipette tip is kept clean by applying just enough positive pressure to maintain a solution flow out of the tip. Also, this avoids contamination of the tip with the bathing solution.

Figure 3 describes the usual sealing procedures used with small cells (panel A) and cells with membrane blebs (panel B). In both cases, it is advantageous to define conditions under which the cells stick loosely to the surface of dishes or chambers employed but under which they can be freed from the surface by a modest flow of solution. In this way, the cell membrane can be approached with the light positive pressure applied to the pipette followed by gentle suction. When the membrane is well positioned within a close distance (5–20 μm), negative pressure is applied to the pipette. Ideally, the membrane or cell is briskly pulled into the pipette tip without rupturing. In our experience, the use of a sensitive acoustic feedback control is highly desirable. We prefer small (0.1–0.3 mV) voltage pulses applied at 100–500 Hz to monitor changes of pipette resistance. After the initial contact, gentle pulsatile pressure is applied. Then, small movements of the pipette tip in all directions often facilitate sealing under the continued application of small negative pressure (1–5 cm H_2O). The voltage pulse is increased to about 10 mV as the seal develops. Usually, the cell is lifted from the surface, and increasingly larger and faster movements of the pipette in each axis often facilitate the final gigohm seal formation (2 to 20 GΩ). The patch is excised by rapid pulses of solution via a solution line placed laterally to the cell. In our case, this is carried out in the recording chamber employed.

Particularly with blebbed cells, membrane vesicles are sometimes formed on patch excision. This is not apparent visually but can be detected as a lack of well-characterized currents during an experiment. The formation of vesicles can be avoided by using sealing conditions that favor a somewhat greater rise of membrane into the pipette tip (i.e.,

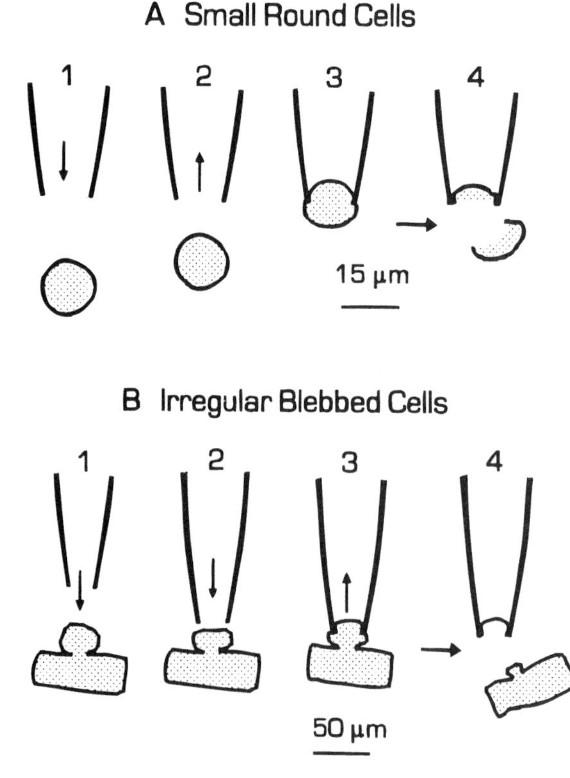

Figure 3. Formation of inside-out giant patches from small, round cells (A) and "blebbed" muscle cells (B). Ideally, cells stick lightly to the chamber surface, so that they can be easily lifted by negative pressure. The pipette tip is brought to a position close to the cell with continuous application of light positive pressure *(1)*. If the cells can be moved easily by negative pressure, the final contact is best made by briskly sucking the cell into the tip *(2–3)*. If the cells do not move easily, the tip is moved to touch the cell in the presence of just enough positive pressure to maintain a solution stream from the tip. In this case, the touch is determined electrically. The formation of gigohm seals is facilitated by pulsatile negative pressure and by movement of the pipette in each axis.

using lower divalent cation concentrations, pipettes with steeper descent to the tip, and greater fire polishing). In our experience, about 50% of vesicles formed can be successfully "popped" to the inside-out configuration by gently touching the pipette tip to a small gas bubble or hydrocarbon (hexane) bead in the recording chamber. Outside-out configurations are sometimes obtained, but conditions to do so routinely have not been established.

It must be mentioned that the ability to obtain stable giant membrane patches routinely with some cell types, in particular blebbed cardiac myocytes, can involve a considerable learning period. In the author's experience with cardiac cells, the ability to obtain very large (25–30 μm diameter), stable patches with a desired membrane configuration continues to grow after 4 years. Seal formation with round cells is less subtle and more quickly learned. Seal formation with *Xenopus* oocytes is far less subtle.

2.5. Seal Formation: Inside-Out Oocyte Patches

Most procedures to form giant inside-out patches from oocytes are the same as for conventional patches (see Fig. 4). Follicles are removed either manually or with a high-phosphate solution. Oocytes are shrunk in two-times hypertonic solution, the vitellin layer is removed mechanically, and the oocytes are returned to isotonic solution. The success rate in obtaining gigohm seals is better, in our experience, when potassium-aspartate is used as the hypertonic agent rather than sucrose. According to Dr. Andreas Jeromin (personal communication), success rate is improved still further if the vitelline layer can be removed with sharpened forceps without injuring unshrunken oocytes.

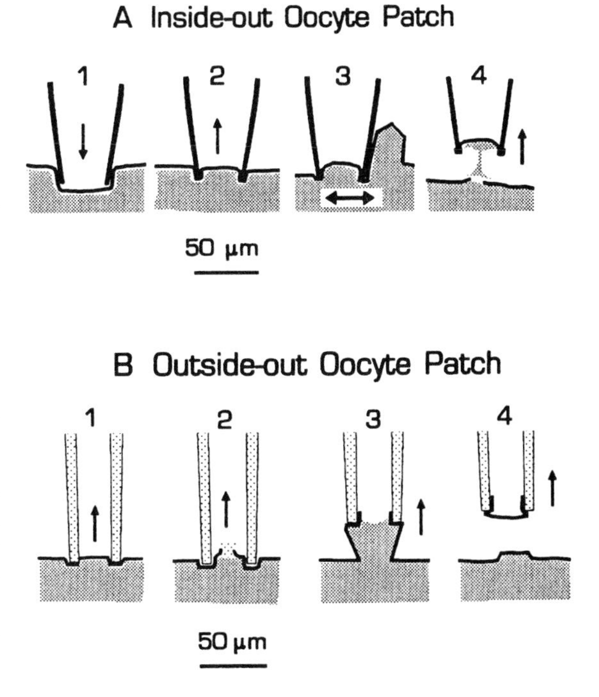

Figure 4. Formation of giant patches from *Xenopus* oocytes. Oocytes are prepared by mechanically removing the vitelline layer. For inside-out patches (A) the pipette is pushed gently into the oocyte in the presence of light positive pressure (Field 1). Then, positive pressure is removed, and light negative pressure is applied, as required to initiate sealing (Field 2). Seal formation can often be facilitated by movements of the pipette in each direction. Patch excision is initiated by increasingly large horizontal movements of the pipette (Field 3). Excision is completed by pulling back the pipette with intermittent horizontal movements to loosen cytoplasmic attachments (Field 4). For outside-out patches (B), relatively straight-ended pipettes are employed, and the seal is made with the tip just touched to the oocyte surface (Field 1), so that rupture of the membrane by negative pressure can be carefully monitored (Field 2). After adjusting pressure so that cytoplasmic contents are immobilized, the pipette is retracted in steps (Field 3) until the membrane seals in an outside-out configuration (Field 4).

For most applications with excised patches, seals should be made in a calcium-free, EGTA-containing solution to avoid activation of the large calcium-dependent chloride conductance of the oocyte membrane on excision of the patch. Accordingly, we usually use bath solutions containing 2–5 mM magnesium as the divalent cation and 1–10 mM EGTA. The monovalent cation employed is not important; 10–20 mM chloride must be present in either the pipette or the bath solution. Our standard bath solution contains 100 mM K-aspartate, 20 mM KCl, 2 mM EGTA, 10 mM HEPES, and 4 mM $MgCl_2$ (pH 7 with KOH). For subsequent seal formation, it is advantageous that the cleaned oocytes stick well to the surface of the petri dishes employed.

If application of a significant free calcium concentration is desired on the cytoplasmic side (e.g., to monitor Na/Ca exchange current or other calcium-activated conductances), then it is essential to perform experiments under completely chloride-free conditions. We find that 2-N-morpholinoethanesulfonic acid (MES) is an appropriate chloride substitute, and a number of other organic anions support significant chloride channel current. To circumvent the requirement for chloride during sealing, a bath solution with 10 to 20 mM chloride can be used. The pipette solution must be chloride-free, and correspondingly a KCl bridge must be used in the pipette. We use routinely a polyethylene tube filled with 50–200 mM agar-agar/KCl that fits snugly over the chlorided silver wire. Alternatively, chloride can be used in the pipette if pipette perfusion techniques are implemented, and the chloride-containing solution is effectively perfused out of the pipette before the beginning of an experiment.

For seal formation, the animal pole of the oocyte is favored, since success on the vegetal pole is less consistent. The pipette tip is brought to the edge of the oocyte in sharp focus, and positive pressure is increased so that a small dent on the side of the oocyte is formed by the solution stream. With continuous positive pressure, the tip is moved into the oocyte until membrane resistance increases by a factor of at least 2, preferably more (Fig. 4A; Field 1). The pipette can be pushed some distance into the oocyte at this time without rupturing the membrane, or the pipette can be left quite superficial. Next, positive pressure is released without application of negative pressure. With favorable oocytes, gigaseals can be obtained simply by small movements of the pipette in each axis and/or application of small negative potential pulses. Depending on the batch of oocytes and the solutions employed, application of negative pressure may be essential (Field 2). The pipette tip can also be pushed deeper into the oocyte, and small movements of the pipette tip in all directions often further facilitate seal formation. The entire sealing procedure can take place in less than 1 min, or it can require up to 5 min to form gigohm seals.

Excision of oocyte patches also requires some practice (Fig. 4; Field 3). The process begins with side-to-side movements of the pipette tip in increasing distances up to nearly the radius of the oocyte (Field 3). In approximately 30-sec intervals, the pipette is retracted about 10 μm from the oocyte, and side-to-side movements are again applied. Toward the end of the excision, strings of cytoplasm are observed to extrude to the patch from the oocyte (Field 4). The great majority of patches finally excise to the inside-out configuration without formation of vesicles. If a vesicle is formed, it usually can be broken by touching the tip against a bubble or bead of hydrocarbon, as with other giant patches. In our procedures, the patch is moved to a temperature-controlled microchamber where solutions applied to the cytoplasmic surface of the patch can be changed in approximately 200 msec (Collins *et al.,* 1992) with no electrical artifacts. A large variety of solution-switching devices have been developed for excised patches, and in principle they should all be appropriate. Inside-out oocyte patches are routinely stable for periods of 15 to 45 min, and they tolerate well holding potentials in the range of −60 to −90 mV.

2.6. Seal Formation: Outside-Out Oocyte Patches

As mentioned in Section 2.1, outside-out giant patches from oocytes can be routinely formed by using pipette tips with a shallow descent to the tip (Rettinger *et al.*, 1994). It is advantageous to position the pipette as nearly horizontal as possible and to view the sealing and excision procedure at a relatively high magnification (400×). Because the pipette solution becomes the cytoplasmic solution, EGTA must be included in the pipette to avoid activation of chloride conductance. The presence of a relatively high divalent ion concentration (2–4 mM) in both the pipette solution (usually magnesium) and the bath solution (magnesium, barium, or calcium) increases success rates. Contact of the pipette tip to the oocyte membrane, monitored electrically, is made at an optical edge of the oocyte in sharp focus (Fig. 4B; Field 1). Seal formation is completed with application of negative pressure. Thereafter, the pipette is retracted until the membrane edge can be clearly observed, and negative pressure is slowly increased (via a water column) to disrupt the membrane with as little extrusion of cytoplasm into the pipette as possible (Fig. 4B, Field 2). Negative pressure is reduced on membrane rupture, such that cytoplasmic material is neither sucked from nor pressed back into the oocyte. The pipette is then slowly retracted (Fig. 4B, Field 3) until the membrane seals back across the tip in the outside-out configuration (Fig. 4B, Field 4). Stability of the giant outside-out patches is similar to that of the inside-out patches.

Figure 5A shows sample current–voltage relationships for the (ouabain-resistant) Na/K pump of the *Torpedo* electroplax, expressed in *Xenopus* oocytes and monitored in an outside-out patch. The results are replotted from Rettinger *et al.* (1994). The endogenous pump current is suppressed with 1 μM ouabain, which does not significantly affect the expressed pump current. The current–voltage relationships were obtained by subtracting voltage pulse results in the absence of extracellular (bath) potassium from results in the presence of the indicated extracellular potassium concentrations [no extracellular sodium; 5 mM cytoplasmic (pipette) ATP and 30 mM cytoplasmic sodium]. The shapes of current–voltage relationships and their dependence on extracellular potassium are very similar to results obtained using the whole-cell, two-microelectrode voltage clamp (not shown). They are also very similar when current–voltage relationships are defined by applying a thousandfold higher ouabain concentration (not shown). The nearly horizontal (flat) current–voltage relationship in the presence of high extracellular potassium and the absence of extracellular sodium is very similar to results obtained for the cardiac Na/K pump in inside-out patches (Fig. 5B). For the inside-out cardiac patch, nearly identical pump current–voltage relationships can be defined by application and removal of either cytoplasmic sodium or cytoplasmic ATP or by pump inhibition with application of cytoplasmic vanadate or extracellular ouabain.

2.7. Pipette Perfusion

While technically demanding, intrapipette perfusion is essential for many purposes and with practice it can be used routinely (H.ilgemann *et al.*, 1002; Matsuoka and

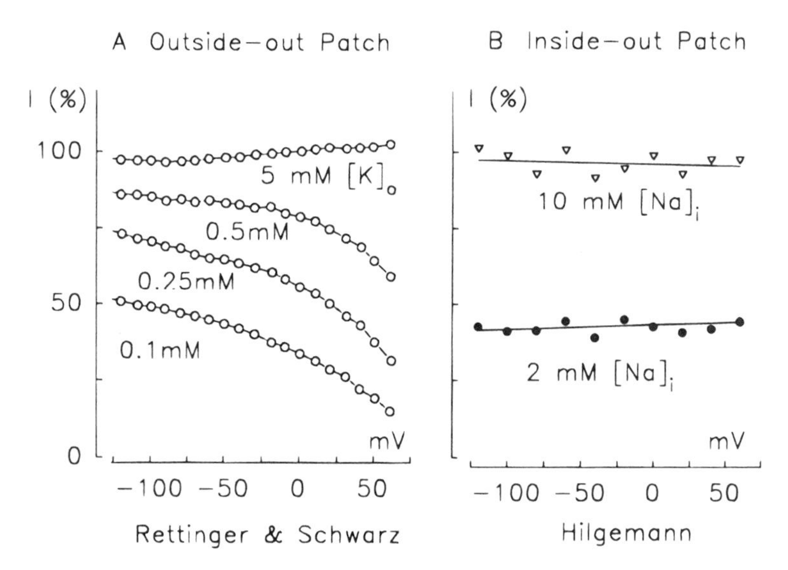

Figure 5. Sodium pump current in outside-out (A) and inside-out (B) giant patches. Panel A shows current–voltage relationships determined for the *Torpedo* electroplax Na/K pump, expressed in *Xenopus* oocyte membrane. These results have been replotted from Rettinger *et al.* (1994). The cytoplasmic (pipette) solution contains 30 mM sodium and 5 mM ATP. The extracellular (bath) solution is sodium-free. Potassium channels are blocked. The current–voltage relationship is defined as the current induced by application of extracellular potassium, and nearly identical current–voltage relationships were obtained by defining pump current as the current blocked by 1 mM ouabain. Note that voltage dependence increases as extracellular (bath) potassium is decreased. Panel B shows current–voltage relationships determined for the native Na/K pump in a giant inside-out membrane patch from a guinea pig myocyte. The extracellular (pipette) solution is sodium-free and contains 10 mM potassium. Potassium channels are blocked. Identical pump currents are defined either by application of cytoplasmic ATP in the presence of cytoplasmic sodium or by application of cytoplasmic sodium in the presence of cytoplasmic ATP. No current is activated in the presence of 200 μM extracellular ouabain. Note the complete voltage independence of the pump current under these conditions.

Hilgemann, 1992). Due to the large diameter of the pipette tips employed with giant patches, the perfusion can be more rapid than with the conventional patch-clamp methods. The negative-pressure method of Soejima and Noma (1984) can be used, with the reservation that pipette perfusion must be established with only 4–8 cm of negative water pressure. The polyethylene tube used for perfusion must be prepared meticulously to insure solution flow.

Recently, pipette perfusion methods using positive pressure have been implemented. These methods are modified from descriptions of Tang *et al.* (1992). As described in Figure 6A, a flexible quartz tubing of 150 μm O.D./70 μm I.D. (Polymicro Technologies, Tuscon, Arizonza) is used to deliver solution to the pipette tip. The tubing is pulled on a gas flame, and the tip is cut to an outer diameter of about 60 μm. Under a microscope, the tip of the perfusion line is placed within 200 μm of the recording pipette tip prior to seal formation. Solution is forced through the quartz tubing to the pipette tip by applying positive air pressure to solution reservoirs with a volume of about 100 μl. The reservoirs are constructed from standard plastic pipette tips. The connections between the reservoirs and final quartz tubing are constructed so as to minimize dead space. A valve is constructed at each reservoir via a short piece of thin silicone tubing. The silicone tubing is squeezed shut by pulling on an elastic tie when the perfusion line is

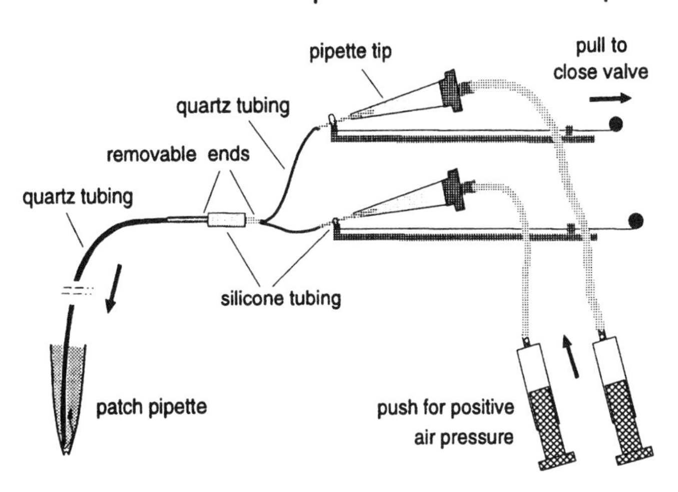

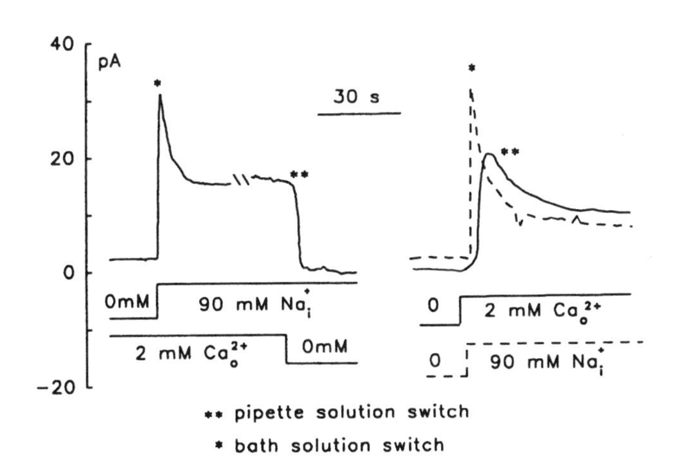

Figure 6. Schematic diagram of a positive pressure pipette perfusion device. A coated quartz tubing (140 μm O.D./70 μm I.D.; Polymicro Technologies; Tuscon, Arizona) is mounted in the pipette tip through a hole sealed with wax. Before seal formation, the quartz line is positioned by hand to within 150 μm of the pipette orifice. The back end of the quartz fiber is glued into a polyethylene (PE) tube, which is inserted into a silicon sleeve. In this way, the quartz tip piece is mated snugly to another PE tube for solution delivery. This end contains two to five quartz lines connected to solution reservoirs. Epoxy glue is used to mount the quartz lines in the PE tube, and the ends are cut with a razor blade. Positive air pressure can be applied from syringes to each of the solution reservoirs, made from plastic pipette tips. To avoid back-flow of solution from a delivery line into other solution lines a valve is constructed at the outlet of each reservoir. To do so, a short stretch of thin silicon tubing is glued between the pipette tip and the quartz tubing line. The silicone tubing can be squeezed shut with an elastic thread when the solution line is not in use.

not in use. This avoids contamination between solutions when positive pressure is applied. Air pressure is applied via syringes to each reservoir to deliver solution to the pipette tip. Such perfusion devices are readily constructed with four or five perfusion reservoirs, as needed. Due to the very small 'dead' volumes, many solution changes can be performed stably with only small changes to solution volume in the patch pipette tip.

Figure 6B illustrates experimental results with the positive pressure pipette perfusion device for outward Na/Ca exchange current in a giant cardiac membrane patch. In the left panel, the outward exchange current was first activated by application of 90 mM sodium from the bath (cytoplasmic) side. The exchange current decayed by about 50% over 40 sec, and the current was then turned off by switching solutions in the pipette tip from one with 2 mM calcium to a calcium-free solution with 100 μM EGTA. Note that the current turns off completely in just a few seconds. This time course approximately reflects the time course of pipette perfusion. In the right panel of Figure 6B, current transients are shown for application of extracellular (pipette) calcium in the presence of cytoplasmic sodium (solid curve), and for application of cytoplasmic sodium in the presence of extracellular calcium (dotted curve). Note that the exchange current decays with a similar time course in these results. This decay reflects an inactivation process (Hilgemann, Matsuoka, and Collins, 1992). That similar results are obtained for these two activation protocols verifies a prediction of our proposed model of inactivation (Hilgemann, Matsuoka and Collins, 1992; Matsuoka and Hilgemann, 1994).

2.8. Measurement of Membrane Patch Capacitance

It is often desirable to determine membrane current densities, using measurementsof membrane capacitance to estimate membrane area. An approximate specific membrane capacitance of 1 μF per cm^2 (= 1 pF per 100 μm^2) is assumed. The patch membrane capacitance must be established as a difference between the capacitance of the pipette tip with intact patch and the capacitance of the tip without the membrane patch (i.e., the glass wall of the tip). To measure the latter, it is essential to seal the pipette orifice shut under a condition comparable to that with the intact patch. Our preferred procedure is as follows: A drop of hydrocarbon mixture (the electrode coating material) is deposited in the recording chamber at an accessible position. At the end of each experiment the pipette tip with intact patch is positioned just next to the hydrocarbon bead. Total pipette capacitance is compensated as completely as possible, and the patch is ruptured by positive pressure. The pipette tip is then sealed to the hydrocarbon bead, the capacitance is again compensated, and the capacitance difference with respect to the measurement with intact patch is the best estimate of the patch membrane capacitance. Care must be taken so that hydrocarbon does not rise up into the pipette tip, as this leads to an underestimation of pipette capacitance.

In our experience, an advantageous hydrocarbon mix for this purpose is hexane with a few percent added phospholipid (95% hexane with 5% phosphatidylcholine), similar to mixtures used to create artificial bilayers. When the pipette tip is touched to the hydrocarbon a tight seal immediately forms. The tip is then retracted from the hydrocarbon, and a stable hydrocarbon film forms across the pipette orifice. The films appear to submicron in thickness. Pipettes with greatly different pipette tip diameters (10–300 μm) differ in their capacitances by only 1 to 4 pF. This indicates that the capacitance of the hydrocarbon film is small and that an artificial bilayer does not form. To measure capacitance as just described, a droplet of hydrocarbon must be kept available in the Petri dishes employed for experiments. To do so, an approximately 0.5 mm hole is drilled in a small piece of a plastic cover slip. This is fixed to the bottom of the recording chamber, and a drop of hydrocarbon is deposited in the hole before adding the experimental solution.

A more convenient estimate of membrane capacitance can be obtained at the end of experiments without sealing the pipette tip shut. Under microscope, the pipette tip is raised as high as possible in the bathing solution, thus reducing as much as possible the capacitance contribution from wetted glass of the pipette tip. The capacitance is compensated as completely as possible, and the pipette is removed from solution (into open circuit configuration with destruction of the membrane patch). The magnitude of the capacitance drop upon removing the pipette from solution is an upper-limit measurement of the membrane patch capacitance, which includes the capacitance of the wetted glass wall of the pipette. An average value of capacitance for the glass wall of pipette tips under this same condition is simply subtracted from the measured capacitance drop to give a reasonable estimate of membrane capacitance. The average value subtracted is obtained in analogous measurements using identically prepared pipettes sealed shut with hydrocarbon film as described above. Again, the tip is raised as high as possible in solution, and the drop of capacitance is monitored upon removing the pipette from solution (1.5–3 pF) to give the pipette tip capacitance. The possible error in these procedures is estimated to be about 1 pF. This is acceptable for most purposes. For cardiac patches, for example, the maximum density of sodium pump current is found to be ~2.2 pA/pF (35 °C) in patches with capacitances ranging from 3 to 12 pF.

3. Giant-Patch Recording

3.1. Patch-Clamp Speed

The low access resistance of pipette tips (<100 kΩ), used with membrane patches with 8–12 pF capacitance, brings with it the potential of voltage-clamping a large membrane area in the megahertz frequency domain. In the long run, this potential should for the first time allow temporal resolution of a variety of important electrogenic events such as (some) ion-binding reactions and fast conformational changes associated with transport function. At present, voltage-clamp speed achieved with our routine instrumentation is limited at about 4 μsec.

With the glass pipettes employed to date, slow charging components can be relatively large, and it is essential to develop reliable current subtraction techniques. In many cases, this will limit resolution of electrogenic processes of interest. The increase of noise with high-frequency recording can of course also be a limiting factor. Remarkably, however, slow membrane components in voltage clamp, because of the membrane "rim" in close apposition to the pipette wall, can be largely avoided when appropriate care is taken during seal formation. Both with oocyte patches and with cardiac membrane patches, it has been possible to select seal conditions so that "rim current" is almost undetectable. For oocyte patches, this simply means that seals are formed with as little negative pressure as possible; similar experience has been reported by Dr. E. Stephani and colleagues (personal communication) in recording of gating and channel currents with giant oocyte patches in the on-cell and inside-out configuration.

Figure 7 illustrates the 4-μsec voltage clamp resolution presently obtained, using the oocyte calcium-activated chloride current as example. Records in Fig. 7A are ten-times data-averaged signals for ±100-mV voltage pulses, with and without 2 μM free calcium on the cytoplasmic side, as indicated. Calcium was buffered with 10 mM EGTA. Two averaged records without calcium are given, one taken before and one after the application of calcium. Slow transient current components are apparent over about 20 μsec, but they are unchanged by activating the Cl conductance and with good certainty reflect charging of the pipette tip. Figure 7B shows the corresponding subtracted record of the actual voltage clamp of the chloride current. The current is stably clamped within 4 μsec.

3.2. Fast Solution Switching

The speeds of solution switching that can be obtained with giant membrane patches will be adequate to resolve the time course of many slow molecular processes. With inside-out patches the speeds obtained will depend on the degree to which membrane rises up in the pipette tip and will be determined by the corresponding diffusion times. With inside-out oocyte patches (about 20–40 μm membrane insertion into the pipette), solutions can be routinely changed at the membrane surface within 100 to 250 msec. For some applications, a much faster apparent solution switch will be obtained if a supersaturating concentration of the desired compound or ion is applied.

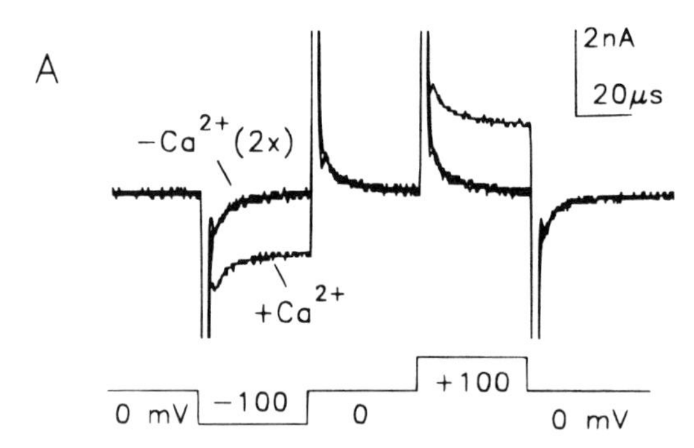

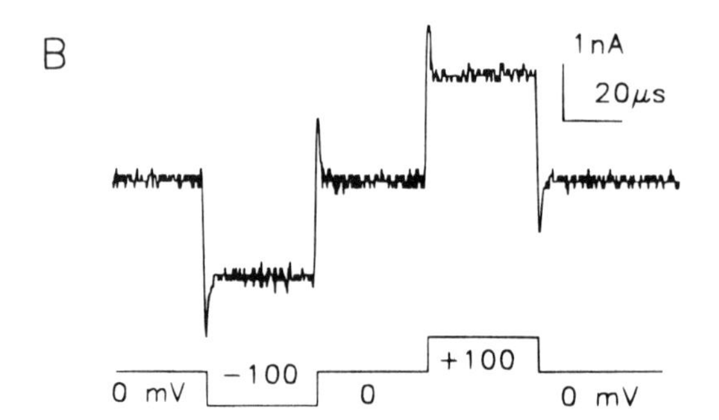

Figure 7. Fast voltage clamp of the calcium-activated chloride current in an inside-out *Xenopus* oocyte giant patch. Panel A shows raw current records with and without 2 μM free cytoplasmic calcium; there was 100 mM chloride on both membrane sides. Records without calcium were taken before and after application of calcium. Panel B shows the subtracted record demonstrating stable voltage clamp within 4 μsec. See text for further details.

Faster solution switches can be obtained with outside-out patches, which are formed essentially at the orifice of the pipette. Figure 8 shows records for the activation by glutamate of the N1/N2A clone of the NMDA channel (P. Ruppersberg, personal communication). As indicated, glutamate was applied and removed together with glycine under conditions selected for inward NMDA current. The solution switch is made by rapidly moving the interface between two solution streams across the pipette. For this, a piezoelectric device is used to move rapidly a pipette with parallel, double (theta) barrels back and forth in front of the pipette. The current rises with a τ of about 18 msec on application of transmitter, and it declines with a τ of about 21 msec on removal of transmitter (dotted lines in Fig. 8). The actual time course of diffusion kinetics to and away from the membrane are thought to be largely complete in 10 msec (P. Ruppersberg, personal communication).

3.3. Single-Channel Recording

It was not expected that the giant patch would be used for single-channel recording. However, Gadsby and colleagues have used cardiac giant patches very successfully to study the modulation of CFTR-like chloride channels by protein kinases and nucleotides (Nagel *et al.*, 1992; Hwang *et al.*, 1994). For the CFTR-like channels, the reasons for success are twofold. First, these channels exist in rather low densities in cardiac membrane, and second, these channels have very long open and closed times. Thus, noise related to the large capacitance of the patch is not limiting.

Figure 9 shows an example of such recordings in an inside-out cardiac patch with at least four channels. In the early portion of the record, the chloride channels are activated by application of protein kinase A (PKA) in the presence of ATP. Then, the nonhydrolyzable ATP analogue AMP-PNP is applied without ATP and PKA. It cannot replace ATP in sustaining channel activity. After reapplication of PKA and ATP, however, AMP-PNP has a stimulating effect that tends to "lock" the channels open. The data are interpreted in terms of multiple nucleotide binding sites and actions modulating CFTR channel activity. As a related application, the giant patch method may prove valuable in recording channel activity during cloning experiments when expression is poor.

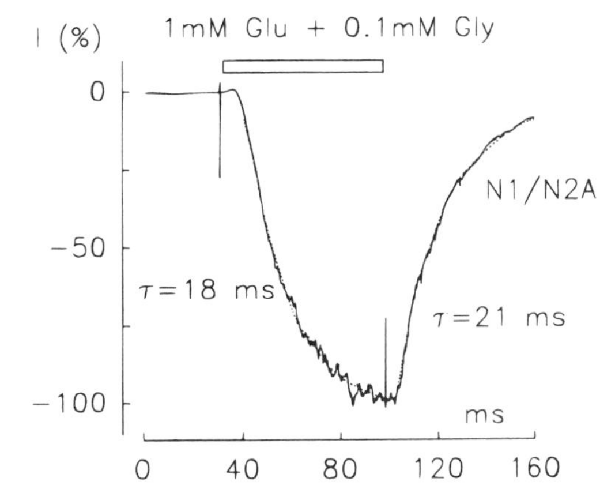

Figure 8. Fast solution switching with an outside-out *Xenopus* oocyte giant patch. The patch is from an oocyte expressing the glutamate-activated NMDA conductance. Conditions were chosen so that application of glutamate with glycine activates a large inward current. Solutions are changed by moving a double-barreled pipette with parallel solution streams rapidly back and forth in front of the patch, such that the interface between the two solutions passes across the pipette. The apparent time constants for activation and relaxation of the current are 18 and 21 msec (see dotted exponentials), respectively, and are thought to reflect largely channel kinetics rather than diffusion times. These results were kindly provided by Dr. P. Ruppersberg, Tübingen.

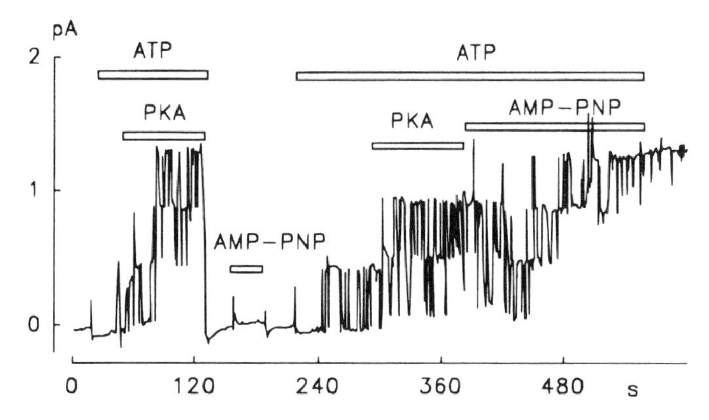

Figure 9. Recording of CFTR Cl channels in giant inside-out guinea pig cardiac membrane patches. The CFTR-like chloride channels of guinea pig cardiac sarcolemma are recorded as an outward current in an excised giant patch. As indicated, applications of both ATP and protein kinase A (PKA) are required to activate the channels. The channels close immediately upon removal of ATP, and the ATP analogue AMP-PNP (0.5 mM) is without effect. After reactivation of the channels with PKA and ATP, however, AMP-PNP can increase channel activity and hold channels open for long periods of time. (This record has been retraced from Hwang *et al.,* 1994, Fig. 3B.)

3.4. Manipulation of Patch-Membrane Composition

The phospholipid composition of a biological membrane is a potentially important determinant of the function of transporters and channels that are embedded within it. We have been particularly interested in the possibility that individual lipids may play second messenger roles in regulatory pathways. We have suggested that modulatory effects of ATP on Na/Ca exchange might be related to changes in negatively charged lipids that strongly stimulate exchange activity (Collins *et al.,* 1992). Possible mechanisms include the fueling of a phospholipid translocase enzyme, which establishes an asymmetric distribution of phosphatidylserine (PS) in favor of the cytoplasmic side, or the fueling of lipid kinases that generate phosphatidylinositols.

In this light, we began to test whether it might be possible to modulate the patch membrane composition by adding phospholipids to the hydrocarbon coat applied to pipette tips. It was found that inclusion of negatively charged phospholipids, such as PS, in the hydrocarbon coating strongly stimulated the exchange current by modulation of an inherent inactivation mechanism of the exchange process. From these results it seems likely that phospholipids can diffuse into and out of the membrane patch along the pipette tip. In subsequent work, we exploited this finding by applying phospholipid mixtures directly to the pipette wall in close proximity to the patch (Collins and Hilgemann, 1993). As predicted, we found that application of negatively charge lipids strongly stimulates the exchange current. At the same time, we tested for effects of the phospholipids on sodium current and found no clear changes. Since there are no apparent changes of membrane capacitance, phospholipids may be exchanging between the pipette wall and the patch.

Next, we tested whether phospholipids could be applied more conveniently as phospholipid vesicles. It is well known that phospholipid vesicles bind to glass and hydrophobic surfaces, including membranes, and we already knew that phospholipids can diffuse along the pipette tip into the patch. In brief, a concentrated vesicle solution is prepared in distilled water via standard sonication techniques. The vesicles are mixed with the cytoplasmic solution just prior to its application to a patch. With this approach, all of the effects observed with direct application of phospholipids have been duplicated and extended.

Figure 10A shows typical effects of PS vesicle application on the Na/Ca exchange current in a cardiac patch. First, ATP was applied to stimulate a rather small exchange current. Then, PS vesicles were applied (0.1 mM apparent concentration). The resulting massive stimulation of exchange current does not reverse on removal of vesicles. The effect can, however, be reversed by application of pure phosphatidylcholine vesicles (not shown); and over time the exchange current can be abolished by application of phosphatidylcholine. In parallel experiments, Na/K pump current was almost unaffected by vesicle application. In general, membrane capacitance does not clearly change or

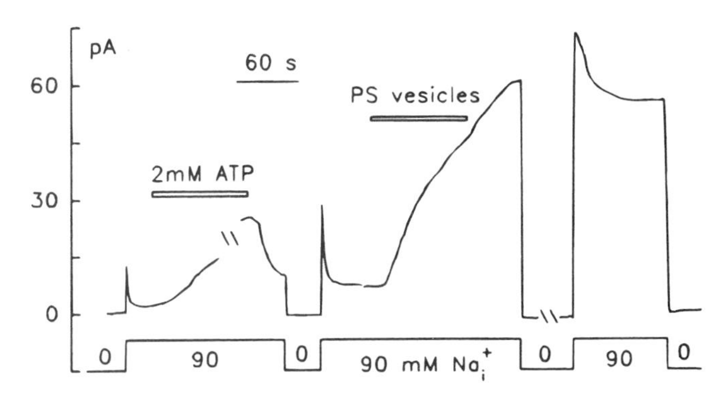

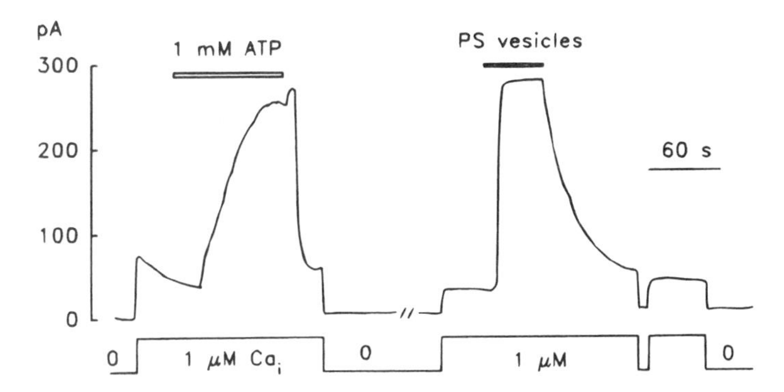

Figure 10. Stimulation of outward cardiac Na/Ca exchange current (A) and *Xenopus* oocyte calcium-activated chloride current (B) by cytoplasmic application of phosphatidylserine (PS) vesicles. As shown in the first part of the records, both currents are stimulated by application of MgATP from the cytoplasmic side. Also, both currents are strongly stimulated by application of PS vesicles (0.1 mM total PS concentration). In cardiac patches, the stimulatory effect of PS does not reverse after removal of vesicles, but it reverses for the most part within 1 min in oocyte patches. Sodium pump current and sodium channel current were unaffected by PS in cardiac membrane.

increases only slightly during application of phospholipids, indicating that the process set in motion has the characteristics of a phospholipid exchange. With phosphatidylcholine, it appears possible that the majority of negatively charged lipids can be exchanged out of the patch.

Figure 10B shows an equivalent experiment with the calcium-activated chloride current in an inside-out oocyte patch. As indicated, the chloride current was activated by application of a solution with 2 μM free calcium (10 mM EGTA) and turned off by removal of calcium. The chloride current is recorded as an outward current (25 mM extracellular chloride; 2 mM cytoplasmic chloride). In the first part of Fig. 10B, 1 mM cytoplasmic ATP is shown to stimulate the chloride current reversibly by about fourfold; this is described here for the first time. The magnitude of the stimulation is very similar to the subsequent stimulatory effect of PS. The stimulatory effect of PS was accompanied by a shift of the calcium dependence of the current to a lower free calcium range. In first approximation, therefore, an increase of negative membrane surface potential might underlie the effect of PS, thereby increasing the effective free cytoplasmic calcium concentration. Preliminary estimates of membrane surface potential, carried out as described previously for cardiac patches (Hilgemann and Collins, 1992), have verified this interpretation (not shown). Remarkably, the stimulation by PS of the chloride current reverses over the course of a minute upon removal of the PS vesicles.

The results with vesicles can tentatively be interpreted as follows: Vesicles presumably bind to the pipette tip and/or the membrane itself. Individual phospholipids subsequently diffuse into the patch, and roughly equal numbers diffuse out of the patch. In cardiac membrane patches, PS incorporated in the membrane is stable. However, it appears that in oocyte membrane PS has a strong tendency to diffuse out of the patch and onto the electrode. It seems possible, therefore, that the formation and excision of membrane patches may in general modify the phospholipid composition of patches with significant consequences for transporter and channel function. Further information is needed, but the evident ability to modify membrane-patch composition opens the way to a wide range of studies of ion-channel and transporter modulation and regulation by membrane lipids.

3.5. Neuronal Calcium Current and the Status of Cytoskeleton in Giant Patches

The cytoskeleton is another potentially important regulatory factor for transporter and channel function that may be altered or disrupted during patch formation and excision. In giant membrane patches from *Lymnaea* (snail) neurons, Johnson and Byerly (1993) have presented evidence that a time-dependent disruption of cytoskeleton after membrane excision may underlie the rundown of calcium channels. As described in Fig. 11, calcium currents can be recorded for several minutes in giant patches from this neuron after excision of the membrane. Upper traces in Fig. 11A are before rundown of the calcium current, and the lower traces in Fig. 11A are after rundown; the remaining voltage-activated current is a proton current. The voltage dependence of the calcium (barium) current is shown in Fig. 11B. The presence of ATP on the cytoplasmic side slows rundown from a T_{50} of about 400 sec to a T_{50} of about 100 sec (see Fig. 11C,D). In the presence of ATP, therefore, channel stability is adequate to study factors such as cytoplasmic free calcium on the calcium current kinetics. The cytoskeleton disrupters cytochalasin B and colchicine were both found to increase the rate of rundown in the presence of ATP to approximately the rate found in the absence of ATP. The cytoskeleton

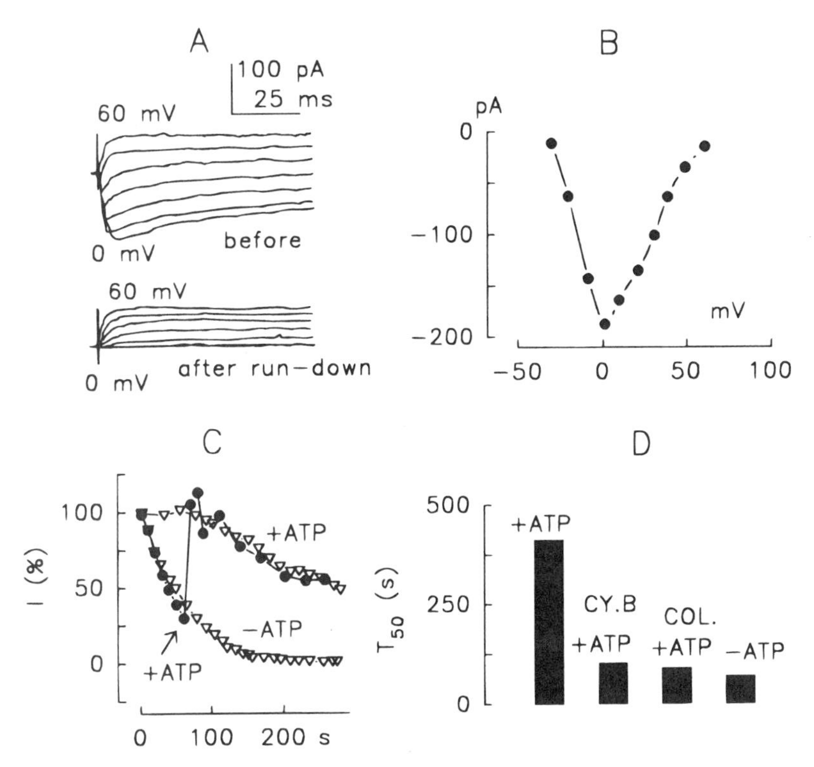

Figure 11. Calcium (barium) current in giant excised inside-out patches from *Lymnaea* neurons. Panel A shows currents recorded just after patch excision (before) and 12 min later after calcium current had run down after rundown. Panel B shows the voltage dependence of barium-carried calcium current. This was obtained by subtracting records after rundown, which are dominated by a voltage-activated proton current, from the records before rundown. Panel C shows the time course of calcium (barium) current rundown in the absence (−ATP) and the presence (+ATP) of 2 mM ATP (open triangles). The filled circles give results in which 2 mM ATP was applied after onset of rundown, whereby ATP increases the calcium current to the +ATP level. Panel D gives the average half-time of calcium (barium) current rundown with 2 mM ATP (+ATP), with 2 mM ATP and 100 μM cytochalasin B (CY.B + ATP), with 2 mM ATP and 100 μM colchicine, and without ATP. Both cytochalasin B and colchicine highly significantly speed current rundown. The results have been replotted from Johnson and Byerly (1993).

stabilizers taxol and phalloidin slow to a lesser extent the rundown of calcium current in the absence of ATP. Although the concentrations of the agents employed are rather high (20–100 μM), the results obtained with the different agents are entirely consistent with the author's interpretation.

In giant membrane patches from cardiac cells and oocytes, we have tested extensively for effects of cytoskeleton disrupters and stabilizers in similar protocols with and without ATP. For Na/Ca exchange current and sodium current in excised cardiac patches, and for the oocyte chloride current in excised patches, we have found no consistent effects of these agents or of specific enzymes such as gelsolin, thought to act on the cytoskeleton. This may mean that with our routine techniques the cytoskeleton is irreversibly disrupted in excised patches. Alternatively, the ion transporters and channels we have examined may not be sensitive to changes in the cytoskeleton.

4. Perspectives

The results outlined in this chapter document a number of applications for the giant membrane patch techniques in the study of ion channels and electrogenic transporters. These examples illustrate considerable potential of the methods in both the biophysical and regulatory areas. In the long run, it is hoped that membrane-linked cell processes such as excitation–secretion coupling and calcium release can be preserved and/or reconstituted in giant excised patches. Free access to the cytoplasmic side in such studies would be highly attractive, for example, to study diffusible macromolecules that modulate such processes.

ACKNOWLEDGMENTS. I am indebted to many colleagues for their comments and input in the form of experimental results. They include J. Rettinger, W. Schwarz, D. C. Gadsby, T. Dousmanis, L. Byerly, P. Ruppersberg, and C.-C. Lu.

References

Collins, A., and Hilgemann, D. W., 1993, A novel method for direct application of phosphlipids to giant excised membrane patches in the study of sodium–calcium exchange and sodium channel currents, *Pflügers Arch.* **423:**347–355.

Collins, A., Somlyo A., and Hilgemann, D. W., 1992, The giant cardiac membrane patch method: Stimulation of outward Na/Ca exchange current by MgATP, *J. Physiol.* **454:**37–57.

Doering, A. E., and Lederer, W. J., 1993, The mechanism by which cytoplasmic protons inhibit the sodium–calcium exchanger in guinea-pig heart cells, *J. Physiol.* **466:**481–499.

Hilgemann, D. W., 1989, Giant excised cardiac sarcolemmal membrane patches: Sodium and sodium–calcium exchange currents, *Pflügers Arch.* **415:**247–249.

Hilgemann, D. W., 1990, Regulation and deregulation of cardiac sodium–calcium exchange in giant excised sarcolemmal patches, *Nature* **334:**242–245.

Hilgemann, D. W., 1992, The giant excised-patch method, *Axobits* **9:**5–6.

Hilgemann, D. W., 1994, Channel-like function of the Na^+,K^+ pump probed at microsecond resolution in giant membrane patches, *Science* **263:**1429–1432.

Hilgemann, D. W., and Collins, A., 1992, The mechanism of sodium–calcium exchange stimulation by ATP in giant cardiac membrane patches: Possible role of aminophospholipid translocase, *J. Physiol.* **454:**59–82.

Hilgemann, D. W., Nicoll, D. A., and Philipson, K. D., 1991, Charge movement during sodium translocation by native and cloned cardiac Na/Ca exchanger in giant excised membrane patches, *Nature* **352:**715–719.

Hilgemann, D. W., Matsuoka, S., and Collins, A., 1992, Dynamic and steady state properties of cardiac sodium–calcium exchange: Sodium-dependent inactivation, *J. Gen. Physiol.* **100:**905–932.

Hwang, T-C, Nagel, G., Nairn, A. C., and Gadsby, D. C., 1994, Regulation of the gating of CFTR Cl channels by phosphorylation and ATP hydrolysis, *Proc. Natl. Acad. Sci. U.S.A.* **210:**4693–4702.

Johnson, B. D., and Byerly, L., 1993, A cytoskeletal mechanism for Ca^{2+} channel metabolic dependence and inactivation by intracellular Ca^{2+}, *Neuron* **10:**797–804.

Matsuoka, S., and Hilgemann, D. W., 1992, Dynamic and steady state properties of cardiac sodium–calcium exchange: Ion- and voltage dependencies of transport cycle, *J. Gen. Physiol.* **100:**962–1001.

Matsuoka, S., and Hilgemann, D. W., 1994, Inactivation of outward Na/Ca exchange current in guinea pig ventricular myocytes, *J. Physiol.* **2380:**443–458.

Matsuoka, S., Nicoll, D. A., Reilly, R. F., Hilgemann, D. W., and Philipson, K. D., 1993, Identification of regulatory regions of the cardiac sarcolemmal Na^+–Ca^{2+} exchanger. *Proc. Natl. Acad. Sci. U.S.A.* **90:**3870–3874.

Nagel, G., Hwang, T.-C., Nastiuk, K. L., Nairn, A. C., and Gadsby, D. C., 1992, The protein kinase A-regulated cardiac Cl^- channel resembles the cytstic fibrosis transmembrane conductance regulator, *Nature* **360:**81–84.

Rettinger, J., Vasilets, L. A., Elsner, S., and Schwarz, W., 1994, Analysing the Na^+/K^+-pump in outside-out giant membrane patches of *Xenopus* oocytes, in: *The Sodium Pump* (E. Bamberg and W. Schoner, eds.), Steikopff Verlag, Darmstadt pp. 553–560.
Soejima, M., and Noma, A., 1984, Mode of regulation of the ACh-sensitive K-channel by the muscarinic receptor in rabbit atrial cells, *Pflügers Arch.* **400:**424–431.
Standen, N. B., Stanfield, P. R., Ward, T. A., and Wilson, S. W., 1984, A new preparation for recording single-channel currents from skeletal muscle, *Proc. R. Soc. Lond.* [*B*] **221:**455–464.
Tang, J. M., Wang, J., and Eisenberg, R. S., 1987, Perfusing patch pipettes, *Methods Enzymol.* **207:**176–180.

Chapter 14

A Fast Pressure-Clamp Technique for Studying Mechanogated Channels

DON W. McBRIDE, JR., and OWEN P. HAMILL

1. Introduction

Mechanosensitive (MS) membrane ion channels provide a means of transducing cell membrane deformation or stretch into an electrical or ionic signal (Howard *et al.,* 1988; Sokabe and Sachs, 1992). They represent the most recently discovered and least understood of the major channel classes. It is only recently that information on their molecular nature has been provided (for references, see Hamill and McBride, 1994a). Yet MS channels are ubiquitous, being found in both eukaryotes and prokaryotes (Martinac, 1993). Although their role in mechanotransduction in sensory cells is evident, in nonsensory cells they have been implicated in diverse mechanosensitive functions (Sachs, 1988). There is a variety of MS channels with different gating (stretch-activated and stretch-inactivated) and ion-selective (Na^+/K^+, K^+, Cl^-, etc) properties (Morris, 1990). There also appear to be two broad mechanisms by which mechanosensitivity can be conferred on a channel. These are direct or indirect, according to the way mechanical energy is coupled to the gating mechanism. In direct coupling, mechanical energy acts directly on the channel molecule, and we refer to these as mechanogated (MG) channels. In indirect coupling the channel itself is not MS but is gated by a second messenger that is regulated by a MS enzyme or process (see Ordway *et al.,* 1992).

A critical feature of gated membrane ion channels, in terms of their signaling function, is their dynamic properties, that is, how fast they open in response to stimulation (i.e., latency), close after removal of stimulation, and inactivate, desensitize, or adapt in response to sustained stimulation. This kinetic information not only is important in the basic characterization of the channel but also can give insight into the underlying molecular mechanisms. For instance, a short latency in response time (i.e., in the submillisecond range) of a channel most likely indicates a direct coupling between the gating stimulus and the channel. On the other hand, longer latencies may indicate intervening steps, possibly involving second messenger systems. Some channel kinetic information can be obtained from studying channels under stationary (i.e., steady-state) stimulating conditions. However, in studying latency, adaptation, inactivation, or desensitization kinetics, it is essential that relaxation or perturbation techniques be used. In particular, the use of the voltage-clamp technique to provide voltage steps proved crucial in the understanding of voltage-gated channels (see Hille,

DON W. McBRIDE, JR., and OWEN P. HAMILL • Department of Physiology and Biophysics, The University of Texas Medical Branch, Galveston, Texas 77555-0641.
Single-Channel Recording, Second Edition, edited by Bert Sakmann and Erwin Neher. Plenum Press, New York, 1995.

1992). Similarly, the recent development of submillisecond concentration-jump techniques has provided new information on the gating kinetics of ligand-gated ion channels (Franke *et al.*, 1991; Maconochie and Knight, 1992). Although voltage- and ligand-gated channels have been well characterized by relaxation methods, the MG channels have usually been studied under stationary conditions, at least in most single-channel/patch-clamp studies (Sachs, 1988; Morris, 1990; Martinac, 1993).

This assumption of stationary behavior of MG channels may be an oversimplification. Certainly, it is well recognized that different mechanoreceptors vary in their dynamic properties, with some showing no adaptation and others displaying rapid adaptation (French, 1992). In the case of mechanoelectric transduction in the vertebrate audiovestibular hair cell, dynamic properties of whole-cell MG currents have been measured using displacement and force perturbation techniques (Hudspeth and Corey, 1977; Howard and Hudspeth, 1987; Jaramillo and Hudspeth, 1993). Although these techniques have provided valuable information on both the activation and adaptation kinetics of macroscopic MG current activity, they are not amenable to the study of single MG channels in patch-clamp experiments. However, recent experiments indicate that single MG channel activity in two previously well-characterized nonsensory cell types (skeletal muscle and *Xenopus* oocytes) displays adaptation in response to sustained stimulation (Hamill and McBride, 1992). In hair cells, adaptation, among other effects, increases the dynamic range of the mechanotransducer. In nonsensory cells, adaptation of MG channel activity may serve to limit Ca^{2+} influx through the MG channel during sustained stimulation caused by, for example, osmotic swelling.

An essential requirement for perturbation studies is the ability to apply a step waveform in stimulation. In the case of MG channels, a pressure-clamp technique (McBride and Hamill, 1992, 1993) has been developed for use in conjunction with patch-clamp techniques to apply suction/pressure steps to membrane patches and whole cells. In its original form the technique could produce pressure/suction steps with transition times of ~10 msec. This response time proved adequate for our initial characterization of adaptation kinetics of MG channels that display decay constants of 100–500 msec at the cell's resting potential (Hamill and McBride, 1994b). However, a 10-msec transition time may represent a limitation for the accurate description of the latency and activation kinetics of MS channels. For example, we have observed that MG channels in oocytes can turn on with a latency of less than 2 msec and a rise time of less than 1 msec (McBride and Hamill, 1993). Thus, MG channel opening occurs during the transition time of the suction/pressure step and before the pressure is clamped to its final value. This prevents a quantitative characterization of the pressure dependence of MG channel activation kinetics.

In this chapter we focus on recent attempts to improve the original pressure clamp. The improvements are centered around reducing the response time to the submillisecond time range and miniaturizing the system to make it more convenient.

2. Summary of Methods for Activating Single MG Channels

Although single MG channels have been activated by osmotic swelling of the cell (Hamill, 1983, Christensen, 1987; Ubl *et al.*, 1988), the most convenient and direct method is to apply suction/pressure to the suction port of the patch pipette holder (Hamill *et al.*, 1981; Guharay and Sachs, 1984). A variety of different means have been used to do this, including the use of mouth, syringe, water aspirator and a thumbwheel-driven piston (see McBride and Hamill, 1992, for references). Although these techniques may be adequate for identifying the channel as MS or studying its steady-state characteristics, they are inadequate in providing rapid and precise pressure/suction waveforms. Furthermore, in many of the early studies, careful attention was not given to the time course of the pressure stimulation, and the output of a pressure transducer, if used, was not shown, and instead stimulation was indicated by arrows or bars. Over the last few years a number of laboratories have reported improved methods for applying suction/pressure to the patch pipette. Unfortunately, published details concerning these methods have been scarce. The first development of a feedback-controlled pressure application system was centered on an oil-based closed system (F. Sachs and D. Borkowski, personal communication; see also Sokabe and Sachs, 1990). The strengths of this oil-based system are its speed (~2 msec) and absence of any external pressure/suction sources. The disadvantages of the system are that it is sensitive to air bubbles and, in terms of routine usage, it is messy and inconvenient (F. Sachs, personal communication). The same laboratory also built an air-coupled pressure-clamp system in which a linear motor from a multiheaded computer disk drive was used to drive a syringe (Sachs, 1987). This system had a rise time of ~30 msec but was abandoned in favor of the oil-filled method (F. Sachs, personal communication). Our laboratory also tried a number of strategies before settling on the air-based pressure-clamp system. These included a microprocessor-controlled piston system that gave reproducible suction/pressure waveforms (Lane *et al.*, 1991), a modified commercially available pressure injection system (Picospritzer, General Valve), and a speaker (woofer) coupled to a syringe.

The basic strategy of the pressure clamp as described in McBride and Hamill (1992) is that the desired pressure applied to the patch is achieved by a balancing of pressure and suction. Central to this balancing is the use of a proportional piezoelectric valve whose opening is proportional to the applied voltage. Through feedback control of this valve, the amount of pressurized N_2 allowed to enter a mixing chamber can be regulated to balance the constant outflow through a continual vacuum efflux. Originally, a Maxtek MV-112 piezoelectric valve was used. This valve has also been employed by Denk and Webb (1992) to stimulate hair cells by a water microjet method and subsequently by Opsahl and Webb (1994) to stimulate alamethicin channels in bilayers formed by the tip-dip method. Although the Maxtek valve does work adequately, we have subsequently found another valve manufactured by the Lee Company (see below) that is overall better suited for the pressure clamp (i.e., in its speed and compactness) and is also less expensive.

3. Mechanical Arrangement of the Improved Pressure Clamp

Figure 1 is a schematic illustrating the mechanical arrangement of the fast pressure clamp. The new elements in this clamp (cf. McBride and Hamill, 1992) are the dual Lee valve arrangement and the compact mixing chamber/transducer. The basic principle of the valve action has also been illustrated. The system volume whose pressure is being controlled is represented by the lightly shaded regions extending from the output ports of both valves into the mixing chamber and the tubing to and including the patch pipette and holder. Simple analysis of the system using the ideal gas law ($PV = nRT$, where V is the system volume, and others terms have their usual meanings; see also McBride and Hamill, 1992) reveals that the speed with which the pressure can be changed is directly proportional to the change in the input or output flux and inversely proportional to the volume of the system ($dP/dt = (RT/V)(dn/dt)$). Therefore, the modifications of the original pressure clamp centered around decreasing the system volume and increasing the speed of the valve.

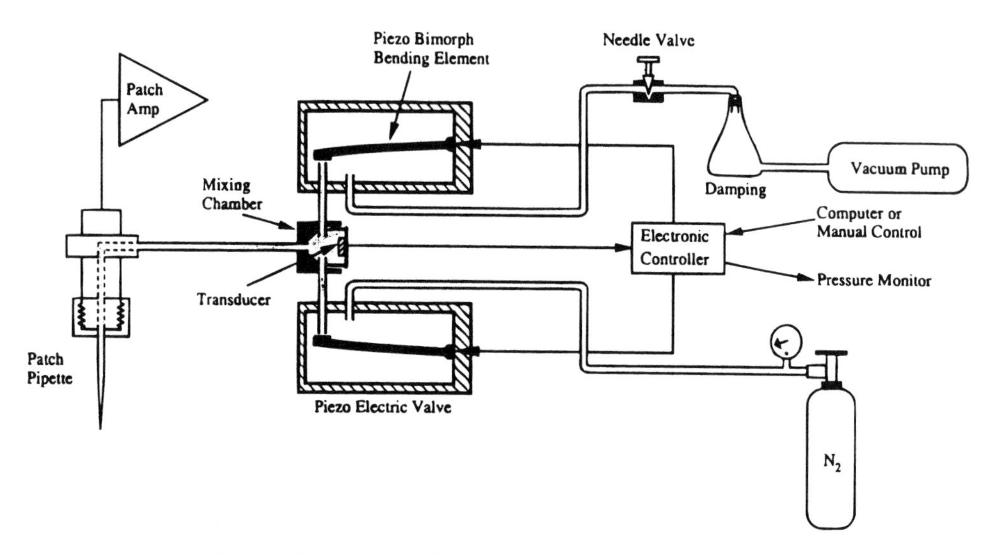

Figure 1. Schematic diagram of the mechanical arrangement of the pressure clamp. The patch-pipette holder and amplifier are shown on the left. The suction port of the pipette holder is connected by a tube to the mixing chamber. The mixing chamber contains the transducer as well as the ports leading directly to each piezoelectric valve. The basic mechanism of the piezoelectric valve is illustrated in which the position of the bimorph is dependent on the applied voltage. The two valves are controlled reciprocally (i.e., as one closes the other opens and vise versa). The upper valve is connected to the vacuum pump through a needle valve (to fine-adjust the suction) and damping flask (necessary to filter out cycle-to-cycle pump fluctuations). The lower valve is directly connected to the regulator on the N_2 tank.

To this end we replaced the Maxtek valve with the LFPA Ultra-Speed Valve from the Lee Company. This valve is considerably smaller (total volume 8 ml versus 170 ml), lighter (13 g versus 800 g), and faster (a response time of ~0.5 msec versus ~2 msec) than the Maxtek. The only disadvantage of the Lee valve is that it is rated for a maximum differential input pressure of 10 psig, whereas the Maxtek valve is rated at 50 psig. We routinely use the Maxtek valve with a N_2 pressure of 60 psig because we found that this pressure is needed to compensate for the high efflux from the mixing chamber provided by the constant vacuum. This high efflux is required to produce a fast suction pulse. However, if the volume, V, is decreased, then the flow rate (dn/dt) can also be decreased without changing dP/dt (see above equation).

Thus, if the system's volume can be reduced sufficiently, flow rate can be reduced, and smaller injection (N_2) pressures can be used. We have done this by (1) using the smaller valve (i.e., the Lee valve) and (2), as described below, incorporating the transducer into the mixing chamber (i.e., the sensor element of the transducer forms one wall of the mixing chamber). These two modifications permit closer proximity of the pressure clamp to the pipette holder and thus minimize the connecting tubing. Elimination of the three-way valve (see McBride and Hamill, 1993) further diminishes the dead space.

In order to consolidate the transducer/mixing chamber, we started with a SenSym transducer, which is enclosed in a metal, transistor-type TO-39 package. The actual metal can has a diameter of 8.25 mm and a height of 7.1 mm. Normally the top of the can has a small-diameter hole that gives access to the sensor. Cutting off the top half of the can exposes the sensor element itself, and the remaining bottom half of the can containing the leads may be conveniently press-fitted into the end of a cylindrically shaped mixing chamber. The mixing chamber has three ports in it. Two ports are diametrically opposed and connect to the piezo valves. The third port is along the axis of the cylinder and opposite the sensor and connects to the pipette holder. With regard to the third port, care must be taken to prevent damage to the sensor element. The total volume of the mixing chamber is about 250 μl.

In the original pressure clamp, only a single piezoelectric valve was used to control the N_2 influx into a constant vacuum. This strategy does work, but it is wasteful and somewhat inconvenient, as one standard-size N_2 tank can be used in a single day's experiments. By including a second piezo valve situated between the mixing chamber and the vacuum source the efflux is controlled electronically, and overall efflux is reduced. To achieve this the electronic controller has been modified to reciprocally control both the efflux and influx of N_2 into the mixing chamber (i.e., one valve opens as the other closes). Therefore, where the original pressure clamp was an open system, the dual-valve arrangement of the new pressure clamp more closely approximates a closed system. Furthermore, the dual-valve system also contributes to the increase in speed of the transition times because the injection of N_2 does not have to overcome the constant N_2 efflux from the mixing chamber as in the old system.

Figure 2A is a photograph of the patch-clamp rig incorporating the fast pressure clamp with the various components as described above. The weight and size of the Lee valves and mixing chamber permit the pressure clamp to be placed on the stage in close proximity to the patch-pipette holder. Figure 2B is a close-up view of an isolated patch-pipette holder connected to the pressure clamp. The modified transducer/mixing chamber with three ports is shown between the two valves.

4. Electronic Control of the Pressure Clamp

The basic design of the electronic controller of the pressure clamp has been retained (McBride and Hamill, 1992). Figure 3 shows the schematic of the electronic controller. It can be divided into four sections. The first section is the voltage command. There are two internal inputs and one external input. The internal control includes both the manual (i.e., potentiometer) and an adjustable test pulse, which can be switched in or out. The test pulse is used during the adjustment procedure for optimizing the step response. This pulse could have been supplied externally. However, because it is frequently used, it has been incorporated into the controller. A combination of an LM555 timer (National Semiconductor, Santa Clara, CA) and an AD633 multiplier chip (Analog Devices, Norwich, MA) was used to generate and scale the test pulse.

The second section of the controller is responsible for pressure measurement. The transducer used is the SCC15GSO (SynSym, Inc., Milpitas, CA). The modifications in incorporating this into the mixing chamber have been discussed above. An LM334 current regulator (National Semiconductor) was used to excite the transducer bridge and was set at a current of ~1 mA. The AD620 (Analog Devices) instrumentation amplifier with a gain of ~500 was used to measure the output of the bridge. The AD620 is an eight-pin chip that is convenient to use. The output of the AD620 is further amplified to give the desired gain of 100 mV/mm Hg. This is followed by an optional filter to remove high-frequency noise. The filter is centered around an LTC1064-3 chip, an eight-order low-pass Bessel filter, along with the 74LS624 (Texas Instruments, Dallas, TX) and necessary op-amps (see application notes for the LTC1064-3 from Linear Technology, Milpitas, CA).

The third section is the feedback control section and is composed of a summation amplifier used to generate the error signal (between the desired and actual pressures), followed

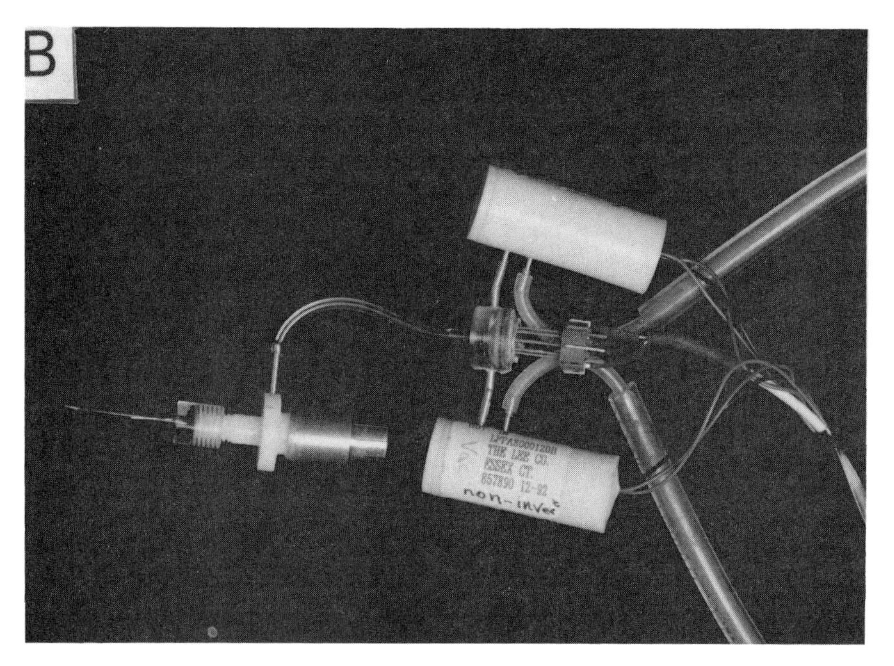

Figure 2. Photographs of the fast pressure-clamp/patch-clamp setup. A: View of the orientation of the pressure-clamp arrangement with respect to the patch-clamp headstage and pipette holder. The small size and weight of the two valves and interposed transducer/mixing chamber that make up the pressure clamp allow the system to be placed on the stage platform. A tube as short as possible yet enabling the pipette to be changed connects the mixing chamber to the suction port of the pipette holder. Tubes to the right of the valves connect to the vacuum and N_2 tank, and single flexible cable connects the transducer and the valves to the pressure-clamp controller. B: Close-up view of the isolated pressure clamp connected to a patch-pipette

by integration or averaging (with compensation) of this error signal. Three adjustable resistors have been included to adjust the response. This may be somewhat redundant, but it does allow flexibility in adjusting the desired waveform. The final stage is the piezo-driver section. This includes two high-voltage op-amps (PA84S from Apex, Tucson, AZ), which are powered at ±80 V. The control signal for the valve controlling the pressure from the N_2 tank has been inverted, resulting in reciprocal control of both valves.

5. Performance of the Improved Clamp

Figure 4 illustrates the fast transition times obtainable with the improved pressure clamp. Whereas with the original clamp transition times as measured by the 20%-to-80% risetime were around 5–7 msec when optimally adjusted, the fast clamp has submillisecond transition times as fast as ~350 μsec with moderate ringing. With less feedback, slower but still submillisecond risetimes can be obtained with no ringing. On the other hand, with more feedback and consequently stronger ringing, transition times as fast as 250 μsec have been observed.

Figure 5A illustrates the operation of the clamp in applying small incrementing suction/pressure steps. In this particular case the pressure increments are 1 mm Hg (10 mm Hg = 1.33 kPa). The pressure rms noise evident in the pressure traces is ~0.1 mm Hg, and this presumably represents the limitation for the minimal distinguishable step size that can be applied by the clamp for the present volume of the system. This sensitivity is adequate for many MG channels, which show half-saturation pressure of 10–15 mm Hg. However, some MG channels have been reported to have half-saturation pressure as low as 1–2 mm Hg (Sackin, 1989; Kim, 1993). For these more sensitive MG channels, the minimal pressure increments could be reduced by increasing the mixing chamber volume (see below).

Figure 5B shows the operation of the clamp at higher pressures. A slight ringing can be seen at a transition time of 1 msec. This ringing at the largest stimulations can be completely eliminated when the clamp is adjusted (see section III of the electronic controller, Fig. 3) to give slower transition times of 1.5–2 msec. At the current gain of 100 mV/mm Hg, the practical pressure limit of the clamp is ±100 mm Hg. In studies using standard-sized patch pipettes (i.e., 1–2 μm), this upper limit value is sufficient for determining stimulus–response relationships of MG channels, since most MG channels studied saturate at <50 mm Hg (Sokabe and Sachs, 1992). Furthermore, this limit exceeds typical patch rupture pressures. However, there are some reports of MG channels that have saturation pressure exceeding ~100 mm Hg (Vandorpe *et al.,* 1994). To achieve higher pressures, the gain of the pressure measurement section could be reduced to, for instance, 50 mV/mm Hg. This would double the practical range to ±200 mm Hg.

---

holder. This enlarged view shows the plastic mixing chamber with three ports into which the pressure transducer has been press fitted. Note that the transducer is plugged into a transistor socket to which controller leads have been soldered. The two valves controlling pressure and vacuum are shown on each side with associated tubing connecting to the N_2 tank and vacuum pump (not shown) as well as the electrical connections to the controller.

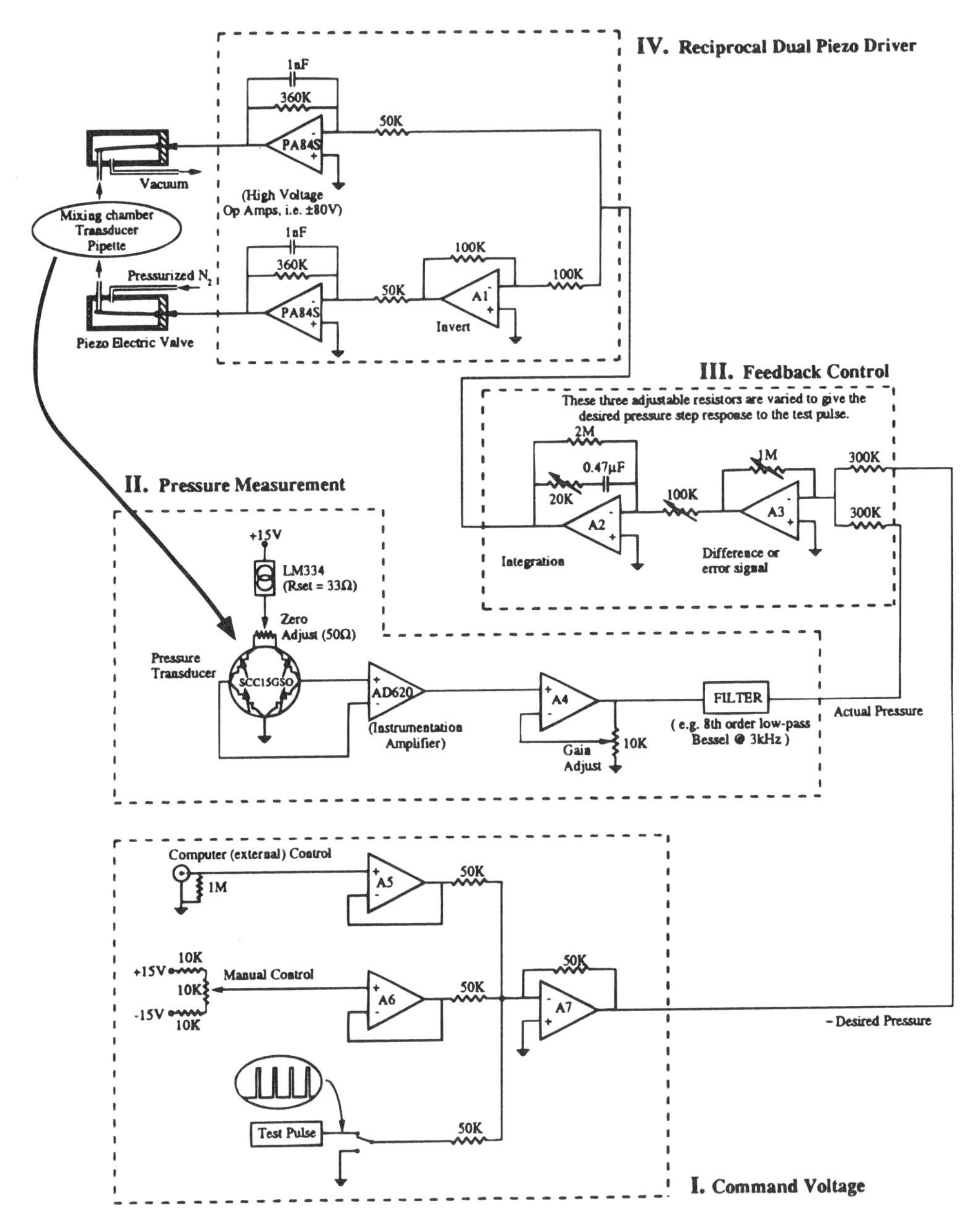

Figure 3. Electronic control of the pressure clamp. The circuit can be divided into four blocks. I: Command voltage. This represents the desired pressure and includes internal and external sources. The internal sources include a manual control for setting the steady-state values and a variable test pulse that can be turned on and off by a toggle switch. The test pulse is used in the step response adjustment procedure. It could have been supplied externally, but, as it is frequently used, for convenience it is incorporated into the circuitry. II: Pressure measurement. This measures the pressure in the mixing chamber. A constant-current source excites the transducer bridge, whose output is measured by the instrumentation amplifier. A voltage follower is included to enable gain adjustment, and the filter is included to remove high-frequency noise. III: Feedback control. This section is composed of an initial summing amplifier that generates an error signal, followed by an integrating amplifier. In this section there are three adjustable resistors. Although to some extent there

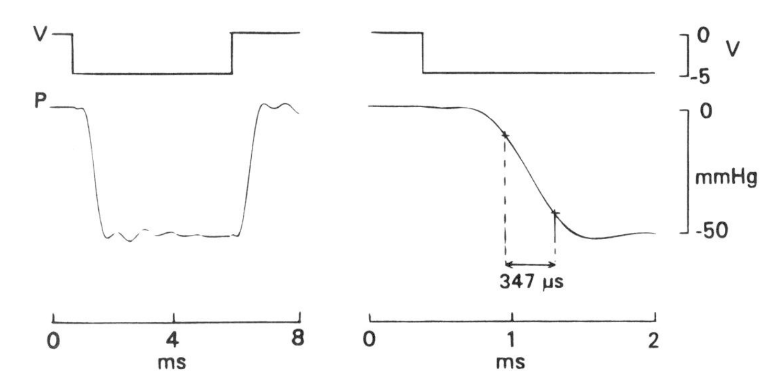

Figure 4. Illustration of the speed of the pressure clamp. In each panel the upper trace represents the command voltage, and the lower trace the pressure waveform. The left hand panel show the full 5-msec duration suction pulse of −50 mm Hg. The right hand panel shows an expanded region of the transition region. The 20% to 80% transition time is ~350μsec. In this case the feedback has been slightly overcompensated to speed up the transition time at the expense of some slight ringing after the transitions.

6. Theoretical Constraints on the Speed and Noise of the Pressure Clamp

Physical factors influencing the speed of the clamp have been previously discussed (McBride and Hamill, 1992). Here we emphasize three constraints on clamp speed. The first is the speed of sound in air (330 m/sec). At this speed a pressure wave would require ~200 μsec to travel 6 cm, which is the approximate length of the tube from the mixing chamber to the pipette tip. Another limitation is the response time of the transducer, which is 100 μsec (from 10% to 90% of full scale). The third limitation is the response time of the piezo valve, which is around 500 μsec for complete opening and 300 μsec for complete closing (*Handbook,* Lee Co., 1991). It appears that the speed of the clamp (300–500 μsec for 20%-to-80% transition) is close to the physical limitations of the system for the present configuration and seems to be limited by the response time of the piezo valve.

In considering noise limitations and sensitivity (i.e., the minimal distinguishable step size), it should be pointed out that there is a reciprocal relationship between these two and the speed of the clamp with regard to the volume of the system. On the one hand, larger volumes decrease the speed (i.e., for a given flux it takes longer to change the pressure in a larger volume). On the other hand, a larger volume stabilizes the pressure of the system, making it less sensitive to fluctuations in input or output flux. This decreases the pressure noise of the system, and also, because it requires a larger change in flux for a given change in pressure, the control can be more sensitive, albeit slower. Our preference has been to optimize the time response of the clamp, since oocyte and muscle MG channels are activated by moderate pressures (~10–20 mm Hg).

may be some redundancy, they do allow flexibility in terms of shaping the desired waveform (i.e., step response). IV: Reciprocal dual piezo driver. This section includes two high-voltage operational amplifiers operated with a gain of about 7 and a low-pass filter. The integrated signal from section III drives one amplifier directly while the other is driven by the inverted integrated signal. Thus giving the reciprocal control.

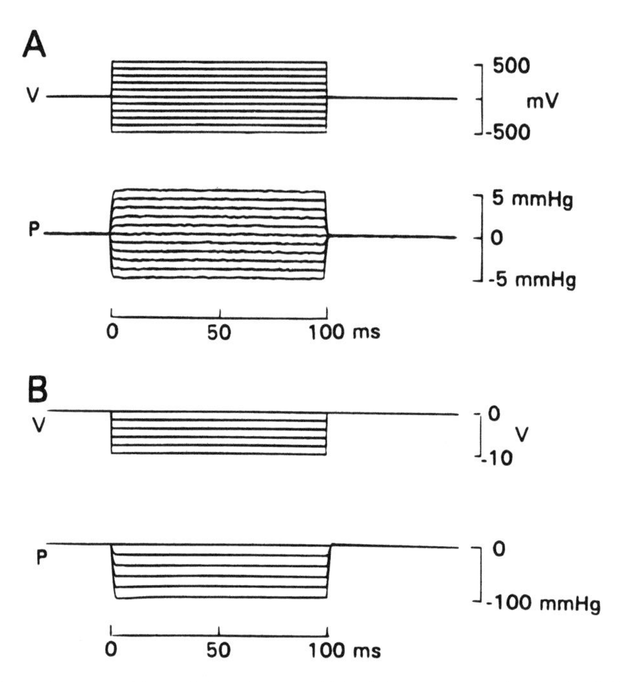

Figure 5. Illustration of the sensitivity (A) and range (B) of the pressure clamp. A: A family of 100-msec pressure/suction steps ranging from 5 to −5 mm Hg in 1-mm Hg increments. The 20%–80% transition time was <1 ms. B: Suction pulses extending up to −100 mm Hg in 20-mm Hg steps. Again a rise time of ~1 msec. Note a slight overshoot after the transition, which can be eliminated by increasing the transition time.

7. Pressure-Clamp Use for Sealing Protocols

Over the last two decades, the patch-clamp technique has been used on a wide range of cell types to address a variety of questions related to membrane transport processes. Common to all patch-clamp experiments is the initial tight-seal formation, which is typically achieved by mouth-applied suction. Although this procedure is adequate in achieving the tight seal, it lacks a certain precision and reproducibility in terms of the mechanical stresses applied to the patch. There have been mixed reports concerning the integrity of the membrane–cytoskeleton complex in sealed patches (Milton and Caldwell, 1990; Sokabe and Sachs, 1990; Ruknudin *et al.*, 1991). However, in the case of voltage- and ligand-gated channels, there has been general agreement (with a few exceptions) between results obtained using whole-cell, single-channel patch-clamp and conventional intracellular voltage-clamp recording techniques. In contrast, with MG channels there are a number of reports that indicate specific changes such as the loss of adaptation and sensitivity in MS channel properties caused by mechanical stresses associated with either sealing and/or mechanical stimulation of the patch (Hamill and McBride, 1992). We have found that using the pressure clamp to apply low (less than 1–2 mm Hg) and reproducible pressure/suction protocols for tight-seal formation leads to more consistent MG channel behavior in terms of adaptation and sensitivity. Perhaps extension of this technique to sealing protocols in studying other channels may remove or reduce some discrepancies and variabilities that have been reported (see Kimitsuki *et al.*, 1990; Johnson and Byerly, 1993; Rosenmund and Westbrook, 1993).

ACKNOWLEDGMENTS. We would like to thank W. B. Davis and W. Drachenburg for their assistance in building the fast pressure clamp. We acknowledge the Muscular Dystrophy Association, the NIH, and the NSF (Instrument Development for Biological Research) for their support.

References

Christensen, O., 1987, Mediation of cell volume regulation by Ca^{2+} influx through stretch-activated channels, *Nature* **330**:66–68.

Denk, W., and Webb, W. W, 1992, Forward and reverse transduction at the limit of sensitivity studied by correlating electrical and mechanical fluctuations in frog saccular hair cells, *Hearing Res.* **60**:89–102.

Franke, C., Hatt, H., and Dudel, J., 1991, Steep concentration dependence and fast desensitization of nicotinic currents elicited by acetylcholine pulses, studied in adult vertebrate muscle, *Pflügers Arch.* **417**:509–516.

French, A. S., 1992, Mechanotransduction, *Annu. Rev. Physiol.* **54**:135–152.

Guharay, F., and Sachs, F., 1984, Stretch-activated single ion channel currents in tissue-cultured embryonic chick skeletal muscle, *J. Physiol.* **352**:685–701.

Hamill, O. P., 1983, Potassium and chloride channels in red blood cells, in: *Single Channel Recordings* (B. Sakmann and E. Neher, eds.), pp. 451–471, Plenum Press, New York.

Hamill, O. P., and McBride, D. W., Jr., 1992, Rapid adaptation of the mechanosensitive channel in *Xenopus* oocytes, *Proc. Natl. Acad. Sci. U.S.A.* **89**:7462–7466.

Hamill, O. P., and McBride, D. W., Jr., 1994a, The cloning of a mechano-gated membrane ion channel, *Trends Neurosci.* **17**:439–443.

Hamill, O. P., and McBride, D. W., Jr., 1994b, Molecular mechanisms of mechanoreceptor adaptation, *News Physiol. Sci.* **9**:53–59.

Hamill, O. P., Marty, A., Neher, E., Sakmann, B., and Sigworth, F. J., 1981, Improved patch clamp techniques for high-resolution recording from cells and cell-free membrane patches, *Pflügers Arch.* **391**:85–100.

Hille, B., 1992, *Ionic Channels of Excitable Membranes,* 2nd ed., Sinauer Associates, Sunderland, MA.

Howard, J., and Hudspeth, A. J., 1987, Mechanical relaxation of the hair bundle mediates adaptation in mechanoelectrical transduction by the bullfrog's saccular hair cell, *Proc. Natl. Acad. Sci. U.S.A.* **84**:3064–3068.

Howard, J., Roberts, W. M., and Hudspeth, A. J., 1988, Mechanoelectrical transduction by hair cells, *Annu. Rev. Biophys. Biophys. Chem.* **17**:99–124.

Hudspeth, A. J., and Corey, D. A., 1977, Sensitivity, polarity and conductance change in the response of the vertebrate hair cells to controlled mechanical stimulation, *Proc. Natl. Acad. Sci. U.S.A.* **74**:2407–2411.

Jaramillo, F., and Hudspeth, A. J., 1993, Displacement-clamp measurement of the forces exerted by gating springs in the hair bundle, *Proc. Natl. Acad. Sci. U.S.A.* **90**:1330–1334.

Johnson, B. D., and Byerly, L., 1993, A cytoskeletal mechanism for Ca^{++} channel metabolic dependence and inactivation by intracellular Ca^{++}, *Neuron* **10**:797–804.

Kim, D., 1993, Novel cation-selective mechanosensitive ion channel in the atrial cell membrane, *Circ. Res.* **72**:225–231.

Kimitsuki, T., Mitsuiye, T., and Noma, A., 1990, Negative shift of cardiac Na^+ channels kinetics in cell-attached patch recordings, *Am. J. Physiol.* **258**:H247–H254.

Lane, J. W., McBride, D. W., Jr., and Hamill, O. P., 1991, Amiloride block of the mechanosensitive cation channel in *Xenopus* oocytes, *J. Physiol.* **441**:347–366.

Lee Company, 1991, *Electro-Fluidic Systems Technical Handbook,* 5th ed., The Lee Company Technical Center, Westbrook, CT.

Maconochie, D. J., and Knight, D. E., 1992, A study of the bovine adrenal chromaffin nicotinic receptor using patch clamp and concentration-jump techniques, *J. Physiol.* **454**:129–153

Martinac, B., 1993, Mechanosensitive ion channels: Biophysics and physiology, in: *Thermodynamics of Cell Surface Receptors* (M. B. Jackson, ed.), pp. 327–352, CRC Press, Boca Raton, FL.

McBride, D. W., Jr., and Hamill, O. P., 1992, Pressure-clamp: A method for rapid step perturbation of mechanosensitive channels, *Pflügers Arch.* **421**:606–612.

McBride, D. W., Jr., and Hamill, O. P., 1993, Pressure-clamp technique for measurement of the relaxation kinetics of mechanosensitive channels, *Trends Neurosci.* **16**:341–345.

Milton, R. L., and Caldwell, J. H., 1990, How do patch clamp seals form? A lipid bleb model, *Pflügers Arch.* **416:**758–765.
Morris, C. E., 1990, Mechanosensitive ion channels, *J. Membr. Biol.* **113:**93–107.
Opsahl, L., and Webb, W. W., 1994, Transduction of membrane tension by the ion channel alamethicin, *Biophys. J.* **66:**71–74.
Ordway, R. W., Petrou, S., Kirber, M. T., Walsh, J. V., and Singer, J. J., 1992, Two distinct mechanisms of ion channel activation by membrane stretch: Evidence that endogenous fatty acids mediate stretch activation of K^+ channels, *Biophys. J.* **61:**A390.
Rosenmund, C., and Westbrook, G. L., 1993, Calcium-induced actin depolymerization reduces NMDA channel activity, *Neuron* **10:**805–814.
Ruknudin, A., Song, M. J., and Sachs, F., 1991, The ultrastructure of patch-clamped membranes: A study using high voltage electron microscopy, *J. Cell. Biol.* **112:**125–134.
Sachs, F., 1987, Baroreceptor mechanisms at the cellular level, *Fed. Proc.* **46:**12–16.
Sachs, F., 1988, Mechanical transduction in biological systems. *CRC Crit. Rev. Biomed. Eng.* **16:**141–169.
Sackin, H., 1989, A stretch-activated K^+ channel sensitive to cell volume, *Proc. Natl. Acad. Sci. U.S.A.* **86:**1731–1735.
Sokabe, M., and Sachs, F., 1990, The structure and dynamics of patch-clamped membranes: A study using differential interference contrast light microscopy, *J. Cell Biol.* **111:**599–606.
Sokabe, M., and Sachs, F., 1992, Towards molecular mechanism of activation in mechanosensitive channels, in: *Advances in Comparative and Environmental Physiology* (F. Ito, ed.), Vol. 10, pp. 55–77, Springer-Verlag, Berlin.
Ubl, J., Murer, H., and Kolb, H-A., 1988, Ion channels activated by osmotic or mechanical stress in membranes of opossum kidney cells, *J. Membr. Biol.* **104:**223–232.
Vandorpe, D. H., Small, D. L., Dabrowski, A. R., and Morris, C. E., 1994, FMRFamide and membrane stretch as activators of the *Aplysia* S-channel, *Biophys. J.* **66:**46–58.

Chapter 15

Electrophysiological Recordings from *Xenopus* Oocytes

WALTER STÜHMER and ANANT B. PAREKH

1. Introduction

1.1. The Oocyte Expression System

A major goal of electrophysiology is to understand, at a molecular level, how an ion channel functions. How does ion permeation occur, how do channels activate and inactivate, how does it sense changes in the electric field, and how is the channel regulated? Our understanding of these fundamental processes has been severely hampered by the lack of a suitable experimental model system. Most cells simultaneously express a plethora of different channels, and it is therefore extremely difficult to study one type of channel in isolation. This is usually achieved by using complex voltage protocols and solutions rich in ion channel blockers and nonpermeant ions, neither of which is likely to be relevant physiologically. Moreover, it is not possible to artificially and systematically manipulate the channel gene in an intact cell, thereby precluding structure–function characterization. What is needed is a cell that has few endogenous ionic conductances and into which the channel under investigation can be exogeneously expressed. These criteria are adequately fulfilled by the *Xenopus* oocyte expression system. Following the key observation of Gurdon *et al.* in 1971 that foreign RNA injected into oocytes could be translated into proteins, Gundersen *et al.* (1983) and Miledi *et al.* (1983) were the first to demonstrate that a variety of receptors and channels from the central nervous system could be functionally expressed in the oocyte. The recent dramatic advances in both molecular biology, where proteins can be routinely cloned and mutated at specific loci, and electrophysiology (predominantly patch clamp) have combined to produce a powerful approach to the study of ion channels.

1.2. Advantages of the Oocyte System

1. Hundreds of viable cells can be isolated from a given donor frog. The cells can be surgically removed without sacrificing the animal, so one frog can be used several times.

WALTER STÜHMER • Max-Planck-Institute for Experimental Medicine, D-37075 Göttingen, Germany. ANANT B. PAREKH • Max-Planck-Institute for Biophysical Chemistry, Am Fassberg, D-37077 Göttingen, Germany.
Single-Channel Recording, Second Edition, edited by Bert Sakmann and Erwin Neher. Plenum Press, New York, 1995.

2. The cells are quite hardy and can survive for up to 2 weeks *in vitro*. The cells can tolerate repeated impalements of microelectrodes and injection pipettes. Moreover, relatively simple facilities are required for maintaining the cells, once isolated.
3. The cells are big (up to 1.3 mm in diameter) and can be easily injected with DNA, RNA, as well as membrane-impermeable drugs.
4. The oocytes faithfully express foreign RNA that has been injected into them.
5. The oocyte has only a few endogeneous channels (the major one being a Ca^{2+}-activated Cl^- channel), which usually carry only a small fraction of the current expressed. This permits a particular channel to be studied in virtual isolation.
6. Macropatch recordings are possible in oocytes. These enable low-noise, fast-clamp patch recordings of many channels that cannot be obtained using other expression systems.
7. Expression cloning into oocytes greatly speeds up the purification and isolation of RNA for a desired protein. Separating total brain RNA, for example, in fractions on the basis of size followed by injection of each fraction enables the rapid location of the protein's RNA to a particular band size.

1.3. Disadvantages of the Oocyte System

1. Because of its large size, whole-cell patch-clamp experiments, where one can control the intracellular ionic composition by dialysis across the patch pipette, are not possible.
2. The endogenous channels, although few, can interfere with current measurements if they are small (e.g., gating charge or mutants having little expression).
3. Posttranslational modifications may be different in the oocyte compared with the native cells. Hence, channels may actually function differently in their native environment.
4. In some laboratories, oocytes exhibit seasonal variation such that channel expression and ability to obtain seals are more difficult in the summer months.
5. It should be borne in mind that *Xenopus* is an amphibian, and the cells should only be studied at room temperature (18–22°C). At higher temperatures, the cells rapidly deteriorate. Most channels and receptors that are expressed in the oocyte are of mammalian origin. Since certain processes depend critically on temperature (oscillations in cytosolic free Ca^{2+}, rates of activation and inactivation of channels), experiments on oocytes may not be of physiological significance with respect to the cells from which the protein is derived.

Despite these limitations, the oocyte continues to be a useful and convenient expression system. Indeed, certain measurements (nonstationary gating analysis) are possible only in the oocyte.

In this chapter, we will first describe procedures for isolating and injecting cells followed by details of the recording techniques. The former have been covered extensively in recent reviews (Methfessel *et al.*, 1986; Stühmer, 1992), and therefore we will give only a brief overview in terms of a general introduction and summary of the techniques involved, extending into more recent developments such as the macropatch, cut-open oocyte, and the giant patch technique (see also Chapter 13, this volume).

2. Procedures and Techniques

2.1. Isolation of Single Oocytes

2.1.1. Stages of Oocytes

Xenopus oocytes can be crudely classified into three developmental stages: first, immature cells; second, mature (arising from immature cells that have re-entered meiosis by lutenizing hormone-induced release of progesterone from the surrounding follicular cells). In the oviduct meiosis is stopped, and the cells are now termed eggs. The final stage is fertilization, occurring when the egg is inseminated.

The immature cells are widely used for expression of channels and receptors, and therefore we will restrict our discussion to this type.

Immature cells have six developmental stages: stages I to VI (Dumont, 1972), the latter stage being the fully grown one (1.2–1.3 mm diameter). Stages I to III are small and lack demarcation between the animal and vegetal pole. At stage IV, separation between the dark brown animal pole (with a high concentration of melanin-containing pigment granules) and the cream/white vegetal pole is clear, although some pigment is still associated with the latter. The poles are well defined in the large stage V and VI cells. The oocytes are surrounded by several structures. From proximal to distal, these are the vitelline membrane (which must be removed for patch-clamp experiments), the follicular layer (in which the follicular cells communicate with the oocyte through gap junctions), and a connective tissue layer (the theca), in which the innervation and vasculature are located. To study implanted channels and receptors or even endogenous currents to the oocyte, it is necessary to remove the surrounding structures. This is because, first, it will provide better access of solutions and drugs to the oocyte itself and, second, it is mechanically difficult to drive low-resistance electrodes through all the intervening cells into the oocyte. Instead, high-resistance (small diameter) electrodes would be needed, and this will increase the membrane time constant for charging (see below). Third, the follicular cells can affect the properties of the oocyte through gap junctions. Ca^{2+} and second messengers readily pass through. Moreover, follicular cells themselves are not inert but contain receptors (e.g., angiotensin II) and channels (delayed-rectifier K type), which will clearly complicate interpretation.

Removal of the follicular layer by enzymatic treatment is described in detail later.

Because of the large size of stage V cells, only simple arrangements are required for injection of RNA, and the cells can tolerate repeated impalements with microelectrodes and patch pipettes. This no doubt explains why this stage is generally chosen for the injection of channel-encoding RNA. One main drawback, however, is that the time constant (t) of charging the membrane capacitance is slow for big cells, which have a large capacitance (C_m). This is because t is related to C_m by the simple relationship: $t = R_s \cdot C_m$ where R_s is the series resistance (determined largely by the electrode resistance). For channels that activate and inactivate very quickly, important kinetic information is therefore lost during the settling time of the clamp. This can be reduced by use of stage III cells, where an eightfold decrease in capacitance and hence speed of the clamp has been reported (Krafte and Lester, 1992). Stage III oocytes are more difficult to handle though, and require specialized techniques for injection of RNA.

One further area of concern with stage VI oocytes is that they possess numerous microvilli and cristae. Dascal (1987) has estimated that the specific membrane capacitance is around 4–7 $\mu F\ cm^{-2}$, if one assumes that the oocyte is a perfect sphere. This is much larger than the usual 1 $\mu F\ cm^{-2}$ estimated for most other cells and would suggest that the oocyte is not a perfect sphere but instead possesses numerous microvilli, which dramatically affect membrane surface area. If channels are expressed on microvilli and other evaginations, and there is no *a priori* reason against this, then not only will the clamp settling speed be slower, and this part of the plasma membrane may not be adequately space clamped, but also kinetic properties of the channel may change in addition to ion accumulation effects affecting the equilibrium potential. In a cell-attached patch on such a membrane, inactivating channels (depending on the location) will inactivate at different rates after a step change in the clamp because of spatial voltage inhomogeneities, and this will manifest itself as a slowly or even partially noninactivating macroscopic patch current. On excision, where contacts with the cytoskeleton and other cytoplasmic components are lost, the microvilli may "open out" and relax such that the clamp speed is faster and more uniform. The macroscopic current will now look rather different and more accurately reflect the properties of the channels themselves. This "microvilli" effect will have a relatively slow onset (probably on a tens of seconds to minutes time scale) and should be borne in mind when slow changes in current are followed in excised patches. On the other hand, interactions with the cytoskeleton might also affect channel performance, and the question arises as to what is the more physiological situation.

Another interesting property of stage VI oocytes relates to ion channel distribution. Ca^{2+}-dependent Cl^- channels, which are endogenous to the oocyte, are found in high density at the animal pole (Lupu-Meiri *et al.,* 1988). Hence, the oocyte has the intrinsic ability to target integral membrane proteins to specific regions. It follows, therefore, that channel density may not be constant but fluctuate. Indeed, following expression of certain K^+ channels, some patches have no current at all, whereas others have large ones, indicating channel confinement to hot spots (Peter *et al.,* 1991). One may have to use several pipettes on an oocyte before locating the region with many channels. Moreover, if channel density is too high, it becomes rather difficult to obtain single-channel recordings. One way around this would be to dilute the RNA so expression is reduced. Not all channels are clustered, though. Na^+ channels seem to be rather uniformly distributed, as may be the Ca^{2+} influx pathway activated by emptying IP_3 stores (Girard and Clapham, 1993).

2.1.2. Maintenance of *Xenopus laevis*

The South African clawed frog, *Xenopus laevis,* can now be obtained commercially from several international companies based in America, Europe, and Japan. On ordering frogs, our experience is that larger ones tend to produce better oocytes than smaller ones. Once obtained, the frogs are relatively simple to keep. The main concerns are temperature and light control as well as treatment of the water. Although frogs can withstand large variations in temperature, the quality of the cells decreases as temperature is raised. Ideally, the frogs should therefore be maintained below 20°C. Furthermore, the frogs should be exposed to a continuous light-dark cycle of 12 hr each. For unknown reasons, this is important for the viability of the cells. Frogs are also sensitive to chlorine and chloramine (a mixture of NH_3 and Cl_2), both present in tap water. These can be removed using high-purity carbon or Barnstead organic cartridge filters.

2.1.3. Operation

Oocytes can be surgically removed from the frog without the animal needing to be killed. First, a healthy-looking frog is selected that has not been operated on for at least 4 weeks. The frog is then placed in a polystyrene box (about 30 × 25 × 20 cm) that contains an ice/water slurry of about 1:1. It is important to have the lid secured tightly (by a heavy object, for example) to prevent the frog from escaping. The cold water slows down the metabolism and makes the frog "drowsy". After 15–20 min, the frog is then transferred to an ice/water slurry containing anesthetic. We routinely use the methane sulfonate salt of 3-aminobenzoic acid ethyl ester (tricaine from Sigma; 1.5 g/l) because it is easily soluble and can be used repeatedly if stored at 4°C between uses. Around 20–30 min exposure to the anesthetic is usually sufficient (depending on the size of the frog). The extent of anaesthesia can be tested in several ways, including the ability to right itself when placed on its back and the reflex dilation of the hind foot on stroking.

Once anethesized, the frog is placed on its back on a tray containing ice, and a tissue paper (immersed in anesthetic) can be draped over the head. This ensures that the frog does not awake during surgery. A small abdominal incisure is made with a sharp razor blade, and the underlying fascia is also cut. Parts of the ovaries can usually be seen, and bundles of oocytes are gently teased out with forceps. As the clumps of cells are extracted, they should be cut into small pieces of around 20–50 cells before being placed into Ca^{2+}-free Barth's medium (composition given in Section 2.2.3). After enough cells have been removed, the remaining ovary is pushed back into the abdomen, and the excision is closed. First the fascia is stitched together, and then the skin is closed. If the fascia is not properly sutured, there is a high chance of infection. The frog is then briefly washed under tap water in order to wash out anesthetic and then transferred to a container containing only water. It is vital that the level of water is not deep enough to drown the frog, since it is slowly recovering from anesthesia, and frogs are air breathers. Once the frog has recovered, it should be transferred to a tank and allowed to recuperate for a couple of days before it is returned to the colony. This is necessary in case the frog becomes infected shortly after operation and could transmit this to the other frogs.

2.1.4. Isolation

The oocytes extracted from the ovary consist of many cells held together by connective tissue and are each surrounded by follicular cells. For both injection and recording, it is necessary to isolate each oocyte. If a small number of cells are required, this can be done manually either by dissecting away the surroundings with microscissors or pulling them off with forceps. This is too laborious if many cells are needed, and enzymatic treatment is therefore preferable. Collagenase is widely used for this purpose. Prolonged exposure is deleterious, and therefore the time in collagenase must be monitored carefully. Moreover, collagenase is usually impure, and it is necessary to test different batches to find the ideal conditions for each one. We use 3 mg/ml type II collagenase (from Sigma) for 3 h in Ca^{2+}-free Barth's medium at 20°C. The oocytes are immersed in this solution as small clumps of cells (20–50) and then placed in a shaking bath at room temperature. The dissociation can be followed by withdrawing cells at different times by means of a blunt Pasteur pipette and observing them under the microscope. Some groups use collagenase in Ca^{2+}-containing Barth's medium. Since Ca^{2+} activates proteases, the time of incubation is usually briefer (around 1 h) and should be carefully monitored. Our impression is that a larger fraction of healthier cells are obtained using Ca^{2+}-free solution. Once the dissociation is complete, the cells should be washed thoroughly in Ca^{2+}-free solution to remove both the collagenase and lysed cells before Ca^{2+} is readmitted as normal Barth's medium. Healthy cells can be observed under a microscope and generally appear as round cells with smooth, homogeneous dark brown and white hemispheres. Collagenase treatment sometimes does not wholly remove the follicular layer, but this can easily be done by means of forceps. For injection, the large stage VI cells are widely used.

2.2. Injection of Oocytes

2.2.1. Sources of RNA

The RNA injected into the oocyte can be total tissue RNA, poly(A) mRNA extracted from tissue samples, or cDNA-derived mRNA (cRNA). Care is also needed in handling and storage of the RNA. It is beyond the scope of this review to describe how such RNA is prepared, but interested readers should consult recent reviews by Snutch and Mandel (1992), Goldin and Sumikawa (1992), and Stühmer, (1992). Instead, we would like to mention briefly a few pertinent points concerning each source.

Although total and poly(A) mRNA isolation requires less preparation than cRNA, it has drawbacks. First, the desired mRNA will be only a small fraction of the total RNA, hence reducing the expression of protein under interest [although size fractionation of poly(A) mRNA can increase the relative amount of desired mRNA]. Second, other proteins will be translated in addition to the protein of interest. For example, suppose we are interested in expressing a Na^+ channel from Purkinje fibers and therefore inject total Purkinje mRNA. On depolarizing the oocyte, the total current will reflect not only the Na^+ current of interest but all other voltage-dependent channels expressed in the Purkinje fiber. One would need appropriate ionic conditions and pharmacological inhibitors to isolate the Na^+ current. Third, structure–function characterization will not be possible, since one is not mutating a specific mRNA.

Injection of cRNA is more elegant, but even this approach is not free from criticism. Most channels do not consist of a simple subunit or multimers of a single component (which would be induced in cRNA experiments) but instead are heteromeric. Dihydropyridine-sensitive Ca^{2+} channels have in $\alpha_1\alpha_2\beta\gamma\delta$ stoichiometry, adult nicotinic ACH receptors $(\alpha)_2\beta\delta e$ and even the Na^+ channel has a β subunit (important for inactivation) in addition to the pore-forming α subunit. Conclusions drawn from studies of cRNA encoding one component of a channel may be of little relevance to the situation in the native cell.

2.2.2. Injection of RNA

In handling mRNA, extreme care must be used at all stages to prevent contamination with RNases and to prevent particles from clogging the injection pipettes. RNase exposure can be reduced by using sterile conditions (e.g., wearing gloves), baking all glassware, and using doubly distilled autoclaved diethyl pyrocarbonate-treated water for all solutions. To precipitate particles present in the RNA solution, the RNA is centrifuged at 25 *g* for 10 min at 4°C. Before spinning, the RNA should be allowed to thaw for a couple of minutes, since it ought to be stored at −80°C in silanized Eppendorf tubes.

Shortly after centrifugation, the RNA should be transferred to the injection pipette. This is a two-step process in which RNA is first taken from the Eppendorf into a capillary and then passed on to the injection pipette. Capillary tubes such as transpipettor tubes (disposable micropipettes, Brand, Malsfeld, Germany) can be used. The tubes should be baked for 2 hr at 180°C to remove RNases, and it is helpful but not essential to silanize them. Once baked, the capillaries should be wrapped in silver foil and stored under sterile conditions until used.

The RNA is taken up into the capillary via a micrometer-controlled syringe containing mineral oil (to reduce the air-filled space in the syringe and associated tubing). Injection pipettes should be prepared shortly before use. We use a two-pull protocol on a standard patch pipette puller (Narashige, Tokyo, Japan) and 2 mm OD borosilicate glass (Hilgenberg, Wertheim, Germany). The current of the second pull should be quite strong so as to pull a finely tipped pipette possessing a long, thin shank. This enables a reasonable estimate of the volume injected, and one can be certain that the cell has indeed been injected by monitoring the displacement of the meniscus in the injection pipette. The tips of the pipettes are broken under a microscope until the opening diameter is around 5–10 μm. Obviously, the broken tip will be jagged and sharp and can easily rupture the oocyte. This is prevented by fire polishing and pulling out a sharp tip. To do this, a microfilament with a droplet of molten glass is brought close to the pipette. The pipette is then allowed to touch the hot microfilament but then is drawn away rapidly. This results in a sharp, needle-like tip and smoothing of the jagged rim by radiation heat. If the tip is too small or big, it can easily be rebroken and the process repeated. If the pipette is prepared shortly before use, the heat of the pulling process as well as that from the microfilament are sufficient to inactivate RNases. The pipettes can be stored in a pipette holder (as for patch pipettes) with a lid to prevent dust from soiling the tips.

The injection pipettes are then affixed to a hand-driven coarse manipulator, allowing movement in three dimensions. The open end of the pipette (opposite to the tip) is connected to a simple syringe by silicone tubing. Filling of the pipette with RNA occurs under a microscope. The pipette tip is maneuvered close to the capillary tube. A droplet of RNA is driven out of the capillary by the micrometer-controlled syringe. The pipette is carefully inserted into this droplet and filled by suction. Only a fraction of the RNA should be drawn out (*ca.* 200 to 500 nl), which is sufficient for injecting around five to ten cells. This is a safety mechanism should the pipette tip clog during injection, which would render the RNA in the pipette unusable. The RNA remaining in the capillary should be drawn back a few millimeters into the tube to reduce evaporation of water from the RNA solution. Following injection of RNA, more RNA is drawn out of the capillary, and the process is repeated.

It is essential to fix the oocytes during injection, because they have a tendency to move away as the pipette touches them. To achieve this, one can carve grooves into a thin perspex plate fitted to the bottom of the injection chamber or use a scratched petri dish or even one covered with silicone or agar. Fixing of the cells occurs by lowering the solution level in the chamber such that surface tension holds the cells in the grooves.

Simple manual syringe systems (which we use) or specially constructed injection machines (e.g., Eppendorf microinjectors or Drummond injectors) can be used. The latter are useful when many cells are to be injected.

For handling of oocytes, Pasteur pipettes are helpful. The tips are broken (to increase tip size) and then fire polished to smooth out the rim.

2.2.3. Incubation

After injection, the oocytes are incubated in Barth's medium of the following composition (mM): 84 NaCl, 1 KCl, 2.4 $NaHCO_3$, 0.82 $MgSO_4$, 0.33 $Ca(NO_3)_2$, 0.41 $CaCl_2$, 7.5 Tris-HCl, pH 7.4. The antibiotics penicillin/streptomycin (100 Uml^{-1}) and gentamycin (50 $mgml^{-1}$) are also added. The cells are stored at 19°C in small petri dishes, and the solution is changed daily. During this latter process, the cells should be inspected daily under a microscope, and unhealthy ones discarded.

Ca^{2+}-free Barth's medium (for collagenase treatment) is simply the above solution without $Ca(NO_3)_2$.

2.3. Recordings

2.3.1. Voltage Clamp

The simplest electrophysiological measurements from oocytes are two-electrode voltage-clamp recordings in which the membrane potential is clamped at a desired value. Following a stepwise change in membrane potential, an initial capacitative current (I_{cap}) will flow that charges the membrane capacitance such that the voltage at time t after the step approaches the desired voltage V_f with an exponential time constant **t:**

$$V_t = V_f\,(1 - e^{-t/t})$$

where **t** = RC. The total current that flows will be determined by I_{cap} and currents associated with voltage-gated proteins (e.g., channels). For ion channels, the current will reflect both gating (from molecular rearrangements in the new electric field) and ionic (from ion permeation across the channel pore) components.

Since $I_{cap} = CdV/dt$, once the potential is clamped, no capacitative current flows. Hence, I_{cap} flows only during the membrane-clamping process. Since many channels (e.g., Na^+) activate and inactivate quickly (100 msec), it is imperative that the clamp settle very quickly. This is the major limitation with voltage-clamp recordings in oocytes. Because of their large size (and therefore capacitance), several milliseconds are required for the membrane to be adequately clamped to the desired value. Ways to increase the speed of the clamp and another recent technical advance (the cut-open oocyte technique) are discussed later.

In two-electrode voltage-clamp recording from oocytes, one intracellular electrode measures the membrane potential (the voltage electrode), and the second (current electrode) passes sufficient current to maintain the desired voltage clamp, using a feedback circuit. The amount of current passed through the current electrode is determined by the discrepancy between the membrane potential and the command potential (the desired value). When these two are equal, no current flows through the current electrode. The current through the current electrode is the measured parameter and can be monitored either as the current flowing to ground through the current-grounding electrode (using a virtual-ground amplifier) or simply as the current flowing through the current electrode. Currents passing to ground through the ground electrode will induce small polarizations in the bath such that it is not actually at ground potential. To compensate for this, it is necessary to have a voltage bath electrode so that the actual membrane potential is taken as the difference between the oocyte potential and the bath potential.

To produce changes in membrane voltage, a pulse generator is necessary. This can be, depending on the need, a simple waveform generator or a computer-controlled pulse generator. The data can be recorded directly onto a simple chart recorder (e.g., Graphtec) if responses are slow (it is also useful to simultaneously video-record the data for analysis later) or on a computer-based data acquisition system for faster events (see Chapter 3, this volume). Electrode pipettes can be easily made from capillary glass containing a thin filament that ensures that solution reaches the tip. Our glass is from Clark (Reading, England) types GC 20TF-10 (thin), GC150F-10 (medium), and GC200F-15 (thick). We use a two-stage pulling process either on a puller that is also used for patch pipettes or on a programmable electrode puller (David Kopf Instruments, Tujunga, CA). It is best to have pipettes with a long, thin shank because this inflicts less damage on impaling the cell. On the other hand, long thin shanks imply a high electrode resistance. Electrodes can be backfilled with 2–3M KCl or 0.5 M K_2SO_4 (or aspartate) containing at least 20 mM Cl^-. Air bubbles lodged near the tip can be removed easily by applying suction through a syringe while observing the tip under a microscope. It is best not to fill the electrode with too much electrolyte because the solution can overflow into the holder when the silver chloride wire (which makes electrical contact with the pipette solution) is inserted. Furthermore, too high a level increases the hydrostatic pressure in the electrode and can result in bulging of the oocyte cytoplasm at the site of electrode insertion as a result of KCl outflow from the electrode. If the level of solution is too low, though, net inflow of cytosol can occur, and this will be seen as an increase in pipette resistance (as 3 M KCl is effectively diluted by cytosol).

During long recordings, it is helpful to wax the end of the pipettes to prevent creep of KCl out of the electrode. This is easily accomplished by heating the open end of the pipette (prior to backfilling with KCl) in a Bunsen flame and then applying a small amount of dental or sticky wax.

The silver chloride electrode that contacts the electrode solution can be a chlorinated silver wire, a silver/silver chloride pellet, or a silver wire immersed in melted silver chloride. In all cases, it is vital that the wire be regularly chlorinated to reduce noise and prevent exposure of the cell to toxic silver ions.

Once the electrodes are backfilled, the resistances will be around 10 MΩ, which is an order of magnitude too high for reasonably fast clamping. The tips are therefore broken either under a microscope or against the bottom of the chamber. Good resistances should be around 0.5–2 MΩ (openings of 1–5 μm). The resistance depends on the nature of the experiment. Obviously, the lower the resistance, the wider the tip, and hence the larger the hole on impaling the cell. This will result in a larger holding current at relatively polarized potentials because of nonselective current across the site of penetration. If one is studying voltage-gated channels that can yield currents of several tens of milliamperes, then a big leak current of even 0.5 μA is tolerable. On the other hand, second-messenger-regulated Ca^{2+} influx is of the order of 200 nA (Ca^{2+}-activated Cl^- current), and leak currents should be no more than 20–30 nA. Most voltage-clamp amplifiers incorporate an electrode resistance measurement. The one from Polder (NPI Electronics, Tamm, Germany) reads out the resistance even while the microelectrodes are still impaled.

To reduce electrode capacitance, it is helpful to Sylgard®-coat the microelectrodes along the shank (silicone curing agent RTV615) and to have a low volume of solution in the bath. This also increases the rate of bath exchange when different solutions are perfused.

Capacitive coupling between the two recording electrodes should also be minimized. This can be achieved by shielding the electrodes from each other (a simple grounded metallic shield around the current electrode up to the solution surface is adequate), reducing electrode capacitance by coating with an insulating material (e.g., Sylgard®), and lowering the bath volume.

Because of their low resistances, the microelectrodes do not clog frequently and hence can be reused for several cells.

Impaling the oocytes is quite straightforward. One way is to carve a groove in the chamber that will hold the oocyte in place. Another way is to touch the cell with the current microelectrode. As the oocyte starts to recoil away, the voltage electrode is inserted. During this process, the oocyte is held in place by the current electrode. Successful impalement of the voltage electrode is easily seen by recording the resting potential. For healthy cells, this is in the range of −40 to −80 mV. Impalement of the current electrode can be easily followed with the Polder amplifier through audio tracking (the voltage electrode can also be followed with this very convenient system). Otherwise, one follows the current electrode in the current-clamp mode. First the voltage electrode is inserted, and the membrane potential is monitored. Small current pulses are applied to the current electrode. On insertion of this electrode, these current pulses will trigger small changes in voltage (the size of which is determined by the input resistance of the oocyte), which will be picked up by the voltage electrode. On insertion of both electrodes, the gain of the feedback loop is set to a low value, and the desired holding potential is set. The clamp is then closed. Small (10-mV) voltage steps are applied at potentials when the channels do not open, and the capacitive currents are observed on an oscilloscope or computer. Compensation of these capacity transients can be achieved by adding the differentiated command voltage step with appropriate amplitude and time constant to the current trace. Better than 90% compensation can be achieved using two components. The Polder amplifier enables compensation to be achieved using three different components. Reasonable holding currents to maintain a clamp potential of −80 mV are around −10 to −100 nA. Higher holding currents indicate leaky cells.

For channels with fast gating kinetics, it is important to have a high feedback gain because it results in a faster settling time of the clamp ($\tau = RC/A$, where A is the gain of the amplifier; see above). However, too high a gain evokes oscillations that can irreversibly damage the cell. The Polder amplifier has a built-automatic oscillation cutoff feature that opens the clamp if the gain is too high.

The two-electrode voltage-clamp recording system is very stable. We can measure currents following repetitive serotonin receptor activation for several hours without the leak current at −80 mV exceeding −40 nA. Also, the cells readily tolerate changes in external perfusion. Solution exchange can be continuous and simply gravity driven, with a suction pipette to control the level, or manual using a Pasteur pipette. If the latter is used, changes in fluid level will change electrode shunt capacitance and hence require readjustment of capacitance compensation and neutralization. If an exact control of fluid level is required, a fluid-level controller (MPCU, Adams and List, Darmstadt, Germany or Westbury, NY) is necessary.

While recording, it is possible to inject various compounds into the cell. Second messengers, ion channel inhibitors, and even proteins have been successfully injected during recording. A third manipulator is necessary for maneuvering the injection pipette (which is made and filled the same way as RNA injection pipettes). It is not always easy to follow impalement and subsequent injection with this third pipette. One way is to see a bulging of the oocyte at the site of penetration following injection. The Polder amplifier is particularly helpful because one can use the audio tracking system. On impaling with the injection pipette, the current electrode normally passes more current, resulting in a change in the pitch of the current electrode audio monitor.

2.3.2. Patch-Clamp Recording

Both cell-attached and excised patches can be easily obtained from oocytes. Before patching, it is necessary to remove the vitelline membrane. This is done manually with fine forceps (see Methfessel *et al.*, 1986). The oocyte is placed in a hyperosmotic medium (mM): 200 potassium aspartate, 20 KCl, 1 $MgCl_2$, 5 EGTA-KOH, 10 HEPES-KOH, pH

7.4. After 3–10 min in this solution, the cell shrivels. The vitelline membrane can now be observed and be gently teased away from the cell using forceps. Once it is removed, the cells are extremely fragile and will easily rupture if mechanically distressed or exposed to air. The cells are also very sticky and will adhere to plastic or glass within a few minutes, and so they should immediately be transferred to the recording chamber.

Patch-clamp recordings on oocytes are essentially the same as for other cells in that the same amplifiers (EPC-7 or Axoclamp) and data acquisition systems can be used (Chapter 3, this volume). One advantage with the oocyte is the ability to obtain macropatches. These enable low-noise, fast-clamp patch recordings of many channels. Patch pipettes for such patches can be pulled from aluminosilicate glass (e.g., Hilgenberg from Malsfeld, Germany) and have opening diameters in the range of 3–8 μm, which is around half the size of most mammalian cells. The pipettes should have as small a taper as possible to reduce access resistance. Gigohm seals using these big pipettes (0.8–2 MΩ depending on shape and pipette solution) can be obtained routinely, but seal formation is generally slow (2–5 min), and less suction is required for seal formation than with smaller pipettes (100 to 200 mm H_2O). Slight depolarization of the patch sometimes assists in seal formation. It is also necessary to have filtered solutions in both the bath and the patch pipette because dust and other suspended particles can hinder seal formation. Slight positive pressure should be applied to the patch pipette (10 to 40 mm H_2O) as it is lowered into the bath solution to prevent both exchange of the bath and pipette solutions and particles from clogging the pipette tip. If seal formation is slow, large omega-shaped patches can form across the pipette tip. These are detected as sudden large increases in patch capacitance that cannot be adequately compensated. Because of the large depth to which the pipettes are immersed in solution, it is necessary to reduce electrode capacitance by coating with a silicone (RTV615) curing agent to within 1–2 diameters of the tip.

Cell-attached macropatches can be easily excised to form inside-out patches. These are obtained by rapid withdrawal of the pipette from the oocyte after seal formation has occurred. In fewer than 20% of trials is a vesicle formed. With macropatches, though, it is usually not possible to rupture a vesicle by brief exposure to air (as can be done for smaller patches). It is necessary that the bath composition mimic the cytosol before excision because the patches are lost if Ca^{2+} is present. Standard bath solution for excised patches is (mM): 100 KCl, 10 EGTA, 10 HEPES; pH 7.4. Inside-out patches can survive for several minutes, but stability can decrease if the bath solution is changed.

One advantage of oocytes (as is the case with large cells) is that the inside-out patch can be reinserted into the cell, a process called patch cramming (Kramer, 1990). This reexposes the inside of the patch to the cytoplasm and is extremely useful when one is studying the regulation of a channel. On excision, suppose the current rapidly runs down, as is the case for the IP_3 store depletion-activated Ca^{2+} current. When the patch is crammed back into the cell, the current rapidly reappears. This therefore suggests that maintenance of the current requires a cytoplasmic factor. Similarly, patches containing RCK4 (Kv1.4) K^+ channels inactivate rapidly on excision. However, the channels reactivate on being crammed back into the cell, demonstrating a role for the cytoplasm in fast inactivation.

Outside-out patches can also be obtained from macropatches using standard techniques as for other cells, but it is somewhat easier if larger-resistance pipettes are used (1.5–2 MΩ). Outside-out patches are extremely stable and readily tolerate extensive changes in the bath solution.

2.3.3. Cut-Open Oocyte

Two-electrode voltage-clamp recordings have the drawbacks that the clamp speed is slow and the intracellular ionic composition cannot be controlled. These problems can be circumvented to some extent by the recently developed cut-open oocyte technique (Perozo *et al.*, 1992). Defolliculated oocytes are placed in a perspex chamber that has three compartments: top (which is recorded from), middle guard pool, and a bottom one for current injection, in which, for internal perfusion, either the cell is exposed to saponin or a hole is made into it. The compartments are isolated from each other by vaseline seals, as used initially for studies on skeletal muscle and nerve fibers. The principle of the recording, described in Perozo *et al.* (1992), is to record the transmembrane potential as the difference between two microelectrodes placed on both sides of the membrane in the top compartment and feed the necessary current to maintain the desired potential through the bottom compartment. Membrane currents are recorded from the top pool and encompass a significant amount of membrane (*ca.* 500–700 μm in diameter). The currents are recorded through a current-to-voltage converter having feedback resistors of 100 kΩ to 1 MΩ. With this clamp, capacity time constants are estimated to be in the range of 50–100 μsec hence enabling rapid settling of the clamp and therefore fast measurements of channel kinetics.

An internal perfusion system running through the bottom pool is effective in dialyzing the cytoplasm. With a perfusion rate of 5–50 ml min^{-1} with a solution lacking K^+, K^+ current is lost within 5 min, demonstrating adequate washout of charge carrier.

Gating currents can also be easily and reliably measured with two-microelectrode recordings in *Xenopus* oocytes as shown by McCormack *et al.* (1994). The use of low-resistance electrodes (0.5 MΩ and below) allows the linear subtraction of capacitive and leak currents using standard *p*/*n* procedures. The procedure is simple, and the recorded current kinetics appear to be similar to those obtained with the cut-open technique.

3. Endogenous Currents to the Oocyte

Although the endogenous currents to the oocyte are of low amplitude, they can assume importance under conditions where exogenously expressed currents are small. It may therefore be fruitful to delineate briefly the endogenous currents and point out ways to eliminate them.

3.1. Ca^{2+}-Activated Cl^- Channel

This is the major current endogenous to the oocyte. Under standard conditions (normal frog Ringer outside and voltage-clamp recording), the reversal potential is around −25 mV (Barish, 1983). The *V–I* relationship shows outward rectification, and single-channel conductance is estimated around 3 pS from noise analysis (Takahashi *et al.*, 1987). In inside-out patches, increasing cytosolic Ca^{2+} increases open probability, demonstrating that an increase in free Ca^{2+} alone is necessary and sufficient for activating the current. In voltage-clamp recordings, the Cl^- current shows voltage dependency, but this is a result of Ca^{2+} influx through endogenous voltage-gated Ca^{2+} channels followed by activation of the Cl^- channels, since intracellular injection of the Ca^{2+} chelators EGTA or BAPTA or perfusion with Ca^{2+}-free solution abolishes this effect.

The Ca^{2+}-activated Cl^- current can be inhibited by intracellular injection (either manually through a syringe or iontophoretically or via a pneumatic pressure injector) of Ca^{2+} chelators (EGTA or BAPTA) or pharmacological blockers (e.g., niflumic or flufenamic acid (White and Aylwin, 1990).

3.2. Voltage-Dependent Ca^{2+} Channel

The endogenous voltage-dependent Ca^{2+} channel is very small and gives rise to about 1 nA whole-cell current. This is almost three orders of magnitude lower than the current achieved following exogenous channel expression. The amplitude of the Ca^{2+} current can be increased by using high Ba^{2+} as the charge carrier and can reach levels of around 50 nA. The channel has an unusual pharmacology in that it is relatively insensitive to even high concentrations of organic Ca^{2+} channel blockers but is potently blocked by Cd^{2+} (IC_{50} = 4 μM; Lory *et al.*, 1990). Although the Ca^{2+} current is small, it is amplified through activation of the Ca^{2+}-dependent Cl^- channel. This can be easily averted either by using external Cl^- channel blockers or injecting Ca^{2+} chelators (see above).

3.3. Pool-Depletion-Activated Current

As in other nonexcitable cells, emptying of agonist-sensitive IP_3 internal stores evokes a Ca^{2+} current in oocytes. The current density of this influx pathway is very low (100–200 fA in a macropatch), and no single-channel events were observed. Although expression of voltage- or ligand-gated channels is unlikely to activate this mechanism, it may manifest itself under certain conditions. An increase in cytosolic Ca^{2+} (by Ca^{2+} influx through voltage-gated channels, for example) can activate phospholipase C, the enzyme that releases IP_3 by hydrolyzing phosphatidylinositol-4,5-bisphosphate. This may result in emptying of stores and activation of the current. Similarly, prolonged exposure to Ca^{2+}-free solution (as occurs in certain experiments) could gradually deplete stores, thereby activating the current. Although no Ca^{2+} current would flow (as there is no Ca^{2+} outside), readmission of Ca^{2+} could result in this current contributing to the total current. The pool-depletion current can be abolished either by injecting the IP_3 receptor antagonist heparin (1 mg ml^{-1} is supramaximal) or extracellular application of Cd^{2+} (200 μM is maximal).

3.4. Stretch-Activated Channels

These channels are revealed following pressure application to the membrane (McBride and Hamill, 1992). They are extremely variable in expression, some batches of oocytes having none which others have many. The channels are nonselective and have a conductance of about 35 pS. The trivalent cation gadolinium blocks the channels at 10 μM (Yang and Sachs, 1989) but may interact with other channels too. If a patch contains stretch-activated channels, it is probably best to discard the patch and try to find a region lacking them.

3.5. Na^+ Channels

Oocytes have a very unusual Na^+ current that slowly manifests itself only after a prolonged depolarization positive to +20 mV for several seconds, suggesting that it is depolarization-induced. The current inactivates only marginally during the depolarization but deactivates rapidly on hyperpolarization. A subsequent depolarization evokes the Na^+ current more quickly and at more negative potentials (>20 mV). If hyperpolarization is maintained for some time between depolarizing pulses, this Na^+ current is lost and requires a priming depolarization once again. The Na^+ channel is largely insensitive to tetrodotoxin, being blocked at 1 mM.

3.6. K^+ Channels

Depolarization positive to −50 mV evokes a slow outward noninactivating K^+ current that is only partially affected by TEA. This current is small, around 30 nA at +10 to +30 mV.

Little is known about the pharmacology of this channel, so it is difficult to abolish it. Many healthy oocytes have resting membrane potentials in the range of −70 mV. It therefore is apparent that a resting K^+ permeability is important for setting the membrane potential. However, the properties of this current have not been investigated.

From the preceding discussion, it is evident that endogenous channel expression varies from frog to frog. It is therefore advisable to keep a few uninjected cells from each batch and examine the endogenous currents. For example, if one has expressed a K^+ channel mutant of low conductance, then the macroscopic current will also be small. If the batch of oocytes also express an endogenous K^+ channel as well, it will be hard to resolve the endogenous current from the exogenous one. It would be advisable to abandon the experiment and take oocytes from another frog.

4. Applications

We would like to end this chapter by briefly describing how a few recent applications of the oocyte expression system has advanced our understanding of certain key physiological issues.

4.1. Properties of Ion Channels

Ten years ago, when the first edition of this volume appeared, no ion channel had been cloned. In the last decade, not only have a whole variety of channels been cloned, but elegant mutation work has revealed the amino acids that are involved in activation, inactivation, and lining the pore. Indeed, it has even been possible to alter the selectivity of several channel types into others by single point mutations. We are beginning to understand, on a molecular level, how a channel functions. This will be of tremendous benefit clinically, since a detailed molecular understanding opens the way for logical pharmacological intervention and hence therapeutic treatment of such debilitating disorders as cystic fibrosis and hyperkalemic periodic paralysis.

4.2. Ca^{2+} Signaling

A variety of hormones and neurotransmitters link to IP_3 production, which subsequently releases Ca^{2+} from internal stores. In most nonexcitable cells, Ca^{2+} release is followed by a plateau of elevated Ca^{2+}, predominantly as a result of Ca^{2+} influx across the plasma membrane through a pathway not gated by voltage yet somehow activated by emptying of IP_3-sensitive stores. One major unresolved question in cellular physiology is the nature of the signal that activates Ca^{2+} influx after emptying stores. Although it is not yet identified, patch-cramming experiments in oocytes show that the signal is a diffusible messenger released after emptying stores. Following depletion of stores, a Ca^{2+} current is observed in cell-attached patches. Excision (now inside-out) results in rapid rundown of this current. If the patch is then crammed back into a different region of the membrane, far from where it was excised, the current rapidly recovers (Parekh *et al.,* 1993). Hence, the signal for activating this current is not localized but diffused below the membrane.

4.3. Secretion

The last year alone has seen a dramatic increase in the number of proteins thought to be involved in secretion. However, few of these molecules have been attributed a function. An elegant approach to address this has been devised by Alder *et al.* (1992). The tactic was to inject total tissue RNA from a secretory cell into the oocyte in the hope of recreating regulated exocytosis. Following injection of total cerebellar mRNA, secretion (monitored through ^{3}H-glutamate release) was evoked in a Ca^{2+}-dependent manner. Having established the oocyte system as a good assay for secretion, they were able to test which proteins were actually involved in secretion by injecting antibodies or antisense oligonucleotides to specific proteins. This approach revealed the key role of synaptophysin, for example. Although such experiments do not reveal the sequence of proteins in the secretory pathway, they convincingly demonstrate whether a protein indeed has a functional role. It is clear that this approach will be of widespread use to a variety of biochemical pathways.

ACKNOWLEDGMENT. Anant B. Parekh was supported, in part, by the Alexander-von-Humboldt-Stiftung.

References

Alder, J., Lu, B., Valtorta, F., Greengard, P., and Poo, M.-M., 1992, Calcium-dependent transmitter secretion reconstituted in *Xenopus* oocytes: requirement for synaptophysin, *Science* **257:**657–661.

Barish, M. E., 1983, A transient calcium-dependent chloride current in the immature *Xenopus* oocyte *J. Physiol.* **342:**309–325.

Dascal, N., 1987, The use of *Xenopus* oocytes for the study of ion channels *Crit. Rev. Biochem.* **22:**317–387.

Dumont, J. M., 1972, Oogenesis in *Xenopus laevis* (Daudin). I. Stages of oocyte development in laboratory maintained animals, *I. Morphol.* **136:**153–179.

Girard, S., and Clapham, D., 1993, Acceleration of intracellular calcium waves in *Xenopus* oocytes by calcium influx, *Science* **260:**229–232.

Goldin, A. L., and Sumikawa, K., 1992, Preparation of RNA for injection into *Xenopus* oocytes, *Methods Enzymol.* **207:**279–297.

Gundersen, C. B., Miledi, R., and Parker, I., 1983, Serotonin receptors induced by exogenous messenger RNA in *Xenopus* oocytes, *Proc. R. Soc. Lond. B.* **219:**103–109.

Gurdon, J. B., Lane, C. D., Woodland, H. R., and Marbaix, G., 1971, Use of frog eggs and oocytes for the study of messenger RNA and its translation in living cells, *Nature* **233:**177–182.

Krafte, D. S., and Lester, H. A., 1992, Use of stage II–III *Xenopus* oocytes to study voltage-dependent ion channels, *Methods Enzymol.* **207:**339–345.

Kramer, R. H., 1990, Patch cramming: Monitoring intracellular messengers in intact cells with membrane patches containing detector ion channels, *Neuron* **2:**335–341.

Lory, P., Rassendren, F. A., Richard, S., Tiaho, F., and Nargeot, J., 1990, Characterization of voltage-dependent calcium channels expressed in *Xenopus* oocytes injected with mRNA from rat heart, *I. Physiol.* **429:**95–112.

Lupu-Meiri, M., Shapira, H., and Oron, Y., 1988, Hemispheric asymmetry of rapid chloride responses to inositol trisphosphate and calcium in *Xenopus* oocytes, *FEBS Lett.* **240:**83–87.

McBride, D. W., and Hamill, O. P., 1992, Pressure-clamp: A method for rapid step perturbation of mechanosensitive channels, *Pflügers Arch.* **421:**606–612.

McCormack, K., Joiner, W. J., and Heinemann, S. H., 1994, A characterization of the activating structural rearrangements in voltage-dependent *Shaker* K^+ channels, *Neuron* **12:**301–315.

Methfessel, C., Witzemann, V., Takahashi, T., Mishina, M., Numa, S., and Sakmann, B., 1986, Patch clamp measurements on *Xenopus laevis* oocytes: Currents through endogenous channels and implanted acetylcholine receptor and sodium channels, *Pflügers Arch.* **407:**577–588.

Miledi, R., Parker, I., and Sumikawa, K., 1983, Recording of single gamma-amino-butyrate- and acetylcholine-activated receptor channels translated by exogenous mRNA in *Xenopus* oocytes, *Proc. R. Soc. Lond.* [*Biol.*] **218:**481–484.

Parekh, A. B., Terlau, H., and Stühmer, W., 1993, Depletion of $InsP_3$ stores activates a Ca^{2+} and K^+ current by means of a phosphatase and a diffusible messenger, *Nature* **364:**814–818.

Perozo, E., Papazian, D. M., Stefani, E., and Bezanilla, F., 1992, Gating currents in *Shaker* K^+ channels. Implications for activation and inactivation models, *Biophys. J.* **62:**160–168.

Peter, A. B., Schittny, J. C., Niggli, V., Reuter, H., and Sigel, E., 1991, The polarized distribution of poly(A+)-mRNA-induced functional ion channels in the *Xenopus* oocyte plasma membrane is prevented by anticytoskeletal drugs in *J. Cell. Biol.* **114:**455–464.

Snutch, T. P., and Mandel, G., 1992, Tissue RNA as source of ion channels and receptors, *Methods Enzymol.* **207:**297–309.

Stühmer, W. 1992 Electrophysiological recordings from *Xenopus* oocytes in *Methods Enzymol.* **207:**319–339.

Takahashi, T., Neher, E., and Sakmann, B., 1987, Rat brain serotonin receptors in *Xenopus* oocytes are coupled by intracellular calcium to endogenous channels in *Proc. Natl. Acad. Sci. U.S.A.* **84:**5063–5067.

White, M. M., and Aylwin, M., 1990, Niflumic and flufenamic acids are potent reversible blockers of Ca^{2+}-activated Cl^- channels in *Xenopus* oocytes, *Mol. Pharmacol.* **37:**720–724.

Yang, X. C., and Sachs, F., 1989, Block of stretch-activated ion channels in *Xenopus* oocytes by gadolinium and calcium ions, *Science* **243:**1068–1071.

Chapter 16

Polymerase Chain Reaction Analysis of Ion Channel Expression in Single Neurons of Brain Slices

HANNAH MONYER and PETER JONAS

1. Introduction

The study of gene expression and regulation in the central nervous system (CNS) is a daunting task because of the diversity of neuronal phenotypes and the complexity of many protein classes. Molecular cloning revealed the presence of a large number of different protein families in the CNS, each comprising several members. Ligand-gated ion channels may serve as an example to illustrate this point (for review, see Unwin, 1993). Heterologous expression combined with electrophysiological analysis suggests that ligand-gated channels are multimeric proteins with functional properties depending on the subunit composition. Very little is known, however, about how the functional properties of the recombinant and native receptors relate to each other. Thus, it is of eminent importance to elucidate the subunit expression profile in different types of neurons in the CNS and to correlate this with the functional properties of the native receptors.

Many of the predictions about basic mechanisms of ligand-gated channel operation are based on the insights that many years of study on the nicotinic acetylcholine receptor of muscle endplate have yielded. The subunit composition of this channel is well defined, consisting of a pentameric arrangement of four different subunits. Although molecular homology points to a similar architecture of other ligand-gated channels, e.g., the $GABA_A$, glycine, and glutamate receptors, the exact subunit stoichiometry of these receptors has remained unknown. Mutational analysis argues for the existence of modules in equivalent positions in each subunit that critically determine functional properties such as conductance, ion selectivity, or gating properties.

As indicated by *in situ* hybridization and immunocytochemistry, different receptor subunits have a distinct yet overlapping expression pattern. These studies permit predictions about the subunit composition of the receptor channels in different neuronal populations. Yet it is at present not feasible to colocalize the large number of receptor subtypes at the level of a single cell by *in situ* hybridization or immunocytochemistry. An additional problem resides in the low expression levels of several mRNAs and the restricted expression in

HANNAH MONYER • Center for Molecular Biology (ZMBH), Im Neuenheimer Feld 282, D-69120 Heidelberg, Germany. PETER JONAS • Department of Cell Physiology, Max-Planck-Institute for Medical Research, D-69120 Heidelberg, Germany.
Single-Channel Recording, Second Edition, edited by Bert Sakmann and Erwin Neher. Plenum Press, New York, 1995.

subpopulations of cells. Technical limitations of the methods employed so far have precluded the direct correlation of the functional properties of receptors expressed by a living neuron with the subunit expression profile of the same cell.

The following recently developed techniques have enabled us to correlate biophysical ion channel characteristics of a defined cell type in a brain slice preparation with specific expression patterns of mRNAs coding for particular channel subunits:

1. Infrared differential interference contrast (IR-DIC) videomicroscopy (Dodt and Zieglgänsberger, 1990; see Chapter 8, this volume) allowing identification of a cell in a slice preparation and visually controlled harvesting of the cellular content.
2. Channel characterization using fast application of agonists to excised membrane patches (for a review on this method, see Chapter 10, this volume).
3. Single-cell polymerase chain reaction (PCR) analysis based on a method described by Lambolez *et al.* (1992) in cultured neurons (see also Bochet *et al.,* 1994).

2. Cell Identification, Recording, Harvesting, and Expelling

The resolution provided by infrared difference interference contrast (IR-DIC) videomicroscopy (see Chapter 8, this volume) allows us to identify visually different types of neurons in brain slices on the basis of location and morphological appearance. For example, interneurons and principal neurons can be easily distinguished in different regions of the CNS. This is illustrated for neocortical nonpyramidal and pyramidal cells in Fig. 1.

2.1. Whole-Cell Configuration

The first step in harvesting the cytoplasm of a single cell for subsequent PCR analysis is to establish a whole-cell recording configuration. Patch pipettes with comparatively large tips should be used to allow aspiration of as much cytoplasm as possible. Depending on the size of the cell type investigated, pipette resistances between 2.5 MΩ (e.g., for comparatively small neocortical nonpyramidal interneurons) and <1 MΩ (e.g., for large neocortical layer V pyramidal cells; Jonas *et al.,* 1994) are appropriate. Thick-walled borosilicate glass tubing is favorable (2 mm outer diameter, 1 mm inner diameter), since gigohm seals form more readily than with thin-walled glass. Pipettes with large tips can be most easily obtained using microprocessor-controlled electrode pullers with horizontal design, e.g., the DMZ puller (Zeitz, Augsburg, Germany) or the Flaming–Brown P-97 puller (Sutter Instruments, Novato, CA).

Several precautions have to be taken to avoid contamination of the harvesting pipette by RNA-degrading enzymes, which could be a source of failure in single-cell PCR experiments:

1. The glass tubing of which the pipettes are fabricated has to be heated prior to pulling (200°C, 4 hr). In our hands, heating is better than autoclaving, since we find it difficult to obtain gigohm seals with autoclaved glass.
2. New pipette solution is made up and autoclaved before every experiment. The intracellular solution we routinely use for single-cell PCR experiments contains (mM): 140 KCl, 5 EGTA, 3 $MgCl_2$, 5 HEPES, pH adjusted to 7.3 with KOH. Solutions containing ATP cannot be used because ATP decomposes during autoclaving. The choice of the predominant monovalent cation in the solution does not appear to be critical: K^+ ions may be replaced by Cs^+ ions if outward currents through K^+ channels

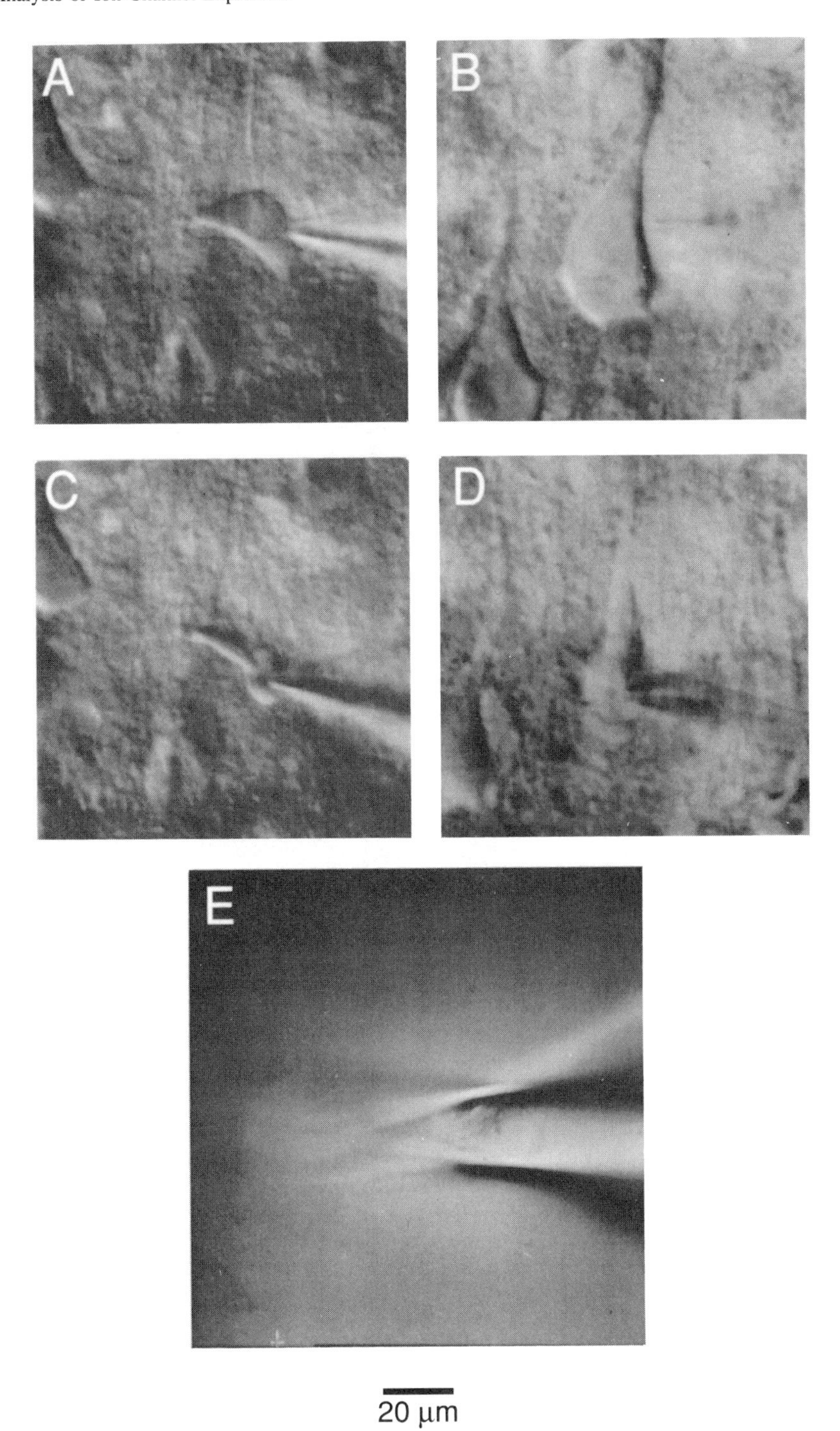

Figure 1. Harvesting the cellular content. A–D: Harvesting of cellular content as observed in the living slice using IR-DIC videomicroscopy. Nonpyramidal cell (layer IV; A,C) and pyramidal neuron (layer V; B,D) in a neocortical brain slice before (A,B) and after harvesting (C,D). E: Patch pipette with harvested material.

need to be suppressed. The pipette solution is filtered through disposable 0.2-μm filters (Schleicher and Schuell). The pipette tip is filled with solution by suction. Subsequently, the end of the pipette is heated using a Bunsen burner (to burn off contaminating material), and the pipette is backfilled with 8 μl solution using a variable microliter pipette and autoclaved tips (GELoader, Eppendorf, Hamburg, Germany).

3. It is recommended that the electrode holders be taken apart and cleaned regularly in ethanol or methanol. Additionally, the silver wire providing the electrical connection to the intracellular solution should be rechlorided electrochemically before each recording.
4. Pipettes, holders, and solutions should be handled only with forceps and gloves (preferentially powder-free).

When the cell body is approached with high positive pressure (about 80 mbar; "blow and seal," see Chapter 8, this volume), the seal forms rapidly when the pressure is released, presumably because the membrane swings back toward the pipette tip. Subsequently, the whole-cell configuration is reached by breaking the patch membrane using a pulse of suction. Apart from being a kind of "precursor" configuration for the harvesting process, a number of electrophysiological measurements can be made before the cytoplasm is aspirated into the patch pipette. For example, it is possible to determine the impulse pattern of a neuron under current-clamp conditions. This has proven to be a useful criterion in identifying GABAergic interneurons in brain slices. Apart from their characteristic morphology, these neurons typically show fast-spiking responses to the injection of sustained depolarizing currents (Jonas *et al.,* 1994). The recording of voltage-activated currents or synaptic currents under voltage-clamp conditions is also possible. These measurements, however, suffer from voltage- and space-clamp problems generally inherent in whole-cell recordings from neurons in brain slices.

To study the kinetic properties of ligand-gated ion channels, fast application of agonists using isolated patches rather than whole cells is required (see Chapter 10, this volume). It is almost impossible, however, to perform fast application and harvesting in one step using the same pipette. With small pipettes ideal for fast application experiments, it is difficult to harvest enough material; with large pipettes ideal for harvesting, it is difficult to obtain stable outside-out patches. In addition, the RNA harvested into the pipette is likely to be degraded during the electrophysiological experiment. We thus prefer a two-step procedure to combine fast application and the PCR technique. First, an outside-out membrane patch is isolated from the cell soma using a 3- to 5-MΩ recording pipette. Different agonists, e.g., glutamate, can be applied rapidly to this patch. Subsequently, a second gigohm seal is formed on the same cell, using a larger (1–3 MΩ) harvesting pipette. So far, these patch and harvesting experiments appear to be restricted to cells with relatively large soma size, e.g., pyramidal cells or basket cells in hippocampal dentate gyrus, which can be patched more than once. The isolation of the outside-out patch prior to harvesting the cytoplasm reduces the success rate of the subsequent PCR amplification only slightly.

2.2. Harvesting of Cell Content

After the whole-cell configuration has been established, the cell content can be harvested into the patch pipette by applying negative pressure (about 50 mbar) to the pipette interior (Fig. 1). It is advantageous to control the aspiration of cytoplasm visually using IR-DIC videomicroscopy and at the same time to monitor the access resistance and the input resistance of the cell electrically. The flow of the cellular contents into the tip of the harvesting pipette is clearly visible in the IR-DIC image. In most cases, however, the cell nucleus soon gets stuck at the patch pipette tip, precluding further harvesting of cytoplasmic material. This can be observed in the IR-DIC image and can be also inferred from a sudden increase of access resistance. To aspirate more cytoplasm, the nucleus has to be harvested into the pipette by increasing the suction up to 200 mbar. The movement of the nucleus needs to be watched carefully because, although initially very slow, it becomes faster with time. When about half of the nucleus is in the harvesting pipette, suction must be released quickly to avoid destroying the cell. Once the nucleus has passed the pipette tip, only very slight suction (about 20 mbar) is sufficient to harvest the remaining cytoplasm. If the harvesting is performed carefully, the input resistance of the cell does not change much; occasionally it is even possible to obtain an outside-out membrane patch when the harvesting pipette is slowly withdrawn from the cell. The entire harvesting procedure takes about 3 min. Debris that occasionally attaches to the outer side of the harvesting pipette should be removed by passing it through an air–fluid interface (repeatedly if necessary).

Cells should be used for subsequent PCR analysis only when the gigohm seal had formed easily and remained intact until the very end of the aspiration procedure. This ensures the harvesting of the cytoplasm of the selected cell, precluding contamination by adjacent material. About 50 control experiments were performed to exclude the possibility of contamination. Pipettes were advanced into the slice and taken out again without obtaining a gigohm seal and without harvesting of cellular content. These control pipettes never gave any PCR amplification product, provided that no material was attached to the outer side of the pipette. In two cases where debris remained attached, however, a false-positive control was obtained. It is therefore strongly recommended that cells be rejected when the outer surface of the harvesting pipette is not clean.

2.3. Expelling

The contents of the patch pipette are expelled into a PCR tube using high positive pressure (4 bar). We use a home-made device (expeller, Fig. 2A) to precisely position the pipette and the reaction tube with respect to each other. The PCR tube is held in a clamp on one side and can be moved up or down using a small translation stage. The harvesting pipette is held in a clamp on the other side and can be moved back and forth as well as in the axial direction. After harvesting of the cellular content, the reaction tube containing 2.5 μl of primer/dNTP mix and dithiothreitol (see below) is mounted onto the expeller. The harvesting pipette is connected to a nitrogen gas tank (via tubing and a gas filter) and is also mounted. A solenoid-driven valve and a footswitch is used to time the pressure application.

Under visual control provided by a 10× binocular microscope, the harvesting pipette tip is placed exactly into the small drop of solution contained in the reaction tube (Fig. 2B). Care should be taken not to break the patch pipette at this stage. Subsequently, pressure is applied to the patch pipette interior. The contents of the tip (which represents only a small fraction of the total volume in the pipette but contains the majority of harvested material) is thereby ejected directly into the reaction tube. After about 10 sec, the harvesting pipette is moved a bit further in the axial direction until it gently touches the bottom of the tube and breaks to a diameter of about 5 μm, and the total volume in the pipette (about 8 μl) is ejected within about 10 sec (Fig. 2C). The procedure described ensures that the harvested material is expelled completely and that it is freely accessible to all reagents in the tube.

A

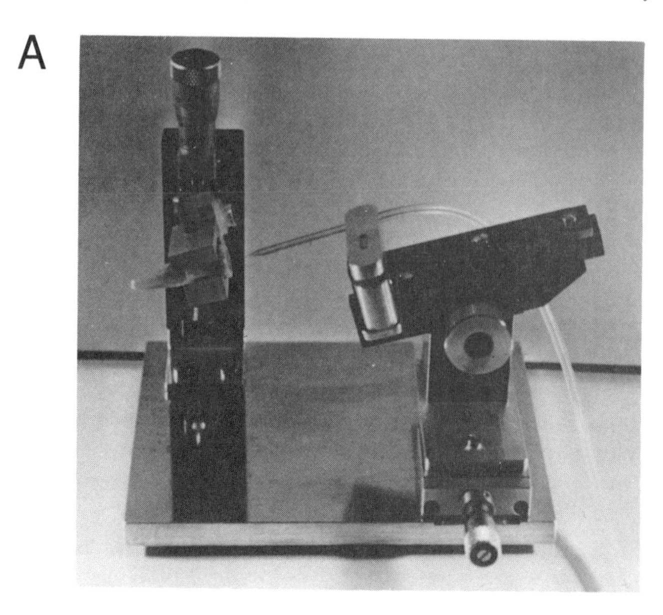

B

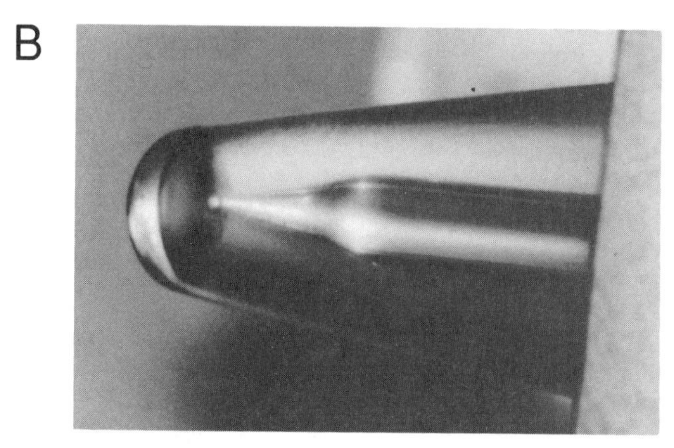

C

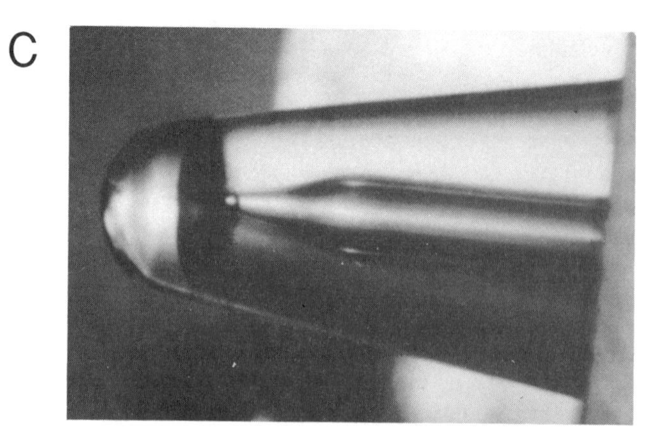

Figure 2. Expelling. A: Device used for expelling the pipette content. The PCR tube is mounted on the left; the harvesting pipette is held in a clamp on the right. The tubing connecting the pipette with the gas tank is also visible. B,C: Harvesting pipette before (B) and after (C) expelling of the cellular content.

Early breaking of the patch pipette is not recommended because the harvested material remains covered by glass and thus is largely inaccessible for the reverse transcriptase.

3. Molecular Analysis of mRNA Expression

3.1. cDNA Synthesis

The handling of the harvested cell material requires the usual precautions when working with RNA (Berger and Kimmel, 1987; Ausubel *et al.*, 1987). An RNase-free environment should be guaranteed. Gloves should be worn during the experiments, and instruments (pipettes, tips, etc.) should be used for these experiments only. It is recommended that the PCR experiments be set up in a room used for that purpose only, particularly in an environment where the plasmids carrying the target sequences for the PCR amplifications are grown.

The reagents required for cDNA synthesis (Fig. 3, step 2) can be prepared in advance and stored in small aliquots at −20°C. A new aliquot is used each day and is kept on ice throughout the experiment. Autoclaved (but not diethylpyrocarbonate-treated) water is used to make up the solutions. A mix is prepared consisting of the hexamer random primers and the four deoxyribonucleoside triphosphates (dNTPs). The antioxidant dithiothreitol (DTT) is also prepared as a stock solution. The enzymes (RNasin and reverse transcriptase) are also stored in 50-μl aliquots.

Although not mandatory, siliconized PCR tubes can be used for these experiments to reduce the loss of cell material through its sticking to the tube. Prior to the mounting of the PCR tube to the expeller, the primer/dNTP mix and DTT are introduced into the tube, which is then placed on ice. The content of the harvesting pipette is then expelled into the PCR tube. After removal of the tube from the expeller, ribonuclease inhibitor and Moloney murine leukemia virus (MMLV) reverse transcriptase are added. The type of reverse transcriptase does not seem to be critical. Similar success rates were obtained in test experiments where Superscript reverse transcriptase (BRL) was used. After adding all components, these are mixed by flicking the tube, followed by centrifugation (PicoFuge centrifuge, Stratagene) for several seconds, and the reaction is incubated at 37°C for 60 min. The tube is transferred onto dry ice and then stored at −20°C until PCR amplification is performed.

3.1.1. Reagents and Solutions

- 5× primer/dNTP mix contains 25 μM hexamer random primers (Boehringer Mannheim) and 2.5 mM of each deoxyribonucleotide (Pharmacia) in Tris-HCl, 10 mM, pH 8.0
- 200 mM DTT (Biomol) dissolved in autoclaved water and filtered (0.2-μm pore size)
- 40 U/μl ribonuclease inhibitor (Promega)
- 200 U/μl MMLV reverse transcriptase (BRL)

3.1.2. cDNA Reaction

- 2 μl primer/dNTP mix
- 0.5 μl DTT
- 6 to 7 μl content of harvesting pipette

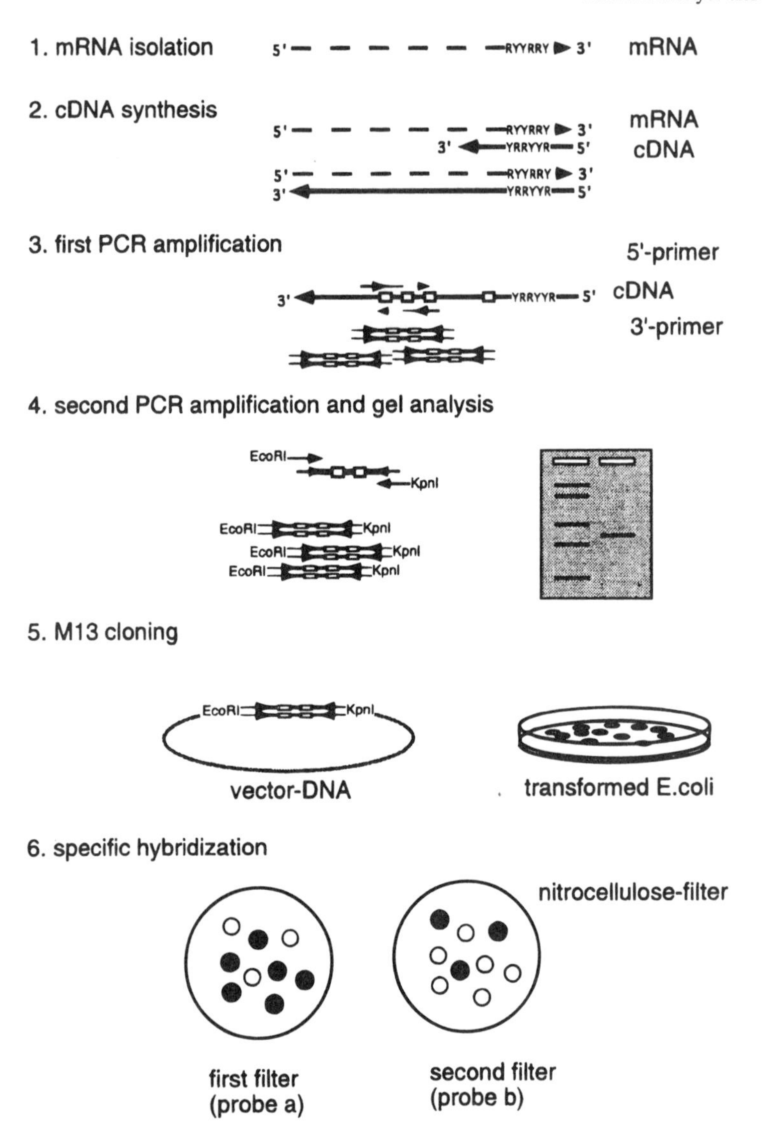

Figure 3. Flow diagram of the steps involved in the single-cell PCR analysis. For details see text.

- 0.5 μl ribonuclease inhibitor
- 0.5 μl MMLV reverse transcriptase

3.2. PCR Amplification

3.2.1. General Considerations

The cDNA reaction is used directly to set up the PCR reaction (Fig. 3, step 3). Hence, 10× buffer, primers, and polymerase are added to the 10 μl of the reverse transcription reaction. The *Taq* polymerase and its buffer from Stratagene were used throughout these experiments with satisfactory results. A test of different commercially available buffers and polymerases has not been performed.

The primer selection for single-cell PCR is based on standard rules for choosing PCR primers (Innis *et al.*, 1990; Ausubel *et al.*, 1987). The primer length is usually 20 to 25 nucleotides, containing around 50% G + C with random distribution. Sequences with secondary structures should be avoided, and so should complementary sequences within primer pairs, particularly at their 3′ ends. Computer programs such as OLIGO (Rychlik and Rhoads, 1989) are of help in designing primers. In case the PCR primers fail to work, a different primer pair may have to be tested. This should not pose a problem when only one mRNA species is to be analyzed. Designing the optimal primer pair may prove more difficult for attempting the amplification of a region of interest in several members of a gene family. In this case, primers are placed in conserved regions, and this limits the primer choice.

The design of the PCR primers requires in addition knowledge about the exon–intron structure of the target molecule in the region to be amplified. The primer localization should be such that the amplified DNA fragment contains at least one intron on the gene level. Thus, amplification of nuclear DNA harvested via nucleus aspiration into the harvesting pipette is precluded.

The successful amplification of the cDNA from a single cell critically depends on the cell size, the expression level of the target molecule, and the amount of harvested cell content. Thus, under optimal conditions, it is possible to visualize the PCR product on an ethidium bromide-stained gel after one round of amplification (40 cycles). The amount of DNA can be estimated by comparing the DNA product with a known amount of DNA size marker, and the specificity of the PCR products can be detected by Southern blot and restriction enzyme analysis.

However, there are numerous instances when the product can be detected by gel analysis only after a second round of PCR amplification (35 cycles). The second PCR amplification (Fig. 3, step 4) serves two purposes: it ensures the acquisition of larger amounts of material for subsequent analyses (agarose gel, Southern blot, specific digests of the PCR product) and permits, in addition, the fractional quantification of the sequence constituents within the obtained product. To achieve the latter, the primers for the second PCR amplification carry restriction sites at their 5′ termini for the directional cloning of the PCR fragments into the appropriate vector. The restriction sites chosen for cloning should preferably not occur as internal sites in the amplified PCR product after the second round of amplification. The yield and specificity of the PCR product are usually increased when at least one nested primer is used in the second amplification. Except for the new primer pair, all reagents are the same as for the first amplification. Optimization of the reamplification reaction comprises the testing of three critical variables: selection of primers, cycle number of the second PCR amplification, and the amount of template used from the first amplification.

Approximately 1 pg of the DNA product from the first PCR reaction is used in the second amplification. When no band can be seen on the gel, 1 μl of the first PCR reaction is used. When all parameters have been optimized, the low DNA target concentration obtained from a single cell often requires too many cycles of amplification. Figure 4A shows the amplified product after a first and second round of PCR amplification using material from three hippocampal basket cells. A visible band can be resolved on an agarose gel in only one of the three cells following a first PCR amplification of 40 cycles. However, a product is obtained in good yield in all three cases following a second PCR amplification (35 cycles) when 1 μl of the first reaction is used. Although it is satisfactory regarding the final DNA yield, this approach harbors the problem of nonspecific background amplification.

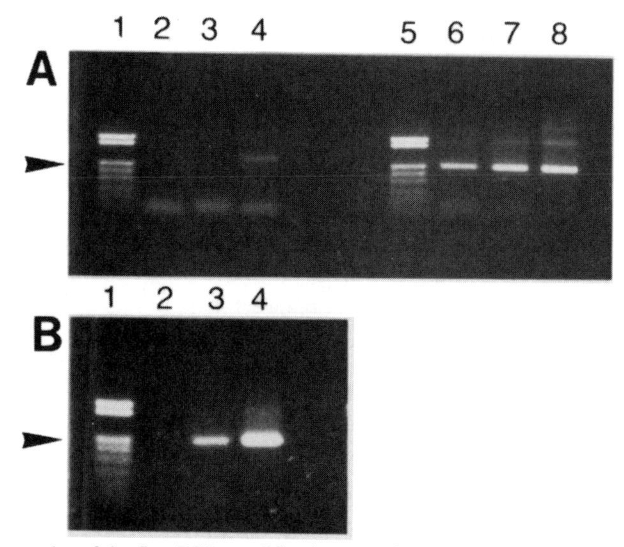

Figure 4. Cycle number of PCR amplification and amount of target DNA critically determine the yield of the product. A: Comparison of PCR product from three hippocampal basket cells after the first and second amplifications. Lanes 1 and 5 contain DNA size marker; 10 μl of the first PCR amplification is loaded into lanes 2 (first cell), 3 (second cell), and 4 (third cell); 5 μl of the second PCR amplification is loaded into lanes 6 (first cell), 7 (second cell), and 8 (third cell). The first PCR amplification consisted of 40 cycles, the second of 35 cycles. B: Comparison of product yield following the second PCR amplification using increasing amounts of the first PCR reaction. No visible DNA band can be resolved on a 1% agarose gel after 20 cycles of the first PCR amplification (not shown). The second PCR amplification (35 cycles) was performed using 1 (lane 2), 5 (lane 3), and 10 μl (lane 4) of the first amplification. Lane 1 contains DNA size marker. Arrows indicate the 635 base pair position.

An alternative protocol can be used that permits keeping the cycle number in the range of standard PCR amplifications. Thus, an overall comparable DNA yield can be obtained by reducing the cycle number of the first PCR amplification and using more material of the first PCR reaction for the second PCR amplification. Figure 4B serves to illustrate this point. In this example, the cycle number of the first PCR amplification is reduced to 20. For the subsequent 35 cycles of the second PCR amplification, 1, 5, and 10 μl, respectively, are used, resulting in an increasing amount of the DNA product. Provided that all steps are optimized, nine out of ten harvested cells have yielded a PCR product.

Detailed descriptions of parameters affecting the efficiency of PCR amplification can be found in laboratory manuals (Innis *et al.*, 1990; McPherson *et al.*, 1991; Ausubel *et al.*, 1987). Different target templates and primers will require optimization of the PCR conditions in order to obtain the desired DNA product in good yield.

3.2.2. First PCR Amplification

- 10 μl reverse transcription reaction
- 10 μl 10× buffer (Stratagene): 100 mM Tris-HCl, pH 8.8, 15 mM $MgCl_2$, 500 mM KCl, 0.01% (w/v) gelatin
- 1 μl 5′ primer (10 μM, made up in 10 mM Tris-HCl, pH 8.0)
- 1 μl 3′ primer (10 μM, made up in 10 mM Tris-HCl, pH 8.0)
- 78 μl water
- 0.5 μl *Taq* polymerase (2.5 U, Stratagene)
- one drop of mineral oil (Sigma, 6-ml bottle), which is added on top of the 100-μl reaction

As for any other PCR amplification, a control reaction containing all reagents except for the DNA is performed in each experiment to monitor for contamination.

The cycling program used for single-cell PCR does not differ from standard PCR amplification protocols and consists of 20 to 40 cycles (see example) using the following temperature profile:

- Denaturation, 94°C, 30 sec
- Primer annealing, 45–49°C (see example), 30 sec
- Primer extension, 72°C, 30 sec to 1 min

A final extension step at 72°C for 10 min is performed at the end, followed by chilling to 4°C. Several trials are often required to choose the optimal annealing temperature after taking into account the T_m values of the primers and the number of mismatches that often can not be avoided when amplifying several members of a family. For certain experiments it may be desirable to perform a hot start to reduce primer dimer formation (see Section 4).

3.2.3. Second PCR Amplification

- A variable amount of first PCR reaction (see above)
- 10 μl 10× buffer (Stratagene)
- 1 μl dNTPs (5 mM each) (Pharmacia)
- 1 μl 5′ primer (10 μM) with restriction site *A*
- 1 μl 3′ primer (10 μM) with restriction site *B*
- Water to 100 μl total volume
- 0.5 μl *Taq* polymerase (2.5 U, Stratagene)

The cycling parameters for the second PCR amplification (35 cycles) are similar to the ones in the first PCR reaction (see example).

The second PCR product can be analyzed in different ways. When the PCR product results from the amplification of several related family members, the first question pertains to the composition of that mixed product. The presence of a particular restriction endonuclease site can be diagnostic for the presence of a certain sequence. Thus, several restriction digests can be performed with the material of the second PCR amplification (Lambolez *et al.*, 1992).

Southern blot analysis is another useful and easy method to detect the presence of a sequence. The signal intensity on a Southern blot, however, depends critically on the specific activity of the labeled probe, the amount of DNA from each cell, and the exposure time of the film. Thus, even fastidious control of all these variables would make precise evaluation of the fractional expression levels of the mRNAs of interest extremely difficult.

3.3. Cloning of the PCR Product

To evaluate constituents numerically, we clone the DNA product of the second PCR amplification into M13mp18 RF (replicative form) DNA (Yanisch-Perron *et al.*, 1985) (Fig. 3, step 5). The PCR product is gel-isolated—standard protocols (Berger and Kimmel, 1987) or a commercially available kit (Jetsorb, Genomed) can be employed—and cleaved with restriction enzymes at the sites contained in the primers for the second PCR amplification. Following passage over a P100 column (Clontech) and ethanol precipitation, an aliquot is visualized on a gel to estimate the amount of the DNA fragment to be cloned into doubly digested M13mp18 RF-DNA. The molar ratio of vector to fragment is 1 : 10. Ligations are carried out as described elsewhere (Berger and Kimmel, 1987; Ausubel *et al.*, 1987). Ligation reactions are used directly to transform competent *E. coli.* We aim at a number of $>$ 100 recombinant plaques per ligation (transformation efficiency 10^6 transformants/μg plasmid DNA). Different insert-to-vector ratios have to be tried at times when the plaque number is too low. Transformation of *E. Coli* by electroporation may be required to obtain desirable plaque yield.

3.3.1. Reagents and Solutions for Ligation

- 10× ligation buffer: 500 mM Tris-HCl (pH 8.0), 100 mM $MgCl_2$ 10 mM DDT, 1 mM EDTA
- 10 mM ATP (Pharmacia)
- T4 DNA ligase (1 U/μl, Boehringer Mannheim)

3.3.2. Ligation Reaction

- 10–20 ng M13 RF-DNA
- 10–20 ng insert
- 1 μl 10× ligation buffer
- 1 μl ATP
- Water to final volume of 10 μl
- 1 μl T4 DNA ligase

The ligation is carried out for 12 to 16 hr at 16°C.

3.4. Plaque Quantification

Dual filter lifts allow for hybridization with specific radiolabeled probes (Fig. 3, step 6). The filters are examined by autoradiography followed by plaque counting. The relative abundance of the mRNAs of interest can thus be assayed.

The most convenient approach entails the use of specific probes for one filter set and a probe designed to recognize all members of a gene family for the second filter set. The length of the oligonucleotide probe should preferably be 30 to 45 nucleotides. The oligonucleotides are 5′-labeled using a standard kinase reaction. Hybridization and washing conditions are chosen based on considerations described in laboratory manuals (Berger and Kimmel, 1987; Ausubel *et al.*, 1987). The same set of filters can be reused and hybridized with other labeled probes after stripping in 80°C hot water if several members of one family are analyzed. To confirm the absence of cross-hybridization of the radiolabeled probes, several hybridizing recombinant phages are grown, and single-stranded DNAs are sequenced.

For each set of primers designed to amplify several members of a gene family, a control experiment should be performed to verify that there is no preferential amplification of any one member. Thus, different ratios of recombinant plasmids containing the target sequences can be mixed and amplified, and the numerical evaluation of the recombinant plaques should reflect the initial plasmid ratios.

For 5′-end labeling of oligonucleotide probe:

- 2 μl 10× buffer (use same buffer as for ligation)
- 4 pmol oligonucleotide
- 8 pmol [γ-^{32}P]ATP (>5000 Ci/mmol, Amersham)

- water to 20 μl total volume
- 1 μl polynucleotide kinase (10 U, New England Biolabs)

The reaction is incubated at 37°C for 30 min. After addition of 30 μl of a 50 mM EDTA solution to stop the reaction, the sample is passed over a Bio-spin 6 column (Bio-rad) to remove unincorporated label.

4. Example

We studied the molecular and functional properties of α-amino-3-hydroxy-5-methyl-4-isoxazolepropionate (AMPA)-type glutamate receptors in different cell types of the CNS. Molecular cloning has revealed the existence of four AMPA receptor subunits—GluR-A to -D (or GluR-1 to −4) that can assemble in homomeric or heteromeric configurations to form recombinant receptor channels with disparate functional properties (for review see Wisden and Seeburg, 1993). The AMPA receptor subunits have a length of approximately 900 amino acids and share an overall amino acid sequence identity of about 70%, which is even higher in the conserved putative transmembrane regions. The AMPA receptor subunits occur in two forms—"flip" or "flop"—with regard to an alternatively spliced exonic sequence of 38 residues.

Recombinant AMPA receptors built from one of the four subunits (GluR-A to -D) alone or assembled from several subunits result in receptors with different gating properties and divalent ion permeabilities. Of special note is the reduced Ca^{2+} permeability of GluR-B containing homomeric channels when compared with channels comprising one of the other three subunits (Hume *et al.,* 1991; Burnashev *et al.,* 1992). Furthermore, the low divalent permeability of the GluR-B subunit is a "dominant trait," as it is imparted onto channels of heteromeric configuration when GluR-B is present. A single amino acid—a glutamine (Q) in GluR-A, -C, and -D versus an arginine (R) in GluR-B—is responsible for this difference in divalent permeabilities. Interestingly, the R in GluR-B is introduced by RNA editing (Sommer *et al.,* 1991).

We addressed the question of whether the expression of different AMPA receptor mRNAs in nonpyramidal and pyramidal neurons of visual cortex brain slices could account for the different electrophysiological properties of AMPA receptors in these cells (Jonas *et al.,* 1994). The remarkable difference in Ca^{2+} permeability of native AMPA receptors in nonpyramidal and pyramidal neurons of the visual cortex rendered these two cell types suitable for elucidating the underlying molecular mechanism.

To amplify all subunits, the sense and antisense primers for PCR amplification were designed to sequences conserved in the four AMPA receptor subunits (Lambolez *et al.,* 1992).

The final reaction conditions for the first PCR amplification in a 100-μl reaction volume were as described in Section 3.2.1. However, because of the substantial primer dimer formation, a hot start protocol was chosen for the amplification of the AMPA subunits. Two solutions, A and B, with the following composition are prepared.

Solution A (60 μl per reaction tube) contains the following components:

- 7 μl of 10× PCR buffer
- 53 μl water

Solution B (30 μl per reaction tube) contains:

- 3 μl 10× PCR buffer
- 1 μl of a 10 μM stock solution of up primer
- 1 μl of a 10 μM stock solution of lo primer
- 24.5 μl water
- 0.5 μl *Taq* polymerase

The PCR amplifications were performed in an automated-thermocycler (Perkin-Elmer/ Cetus) in the following way. First the 60 μl of solution A and then a drop of mineral oil were added to the 10 μl of the reverse transcription reaction. The tubes were then placed in the PCR machine, and after a hot start of 3 min, the 30 μl of solution B was added to the PCR tube. The temperature profile for amplification of AMPA receptor subunits was:

- Denaturation at 94°C for 3 min
- Five cycles at 94°C for 30 sec, 45°C for 30 sec, 72°C for 30 sec
- 35 cycles at 94°C for 30 sec, 49°C for 30 sec, 72°C for 30 sec
- 10 min at 72°C to permit completion of final extensions

For the second PCR amplification we used a sense primer (upEco) containing the *EcoRI* restriction site followed by the sequence of the sense primer utilized in the first PCR amplification (Table I). The antisense primer (loKpn1) was a nested primer containing the *KpnI* site and was located 134 nucleotides upstream of the antisense primer used in the first PCR amplification (Table I, Fig. 5). The compositions of solution A and solution B were similar to those used for the first PCR amplification except for the following changes: solution A contained in addition 1 μl of the 100× dNTP solution, and solution B contained the primer pair with the restriction site. A total of 35 cycles were run for the second PCR reaction at an annealing temperature of 49°C for all cycles.

The first and second PCR reactions resulted in DNA fragments of approximately 750 bp and 620 bp, respectively, in length. Following the second PCR amplification, 10 μl of the reaction was resolved on a 1% agarose gel, and Southern blot analysis was performed after transfer onto nylon membranes (0.2 mm, Schleicher & Schuell) according to standard

Table I. Sequences of PCR Primers and Specific Probes Used for AMPA Receptor Subunit Analysis

First PCR amplification
up primer: 5′-CCTTTGGCCTATGAGATCTGGATGTG-3′
lo primer: 5′-TCGTACCACCATTTGTTTTTCA-3′
Second PCR amplification
up Eco primer: 5′-GCGAATTCTTTGGC(CT)TATGA(GA)ATCTGGATGTG-3′
lo Kpn1 primer: 5′-GCGGTACCAAGTTTCC(AT)CC(CA)ACTTTCAT(GC)GT-3′
Oligonucleotides for Southern analyses and quantification
pan A: 5′-GTCACTGGTTGTCTGGTCTCGTCCCTCTTCAAACTCTTCGCTGTG-3′
pan B: 5′-TTCACTACTTTGTGTTTCTCTTCCATCTTCAAATTCCTCAGTGTG-3′
pan C: 5′-AGGGCTTTGTGGGTCACGAGGTTCTTCATTGTTTTCTTCCAAGTG-3′
pan D: 5′-CTGGTCACTGTGTCCTTCCTTCCCATCCTCAGGTTCTTCTGTGTG-3′
AMPA pan: TA(CT)TC(AG)TTCAT(GCT)GT(CG)GACTCCAG
Subunit-specific primers for flip-flop analyses.
A5: 5′-GCGAATTCGAGGGACGAGACCAGACAACC-3′
B5: 5′-GCGAATTCACACAAAGTAGTGAATCAACT-3′
C5: 5′-CCGAATTCACAAAGCCCTCCTGATCCTC-3′
D5: 5′-GCGAATTCCTGAGGATGGGAAGGAAGG-3′
lo Kpn2: 5′-GCGGTACCTCGTACCACCATTTG(TC)TTTTCA-3′

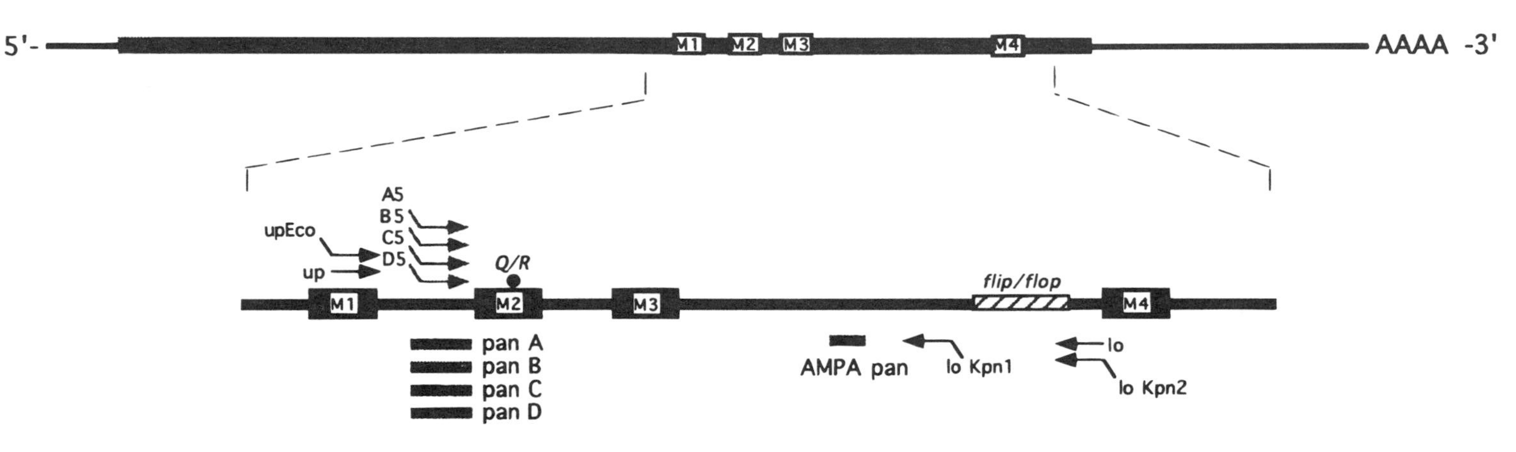

Figure 5. Location of PCR primers and probes on cDNA encoding AMPA receptor subunits. M1 to M4 denote the putative transmembrane regions. The flip-flop module between M3 and M4 is generated by alternative splicing (Sommer *et al.*, 1990). The *Q/R* site indicates the position within M2 where an arginine codon is generated by RNA editing (Sommer *et al.*, 1991). Arrows indicate primers; bars indicate probes used for hybridization.

procedures (Berger and Kimmel, 1987; Ausubel *et al.,* 1987). In our example four blots were hybridized with four specific probes (Table I, Fig. 5). The subunit-specific 45-mer oligonucleotides located between M1 and M2 were 5′-end-labeled with ^{32}P to a specific activity of approximately 10^7 cpm/μg. Hybridizations were carried out in 5×SSC (1×SSC = 0.15 M NaCl, 0.015 M sodium citrate), 50% formamide at 37°C. The washing conditions were 1×SSC, 60°C.

Dual filter lifts (nitrocellulose filters, 0.45 mm; Schleicher & Schuell) from plaque-containing plates were hybridized with the GluR-B specific oligonucleotide (panB) and an oligonucleotide (AMPA pan) designed to recognize all four AMPA receptor subunits. Hybridization and washing conditions were as specified above for the subunit-specific probe. For the 21-mer AMPA pan probe, hybridization was carried out in 5×SSC, 50% formamide at room temperature, and washing was performed in 1×SSC at 40°C. The filters were exposed to x-ray film for 3 to 5 hr. The number of hybridizing phage plaques permitted the evaluation of the abundance of GluR-B mRNA relative to that of the other AMPA receptor mRNA species (Jonas *et al.,* 1994).

Control experiments showed that PCR amplifications as used in these experiment do not lead to a preferential amplification of one subunit. GluR-A, -B, and -C plasmid DNAs were mixed using following amounts: 1:1:10 pg; 1:10:1 pg; 10:1:1 pg; 1:1:1 pg. After DNA amplification and cloning, the numbers of GluR-A-, -B-, and -C-specific recombinant phage plaques were, respectively, 144,98,417; 103,456,25; 446,72,24; 167,158,88. This indicates that PCR amplification followed by subcloning represents a suitable assay to semiquantitatively report the abundance of AMPA GluR-specific mRNAs in a single cell.

The Southern blot analysis revealed an unforeseen diversity of expression profiles of AMPA receptor subunits in the analyzed cells. Both nonpyramidal and pyramidal neurons expressed the GluR-B subunit. The cloning of the PCR fragment obtained from nonpyramidal cells resulted in a significantly lower number of GluR-B-specific plaques compared with the number obtained from pyramidal neurons. This result correlates well with the different Ca^{2+} permeabilities of AMPA receptors in both cell types (Jonas *et al.,* 1994). It should be emphasized, however, that the ratios of AMPA receptor mRNAs in a cell do not necessarily reflect the subunit stoichiometry of native receptors, e.g., because of differences in translation efficacy between different subunits.

Different sets of second primer pairs can be designed to permit detailed analysis of functionally critical regions. The primer pair upEco and loKpn1 allowed analysis of AMPA receptor-specific sequences in a region including the transmembrane segments M1 to M3. Analysis of yet another site of interest—the flip/flop region (Sommer *et al.,* 1990)—was not possible because the location of this site is downstream of the loKpn1 primer. Thus, we reamplified the first PCR product with subunit-specific nested sense primers (A5, B5, C5, D5) located downstream of the up primer and the loKpn2 primer (Table I, Fig. 5). The use of subunit-specific primers for the second PCR amplification for this experiment requires a higher annealing temperature (55°C) to avoid amplification of related subunits. Any further question can refer to AMPA receptors only and is limited to the sequences amplified in the first PCR reaction, as the entire reverse transcription reaction of each cell is used for the subsequent first PCR amplification.

Another approach to analyze gene expression in single neurons is based on RNA amplification and is devised to produce sufficient amplified RNA for the analysis of any expressed gene (van Gelder *et al.,* 1990; Eberwine *et al.,* 1992; Mackler *et al.,* 1992).

It remains to be shown whether a combination of both methods might be advantageous. Thus, an initial step of nucleic acid enrichment via either RNA or DNA amplification with random primers could help in providing sufficient material to permit splitting the material so that PCR reactions with different specific primers could be performed.

ACKNOWLEDGMENTS. We thank J. Geiger and U. Keller for their help with the experiments and Profs. B. Sakmann and P. H. Seeburg for helpful discussions and critically reading the manuscript. We also thank Drs. J. G. G. Borst and M. Koenen for reading the manuscript and Drs. B. Lambolez and J. Rossier for helping to establish the method of single-cell PCR.

References

Ausubel, F. M., Brent, R., Kingston, R. E., Moore, D. D., Seidman, J. G., Smith, J. A., and Struhl, K. (eds.), 1987, *Current Protocols in Molecular Biology,* John Wiley & Sons, New York.

Berger, S. L., and Kimmel, A. R. (eds.), 1987, *Guide to Molecular Cloning Techniques,* Academic Press, San Diego.

Bochet, P., Audinat, E., Lambolez, B., Crépel, F., Rossier, J., Iino, M., Tsuzuki, K., and Ozawa, S., 1994, Subunit composition at the single-cell level explains functional properties of a glutamate-gated channel, *Neuron* **12:**383–388.

Burnashev, N., Monyer, H., Seeburg, P. H. and Sakmann, B., 1992, Divalent ion permeability of AMPA receptor channels is dominated by the edited form of a single subunit, *Neuron* **8:**189–198.

Dodt, H.-U., and Zieglgänsberger, W., 1990, Visualizing unstained neurons in living brain slices by infrared DIC-videomicroscopy, *Brain Res.* **537:**333–336.

Eberwine, J., Yeh, H., Miyashiro, K., Cao, Y., Nair, S., Finnell, R., Zettel, M., and Coleman, P., 1992, Analysis of gene expression in single line neurons, *Proc. Natl. Acad. Sci. USA,* **89:**3010–3014.

Hume, R. I., Dingledine, R., and Heinemann, S. F., 1991, Identification of a site in glutamate receptor subunits that controls calcium permeability, *Science* **253:**1028–1031.

Innis, M. A., Gelfand, D. H., Sninsky, J. J., and White, T. J. (eds.), 1990, *PCR Protocols,* Academic Press, San Diego.

Jonas, P., Racca, C., Sakmann, B., Seeburg, P. H., and Monyer, H., 1994, Differences in Ca^{2+} permeability of AMPA-type glutamate receptor channels in neocortical neurons caused by differential GluR-B subunit expression, *Neuron* **12:**1281–1289.

Lambolez, B., Audinat, E., Bochet, P., Crépel, F., and Rossier, J., 1992, AMPA receptor subunits expressed by single Purkinje cells. *Neuron* **9:**247–258.

Mackler, S. A., Brooks, B. P., and Eberwine, J. H., 1992, Stimulus-induced coordinate changes in mRNA abundance in single postsynaptic hippocampal CA1 neurons, *Neuron* **9:**539–548.

McPherson, M. J., Quirke, P., and Taylor, G. R. (eds.), 1991, *PCR, A Practical Approach,* Oxford University Press, Oxford.

Rychlik, W., and Rhoads, R. E., 1989, A computer program for choosing optimal oligonucleotides for filter hybridization, sequencing and *in vitro* amplification of DNA, *Nucl. Acids Res.* **17:**8543–8551.

Sommer, B., Keinänen, K., Verdoorn, T. A., Wisden, W., Burnashev, N., Herb, A., Köhler, M., Takagi, T., Sakmann, B., and Seeburg, P. H., 1990, Flip and flop: A cell-specific functional switch in glutamate-operated channels of the CNS, *Science* **249:**1580–1585.

Sommer, B., Köhler, M., Sprengel, R., and Seeburg, P. H., 1991, RNA editing in brain controls a determinant of ion flow in glutamate-gated channels, *Cell* **67:**11–19.

Unwin, N., 1993, Neurotransmitter action: Opening of ligand-gated ion channels, Review Supplement to *Cell* **72** and *Neuron* **10:**31–41.

van Gelder, R. N., von Zastro, M. E., Yool, A., Dement, W. C., Barchas, J. D., and Eberwine, J. H., 1990, Amplified RNA synthesized from limited quantities of heterogenous cDNA, *Proc. Natl. Acad. Sci., U.S.A.* **87:**1663–1667.

Wisden, W., and Seeburg, P. H., 1993, Mammalian ionotropic glutamate receptors. *Curr. Opin. Neurobiol.* **3:**291–298.

Yanisch-Perron, C., Vieira, J., and Messing, J., 1985, Improved M13 phage cloning vectors and host strains: Nucleotide sequences of M13mp18 and pUC19 vectors, *Gene* **33:**103–119.

Chapter 17

Force Microscopy on Membrane Patches
A Perspective

J. K. HEINRICH HÖRBER, JOHANNES MOSBACHER, and WALTER HÄBERLE

1. Introduction

Force microscopy, originally named atomic force microscopy by the inventors (Binnig *et al.*, 1986), is a new type of measuring instrument in the nanometer range, down to the size of single atoms, developed after the invention of the scanning tunneling microscope (Binnig *et al.*, 1982). These microscopes represent not only new types of imaging instruments, hence the term "microscopes," but, more importantly, a new technology called scanning probe techniques (SPT). With these methods, one can measure and manipulate in the nanometer range in any environment, opening up a new field of interdisciplinary experimental science. The common features of these microscopes are a measurable and strongly distance-dependent parameter, a probe small enough for the desired spatial resolution, and a mechanism to scan over a surface with the necessary stability.

Different SPT instruments can be grouped on the basis of the type of interaction between probe and sample they use to determine a surface structure. One group uses the exponentially decaying nearfield effect, measuring either photons, as does the scanning nearfield optical microscope (SNOM), or electrons, as does the scanning tunneling microscope (STM). Others measure, for instance, van der Waals forces, such as the atomic force microscope (AFM), or forces acting between surfaces in solution. The latter sometimes is called a scanning force microscope (SFM), as it does not work on the atomic scale. Finally, some instruments such as the thermocouple microscope (TCM) measure temperature in the nanometer range, or the scanning ion conductance microscope (SICM), measuring ion currents with small pipette tips of 100 nm. These are summarized in Table I.

Applications of these instruments to biological material started immediately after the invention of STM: by imaging DNA samples in a vacuum chamber (Binnig *et al.*, 1982). After it became clear that samples can be probed with SPT in air and even in solutions, the study of membranes began on artificial lipid films with STM (Fuchs *et al.*, 1987; Hörber *et al.*, 1988; Smith *et al.*, 1987) and AFM (Egger *et al.*, 1990; Meyer *et al.*, 1989). First attempts to study natural membranes with AFM (Worcester *et al.*, 1988) and STM (Guckenberger *et*

J. K. HEINRICH HÖRBER • Department of Cell Biophysics, European Molecular Biology Laboratory (EMBL), D-69117 Heidelberg, Germany. JOHANNES MOSBACHER • Department of Cell Physiology, Max-Planck Institute for Medical Research, D-69120 Heidelberg, Germany. WALTER HÄBERLE • IBM Physics Group Munich, D-80799 Munich, Germany.

Single-Channel Recording, Second Edition, edited by Bert Sakmann and Erwin Neher. Plenum Press, New York, 1995.

Table I. Types of scanning probe microscopes and their applications

SPM	Interactions	Resolution	Application	Citations[a]
STM	Electron–electron	10^{-11} m	Conducting substrates	(Binnig *et al.*, 1982)
SNOM	Electrodynamic	5×10^{-8} m	Spectroscopy	(Pohl *et al.*, 1988)
AFM, SFM	Van der Waal, electrostatic, friction	10^{-10} m	Conducting and nonconducting substrates	(Binnig *et al.*, 1986, Martin and Wickramasinghe, 1988)
SCM	Electrostatic	10^{-9} m	Capacitance	(Matey and Blanc, 1985)
MFM	Magnetic	2.5×10^{-8} m	Magnetic storage devices	(Sáenz *et al.*, 1987)
TCM	Phonons (temperature)	5×10^{-8} m (10^{-4} degrees)	Photothermal adsorption spectroscopy	(Williams and Wickramasinghe, 1988)
SICM	Ion currents	$\approx 10^{-7}$ m	Electrochemistry, ion channels	(Hansma *et al.*, 1989)

[a]Citations are those that gave the first descriptions of each technique.

to study natural membranes with AFM (Worcester *et al.*, 1988) and STM (Guckenberger *et al.*, 1989; Hörber *et al.*, 1991; Jericho *et al.*, 1990) followed, but these showed that reliable preparations of such membranes on substrates are quite difficult to obtain because numerous surface effects influence the preparation and the imaging and introduce considerable difficulty in identifying structures.

Therefore, it is tempting to combine SPT with patch-clamp techniques to improve the preparation of membranes by holding whole cells or membrane patches at the tip of a pipette. In this way the membrane is stabilized by cellular structures and by the attachment to the supporting glass rim of the pipette which can be verified by measuring whether a gigaseal is formed.

This approach to investigating cell membranes emerged in 1989, when the first reproducible images were obtained of the plasma membrane of a living cell (Häberle *et al.*, 1989). In this setup, the cell, fixed by a pipette in its normal growth medium, could be kept alive for several days, and changes in the shape of the plasma membrane were recorded in a video movie. Furthermore, all electrophysiological techniques can be used simultaneously in such experiments.

With this step in the development of scanning probe instruments, our ability to investigate the dynamics of biological processes of cell membranes under physiological conditions could be extended into the nanometer range. Presently, structures as small as about 10–20 nm can be resolved, which gives access to processes such as the binding of labeled antibodies, endo- and exocytosis, and pore formation.

With the probe of an SFM, forces are applied to the investigated plasma membrane, and so the mechanical properties of cell surface structures become involved in the imaging process. On one hand, this fact mixes topographic and elastic properties of the sample in the images; on the other hand, it provides additional information about cell membranes and their dynamics in various situations during the lifetime of the cell. But the two aspects in the SFM images have to be separated. Thus, independent data are needed, such as topographic data obtained by electron microscopy or by modulation techniques of the SFM.

The ability of the SFM to apply modulated forces very locally, down to the scale of piconewtons, onto the cell membrane gives access to the mechanical properties of the plasma membrane and the processes of cellular reactions on mechanical stress at a molecular level. The structures controlling such processes are probably ion channels, which can be studied by the combination of SFM and the patch-clamp technique. Furthermore, the combination can help to improve the preparative methods of cell membranes on solid supports for SPT in general, as the attachment to the glass rim of the pipette can be controlled by measuring the resistance over the membrane, which reflects the contact between membrane and glass surface. The structure of the glass surface in solution can be characterized by the SFM with nanometer resolution. If this is done before the pipette is used the surface structures involved in forming a gigaseal can possibly be analyzed.

The thickness of the glass wall of the pipette (0.5–1 μm) is large enough to be used as a solid support for SFM investigations of the membrane, which should give the same high resolution of about 1 nm as gained on molecular structures fixed on other solid supports. In addition, with such preparation processes, the membrane spanning the pipette opening can be studied if manipulations of the environment on both sides of the membrane or the undisturbed transport through the membrane is required, although in this case the resolution might be lower.

2. Force Microscopy

An important part of scientific development is linked to extending the range of our senses. Microscopes and telescopes have improved our sight, analytical chemistry our sense of smell and taste, and electronics our hearing. With SFM we have now started to extend the range of our sense of touch into the dimensions of single molecules. Learning about our world by touching things is quite fundamental, and the idea of building instruments for this purpose is quite old. The most important aspect is the dependence on the size of the probe used. There is surely a dependence on the material of the tip used not only because the tip material determines the forces acting between tip and surface, but also because the tip geometry alone makes quite a difference.

On a very small scale, with a single atom as the final tip, only really flat surfaces on an atomic scale of Ångstroms can be investigated for geometric reasons, and what is learned about the surface depends on the interaction between the tip atom and the single atoms of the surface. Such a situation, however, can only be prepared under extremely clean conditions.

In the nanometer range, the tip, depending on the way it is produced, can look rather rough but can still have a single atom as the final tip. On this scale, objects of interest can also have quite rough surfaces, and one has to take into account that the images are a superimposition of the tip geometry and the surface geometry. The effects occurring in the nanometer range, especially in solutions, are quite complex (Israelachvili, 1992) and may vary significantly over a surface. If the tip radius gets into the micrometer range or even larger, then these effects are averaged, and everybody knows what happens when a needle moves over a record surface.

The tips usually used for an SFM are etched silicon structures made with microfabrication techniques, together with the whole cantilever used to measure the applied force (Fig. 1). Microfabrication has to be used because the lever should be sensitive in the nanonewton range, which requires a small spring constant ($k \leq 0.1$ N/m), and the resonance frequency should be high ($f \approx 1$ kHz) to allow scanning at an acceptable speed. Because the resonance frequency f is coupled to the spring constant k and the effective mass m of the lever [$f = (1/2\pi)\,(k/m)^{-1/2}$], the latter has to be in the microgram range to satisfy the above mentioned

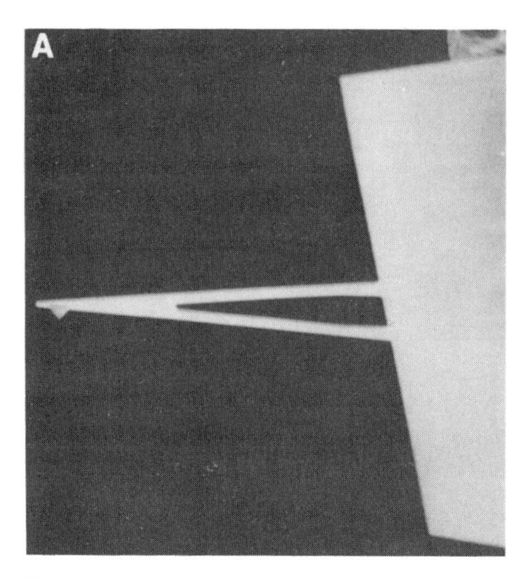

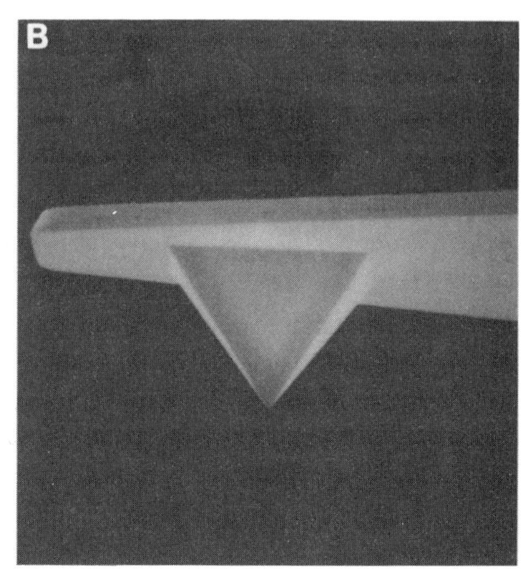

Figure 1. Cantilever and tip of a scanning force microscope imaged by a scanning electron microscope with 700 times and 7000 times magnification on left and right sides, respectively. The lever has a length of about 100 μm, and the pyramidal-shaped probe has a height of about 3 μm. The images are provided courtesy of Park Scientific Instruments.

conditions. For such a lever fixed at one end, the movement with the lowest resonance frequency is bending up and down. For torsions, the levers are stiffer and therefore have higher resonance frequencies. For other degrees of freedom they are even stiffer. Thus, in the case of small forces, the spring constant for bending up and down plays the principal role.

This z movement of the lever, which is proportional to the force between tip and sample perpendicular to the surface, can be detected either by an STM tip at the rear of the SFM cantilever (Binnig *et al.,* 1986) or by optical detection systems (Meyer and Am, 1988). The method used in most instruments is the deflection of a laser beam focused on the gold-coated back of the end of the lever. If the lever is bent, this laser beam changes its direction slightly. This can be detected by a bicell photo diode, which gives a voltage signal proportional to the differences of light intensities detected by its two sides.

With this detection of the force acting on the tip, there are two possible ways to produce an image. One is to hold the force constant with a feedback circuit and detect the z movement of the sample necessary to do so; the other is to hold the z position constant and detect the deflection of the lever.

Further information can be obtained by analyzing friction forces acting lateral to the scanning direction if they are strong enough to introduce torsion of the lever, which also leads to a change in the deflection of the laser beam. These two different types of interactions between tip and sample can be separated if a quadrant photo detector is used and the laser beam runs, for example, perpendicular to the length axis of the lever. In this way, torsion changes the difference between the upper half and the lower half of the detector, and the normal bending produces changes in the difference between the left and right halves of the detector.

The forces measured by such tips on surfaces are the result of electromagnetic interaction, which becomes rather complex because more than two charges are involved. Especially in ionic solutions, the ambient conditions have to be taken into account, which can add many inducible and rearrangeable dipole moments. What is measured by the deflection of the cantilever in this case is the sum of all attractive and repulsive forces. Therefore, the topography determined corresponds to the forces involved in the measuring process and is changed if these forces change, which can occur, for example, with changes in the environment. Also, it should be always kept in mind that the surface of the sample and the surface of the tip are influenced by the environmental conditions. Under ambient conditions all solids are covered by thin layers of water. The final thickness of the water layer depends on the humidity and the temperature difference between air and solid. If a typical SFM tip approaches a surface, these water layers can give rise to very strong capillary forces ($\approx 10^{-6}$ N) when the gap is closed by a water bridge. This effect can be avoided if the whole sample and the probe are kept in water. In this case, smaller forces between the tip and the surface can be detected much more easily when the tip approaches the surface. For instance, the van der Waals force is attractive at a distance of about 2 nm ($\approx 10^{-10}$ N) but is canceled by a repulsive force below 1 nm; the equilibrium point is called the van der Waals radius. The repulsive force rapidly gets stronger with decreasing distance, which is because of the Pauli principle not allowing two electrons to be in the same quantum-mechanical state.

If additional molecules are dissolved in the surrounding water, other types of interactions may occur. One is a hydrophobic force keeping nonpolar molecules together, even if they are already adsorbed on a surface in water. Another force is produced by concentration gradients; for instance if a surface releases ions into the water, a layer of higher concentration is formed above the surface. A very important force, especially for larger biological molecules, is the hydration force. It is established by the much tighter binding of the last, or the last few, layers of water molecules. This force is affected quite strongly by ions and charged parts of molecules. The forces involved are often of the same magnitude as some internal molecular forces that lead to the overall shape of the molecules. These forces play an important role in the final folding of larger molecules and especially in their interaction. These layers may also be responsible for structures seen with the SFM, for example, with DNA preparations, which differ strongly from the expected structures of molecular computer models, showing clearly that on this nanometer scale the type of interaction used for measuring determines the topography of the surface detected. Strongly bound adsorbates to molecular surface structures such as water and ions are part of this structure and are therefore measured by the SFM as the surface contour, determined by the balance of the forces applied by the lever and the binding forces of the adsorbates in the environment.

To interpret the results obtained by SFM and scanning probe instruments in general, it is therefore necessary to take into account the environment and the type of interaction involved. If the resolution is at the atomic scale, the electrostatic interaction occurring between atoms plays the most important role. On a scale of 1–10 nm, effects like the hydration force are more important. At resolutions of less than 10 nm, the shape of the tip normally plays the most important role, as there is considerable averaging over various types of interaction. The resolution below 10 nm can change rapidly with changes in the tip structure. This shows that the final structure of the tip often is determined in solution by adsorbates that can suddenly appear or disappear.

In summary, images with high resolution are made mostly by luck when the environment fits to the tip and surface structure in a way that the interactions between tip and surface decay very rapidly with distance. Images with a resolution in the 10-nm range are made more regularly with tips sharp enough to give this resolution.

3. Force Microscopy on Whole Cells

Figure 2 shows the schematic arrangement of the SFM above the objective of an inverted optical microscope. The sample area is observed from below through a planar surface defined by a glass plate at a magnification of 600–1200× and from above by a stereomicroscope with a magnification of 40–200×. The illumination is from the top through the less-well-defined surface of the aqueous solution. In order not to block the illumination, the manipulator for the optical fiber and for the micro pipette point toward the focal plane at an angle of 45°. The lever is mounted in a fixed position within the

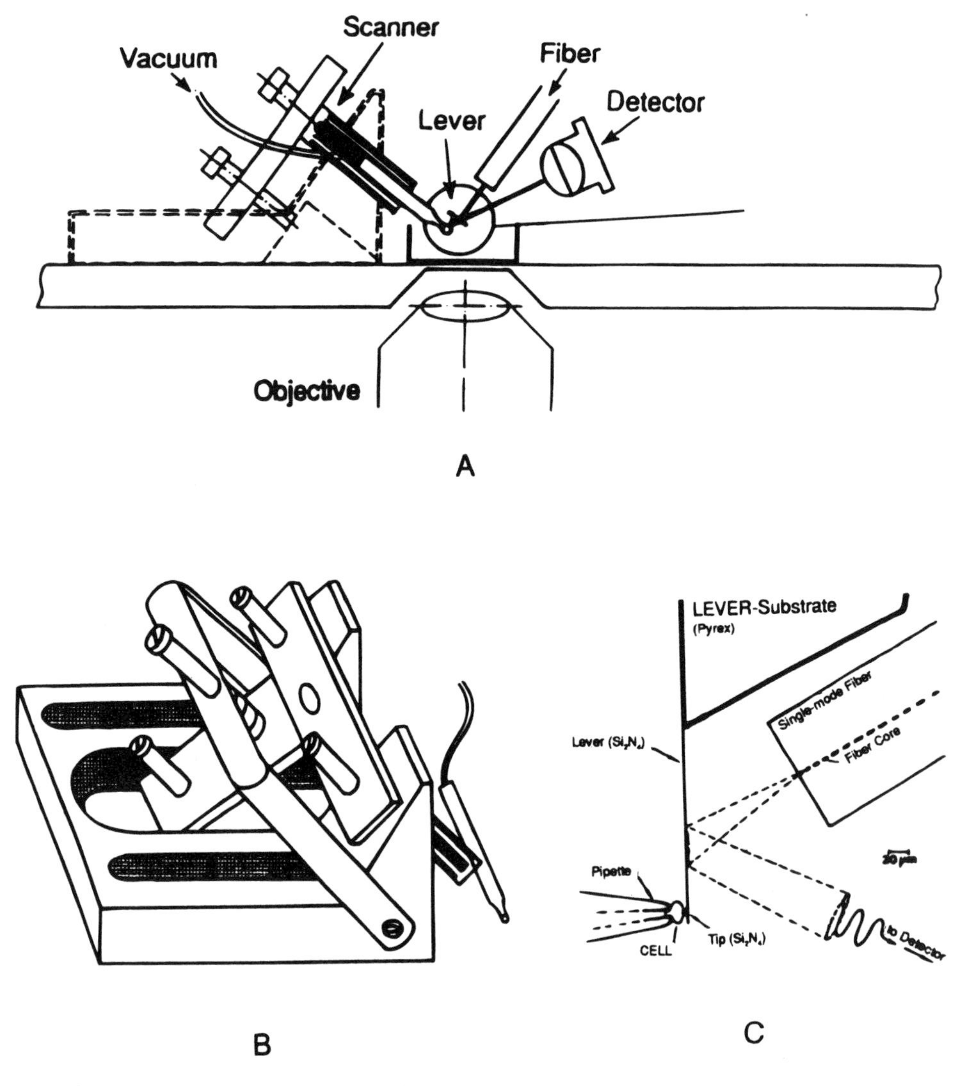

Figure 2. Scheme of the setup used for whole-cell imaging mounted on a inverted optical microscope. The mechanical part of the piezo tube holder is shown in B. C shows the geometry of the optical system in a magnification of 500 : 1.

liquid slightly above the glass plate tilted by 45°. In this way the lever is deflected in a direction parallel to the optical fiber. The single-mode optic fiber is used to reduce the necessary optical components by bringing the end as close as possible to the lever. For this reason several millimeters of the fiber's protective jacket are removed. The minimum distance is determined by the diameter of the fiber cladding and the geometry of the lever. For 633-nm light and levers that are 100 μm long, we normally use a fiber that has a nominal cladding diameter of 125 μm. Holding the fiber at an angle of 45° with respect to the lever means that one can safely bring the fiber core to within about 150 μm of the desired spot on the lever. The 4-μm-diameter core has a numerical aperture of 0.1, and the light emerging from the fiber therefore expands with an apex angle of 6°. For the geometry given above, the smallest spot size achievable is 50 μm, approximately the size of the triangular region at the end of the cantilever. Because of the construction of the mechanical pieces holding the pipette and fiber positioner, the closest one may bring the position-sensitive quadrant detector to the lever is approximately 2 cm. This implies a minimum spot size of 2.1 mm within the 3-mm × 3-mm borders of the detector. With a 2-mW HeNe laser under normal operational conditions, the displacement sensitivity is below 0.01 nm with a signal-to-noise ratio of 10 and a bandwidth of 1 Hz, comparable to the sensitivity of other optical detection techniques. The advantage of this method is that there are no lenses, and there is no air/liquid or air/solid interface across which the incoming light beam must travel. Only the outgoing light has to cross the water–air interface, as the detector cannot be immersed in water.

The pipettes used for holding the cell are made out of an 0.8-mm borosilicate glass capillary that is pulled to about 2–4 μm in three pulling steps. It is mounted on the piezotube scanner and coupled to a fine and flexible Teflon® tube through which the pressure in the pipette can be adjusted by a piston or water pump. The pipette is fixed at an angle that allows imaging of the cell without the danger of touching the pipette with the lever. These components are located in a container of 50-μl volume. The glass plate above the objective of the optical microscope forms the bottom of this container.

After several microliters of a cell suspension are added, a single cell is sucked onto the tip of the pipette and fixed there by a low negative pressure in the pipette. The fixed cell is placed close to the SFM lever by a rough approach with screws and finally positioned by the piezo scanner. When the cell is in close contact with the tip of the lever, scanning of the capillary with the cell attached leads to position-dependent deflections of the lever. The levers used are microfabricated silicon and silicon nitrite triangles with 100-μm length and with a spring constant of 0.12 N/m. They were provided courtesy of Shinya Akamine (Stanford University) and Mike-Kirk (Park Scientific Instruments). The forces that can be applied with these levers and the sensitivity of the detection system can be as low as 0.1 nN.

With such forces applied to the plasma membrane of a cell, the stiffness of the membrane structures is important. The scaffold of the cortical layer of actin filaments and actin-binding proteins that are cross-linked into a three-dimensional network and closely connected to the surface membrane may be resolved best. These filaments have a structural role: they may pull on the membrane and create the changes seen in the SFM by means of the dynamics of the surface protrusions (Fig. 3).

The stable arrangements of the actin filaments are responsible for their relatively persistent structure. During phagocytosis or cell movement, however, rapid changes of shape occur at the cell surface. These changes depend on the transient and regulated polymerization of cytoplasmic free actin or the depolarization during the breakdown of

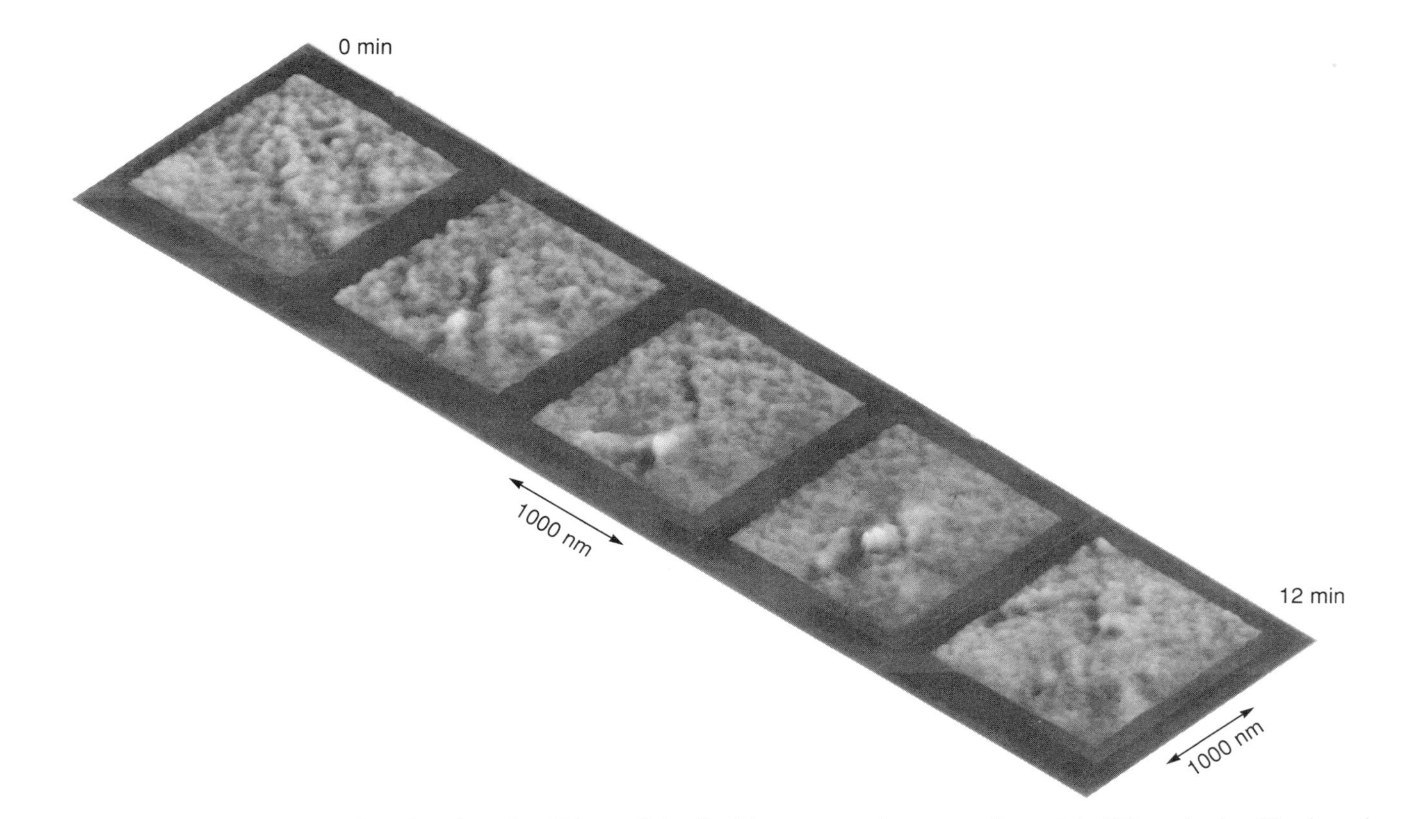

Figure 3. Plasma membrane structures of a cultured monkey kidney cell involved in an exocytotic process of a particle 100 μm in size. The dynamic process lasting several minutes could be documented using real-time video images, which are represented here by several single images.

the actin filament complex. The time scale of cell surface changes on a larger scale observed with the SFM was about 1–2 hr at 24°C.

The method of fixing a cell to a pipette tip is flexible enough to allow integration of and combination with well-established manipulations used in investigations of single-cell properties. The structures observed might be partially related to known features of membranes, but detailed structural analysis has to be left to further investigation. More important than the observations of structures with this technique is the possibility of conducting dynamic studies of living organisms on this scale (Hörber *et al.*, 1992). In general, this technique makes studies of the evolution of cell membrane structures possible and provides information that brings us closer to understanding not only the "being" of these structures but also their "becoming."

4. Force Microscopy on Excised Patches

One obvious next step in the development of a force microscope for biological membranes is the integration of the SFM in a patch-clamp setup to scan excised membranes under controlled physiological conditions, such as ion gradients and transmembrane voltage. This section describes first attempts with such a device and is divided into three parts. Construction details for do-it-yourself researchers are discussed first. Thereafter, technical details concerning experimental procedure and the expected resolution of images from "soft" biological samples based on experience with our first trials are given. At the end of this chapter we give an outlook on further developments and applications of the SPT on cells and membranes.

4.1. Construction of the Setup

Several conditions must be fulfilled to facilitate the implementation of an SFM into a patch-clamp setup: the tip of the cantilever has to be placed almost directly opposite to the patch pipette; there has to be enough space for the optical detection system, laser, mirrors, and photodiodes; and, finally, it must be taken into consideration that the stability of the combination lever–chamber–patch pipette has to allow imaging reproducible in the range of nanometers.

A device built with these constraints in mind is shown schematically in Fig. 4. The top and bottom of the bath chamber consist of two coverslips holding the bath solution by surface tension. The dimensions of the chamber are about 15 mm (l), 8 mm (w), and 5 mm (h). The two sides of the chamber are open, which allows the patch pipette to come from one side and the application pipette from the other side. The SFM cantilever is fixed by a steel spring that is pressed against the wall of the chamber by a screw. It is located on the side of the application pipette. The laser beam (1-mW laser diode, $\lambda = 670$ nm, collimator lenses included in the mounting, smallest focus distance about 10 mm, Laser 2000, Weβling, Germany) is set parallel to the length axis of the cantilever. It is focused by the lenses of its collimator on the triangular-shaped end of the cantilever. The quadrant photodetector (5 mm^2 active surface, AMS Optotech, Munich, Germany) is covered by a 670-nm band-pass filter (e.g., L.O.T., Darmstadt, Germany) to reduce influences of ambient light sources. About half of the surface of the detector is illuminated by the laser beam. Both laser and detector are at a distance of about 50 mm from the

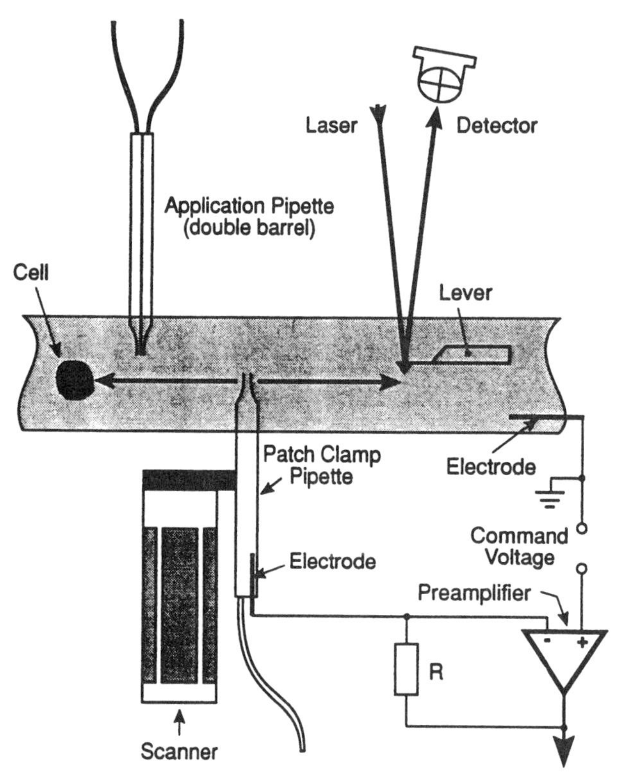

Figure 4. Schematic top view of the bath chamber used in the SFM/patch-clamp setup. Two coverslips as top and bottom hold the solution in the chamber, which is perfused from left to right. Patch and application pipette enter the chamber from the side. The whole chamber and the optical system are mounted on an *xyz* manipulator, whereas the patch-pipette holder is glued on the piezo scanner, which is fixed on a piece of macor and screwed on the mounting plate. The SFM cantilever is held by a steel spring.

cantilever. Thus, the magnification of the movement of the lever can be calculated to be about 5×10^3 (distance lever–detector divided by the length of the cantilever).

Laser and detector are adjustable by microscrews in three dimensions. We took parts of the Microbench provided by Spindler and Hoyer (Göttingen, Germany) to construct the optical system. A glass coverslip is placed at the positions where the laser beam crosses the water–air interface to suppress changes of beam direction from fluctuations of the water surface. The chamber and the double-barrel pipette for fast application of substances that may change the channel structure, as well as the optical detection system are mounted on a motor-controlled *xyz* manipulator with a step size of 10 nm (Marzhäuser, Wetzlar, Germany). All parts except those that have direct contact to the bath solution are made from steel. The quadrant piezo tube with a diameter of 10 mm and a wall thickness of 0.6 mm (Stavely Sensors, East Hartford, CT) is glued on a piece of macor or zerodur (Schott Glaswerke, Mainz, Germany) by ceramic glue (e.g., Torr Seal, Varian Vacuum Products, Lexington, MA). Because high voltages up to 500 V are applied to the piezo, it should be well isolated from metal parts. On the other side, the piezo should be fixed by very stiff material to provide a high resonance frequency.

The pipette is held by the holder fixed on the other end of the piezo tube. This pipette holder was made of steel because of its mechanical properties and its ability to shield the piezo by grounding the holder. Otherwise, the noise induced by the high voltage applied on the piezo for scanning would disturb the patch-clamp current measurements. A glass coverslip between pipette holder and piezo tube was used to insulate the two parts.

The microscope can be moved by an *xy* table relative to the 30-mm-thick steel plate on which the *xyz* manipulator holding the chamber and the optical system is fixed. Because the patch pipette has to be moved for scanning in the SFM application, it cannot be fixed on the headstage by standard pipette holders. We added a short silicon tube between the pipette holder and the pipette to apply suction or pressure.

The scan direction was parallel to the axis of the lever. In this way frictional forces can be separated from topological data either by the information from the quadrant detector or by subtracting the backward and forward scans. The feedback electronics used were essentially similar to that developed at Stanford University for an STM (Bryant *et al.*, 1986). A schematic drawing of the SFM electronics is shown in Fig. 5A. The asymmetric HV amplifier and the feedback circuit are shown in diagrams in Figs. 5B and 5C. The quadrant photo diode was read out as shown in Fig. 5D and connected to the feedback circuit and the video card.

A fast and feasible way to build an SFM integrated in a patch-clamp setup would be the construction of the mechanical parts and the integration of commercially available electronic hard- and software. Some companies supply these parts from their SFM or STM devices (e.g., RHK/USA), but complete commercial instruments at present are not flexible enough to be integrated into a patch-clamp setup.

4.2. Experimental Procedures

It is essential to adjust the laser beam accurately and to correlate the voltage applied to the piezo with the lateral (*xy*) amplitude as well as the *z* variation of the pipette tip. A piece of mica, or HOPG, or other known structures such as a diffraction grating glued on a glass capillary tube can be used to calibrate the scan width of the piezo. Michelson interferometry can be used to calibrate the *z* movement of the piezo in an external device. In instruments like the one described here, the length of the pipette influences the scan width. Therefore, it is always recommended to position the pipette tip at the same place, for example, by using a marking fixed relative to the pipette holder.

A fast rough calibration is obtained by scanning a pipette with a known tip resistance in physiological solution and correlating the image with the resultant estimated tip diameter. We used pipettes pulled from thick-wall borosilicate glass with inner tip diameters of 1–3 μm. The tip surfaces never appear as perfect as the images obtained by an SEM. Metal coating and the vacuum environment of the SEM seem to smooth out the structure of the glass and prevent adsorbents on the tip. We found that with standard heat polishing little effect on the shape of the glass surface was seen, and it seems to be useful mainly for removing dust and adsorbents from pipettes that have not been freshly pulled.

Steep, rigid structures such as an open pipette are, however, ill-suited for the SFM because they can destroy the tip of the probe if the feedback circuit of the force microscope is too slow to pull back fast enough during the approach. Therefore, the approach of the pipette should be done under visual control (Fig. 6) with a slow scan rate and a scan width just large enough ($\approx$ 1 μm) to find the pipette in an acceptable time. The approach can take several minutes before a satisfying image is obtained. It is recommended to find and store the position of the lever tip before isolating a patch. In this way, the pipette can be moved rapidly to the lever tip after formation of the membrane patch. While the SFM is used, the motors of the *xyz* stage have to be turned off, as some motor-driven manipulators may oscillate with an amplitude in the range of the specified precision.

In principle, it would be possible to isolate membrane patches from preparations

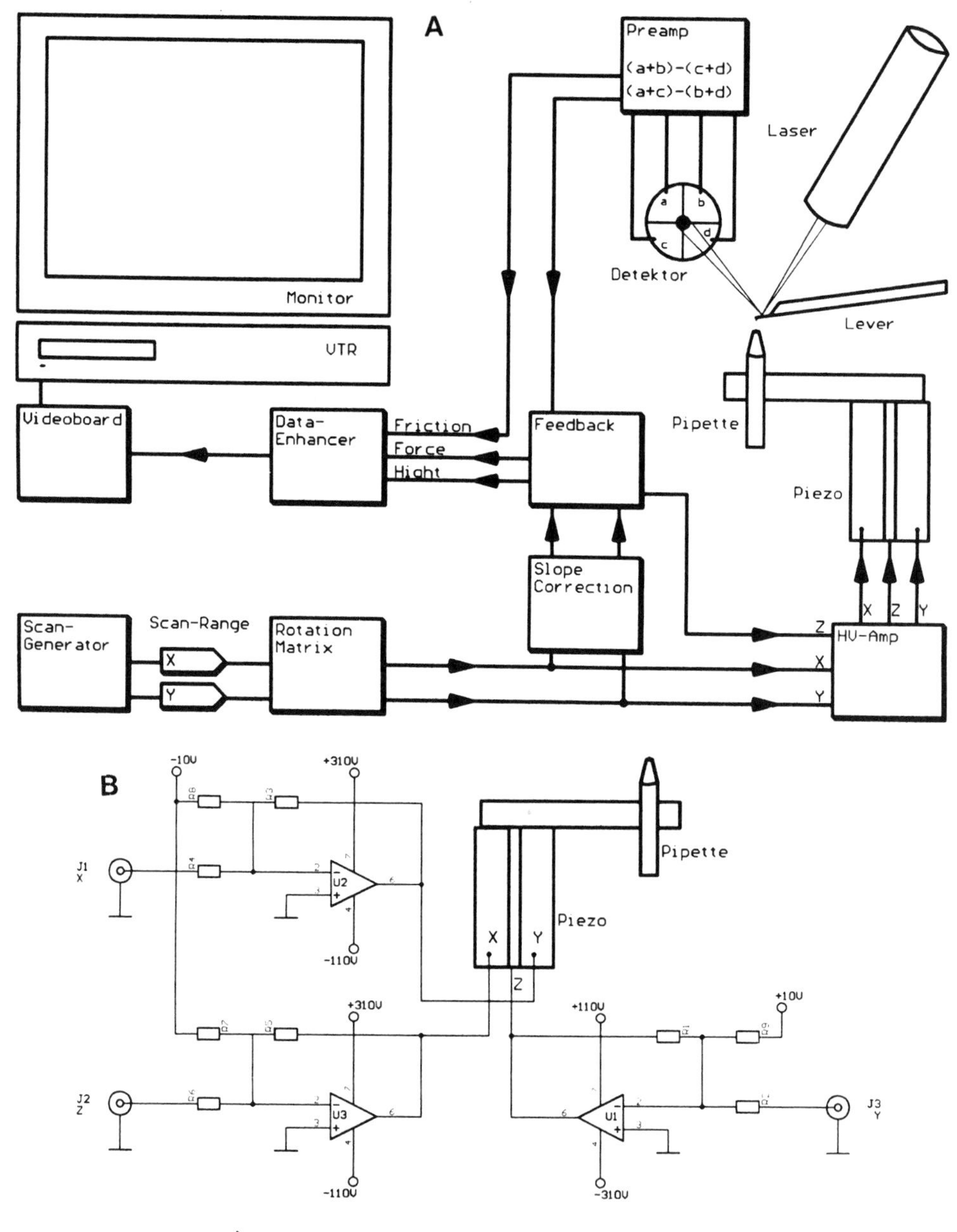

Figure 5. Instrumentation and electronics used for the SFM. (A) Schematic overview of the components with their connections. Descriptions of single parts are given in the text and in the captions to B–D. (B) High-voltage amplifier. The specified voltage range of piezo tubes depends on the wall thickness, with a reverse voltage of only about half of the maximum forward voltage. If a symmetrical ± 150-V high-voltage amplifier is used, the wall thickness has to be chosen in such a way that the reversed voltage does not damage the tube. Therefore, the full voltage range of the piezo cannot be used. To achieve the maximal scanning range, we used an asymmetric high-voltage amplifier applying -100 V to $+200$ V to the z electrode and -200 V to $+100$ V to the four outer electrodes. In this way the total voltage ranges from -200 V to $+400$ V seen from the inner electrode and makes full use of the possible scanning range of the piezo. (C) Feedback circuit. The task of the feedback circuit is to keep the force between the lever and the surface constant. To accomplish this, the force detected by the preamplifier is subtracted from the set point given by a potentiometer in the feedback circuit. This difference is integrated and fed into the z part of the

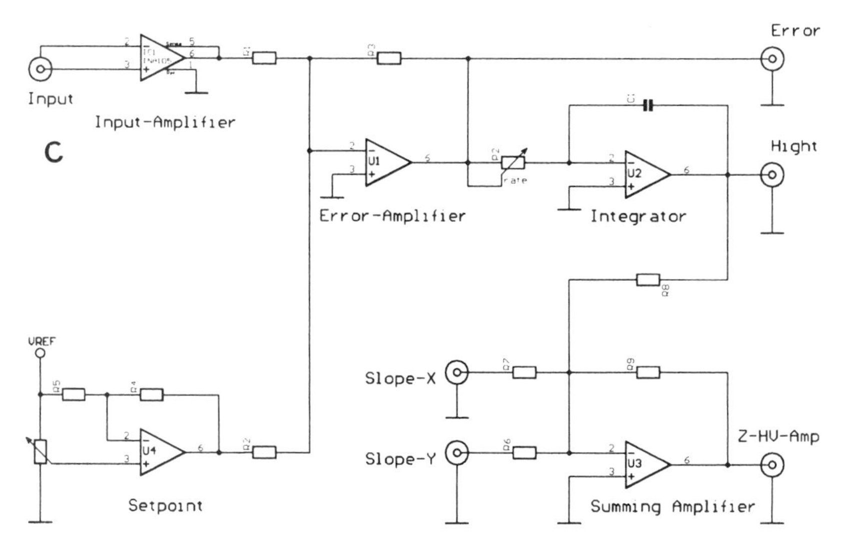

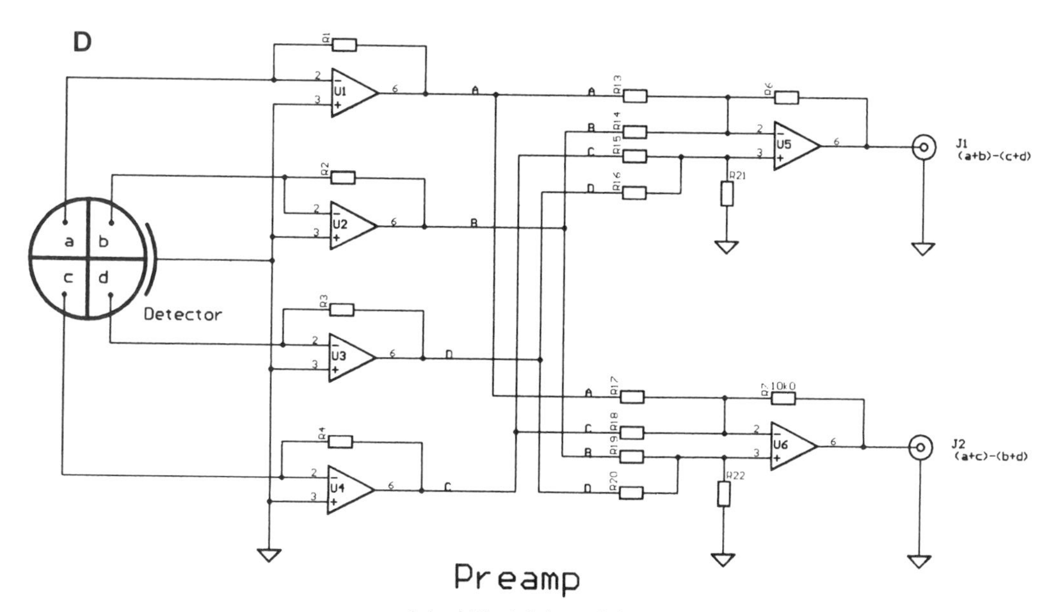

high-voltage amplifier. If the force is too weak, the piezo approaches; if it is too strong, the piezo retracts until the force is equal to the set point. Some additional inputs allow the correction of a slope of the surface by adding adjustable parts of the scan voltages to z (slope correction). In this way there are two possible ways to form an image. (1) The z signal can be used to obtain information about the height differences on the surface. To do so, the scan has to be slow enough, using a small time constant of the integrator, so that the z voltage and the piezo can follow the surface immediately, keeping the force constant (constant-force mode). (2) The deviation of the force while scanning over the surface can be used at relatively high speeds, while the average force is kept constant (constant-height mode). In this way, the z voltage is almost constant, and height differences of the surface cause a deviation of the force between surface and lever tip. (D) Preamplifier for laser detection. The deflection of the laser beam is measured by a quadrant detector, giving three different signals: the sum of all four elements, to measure the total intensity and get the maximum signal from the sensor; the difference between the upper and lower elements, to get the force between the lever tip and the surface; and the difference between the left and right elements, which gives the friction between the lever and the surface. Since the elements of a quadrant detector deliver a current proportional to the intensity of the incoming light, this has to be converted into a voltage signal by the current-to-voltage converters on the left side of the scheme. These signals are finally added and subtracted by an operational amplifiers to achieve the desired three signals.

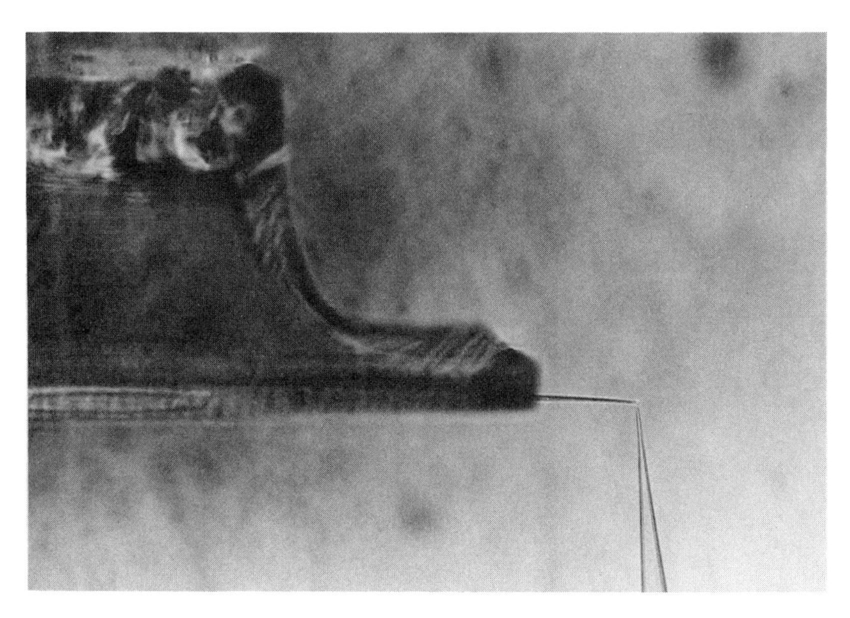

Figure 6. Image of a patch pipette in front of the SFM lever. The silicon substrate of the lever is visible on the upper left corner of the image. At the end of the lever, the tip is detectable as a small point directed to the tip of the pipette. A 10× objective magnification was used.

like brain slices or cultured cells in the setup described above. To do so, however, it is necessary to position the patch-clamp pipette in an angle of at least 15° relative to the chamber to excise the patch and then turn back to the horizontal plane for scanning the patch. This would be possible with a $xyz\phi$ manipulator. Our experiments were made with *Xenopus* oocytes to circumvent this problem. These large cells (diameter is about 1 mm) are easily accessible from the side when they are attached to the bottom of the chamber. We prepared oocytes as usual (Methfessel *et al.*, 1986) and positioned one of them on the bottom of the chamber.

A further reason for using oocytes is the fact that stretch-activated channels are present in their membrane and characterization of mechanosensitive (MS) channels is a promising task for experiments with a SFM/patch-clamp setup. With the probe of the SFM, well-defined forces down to the piconewton range can be applied. Furthermore, heterologously expressed recombinant MS channels (e.g., Hong and Driscoll, 1994; Huang and Chalfie, 1994) can be investigated in wild-type as well as mutated forms. It should be possible to activate single MS channels and verify this by measurements of membrane currents. Ultimately, it may be possible to estimate the force needed to gate a single MS channel. This experiment, however, exceeds the current limit of resolution of an SFM: a single MS channel of a stereocilium coupled to a tip link of a hair cell can be activated by forces of some piconewtons (for review, see Hudspeth and Gillespie, 1994). The minimal force that can be applied perpendicular to the membrane by a cantilever tip in the contact mode under water is about 10 pN, and the extent to which this direction of the force is transformed into the direction necessary to open these channels will depend on the elastoviscosity of the membrane.

At the moment, before experiments like the activation of a single MS channel can be performed, some problems in the approach to an outside-out patch still have to be solved. We have not been able, until now, to scan an outside-out patch. When the tip of the patch pipette approached the SFM pyramidal tip, the patch broke, which was detectable by the sudden resistance decrease simultaneous with the first signal from the SFM. Assuming an omega-shaped outside-out patch, it seems possible that the tip cuts the patch when approaching it from the side. Therefore, the exact approach of the pipette tip toward the lever tip has to be improved.

The external face of the plasma membrane can be studied at present only if vesicles are formed at the tip of the pipette. Vesicles were obtained by pulling back the patch pipette after forming a gigaseal between pipette tip and cell membrane. Vesicles seem to be attached much deeper in the pipette, and the membrane on the pipette tip is attached inside the surrounding glass walls. An example of an image of a pipette tip holding a membrane vesicle is shown in Fig. 7. Part of the vesicle membrane is visible as a smooth structure in the middle of the pipette opening, with the resolution in some areas as good as about 50 nm. It should be possible to improve the setup in a way to increase the resolution to 10–20 nm on membrane patches, similar to the resolution obtained for whole cells. A higher resolution (1–2 nm) should be addressed by scanning the membrane on the rim of the pipette where the membrane is stabilized by the glass surface.

Despite the lower lateral resolution, changes in the height (z direction) are accurately measurable ($\Delta z \approx 2$ nm). This enables investigations of membrane changes to the application of suction or pressure or to voltage pulses. Two examples are presented in Figs. 8A and B. We applied voltage pulses of ± 100 mV for 1 sec or 2 sec across the vesicle. Some vesicles changed the position of their membrane simultaneously with the voltage pulse, and it seems promising to investigate the behavior of membranes when

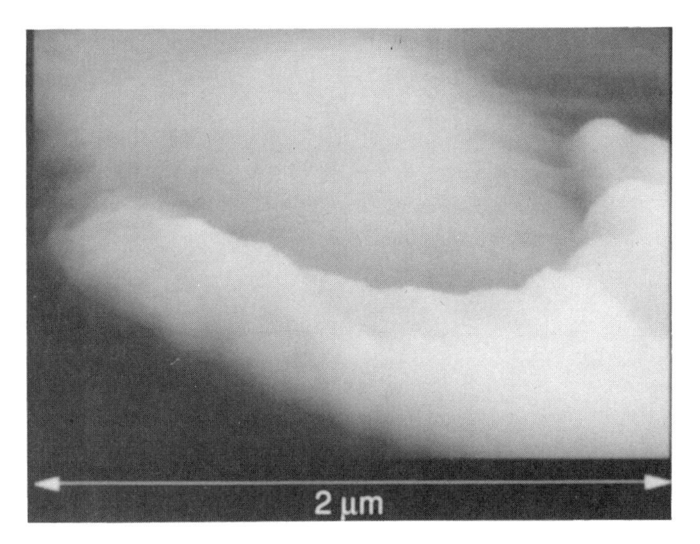

Figure 7. Tip of the patch pipette holding a vesicle. The vesicle is detectable as an unsharp structure in the middle of the pipette tip. On the right side of the pipette, the bending of the vesicle on the glass wall is detectable. SFM images of soft samples often show a lower resolution than those of more rigid structures. Additionally, the higher frictional forces distort the image because these forces produce pseudoheights or -depths depending on the scan direction. Scan frequency was 40 Hz, and scan size was 2 µm. The image was obtained in the constant-force mode, and the height differences between the white and dark parts of the image are about 1 µm. Raw data of the scan were filtered and displayed as a side-view image.

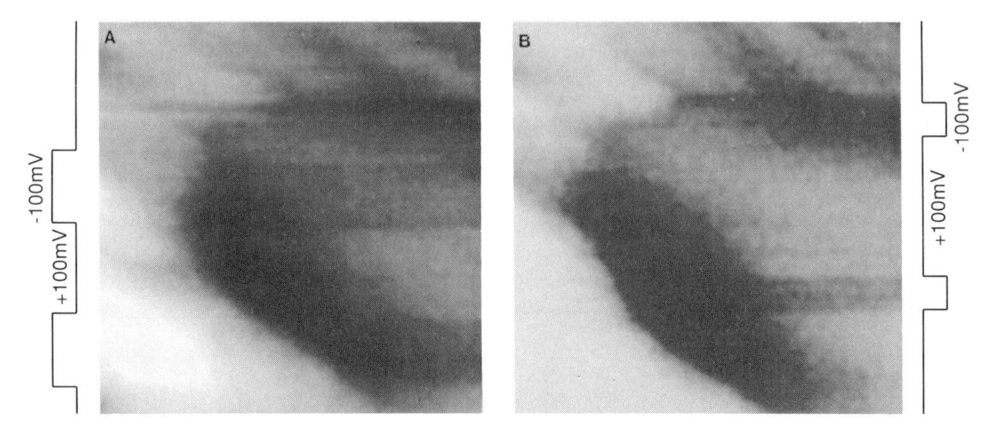

Figure 8. Scanning force microscope images of a vesicle to which voltage steps are applied from +100 mV to −100 mV for 1 sec (right) and 2 sec (left). It produces a movement of the outer membrane of about 10 nm, detectable as darker bars in the image (i.e., moving deeper in the pipette). These bars are visible only on the membrane but not on the glass wall, indicating that they are induced only by a movement of the membrane itself. On the left and right sides, the time courses of the applied voltage are displayed. The vesicle was scanned from above with a scan frequency of 20 Hz. Raw data of the whole image are shown.

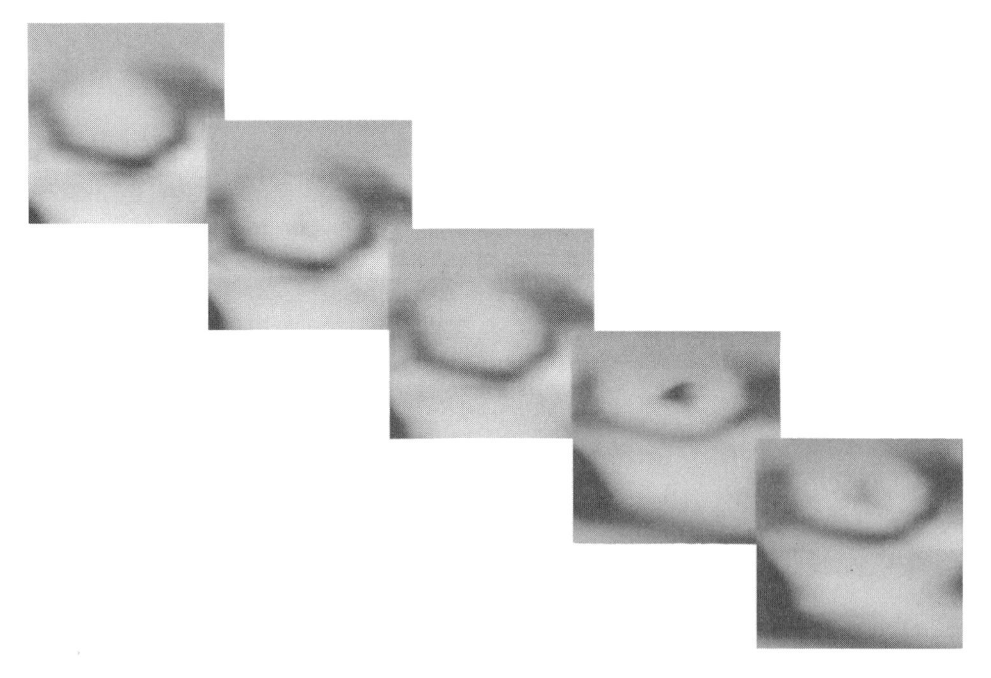

Figure 9. Application of suction to an inside-out patch is demonstrated by a series of pictures of inside-out patches. Time increases from left to right. Raw data (scan frequency ≈60 Hz, and scan size ≈3 μm) are shown. The inside-out patch was obtained without suction while forming the cell-attached gigaseal and is attached close to the end of the pipette. Suction tears the middle of the patch into the pipette. This effect could be reversed by reducing the suction or applying pressure.

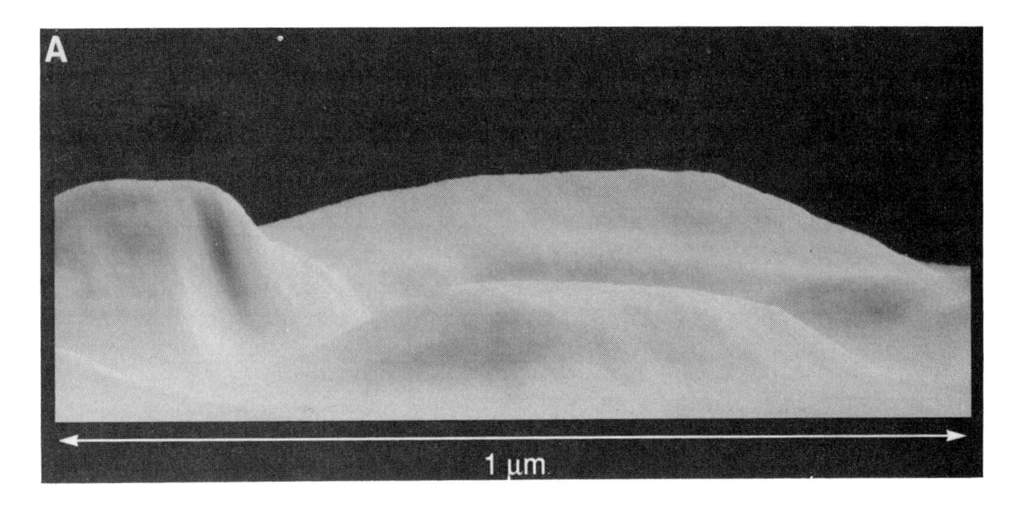

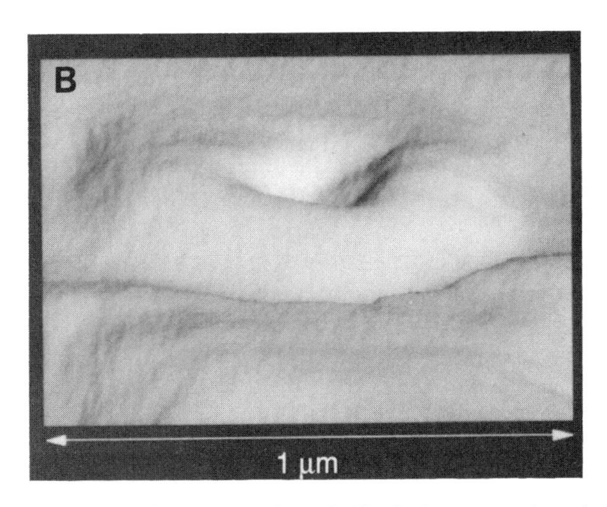

Figure 10. Three-dimensional plots of the images shown in Fig. 9. A corresponds to the first image on the left, and B to the last image from the left. Two different image-processing procedures were used. In A, the membrane attachment to the glass wall (the higher parts in the background) is clearly visible. The center of the patch on the image of A is coming out of the pipette while the surrounding part is sagging into the tip opening. In B, the applied suction tears only the middle of the patch into the opening of the pipette tip. Here, the high structures in the foreground of the image are parts of the glass wall. Image sizes are about 1 μm in A and in B. Height differences between "mountains" and "valleys" are about 200 nm in both images.

voltage-gated ion channels are present in the membrane, as dipoles in channel proteins could be rearranged by an external electric field inducing a tension on the membrane patch.

Inside-out patches were considerably more stable than outside-out patches, perhaps because they are located deeper in the pipette tip and have more contact with the glass. Inside-out patches formed without previous application of suction seemed to be folded into the pipette tip by not more than about 2 μm, since this is the estimated depth that the lever can move into an opening of 2-μm diameter without knocking the substrate of the probe on the edges of the opening. It seems possible that parts of the cytoskeleton are still attached to the inside-out membrane patch, and therefore structural differences may be detectable between the inner and the outer sides of the membrane. Figure 9 shows several images of an inside-out patch spanning the tip of the pipette. Positive or negative pressure applied to the pipette produced large changes in the shape of the patch (see also Fig. 10). The seal resistance did not change when the patch was exposed to the pressure changes. It should be possible to measure the elasticity of different membranes by applying a defined pressure (see Chapter 14, this volume). In this way, the pressure can be correlated with the detected changes in z direction in the patch.

4.3. Outlook

The experiments described above are still in progress and suggest that the combination of SPT with the patch-clamp technique is potentially useful in characterizing cell membranes. Further experiments are planned such as real-time resolution of dynamic processes in the range of 100 msec for complete images and in the range of a few microseconds for small areas of the membrane surface. This might enable simultaneous measurement of capacitance and surface structure of cells undergoing exo- and endocytosis or the direct excitation of mechanical sensors like the hair cells of the inner ear.

Recent progress in analyzing MS channels (Sukharev *et al.*, 1994) as well as cloning of a MS channel from *Caenorhabditis elegans* (Hong and Driscoll, 1994; Huang and Chalfie, 1994), which is thought to be involved in touch sensitivity make combined studies employing SPT and patch-clamp methods even more exciting. Expression of cloned MS channels opens possibilities such as measures of the effects of site-directed mutagenesis for structure–function analysis.

Technically, recent advances in the development of the SPT have extended the possibilities of the SFM. For instance, the development of cantilevers with piezoresistant thin layers can circumvent an optical detection system. The bending of the cantilever in this case is proportional to the resistance of the thin layer on top of the cantilever. Also, the development of procedures for preparing tips coated with antibodies, agonists, or drugs continue, which will facilitate investigations of interactions between specific molecules. Thus, the major advantage in the future of introducing the SPT may not be the suboptical resolution on surfaces of cell membranes but the possibility of probing mechanical and chemical properties in the nanometer range.

ACKNOWLEDGMENT. We thank Dr. J. P. Ruppersberg for his help with the development of the SFM/patch-clamp setup and Dr. L. P. Wollmuth for reading the manuscript.

References

Binnig, G., Rohrer, H., Gerber, C., and Weibel, E., 1982, Surface study by scanning tunneling microscopy, *Phys. Rev. Lett.* **49**:57–60.

Binnig, G., Quate, C. F., and Gerber, C., 1986, Atomic force microscope, *Phys. Rev. Lett.* **56**:930–933.

Bryant, A., Smith, D. P. E., Ohnesorge, F., Weisenhorn, A. J., Heyn, S. P., Drake, B., Prater, C. B., Gould, S. A., Hansma, P. A., and Gaub, H., and Quate, C. F., 1986, *Appl. Phys. Lett.* **48**:832–834.

Egger, M., Ohnesorge, F., Weisenhorn, A. J., Heyn, S. P., Drake, B., Prater, C. B., Gould, S. A.,

Hansma, P. A., and Gaub, H., 1990, Wet lipid–protein membranes imaged at submolecular resolution by atomic force microscopy, *J. Struct. Biol.* **103:**89–94.

Fuchs, H., Schrepp, W., and Rohrer, H., 1987, STM investigations of Langmuir–Blodgett films, *Surf. Sci.* **181:**391–393.

Guckenberger, R., Weigrabe, W., Hillebrand, A., Hartmann, T., Wang, Z., and Baumeister, W., 1989, Scanning tunneling microscopy of a hydrated bacterial surface protein, *Ultramicroscopy* **31:**327–331.

Häberle, W., Hörber, J. K. H., and Binnig, G., 1989, Force microscopy on living cells, *J. Vac. Sci. Technol.* **B9:**1210–1212.

Hansma, P. K., Drake, B., Marti, O., Gould, S. A. C., and Prater, C. B., 1989, The scanning ion conductance microscope, *Science* **243:**641–643.

Hong, K., and Driscoll, M., 1994, A transmembrane domain of the putative channel subunit MEC-4 influences mechanotransduction and neurodegeneration in *C. elegans*, *Nature* **367:**470–474.

Hörber J. K. H., Lang, C. A., Hänsch, T. W., Heckl, W. M., and Möhwald, H., 1988, Scanning tunneling microscopy of lipid films and embedded biomolecules, *Chem. Phys. Lett.* **145:**151–154.

Hörber, J. K. H., Schuler, F. M., Witzemann, W., Schröder, K. H., and Müller, H., 1991, Imaging of cell membrane proteins with a scanning tunneling microscope, *J. Vac. Sci. Technol.* **B9:**1214–1218.

Hörber, J. K. H., Häberle, W., Ohnesorge, F., Binnig, G., Liebich, H. G., Czerny, C. P., Mahnel, H., and Mayr, A., 1992, Investigation of living cells in the nanometer regime with the scanning force microscope, *Scan. Microsc.* **6:**919–929.

Huang, M., and Chalfie, M., 1994, Gene interactions affecting mechanosensory transduction in *Caenorhabditis elegans*, *Nature* **367:**467–470.

Hudspeth, A. J., and Gillespie, P. G., 1994, Pulling springs to tune transduction: Adaption by hair cells, *Neuron* **12:**1–9.

Israelachvili, J. N., 1992, *Intermolecular and Surface Forces*, 2nd. ed., Academic Press, San Diego.

Jericho, M. H., Blackford, B. D., Dahn, D. C., Frame, C., and Maclean, D., 1990, Scanning tunneling microscopy imaging of uncoated biological material, *J. Vac. Sci. Technol.* **88:**661–666.

Martin, Y., and Wickramasinghe, H. K., 1988, High resolution capacitance measurement and potentiometry by force microscopy, *Appl. Phys. Lett.* **52:**1103.

Matey, J. R., and Blanc, J., 1985, Scanning capacitance microscopy *J. Appl. Phys.* **57:**1437–1439.

Methfessel, C., Witzemann, V., Takahashi, T., Mishima, M., Numa, S., and Sakmann, B., 1986, Patch clamp measurements on *Xenopus laevis* oocytes: Currents through endogenous channels and implanted acetylcholine receptor and sodium channels, *Pflügers Arch.* **407:**577–588.

Meyer, E., Howald, L., Overney, R. M., Heinzelmann, H., Frommer, J., Guntherodt, H.-J., Wagner, T., Schier, H., and Roth, S., 1989, Molecular-resolution images of Langmuir–Blodgett films using atomic force microscopy, *Nature* **349:**398–399.

Meyer, G., and Am, N. M., 1988, Novel optical approach to atomic force microscopy, *Appl. Phys. Lett.* **53:**1045–1047.

Pohl, D. W., Fischer, U. C., and Dürig, U. T., 1988, Scanning near-field optical microscopy (SNOM), *J. Microsc.* **152:**853–861.

Sáenz, J. J., Garcia, N., Grutter, P., Meyer, E., Heinzelmann, H., Wiesendanger, R., Rosenthaler, L., Hidber, H. R., and Guntherodt, H.-J., 1987, Observation of magnetic forces by the atomic force microscope, *J. Appl. Phys.* **62:**4293–4295.

Smith, D. P. E., Bryant, A., Quade, C. F., Rabe, J. P., Berger, Ch., and Swalen, J. D., 1987, Images of a lipid bilayer at molecular resolution by scanning tunneling microscopy, *Proc. Natl. Acad. Sci. U.S.A.* **84:**969–972.

Sukharev, S. I., Blount, P., Martinac, B., Blattner, F. R., and Kung, C., 1994, A large-conductance mechanosensitive channel in *E. coli* encoded by *mscL* alone, *Nature* **368:**265–268.

Williams, C. C., and Wickramasinghe, H. K., 1988, Thermal and photo thermal imaging on a sub 100 nm scale, *Proc. SPIE* **897:**129.

Worcester, D. L., Miller, R. G., and Bryant, P. J., 1988, Atomic force microscopy of purple membranes, *J. Microsc.* **152:**817–821.

Part III

ANALYSIS

Chapter 18

The Principles of the Stochastic Interpretation of Ion-Channel Mechanisms

DAVID COLQUHOUN and ALAN G. HAWKES

1. The Nature of the Problem

1.1. Reaction Mechanisms and Rates

Most mechanisms that are considered for ion channels (as for any other sort of chemical reaction) involve reversible transitions among the various possible discrete chemical states in which the system can exist. Other sorts of mechanisms may, of course, exist; for example, there may be an irreversible reaction step, a problem that is considered in Section 7. Our primary aim is to gain insight about the nature of the reaction mechanism from experimental observations. In this process, we may also obtain estimates of numerical values for the rate constants in the mechanism.

The sort of mechanism that is commonly considered can be illustrated by the following examples. In each case, the symbol that denotes the rate constant for a transition is appended to the arrow that represents the transition; the interpretation of these rate constants is considered below. The simplest reaction mechanism consists of a transition between a single shut state of the ion channel and a single open state:

$$\text{Shut} \underset{\alpha}{\overset{\beta'}{\rightleftharpoons}} \text{Open} \tag{1}$$

There are two states altogether. If a ligand must be bound before the ion channel can open, at least three discrete states are needed to describe the mechanism. The mechanism of Castillo and Katz (1957) has two shut states and one open state; this is usually represented as

$$\overbrace{\text{R} \underset{k_{-1}}{\overset{k_{+1}}{\rightleftharpoons}} \text{AR}}^{\text{shut}} \underset{\alpha}{\overset{\beta}{\rightleftharpoons}} \overbrace{\text{AR}^*}^{\text{open}} \qquad \underbrace{\text{R}}_{\text{vacant}} \quad \underbrace{\text{AR}, \text{AR}^*}_{\text{occupied}} \tag{2}$$

where R represents a shut channel, R* an open channel, and A represents the agonist molecule.

Note to the reader: At the author's request this chapter will use British spelling and the abbreviations ms and μs instead of msec and μsec.

DAVID COLQUHOUN • Department of Pharmacology, University College London, London WC1E 6BT, England, and ALAN G. HAWKES, European Business Management School, University of Wales Swansea, Swansea SA2 8PP, Wales.

Single-Channel Recording, Second Edition, edited by Bert Sakmann and Erwin Neher. Plenum Press, New York, 1995.

The simplified mechanism of the axonal sodium channel also has two discrete shut states and one open state; if inactivation of the channel can occur without channel opening, the reaction must be written in a cyclical form:

$$\begin{array}{ccc} & & \text{Open} \\ & \nearrow\!\!\swarrow & \updownarrow \\ \text{Shut} & & \\ & \nwarrow\!\!\searrow & \\ & & \text{Inactivated} \end{array} \tag{3}$$

Some other examples are considered below (see Sections 4 and 13).

The usual procedure would be first to postulate a plausible mechanism, then to use the law of mass action to predict its expected kinetic and equilibrium behaviour, and finally to compare these predictions with experimental observations. Such predictions concern, of course, the *average* behaviour of the system.

If we are recording from a large number of molecules (ion channels, in the present case), then it is only the average behaviour that can be observed. For example, if the transition rates between the various states are constant (do not vary with time), then the time course of the mean current, $I(t)$, through the ion channels will be described by the sum of $k - 1$ exponential terms, where k is the number of states in the system (see examples above). Thus,

$$I(t) = I(\infty) + w_1 e^{-t/\tau_1} + w_2 e^{-t/\tau_2} + \cdots. \tag{4}$$

[Note that $\exp(-t/\tau)$ is often used as an alternative way of writing $e^{-t/\tau}$.] For any specified mechanism, the amplitudes w_i and the time constants τ_i can be calculated by the methods given, for example, by Colquhoun and Hawkes (1977), as can the predicted noise spectrum. A 'cookbook' approach to programming such calculations is provided in Chapter 20 (this volume). The values of τ_i each depend on *all* of the rate constants in the mechanism, and they have, in general, no simple physical significance (although in particular cases they may approximate some physical quantity such as mean open lifetime or mean burst length).

If, on the other hand, we record from a fairly small number of ion channels, the fluctuations about the average behaviour become large enough to measure, and Katz and Miledi (1970, 1972) showed how these fluctuations (or 'noise') could be interpreted in terms of the ion channel mechanism. Suppose, for example, that there are $N = 10^6$ ion channels and that, at equilibrium, there are 1000 channels open on average. The probability that an individual channel is open at a given moment is $p = 1000/10^6 = 0.001$, so the standard deviation of the number of open channels is given by the binomial distribution as $[Np(1 - p)]^{1/2} = 31.6$. The number of channels that are open at equilibrium is therefore not constant at 1000, but is 1000 ± 31.6, where the standard deviation reflects the random fluctuations in the number of open channels from moment to moment (see examples in Colquhoun, 1981, for an elementary discussion).

The law of mass action states that the rate of any reaction is proportional to the product of the reactant concentrations. The proportionality constant is described as a 'rate constant' (α, β, k_{-1}, etc.) and is supposed to be a genuine constant, i.e., not to vary with time. This is not necessarily true, however; for example, the channel-shutting rate constant, α, is known to be dependent on membrane potential (for muscle-type nicotinic receptors), so it will stay constant only if the membrane potential stays constant (i.e., only as long as we have an effective voltage clamp). Furthermore, for an association reaction with rate constant k_{+1} (dimension $M^{-1}s^{-1}$), the transition rate (dimensions s^{-1}) will be $k_{+1}x_A$ where x_A is the free ligand concentration; this will be constant only if neither the ligand concentration nor the rate constant varies with time. In many sorts of experiments there is a considerable risk of transient concentration changes that would violate this condition. Of course, it is still possible to solve the kinetic equations even if the transition rates are not constant, as long as their time course is known, but this adds considerably to the complexity (we no longer get sums of exponential terms) and is not considered here.

We have already mentioned that the current, $I(t)$, is only an average value; there will be fluctuations about this average as a result of moment-to-moment random variations in the number of ion channels that are open. The smaller the number of ion channels that we record from, the larger (relative to the mean current) the fluctuations will be. When we record from a single ion channel, the current varies in a step-like fashion between (in the simplest cases) two values, fully open and fully shut (Neher and Sakmann, 1976; Hamill *et al.*, 1981). The current is effectively *never* equal to its equilibrium value; it is always zero or 100%. Equilibrium can be defined only over a long time period; the term "fraction of channels open at equilibrium" must be replaced by "the fraction of time for which the single channel stays open," a quantity that can be measured accurately only over a period of observation long enough to contain many open and shut intervals. A long stretch of record is needed because we are looking at a single molecule, and its behaviour is, of course, random.

1.2. Rate Constants and Probabilities

In ordinary chemical kinetics, a rate constant describes the rate of reaction; for example, α in equation 1 or 2 describes the rate of the channel-shutting reaction. The transition rates (e.g., k_{-1} or $k_{+1}x_A$) have the dimensions of frequency (s^{-1}), and they can be interpreted as frequencies. For example, in equation 1, the number of shuttings that occur per second (of individual molecules) is simply α multiplied by the fraction of channels that are in the open state. At equilibrium, the number of shuttings per second (α times the fraction of channels that are open) will be equal, on average, to the number of openings per second (β' times the fraction of channels that are shut). This frequency interpretation of rate constants is described in more detail by Colquhoun and Hawkes (1994) and is illustrated in Sections 4.6 (Fig. 6) and 9.1.

However, when we look at a single ion channel, we see that the shutting takes place at random, so the rate constant must be interpreted in a probabilistic way: α is a measure of the probability that an open channel will shut in unit time (though, because α has dimensions of s^{-1}, it is clearly not an ordinary probability, which must be dimensionless). Roughly, we can say that for a time interval Δt,

$$\text{Prob(open channel shuts during } \Delta t) \cong \alpha\Delta t$$

This is dimensionless, but it is still not a proper probability because it can be greater than unity. Also, this definition does not make clear whether or not several openings and shuttings are allowed to occur during the time interval Δt. It turns out that the proper way to write this definition can be arrived at by introducing a 'remainder term,' which we do not specify in detail but which has the property that it disappears (relative to Δt) as Δt becomes very small. This term would describe, for example, the possibilities of several transitions occurring during Δt, which clearly becomes negligible for small Δt. This remainder term is written as $o(\Delta t)$ (further discussion of such terms is found, for example, in Colquhoun, 1971, Appendix 2). Thus, we can now write:

$$\text{Prob(open channel shut during } \Delta t) = \alpha \Delta t + o(\Delta t)$$

More properly, the left-hand side should be written as a conditional probability, the probability that a channel shuts during Δt *given* that it was open at the start of this period:

$$\text{Prob(channel shuts between } t \text{ and } t + \Delta t \,|\, \text{channel was open at } t) = \alpha \Delta t + o(\Delta t) \quad (5)$$

(the vertical bar is read as "given"; see Section 2).

Notice that this is supposed to be the same at whatever time t we start timing our interval, and also to be independent of what has happened earlier, i.e., it depends only on the present (time t) state of the channel. This is a fundamental characteristic of our type of random process (a homogeneous Markov process). Further progress appears to be prevented by the unspecified term $o(\Delta t)$ in equation 5. This term can, however, always be eliminated by dividing by Δt and then letting Δt tend to zero. In this way, the $o(\Delta t)$ terms disappear, and we obtain a differential equation that can be solved in the normal way. This procedure is illustrated in Section 3.

We have defined a probability in terms of a rate constant in equation 5; rearrangement of this gives a definition of the rate constant in probabilistic terms. Roughly,

$$\alpha \approx \text{(Probability of shutting in } \Delta t)/\Delta t$$

or, more properly, from equation 5,

$$\alpha = \lim_{\Delta t \to 0} [\text{Prob(channel shuts between } t \text{ and } t + \Delta t \,|\, \text{open at } t)/\Delta t]$$

$$= \lim_{\Delta t \to 0} [\text{Prob(channel shut at } t + \Delta t \,|\, \text{open at } t)/\Delta t]. \quad (6)$$

More generally, we can define any transition rate in this way. Denote as q_{ij} the transition rate from state i to state j. Then

$$q_{ij} = \lim_{\Delta t \to 0} [\text{Prob(state } j \text{ at time } t + \Delta t \,|\, \text{state } i \text{ at } t)/\Delta t] \qquad i \neq j$$

$$= \lim_{\Delta t \to 0} [P_{ij}(\Delta t)/\Delta t] \quad (7)$$

where

$$P_{ij}(t) = \text{Prob(state } j \text{ at time } t \,|\, \text{state } i \text{ at time 0)} \quad (8)$$

Notice that a transition rate must always have dimensions of s^{-1}.

1.3. Fractal and Diffusion Models

Classical chemical kinetics, based on the law of mass action, has always entailed the assumption that the system can exist in a small number of discrete states, as in the examples given above. The advent of single-channel recording has provided perhaps the strongest and most direct evidence for the existence of discrete states of large protein molecules. The switch between shut and open states, or between open states of different conductance, is very fast, with no detectable intermediate states; this is exactly what is expected on the basis of the classical postulates of chemical kinetics. Yet, ironically, it is these same observations that have caused these postulates to be questioned.

It has been suggested that, because proteins can exist in an essentially infinite number of conformations, it is inappropriate to postulate a small number of discrete states, and some sort of fractal or diffusion model should be preferred (Liebovitch *et al.*, 1987; Läuger, 1988; Milhauser *et al.*, 1988; Liebovitch, 1989). Such models usually predict that the probability of a transition occurring in unit time will not be constant and so differ fundamentally from the Markov model. There are several reasons to think that such approaches are not, at present, likely to be very helpful.

The most important reason is that the experimental evidence shows Markov models to fit the data better than the alternatives (as formulated up to now); see, for example Korn and Horn (1988), McManus *et al.* (1988), Sansom *et al.* (1989), McManus and Magleby (1989), Petracchi *et al.* (1991), and Gibb and Colquhoun (1992). An example of this evidence is given later (see Section 10.3).

A second reason is that the theoretical argument is not entirely convincing. Fractals stem from mathematics rather than physics, so it is far from clear what they can tell us about the real world (the same comment applies to catastrophe theory, which, at the height of its fashion, was said to "explain" almost every biological phenomenon from riots to action potentials but is now almost forgotten). Diffusion theory, on the other hand, has a sound physical basis and must be taken more seriously. Clearly, a protein (or, indeed, much smaller molecules) can exist in an infinite number of conformations, but this does not preclude the existence of a limited number of states, or conformations, that are much more stable than the others (e.g., Läuger, 1985, 1988). Such "discrete" states are not, of course, fixed and stationary. All the parts of the molecule have thermal motion, much of it very rapid (on a picosecond time scale), so there is continuous fluctuation around the average structure of the "discrete" state, but if these fluctuations are of no great functional significance (e.g., have only a small effect on channel conductance), then there is no need to incorporate them into the model. To attempt to do so merely increases vastly the number of parameters to be estimated without contribution to the usefulness of the model. In fact, the fractal formulation does not attempt the impossible task of estimating all the relevant parameters but, on the contrary, attempts to describe the data with only two, neither of which has any obvious physical significance.

In summary, the simple forms of the fractal argument that have been used fail to fit the data adequately in many cases. Furthermore, even if it were true that an infinite (or at least very large) number of states should be considered, it would be impossible to estimate from experimental data parameters with genuine physical significance. Just as in all other science, the mechanisms with a few discrete states that we use are undoubtedly approximations, but they have parameters that can be estimated, have real physical significance, and they have proven predictive value (see Colquhoun and Ogden, 1986; Horn and Korn, 1989).

2. Probabilities and Conditional Probabilities

The definitions just given are in terms of probabilities and conditional probabilities. It may be useful at this stage to illustrate exactly what these terms mean. Consider the behavior

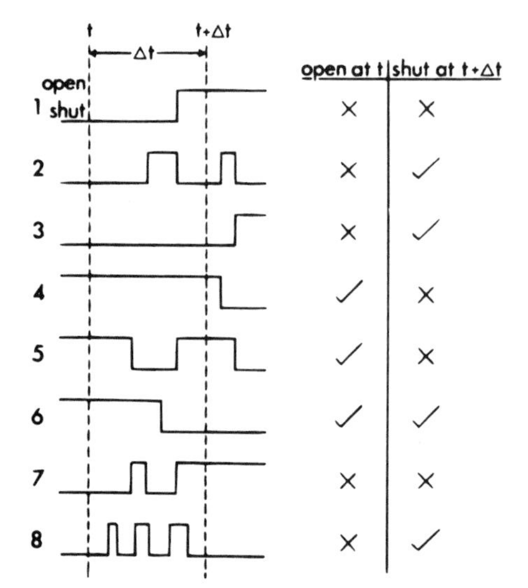

Figure 1. Illustration of the meaning of probabilities and conditional probabilities (see text).

of eight individual ion channels illustrated in Fig. 1. Imagine that these eight channels behave in a manner typical of a much larger number of channels so that the ratios given are good estimates of the true or long-term average values of the probabilities. One possible example of their behaviour is shown in Fig. 1.

Only one channel of the eight is both open at t and shut at $t + \Delta t$, so

$$\begin{aligned}\text{Prob(open at } t \textit{ and } \text{shut at } t + \Delta t) &= (\text{number open at } t \textit{ and } \text{shut at } t + \Delta t)/(\text{total number}) \\ &= 1/8.\end{aligned}$$

However, the conditional probability, Prob(shut at $t + \Delta t \mid$ open at t), although it has the same numerator, has a different denominator; it is defined with respect to the population of channels that obey the prior condition, i.e., those that were open at t. These are three in number (channels 4, 5, and 6), so the conditional probability is

$$\begin{aligned}\text{Prob(shut at } t + \Delta t \mid \text{open at } t) &= (\text{number open at } t \textit{ and } \text{shut at } t + \Delta t)/(\text{number open at } t) \\ &= 1/3.\end{aligned}$$

This is an example of the general rule of probability that for any events A and B,

$$\text{Prob(B} \mid \text{A)} = \text{Prob(A } \textit{and } \text{B)/Prob(A)}$$
$$\text{Prob(A } \textit{and } \text{B)} = \text{Prob(A) Prob(B} \mid \text{A)} \tag{9}$$

In this case, A is 'open at t'; B is 'shut at $t + \Delta t$.' So in this example,

$$\text{Prob(B} \mid \text{A)} = (1/8)/(3/8) = 1/3$$

In general, if A and B are *independent*, then the probabilities of B cannot depend on whether A has occurred or not. Therefore, Prob(B | A) can be written simply as Prob(B), and equation 9 reduces to the simple multiplication rule:

$$\text{Prob(A } \textit{and} \text{ B)} = \text{Prob(A) Prob(B)} \tag{10}$$

3. The Distribution of Random Time Intervals

3.1. The Lifetime in an Individual State

We are interested in the length of time for which the system stays in a particular state, for example, the open state. These lengths of time are random variables, and the form of their variability can be described by a probability distribution. Time is a continuous variable, so we wish to find the probability density function (pdf) of, for example, open lifetimes. This is a function $f(t)$, defined so that the area under the curve up to a particular time t represents the probability that the lifetime is equal to or less than t. Thus, the pdf can be found by differentiating the cumulative distribution (or *distribution function*) Prob(open lifetime $\leq t$), which is usually denoted $F(t)$. The pdf is thus

$$f(t) = \lim_{\Delta t \to 0} [\text{Prob(lifetime is between } t \text{ and } t + \Delta t)/\Delta t] \tag{11}$$

A number of approaches to the derivation of this distribution are possible. We shall first derive it directly and then mention some other approaches. Take, as an example, the length of time for which a channel stays open. First define a probability, which we shall denote $R(t)$, as

$$R(t) = \text{Prob(channel stays open } \textit{throughout} \text{ the time from 0 to } t) \tag{12}$$

It is worth noting that this is a rather different sort of probability from that used in analyzing relaxations or noise. In these cases, we are interested, for example, in the probability that a channel is open at time t, given that it was open at $t = 0$, regardless of whether the channel may have shut one or more times in between. In reliability theory, the sort of probability defined in equation 12 is known as the *reliability function;* it represents the probability that a system remains operational throughout the period 0 to t.

Now, from equation 5 we know that

$$\text{Prob(shut at } t + \Delta t | \text{open at } t) = \alpha\Delta t + o(\Delta t), \tag{13}$$

where α is the ordinary rate constant for the shutting reaction (or, more generally, the sum of the rate constants for all routes by which the open channel can shut). The channel obviously must either shut or not shut during Δt, so the probabilities for these two alternatives must add to unity. Hence,

$$\text{Prob(channel does not shut between } t \text{ and } t + \Delta t | \text{open at } t) = 1 - \alpha\Delta t - o(\Delta t), \tag{14}$$

From equations 9 and 12, we can now define

$$\begin{aligned} R(t + \Delta t) &= \text{Prob(channel stays open throughout the time from 0 to } t + \Delta t) \\ &= \text{Prob(open throughout } 0, t) \cdot \text{Prob(open throughout } t, t + \Delta t \,|\, \text{open} \\ &\qquad \text{throughout } 0, t) \end{aligned} \tag{15}$$

Now the crucial Markov assumption, discussed following equation 5, is that the last probability does not depend on the whole history from 0 to t but only on the fact that the channel is open at time t. Thus,

$$\begin{aligned} &\text{Prob(open throughout } t, t + \Delta t \,|\, \text{open throughout } 0, t) \\ &\quad = \text{Prob(open throughout } t, t + \Delta t \,|\, \text{open at } t) \end{aligned} \tag{16}$$

But equation 16 is just the probability that was derived in equation 14, and the first probability in equation 15 is simply $R(t)$, so equation 15 can be written as

$$R(t + \Delta t) = R(t)[1 - \alpha\Delta t - o(\Delta t)] \tag{17}$$

Thus,

$$\frac{R(t + \Delta t) - R(t)}{\Delta t} = -R(t)\left[\alpha + \frac{o(\Delta t)}{\Delta t}\right]$$

If we now let $\Delta t \rightarrow 0$, the left-hand side becomes the first derivative of $R(t)$, and the remainder term disappears, so

$$\frac{dR(t)}{dt} = -\alpha R(t). \tag{18}$$

As long as α is a constant (not time dependent), the solution of this equation is

$$R(t) = e^{-\alpha t} \tag{19}$$

because $R(0) = 1$ (i.e., channel cannot move out of the open state in zero time).

Next, we notice that if the channel stays open throughout the time from 0 to t, its open lifetime must be at least t. This is the crucial step that relates the argument to the distribution of open times. We can therefore write

$$\begin{aligned} R(t) = e^{-\alpha t} &= \text{Prob(channel stays open throughout time from 0 to } t) \\ &= \text{Prob(open lifetime} > t) \end{aligned} \tag{20}$$

and therefore,

$$\text{Prob(open lifetime} \leq t) = 1 - R(t) = 1 - e^{-\alpha t} = F(t) \tag{21}$$

This defines the cumulative distribution, $F(t)$, of open-channel lifetimes. The required pdf for the open-channel lifetime is the first derivative of this, i.e.,

$$f(t) = dF(t)/dt = -\frac{dR(t)}{dt} = \alpha e^{-\alpha t} \qquad (t \geq 0) \tag{22}$$

[for times less than zero, $f(t) = 0$].

This pdf is described as an *exponential distribution,* or exponential density, with mean $1/\alpha$. It is a simple exponentially decaying curve. This is quite different in shape from the well-known Gaussian or "normal" distribution: rather than being a symmetrical bell-shaped curve, it is an extreme example of a positively skewed distribution with the mode (maximum) at $t = 0$ (compare with the Gaussian curve for which the mode is the same as the mean). The exponential distribution has the same sort of central role in stochastic processes as the Gaussian distribution has in large areas of classical statistics.

For any pdf, $f(t)$, the mean is given by

$$\text{mean} = \int_{-\infty}^{\infty} tf(t)dt$$

For nonnegative random variables, $f(t) = 0$ when $t < 0$, and the lower limit of this integral can be taken as zero; then integration by parts yields a useful alternative formula that is sometimes easier to calculate. Thus,

$$\text{mean} = \int_{0}^{\infty} tf(t)dt = \int_{0}^{\infty} R(t)dt \tag{23}$$

which, in the present example, is $1/\alpha$, the mean open lifetime.

3.2. Another Approach to the Exponential Distribution

An open channel must overcome a certain energy barrier before it can flip to a shut conformation. The energy needed for this purpose comes from the random thermal energy of the system. The bonds of the channel protein will be vibrating, bending, and stretching, and much of this motion will be very rapid, on a picosecond time scale. One can imagine that each time the molecule stretches, it has a chance to surmount the energy barrier and flip shut. Each "stretch" is like a binomial trial with a certain probability, p, of success (i.e., shutting) at each trial. Since the stretching is on a picosecond time scale, but the channel stays open for milliseconds, clearly, the chance (p) of success at each trial must be small, and a large number (N) of trials will be needed before the channel shuts. Now, when N is large and p is small, the binomial distribution approaches the Poisson distribution. The Poisson distribution gives the probability of there being no successes (i.e., no shutting) in time t as $e^{-\alpha t}$, where α is the mean frequency of successes in unit time. If there is no success at shutting in time t, then the open lifetime must be greater than t, and, since we have found that the probability of this is $e^{-\alpha t}$, we are led directly to equation 20 and hence to the exponential pdf.

3.3. Generalizations

The above argument was mostly concerned with the open time, but clearly the same argument applies to the time spent in any single specified state. For the simplest mechanism (equation 1), therefore, the open time is exponentially distributed with mean $1/\alpha$, and the shut time has pdf $g(t) = \beta' e^{-\beta' t}$, i.e., it is exponentially distributed with mean $1/\beta'$ for a single channel. If more than one channel contributes to the observations, the shut times will appear to be shorter than this, of course. In general, we can, by a similar argument, give the following rule:

$$\text{Lifetime in any single state is exponentially distributed with mean} \\ = 1/(\text{sum of transition rates that lead away from the state}) \tag{24}$$

In general, for mechanisms with many states, we expect that all of the distributions of quantities such as open times, shut times, burst lengths, and so on will be mixtures of exponentials. This will be the case when the transition rates are constant (see above), and we therefore expect Markov behaviour. Such distributions are often described has being a *sum of exponential components,* just like the macroscopic current in equation 4 (except that there will usually be fewer components in the single-channel distributions). It is actually preferable to refer to such distributions as having the form of a *mixture of exponential distributions* (or of exponential densities). Each component can be written in the form of a simple exponential distribution, i.e., $\lambda_t \exp(-\lambda_i t)$, where λ_i is the reciprocal of the time constant, or mean, for the *i*th component, $\tau_i = 1/\lambda_i$. Each such distribution has unit area, and to ensure that the final distribution also has unit area, each component is multiplied by a fraction area, a_i, the relative area occupied by the *i*th component; these are such that the sum of the areas is unity. Thus, the general form for a pdf that is a mixture of exponentials is

$$f(t) = a_1\lambda_1 e^{-\lambda_1 t} + a_2\lambda_2 e^{-\lambda_2 t} + \cdots,$$

or, for *n* components,

$$f(t) = \sum_{i=1}^{i=n} a_i\lambda_i e^{-\lambda_i t} \tag{25}$$

where

$$\sum a_i = 1.$$

The question of the number of components that would be expected in particular cases is addressed below and discussed more generally in Section 13.5.

It is often of interest to know the distribution of the time spent within any specified set of states (e.g., all shut states) rather than in a single state. In this case the system can oscillate among the various states within the set in a random way; the time that elapses before the set of states is eventually left will (under our usual assumptions) be described by a mixture of exponential distributions. The derivation of such distributions is exemplified by the burst-length distribution discussed in Section 4.7 and by the derivation of the shut-time distribution given by Colquhoun and Hawkes (1994). The general solution is given by Colquhoun and Hawkes (1982), and this is discussed briefly in Section 13.3 and Chapter 20 (this volume).

3.4. Relationship between Single-Channel Events and Whole-Cell Currents

It is, of course, no coincidence that, on one hand, the mean current through a large number of ion channels follows an exponential time course (see equation 4) and, on the other hand, the random lifetimes of elementary events are described by exponential distributions. This can be illustrated schematically for the case of the decay phase of a miniature end-plate current. According to Anderson and Stevens (1973), the decay phase, which follows a simple exponential time course, is determined by the lifetime of individual open channels. This is illustrated in Fig. 2*b,c*. At zero time, a number of ion channels are opened, almost

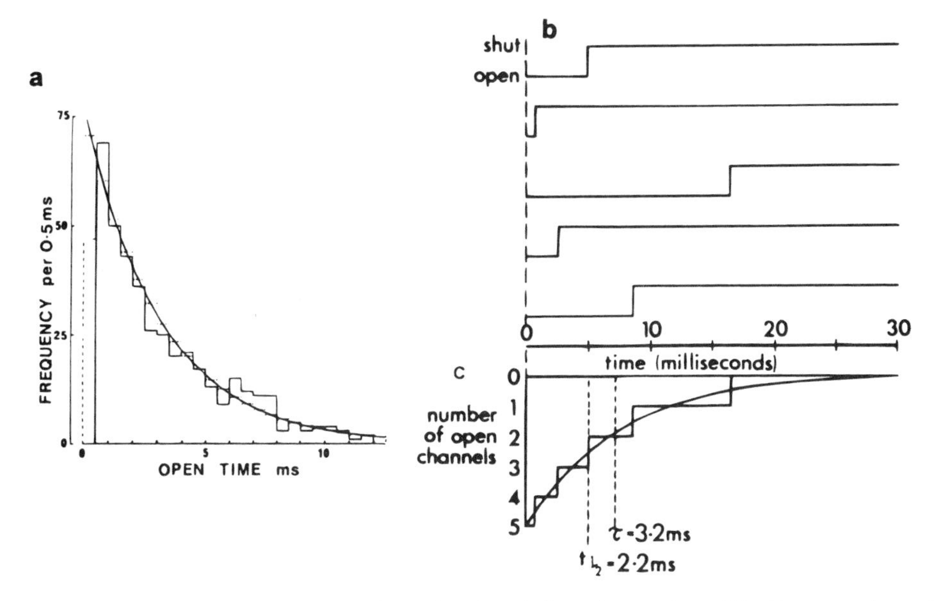

Figure 2. a: An exponential distribution of the duration of channel-open times. The histogram shows the number of openings per bin of 0.5-msec width (*R. temporaria,* synaptic channels, 50 nM acetylcholine, −80 mV, 8°C; D. C. Ogden, D. J. Adams, and D. Colquhoun, unpublished data). The continuous line shows an exponential probability density function that has been fitted to the observations (above 0.5 ms) by the method of maximum likelihood (see Chapter 19, this volume). It has a time constant of $\tau = 3.2$ ms, i.e., a rate constant of $\lambda = 1/\tau = 312.5\ s^{-1}$. The estimated exponential probability density function is therefore (see equation 22) $f(t) = \lambda e^{-\lambda t} = 312.5e^{-312.5t}\ s^{-1}$, the area under this curve is, as for any probability density function, unity. In the figure, $f(t)$ has been multiplied by the number of observations that lie under the fitted curve (480) and expressed in units of $(0.5\ ms)^{-1}$ rather than s^{-1}, so the continuous curve can be superimposed on the histogram (see Section 5.1.5 of Chapter 19, this volume). Thus, for example, the intercept at $t = 0$ is plotted not as $f(0) = 312.5\ s^{-1}$ but as $312.5\ s^{-1} \times 480/2000 = 75\ (0.5\ ms)^{-1}$, where the factor 2000 is 0.5 ms/1s. The horizontal dashed lines show the frequency in each bin as calculated from the continuous curve. b: Simulated behaviour of five individual channels that were open at the time ($t = 0$) at which the acetylcholine concentration had fallen to zero. Opening is plotted downward. The channels stay open for a random (exponentially distributed) length of time with a mean of 3.2 ms. c: Sum of the five records shown in b. The total number of open channels decays exponentially (as illustrated in Fig. 3c) with a time constant of 3.2 ms in this example. Reproduced from Colquhoun (1981), with permission.

synchronously, by a quantum of acetylcholine; the acetylcholine then rapidly disappears so that a channel, once it has shut, cannot reopen. The length of time for which each channel stays open is described by an exponential distribution (Fig. 2*a,b*), which ensures that the total current through a large number of such channels will decay along an exponential time course (Fig. 2c).

This simple argument works only because the channels were supposed to open almost simultaneously. This is true, to a good approximation, for synaptic transmission mediated by nicotinic receptors, but it is far from true for NMDA-type glutamate receptors (Edmonds and Colquhoun, 1992). In such cases we need also to consider the distribution of the time (first latency) from the application of the stimulus (e.g., synaptic release of transmitter) to the time when the channel first opens. The complications that arise in such cases will be considered in Sections 9–11.

3.5. Pooling States That Equilibrate Rapidly

If, in mechanism 2, the binding step were very fast compared with the subsequent conformation change, and so fast that it was beyond the resolution of the experiment, then the vacant and occupied states would behave, experimentally, as a single (shut) state. This may be represented diagrammatically by enclosing the two states in a box, thus:

$$\boxed{\mathrm{R} \underset{}{\overset{K_\mathrm{A}}{\rightleftharpoons}} \mathrm{AR}} \underset{\alpha}{\overset{\beta'}{\rightleftharpoons}} \mathrm{AR}^* \tag{26}$$

shut 'state'

If the binding and dissociation are fast enough, the vacant and occupied states will be close to equilibrium at all times (even if the system as a whole is not). Therefore, the transition between them has been labelled only with the equilibrium constant, $K_\mathrm{A} = k_{-1}/k_{+1}$, rather than with the separate rate constants. This procedure has reduced the effective number of states in the mechanism from three to two (just shut and open). This does not affect the way we look at the shutting reaction, with rate constant α. However, we have to be more careful about how we treat the opening reaction. The transition rate from shut to open can no longer be taken as β, because the "shut state" spends part of its time without ligand bound (R), and while the receptor is not occupied, opening is impossible. The fraction of time for which the "shut state" is occupied (in AR) and so capable of opening is simply the equilibrium fraction of shut states that are occupied, i.e., $x_\mathrm{A}/(x_\mathrm{A} + K_\mathrm{A})$. Thus, the effective opening rate constant is

$$\beta' = \beta\left(\frac{x_\mathrm{A}}{x_\mathrm{A} + K_\mathrm{A}}\right) \tag{27}$$

When β' is so defined, the three-state mechanism in equation 2 becomes formally identical to the two-state mechanism in equation 1.

The argument just presented is related to the discussion of 'discrete states' in Section 1.3. What we refer to here as a discrete state must, since the protein is not stationary, consist of many conformations (or substates) that interchange rapidly. But if the lifetime of each individual substate is exponentially distributed, we expect that the overall lifetime will itself be a mixture of exponentials with Markovian behaviour. If most of the time is spent in one substate, which may be visited many times, then we expect that the distribution of the overall lifetime will itself be essentially a simple exponential. This is a result of the following fact, which is derived later, in Section 9.3:

The sum of a random number of exponentially distributed time intervals is itself exponentially distributed. (28)

What we mean by a "fast" reaction depends entirely on the time resolution of the experiment as well as on the rates of other steps in the mechanism. What is fast in one context may be slow in another; further examples are given by Colquhoun and Hawkes (1994).

4. A Mechanism with More Than One Shut State: The Simple Open Ion Channel-Block Mechanism

The two-state mechanism in equation 1 is simple because it is possible to tell which of the two states the system is in at any moment simply by inspecting the experimental record (though, in practice, complications would arise if more than one channel were contributing to the recording; see Section 8). In most cases of practical interest, there are likely to be several (experimentally indistinguishable) shut states, and possibly more than one open state too (see, for example, Section 13). It may be noted that, insofar as there will usually be more shut states than open ones, the distributions of shut periods are potentially far more informative than the distributions of open times. A simple example of a mechanism with two shut states is now considered in some detail.

4.1. A Simple Ion Channel-Block Mechanism

Consider the following simple mechanism (Armstrong, 1971; Adams, 1976) for ion channel block, which assumes that agonist binding is much faster than the open–shut reaction, as discussed in Section 3.5 (this is unlikely to be true, at least for the muscle nicotinic receptor).

$$\text{Shut} \underset{\alpha}{\overset{\beta'}{\rightleftharpoons}} \text{Open} \underset{k_{-\text{B}}}{\overset{k_{+\text{B}}}{\rightleftharpoons}} \text{Blocked.} \tag{29}$$

State number: 3 1 2

In this mechanism, the transition rate from open to blocked states is $k_{+\text{B}}x_\text{B}$, where x_B is the concentration of the blocking molecule. In this example, there are two shut states (shut and blocked). Neither of the shut states conducts any current, so it is not possible to tell for certain which of the two shut states the system is in at any moment simply by looking at the experimental record. This makes matters more complicated.

4.2. Relaxation and Noise

In this example, there are $k = 3$ states, so it would be expected that relaxations and noise experiments would be described by the sum of two components (exponential or Lorentzian, respectively) with rate constants denoted λ_1 and λ_2. The following results can be derived as described by Colquhoun and Hawkes (1977).

Although it is often convenient to derive results in terms of the rate constants λ_1 and λ_2, it is preferable, whenever possible, to refer to the time constants, $\tau_1 = 1/\lambda_1$ and $\tau_2 = 1/\lambda_2$ (as used in equation 4). There are two reasons for this. First, it avoids confusion between the *fundamental rate constants* in the mechanism (k_{-1}, etc), and the derived or *observed rate constants,* λ. Each of the observed rate constants depends on *all* of the fundamental rate constants. These components are easy to observe in the case of some channel-blocking drugs, as illustrated in Fig. 3. Second, it is easier to think in terms of time rather than frequency.

In this case, we find the two rate constants to be the solutions of the quadratic equation

$$\lambda^2 + b\lambda + c = 0 \tag{30}$$

where

$$-b = \lambda_1 + \lambda_2 = \alpha + \beta' + k_{+B}x_B + k_{-B}$$

$$c = \lambda_1\lambda_2 = \alpha k_{-B}\left[1 + \frac{\beta'}{\alpha}\left(1 + \frac{x_B}{K_B}\right)\right] \tag{31}$$

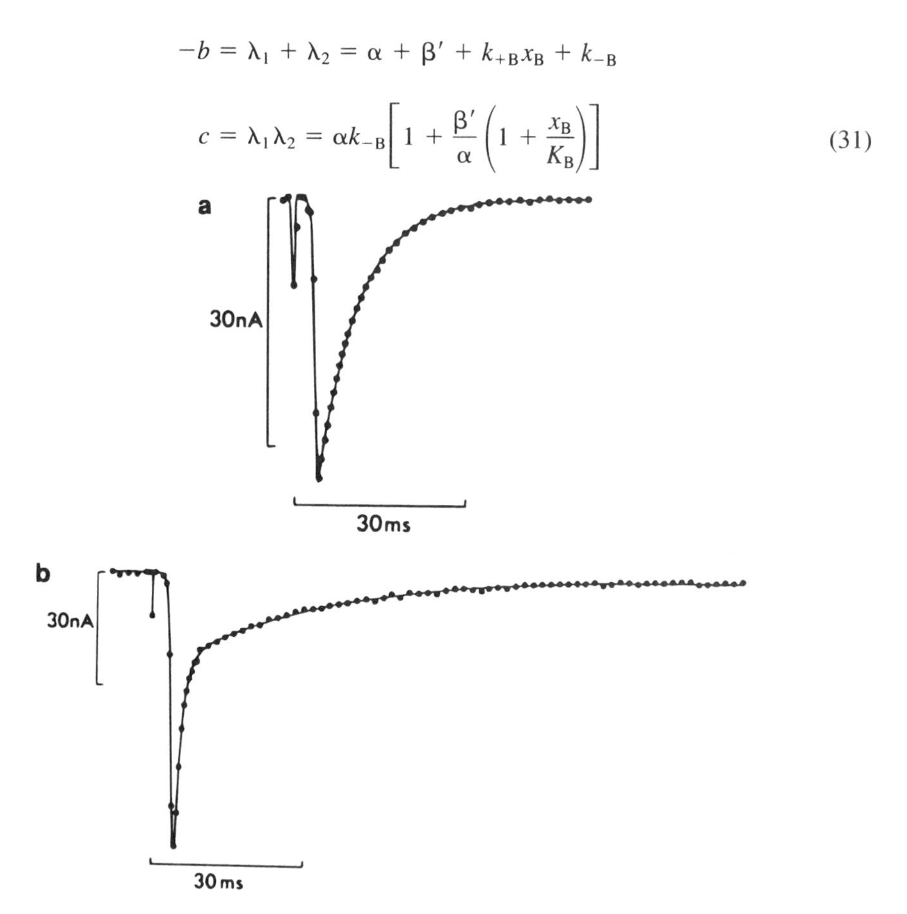

Figure 3. a and b: Endplate currents at -130 mV (dots) evoked by nerve stimulation (inward current is shown downward). a: Control, fitted with single exponential (τ = 7.1 ms). b: In presence of 5 μM gallamine, fitted with sum of two exponentials (τ = 1.37 and 28.1 ms). c and d: Spectral density (dots) of noise (at -100 mV) induced by carbachol. c: Carbachol (20 μM), fitted with single Lorentzian (τ = 3.47 ms) d: Carbachol (100 μM) in presence of gallamine (20 μM), fitted with sum of two Lorentzians (τ = 0.65 ms and 7.28 ms). Reproduced from Colquhoun and Sheridan (1981), with permission.

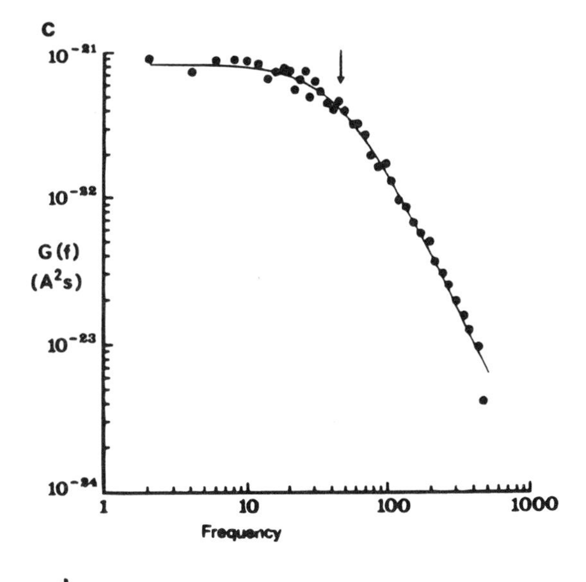

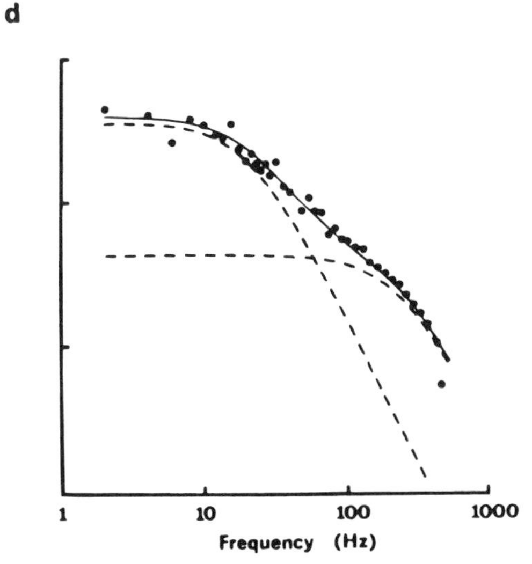

Figure 3. *Continued.*

and $K_B = k_{-B}/k_{+B}$ is the equilibrium constant for blocker binding. The relative amplitudes of the two components are also related to the reaction rate constants, though in a rather complicated way (see Colquhoun and Hawkes, 1977; Chapter 20, this volume).

4.3. Open Lifetimes of Single Channels

By contrast with noise or relaxation, the analysis of open times for single channels is very simple in this case. There is only one open state, and it is identifiable on the experimental record. By virtue of the rule given in expression 24, the open lifetime must therefore be distributed exponentially with mean $1/(\alpha + k_{+B}x_B)$. This follows from expression 24 because there are two ways out of the open state (shutting or blocking) with transition rates α and $k_{+B}x_B$, respectively.

4.4. Shut Lifetimes of Single Channels

Because there are two indistinguishable shut states, this is not as simple as previous cases. However, in this particular mechanism, the two shut states cannot intercommunicate directly but only by going through the open state. This makes matters much simpler than they would otherwise be, because each period for which the channel is shut must consist *either* of a single sojourn in the shut state (exponentially distributed with mean $1/\beta'$) *or* of a single sojourn in the blocked state (exponentially distributed with mean $1/k_{-B}$). The overall distribution of shut times is therefore simply a mixture of these two distributions in proportions dictated by the relative frequency of sojourns in the shut and blocked states (as long as only one channel contributes to the observations; see Section 8). These frequencies will be proportional to α and $k_{+B}x_B$, respectively, because these rate constants give the relative frequencies (probabilities) with which each of the two shut states is entered, starting from the open state. Thus, the pdf of all shut times can be put into the general form of a mixture of (in this case) two exponentials (see equation 25), as

$$f(t) = a_1\beta' e^{-\beta' t} + a_2 k_{-B} e^{-k_{-B}t}$$

where the areas of the two components are

$$a_1 = \left(\frac{\alpha}{\alpha + k_{+B}x_B}\right) \quad \text{and} \quad a_2 = \left(\frac{k_{+B}x_B}{\alpha + k_{+B}x_B}\right) \tag{32}$$

4.5. Bursts of Openings

If the agonist concentration is low (β' is low), openings are infrequent, and if the blocker dissociates quite rapidly from the open channel (k_{-B} is large), blockages are brief. In this case, openings would be expected to occur in bursts as the channel blocks and unblocks several times in quick succession before entering a long shut period. This has been observed in many cases (e.g., Neher and Steinbach, 1978; Ogden *et al.,* 1981) and is illustrated in Fig. 4.

The burst-like appearance is just the single-channel equivalent of a double-exponential relaxation, as illustrated in Fig. 5 (compare Fig. 2, in which a simple exponentially distributed open lifetime gave rise to a simple exponential relaxation). But not all channel blockers will produce such obvious effects, as discussed next.

4.5.1. Fast Channel Blockers

Some low-affinity blockers produce very brief blockages (e.g., about 20 μs for acetylcholine itself on nicotinic receptors; Ogden and Colquhoun, 1985). They therefore have noticeable effects only at high concentrations, at which blockages are very frequent. Bursts will consist

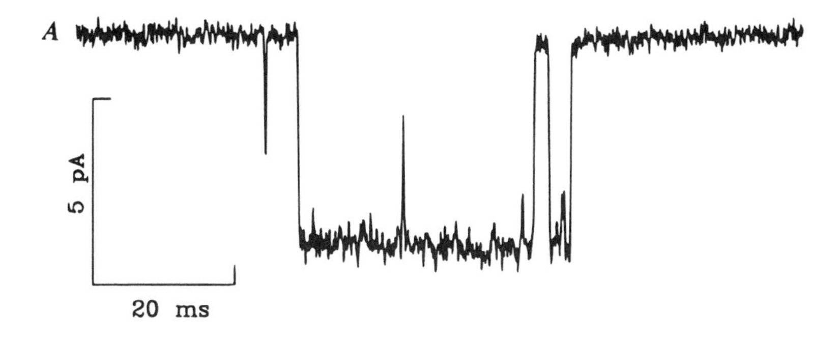

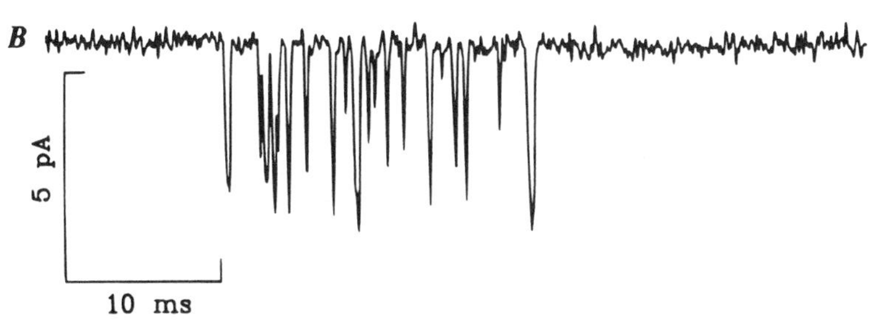

Figure 4. Channel blockage at the single-channel level, illustrated by the block of NMDA-type glutamate receptor channels by magnesium ions. The control trace in *A* shows openings of an NMDA receptor in the absence of magnesium (30 nM glutamate and 1 μM glycine in calcium-free EDTA-buffered solution, with free Mg^{2+} about 0.2 μM; see Gibb and Colquhoun, 1992). The trace in *B* was obtained in the presence of 25 μM free Mg^{2+} (EDTA-buffered solution, 100 nM glutamate). Despite the complexity of the shut-time structure even in the absence of magnesium, the effect of channel blockage is very obvious. The channel openings in *B* are much shorter on average (so filtering prevents most of them from attaining full amplitude) and are frequently interrupted by brief blockages. Filtered at 3 kHz (−3 dB). (Unpublished results of A. J. Gibb on dissociated adult rat hippocampal CA1 cells.)

of a large number of very short openings ($k_{+B}x_B$ is large) separated by very short blockages (k_{-B} is large). Most of these are too brief to be clearly resolved, so the bursts look like single noisy openings of reduced amplitude. According to mechanism 29, we would expect to see only the slow component of the biphasic relaxation (see Fig. 5), which corresponds roughly to the burst length (see below). Such blockers would appear to slow down the relaxation. (In the case of very brief interruptions that occur with agonist alone, it is also true that the relaxation will be approximately exponential, with a time constant similar to the mean burst length; see Section 5 and Fig. 8.)

4.5.2. Slow Blockers

At the other extreme, some blockers produce very long blockages, e.g., tubocurarine on nicotinic receptors, with mean blockage durations of seconds (Colquhoun *et al.*, 1979). Such blockers will appear to speed up the relaxation. At the single-channel level individual openings will be shortened on average, as described below, but if, following the opening, there is a blockage that lasts for 3 s, it would be impossible to tell by looking at the record that the opening before the blockage and that after it were both part of one very long burst.

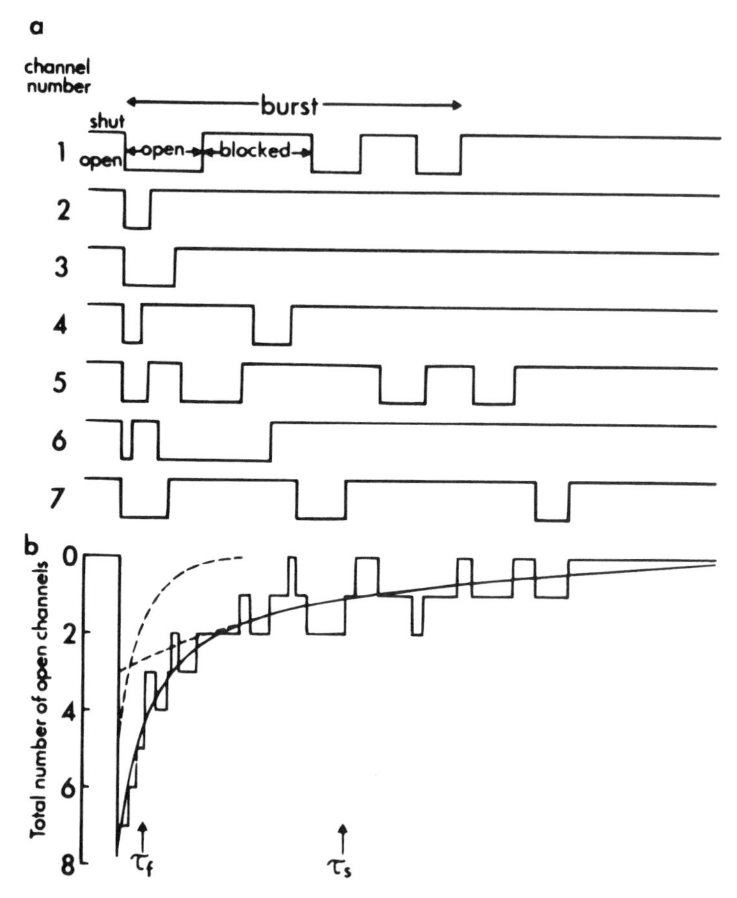

Figure 5. Schematic illustration to show why the occurrence of channel openings in bursts may result in biphasic relaxations (like, for example, that illustrated in Fig. 3b). The open state is shown as a downward deflection. a: Simulated behavior of seven individual ion channels in the presence of an ion-channel-blocking drug. Channels are supposed to be opened nearly synchronously at time zero by a quantum of acetylcholine, and the acetylcholine is supposed to disappear rapidly from the synaptic cleft. Thus, each channel produced only one burst of openings before it finally shuts (as marked on channel 1, which has two blockages and therefore three openings before it shuts). b: Sum of all seven records shown in a. The initial decline is rapid (time constant τ_f) as open channels become blocked, but the current thereafter declines more slowly (time constant τ_s). The continuous line is the sum of two exponential curves (shown separately as dashed lines) with time constants τ_f and τ_s. The slow time constant, under these conditions, reflects primarily the burst length rather than the length of an individual opening. (See also Neher and Steinbach, 1978.)

At the macroscopic level, the relaxation would reflect only the shortened openings; there would in fact be a very slow component too, but it would have such small amplitude that it would be undetectable.

4.6. The Number of Openings per Burst

The number of openings per burst will, of course, be a random variable. Its distribution can be found as follows. Define as π_{12} the probability that an open channel (state 1) will, as its next transition, become blocked (state 2). This probability takes no account of how much time elapses before the transition occurs but only of where the transition leads when it eventually does occur. It therefore depends simply on the rate of transition from state 1 to state 2, $k_{+B}x_B$; this rate must be divided by the sum of all rates for leaving state 1 so that the probabilities add to unity. Thus, we obtain

$$\pi_{12} = \frac{k_{+B}x_B}{\alpha + k_{+B}x_B} \tag{33}$$

which is precisely what was used in equation 32 to define the relative frequency of entry into each shut state. If the open channel does not block next, the only other possibility is that it shuts next, so the probability that the next transition of the open channel is to the shut state (state 3) is

$$\pi_{13} = 1 - \pi_{12} = \frac{\alpha}{\alpha + k_{+B}x_B} \tag{34}$$

We shall also need the probability that the next transition of the blocked channel is to the open state. In this particular mechanism, equation 29, there is nowhere else the blocked channel can go, so

$$\pi_{21} = 1 \tag{35}$$

The probability that a burst has only one opening is simply the probability that the channel, once open, then shuts, i.e., π_{13}. If the burst has two openings (and therefore one blockage), the open channel first blocks (probability π_{12}), then reopens (probability π_{21}), and finally shuts (probability π_{13}). So the overall probability of seeing two openings is the product of these three probabilities, i.e., $(\pi_{12}\pi_{21})\pi_{13}$. Extension of this argument gives the probability of a burst having r openings (and $r - 1$ blockages) as

$$P(r) = (\pi_{12}\pi_{21})^{r-1}\pi_{13} = (\pi_{12}\pi_{21})^{r-1}(1 - \pi_{12}). \qquad (r = 1, 2, \ldots, \infty) \tag{36}$$

This form of distribution is called a *geometric distribution*. The cumulative form of this distribution, the probability that we observe *n or more* openings per burst, is

$$P(r \geq n) = (\pi_{12}\pi_{21})^{n-1} \tag{37}$$

The mean number of openings per burst (denoted m_r) is

$$m_r = \sum_{r=1}^{\infty} rP(r) = \frac{1}{1 - \pi_{12}} = 1 + \frac{k_{+B}x_B}{\alpha} \tag{38}$$

This last result predicts that the mean number of openings per burst should increase linearly with the blocker concentration, with slope k_{+B}/α.

The number of blockages per unit open time is predicted to be simply $k_{+B}x_B$, so a plot of the observed blockage frequency against x_B should go through the origin and have a slope of k_{+B}. This behaviour has been observed directly in some cases, as illustrated in Fig. 6.

The geometric distribution is the discrete equivalent of the exponential distribution. It has the characteristic that a given increment in r reduces $P(r)$ by a constant factor (*viz.* $\pi_{12}\pi_{21}$), which is analogous to the behaviour of an exponential. And when m_r is large, the geometric distribution is well approximated by an exponential distribution with mean m_r. More generally, we expect that, under conditions where distributions of time intervals are described by a mixture of exponentials (see equation 25), the distributions of quantities such as the number of openings per burst (which can adopt only discrete integer values) will be described by a mixture of geometric distributions (see also Chapter 19, this volume). The number of geometric components in the distribution of the number of openings per burst is determined by the number of routes between open states and short-lived shut states (see Section 13.4) and is therefore not more than the number of open states (see also Chapter 20, this volume; Colquhoun and Hawkes, 1982, 1987).

4.7. Lifetime of Various States and Compound States

From the rule obtained earlier, in equation 24, we can immediately obtain the distribution of lifetimes in the three individual states. These will be exponentially distributed with

$$\text{Mean open lifetime: } m_0 = 1/(\alpha + k_{+B}x_B) \qquad (39)$$

$$\text{Mean blocked lifetime (gap within a burst): } m_w = 1/k_{-B} \qquad (40)$$

$$\text{Mean shut lifetime (gap between bursts): } m_b = 1/\beta' \qquad (41)$$

Thus, if bursts can clearly be distinguished in the observed record, we can, for example, obtain an estimate of k_{-B} simply by measuring the mean length of gaps within bursts. The

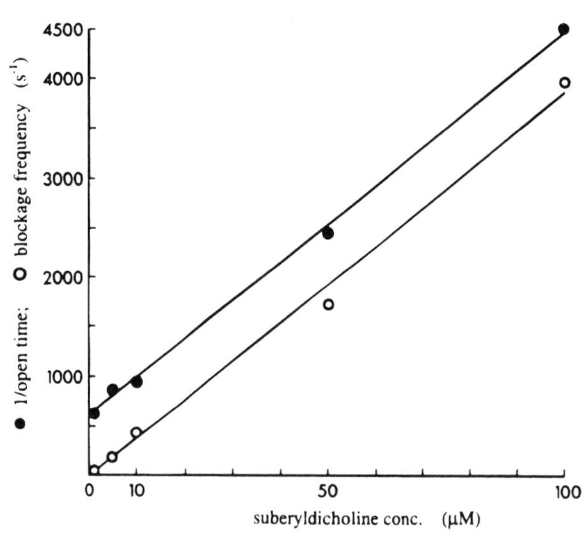

Figure 6. Block of suberyldicholine-activated endplate channels by the agonist itself at high concentrations. The blockages produce a characteristic component with a mean of about 5 ms in the distribution of shut times (so $k_{-B} \approx 1/5$ ms $= 200$ s^{-1}). Blockages become more frequent (the relative area of the 5-ms component increases) with concentration. *Closed circles:* reciprocal of the corrected mean open time, plotted against concentration (slope 3.9×10^7 M^{-1}s^{-1} intercept 594 s^{-1}). *Open circles:* the 'blockage frequency plot'—the frequency of blockages per unit of open time, plotted against concentration (slope $= 3.8 \times 10^7$ M^{-1}s^{-1}, intercept 4 s^{-1}). Reproduced with permission from Ogden and Colquhoun (1985).

division of openings into bursts may be a particularly useful procedure if more than one channel is contributing to the experimental record (see Section 8). In this case, it will not be known whether a burst originates from the same channel as the previous burst, so the lengths of shut times (gaps) *between* bursts will not be interpretable. Usually, however, it will be likely that all openings within a particular burst originate from the same channel, so the lengths of gaps within bursts will be interpretable and useful.

By use of the means in equations 39–41, we can immediately obtain the average value for the total open time during a whole burst. It will be the mean number of openings per burst, m_r from equation 38, multiplied by the mean length of an opening, m_0 from equation 39. Thus,

$$\text{Mean open time per burst} = m_r m_0 = (1 + k_{+\mathrm{B}}x_\mathrm{B}/\alpha)\left(\frac{1}{\alpha + k_{+\mathrm{B}}x_\mathrm{B}}\right)$$
$$= 1/\alpha \tag{42}$$

Thus, the mean open time per burst is exactly what the mean length of an opening would have been if no blocker were present, as was first pointed out by Neher and Steinbach (1978). This result seems surprising at first, and it will be discussed again in Section 6. Similarly, the total length of time spent in the blocked state per burst (total shut time per burst) is, on average,

$$\text{Mean shut time per burst} = (m_r - 1)m_w = \frac{k_{+\mathrm{B}}x_\mathrm{B}}{\alpha}(1/k_{-\mathrm{B}}) = c_\mathrm{B}/\alpha \tag{43}$$

where we denote the blocker concentration, normalized with respect to its equilibrium constant ($K_\mathrm{B} = k_{-\mathrm{B}}/k_{+\mathrm{B}}$), as

$$c_\mathrm{B} = x_\mathrm{B}/K_\mathrm{B} \tag{44}$$

Addition of equations 42 and 43 gives the mean burst length as

$$\text{Mean burst length} = \frac{1 + c_\mathrm{B}}{\alpha} \tag{45}$$

as derived by Neher and Steinbach (1978). This is predicted to increase linearly with the concentration of blocker.

We have just obtained means for the durations of various quantities characteristic of the burst, but so far we have not mentioned the distribution of these variables. It can be shown (see below; Colquhoun and Hawkes, 1982) that the fact that there is only one open state implies that the total open time per burst has a simple exponential distribution (with mean $1/\alpha$ as found above). Similarly, the fact that the gaps within bursts are spent in a single state (state 2, the blocked state) implies that the total shut time per burst (excluding bursts that have no blockages in them) will also have a simple exponential distribution, with the overall mean derived in equation 43 divided by the probability that there is at least one blockage, which, from equation 37, is $P(r \geq 2) = \pi_{12}\pi_{21}$.

The distribution of the number of openings per burst had one (geometric) component because there is only one open state in this example. However, the distribution of the burst *length* will be described by the sum of two exponential terms (because the burst is a period of time spent in either of two states, open or blocked). This distribution can be derived as follows.

4.8. Derivation of Burst Length Distribution for the Channel-Block Mechanism

We note that a burst consists of a sojourn in either of two states, open or blocked. As soon as the shut state (see equation 29) is entered, the burst ends. Thus, we want to find the distribution of the time spent oscillating within the pair of burst states (open $\rightleftharpoons$ blocked) without leaving this pair for the shut state.

We have already considered one example, the distribution of all shut times, that involved a sojourn in a pair of states, but this was unusually simple to deal with because the two shut states in question (shut and blocked) could not intercommunicate. In this case, the two states of interest (open and blocked) can intercommunicate, so a more general approach is needed; a similar approach can be used for many problems that involve a sojourn in a set of two or more states.

The burst starts at the beginning of the first opening and ends at the end of the last opening; the channel is open at the start and end of the burst. We have already considered in equation 8 a probability defined as

$$P_{11}(t) = \text{Prob}(\text{open at time } t | \text{open at time } 0) \tag{46}$$

This is what is needed for derivation of the time course of the macroscopic current or for the noise spectrum. However, it is not quite what we need now; this probability allows for the possibility that the system may enter any of the other states between 0 and t, but if the shut state is entered, the burst is ended, and we are no longer interested. What we need is a modified version of this that restricts the system to staying in the burst (i.e., in either open or blocked states) *throughout* the time between 0 and t. This sort of probability will be denoted by a prime. Thus,

$$P'_{11}(t) = \text{Prob}(\text{stays in burst throughout } 0, t \textit{ and } \text{open at } t | \text{open at } 0) \tag{47}$$

By analogy with the procedure in Section 3, we start by obtaining an expression for $P'_{11}(t + \Delta t)$, the probability that the channel stays within a burst for the whole time from 0 to $t + \Delta t$ and is open at $t + \Delta t$, given that it was open at $t = 0$. This can happen in either of two ways (the probabilities of which must be added):

1. The channel is open at t [with probability $P'_{11}(t)$] *and* stays open during the interval Δt between t and $t + \Delta t$. The probability that the channel does *not* stay open during Δt is $(\alpha + k_{+B}x_B)\Delta t + o(\Delta t)$, so the probability that it does stay open during Δt is $1 - (\alpha + k_{+B}x_B)\Delta t - o(\Delta t)$.

2. The channel is blocked (state 2) at time t [the probability of this, following the notation in equation 8 is $P'_{12}(t)$], *and* the channel unblocks during Δt [with probability $k_{-B}\Delta t + o(\Delta t)$]. Assembling these values gives our required result as

$$P'_{11}(t + \Delta t) = P'_{11}(t)[1 - (\alpha + k_{+B}x_B)\Delta t - o(\Delta t)] + P'_{12}(t)[k_{-B}\Delta t + o(\Delta t)] \tag{48}$$

Rearrangement of this, followed by allowing Δt to tend to zero, gives (by the method used in equations 17 and 18)

$$\frac{dP'_{11}(t)}{dt} = -(\alpha + k_{+B}x_B)P'_{11}(t) + k_{-B}P'_{12}(t) \tag{49}$$

This cannot be solved as it stands because there are two unknowns, $P'_{11}(t)$ and $P'_{12}(t)$. However, if an exactly analogous argument to that just given for $P'_{11}(t)$ is followed for $P'_{12}(t)$, we obtain another differential equation,

$$\frac{dP'_{12}(t)}{dt} = k_{+B}x_B P'_{11}(t) - k_{-B}P'_{12}(t) \tag{50}$$

We now have two simultaneous equations in two unknowns, which can be solved. For example, equation 49 can be rearranged to give an expression for $P'_{12}(t)$, which is substituted into equation 50. In this way, $P'_{12}(t)$ is eliminated, and we obtain a single (second-order) equation in $P'_{11}(t)$ only:

$$\frac{d^2P'_{11}(t)}{dt^2} + \frac{dP'_{11}(t)}{dt}(\alpha + k_{+B}x_B + k_{-B}) + \alpha k_{-B}P'_{11}(t) = 0 \tag{51}$$

Standard methods give the solution of this as the sum of two exponential terms with rate constants λ_1 and λ_2:

$$P'_{11}(t) = \frac{1}{\lambda_2 - \lambda_1}[(k_{-B} - \lambda_1)\exp(-\lambda_1 t) + (\lambda_2 - k_{-B})\exp(-\lambda_2 t)] \tag{52}$$

The two rate constants, λ_1 and λ_2, are found by solution of the quadratic equation,

$$\lambda^2 + b\lambda + c = 0$$

where

$$-b = \lambda_1 + \lambda_2 = \alpha + k_{+B}x_B + k_{-B}$$

$$c = \lambda_1\lambda_2 = \alpha k_{-B} \tag{53}$$

The pdf for the burst length follows directly from this. It is defined as

$$f(t) = \lim_{\Delta t \to 0}[\text{Prob(burst lasts from 0 to } t \textit{ and } \text{leaves burst in } t, t + \Delta t)/\Delta t] \tag{54}$$

In this case, the burst can be left only by direct transition from the open state to the shut state; the blocked state cannot shut directly, so there is no possibility of an experimentally invisible period in the blocked state following the last opening (cf. the example in Section 5). The probability that the burst is left in $t, t + \Delta t$ is simply $\alpha\Delta t + o(\Delta t)$, conditional on the burst lasting from 0 to t and being in open state 1 at t. Insertion of this into equation 54 gives

$$f(t) = \lim_{\Delta t \to 0} \{P'_{11}(t)[\alpha\Delta t + o(\Delta t)]/\Delta t\} = \alpha P'_{11}(t) \tag{55}$$

where $P'_{11}(t)$ is given by equation 52. Thus, the final form of the distribution of the burst length is, in the standard form specified in equation 25,

$$f(t) = a_1\lambda_1 e^{-\lambda_1 t} + a_2\lambda_2 e^{-\lambda_2 t} \tag{56}$$

where the areas of the two components are

$$a_1 = \frac{\alpha(k_{-B} - \lambda_1)}{\lambda_1(\lambda_2 - \lambda_1)} \quad \text{and} \quad a_2 = \frac{\alpha(\lambda_2 - k_{-B})}{\lambda_2(\lambda_2 - \lambda_1)} \tag{57}$$

The mean burst length follows from

$$m = \int_0^\infty tf(t)dt = \frac{a_1}{\lambda_1} + \frac{a_2}{\lambda_2} = \frac{(1 + c_B)}{\alpha} \tag{58}$$

which agrees with the result already found in equation 45 by a different route.

Two things are noteworthy about this distribution. (1) Unlike the simple case in which states do not intercommunicate, which was exemplified in equation 32, the two rate constants defined by equation 53 are compound quantities with no exact physical significance. (2) The two rate constants found here are not the same as those found for noise and relaxation experiments, as given in equations 30 and 31. The present versions are simpler because they do not involve rate constants that are concerned only with transitions from states outside the burst; i.e., they do not involve β' in this case. However, if few channels are open (β' is small), the rate constants for noise and relaxation, from equation 31, will become similar to those for the burst length distribution, from equation 53.

5. A Simple Agonist Mechanism

The mechanism of Castillo and Katz (1957), which has already been discussed in Sections 1.1 and 3.5, also has, like that just discussed, two shut states and one open state. However, in this case, the two shut states can intercommunicate directly. It will be convenient to number the three states thus

$$\mathrm{R} \underset{k_{-1}}{\overset{k_{+1}}{\rightleftharpoons}} \mathrm{AR} \underset{\alpha}{\overset{\beta}{\rightleftharpoons}} \mathrm{AR}^* \tag{59}$$

State Number 3 2 1

5.1. Shut Times

This mechanism can be analyzed in much the same way as the channel-block mechanism. In this case, because the two shut states intercommunicate, the distribution of all shut periods, although it will still have two exponential terms, will not have rate constants that have a simple physical significance. The rate constants must be found by solving a quadratic as in equation 53. The derivation follows the same lines as that for the distribution of the burst length and is given in full by Colquhoun and Hawkes (1994), so it will not be repeated here.

5.2. Bursts of Openings

Again, openings are predicted to occur in bursts, in this case bursts of several openings during a single occupancy (i.e., oscillation between AR and AR* before final dissociation). The bursts will be obvious as long as the time spent in AR, on average $1/(\beta + k_{-1})$ from rule 24, is short compared with the time between bursts. The distribution of the gap between bursts will be complicated by the fact that repeated occupancies ($R \rightleftharpoons AR$) may take place before a burst starts, and the gap between bursts will also include the time spent in AR immediately before the first opening of the burst and immediately after the last opening, as illustrated in Fig. 7.

The distribution of the number of openings per burst is geometric (as in Section 4.6), with mean, m_r, given by

$$m_r = 1 + (\beta/k_{-1}) \tag{60}$$

The openings have mean length $1/\alpha$, so the mean open time per burst is therefore m_r/α. Each burst will contain, on average, $(m_r - 1)$ brief shuttings, each of mean length $1/(\beta + k_{-1})$, giving a mean total shut time per burst of $(m_r - 1)/(\beta + k_{-1})$. The mean burst length will be the sum of these two quantities.

The *distribution* of the burst length can be found, much as in the channel-block example (Section 4.8), by deriving an expression for $P'_{11}(t)$. The way that the burst ends is rather different in this case, however; it cannot end (reach state 3) *directly* from the open state but only via AR. Therefore $P'_{11}(t)$ must be multiplied not only by the transition rate from AR* to AR, i.e., by α, as in equation 55, but also by the probability that, once in AR, the burst ends rather than continues, i.e., $\pi_{23} = k_{-1}/(\beta + k_{-1})$.

The openings of many sorts of ion channel are observed to occur in bursts, as illustrated

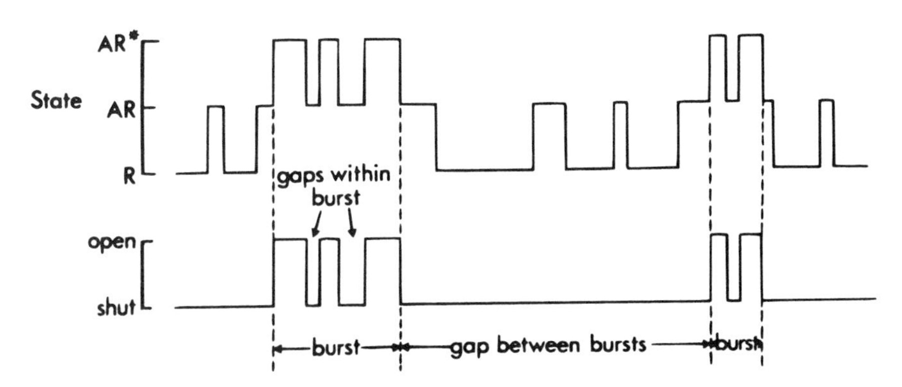

Figure 7. Schematic illustration of transitions between various states (top) and observed single-channel currents (bottom) for the simple agonist mechanism in equation 59. This illustrates the molecular events that underlie a burst of openings.

in Fig. 8 for the nicotinic acetylcholine receptor. Such observations have been interpreted along the lines suggested above, though a somewhat more complex mechanism than 59 is needed, as discussed in Sections 11 and 13; Chapter 20, this volume; Colquhoun and Hawkes, 1994; see, for example, Colquhoun and Sakmann, 1985; Sine and Steinbach, 1986).

5.3. Effective Openings

If the resolution of the experiment is poor, few of the brief shuttings, of the sort shown in Fig. 8 will be detected, and the bursts will appear to be single openings (see Section 12), with mean length equal to the burst length. When the shut times within bursts are short, the mean length of this "effective opening" will be little different from the total open time per burst, m_r/α, i.e., from equation 60,

$$\text{Mean open time per burst} = \frac{1}{\alpha}\left(1 + \frac{\beta}{k_{-1}}\right) \tag{61}$$

Furthermore, by virtue of equation 28, the duration of the "effective opening" will be approximately a single exponential with this mean.

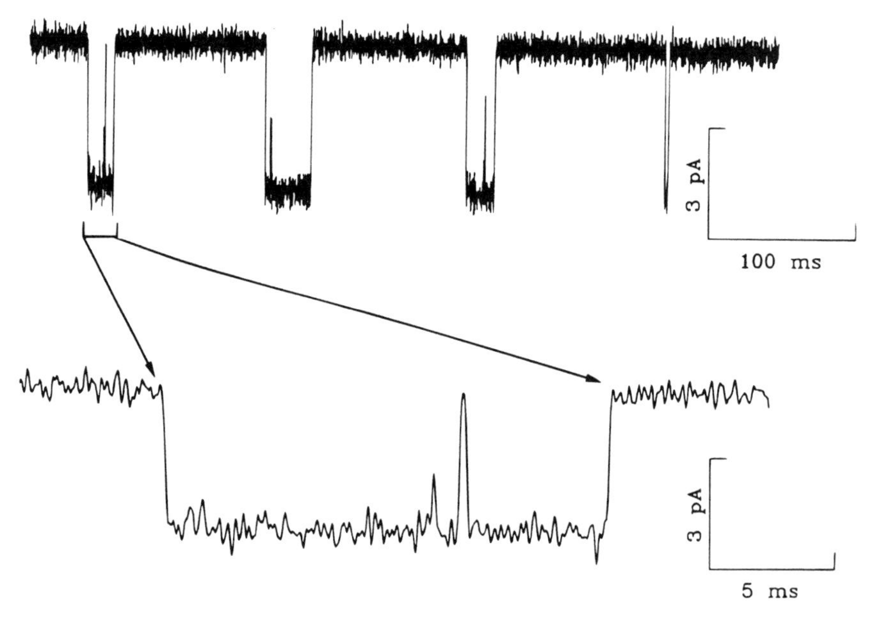

Figure 8. Example of bursts of channel openings elicited by an agonist. The upper trace shows four bursts of openings elicited by acetylcholine (100 nM, adult frog endplate, filter 2.5 kHz −3 dB; unpublished data of D. Colquhoun and B. Sakmann, methods as in Colquhoun and Sakmann, 1985). The last burst appears to consist of a single short opening, but the other three contain at least two or three openings separated by short shut periods. The lower section shows the first burst on an expanded time scale. It contains one fully resolved shut period and a partially resolved shutting. On the assumption that the partially resolved event is indeed a complete closure, time-course fitting (see Chapter 19, this volume) suggests that the burst contains three openings (durations 10.7 ms 1.0 ms and 5.7 ms) separated by two closed periods (durations 61 μs and 289 μs).

If a channel-blocking agent is added to the mechanism in equation 59, there will now be three shut states rather than two. If the channel blockages are, on average, much longer than the spontaneous brief shuttings just discussed, then each activation of the channel can be considered as a *cluster* of openings, the bursts within a cluster being separated by channel blockages, and the openings within a burst being separated by spontaneous brief shuttings. The formal theory of clusters of bursts was presented by Colquhoun and Hawkes (1982). This theory was used by Ogden and Colquhoun (1985) to show that, in the case where the spontaneous brief shuttings cannot be resolved, the whole burst will behave approximately like a single "effective opening," and application of the simple channel-block theory given in Section 4 will not give rise to serious errors. For example, the mean length of the effective opening will be reduced by the presence of a channel blocker in the same way as the mean length of the actual openings is reduced.

5.4. Macroscopic Currents

When the gaps within bursts are brief, noise and relaxation experiments will give a time constant that corresponds approximately to the mean burst length (rather than the mean open time). This can be shown as follows.

The two macroscopic rate constants are found, as usual, by solving a quadratic equation,

$$\lambda^2 + b\lambda + c = 0.$$

The well-known solution of this quadratic is

$$\lambda_1, \lambda_2 = 0.5(-b \pm \sqrt{b^2 - 4c}) \tag{62}$$

A less well-known alternative is

$$\lambda_1, \lambda_2 = \frac{2c}{-b \mp \sqrt{b^2 - 4c}} \tag{63}$$

where, as before,

$$-b = \lambda_1 + \lambda_2, \qquad c = \lambda_1\lambda_2 \tag{64}$$

When one of the rate constants is much bigger than the other (say $\lambda_f \gg \lambda_s$, where the subscripts denote *fast* and *slow*), i.e., when $b^2 \gg c$, then the faster rate constant, λ_f, is approximately

$$\lambda_f \approx -b \tag{65}$$

and, from the alternative form, equation 63, the slower rate constant, λ_s, is approximately

$$\lambda_s \approx -c/b \tag{66}$$

In this case the coefficients are given (e.g., Colquhoun and Hawkes, 1977) by

$$-b = \lambda_s + \lambda_f = \alpha + \beta + k_{+1}x_A + k_{-1} \tag{67}$$

$$c = \lambda_s\lambda_f = \alpha k_{-1}\left[1 + \frac{x_A}{K_A}\frac{(\alpha + \beta)}{\alpha}\right]$$

$$= \alpha k_{-1} + \alpha k_{+1}x_A + \beta k_{+1}x_A \quad (68)$$

where K_A is the microscopic equilibrium constant for the binding reaction, i.e., k_{-1}/k_{+1}. When the shut times within bursts are very short, only the slower of these components will be detectable, and the time constant for this component will, from equation 66, be approximately

$$\tau_s = 1/\lambda_s \approx \frac{\alpha + \beta + k_{+1}x_A + k_{-1}}{\alpha k_{-1} + \alpha k_{+1}x_A + \beta k_{+1}x_A}$$

$$\approx \frac{1}{\alpha}\left(1 + \frac{\beta}{k_{-1}}\right) \quad (69)$$

The second approximation is valid (1) when the agonist concentration, x_A, is sufficiently low, and (2) when the shut times within bursts, with mean length $1/(\beta + k_{-1})$, are short enough that $(\beta + k_{-1}) \gg \alpha$. The result in equation 69 is seen to be the mean open time per burst, as found in equation 61, i.e., approximately the mean burst length. Several approximations had to be made to get this result; this illustrates the general principle that there is usually no simple correspondence between the time constants for macroscopic currents and the time constants for the single-channel distributions.

This topic is discussed further in Sections 11 and 13 (see also Chapter 20, this volume; Colquhoun and Hawkes, 1981, 1982, 1994).

6. Some Fallacies and Paradoxes

The random nature of single-channel events leads to behaviour that is often not what might, at first sight, be expected intuitively. Some examples of apparently paradoxical behaviour and of common fallacies are now discussed.

6.1. The Waiting Time Paradox

This is most easily illustrated by consideration of a simple binding reaction

$$\mathrm{R} \underset{k_{-1}}{\overset{k_{+1}}{\rightleftharpoons}} \mathrm{AR}$$

State number: 2 1

Imagine that the receptors (R) have attained equilibrium with a concentration x_A of the ligand (A). The fraction of receptors that are occupied (or the fraction of time for which a particular receptor is occupied) will be $p_1(\infty) = x_A/(x_A + K_A)$, where $K_A = k_{-1}/k_{+1}$. Suppose that at an arbitrary moment, $t = 0$, the ligand concentration is reduced to zero. We should expect to see a simple exponential decline in receptor occupancy with a time constant $1/k_{-1}$ (though in practice diffusion problems usually preclude such simplicity). This, of course, would be interpreted (see Fig. 2) in stochastic terms by pointing out that the length of time for which a particular receptor remains occupied is exponentially distributed with a mean of $1/k_{-1}$, from rule 24.

However, it might be objected that a receptor that is occupied at the arbitrary moment $t = 0$ must already have been occupied for some time before $t = 0$, and what we measure in the experiment is the residual lifetime of the occupied state from $t = 0$ until dissociation eventually occurs, as illustrated in Fig. 9. Because the mean lifetime of the entire occupancy, measured from the moment the receptor becomes occupied to the moment of dissociation, is on average $1/k_{-1}$, surely this residual lifetime should be shorter! On the other hand, since the drug-receptor complex does not 'age'—i.e., it has no knowledge of how long it has already existed—the mean lifetime measured from any arbitrary moment must always be $1/k_{-1}$. Both of these arguments sound quite convincing, but the latter argument is the correct one.

The resolution of the paradox lies in the fact that we are looking, in the experiment, only at those drug–receptor complexes that happened to exist at the moment, $t = 0$, when we chose suddenly to reduce the ligand concentration to zero (no more complexes can form after this moment). These particular complexes will not be typical of all drug–receptor complexes: we have a greater chance of catching in existence long-lived complexes than short-lived ones. This happens because of a phenomenon known as length-biased sampling. Although complexes with above-average lifetimes are *fewer* in number than those with below-average lifetimes (because of the positive skew of the exponential distribution), the former actually occupy a *greater* proportion of the total time than the latter. The above-average lifetimes have, therefore, a greater probability of being caught in existence at an arbitrary moment. Although the mean length of all occupancies is $1/k_{-1}$, the mean lifetime of the particular complexes that are in existence at $t = 0$ is twice as long, $2/k_{-1}$. These complexes will, on average, have been in existence for a time $1/k_{-1}$ before $t = 0$ and for a time $1/k_{-1}$ (the residual lifetime) after $t = 0$. The paradox is resolved. Further details can be found, for example, in Feller (1966) or Colquhoun (1971, Chapter 5 and Appendix 2).

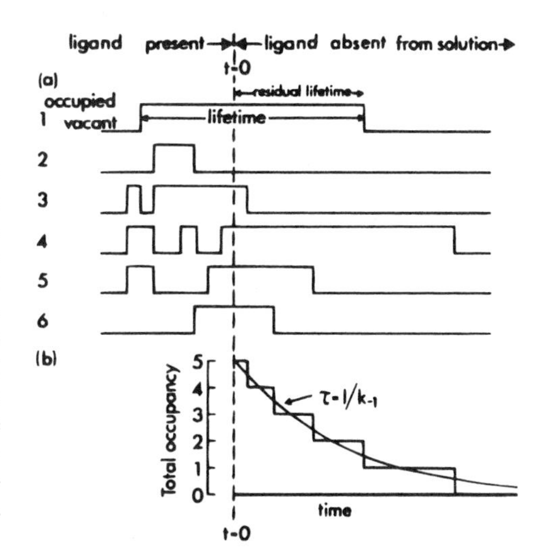

Figure 9. Illustration of the waiting time paradox. (a) Simulated behavior of six individual receptors. Before $t = 0$, ligand is present, and the receptor becomes occupied and vacant at random. The average lifetime of an occupancy is $1/k_{-1}$ (where k_{-1} is the dissociation rate constant). At $t = 0$, the ligand is removed from solution, and receptors that were occupied at $t = 0$ dissociate after a variable length of time. (b) The total of the records in a, showing the time course of decline of occupancy. The time course clearly reflects the distribution of the residual lifetime (defined on channel 1), which turns out to be identical with the distribution of the total lifetime (also defined on channel 1). Both are exponentially distributed with mean $1/k_{-1}$.

6.2. The Unblocked Channel Fallacy

Consider the simple channel-block mechanism of equation 29. In the absence of the blocking drug, the mean length of an opening would be $1/\alpha$. It was found above that in the presence of the blocker in concentration x_B, the mean length of an individual opening is reduced to $1/(\alpha + k_{+B}x_B)$. The easiest way to imagine why the opening is, on average, shorter is to suppose that its normal lifetime is cut short by a blocking molecule, which causes it to cease conducting prematurely (before it would otherwise have shut). But not every opening is ended by a blockage. The number of blockages per burst is a random variable, and a certain number of openings will end in the normal way, by transition to the shut state, rather than by the channel being blocked. This will be true of openings that have no blockage, so there is only one opening in the burst (and, more generally, for the last opening in any burst). Surely, these openings, which have not been cut short by a blockage, must be perfectly normal, with a mean lifetime $1/\alpha$.

On this basis, it is sometimes suggested, for example, that the noise spectrum should contain a component with the normal time constant $(1/\alpha)$, which corresponds to those channels that do not block. However, this is quite inconsistent with rule 24, which states that because there is only one open state, its lifetime must follow a simple exponential distribution with a mean, in this example, of $1/(\alpha + k_{+B}x_B)$. There should be no component with mean $1/\alpha$. In fact, if openings that end by shutting in the normal way rather than by blocking (e.g., bursts with only one opening) were measured separately from all other openings, it would be found that their duration was a simple exponential with mean $1/(\alpha + k_{+B}x_B)$; they *are* shorter than "normal" even though no blockage has occurred. The reason is again connected with length-biased sampling. Openings that happen to be very long will tend to get blocked before they shut, so, conversely, the openings that happen to be short (less than $1/\alpha$) will predominate among those that have no blockage. The extent to which these are shorter than $1/\alpha$ turns out, with great elegance, to be precisely sufficient to make their mean lifetime $1/(\alpha + k_{+B}x_B)$, exactly the same as that for openings that *are* terminated by being blocked.

6.3. The Last Opening of a Burst Fallacy

There are a number of other fallacies that can be disposed of easily by rule 24, which gives the distribution of the length of time spent in a single state. The explanation is, as in the last example, usually based on length-biased sampling. For example, the simple agonist mechanism, equation 59 predicts that openings should occur in bursts. The average length of an opening should be $1/\alpha$ regardless of where it occurs in the burst as long as there is only one open state (though if there is more than one open state, this may no longer be true; see below). According to mechanism 59, the agonist cannot dissociate from the open channel. If it were able to, it might be thought that this dissociation would end the burst and would cut short the lifetime of the last opening in the burst. Thus, might it be possible to test the hypothesis that the agonist can dissociate from the open state by seeing whether the last opening of the burst has a different distribution from the others? If there is only one open state, clearly this would *not* be possible. It is true that if a channel could shut by another route as well as that shown in equation 59, the mean lifetime of the open state would be reduced to something less than $1/\alpha$. But all openings regardless of position in the burst would have, on average, this same reduced lifetime.

6.4. The Total Open Time per Burst Paradox

It was pointed out earlier, in equation 42, that for the simple channel-block mechanism, the total time per burst that is spent in the open state will be, on average, $1/\alpha$. This is exactly what the mean open time would be in the absence of a blocker (a fact that, incidentally, explains the inefficiency of channel block in reducing the equilibrium current when the agonist concentration is low). How can this happen? The channel cannot 'know' how long it has been open earlier in the burst and so make up the total open time to $1/\alpha$. After a blockage, the channel is not continuing a normal open time (mean length $1/\alpha$) but starting a new open time [with mean length $1/(\alpha + k_{+B}x_B)$]. Clearly, since the mean length of a single opening is $1/(\alpha + k_{+B}x_B)$, it follows at once that the mean open time per burst for bursts with r openings must simply be $r/(\alpha + k_{+B}x_B)$. The relative proportions of bursts with $r = 1, 2.\ldots\ldots$ openings, given by equation 36, must be such that, on average, the total open time per burst is $1/\alpha$.

One way of understanding this is as follows (see Colquhoun and Hawkes, 1982). Imagine that a clock is started at the beginning of the first opening of a burst; the clock is stopped when the channel blocks and restarted when the channel reopens. It is finally stopped at the end of the burst, i.e., as soon as the channel shuts (as opposed to blocking). Thus, the clock runs only while the channel is open, and when finally stopped, it shows the total open time per burst. While the channel is open, the probability that it will leave the open state in Δt is $(\alpha + k_{+B}x_B)\Delta t + o(\Delta t)$, but if it leaves for the blocked state, the clock is stopped only temporarily. For the whole time that the clock is running, the probability that the clock is stopped finally, i.e., that the channel shuts (as opposed to blocking), in Δt is $\alpha\Delta t + o(\Delta t)$. This fact is sufficient to ensure that the time shown when the clock stops finally, the total open time per burst, has a simple exponential distribution with mean $1/\alpha$; this follows from the derivation of the exponential distribution given in Section 3.

A more general treatment (see Section 13; Chapter 20, this volume; Colquhoun and Hawkes, 1982; Neher, 1983) shows that the total open time per burst will be $1/\alpha$ for any mechanism with one open state as long as it fulfills the following condition. Suppose that there are any number of short-lived shut states ($\mathcal{B}$ states, say) in which the system stays during a gap within a burst, and that there are any number of long-lived shut states ($\mathcal{C}$ states, say) in which the system stays during a gap between bursts. If the only route from the former set of states ($\mathcal{B}$) to the latter ($\mathcal{C}$) is via the open state, and the total transition rate from the open state to the $\mathcal{C}$ states is α, then the total open time per burst must be, on average, $1/\alpha$. If, on the other hand, there are routes from $\mathcal{B}$ states to $\mathcal{C}$ states that do not go through the open state (e.g., if the blocked channel can shut without reopening in the channel block example), the mean open time per burst must be less than $1/\alpha$.

7. Reversible and Irreversible Mechanisms

Most reaction mechanisms are such that the system, left to itself, will move spontaneously towards a true thermodynamic equilibrium. All the reaction steps in such mechanisms will usually be reversible, and the mechanism must obey the principle of microscopic reversibility or detailed balance (see Denbigh, 1951). This principle states that at equilibrium, each individual reaction step will proceed, on average, at the same rate in each direction. This means, for example, that a cyclic reaction mechanism cannot have, at equilibrium, any tendency to move predominantly in one direction around the cycle; this has implications for the form of the distribution of open times (see example below). A slightly more subtle consequence of the principle of microscopic reversibility is that it implies that the stochastic properties of the mechanism must show time reversibility; they must be, on average, the same whether the record is read from left to right or from right to left (see Kelly, 1979). One example of such time symmetry is given by Colquhoun and Hawkes (1982), who discuss a mechanism in which all openings in a burst have not got the same distribution; these distributions are, however, the same for the first and last openings, and for the second and next-to-last openings, and so on. Another example is discussed below.

Reaction mechanisms with irreversible steps, such as that in equation 74 below, do not obey the principle of microscopic reversibility and do not tend spontaneously towards equilibrium. Such reactions may, however, be maintained in a steady state if they are coupled to a source of energy. If a steady state is attained, then all of the distributions derived by Colquhoun and Hawkes (1982) are still valid, although time symmetry is not, of course, expected.

7.1. A Simple Example

Some of the consequences of reversibility and irreversibility can be illustrated by a simple example, a cyclic mechanism that has one shut state (C) and two open states (O_1 and O_2). First consider the possibility that there might be a net clockwise circulation around the cycle. To achieve this, we might assign rate constants (all with dimensions s^{-1} as follows:

$$\begin{array}{l} C \xrightarrow{98} O_1,\quad O_1 \xrightarrow{100} C \\ O_1 \xrightarrow{100} O_2 \\ C \xrightarrow{2} O_2,\quad O_2 \xrightarrow{50} C \end{array} \tag{70}$$

For reversible reactions, however, the principle of microscopic reversibility implies that the product of the rate constants going one way around the cycle is the same as the product going the other way around. The rate constant for $O_2 \rightarrow O_1$, which has been omitted from scheme 70, must therefore be 2450 s^{-1}. The complete mechanism is thus

$$\begin{array}{l} C \xrightarrow{98} O_1,\quad O_1 \xrightarrow{100} C \\ O_1 \xrightarrow{100} O_2,\quad O_2 \xrightarrow{2450} O_1 \\ C \xrightarrow{2} O_2,\quad O_2 \xrightarrow{50} C \end{array} \tag{71}$$

Denote the closed state as state 3 and the open states O_1 and O_2 as states 1 and 2, respectively. At equilibrium, the occupancy of each state is

$$\begin{aligned} p_1(\infty) &= 0.4851 \\ p_2(\infty) &= 0.0198 \\ p_3(\infty) &= 0.4951 \end{aligned} \tag{72}$$

Thus, the mean frequency of the $O_1 \rightarrow O_2$ transitions, $0.4851 \times 100 = 48.51$ per second, is the same as the mean frequency of the $O_2 \rightarrow O_1$ transitions, $0.0198 \times 2450 = 48.51$ per second. The same applies to the other two reaction steps. There is no net circulation.

The rule given in statement 24 shows that the mean lifetime, m, of a sojourn in each of the states for reaction 71 is

$$m_1 = 5 \text{ ms}$$

$$m_2 = 0.4 \text{ ms}$$

$$m_3 = 10 \text{ ms} \tag{73}$$

In order to provide a contrast to the reversible scheme in reaction 71, consider the case in which all transitions are irreversible, and reaction can proceed only clockwise around the cycle. Suppose this mechanism is maintained in a steady state by coupling to an energy supply, and we choose rate constants for the transitions such that the steady-state occupancies are the same as for reaction 71; these are given in equations 72. In this example, the occupancies must be proportional to the mean lifetime of each state, so a suitable choice of rate constants would be

There are two interesting respects in which the reversible (71) and irreversible (74) reactions can be compared.

7.2. Distribution of the Lifetime of an Opening

Suppose that the conductance of the two open states is identical, so they cannot be distinguished. The distribution of the duration of an opening for the reversible mechanism (71) can be shown (e.g., from equation 3.64 of Colquhoun and Hawkes, 1982) to have a probability density function

$$f(t) = 97.962e^{-t/\tau_1} + 1.037e^{-t/\tau_2} \tag{75}$$

where the time constants of the two exponential components are $\tau_1 = 10.204$ ms and $\tau_2 = 0.384$ ms. Notice that the coefficients of both terms are positive so the distribution is a monotonically decreasing curve. It has not got a maximum or even a point of inflection (see

Fig. 10*a*). It can be shown that this must always be true, whatever rate constants are inserted in equation 71. This result is interesting in connection with observations by Gration *et al.* (1982) of an open-time distribution that appeared to go through a maximum; their result seems to be incompatible with any equilibrium reversible reaction mechanism.

In contrast, the irreversible mechanism (74) must give a steady-state open-time distribution with a maximum. The opening must always start in state 1 and then proceed through state 2 before shutting can occur, so there are few very short openings. The pdf can again be found from equation 3.64 of Colquhoun and Hawkes (1982) or, in this case, by analogy with equation 91 below. The result for scheme 74 is

$$f(t) = 106.34(e^{-t/\tau_1} - e^{-t/\tau_2}) \tag{76}$$

where the time constants, in this case, are simply the mean lifetimes of open states 1 and 2, i.e., 9.804 ms and 0.4 ms, respectively. This distribution has a term with a negative sign and must go through a maximum (see Fig. 10a), whatever the particular values of rate constants. In these examples, the mean open lifetime is 10.2 ms for both reversible (71 and 75) and irreversible (74 and 76) cases.

7.3. Probabilities of Particular Sequences of Transitions when the Open States Are Distinguishable

Let us suppose now that open state 2 has a lower conductance than open state 1, so the two states can be distinguished on the experimental record. Such conductance substates have

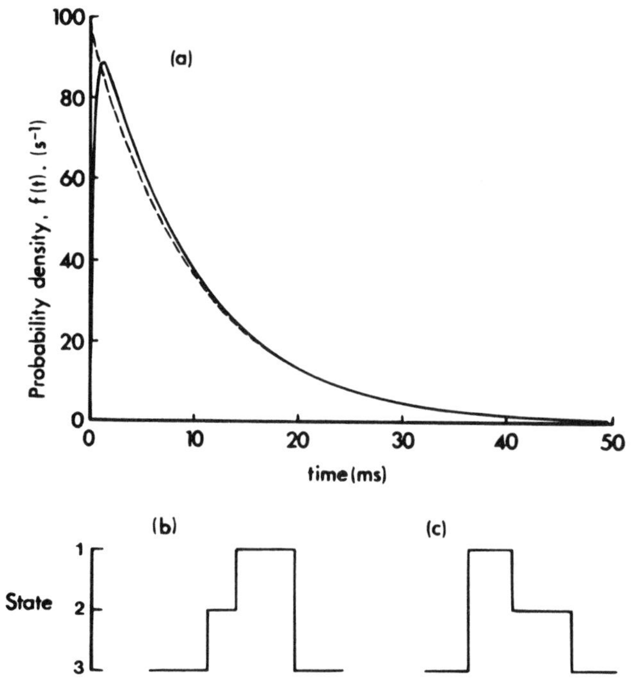

Figure 10. Reversible and irreversible mechanisms. a: The dashed line shows the pdf of the lifetime of the open (O_1 *or* O_2) state for the reversible mechanism in equation 71. The equation for this curve is given as equation 75. The fast component ($\tau = 0.384$ ms) of the distribution is too small in amplitude (only 1.04 s^{-1}) to be easily visible. The continuous line shows the pdf of the lifetime of the open state for the irreversible mechanism in equation 74 which is specified in equation 76. In this case, the fast component ($\tau = 0.4$ ms) has a negative sign, and the pdf goes through a maximum (i.e., very short open times will rarely be seen). b and c: Two possible types of opening for the reversible mechanism in equation 71; these will be observable only if the two open states, O_1 and O_2, differ in conductance.

now been observed in many sorts of ion channel. Consider, for example, an opening in which state 1 is entered first, then state 2, so the sequence of transitions is $3 \rightarrow 1 \rightarrow 2 \rightarrow 3$ (see Fig. 10c). Contrast this with the sort of openings in which state 2 is entered first, then state 1, i.e., the sequence $3 \rightarrow 2 \rightarrow 1 \rightarrow 3$ (see Fig. 10b). These sequences are mirror images of each other in time, so the principle of time symmetry discussed above suggests that for any reversible mechanism at equilibrium, they should occur equally frequently.

At first sight, it is not obvious how this can happen for the mechanism in equation 71. For this mechanism, it is fairly obvious that any open period has a 98% chance of starting in state 1 and only a 2% chance of starting in state 2, because the opening rate constants are 98 s^{-1} and 2 s^{-1} (in general, such probabilities can be found from equation 3.63 of Colquhoun and Hawkes, 1982). It is nevertheless true that the above sequences are equally probable. This can easily be shown by calculation of the probabilities (π_{ij} values) that a channel in one state (i) will move next to another (j); this sort of argument has already been illustrated in Section 4, equations 33–37.

Consider first the $1 \rightarrow 2 \rightarrow 3$ transition. For the values in equation 71, the probability that a channel in state 1 will next move to state 2 is $\pi_{12} = 100/(100 + 100) = 0.5$, and the probability that once in state 2 it will move to state 3 is $\pi_{23} = 50/(50 + 2450) = 0.02$ The probability of the $1 \rightarrow 2 \rightarrow 3$ sequence is therefore $\pi_{12}\pi_{23} = 0.5 \times 0.02 = 0.01$. Now the probability that the opening starts in state 1 in the first place is 0.98, so a fraction 0.98×0.01, i.e., 0.98%, of all openings will be of the $3 \rightarrow 1 \rightarrow 2 \rightarrow 3$ type shown in Fig. 10c. Similarly, the probability of a $2 \rightarrow 1 \rightarrow 3$ sequence is $\pi_{21}\pi_{13} = 0.98 \times 0.5 = 0.49$. However, only 2% of openings start in state 2, so a fraction 0.49×0.02, i.e., 0.98%, of all openings should be of the $3 \rightarrow 2 \rightarrow 1 \rightarrow 3$ type shown in Fig. 10b. This is exactly the same fraction as for the mirror-image sequence, as predicted.

There have been some reports of asymmetry in sublevel structure (e.g., Hamill and Sakmann, 1981; Cull-Candy and Usowicz, 1987), though the majority of cases where the question has been inspected show no sign of asymmetry (e.g., Howe *et al.*, 1991; Gibb and Colquhoun, 1992; Stern *et al.*, 1992). The finding of asymmetry suggests that either the reaction mechanism is not reversible or that it is not at equilibrium (see, for example, Läuger, 1985; Chapter 23 this volume). Clearly, the flow of ions through an open channel is far from equilibrium, so any coupling between ion flow and channel gating could, in principle, give rise to asymmetry.

The conclusion that has just been illustrated is quite general. For a reversible mechanism at equilibrium, any sequences that are mirror images in time should be equally frequent, and the length of time spent in each of the states should have the same distribution whether the record is read from left to right or from right to left. Furthermore, this remains true even if each of the experimentally distinguishable states actually consists of any number of indistinguishable (equal-conductance) states.

None of this is, of course, true for an irreversible mechanism. For that shown in reaction 74, it is clear that every opening will consist of a $3 \rightarrow 1 \rightarrow 2 \rightarrow 3$ sequence of transitions.

8. The Problem of the Number of Channels

It is clear that, in general, there may well be more than one ion channel in the patch of membrane from which a recording is made. This means that one cannot, in general, be sure that a particular single channel current in the recording originates from the same individual ion channel that produced the previous current pulse. This, in turn, means that the distribution of the length of the shut periods between openings cannot be interpreted without knowledge of the number of ion channels that are present. This is very unfortunate because, insofar as there will usually be more shut states than open states, the distribution of shut times is potentially more informative than the distribution of open times.

There are at least three things that can be done about this problem: (1) make an estimate of the number of channels present and make appropriate allowance if there is more than one; (2) use recordings only from patches that have one channel (evidence for this is considered below); (3) use only the brief shut periods within a burst of openings, which may be interpretable even if interburst intervals are not. It must be said, however, that quite often none of these procedures proves to be entirely satisfactory, and lack of knowledge of the number of channels continues to be a serious problem. The procedures are now discussed in a bit more detail.

8.1. Estimation of the Number of Channels

Suppose that there are N independent channels present. The probability that r of those channels are open simultaneously should be given by the binomial distribution as

$$P(r) = \frac{N!}{r!(N-r)!} p_0^r(1 - p_0)^{N-r} \qquad (r = 0, 1 \ldots, N) \tag{77}$$

where p_0 is the probability that an individual channel is open (this will, of course, be unknown and will be lower than the *observed* probability of being open, if more than one channel is present). In principle, the value of N can be estimated from data by comparing the distribution of simultaneously open channels with the predictions of the binomial distribution. The estimation of the binomial parameter N is, however, a problem with a notorious reputation among statisticians (see Olkin *et al.*, 1981). The problem is discussed critically, in the single-channel context, by Horn (1991). He compares seven different ways of estimating N on a series of simulated data sets with a range of parameter values.

The simplest estimate of N is just the largest number of simultaneously open channels that is seen in the record. Although this sounds crude (it *is*), other methods that might be thought of as more subtle (such as maximum-likelihood estimation of N) will often produce much the same answer. The fact is that many sorts of record contain very limited information about the size of N, so no method can extract much from them. It is obvious, for example, that when a very low agonist concentration is used on a muscle endplate, long records can be obtained without any double openings at all despite the fact that the patch contains hundreds of channels. In general, it will never be possible to estimate N when the number of channels is large and the probability of each being open is small. In this case, the binomial distribution approaches a Poisson distribution, and N becomes indeterminate (only the mean, Np_0, can de determined). (Exactly the same problem arises in the study of quantal transmitter release.) In order to have any hope of estimating N, the experiment must be done under conditions where p_0 is as high as possible (see Horn, 1991). The problem, however, remains that p_0 is the probability of being open for *one* channel, and so it cannot be inferred directly from a record derived from an unknown number of channels.

A further problem is that it is possible that the assumptions of the binomial analysis are not met. Receptor heterogeneity is a real problem (especially in the central nervous system) for this analysis (as well as for many others). There is also a possibility that receptors may not always be independent; e.g., the opening of one receptor might influence the opening of adjacent receptors. There have been reports of such interactions, for example, Yeramian *et al.* (1986), but most are not as convincing as this one.

8.2. Evidence for the Presence of Only One Channel

Obviously, if one or more double openings are seen, there must be more than one channel. If, on the other hand, the observed record consists entirely of periods with either zero or one channel open, then there may be only one channel present. If there is a channel open for most of the time, and yet no double openings are seen, then it is fairly obvious that all the openings must come from the same ion channel. This is the basis for determining the fraction of time for which an individual channel is open by looking at clusters of channel openings at high agonist concentrations (see, for example, Sakmann *et al.,* 1980; Sine and Steinbach, 1987; Colquhoun and Ogden, 1988). If, however, much of the time is spent with no channels open, it will not be obvious how many channels are present, and some sort of statistical test is desirable. Horn (1991) suggests, on the basis of his binomial simulations, that if no double openings are seen, and the channel is open for more than about 50% of the time, then it is very likely that one channel is present. Some variants on this approach will now be discussed.

8.2.1. A Simple Approximation

Suppose that (1) channels can exist in two states only, open and shut, as in equation 1, that (2) we observe n_o single openings but no double openings, and that (3) for most of the time no channel is open; i.e., if we denote the observed mean (singly) open time m_o, and the observed mean shut time as m_s, then we assume $m_s \gg m_o$. How probable is this observation if there are actually N independent channels present? If we start with one channel open, the probability, π, that the next transition is the shutting of this one channel, with rate α, rather than a second channel opening, with rate $(N - 1)\beta'$, is

$$\pi = \frac{\alpha}{\alpha + (N - 1)\beta'} \tag{78}$$

The observed probability of being open in the experimental record, P_{ON} say, is

$$P_{ON} = \frac{m_O}{m_O + m_S} \tag{79}$$

Furthermore, given our assumption that $m_s \gg m_o$, the rate constants in this can be estimated from the data as

$$\hat{\alpha} \approx 1/m_O$$

$$\hat{\beta}' \approx 1/Nm_S \tag{80}$$

so we can estimate π from the observations as

$$\pi \approx \frac{1}{1 + \left(\frac{N-1}{N}\right)\frac{m_O}{m_S}}$$

$$= \frac{1 - P_{ON}}{1 - P_{ON}/N}. \tag{81}$$

We can now ask how many consecutive single openings are likely to be seen when there is actually more than one channel present. If we note that the probability that the singly open channel is followed by a transition to a doubly open channel is $(1 - \pi)$, then the probability of getting r single openings before the first multiple opening occurs (given that there is at least one single opening) is

$$P(r) = \pi^{r-1}(1 - \pi) \tag{82}$$

This is a geometric distribution of the sort already encountered in equation 36. The mean number of consecutive single openings, m, is, as in equation 38, thus

$$m = \frac{1}{1-\pi} = \frac{1}{P_{ON}}\left(\frac{N}{N-1}\right)(1 - P_{ON}/N) \tag{83}$$

We have observed n_o consecutive single openings, so the run of single openings must be at least n_o in length. The probability of observing n_o or more single openings is, as in equation 37,

$$P(r \geq n_O) = \pi^{n_O - 1} \tag{84}$$

This result can also be derived, under the above assumptions, as the approximate probability that the waiting time until the first double opening is greater than the length, T, of the observed record, given that N channels are present.

Consider, for example, a record consisting of single openings of mean length $m_o = 1$ ms and mean shut time $m_s = 99$ ms, so $m_o/m_s = 0.0101$, and $P_{ON} = 0.01$. On the hypothesis that there are actually $N = 2$ channels present, equation 81, gives $\pi = 0.9949749$. If we observe $n_o = 300$ openings (i.e., about a 30-s record) with no double openings, then equation (84) gives the probability of a run at least this long as 0.222 (or 0.134 if $N = 3$). The observation would not be surprising if there were actually two or three channels present, even though no multiple openings have been observed, so the data are insufficient to provide good evidence for the hypothesis that there is only one channel present. On the other hand, if $n_o = 1200$ single openings were observed (2 min with no double openings), equation (84) would give the probability of a run at least this long as 0.0024 if there were two channels present (or 0.0003 if three channels were present), so it is unlikely that more than one channel is functioning.

8.2.2. A Better Approximation

The case of $N = 2$ channels is discussed in detail by Colquhoun and Hawkes (1990). They give an approximation for the mean number of openings in a run of single openings as

$$m \approx \frac{2}{P_{O_2}}(1 - 0.5P_{O_2} - 0.75P_{O_2}^2) \tag{85}$$

which differs from the result in equation (83) only by virtue of the $0.75{P_{O_2}}^2$ term. In either case, the result approaches $2/P_{O_2}$ when the observed probability of being open, P_{O_2}, is sufficiently small. The result in equation (85) is plotted in Fig. 11, together with approximate upper confidence limits for the number of openings per run.

8.2.3. Exact Solutions

Colquhoun and Hawkes (1990) also present exact calculations concerning the lengths of runs of single openings (and bursts) in the case where there are $N = 2$ identical independent channels. Such calculations are needed to explore the conditions under which the approximations are adequate, though in order to obtain exact results it is necessary to specify the channel mechanism (whereas the above approximations have the virtue that this is not necessary). The approximation works well when (1) the openings occur singly and are well separated from each other, and (2) the openings occur in compact, well-separated bursts, the shut times within the bursts being brief relative to the openings. In the latter case, the word "opening" in the approximate argument should be replaced by "burst"; the burst is open for a large proportion of the time (so that two overlapping bursts will certainly produce a double-amplitude event), and for the purposes of the present argument (as well as for physiological purposes) the burst is the "effective opening." However, in cases where the shut times within bursts are of the same order of magnitude as the open times, the approximation may be poor.

8.2.4. Problems of Prolonged Bursts, Desensitization, and Sleepy Channels

The approximation presented above works well when openings occur singly or in compact bursts, but it would probably not be very good for channels such as the NMDA-

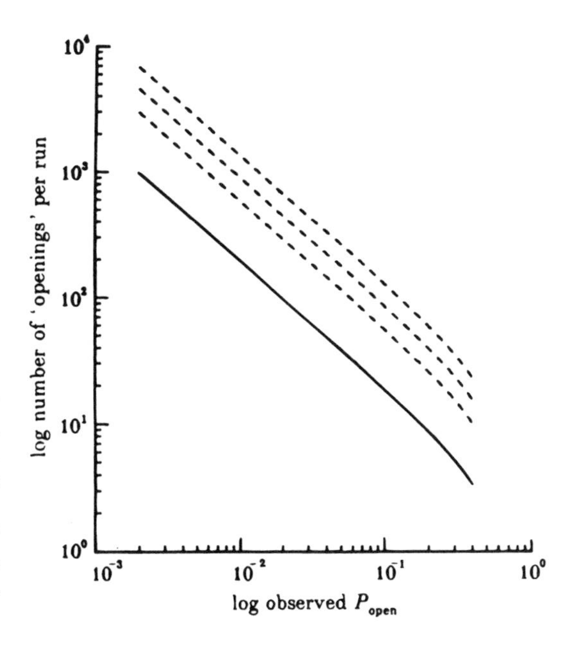

Figure 11. The mean number of openings per run of single openings in a membrane patch that contains two channels, as calculated from the approximation given in equation 85. The dashed lines show the approximate upper confidence limits for the number of openings per run, for $P = 0.05, 0.01$, and 0.001. Reproduced with permission from Colquhoun and Hawkes (1990).

type glutamate receptor, which produce complex and prolonged burst-like channel activations containing shut periods some of which are considerably longer than the openings.

Furthermore, most channels show desensitization, inactivation, or 'sleeping' phenomena, which involve entry into shut states that may have very long lifetimes. If, for example, a patch contains two channels, but one of them is desensitized at the beginning of the recording, it will appear that only one channel is present. If the channel is open for much of the time, then it will be obvious when the second channel emerges from its desensitized state, because double openings will be seen straight away, but if the probability of being open is low, it may not be at all obvious that a second channel has appeared. Clearly, though, any method for trying to estimate N will not work well if N is effectively changing during the recording. This is probably one of the most serious problems in practice.

8.2.5. Fitting with a Known Number of Channels

If an estimate of the number of channels can be made, then it is possible to fit some sorts of distribution even when records contain more than one channel open at the same time (Jackson, 1985; Horn and Lange, 1983). These methods are discussed in Chapter 19 (this volume).

8.3. Use of Shut Periods within Bursts

Most channels seem to produce openings in bursts rather than singly. This observation implies only that there is more than one shut state (see Sections 4, 5, and 13; Colquhoun and Hawkes, 1982). Regardless of the mechanism, it is likely, if the gaps within a burst are short, that all of the openings in one burst originate from the same individual channel, even if there are several channels present so the next burst may originate from a different channel. In this case, the distribution of the lengths of shut periods within (but not between) bursts can be interpreted in terms of mechanism as though only one channel was present, even when it is not known how many channels are actually present. This procedure was employed, for example, by Colquhoun and Sakmann (1985) and Sine and Steinbach (1986).

9. Distribution of the Sum of Several Random Intervals

Many problems involve finding the distribution of the sum of two or more random intervals, for example, the durations of the sojourns in the various states that constitute a burst of openings. This sort of problem also arises when we consider the relationship between single-channel currents and macroscopic currents (see Section 11). Some useful examples will be discussed in this section.

9.1. The Sum of Two Different Exponentially Distributed Intervals

By way of an example, consider again the simple two-state mechanism specified in equation 1, namely:

$$\text{Shut} \underset{\alpha}{\overset{\beta'}{\rightleftharpoons}} \text{Open} \tag{86}$$

What is the distribution of the time interval between two successive openings? This time interval consists of one open time plus one shut time. The pdf of the open time is $f_1(t) = \alpha e^{-\alpha t}$, and that of the shut time is $f_2(t) = \beta' e^{-\beta' t}$. We wish to know the pdf, $f(t)$, for one open time plus one shut time. Suppose that the open time is of length τ; if the total length of the gap between openings is t, then the length of the following shut period must be $t - \tau$. Since it is a basic characteristic of our random process that events occurring in nonoverlapping time intervals are independent, we can simply multiply the corresponding probability densities, which gives $f_1(\tau)f_2(t - \tau)$. However, the length, τ, of the opening may have any value from 0 to t, so to obtain the pdf, we must sum over these possibilities. This summation, because τ is a continuous variable, must be written as an integral, so we obtain the pdf of the time between openings as

$$f(t) = \int_{\tau=0}^{\tau=t} f_1(\tau)f_2(t-\tau)d\tau. \tag{87}$$

This form of integral is called a *convolution* (of f_1 and f_2).*

This argument leads to the general rule that the pdf of a sum of random intervals is the convolution of their individual pdfs. In this case, with simple exponential pdfs, the convolution (equation 87) can easily be integrated directly. In general, however, it is much easier to solve this sort of problem by use of the Laplace transform of the pdfs, because simple multiplication of the transforms corresponds with convolution in the time domain. This is the method that must be used for a more general treatment (see Section 13; Colquhoun and Hawkes, 1982), which provides another reason to discuss a simple example now.

We shall denote the Laplace transform of $f(t)$ as $f^*(s)$. In this example, we have

$$f_1(t) = \alpha e^{-\alpha t} \quad \text{and} \quad f_2(t) = \beta' e^{-\beta' t} \tag{88}$$

so their Laplace transforms, which can be obtained from tables (e.g. Spiegel, 1965), are

$$f_1^*(s) = \alpha/(s + \alpha) \qquad f_2^*(s) = \beta'/(s + \beta'). \tag{89}$$

The Laplace transform of the required pdf (equation 87) is therefore

$$f^*(s) = f_1^*(s)f_2^*(s) = \frac{\alpha\beta'}{(s + \alpha)(s + \beta')} = \frac{\alpha\beta'}{\alpha - \beta'}\left(\frac{1}{s + \beta'} - \frac{1}{s + \alpha}\right) \tag{90}$$

Inversion of this transform gives the required pdf as

$$f(t) = \frac{\alpha\beta'}{\alpha - \beta'}(e^{-\beta' t} - e^{-\alpha t}) \tag{91}$$

*In general, the integral for a convolution would be from $-\infty$ to $+\infty$, but in this case the pdfs are zero for times less than zero.

Notice that this pdf is the difference between two exponential terms and therefore, unlike the simple exponential, goes through a maximum (as already illustrated; see equation 76 and Fig. 10a). This shape indicates a deficiency of very short values (compared with a simple exponential distribution), and this is what would be expected intuitively, because in order to get a short interval *both* the open and shut times must be very short, and this is relatively unlikely to happen. This characteristic shape is illustrated again in Section 11, Fig. 16, when the relationship between single-channel currents and macroscopic currents is discussed. The mean of this pdf, the mean time between openings, is, from equation 23,

$$\text{mean} = \int_0^\infty tf(t)dt = \frac{1}{\alpha} + \frac{1}{\beta'}. \tag{92}$$

As expected, this is merely the sum of the mean open time and the mean shut time. The mean opening frequency is the reciprocal of this, i.e.,

$$\frac{\alpha\beta'}{\alpha + \beta'} = \alpha p_1(\infty) = \beta' p_2(\infty) \tag{93}$$

where $p_1(\infty)$ and $p_2(\infty)$ are the equilibrium probabilities (or fractions) of open and shut channels, respectively. In other words, the mean opening frequency is the opening transition rate, β', multiplied by the probability, $p_2(\infty)$, that a channel is shut (i.e., available to open). It is, of course, equal to the mean equilibrium shutting frequency, $\alpha p_1(\infty)$. This provides another way of interpreting rate constants in terms of the frequency with which transitions occur (see also Sections 1.2 and 4.6 and Fig. 6).

9.2. The Distribution of the Sum of *n* Exponentially Distributed Intervals

As an example, consider the case where we wish to know the distribution of the total open time in a burst of openings that contains exactly n openings (this will be close to the burst length if the shut periods are short). Suppose that each of the openings has the same exponentially distributed length with mean $1/\alpha$, i.e., they have pdf $f_1(t) = \alpha e^{-\alpha t}$, as above. We need, according to the argument in the previous section, the n-fold convolution of $f_1(t)$ with itself. This is made easy by using Laplace transforms as in equations 89 and 90. The Laplace transform of the required result is

$$f^*(s) = [f_1^*(s)]^n = \left(\frac{\alpha}{s + \alpha}\right)^n \tag{94}$$

Inversion of this transform gives the required pdf as

$$f(t) = \frac{\alpha(\alpha t)^{n-1} e^{-\alpha t}}{(n - 1)!}. \tag{95}$$

This is known as a *gamma distribution.* It has a mean n/α, simply n times the lifetime of an individual interval, as expected, and a variance n/α^2. Like the result in equation 91, it is zero at $t = 0$ and goes through a maximum at $t = t_{\max} = (n - 1)/\alpha$. For $n = 1$ it reduces to a simple exponential, but as n gets larger, the pdf becomes more and more symmetrical, eventually approaching a Gaussian shape. The cumulative form of this distribution is given, for example, by Mood and Graybill (1963):

$$F(t) = 1 - \sum_{r=0}^{n-1} \frac{(\alpha t)^r}{r!} e^{-\alpha t} \tag{96}$$

9.3. The Distribution of a Random Number of Exponentially Distributed Intervals

The results in the last section referred to the sum of a *fixed* number of exponentially distributed values. In the case of, for example, a burst of channel openings, the number of openings is not fixed but random. In the simplest cases the number of openings per burst will follow a geometric distribution, as exemplified in Sections 4.6 and 8.2. If we write the geometric distribution in the form already used in equation (82), the probability of there being r intervals (e.g., r openings per burst) is

$$P(r) = \pi^{r-1}(1 - \pi) \tag{97}$$

with mean

$$m_r = 1/(1 - \pi). \tag{98}$$

The required pdf can be found by weighting the pdf for r openings, with Laplace transform $f_1^*(s)^r$, as in equation 94, with $P(r)$ from equation 97. This gives

$$\begin{aligned} f^*(s) &= \sum_{r=1}^{r=\infty} P(r)[f_1^*(s)]^r \\ &= \frac{1 - \pi}{\pi} \sum_{r=1}^{r=\infty} \left(\frac{\pi\alpha}{s + \alpha}\right)^r \\ &= \frac{\alpha(1 - \pi)}{s + \alpha(1 - \pi)} \end{aligned} \tag{99}$$

Comparison of this result with that in equation 89 shows that its inverse transform is a simple exponential with mean $1/\alpha(1 - \pi) = m_r/\alpha$, i.e., simply the mean number of intervals, m_r from equation 98, times the mean length of one interval; thus,

$$f(t) = (\alpha/m_r) \exp(-\alpha t/m_r) \tag{100}$$

This completes the derivation of the result already given in equation 28.

10. Correlations and Connectivity

It seems surprising, at first sight, that a memoryless process can show a correlation between the length of one opening and the length of the next. Nevertheless, this is the case, as was first pointed out by Fredkin *et al.* (1985). The existence of such correlations is of importance in two main respects. First, the behaviour of channels after a perturbation (e.g., a voltage jump or concentration jump) depends on the nature of correlations (see also Section 11). And second, correlation phenomena can potentially give information about the way that the various states in the mechanism are connected. This latter ability is of considerable interest for the investigation of mechanisms, though its full potential has yet to be exploited experimentally. Both macroscopic and, to a greater extent, single-channel experiments can give information about the *number* of states that exist, but it is much harder to discover how these states are connected to each other, and the ability of correlation measurements to provide such information is a unique advantage of being able to measure the behaviour of single molecules.

10.1. Origins of Correlations

According to our (Markov) assumptions, the duration of a sojourn in any individual state must be independent of (and therefore not correlated with) the length of the sojourn in the previous state. It is for this reason that no correlations between open or shut times would be expected for the simple two-state mechanism in equation 1 or, indeed, for any of the mechanisms that have been discussed so far. In fact, correlations can arise only if there are at least two indistinguishable shut states and at least two *indistinguishable* open states (i.e., at least two open states with the same conductance). Furthermore, there must be at least two routes from open states to shut states before correlations are expected (Fredkin *et al.*, 1985; Colquhoun and Hawkes, 1987; Ball and Sansom, 1988a). More precisely, correlations will be found if there is no single state, deletion of which totally separates the open states from the shut states. The number of states that must be deleted to achieve such a separation is the *connectivity* of open and shut states, so correlations will be seen if the connectivity is greater than 1. The mechanisms in schemes 101 each have two open states (denoted O) and three shut states (denoted C).

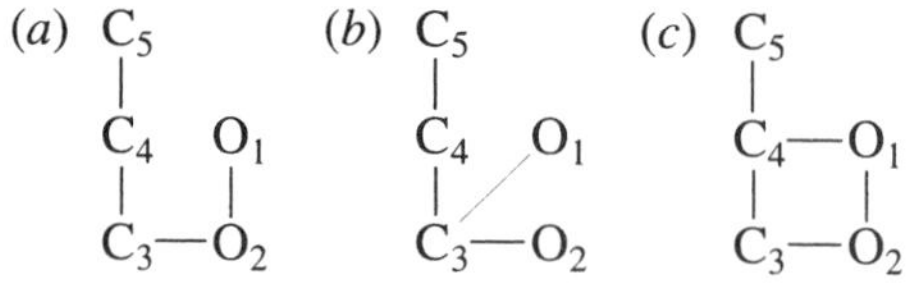

$$
\begin{array}{lll}
(a)\ \begin{array}{ccc} C_5 & & \\ | & & \\ C_4 & & O_1 \\ | & & | \\ C_3 & — & O_2 \end{array}
&
(b)\ \begin{array}{ccc} C_5 & & \\ | & & \\ C_4 & & O_1 \\ | & \diagup & \\ C_3 & — & O_2 \end{array}
&
(c)\ \begin{array}{ccc} C_5 & & \\ | & & \\ C_4 & — & O_1 \\ | & & | \\ C_3 & — & O_2 \end{array}
\end{array}
\qquad (101)
$$

In schemes *a* and *b* there will be no correlations; deletion of state C_3 (or of state O_2) in *a* separates the open and shut states, as does deletion of C_3 in *b*. In *c*, on the other hand, the connectivity is 2 (e.g., deletion of C_3 *and* C_4 will separate open and shut states), so correlations between open times may be seen. Even in this case, correlations between successive open times will be seen only if the two open states, O_1 and O_2 have different mean lifetimes. The correlations result simply from the occurrence of several $C_4 \rightleftharpoons O_1$ oscillations followed by a $C_4 \rightarrow C_3$ transition and then several $C_3 \rightleftharpoons O_2$ oscillations, so runs of O_1 and runs of O_2

openings occur. The effect will clearly be most pronounced if the $C_4 \rightleftharpoons C_3$ reaction is relatively slow.

For example, most of the properties of the nicotinic receptor are predicted well by *c:* in this case O_2 has a long mean lifetime compared with O_1 (but it has the same conductance), whereas C_3 has a very short lifetime. Thus, long open times tend to occur in runs (so there is a positive correlation between the length of one opening and the next), but long openings tend to occur adjacent to short shuttings, giving a negative correlation between open time and subsequent shut time (Colquhoun and Sakmann, 1985).

These results can be extended to correlations between the lengths of bursts of openings and between the lengths of openings within a burst (Colquhoun and Hawkes, 1987). There will be correlations between bursts when the connectivity (as defined above) between open states and *long-lived* shut states is greater than 1. There will be correlations between openings within a burst when the *direct* connectivity between open states and *short-lived* shut states is greater than 1 (the term *direct connectivity* refers only to routes that connect open and short-lived shut states directly, not including routes that connect them indirectly via a long-lived shut state, entry into which would signal the end of a burst). Thus, for the examples in scheme 101, taking C_5 to be the long-lived shut state, neither *a* nor *b* would show any such correlations, whereas *c* would show correlations within bursts but no correlations between bursts (as observed experimentally by Colquhoun and Sakmann, 1985). The following scheme (in which C_5 and C_6 both represent long-lived shut states), on the other hand, would show all three types of correlation.

$$\begin{array}{ccc} C_5 & - & C_6 \\ | & & | \\ C_4 & - & O_1 \\ | & & | \\ C_3 & - & O_2 \end{array} \qquad (102)$$

10.2. Measurement and Display of Correlations

Correlations of this sort have been reported for many other ion channels, by, for example, Jackson *et al.* (1983), Labarca *et al.* (1985), Ball *et al.* (1988), McManus *et al.* (1985), Blatz and Magleby (1989), Magleby and Weiss (1990b), and Gibb and Colquhoun (1992).

In the earlier work in this field, it was usual to measure correlation coefficients from the experimental record. However, it is visually more attractive, and in some respects more informative, to present the results as graphs, as suggested by McManus *et al.* (1985), Blatz and Magleby (1989), Magleby and Weiss (1990b), and Magleby and Song (1992). An example of such a plot is shown in Fig. 12. This graph illustrates correlations found for the NMDA-type glutamate receptor (Gibb and Colquhoun, 1992). To construct this graph, five contiguous shut-time ranges were defined (each centered around the time constant of a component of the shut-time distribution). Then, for each range, the average of the open times was calculated for all openings that were adjacent to shut times in this range, and this average open time was plotted against the mean of the shut times in the range. The graph in Fig. 12 shows a continuous decline, so it is clear that long open times tend to be adjacent to short shut times, and vice versa.

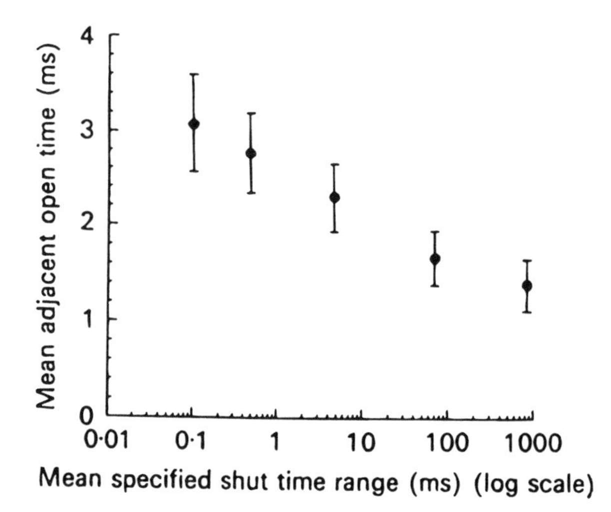

Figure 12. Relationship between the mean durations of adjacent open and shut intervals. The graph shows the mean (± standard deviation of mean) of the open times in 16 different patches plotted against the average of the mean adjacent shut time ranges used in each patch. Long open times tend to be next to short shut times. These results are for NMDA-type glutamate receptors in dissociated cells of adult rat hippocampus (CA1 region), activated by low glutamate concentrations. Reproduced from Gibb and Colquhoun, 1992.

10.3. Correlations as a Test of Markov Assumptions

The reason the openings that are adjacent to short closings tend to be long was investigated further, with the results shown in Fig. 13 (Gibb and Colquhoun, 1992). Figure 13A,B shows the conditional distributions of open times for openings that occur adjacent to the shortest closings (in A) and for openings that occur adjacent to the longest closings (in B). (The means from these distributions contribute points to Fig. 12.) The distributions are displayed as distributions of log(duration), as explained in Chapter 19 (Section 5.1.2) (this volume). The dashed line in A shows the (scaled) fit from B, and the dashed line in B shows the (scaled) fit from A. It can be seen that there is an excess of long openings in A, and an excess of short openings in B. This is shown quantitatively in Fig. 13C,D; it is clear from Fig. 13C that the time constants for the open-time distribution are much the same for all openings, regardless of whether they are adjacent to short or long shuttings. The mean open times differ only because the areas attached to each time constant differ, as shown in Fig. 13D. Similar observations were made by McManus and Magleby (1989) for the large-conductance calcium-activated potassium channel; they pointed out that this behaviour is a clear prediction of the Markov assumptions, whereas at least some non-Markov models do not predict such behaviour and can therefore be rejected on the basis of these observations (see Section 1.3).

10.4. Two-Dimensional Distributions

In order to extract all the information from the experimental record, it is necessary, if correlations are present, to consider two-dimensional distributions rather than the one-dimensional distributions considered so far (Fredkin *et al.*, 1985). An example of a two-dimensional distribution is shown in Fig. 14A (Magleby and Song, 1992). This distribution shows open time on one coordinate and shut time on the other. It was constructed from simulated data that were derived from the mechanism shown in equations 101*c* and 110, so there are two components in the open-time distributions and three components in the shut-time distributions (which resemble qualitatively the distribution shown in Chapter 19, this

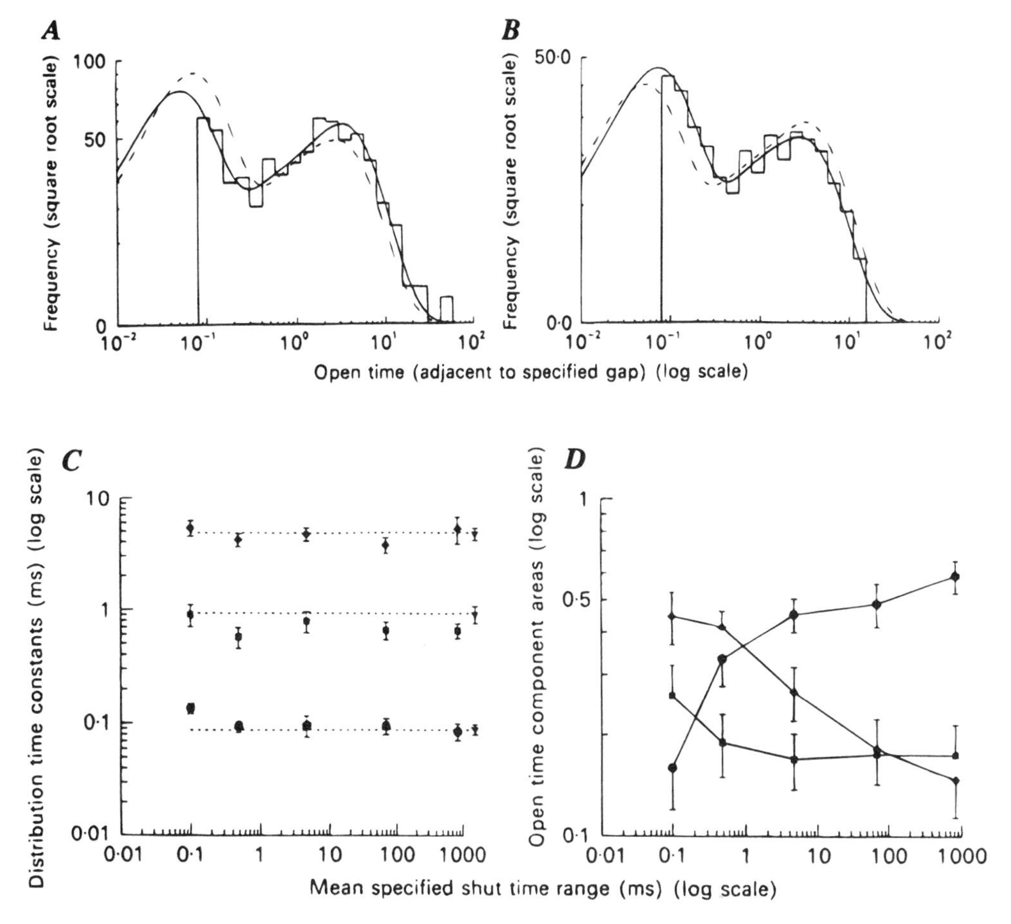

Figure 13. Conditional distributions of apparent open times adjacent to brief (*A*) and long (*B*) shut times. Data were as described in Fig. 12. *A;* From a total of 1206 apparent open times, 640 were identified as adjacent to shut times in the range 50 μs to 0.3 ms. These were fitted with the sum of three exponential components (solid curve) with time constants (areas in parentheses) of 48 μs (52%), 0.36 ms (8%), and 3.21 ms (40%). The fit predicted a total of 1154 open times. The dashed line in *A* shows the fit from *B* scaled to contain the same number of openings as the solid line. *B:* A total of 335 open times were identified as adjacent to shut times in the range 50–5000 ms. These were fitted with the sum of three exponentials (solid curve) with time constants (areas in parentheses) of 68 μs (60%), 0.46 ms (5.8%), and 2.79 ms (35%). The dashed line shows the fit from *A* scaled to contain the same number of openings as in the solid line. The difference between dashed and solid lines indicates that, relative to openings adjacent to long shut periods, there are more long openings and fewer short openings adjacent to short shut periods. *C* and *D:* Mean time constants (*C*) and areas (*D*) of the exponential components describing conditional open-time distributions from 16 different patches. The mean (and its standard deviation) for each fitted parameter is shown plotted against the average of the mean adjacent shut-time ranges used in each patch. The inverted triangles to the right of the data values in *C* show the mean (and its standard deviation) for the time constants from the unconditional open-time distributions, and the dashed lines show the position of each mean across the plot. In *D,* the filled circles, filled squares, and filled diamonds refer to the area of the fast, intermediate, and slow components of the open-time distributions. The lines drawn in *D* have experimental values only at the data points and are drawn only so that the data values for each open time component can be clearly identified. Reproduced from Gibb and Colquhoun, 1992.

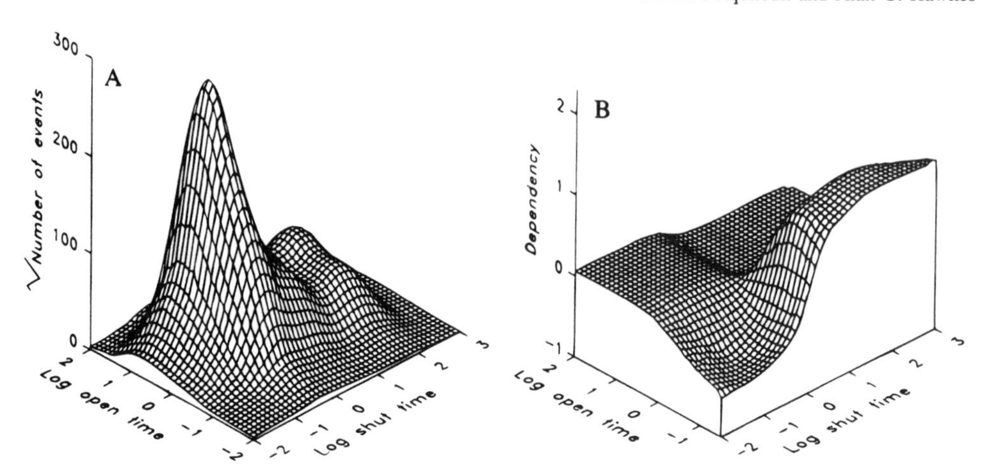

Figure 14. Illustrations of correlations based on 10^7 simulated observations (reproduced from Magleby and Song, 1990, with permission). *A:* A bivariate distribution of open time on one axis and shut time on the other. The distribution of log(duration) is shown, so that peaks in the distribution occur at times corresponding to the time constants of the exponential components (see Chapter 19, this volume, Section 5.1.2). *B:* An example of the dependency plot (see text) for the same simulated data.

volume, Fig. 15). The distributions in Fig. 14 are displayed as distributions of log(duration), as explained in Chapter 19 (this volume) (Section 5.1.2).

The two conditional open-time distributions that were shown in Fig. 13A,B are simply sections (at two particular fixed shut times) across the two-dimensional distribution in Fig. 14A. In practice, in order to construct the conditional distributions from experimental data, it is necessary to use a range of shut times (i.e., a shut-time bin) rather than a single exact value.

The fact that the open-time distribution differs according to the adjacent shut time (as in Fig. 13A,B) is visible in the two-dimensional distribution, but it is not very prominent. It was therefore suggested by Magleby and Song (1992) that the correlations could be made more obvious by displaying the data in the form of a *dependency plot.* They define *dependency* as the (normalized) difference between the actual frequency of particular shut–open time pairs and the frequency that would be expected if openings and shuttings were independent. Define $f_O(t_O)$ and $f_S(t_S)$ as the unconditional probability density functions for open times and shut times, respectively, and $f(t_O,t_S)$ as the two-dimensional distribution. If there were no correlations, then the two-dimensional distribution would simply be the product of the separate distributions, $f_O(t_O)f_S(t_S)$. Thus, dependency, $d(t_O,t_S)$, was defined as

$$d(t_O,t_S) = \frac{f(t_O,t_S) - f_O(t_O)f_S(t_S)}{f_O(t_O)f_S(t_S)} .$$

This will be zero for independent intervals, and a value of +0.5 would indicate that there are 50% more observed interval pairs than would be expected in the case of independent adjacent intervals. A description of how to calculate the plot from experimental values is given by Magleby and Song (1992). An example is shown in Fig. 14B for the data shown in Fig. 14A. The dependency plot clearly shows the excess of short open times adjacent to long shut times, and the deficiency of short open times adjacent to short shut-times.

Plots of the sort shown in Fig. 14 can be used to distinguish between different kinetic mechanisms and as an aid in fitting. It may be mentioned here that full maximum-likelihood fit of the entire idealized data record, not of separate distributions, with a particular mechanism (see Sections 12.5 and 13.7) is probably the best way of extracting all the information from an experimental record. But plots like those in Figs. 13 and 14 are still good ways to display the quality of the fit so obtained, even though they are not used for the fitting process itself.

10.5. The Decay of Correlations

In a record at equilibrium, the correlation between, for example, an open time and the nth subsequent open time (for a single channel) will decay towards zero with increasing lag (n). Likewise, the distribution of open times following a jump will, after sufficient time, eventually become the same as the equilibrium distribution of all open times (see Section 11).

In principle, the connectivity between open and shut states can be measured experimentally, because the decay of the correlation coefficient with increasing lag (n) should be described by the sum of m geometric terms, where m is the connectivity minus one (Fredkin *et al.* 1985; Colquhoun and Hawkes, 1987). A similar decay should be seen in the mean lifetimes of events following a jump (Ball *et al.,* 1989). The full potential of measurements of this sort has yet to be achieved in practice.

10.6. Spurious Correlations

It was pointed out in Section 10.1 that the correlations will be strongest for the mechanism in equation 101c when the $C_4 \rightleftharpoons C_3$ reaction is relatively slow. At the extreme case, when this rate is zero, we are left with two separate channels with different mean open and shut times. Furthermore, neither of these channels would, by itself, show any correlations. Clearly, it is quite possible for spurious correlations to arise as a result of receptor heterogeneity (which is a major problem in many studies). In fact, it is even possible in principle for spurious correlations to arise even when there is more than one *identical* channel in the membrane patch (Colquhoun and Hawkes, 1987), though the importance of this has not yet been investigated. Furthermore, the inability to detect brief events may give rise to strong correlation in the observed record when there is actually little or no correlation (as exemplified in Section 12.4). In other cases imperfect resolution may attenuate real correlations (Ball and Sansom, 1988a).

11. Single Channels and Macroscopic Currents after a Jump

Essentially everything that has been said so far concerns single-channel records that are in a steady state (see Section 7). However, synapses and action potentials do not function in a steady state; they operate far from equilibrium, and so it is important to understand single-channel behaviour in the transient state before equilibrium is attained.

We shall discuss here only the case where the transition rates between states are constant, i.e., do not vary with time. This means, for example, that membrane potential and/or ligand concentration must be constant (see Section 1.1). We therefore consider only the cases where membrane potential or ligand concentration are changed in a stepwise fashion from one constant value to another. Such experiments are usually referred to as *voltage jumps* and *concentration jumps,* respectively. It is, for example, common to mimic a synaptic current by applying a very brief rectangular pulse of agonist (to an outside-out membrane patch). We shall not discuss here the practical problems that often arise in achieving sufficiently rapid changes in potential or concentration to fulfill the assumptions. Some of the practical aspects are discussed in Chapter 19.

11.1. Single Channels after a Jump in the Absence of Correlations

The behaviour of single channels following a sudden change in membrane potential or ligand concentration is not necessarily the same as that in the steady state. The differences depend primarily on two things. First, they depend on the state of the system at the moment the jump was applied ($t = 0$). Second, they depend on whether the channels show correlations of the sort discussed in Section 10.

The simplest case occurs when channel openings are uncorrelated (see Section 10). This will, for example, always be the case if there is only one open state. In this case, all the openings and shuttings that follow the jump will, with one exception, have exactly the same distributions as in the steady state.

The one exception is the *first latency.* Consider, as an example, a membrane patch that is initially bathed in an agonist-free solution, so the channel(s) in it are shut. At $t = 0$ the agonist concentration is suddenly increased from zero to a finite value. The time that elapses before the first channel opening occurs is defined as the first latency, and its distribution depends on the fraction of receptors that are in each of the different shut states at $t = 0$.

Consider, for example, the simple agonist mechanism that was discussed in Section 5 and is shown again in equation 103. The channels would all be in state 3 (R, the resting state) in the absence of agonist. Compare this situation with that which obtains during a steady-state record with a constant agonist concentration: in this case, the shut channels would not all be in state 3 (R) but would be divided between state 3 and state 2 (AR), according to the value for the equilibrium constant for binding. The *initial condition* from which an opening occurs differs in these two cases, so the distribution of the shut times that precede openings will differ accordingly. This is intuitively very reasonable. At equilibrium, every shut period is preceded by an opening, so the shut period must always start in state 2 (AR), and similarly, every shut period must end in state 2 [this is why the probability $P_{22}(t)$ is needed for the derivation of the shut-time distribution given by Colquhoun and Hawkes, 1994, Appendix 1]. Because opening can occur directly from state 2, it is easy to see that the shut state preceding the next opening may be quite short; there *may* be no sjourn in state 3 before the next opening. When on the other hand, the channels are all initially in state 3 (R), the channel *must* spend time both in state 3 and in state 2 before opening is possible, and so a longer time is likely to elapse before an opening occurs.

As usual for a Markov process, these differences in the distributions depend entirely on differences in areas rather than time constants. In the case of mechanism 59, there are two shut states, so distributions of shut times are therefore a mixture of two exponentials. The time constants for the two components are the same for all shut-time distributions, including that for the first latency after a jump; but the area of the faster component will be larger for channels that were initially in state 2 (AR) than it is for channels that were initially in state 3 (R). The steady-state equivalent of this phenomenon has already been illustrated in Section 10.3 and Figs. 13 and 14.

As an example, consider the mechanism in equation 59 with the following transition rates (all in s^{-1}):

$$\underset{3}{\text{R}} \underset{1000}{\overset{100}{\rightleftharpoons}} \underset{2}{\text{AR}} \underset{1000}{\overset{1000}{\rightleftharpoons}} \underset{1}{\text{AR*}} \tag{103}$$

(for example, we could have $k_{+1} = 10^7\text{M}^{-1}\text{s}^{-1}$ and a concentration of 10 μM, giving the binding rate as 100 s^{-1}). The equilibrium shut-time distribution, calculated as described in Colquhoun and Hawkes (1994) (see also Section 13; Chapter 20, this volume) is, in the standard form of equation 25,

$$f(t) = a_1\lambda_1 e^{-\lambda_1 t} + a_2\lambda_2 e^{-\lambda_2 t} \tag{104}$$

where the time constants are

$$\tau_1 = 1/\lambda_1 = 0.4875 \text{ ms} \quad \text{and} \quad \tau_2 = 1/\lambda_2 = 20.51 \text{ ms} \tag{105}$$

and the areas of the components are

$$a_1 = 0.4750 \quad \text{and} \quad a_2 = 0.5250 \tag{106}$$

The mean length of a shut period at equilibrium is therefore

$$\text{mean shut time} = a_1\tau_1 + a_2\tau_2 = 11.00 \text{ ms} \tag{107}$$

Since, from equation 24, the mean lifetime of state 3 is 10 ms, and the mean lifetime of state 2 is 0.5 ms, it is clear that shut times consist, on average, of a $2 \rightarrow 3 \rightarrow 2$ transition (the rate constants show that it is equally likely that a channel in state 2 will, at its next transition, move to state 3 or state 1). This is for shut periods that start in state 2 and end in state 2. However, if we consider a concentration jump from zero concentration to 10 μM, the channels are initially all in state 3. The distribution of the latency until the first opening will therefore be the distribution of shut times conditional on starting in state 3. This can be found in the way given by Colquhoun and Hawkes (1994), but in this case the appropriate probability would be $P_{32}(t)$ rather than $P_{22}(t)$. The result is a distribution like that in equation 104 with the same rate constants, as given in equation 105, but with areas

$$a_1 = -0.02435 \quad \text{and} \quad a_2 = 1.02435 \tag{108}$$

and

$$\text{mean shut time} = a_1\tau_1 + a_2\tau_2 = 21.00 \text{ ms} \tag{109}$$

In this case one of the areas is negative, which means that the distribution goes through a maximum, as illustrated earlier; in other words, very short latencies are unlikely. Correspondingly, the macroscopic current for such a jump would have a sigmoid start; the time constants for the relaxation would be, from equation 62–64, 0.3778 ms and 2.205 ms, the former having a negative amplitude.

After the first latency has elapsed, all openings and shuttings have exactly the same distributions as at equilibrium (mean open time 1.00 ms; mean shut time 11 ms). This is a consequence of the absence of correlations in this mechanism.

11.2. Single Channels after a Jump in the Presence of Correlations

The characteristics of single-channel openings after a jump when correlations are present have been considered by Colquhoun and Hawkes (1987) and by Ball *et al.* (1989). As an example, consider the mechanism in equation (101c). This has been used by several authors (e.g., Colquhoun and Sakmann, 1985) to describe the nicotinic acetylcholine receptor (see also Section 13; Chapter 20, this volume; Colquhoun and Hawkes, 1982). In this context, the mechanism may be written to show the binding of two agonist molecules (A) to the shut (R) and open (R*) receptor, thus:

$$
\begin{array}{ccccc}
\text{State number} & & & & \text{State number} \\
5 & \mathrm{R} & & & \\
 & k_{-1} \Big\uparrow\Big\downarrow 2k_{+1} & & & \\
4 & \mathrm{AR} & \underset{\alpha_1}{\overset{\beta_1}{\rightleftharpoons}} & \mathrm{AR}^* & 1 \\
 & 2k_{-2} \Big\uparrow\Big\downarrow k_{+2} & & 2k_{-2}^* \Big\uparrow\Big\downarrow k_{+2}^* & \\
3 & \mathrm{A_2R} & \underset{\alpha_2}{\overset{\beta_2}{\rightleftharpoons}} & \mathrm{A_2R}^* & 2
\end{array}
\qquad (110)
$$

The experimental evidence suggests that the mean lifetime of open state 1 (the singly liganded open state) is considerably shorter than that of open state 2 and that the mean lifetimes of shut states 3 and 4 are short. This would account for the observed correlations, as explained in Section 10.1. An example of the calculation of single-channel properties after a jump is given by Colquhoun and Hawkes (1987) for this mechanism. They used the rate constants that were found by Colquhoun and Sakmann (1985) to provide a fair description of nicotinic receptor behaviour and used these values to predict the behaviour of channels following a concentration jump from zero to 4 nM. The time constants were, of course, the same for all distributions, but the areas changed such that the mean shut times were as follows:

- Mean latency to first opening 1539 s
- Mean shut time between first and second openings 1038 s
- Mean shut time between second and third openings 806.1 s
- Mean shut time between third and fourth openings 698.6 s

and so on, until the equilibrium mean shut time of 605.7 s is reached.

Similarly, mean lengths of the first, second, etc. openings following the jump were 0.754 ms, 1.029 ms, 1.156 ms, 1.215 ms and so on until the equilibrium mean open time of 1.267 ms was attained. The calculation of these values, for a mechanism as complex as that in equation 110, cannot be written explicitly but requires the use of matrix methods (see Section 13 below; Chapter 20, this volume; Colquhoun and Hawkes 1982, 1987). It has been shown by Ball *et al.* (1989) that such measurements can be used to provide information about mechanisms.

11.3. The Relationship between Single-Channel Currents and Macroscopic Currents

From the experimental point of view, the relationship is simple: the macroscopic current is just the sum or average of a set of single-channel records. Two schematic examples have already been shown, in Figs. 2 and 5, of the relationship between single-channel currents and macroscopic currents. In both of these it was supposed for simplicity that the channels open synchronously at $t = 0$ or, in other words, that the first latency was negligible. This is not always true. An example of experimental measurements in which it is certainly not true is shown in Fig. 15. This shows responses of a membrane patch that contained NMDA-type glutamate receptors to application of the agonist (glutamate) for 1 ms. The patch probably contained only one channel (see Section 8), and six individual responses are shown. The average of these responses is shown at the top of the figure and is seen to follow a time course that is typical of NMDA receptors, with a relatively slow rise time followed by a slow double-exponential decay, with time constants, in this case, of $\tau = 61.5$ ms and 208 ms. In this experiment it is clear that the latency until the first opening occurs is sometimes very long indeed, and this will have a profound effect on the time course of the macroscopic current. For example, in the third trace from the top in Fig. 15, the first opening occurs about 860 ms after the 1-ms pulse, and in the fifth trace the latency is about 1340 ms.

In order to predict, from some specified mechanism, the results of an experiment like that shown in Fig. 15, we first note that the experiment involves *two* concentration jumps. First, there is a jump from zero concentration to 1 mM, the channels being initially in their resting state. This is followed, 1 ms later, by a jump from 1 mM to zero. The initial condition (i.e., the fraction of channels in each state) for the second jump is found during the calculation of the response to the first jump; it is simply the fraction of channels in each state, $p_i(t)$, at $t = 1$ ms. The methods for calculating the macroscopic (average) current have been mentioned above and are described in Chapter 20 (this volume).

11.3.1. The Simplest Example of the Effect of First Latency

In order to investigate the effect of nonsynchronous channel opening, it will be useful to consider first the simplest possible case. This case concerns a hypothetical channel that, after brief agonist application, produces an activation consisting of a single opening, after the first latency has elapsed (for the NMDA receptor, the activation is actually a great deal more complicated than a single opening). In Fig. 16A, nine examples are shown of simulated channels with a mean first latency of 1 ms and a mean open time of 10 ms (the variability of both being described by simple exponential distributions). The average current (shown at the top) is seen, not surprisingly, to have a rising phase that can be fitted with an exponential with a time constant of about 1 ms, and the decay phase has a time constant of about 10 ms. Apart from being about ten times too slow, this example is similar to what happens at a neuromuscular junction.

More surprising, perhaps, are the results shown in Fig. 16B, in which the numbers are reversed, and simulated channels have a mean first latency of 10 ms and a mean open time of 1 ms. The averaged current shown at the top is seen to have essentially the same shape as in Fig. 16A (though it is ten times smaller and considerably noisier relative to its amplitude). Thus, in this latter case, the rate of decay reflects the duration of the first latency, whereas the rate of rise represents the mean channel-open time. The reason for this result, which seems paradoxical at first sight, can be seen from the simulations (e.g., the exponential distribution of first latencies means that short latencies are more common than long ones) and from the relevant theory, which was outlined in Section 9.1. The distribution of the time from the stimulus until the channel shuts finally is simply the distribution of the sum of (1) the first latency (mean length $1/\beta'$, say), and (2) the length of the channel opening (mean length $1/\alpha$, say). This distribution has already been found, as the convolution in equation

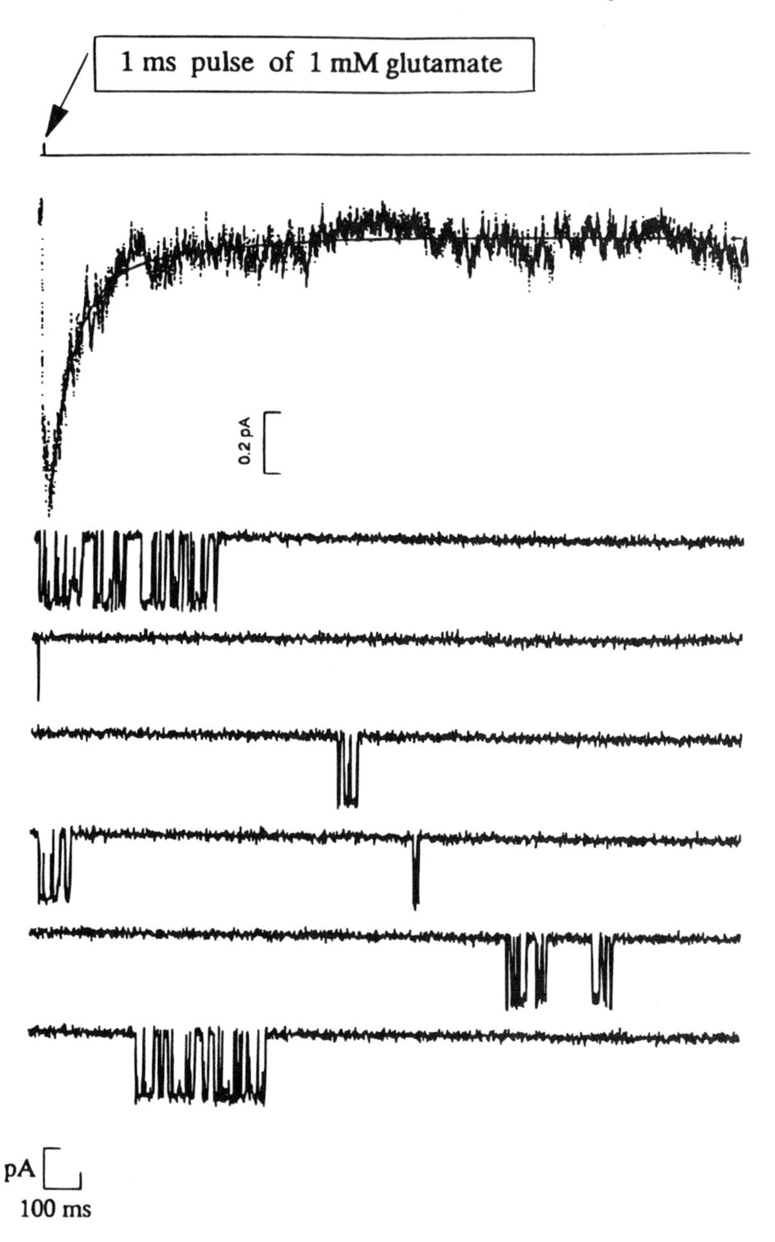

Figure 15. Illustration of first latency measurement. The lower section shows six individual responses, each 2000 ms in duration, to a 1-ms pulse of 1 mM glutamate. The time at which the command signal for the pulse was applied is shown in the topmost trace (the actual concentration change at the patch started about 1 ms later). This membrane patch contained, almost certainly, only one active channel, and it is clear that the latency before the first opening is often long (see text). The average of 122 such records is shown at the top. The decay phase of the average (starting from $t = 37$ ms) was fitted with two exponentials. Their time constants were 61.5 ms and 208 ms (the latter accounts for 23.6% of the amplitude at the starting point for the fit). (Data of B. Edmonds; outside-out patch from rat dentate gyrus granule cell at -60 mV, in solution containing 5 μM glycine and 5 μM CNQX. Methods as in Edmonds and Colquhoun, 1992.)

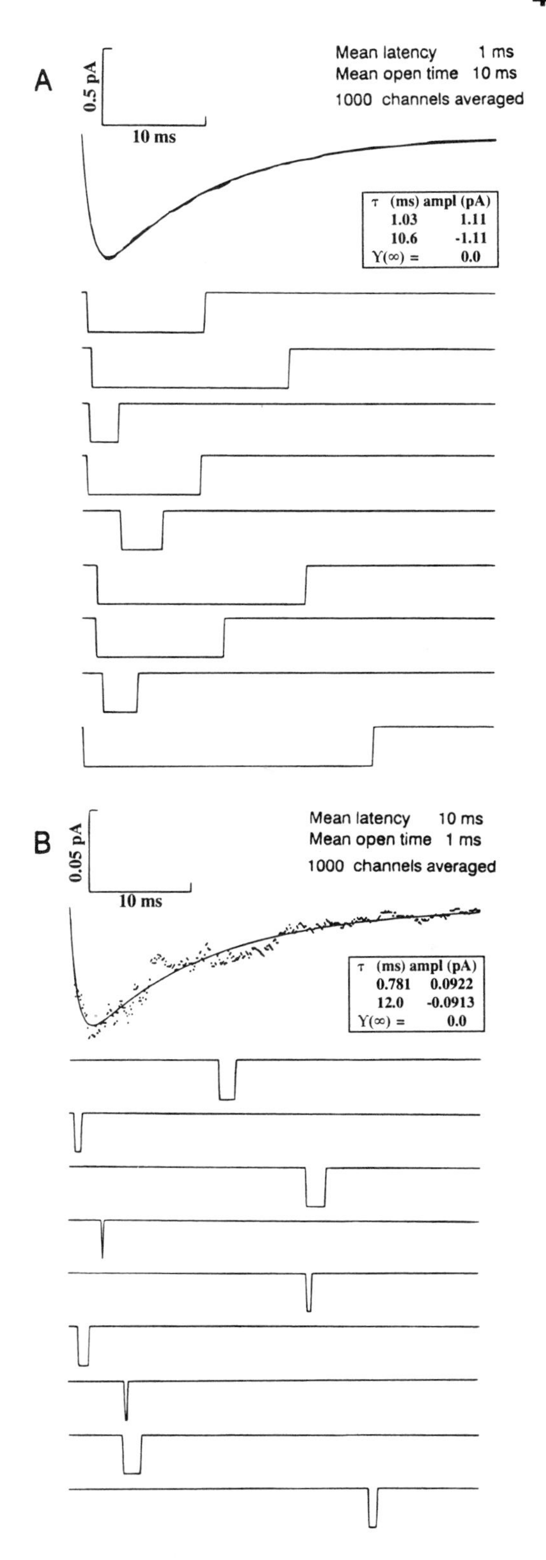

Figure 16. A simple simulation of a synaptic current in which each channel is supposed to produce only a single opening after an exponentially distributed latency. A: Mean latency 1 ms, mean open time 10 ms. The lower part shows nine examples of simulated channels. The top trace is the average of 1000 such channels; the double-exponential curve fitted to the average has $\tau = 1.03$ ms (amplitude 1.11 pA), and $\tau = 10.6$ ms (amplitude -1.11 pA). B: Similar, but with a mean latency of 10 ms and a mean open time of 1 ms. The double-exponential curve fitted to the average has $\tau = 0.76$ ms (amplitude 0.092 pA) and $\tau = 12.0$ ms (amplitude -0.091 pA).

87. Since both intervals have been taken to be simple exponentials, $f_1(t) = \alpha e^{-\alpha t}$ and $f_2(t) = \beta' e^{-\beta' t}$, the result, $f(t)$, is exactly as has already been given in equation 91. It has the form of the difference between two exponentials, and it is the curve that has been fitted to the averages in Fig. 16.

In this particular simple case, though not in general, there is a very simple relationship between the distribution, $f(t)$, of the total event length, and the shape of the averaged current. The time course of the current is given, apart from a scale factor, by the probability that a channel is open at time t. This we shall denote $P_{\text{open}}(t)$, and it can be found as follows. A channel will be open at time t if (1) the first latency is of length u, *and* (2) the channel stays open for a time *equal to or greater than* $t - u$. The probability that a channel stays open for a time $t - u$ *or longer* is, from equation 21 the cumulative distribution

$$R_1(t - u) \equiv e^{-\alpha(t-u)} \tag{111}$$

so, by an argument exactly like that used to arrive at equation 87, the probability that a channel is open at time t is

$$P_{\text{open}}(t) = \int_{u=0}^{u=t} f_2(u) R_1(t - u)\, du \tag{112}$$

This differs from equation 91 only by a factor of $1/\alpha$, the mean open lifetime, so

$$P_{\text{open}}(t) = f(t)/\alpha = \frac{\beta'}{\alpha - \beta'}(e^{-\beta' t} - e^{-\alpha t}) \tag{113}$$

which is, apart from its amplitude, unchanged when α and β' are interchanged. The amplitudes of the two exponential components are equal and opposite, being, from equation 113,

$$a = \frac{\beta'}{(\alpha - \beta')} \tag{114}$$

with a maximum at t_{max}, which is given by

$$t_{\text{max}} = \frac{\ln(\beta'/\alpha)}{\beta' - \alpha}. \tag{115}$$

The simulated average currents in Fig. 16 are indeed well fitted by these values.

11.3.2. The Effect of First Latency in General

If there is more than one sort of shut state (which there invariably is for real channels), the possibility arises that the channel may open more than once after a pulse of agonist (or of membrane potential) is applied. This would, for example, be the case for the channel-block mechanism discussed in Section 4 (see Fig. 5). It is also clearly the case for the NMDA receptor as shown by the experiment in Fig. 15. If there is only one open state, then the result given above can be generalized as follows. The probability of the channel being open at time t (and hence the macroscopic current at time t) is given by the convolution of the first latency distribution with $P_{11}(t)$, where the latter was defined in equation 46 as

$$P_{11}(t) = \text{Prob(open at time } t \,|\, \text{open at time 0)}. \tag{116}$$

This result has been used, for example, by Aldrich et al. (1983) and by Horn and Vandenberg (1984) for the interpretation of experiments on sodium channels, in which measurements of first latency turned out to be important for investigations of the channel mechanism.

It is important to note that $P_{11}(t)$ is the sort of probability that is used in the calculation of macroscopic currents or noise; it does *not* specify that the channel should be open *throughout* the time from 0 to t (as would be the case for analysis of single channels; e.g., see Section 4.8) but merely that it was open at 0 and at t, regardless of what happens in between. In fact, $P_{11}(t)$ describes the time course of the current that would be found by averaging single-channel records after aligning the starting points of the first opening in each record.

In general, the expression for $P_{11}(t)$ will be given by the sum of $k - 1$ exponentials that have the time constants found for macroscopic relaxations (as in equation 4) (they will be the eigenvalues of $-\mathbf{Q}$; see Section 13 and Chapter 20, this volume). These time constants will not, in general, be the same as those for any of the single-channel distributions. Thus, although the first latency distribution is a 'single-channel quantity', $P_{11}(t)$ is not, and there is, therefore, in general, no simple relationship between single-channel distributions and macroscopic currents.

In particular, *it is impossible to predict the response to a jump from measurements of steady-state single-channel recordings.* This is generally true, though if the single channel recordings were made under a range of conditions and were detailed enough to allow complete identification of the mechanism and all its rate constants (see Section 12), then it would of course be possible to predict the time course of macroscopic currents. This was illustrated by Edmonds and Colquhoun (1992), who show that simple averaging of aligned channel activations (measured in steady-state records) does not reproduce the shape of the macroscopic currents. However, this procedure would work, to a good approximation, for muscle-type nicotinic receptors, which produce compact bursts of openings with a very short first latency and are therefore close to the situation illustrated in Fig. 2.

If there is more than one open state, then the result stated above can be further generalized, using matrix methods, by what amounts to using a separate first-latency distribution for entry into each of the open states (see Section 13 and Chapter 20, this volume).

12. The Time Interval Omission Problem

The filtering effect of the recording apparatus, together with noise and sampling the signal at regularly spaced points in time, means that brief openings or shuttings of the ion channel will not be detectable. This will cause a distortion of the histograms of the distributions of open times and shut times that can be quite serious (see example below).

12.1. Definition of the Problem

We suppose in what follows that all events that are shorter than some fixed *resolution* or *dead time* (denote ξ_o for open times, ξ_s for shut times) are *not* detected, whereas all events longer than this are detected and measured accurately. The resolution is usually not well defined, but may be imposed retrospectively on the measurements by concatenating any observed shut time below ξ_s with the open times on each side of it to produce one long "apparent opening." The procedures necessary for imposition of a fixed dead time are discussed in Chapter 19 (this volume, Section 5.2). The effect on the distribution of open times that is caused by missing open times that are shorter than ξ_o is easily allowed for (see Section 6.8.1 of Chapter 19, this volume). But the concatenation of adjacent open times that occurs when the short shut time separating them is missed is potentially far more serious and may cause openings to appear to be far longer than they really are.

12.1.1. Dependence on the Method of Analysis of Experimental Records

Before any attempt can be made to make allowance for missed events, the problem must be formulated precisely. The problem is to decide how to define what it is that is *actually* measured when an experimental record is analyzed. The answer to this question will depend, to some extent, on the method that is used for the analysis. If a threshold-crossing method is used, it seems natural to define the dead time as the duration of an event that is just long enough for the signal to reach the threshold (in the absence of noise). There are two problems with this definition. First, the universal presence of noise will mean that some events that are longer than the dead time will be missed, and some events that are shorter than the dead time will be detected (see Chapter 19, this volume). Second, as pointed out by Magleby and Weiss (1990a), events that are both shorter than the dead time but are close together may sum to produce a signal that crosses the threshold. Both of these problems are less severe if the record is fitted by time-course fitting with subsequent imposition of a fixed dead time (see Chapter 19, this volume, Section 5.2).

Ideally, the method used for missed-event correction should take into account the actual properties of the method used for analysis. Draber and Schultze (1994) have made an attempt to do this (though for an analysis method that has not yet been much used in practice). The only realistic method for doing this is the (very slow) repeated simulation of the entire analysis, as proposed by Magleby and Weiss (1990a).

We now define a theoretical quantity, the *apparent* open time. This quantity is intended to be, as far as possible, what would actually be measured from an experimental record, the *observed* open time. In fact, this distinction will often be neglected, and both quantities referred to as *apparent.* The mean length of apparent openings will be denoted ${}^{e}\mu_o$ (where the superscript e stands for *effective*). For the purpose of the theory, an apparent opening is defined as starting with an opening longer than ξ_o (which is therefore visible); this is followed by any number of openings, which may be of any length but are separated by gaps that are all shorter than ξ_s and are therefore not detected; this process is ended when a shut time in excess of ξ_s is observed. Short openings, less than ξ_o are similarly treated to obtain 'apparent shut times'. The extent to which this definition mimics reality will, as mentioned above, depend on the method used to analyze experimental records. A run of short random openings and shuttings will, from time to time, produce signals of quite unrecognizable shape, so it is impossible to anticipate all possibilities. At least it is impossible to do so in any analysis program that allows the operator to approve or disapprove the fitted durations, and, as explained elsewhere, there are good reasons, unconnected with the missed-events problem, why it is always desirable to inspect what the computer is doing to your data. There will inevitably (and probably quite rightly) be a subjective element in the operator's response to oddly shaped signals. Fortunately, such oddities are rare in most data and so should not give rise to serious errors.

12.1.2. Dependence on the Channel Mechanism

It is an unfortunate fact that in order to make proper allowance for missed events, it is necessary to postulate a kinetic mechanism for the operation of the ion channel. When substantial numbers of both brief openings and brief shuttings are missed, very little can be done without a realistic knowledge of the mechanism, as is made clear by the discussion below. However, it is quite often the case, to a first approximation at least, that most openings are detected but many short gaps are missed (or, more rarely, the other way round). In this case, corrections can be made without detailed knowledge of the mechanism. When most openings are detected, the shut-time distribution will (apart from the lack of values below ξ_s) be quite accurate; i.e., it will have approximately the correct time constants (see Fig. 18, for example). We can, therefore, obtain a realistic estimate of the number (and duration) of missed shut times simply by extrapolating the fitted shut-time distribution to $t = 0$. This is essentially the procedure used by Colquhoun and Sakmann (1985), and it is given below (see equation 124). Even in this case, however, it was necessary to assume something about mechanisms in order to do the correction. The reason for this is that, in their data, the distribution of (apparent) open times or of burst lengths had two exponential components, so, although an estimate could be made of the number of brief shuttings that were missed, there was no way of knowing whether they were missed from 'long bursts' or from 'short bursts'. The data suggested that short bursts contained few short gaps, so, in order to perform the correction, it was assumed that *all* the missed gaps were missed from long bursts. This procedure was subsequently shown to behave quite well when tested by the exact procedures discussed below, but there can be no guarantee that it will always do so.

We shall first discuss the (oversimplified) case in which the system has only one shut state and one open state.

12.2. The Two-State Case

Suppose the true open times and shut times both follow simple exponential distributions with means μ_o and μ_s, respectively. Then we have

$$P(\text{shut time} > \xi_s) = e^{-\xi_s/\mu_s} \tag{117}$$

and so

$$\text{Mean number of openings per apparent opening} = 1/e^{-\xi_s//\mu_s} = e^{\xi_s/\mu_s} \tag{118}$$

It is well known (see Chapter 19, this volume, Sections 6.6 and 6.8) that for an exponential distribution

$$\text{Mean length of shuttings longer than } \xi_s = \xi_s + \mu_s \tag{119}$$

then

$$\text{Mean of shuttings less than } \xi_s = [\mu_s - (\xi_s + \mu_s)e^{-\xi_s/\mu_s}]/(1 - e^{-\xi_s/\mu_s}) \tag{120}$$

because $e^{-\xi_s/\mu_s} \times$ (expression 119) + $(1 - e^{-\xi_s/\mu_s}) \times$ (expression 120) must equal μ_s, the overall mean shut time. The mean apparent open time, denoted ${}^e\mu_o$, is therefore $\xi_o + \mu_o e^{\xi_s/\mu_s} + (e^{\xi_s/\mu_s} - 1) \times$ (expression 120), as there is one less shutting than opening contributing to the apparent open time. Thus,

$$^{e}\mu_0 = \xi_0 + (\mu_0 + \mu_s)e^{\xi_s/\mu_s} - (\xi_s + \mu_s) \tag{121}$$

Similarly, the mean apparent shut time is

$$^{e}\mu_s = \xi_s + (\mu_o + \mu_s)e^{\xi_o/\mu_o} - (\xi_o + \mu_o) \tag{122}$$

The values of ${}^e\mu_o$ and ${}^e\mu_s$ can be estimated from the data by averaging the observed open and shut times; ξ_o and ξ_s are known, so that equations 121 and 122 are a pair of nonlinear simultaneous equations that can be solved numerically for the true means μ_o and μ_s. For example, obtain an expression for μ_s from equation 122 and substitute it into equation 121 to obtain an equation in μ_o only, that can be solved by bisection. It turns out that these equations usually have two pairs of solutions. Suppose, for example, that $\xi_o = \xi_s = 200$ μs, ${}^e\mu_o = 0.6$ ms, and ${}^e\mu_s = 2.0$ ms. Then, there is a 'slow' solution ($\mu_o = 299.0$ μs, $\mu_s = 878.7$ μs) and a 'fast' solution ($\mu_o = 106.3$ μs, $\mu_s = 214.8$ μs). The slow solution implies, for example, that on average an observed shut time comprises 1.95 shut times separated by 0.95 (short) open times, whereas the equivalent figures for the fast case are 6.56 and 5.56. In principle, the ambiguity is not quite complete because the forms of the distributions of observed times are predicted to be different (though they have the same means) for these two solutions, but in practice the difference may be too small to be useful (Hawkes *et al.*, 1990). Furthermore with the fast solution consisting of rapid alternation of openings and shuttings of duration comparable to the resolution, the apparent openings would have the appearance of a noisy opening of reduced amplitude.

This problem has been further studied using an approximate likelihood method by Yeo *et al.* (1988), Milne *et al.* (1989), and Ball *et al.* (1990), yielding a likelihood with two almost equally high peaks. They showed that the two solutions could be resolved by making additional analyses in which ξ_o and ξ_s are changed, the real solution remains the same, and the false one is altered.

The above model, assuming fixed resolution, is used throughout this section, but Draber and Schultze (1994), following Magleby and Weiss (1990a), used a (theoretically) specified model of a detector (see above), and in the two-state problem they obtain the alternative result, in the case $\xi_o = \xi_s = \xi$,

$$^{e}\mu_o = \frac{1}{(\mu_o - \mu_s)} \times \left\{ -\frac{2\mu_o^2\mu_s}{(\mu_o - \mu_s)}[1 - e^{(\xi/\mu_s - \xi/\mu_o)}] + \xi(\mu_o + \mu_s) - \mu_o^2 e^{(\xi/\mu_s - \xi/\mu_o)} - \mu_o\mu_s \right\} \tag{123}$$

with a similar result for ${}^e\mu_s$, the subscripts *o* and *s* being interchanged. These results are close to those given by equations 121 and 122 if μ_o and μ_s are greater than about 2ξ.

12.2.1. The Case when Only Gaps Are Missed

If openings are long enough that very few are missed, then the results simplify. Thus, if $\xi_o \ll \mu_o$, equation 122 reduces to ${}^e\mu_s \simeq \xi_s + \mu_s$. The openings, however, are still extended by missing gaps, but, as μ_s is now known from this result, equation 121 can be solved for μ_o as

$$\mu_o = e^{-\xi_s/\mu_s}[({}^e\mu_o - \xi_o) + (\xi_s + \mu_s)] - \mu_s \tag{124}$$

Analogous results, interchanging *o* and *s*, can be obtained if shut times are long.

12.2.2. Bursts of Openings

We consider here only the case in which most openings are detected but many shut times within a burst are undetected. As before, this implies that gaps will rarely be extended by undetected openings, so if μ_g now denotes the true mean length of gaps within bursts, the observed mean length of such gaps will again be $\xi_s + \mu_g$. Since we have assumed that most openings are detectable, the mean length of the observed burst will be close to the true mean burst length. Thus, both intraburst gap lengths and burst lengths can be estimated from the data. However, the apparent openings will be longer than the true openings, and the observed number of openings per burst will be correspondingly too small. It is for this reason that Colquhoun and Sakmann (1985) presented primarily distributions of gap lengths and burst lengths but not those of apparent open times or of the number of apparent openings per burst.

Corrected means for the last two distributions can be obtained as follows for the case in which the true openings and the true gaps within a burst each have simple exponential distributions. The burst distribution is fitted to give an estimate of the mean burst length, μ_{bst}, and the number of bursts, N_{bst}, each of which should be close to the true values. The distribution of lengths of gaps within bursts is fitted to give estimates of their true length, μ_g, and of their true number, N_g, which may be considerably greater than the observed number, $n_s = N_g e^{-\xi_s/\mu_g}$. The true number of gaps per burst, μ_r, is estimated as the total number of gaps divided by the total number of bursts, so $\mu_r = N_g/N_{bst}$. The true mean open time can be estimated by noting that the mean total shut time per burst including undetected gaps, is $\mu_g\mu_r$, so

$$\mu_o = \frac{\text{mean open time per burst}}{\text{mean number of openings per burst}} = (\mu_{bst} - \mu_g\mu_r)/(\mu_r + 1) \tag{125}$$

This is essentially the correction employed by Colquhoun and Sakmann (1985).

12.3. The General Markov Model

The previous section discussed only the two-state case and was concerned only with the *means* of the apparent observed open times and of observed shut times, according to particular assumptions about how these arise from the inability to observe small intervals. We need to extend this to models of channel action with any number of shut states and open states; we also need to predict the distributions of observed quantities, not only their means. So far, this has been achieved only in the case where all open states have the same conductance. Several approximate methods have been described, for example by Blatz and Magleby (1986), Yeo *et al.* (1988), and Crouzy and Sigworth (1990), in each case approximating the distributions by mixtures of exponential distributions. An exact solution in terms of Laplace transforms was obtained by Ball and Sansom (1988b) following earlier work by Roux and Sauvé (1985).

Hawkes *et al.* (1990) obtained the exact algebraic forms of these probability density functions in the case $\xi_o = \xi_s$; they presented some numerical examples that suggested that the best of the mixed-exponential approximations was that of Crouzy and Sigworth (1990). Unlike the distributions of true open times or shut times, they are not mixtures of exponentials but are sums of exponentials multiplied by polynomials in *t;* a different form holds over different ranges of length ξ, so that over the interval $[\xi r < t < \xi(r + 1)]$ the multiplying polynomials are of degree $(r - 1)$; the density is, of course, zero for $t < \xi$. These distributions are reasonably easy to compute for small t but get progressively more complicated as t increases and eventually become numerically unstable. An alternative approach by Ball and Yeo (1994) is based on numerical solution of a system of integral equations. Ball *et al.* (1991, 1993b) obtained a solution, in terms of Laplace transforms, in the more general setting of semi-Markov processes (which includes fractal and diffusion models as well as the Markov model discussed here). Ball *et al.* (1993a) showed that a general result of Hawkes *et al.* (1990), from which the above result specific to Markov models was obtained, can be extended into this more general setting.

Jalali and Hawkes (1992a,b) (see also Hawkes *et al.,* 1992) obtained asymptotic forms for these probability densities that are extremely accurate except possibly for quite small values of t. They recommend using the exact form for $t < 3\xi$ and the asymptotic form for $t > 3\xi$. Brief details are given in Section 13.7. The asymptotic distribution not only has the form of a mixture of exponentials, but it also has the same number of exponential components as the true distribution (that which would be found if no intervals were missed). However, the values of the time constants and of their associated areas may be quite different. It is this asymptotic form that would be estimated when fitting a mixture of exponentials to experimentally observed time intervals using the methods described in Chapter 19 (this volume, Section 6.8).

We consider the mechanism of scheme 110 (see also equation 127), which has two open states and three shut states, with parameter values $\alpha_1 = 3000\ s^{-1}$, $\alpha_2 = 500\ s^{-1}$, $\beta_1 = 15\ s^{-1}$, $\beta_2 = 15{,}000\ s^{-1}$, $k_{+1} = 5 \times 10^7\ M^{-1}s^{-1}$, $k_{+2} = k^*_{+2} = 5 \times 10^8\ M^{-1}s^{-1}$, $k_{-1} = k_{-2} = 2000\ s^{-1}$, $k^*_{-2} = (1/3)\ s^{-1}$, and agonist concentration $x_A = 0.1\ \mu M$. A set of data, in the form of a sequence of open and shut times, was simulated from this model; a resolution of 50 μs (for both open and shut times) was then imposed on the record (see Chapter 19, this volume, Section 5.2) to produce a sequence of 10,240 *apparent* open times alternating with 10,240 *apparent* shut times. We will refer to this as the *simulation model* for the remainder of this section.

The true open time distribution has two exponential components with time constants 2.00 ms and 0.328 ms with corresponding areas of 0.928 and 0.072, giving an overall mean of 1.88 ms. Figure 17 shows the theoretical distribution of the logarithm of apparent open times (see Section 5.1.2 of Chapter 19, this volume), and this compares well with a histogram arising from the simulation. The true distribution of open times is shown for comparison. Compared with the true distribution, the distribution of apparent open times has been shifted to the right, having a mean of 3.52 ms rather than 1.88 ms; this shift is not caused by missing the short open times but results from missing the short shut times.

The distribution of apparent open times does not have a mixed exponential form for small t, but for $t > 3\xi$ it is very well approximated by the asymptotic distribution, which is a mixture of exponentials. This gives us another way of comparing the true and apparent distributions. First note that the pdf of apparent open times is zero below the dead time, $t = \xi$, whereas the true open time pdf starts at $t = 0$; thus, in order to compare the relative

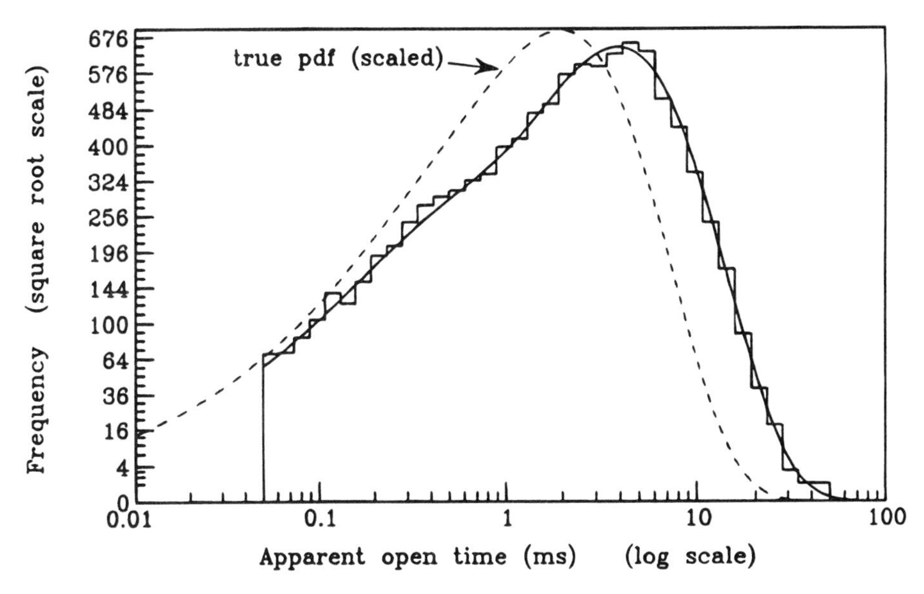

Figure 17. Probability density of the logarithm of open times for the model defined in the text. The solid line is the theoretical distribution of apparent open times for the case when the dead time is $\xi = 50$ µs. The histogram shows the distribution of 10,240 open times, which were simulated on the basis of this model and then subsequently had a resolution of 50 µs imposed, as described in Section 5.2 of Chapter 19 (this volume). The dashed line shows the true distribution of open times, scaled to predict the correct number of observations greater than $t = \xi$.

areas of components of the true distribution with those of the asymptotic distribution of apparent open times, it is necessary to project the exponentials of the asymptotic distribution back to $t = 0$. When this is done, we obtain an approximate distribution with time constants of 3.89 ms and 0.328 ms with corresponding areas of 0.869 and 0.131 (so the overall mean is 3.42 ms, which is close to that of the exact distribution of apparent open times). Comparing this with the true distribution, we see that the short time constant remains virtually the same but the longer one has almost doubled. This happens because, according to the mechanism used for the example, most of the short shut times occur in (and are therefore missed from) the 'long bursts' (see also Section 12.1).

Let us turn now to the distribution of shut times. The true distribution of all shut times in this example is a mixture of three exponentials with time constants of 3789 ms, 0.485 ms, and 53 µs, with areas of 0.262, 0.008, and 0.730, respectively. The distribution of apparent shut times, for $t > 3\xi$, is well approximated by the asymptotic form of this distribution, which is a mixture of three exponentials with time constants of 3952 ms, 0.485 ms and 54 µs; the areas are 0.263, 0.008, and 0.729, respectively (when the asymptotic distribution is projected back to $t = 0$, as above). Apart from a slightly increased long time constant, this is almost identical with the true distribution. This is illustrated in Fig. 18, which shows the logarithmic plots of apparent shut times and true shut times. When the latter are scaled to consider only intervals greater than ξ, they are almost identical.

The reason for these two different types of behaviour is that the true shut time distribution has an important component, area 0.730, with a time constant of 53 µs, which is almost the same length as the dead time; many of these will be missed, leading to concatenated open times. In contrast, the shortest open time constant is more than six times the dead time, so

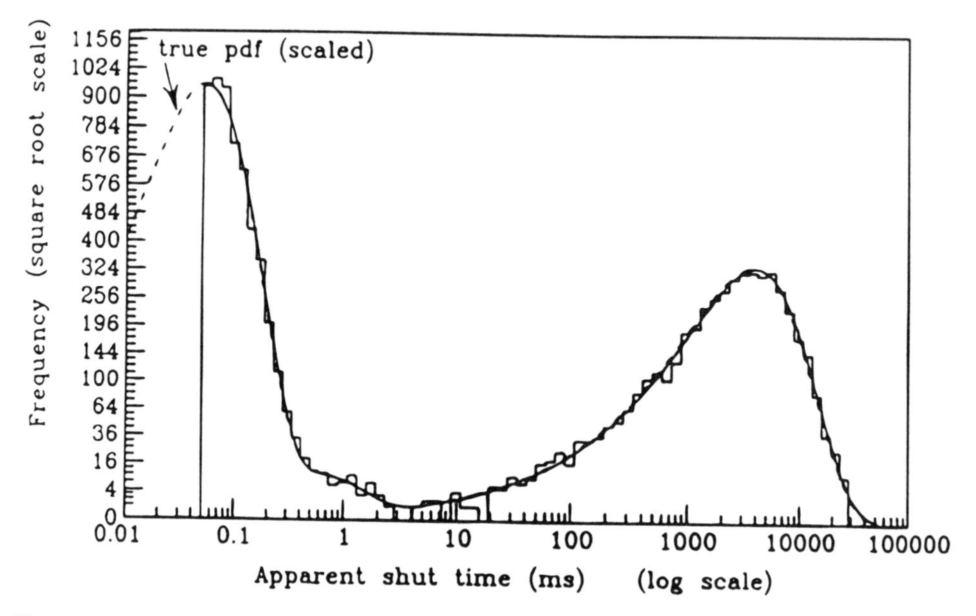

Figure 18. Probability density of the logarithm of shut times for the model defined in the text. The solid line is the theoretical distribution of apparent shut times when the deadtime $\xi = 50$ μs. The histogram shows the distribution of 10,240 shut times, which were simulated on the basis of this model after imposition of a 50-μs resolution. The dashed line showing the true distribution of shut times, scaled as described for Fig. 17, is virtually indistinguishable from the distribution of apparent shut times when $t > \xi$.

few open times will be missed. Thus, the distribution of shut times is almost undistorted in the sense that fitting the observed values with exponentials will give something close to the true time constants and areas (though the overall average of the observed values would be considerably increased, from 993 ms to 1855 ms, because of missing short shut times).

12.4. Joint Distributions of Adjacent Intervals

Magleby and co-workers (Blatz and Magleby, 1989; Weiss and Magleby, 1989; McManus and Magleby, 1989; Magleby and Weiss, 1990a,b) have used extensive simulation to show that the joint distribution of adjacent apparent open times and shut times can be very useful in distinguishing between different mechanisms that have very similar overall distributions of these variables when considered separately. They also use them for parameter estimation. The extra information concerning the relationship between the durations of neighbouring intervals is very valuable (see Sections 10 and 11).

The methods of Hawkes *et al.* (1992) described above can also be used to obtain the theoretical joint distributions of the adjacent apparent open and closed intervals, allowing for time interval omission. The appropriate formulas are outlined in Section 13.7; more detailed formulas and software to calculate and display these distributions are given by Srodzinski (1994).

Figure 19A shows the distribution of apparent open times that are adjacent to short shut times (i.e., those less than 150 μs) for our simulation example. Note again the good correspondence between theory and simulation. Compared with the overall distribution of apparent open times there are relatively few short open times. Figure 19B shows equivalent results for apparent open intervals adjacent to long apparent shut times (greater than 10 ms). This time we see an excess of short open times. The complementary features of these two graphs are a representation of a negative correlation between adjacent apparent open times and apparent shut times. This is illustrated another way in Fig. 19C, which shows how the mean apparent open time, calculated from the above theory, decreases for intervals adjacent to larger apparent shut times. This graph shows a continuous but very nonlinear decline. In practice, shut-time ranges must be used, as in the experimental example in Fig. 12, so the graph is not continuous. The model and parameter values used for Figs. 17 to 19 are based on observations for the frog muscle nicotinic receptor; the form of the decline for the NMDA receptor, illustrated in Fig. 12, is more linear than that plotted in Fig. 19C.

It is worth noting that time interval omission can induce a correlation not present in the true record. For example, model I of Blatz and Magleby (1989) has two open states whose mean lifetimes are almost the same; the discussion in Section 10.1 implies that there should be little correlation between adjacent *true* intervals (none at all if the means are identical). Nevertheless, there is quite strong negative correlation between *observed* open times and adjacent *observed* shut times. The reason for this can be explained with respect to a modified version of our simulation model, which then becomes a simpler version of Blatz and Magleby's model. Modify our model so that direct interchange between the two open states is impossible, and make $\alpha_1 = \alpha_2$, so the mean lifetimes of the two open states are the same. Now open state 2 is next to shut state 3, which has a mean life of 53 μs (just larger than the dead time of 50 μs) with a high probability of returning to state 2, so that successive open sojourns in state 2 are likely to be concatenated with short sojourns in state 3 to form long apparent open times, which are likely to be adjacent to short apparent shut times. State 1, however, is next to state 4, which not only has a mean life of 455 μs but is highly likely to result in a subsequent visit to the very long-lived shut state 5. Thus, open sojourns in state 1 are likely to be isolated and therefore constitute relatively short apparent open times adjacent to quite long apparent shut times. A negative correlation therefore appears between the adjacent *apparent* times, although there is none between *true* adjacent times.

12.5. Maximum-Likelihood Fitting

The ability to calculate the theoretical distributions of things that are actually observed (rather than of what would be observed if the resolution was perfect) opens the way to fitting a specified mechanism directly to the data. Previously, one could only fit empirical mixtures of exponentials separately to open times, shut times, bursts lengths, etc., but to interpret these results in terms of a mechanism and to estimate from them the values of the underlying mass-action rate constants are feasible only approximately and in simple cases. In any case, such methods are very *ad hoc* and almost certainly inefficient. It is, for example, far from obvious how to combine the (often overlapping) information from fitting various different sorts of distribution. For example, the distributions of burst length and of total open time per burst contain different but overlapping information about the burst structure in the data.

However, by using the above distributions for *apparent* open and shut times, i.e., the distributions of what is actually *observed,* it is possible to calculate the *likelihood* for an entire single-channel record, represented as an alternating sequence of open and shut times.

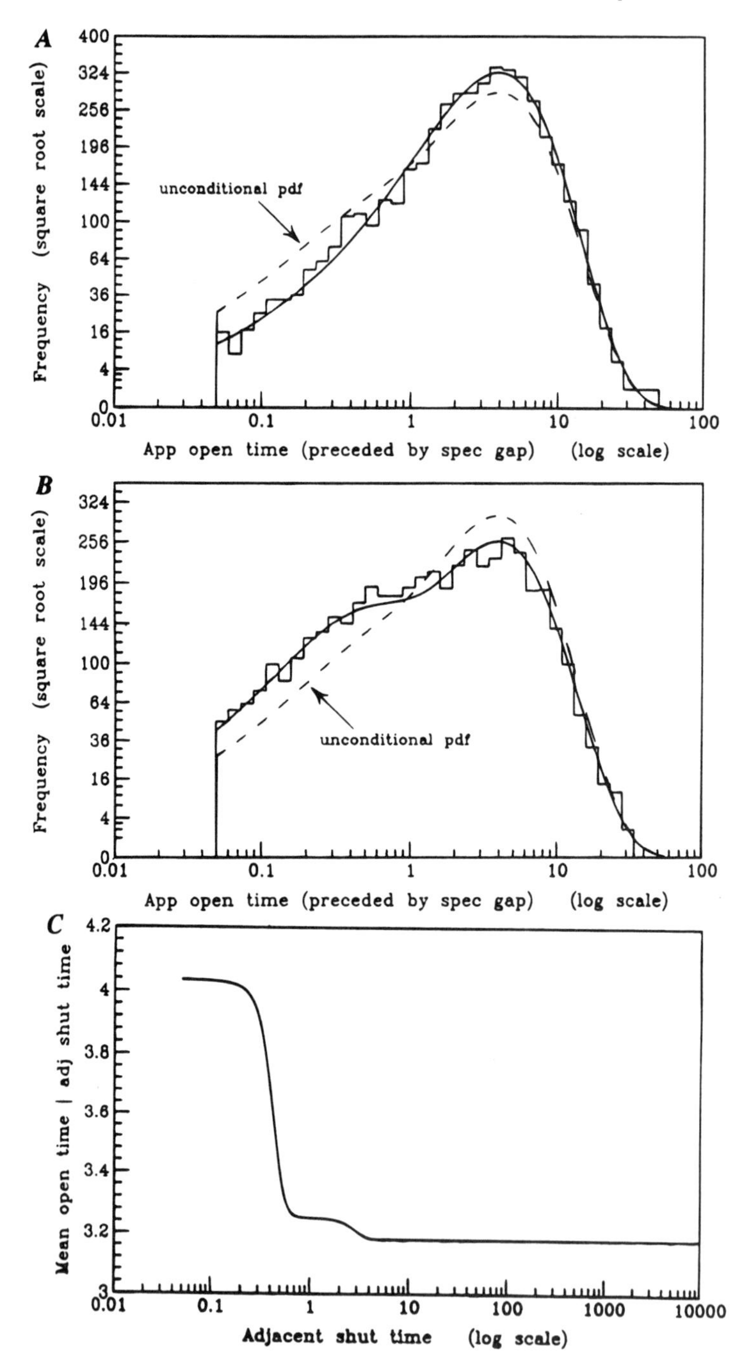

Figure 19. *A* shows the theoretical distribution (solid line) of apparent open times that are adjacent to short apparent shut times (less than 150 μs). The histogram shows the distribution of simulated (see text) values of the same quantity. The dashed line corresponds to the overall theoretical distribution of apparent open times, given in Fig. 17. *B* shows similar results for openings adjacent to long apparent shut times (greater than 10 ms). *C* shows the plot of the theoretical mean of apparent open times that are adjacent to apparent shut times of a given duration, against the logarithm of the shut time.

This calculation takes the form of an enormous number of matrix multiplications, the calculation for the first opening provides the appropriate initial condition for the calculation for the subsequent shut time, which in turn provides an appropriate initial condition for the next opening, and so on up to the end of the data (see Section 13.7). Thus, the *order* in which openings and shuttings occur, and so information about correlations between the durations of neighbouring intervals, is taken into account correctly; in contrast, the separate distributions of apparent open and shut times, described above, lose this information. The calculations are done in a manner similar to that used by Ball and Sansom (1989), following earlier work by Horn and Lange (1983), assuming ideal data ($\xi = 0$).

With this approach, the parameters to be fitted are the actual mass-action rate constants in the reaction mechanism (*not* empirical time constants and areas). The values of these parameters are adjusted by a suitable search routine so as to maximize the likelihood for the entire record. We have found this to be quite feasible on a fast PC for a record consisting of several thousand intervals. Furthermore, it is possible to fit simultaneously data from several different sorts of measurement, for example, recordings made with different concentrations of agonist.

Thus, we do not have to fit separately all of the sorts of distribution mentioned above. However, it will be useful for model validation to compare observed histograms with theoretical distributions calculated from the model, especially the joint distributions of adjacent open and shut times, using values of the parameters fitted by the maximum-likelihood method.

The likelihood itself can be used to judge the relative merits of alternative postulated mechanisms. If each of the proposed mechanisms is fitted to the same data, the relative plausibility of each mechanism can be assessed from how large its maximised likelihood is; this was done, for example, by Horn and Vandenberg (1984) (without missed-event correction).

On the basis of the data from our simulation example above, the free parameters were adjusted (by a simplex method) to maximise the likelihood of the sequence. A comparison of the true and estimated parameter values is given in Table I. These agree very well, but in some cases, especially with less data, one would expect that some parameters in a mechanism would be estimated quite well and others poorly. This feature depends on the nature of the mechanism (e.g., rates leading from a state that is rarely visited will be poorly estimated) and is found in other methods of estimation (see Fredkin and Rice, 1991).

Table I. Comparison of True and Estimated Parameters from a Simulation of Mechanism 110

Parameter	True value	Estimated value[a]	Units
α_1	3000	2848	s^{-1}
α_2	500	521.4	s^{-1}
β_1	15	15.74	s^{-1}
β_2	15,000	15,592	s^{-1}
$2k_{+1}$	1×10^8	9.529×10^7	$M^{-1}s^{-1}$
k_{+2}	5×10^8	5.103×10^8	$M^{-1}s^{-1}$
k^*_{+2}	5×10^8	5.103×10^8	$M^{-1}s^{-1}$
k_{-1}	2000	1960	s^{-1}
$2k_{-2}$	4000	3919	s^{-1}
$2k^*_{-2}$	0.666667	0.7243	s^{-1}

[a]Note that parameter estimates have been constrained so that $k_{-1} = k_{-2}$ and $k_{+2} = k^*_{+2}$ and microscopic reversibility is preserved.

13. A More General Approach to the Analysis of Single-Channel Behaviour

It would involve a great deal of work if the sort of analysis given for the channel-block mechanism (Section 4) had to be repeated for every type of mechanism that one wished to consider. Furthermore, it is found that the approach given above is not sufficiently general to allow analysis of some mechanisms that are of direct experimental interest. In particular, mechanisms with more than one open state and/or cyclic reactions cannot be analyzed by the relatively simple methods used so far. Consider, for example, the mechanism in equation 110, which has two open states (labelled 1 and 2) and three shut states.

We shall assume that the conductance of the two open states is the same, so, during a single opening, there may be any number of oscillations between them: $AR^* \rightleftharpoons A_2R^*$. Similarly a gap within a burst may involve any number of oscillations between $AR \rightleftharpoons A_2R$. The analysis is further complicated by the fact that there are two different ways in which the opening may start (via $AR \rightarrow AR^*$ or $A_2R \rightarrow A_2R^*$) and, correspondingly, two routes by which the opening may end. Clearly, the probability that an opening starts by one of these routes rather than the other will depend on how the previous opening ended. One would expect, for example, that the first opening in a burst is more likely to start via $AR \rightarrow AR^*$ than subsequent openings because the start of a burst must involve passage through AR, whereas a gap within a burst may be spent entirely in A_2R.

In Section 4.6, the distribution of the number of openings per burst was found by simple multiplication of probabilities for the routes through the burst. In the present example there are many different possible routes through a burst, and the only way in which it is practicable to find the appropriate combination of probabilities is to describe them by matrix multiplication. It turns out that matrix notation is very convenient for this sort of problem. By its use one can write down just a few equations for equilibrium single-channel behaviour (Colquhoun and Hawkes, 1982). This enables a single computer program to be written that will evaluate numerically the predicted behaviour of any mechanism, given only the transition rates between the various states. Chapter 20 (this volume) contains details of various matrix results and methods of computation; we suggest that it be read in conjunction with this section.

13.1. Specification of Transition Rates

The transition rates are most conveniently specified in a table or matrix (denoted $\mathbf{Q}$), with the entry in the ith row and jth column (denoted q_{ij}) representing the transition rate from state i to state j (as already defined in equation 7). This fills the whole table except for the diagonal elements ($i = j$). These, it turns out, are most conveniently filled with a number such that the sum of the entries in each row is zero. Thus, from rule 24, $-1/q_{ii}$ is the mean lifetime of a sojourn in the ith state, as is clear from the following examples. For the simple channel-block mechanism (equation 29) with $k = 3$ states, we have

$$\mathbf{Q} = \begin{array}{c} \\ 1 \\ 2 \\ 3 \end{array}\begin{array}{c} \begin{array}{ccc} 1 & 2 & 3 \end{array} \\ \begin{bmatrix} -(\alpha + k_{+B}x_B) & k_{+B}x_B & \alpha \\ k_{-B} & -k_{-B} & 0 \\ \beta' & 0 & -\beta' \end{bmatrix} \end{array} \qquad (126)$$

Similarly, for the more complex agonist mechanism in scheme 110, with $k = 5$ states, we have

$$
\mathbf{Q} = \begin{array}{c} \\ 1 \\ 2 \\ 3 \\ 4 \\ 5 \end{array}
\begin{array}{c}
\begin{array}{ccccc} 1 & 2 & 3 & 4 & 5 \end{array} \\
\begin{bmatrix}
-(\alpha_1 + k^*_{+2}x_A) & k^*_{+2}x_A & 0 & \alpha_1 & 0 \\
2k^*_{-2} & -(\alpha_2 + 2k^*_{-2}) & \alpha_2 & 0 & 0 \\
0 & \beta_2 & -(\beta_2 + 2k_{-2}) & 2k_{-2} & 0 \\
\beta_1 & 0 & k_{+2}x_A & -(\beta_1 + k_{+2}x_A + k_{-1}) & k_{-1} \\
0 & 0 & 0 & 2k_{+1}x_A & -2k_{+1}x_A
\end{bmatrix}
\end{array}
\tag{127}
$$

where x_A is the agonist concentration. This matrix, with some specific parameter values, was used for some numerical examples in Section 12 and in Chapter 20 (this volume). Notice that these two examples illustrate the convenient numbering convention for the states that underlies the notation introduced by Colquhoun and Hawkes (1982). The open states have the lowest numbers ($1, \ldots, k_{\mathcal{A}}$), and shut states have the higher numbers. For the purpose of analysis of bursts, short-lived shut states are given lower numbers than long-lived shut states. This convention allows convenient partitioning of the **Q** matrix into subsections. This partitioning is shown explicitly in Section 2 of Chapter 20 (this volume), and is used throughout this section.

13.2. Derivation of Probabilities

The probabilities that are needed for noise and relaxation analysis, which were defined as $P_{ij}(t)$ in equation 8, can be considered as elements of a matrix, which we shall denote $\mathbf{P}(t)$. It can be found by solution of a differential equation:

$$d\mathbf{P}(t)/dt = \mathbf{P}(t)\mathbf{Q} \tag{128}$$

The solution is, quite generally,

$$\mathbf{P}(t) = e^{\mathbf{Q}t} \tag{129}$$

This has a matrix in the exponent, but its evaluation requires only operations of matrix addition and multiplication, because the exponential is defined in terms of its series expansion:

$$e^{\mathbf{Q}t} = \mathbf{I} + \mathbf{Q}t + (\mathbf{Q}t)^2/2! + \cdots \tag{130}$$

where **I** is a unit matrix (unit diagonals, zeroes elsewhere). In practice, this is not the most convenient way to evaluate the exponential term (see Chapter 20, this volume); in fact, each element of $\mathbf{P}(t)$ (and hence the relaxation or the autocovariance function of noise) can be written in terms of the sum of $k - 1$ exponential terms of the form

$$P_{ij}(t) = p_j(\infty) + w_1 e^{-\lambda_1 t} + w_2 e^{-\lambda_2 t} + \cdots \tag{131}$$

In this expression $p_j(\infty)$ is the equilibrium probability that the system is in state j, which $P_{ij}(t)$ must approach after a long time ($t \rightarrow \infty$). The coefficients w_i can be determined from **Q** by the methods described in Chapter 20 (this volume; see Colquhoun and Hawkes, 1977,

1981, 1982, for details). The rate constants λ_i, which were found by solution of a quadratic in equations 30 and 53, are found, in general, as the solution of a polynomial of degree $k - 1$ that can be derived from **Q.** They are known as the eigenvalues of **Q** (actually, the eigenvalues of $-\mathbf{Q}$). One of the eigenvalues is zero, as **Q** is singular, leaving only $k - 1$ further eigenvalues to find. Standard computer subroutines exist for finding them. These methods are discussed in Chapter 20 (this volume) together with the use of $e^{\mathbf{Q}t}$ in describing the relaxation of the macroscopic current toward equilibrium following a jump. General methods for the calculation of the equilibrium occupancies directly from the **Q** matrix are given in Section 3 of Chapter 20 (this volume).

For the analysis of single channels, however, we usually need a different sort of probability, one that requires that we stay within a specific subset of states throughout the whole time from 0 to t. An example of such a probability was defined in equation 47 and explicitly derived in equations 48–54 when the distribution of the burst length for the channel block mechanism (equation 29) was considered. In that case, we specified in equation 47 that we stayed within the burst (i.e., in state 1 or 2) from 0 to t. It will be convenient to give a symbol $\mathscr{E}$, say, to this set of 'burst states' and to denote the number of such states as $k_{\mathscr{E}}$ ($k_{\mathscr{E}} = 2$ in this case). Similarly, in the case of the more complex agonist mechanism of equation 110, $\mathscr{E}$ would consist of states 1, 2, 3, and 4, and $k_{\mathscr{E}} = 4$. Probabilities such as that in equation 47 will be denoted, by analogy with equation 8, as $P'_{ij}(t)$, in which the subscripts i and j can stand for any of the $\mathscr{E}$ states. In the case of burst length, we can appropriately denote the $(k_{\mathscr{E}} \times k_{\mathscr{E}})$ matrix of such quantities as $\mathbf{P}_{\mathscr{E}\mathscr{E}}(t)$, and it is given quite generally by

$$\mathbf{P}_{\mathscr{E}\mathscr{E}}(t) = e^{\mathbf{Q}_{\mathscr{E}\mathscr{E}}t} \tag{132}$$

where $\mathbf{Q}_{\mathscr{E}\mathscr{E}}$ is the submatrix of **Q** relevant to the burst states. In the case of the simple channel-block mechanism, for example, this is the top left-hand corner of expression 126:

$$\mathbf{Q}_{\mathscr{E}\mathscr{E}} = \begin{bmatrix} -(\alpha + k_{+\mathrm{B}}x_{\mathrm{B}}) & k_{+\mathrm{B}}x_{\mathrm{B}} \\ k_{-\mathrm{B}} & -k_{-\mathrm{B}} \end{bmatrix} \tag{133}$$

Notice that equation 132 is analogous to 129, although it is rather simpler because it involves a smaller matrix. The upper left-hand element of $\mathbf{P}_{\mathscr{E}\mathscr{E}}(t)$ is $P'_{11}(t)$, which has already been derived in equation 52. In general, the elements of $\mathbf{P}_{\mathscr{E}\mathscr{E}}(t)$ can be expressed as the sum of $k_{\mathscr{E}}$ exponential terms; the rate constants for these terms (e.g., those given in equation 53 for simple channel block) are given by the eigenvalues of $-\mathbf{Q}_{\mathscr{E}\mathscr{E}}$ (which are $k_{\mathscr{E}}$ in number, not $k_{\mathscr{E}} - 1$, because $\mathbf{Q}_{\mathscr{E}\mathscr{E}}$, unlike **Q,** is not singular).

13.3. The Open-Time and Shut-Time Distributions

A similar procedure can be followed for any other specified subset of states. The result will always involve a sum of exponential terms, the number of terms being equal to the number of states. For example, let us denote the set of open states as $\mathscr{A}$; this would contain state 1 only for the simple mechanisms in equations 1 and 59, but it would contain states 1 and 2 for the more complex mechanism in equation 110. Again, we can define the subsection of $\mathbf{Q}$ that concerns transitions within $\mathcal{A}$ states; for the mechanism in equation 110, this consists of the top left-hand 2 × 2 section of matrix 127:

$$\mathbf{Q}_{\mathcal{A}\mathcal{A}} = \begin{bmatrix} -(\alpha_1 + k^*_{+2}x_A) & k^*_{+2}x_A \\ 2k^*_{-2} & -(\alpha_2 + 2k^*_{-2}) \end{bmatrix} \quad (134)$$

whereas for the channel-block mechanism 29, with only one open state, we have simply, from equation 126:

$$\mathbf{Q}_{\mathcal{A}\mathcal{A}} = -(\alpha + k_{+B}x_B). \quad (135)$$

In general, we can write (see Colquhoun and Hawkes 1977, 1981, 1982) the distribution of open times as

$$f(t) = \phi e^{\mathbf{Q}_{\mathcal{A}\mathcal{A}}t}(-\mathbf{Q}_{\mathcal{A}\mathcal{A}})\mathbf{u}_{\mathcal{A}} \quad (136)$$

with mean

$$m = \phi(-\mathbf{Q}^{-1}_{\mathcal{A}\mathcal{A}})\mathbf{u}_{\mathcal{A}}$$

An alternative way to write the same thing is

$$f(t) = \phi\mathbf{G}_{\mathcal{A}\mathcal{F}}(t)\mathbf{u}_{\mathcal{F}} \quad (137)$$

where $\mathcal{F}$ represents the set of shut states, and we define (as in equation 142 below)

$$\mathbf{G}_{\mathcal{A}\mathcal{F}}(t) = e^{\mathbf{Q}_{\mathcal{A}\mathcal{A}}t}\mathbf{Q}_{\mathcal{A}\mathcal{F}}$$

The result in equation 136 is an exact matrix analogue of the simple exponential distribution in equation 22 with $-\mathbf{Q}_{\mathcal{A}\mathcal{A}}$ replacing α. All that has been added are an initial vector ϕ, which specifies the relative probabilities of an opening starting in each of the open states, and a final vector, $\mathbf{u}_{\mathcal{A}}$, with $k_{\mathcal{A}}$ (the number of open states) unit elements. Despite the simple appearance of equation 136, it is perfectly general; it works for *any* mechanism, however complex.

In general, different classes of open times will have different distributions, determined by supplying an appropriate initial vector ϕ. For example, the distribution of all open times in a steady-state record is found by using ϕ_0, defined in Chapter 20 (this volume, equation 42). The appropriate ϕ for open times after a jump are considered below, and cases such as the first or last opening in a burst are given by Colquhoun and Hawkes (1982). In some cases, when considering certain specified open times in the middle of a burst, the vector $\mathbf{u}_{\mathcal{A}}$ must be replaced by another vector that describes the way in which a burst ends (see Colquhoun and Hawkes, 1982).

If there is only one open state, both ϕ and $\mathbf{u}_{\mathcal{A}}$ are unity and so can be omitted, and in this case all classes of open times have the same distribution: for example, insertion of equation 135 into 136 gives the result, already derived (see Section 4.3), that the open time is described by a simple exponential distribution with mean $1/(\alpha + k_{+B}x_B)$.

More generally, equation 136 can be expressed without use of matrices as a sum of exponential terms, the number of terms being equal to the number of open states and the rate constants being the eigenvalues of $-\mathbf{Q}_{\mathcal{A}\mathcal{A}}$, as described in Chapter 20 (this volume).

The distribution of shut times is exactly equivalent to that for open times given above, but we use the matrix $\mathbf{Q}_{\mathcal{F}\mathcal{F}}$ instead of $\mathbf{Q}_{\mathcal{A}\mathcal{A}}$ and replace $\mathbf{u}_{\mathcal{A}}$ by a vector $\mathbf{u}_{\mathcal{F}}$ containing $k_{\mathcal{F}}$ unit elements, where $\mathcal{F}$ denotes the set of all shut states ($k_{\mathcal{F}}$ in number). The initial $(1 \times k_{\mathcal{F}})$ vector ϕ now gives the probability that a shut period starts in each of the shut states (in the steady state this would be given by ϕ_s; see equation 50 of Chapter 20, this volume). Thus, the probability density is

$$f(t) = \phi e^{\mathbf{Q}_{\mathcal{F}\mathcal{F}}t}(-\mathbf{Q}_{\mathcal{F}\mathcal{F}})\mathbf{u}_{\mathcal{F}} \tag{138}$$

with mean

$$m = \phi(-\mathbf{Q}_{\mathcal{F}\mathcal{F}}^{-1})\mathbf{u}_{\mathcal{F}} \tag{139}$$

This distribution can be expressed as a sum of exponential terms, the number of terms being equal to the number of shut states, and the rate constants being the eigenvalues of $-Q_{\mathcal{F}\mathcal{F}}$, as described in Chapter 20 (this volume).

We can now see, in matrix terms, why the distribution of all shut periods was so simple for the simple channel-block mechanism. The shut states are states 2 and 3 in this case, so $\mathbf{Q}_{\mathcal{F}\mathcal{F}}$ consists of the lower right-hand 2×2 section of equation 126. The lack of intercommunication between the shut states in this mechanism is reflected by the fact that this submatrix is diagonal (elements not on the diagonal are zero); consequently, the eigenvalues of $-\mathbf{Q}_{\mathcal{F}\mathcal{F}}$ are simply its diagonal elements, $k_{-\mathrm{B}}$ and β'.

13.4. A General Approach to Bursts of Ion-Channel Openings

The analysis of bursts of openings can be approached in a way that is valid for any mechanism of the sort discussed above. The analysis given by Colquhoun and Hawkes (1982) starts by dividing the k states of the system into three subsets defined as follows: (1) open states, denoted $\mathcal{A}$ ($k_{\mathcal{A}}$ in number), (2) short-lived shut states, denoted $\mathcal{B}$ ($k_{\mathcal{B}}$ in number), and (3) long-lived shut states, denoted $\mathcal{C}$ ($k_{\mathcal{C}}$ in number). The short-lived shut states ($\mathcal{B}$) are defined such that any sojourn in this set of states is brief enough to be deemed a gap within a burst, whereas a sojourn in $\mathcal{C}$ would be deemed a gap between bursts. This is illustrated schematically in Fig. 20. The division into subsets is, of course, arbitrary; it is part of our hypothesis about how the observations should be interpreted. Furthermore, the division may depend on the conditions of the experiment (e.g., ligand concentrations) as well as on the mechanism itself.

Take, as an example, the agonist mechanism in equation 110. The set of open states, $\mathcal{A}$, is made up of states 1 and 2. For most plausible values of the rate constants, the lifetimes of shut states 3 and 4 will be short, so they constitute set $\mathcal{B}$. At low agonist concentration (but not otherwise), the lifetimes of the vacant state, 5, will be long, so it is the sole member of set $\mathcal{C}$. The transition rates for the mechanism, which are tabulated in matrix 127, can now be divided up according to this subdivision of states. For example, transition rates among open states are in the $k_{\mathcal{A}} \times k_{\mathcal{A}}$ matrix $\mathbf{Q}_{\mathcal{A}\mathcal{A}}$ that has already been defined in equation 134.

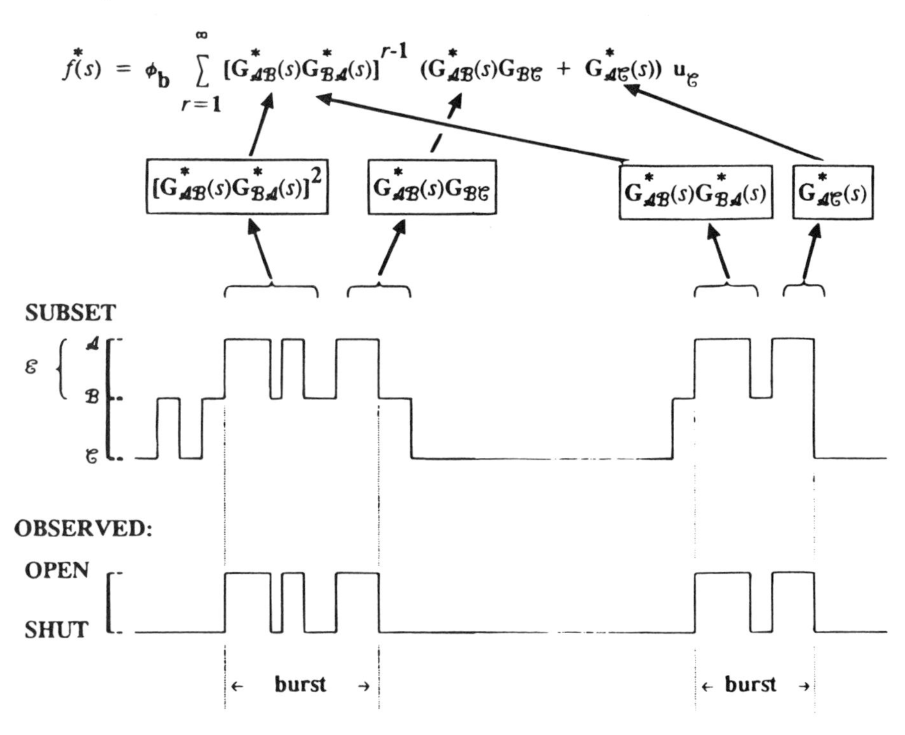

Figure 20. A more general definition of bursts of channel openings. The diagram shows two bursts, the first with $r = 3$ openings and the second with $r = 2$ openings. The two bursts shut by different routes, though this would not be visible on the experimental record; the second burst ends *via* a direct $\mathscr{A} \rightarrow \mathscr{C}$ transition, whereas the first gets to $\mathscr{C}$ *via* $\mathscr{B}$. The lower part of the diagram shows the current that would be observed (if all the open states in set $\mathscr{A}$ have the same conductance). At first, this diagram looks very like Fig. 7 (except that direct $3 \rightleftharpoons 1$ transitions were not allowed in mechanism 59). In fact, it is much more general, because the three levels in the upper diagram no longer represent three discrete states in a particular specified mechanism; they now represent three sets of states (each of which may contain any number of discrete states) that can be defined for any mechanism that results in the occurrence of channel openings in bursts. At the top, the expression for the Laplace transform of the burst length distribution (equation 149) is reproduced, and arrows show the terms in the equation that correspond to the events depicted in the diagram.

Similarly, the transition rates from $\mathscr{A}$ states to $\mathscr{B}$ states are in the $k_{\mathscr{A}} \times k_{\mathscr{B}}$ matrix defined, from matrix 127, as

$$\mathbf{Q}_{\mathscr{A}\mathscr{B}} = \begin{bmatrix} 0 & \alpha_1 \\ \alpha_2 & 0 \end{bmatrix} \tag{140}$$

We can define a probability density that describes the probability of staying within a particular subset, say $\mathscr{A}$ (the open states), throughout the time from 0 to t and then leaving $\mathscr{A}$ for a shut state in $\mathscr{B}$, say. For any state i that is open (in $\mathscr{A}$) and any state j in $\mathscr{B}$, this density is defined as

$$g_{ij}(t) = \lim_{\Delta t \to 0} [\text{Prob(stay within } \mathscr{A} \text{ from 0 to } t \text{ and leave } \mathscr{A} \text{ for state } j \text{ between } t \text{ and } t + \Delta t \,|\, \text{in state } i \text{ at time 0)}/\Delta t] \tag{141}$$

The $k_{\mathcal{A}} \times k_{\mathcal{B}}$ matrix of such quantities we denote $\mathbf{G}_{\mathcal{AB}}(t)$. It can be calculated simply as

$$\mathbf{G}_{\mathcal{AB}}(t) = e^{\mathbf{Q}_{\mathcal{AA}}t}\mathbf{Q}_{\mathcal{AB}} \tag{142}$$

The $k_{\mathcal{A}} \times k_{\mathcal{A}}$ matrix $\mathbf{P}_{\mathcal{AA}}(t) = e^{\mathbf{Q}_{\mathcal{AA}}t}$ analogous to equation 132, can be expressed as the sum of $k_{\mathcal{A}}$ matrices multiplied by scalar exponential terms, as described in Chapter 20 (this volume).

13.4.1. The Number of Openings per Burst

In Section 4.6, the distribution of the number of openings per burst was derived for a simple mechanism. For more complex mechanisms, quantities like the π_{12} used there are no longer convenient. We wish to know the probabilities for transitions from, for example, $\mathcal{A}$ states to $\mathcal{B}$ states regardless of when this transition occurs. The simple quantity π_{12} is replaced by a matrix of transition probabilities, denoted simply $\mathbf{G}_{\mathcal{AB}}$ (the argument, t, is omitted to indicate that this now contains simple probabilities that do not depend on time). Its elements give the probabilities that the system exits from $\mathcal{A}$ (after any number of transitions within $\mathcal{A}$ states) to a particular state j in $\mathcal{B}$, given that it started in state i in $\mathcal{A}$. It can be calculated as

$$\mathbf{G}_{\mathcal{AB}} = \int_0^\infty \mathbf{G}_{\mathcal{AB}}(t)dt = -(\mathbf{Q}_{\mathcal{AA}})^{-1}\mathbf{Q}_{\mathcal{AB}} \tag{143}$$

which can be found directly from the relevant subsections of $\mathbf{Q}$ defined above. Alternatively, we can take the Laplace transform of equation 142; the result for matrices is exactly analogous to that given for the simple exponential in equation 89:

$$\mathcal{L}[\mathbf{G}_{\mathcal{AB}}(t)] = \mathbf{G}^*_{\mathcal{AB}}(s) = (s\mathbf{I} - \mathbf{Q}_{\mathcal{AA}})^{-1}\mathbf{Q}_{\mathcal{AB}} \tag{144}$$

The integration in equation 143 is equivalent to setting $s = 0$ in the Laplace transform, which gives the same result as in 143. Thus, we can also define $\mathbf{G}_{\mathcal{AB}}$ as $\mathbf{G}^*_{\mathcal{AB}}(0)$. Equivalent distributions involving transition from the $\mathcal{B}$ states to the $\mathcal{A}$ states are given by

$$\mathbf{G}_{\mathcal{BA}}(t) = e^{\mathbf{Q}_{\mathcal{BB}}t}\mathbf{Q}_{\mathcal{BA}} \tag{145}$$

and

$$\mathbf{G}_{\mathcal{BA}} = (-\mathbf{Q}_{\mathcal{BB}})^{-1}\mathbf{Q}_{\mathcal{BA}} \tag{146}$$

With the help of expressions such as this, we can, for example, write quite generally, for any mechanism, the probability that a burst contains r openings as

$$P(r) = \phi_{\mathrm{b}}(\mathbf{G}_{\mathcal{AB}}\mathbf{G}_{\mathcal{BA}})^{r-1}(\mathbf{I} - \mathbf{G}_{\mathcal{AB}}\mathbf{G}_{\mathcal{BA}})\mathbf{u}_{\mathcal{A}} \tag{147}$$

with mean

$$\mathbf{m} = \phi_{\mathrm{b}}(\mathbf{I} - \mathbf{G}_{\mathcal{AB}}\mathbf{G}_{\mathcal{BA}})^{-1}\mathbf{u}_{\mathcal{A}} \tag{148}$$

These are matrix analogues of the simple expressions given in equations 36 and 38, with the matrix $\mathbf{G}_{\mathcal{AB}}\mathbf{G}_{\mathcal{BA}}$ describing the transitions from open to brief-shut and back to open, rather than $\pi_{12}\pi_{21}$. The only extra features that are needed (and only if there is more than one open state) are the initial vector ϕ_b, which is introduced to give the relative probabilities of a burst starting in each of the open states (see Section 6 of Chapter 20, this volume), and the usual vector of unit values, $\mathbf{u}_{\mathcal{A}}$.

As described in Chapter 20 (this volume), the distribution in equation 147 can be expressed, without use of matrices, as a mixture of geometric distributions (as in equation 58 of Chapter 19)

$$P(r) = \sum_i a_i(1 - \rho_i)\rho_i^{r-1}$$

where the ρ_i values are given by the eigenvalues of $\mathbf{G}_{\mathcal{AB}}\mathbf{G}_{\mathcal{BA}}$, and a_i is the area of the ith component. The number of *proper* ($\rho_i \neq 0$) geometric components is the rank of the matrix $\mathbf{G}_{\mathcal{AB}}\mathbf{G}_{\mathcal{BA}}$, which is at most (almost always equal to) the direct connectivity between $\mathcal{A}$ and $\mathcal{B}$ (see Section 10) and therefore does not exceed the smaller of $k_{\mathcal{A}}$ and $k_{\mathcal{B}}$. If the number of proper components is less than $k_{\mathcal{A}}$, there is also a component corresponding to zero eigenvalues, $\rho = 0$, which contributes to $P(1)$ but not to any other $P(r)$ (because 0^{r-1} is zero for $r > 1$, but 0^0 is taken as 1). This component is trivial in a mathematical sense, because it corresponds to the probability distribution of a random variable that can take only the value 1 but is of great practical interest because it corresponds to an excess of bursts that consist of a single opening. However, this component does not always exist even when $\mathbf{G}_{\mathcal{AB}}\mathbf{G}_{\mathcal{BA}}$ does have zero eigenvalues, because the area a_i attached to it may be zero. This typically happens when the connectivity between the set of open states, $\mathcal{A}$, and the complete set of shut states, $\mathcal{F}$, is the same as the direct connectivity between $\mathcal{A}$ and $\mathcal{B}$; this will be true, for example, if there is no direct connection between $\mathcal{A}$ and $\mathcal{C}$, only indirect links via $\mathcal{B}$. Examples and further discussion of this complex point are to be found in Colquhoun and Hawkes (1987).

13.4.2. Distribution of the Burst Length

A burst starts in an open state (one of the $\mathcal{A}$ states) and then may oscillate any number of times (0, 1, . . . ∞) to the short-lived shut states ($\mathcal{B}$ states) and back to $\mathcal{A}$. The probability (densities) for all possible numbers of oscillations must be added (hence the summation sign in equation 149 below). Such oscillations are illustrated in Fig. 20 for bursts with three and two openings. The burst ends at the end of the last opening, before the long-lived shut states (set $\mathcal{C}$) are reached. This may happen by direct transition from $\mathcal{A}$ to $\mathcal{C}$ (as in the second burst in Fig. 20), or it may occur *via* an intermediate sojourn in $\mathcal{B}$ (as in the first burst in Fig. 20). In the latter case, the final sojourn in $\mathcal{B}$ is invisible to the observer, so its duration must not be counted as part of the burst length. It is at this point that we see the great power of working with Laplace transforms. The burst length, t, consists of the sum of the lengths of many individual sojourns in different states; these may be of any length, but they add up to t. The problem is, therefore, a more complicated version of the convolution problem described in Section 9. As in Section 9, it can be solved most conveniently by multiplying the Laplace transforms of the individual distributions. Hence, we obtain a term

$\mathbf{G}^*_{\mathcal{AB}}(s)\mathbf{G}^*_{\mathcal{BA}}(s)$ to describe an oscillation from $\mathcal{A}$ to $\mathcal{B}$ and back. On the basis of this argument, we find the Laplace transform, $f^*(s)$, of the burst length distribution as

$$f^*(s) = \phi_b \sum_{r=1}^{\infty} [\mathbf{G}^*_{\mathcal{AB}}(s)\mathbf{G}^*_{\mathcal{BA}}(s)]^{r-1}[\mathbf{G}^*_{\mathcal{AB}}(s)\mathbf{G}_{\mathcal{BC}} + \mathbf{G}^*_{\mathcal{AC}}(s)]\mathbf{u}_{\mathcal{C}} \tag{149}$$

The final term in this describes the end of the burst. The last periods in $\mathcal{A}$ count as part of the burst duration (as illustrated in Fig. 20); hence, the terms $\mathbf{G}^*_{\mathcal{AB}}(s)$ and $\mathbf{G}^*_{\mathcal{AC}}(s)$. The silent final sojourn in $\mathcal{B}$ (see first burst in Fig. 20) is dealt with very elegantly simply by setting $s = 0$, so the last term in equation 149 contains $\mathbf{G}_{\mathcal{BC}}$, i.e., $\mathbf{G}^*_{\mathcal{BC}}(0)$, rather than $\mathbf{G}^*_{\mathcal{BC}}(s)$. This notation, introduced by Colquhoun and Hawkes (1982), removes the need for the clumsier deconvolution procedures used by Colquhoun and Hawkes (1981).

The final part of the problem is to invert the Laplace transform in equation 149 to find the burst length distribution itself. This is a somewhat lengthy procedure (see Colquhoun and Hawkes, 1982), but the result is very simple. It is

$$f(t) = \phi_b[e^{\mathbf{Q}_{\mathcal{EE}}t}]_{\mathcal{AA}}(-\mathbf{Q}_{\mathcal{AA}})e_b \tag{150}$$

where the $(k_{\mathcal{A}} \times 1)$ vector $e_b = (\mathbf{G}_{\mathcal{AB}}\mathbf{G}_{\mathcal{BC}} + \mathbf{G}_{\mathcal{AC}})\mathbf{u}_{\mathcal{C}}$ replaces the usual unit vector; it describes the paths by which the burst can end. The result in equation 150, although perfectly general, looks hardly any more complicated than the general open time distribution given in equation 136. The subscript $\mathcal{AA}$ means that the calculation is done using only the upper $k_{\mathcal{A}} \times k_{\mathcal{A}}$ section of $e^{\mathbf{Q}_{\mathcal{EE}}t}$ (which is a $k_{\mathcal{E}} \times k_{\mathcal{E}}$ matrix). The form of this result is intuitively appealing: it describes a sojourn in the burst states (set $\mathcal{E}$) that starts and ends in an open state (set $\mathcal{A}$). It can be expressed in scalar form, as a mixture of $k_{\mathcal{E}}$ exponentials with rates that are the eigenvalues of $-\mathbf{Q}_{\mathcal{EE}}$, as described in Chapter 20 (Section 7, this volume).

13.4.3. Distribution of the Total Open Time per Burst

In Section 6.4 we discussed the fact that, under certain circumstances, if there is only one open state, then the total time for which a channel is open within a burst has an exponential distribution. Having got as far as writing equation 149 for the Laplace transform of the burst length distribution, it is very easy to obtain various related distributions, such as that for the total open time per burst. The various possible routes through the burst are described by equation 149, but now we are not interested in the time spent in the shut states, so we merely set $s = 0$ in all the $\mathbf{G}^*_{\mathcal{BA}}(s)$; i.e., we replace them with $\mathbf{G}_{\mathcal{BA}}$. Inversion of the result gives, again for any mechanism, a probability density with a form that is very similar to that for single open times in equation 136. It is given by

$$f(t) = \phi_b e^{\mathbf{V}_{\mathcal{AA}}t}(-\mathbf{V}_{\mathcal{AA}})\mathbf{u}_{\mathcal{A}} \tag{151}$$

where ϕ_b is as above. In this result, $\mathbf{V}_{\mathcal{AA}}$ is a $k_{\mathcal{A}} \times k_{\mathcal{A}}$ matrix of transition rates between the set of $\mathcal{A}$ states that takes into account the possibility of going from one to another via a sojourn in $\mathcal{B}$ but not how long it takes to make the sojourn (because any time spent in a gap does not contribute to the total open time); thus,

$$\mathbf{V}_{\mathcal{AA}} = \mathbf{Q}_{\mathcal{AA}} + \mathbf{Q}_{\mathcal{AB}}\mathbf{G}_{\mathcal{BA}} \tag{152}$$

It follows that this distribution is also a sum of $k_{\mathscr{A}}$ exponentials whose rate constants are the eigenvalues of $-\mathbf{V}_{\mathscr{A}\mathscr{A}}$. In particular, it is a simple exponential distribution if there is only one open state.

13.5. Some Conclusions from the General Treatment

A number of general conclusions can be drawn for single-channel observations in the steady state from the analysis of Colquhoun and Hawkes (1982). For example, we can make the following statements:

1. The analysis of single-channel observations depends on submatrices of $\mathbf{Q}$ that correspond to observable sets of states. Insofar as these are smaller than $\mathbf{Q}$ itself, the analysis will be simpler than that of noise and relaxation experiments.
2. The number of exponential components in the distributions of various open lifetimes and of the total open time per burst should be equal to the number of open states. In practice, of course, some components may be too small to observe. In mechanisms with more than one open state, the distribution of open times will not generally be the same for all of the openings in a burst (and similarly for gaps within a burst). The distributions of durations of other intervals of interest also have distributions that are sums of exponentials. The numbers of components in these distributions are summarized in Table II.
3. In general, if a distribution contains more than one exponential component, the time constants for these components cannot be interpreted simply as the mean lifetimes of particular species, and the areas under the individual components cannot be interpreted as the number of sojourns in a particular state. Nevertheless, in particular cases, such interpretations may be approximately valid.
4. The distribution of the number of openings per burst should consist of a mixture of a number of geometric distributions; the number of components is determined by the direct connectivity of the open states, $\mathscr{A}$, and the short-lived shut states, $\mathscr{B}$. In some circumstances there may be an additional component that modifies the probability of a burst consisting of a single open time.
5. It is, for all practical purposes, not possible to analyze mechanisms such as equation 110 without the help of matrix notation. Use of this notation allows a single computer program to be written that can calculate numerically the single-channel, noise, and relaxation behavior of any specified mechanism (see Chapter 20, this volume).

Table II. Numbers of Exponential Components in Various Distributions

Type of interval	Number of components
Open times	$k_{\mathscr{A}}$
Shut times	$k_{\mathscr{F}} = k_{\mathscr{B}} + k_{\mathscr{C}}$
Burst length	$k_{\mathscr{E}} = k_{\mathscr{A}} + k_{\mathscr{B}}$
Total open time per burst	$k_{\mathscr{A}}$
Total shut time per burst	$k_{\mathscr{B}}$
Gaps within bursts	$k_{\mathscr{B}}$
Gaps between bursts	$k_{\mathscr{F}} + k_{\mathscr{B}}$

13.6. Distributions following a Jump

Suppose there is a single jump of agonist concentration or voltage applied at time zero. Certain complications occur in the case where there is zero agonist concentration after the jump; there will be only a finite number of subsequent openings, and there may be none at all. We will not consider such cases here.

The basic results already given for open- and shut-time distributions still hold after a jump; the only thing that is different is the initial vector (denoted ϕ above) that describes the relative probabilities of starting in each of the open states or shut states. If, for example, the channel is shut at the moment the jump occurs, $t = 0$, then the distribution of the subsequent shut time (the *first latency,* see Section 11) is described by exactly the same expression as has already been given, but now ϕ must give the relative probabilities that the channel is in each of the shut states at $t = 0$. We denote the occupancies at time t as $\mathbf{p}(t)$ and partition this vector into the occupancies of open states $\mathbf{p}_{\mathcal{A}}(t)$ (a $1 \times k_{\mathcal{A}}$ vector) and the occupancies of shut states $\mathbf{p}_{\mathcal{F}}(t)$ (a $1 \times k_{\mathcal{F}}$ vector), as described in Chapter 20 (this volume). The relative probability of being in each shut state at $t = 0$ is therefore

$$\phi(0) = \mathbf{p}_{\mathcal{F}}(0)/\mathbf{p}_{\mathcal{F}}(0)\mathbf{u}_{\mathcal{F}}$$

where the denominator is merely the sum of the terms in the numerator, which is included to make the elements of $\phi(0)$ add up to 1. Using $\phi(0)$ in equation 138 immediately gives the distribution of first latencies. In order to use this result, we must be able to postulate appropriate values for $\mathbf{p}_{\mathcal{F}}(0)$. An example is given in Chapter 20 (this volume, Section 8). If the channel has come to equilibrium before $t = 0$, the equilibrium occupancies (under prejump conditions), calculated as in Chapter 20 (this volume, Section 3) can be used. If the channel is not at equilibrium at $t = 0$, e.g., because there was another jump just before $t = 0$, then the occupancies at $t = 0$ can be calculated as described in Chapter 20 (this volume Section 4).

More generally, when we allow for the possibility that the channel may be open at $t = 0$, we can calculate the first latency as follows. If the channel is open at time zero, the first latency is defined to be zero. Let $f_j(t)$ denote the probability density that the first latency has duration t and that when it ends, the channel enters open state j; let $\mathbf{f}(t)$ be the row vector with elements $f_j(t)$. Then

$$\mathbf{f}(t) = p_{\mathcal{A}}(0)\delta(t) + p_{\mathcal{F}}(0)e^{\mathbf{Q}_{\mathcal{F}\mathcal{F}}t}\mathbf{Q}_{\mathcal{F}\mathcal{A}} \tag{153}$$

where $\delta(t)$ is the Dirac delta function. The first term represents the 'lump' of probability at $t = 0$ that results from channels that were open at $t = 0$. The second term, which is of the form described above, gives the distributions of first latencies for channels that were shut at $t = 0$. The overall density of the first latency, $f_{s1}(t)$ say, is obtained by summing over j, so it can be written as

$$f_{s1}(t) = \mathbf{f}(t)\mathbf{u}_{\mathcal{A}} \tag{154}$$

with mean

$$m = \mathbf{p}_{\mathcal{F}}(0)(-\mathbf{Q}_{\mathcal{F}\mathcal{F}})^{-1}\mathbf{u}_{\mathcal{F}} \tag{155}$$

After this first, rather special, shut time, all subsequent open and shut times have the standard distributions given by equations 136 to 139, provided we supply the appropriate initial probability vector, ϕ. The vector of entry probabilities, for the state in which the first open time begins, is

$$\phi_{o1} = \int_0^\infty \mathbf{f}(t)dt = p_{\mathcal{A}}(0) + p_{\mathcal{F}}(0)(-\mathbf{Q}_{\mathcal{F}\mathcal{F}})^{-1}\mathbf{Q}_{\mathcal{F}\mathcal{A}} = p_{\mathcal{A}}(0) + p_{\mathcal{F}}(0)\mathbf{G}_{\mathcal{F}\mathcal{A}} \tag{156}$$

The rth open time following the jump has a probability density given by the standard result, equation 136, but with initial vector, ϕ_{or}, given by

$$\phi_{or} = \phi_{o1}(\mathbf{G}_{\mathcal{A}\mathcal{F}}\mathbf{G}_{\mathcal{F}\mathcal{A}})^{r-1} \qquad r \geq 2 \tag{157}$$

and the mean for the rth open time is

$$m = \phi_{or}(-\mathbf{Q}_{\mathcal{A}\mathcal{A}})^{-1}\mathbf{u}_{\mathcal{A}}\,. \tag{158}$$

Similarly, the rth shut time, for $r \geq 2$, has probability density given by equation 138 with the initial vector, ϕ, defined as

$$\phi_{sr} = \phi_{o1}(\mathbf{G}_{\mathcal{A}\mathcal{F}}\mathbf{G}_{\mathcal{F}\mathcal{A}})^{r-2}\mathbf{G}_{\mathcal{A}\mathcal{F}} \qquad r \geq 2 \tag{159}$$

The mean for the rth shut time is

$$m = \phi_{sr}(-\mathbf{Q}_{\mathcal{F}\mathcal{F}})^{-1}\mathbf{u}_{\mathcal{F}} \tag{160}$$

The results at the end of Section 11.2 can be obtained from these formulas.

Equation 135 generalizes by use of the total probability theorem and the (strong) Markov property to

$$P_{\text{open}}(t) = \int_0^t \Sigma_j f_j(u)P(\text{open at } t \,|\, \text{in open state } j \text{ at time } u)du$$

i.e.,

$$P_{\text{open}}(t) = \int_0^t \mathbf{f}(u)[e^{\mathbf{Q}(t-u)}]_{\mathcal{A}\mathcal{A}}\mathbf{u}_{\mathcal{A}}du \tag{161}$$

where $[e^{\mathbf{Q}t}]_{\mathcal{A}\mathcal{A}}$ stands for that part of the matrix $e^{\mathbf{Q}t}$ obtained by choosing only those rows and columns corresponding to open states. If there is only one open state, this is just the element $P_{11}(t)$ discussed in Section 11.3. But when there is more than one open state, we see that equation 161 does not contain the first latency distribution [i.e., $\mathbf{f}(t)\mathbf{u}_{\mathcal{A}}$ from 154] as such. We thus see that there is not a simple direct relationship between the macroscopic current and the first latency distribution; rather, both can be obtained from the vector $\mathbf{f}(t)$. The macroscopic time course, $P_{\text{open}}(t)$, can be calculated as described in Chapter 20 (this volume), it will have the form of a sum of $k - 1$ exponentials with rate constants that are the eigenvalues of $-\mathbf{Q}$.

The fact that equation 161 is in the form of a convolution implies that it can be expressed simply in terms of Laplace transforms as

$$P^*_{\text{open}}(s) = \mathbf{f}^*(s)[s\mathbf{I} - \mathbf{Q}]^{-1}_{\mathcal{A}\mathcal{A}}\mathbf{u}_{\mathcal{A}} \tag{162}$$

where, by taking the Laplace transform of equation 153,

$$\mathbf{f}^*(s) = p_{\mathcal{A}}(0) + p_{\mathcal{F}}(0)(s\mathbf{I} - \mathbf{Q}_{\mathcal{F}\mathcal{F}})^{-1}\mathbf{Q}_{\mathcal{F}\mathcal{A}}. \tag{163}$$

13.7. Time Interval Omission and Maximum-Likelihood Fitting

The general nature of the problem that arises was discussed in Section 12. Here we give a brief outline of the matrix approach for general Markov mechanisms as developed in Hawkes *et al.* (1990) and Jalali and Hawkes (1992a,b), assuming a constant dead time ξ for both open and closed intervals. This follows the basic method of Ball and Sansom (1988b) of defining *e*-open times and *e*-shut times, which begin and end at time ξ after the start of the observed open and shut times (see Fig. 21), and we say that an event of type j occurs at such a point if the channel is in state j at that instant. An alternative approach (Ball *et al.*, 1991, 1993b) that concentrates on the beginnings of the observed intervals is less mathematically elegant but more physically natural, more general, and likely to be a more fruitful approach to future problems; however, in this brief outline it is simpler to use the first approach.

The key to the problem is a matrix function ${}^{\mathcal{A}}\mathbf{R}(t)$ whose ijth element (i, j being open states) is

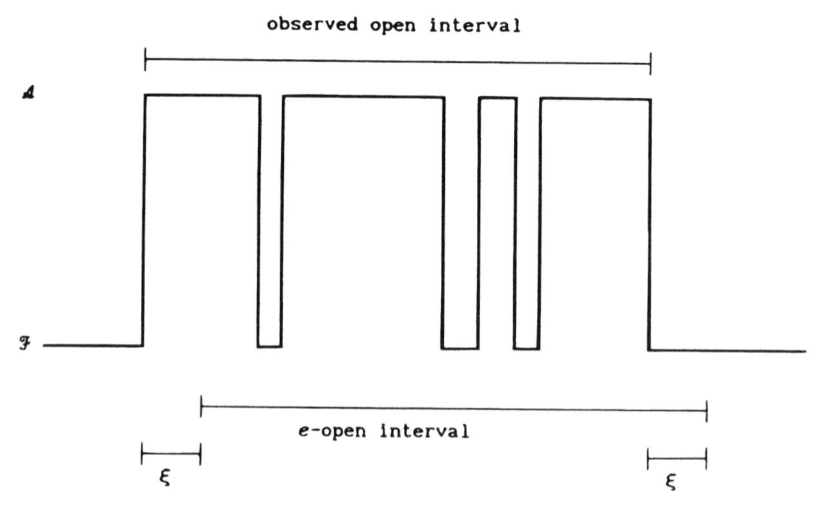

Figure 21. Illustration of the definition of an observed or apparent open interval that begins with an open time greater than ξ. An *e*-open interval has the same duration but begins time ξ after the start of the observed open interval and ends at time ξ after the start of the following observed shut interval. The events of the semi-Markov process discussed in the text occur at these points.

$${}^{\mathcal{A}}R_{ij}(t) = P[X(t) = j \text{ and no shut time is detected over } (0,t) | X(0) = i]$$

where $X(t)$ is the state of the channel at time t, and a detectable shut time is a sojourn in $\mathcal{F}$ of duration greater than ξ. A similar function ${}^{\mathcal{F}}\mathbf{R}(t)$ is defined for shut times. Hawkes *et al.* (1990) showed how to compute ${}^{\mathcal{A}}\mathbf{R}(t)$ by a method that is quite simple for small t but becomes more complicated and numerically unstable for large t. However, Jalali and Hawkes (1992a,b) showed that it could be extremely well approximated, for all except quite small values of t, by a sum of $k_{\mathcal{A}}$ exponentials:

$${}^{\mathcal{A}}\mathbf{R}(t) \simeq \sum_{i=1}^{k_{\mathcal{A}}} \mathbf{M}_i e^{-\lambda_i' t} \tag{164}$$

where the $\mathbf{M}_i$ are $k_{\mathcal{A}} \times k_{\mathcal{A}}$ matrices, and $-\lambda_i'$ are some kind of generalized eigenvalues. They recommend using this for $t > 3\xi$ and the exact result for $t \leq 3\xi$.

Now let ${}^{e}\mathbf{G}_{\mathcal{A}\mathcal{F}}(t)$ denote a semi-Markov matrix whose ijth element (i in $\mathcal{A}$ and j in $\mathcal{F}$) gives the probability density of an e-open interval being of length t and the probability that it ends in shut state j, given that it began in open state i. It is given by

$${}^{e}\mathbf{G}_{\mathcal{A}\mathcal{F}}(t) = {}^{\mathcal{A}}\mathbf{R}(t - \xi)\mathbf{Q}_{\mathcal{A}\mathcal{F}} \exp(\mathbf{Q}_{\mathcal{F}\mathcal{F}}\xi) \tag{165}$$

because, for the e-open interval to end at time t, there must be a transition from $\mathcal{A}$ to $\mathcal{F}$ at time $t - \xi$ (with no detectable sojourn in $\mathcal{F}$ up to then), followed by a sojourn of at least ξ in $\mathcal{F}$. ${}^{e}\mathbf{G}_{\mathcal{A}\mathcal{F}}(t)$ replaces the matrix function $\mathbf{G}_{\mathcal{A}\mathcal{F}}(t)$ that occurs in the ideal ($\xi = 0$) theory. A similar function for shut times is

$${}^{e}\mathbf{G}_{\mathcal{F}\mathcal{A}}(t) = {}^{\mathcal{F}}\mathbf{R}(t - \xi)\mathbf{Q}_{\mathcal{F}\mathcal{A}} \exp(\mathbf{Q}_{\mathcal{A}\mathcal{A}}\xi) \tag{166}$$

These functions enable us to obtain many results of interest in a form that superficially resembles those found in the ideal case.

13.7.1. Distributions of Observed Open Times and Shut Times

The probability density of observed open times is

$$f(t) = \phi_{\mathcal{A}}{}^{e}\mathbf{G}_{\mathcal{A}\mathcal{F}}(t)u_{\mathcal{F}} \tag{167}$$

which may be compared with the ideal form as given in equation 137. The probability density of observed shut times is given by the similar expression:

$$f(t) = \phi_{\mathcal{F}}{}^{e}\mathbf{G}_{\mathcal{F}\mathcal{A}}(t)u_{\mathcal{A}} \tag{168}$$

In these results $\phi_{\mathcal{A}}$ and $\phi_{\mathcal{F}}$ are equilibrium probability vectors for the states occupied at the start of e-open intervals or e-shut intervals, respectively. Formulas for calculating them are given by Hawkes *et al.* (1990).

The importance of result 164 is that, apart from very short times ($t < 3\xi$), the density, obtained by substituting equation 165 into expression 167, is very well approximated by a mixture of $k_{\mathscr{A}}$ exponentials with rate constants λ_i'; in other words, it behaves much like the true distribution of open times, having the same number of components but with modified time constants (see Section 12.3 for a numerical example). Similar results apply to equation 168, resulting in a distribution approximated by a mixture of $k_{\mathscr{F}}$ exponentials.

13.7.2. Joint Distributions of Adjacent Observed Intervals

The above results easily generalize: for example, the joint probability density of an observed shut time followed by an observed open time is given by

$$f(t_s,t_o) = \boldsymbol{\phi}_{\mathscr{F}}{}^e\mathbf{G}_{\mathscr{F}\mathscr{A}}(t_s)\ {}^e\mathbf{G}_{\mathscr{A}\mathscr{F}}(t_o)u_{\mathscr{F}} \tag{169}$$

By further operations on this joint distribution, we can obtain various conditional distributions and conditional means, examples of which are given in Section 12.4. Details of these procedures are given in Srodzinski (1994).

By interchanging $\mathscr{A}$ and $\mathscr{F}$ in the above formula we get the joint distribution of an observed open time followed by an observed shut time. For a reversible process, these two distributions are identical when dealing with *true* open and shut times. However, the method used for defining *observed* intervals is not symmetrical in time; consequently, we have found in numerical examples that these two distributions are not actually identical, though they are so close that the difference would not be detected in practice.

13.7.3. Likelihood of a Complete Record

The formula 169 is easily extended to an entire record. If, for example, we have a sequence of $2n$ intervals that starts with a shut time and ends with an open time, and if the ith pair of adjacent observed shut and open times are denoted by t_{si}, t_{oi}, then the likelihood of the entire record is given by multiplying together all the appropriate matrices. Thus, the likelihood is given by

$$\boldsymbol{\phi}_{\mathscr{F}}\ {}^e\mathbf{G}_{\mathscr{F}\mathscr{A}}(t_{s1})\ {}^e\mathbf{G}_{\mathscr{F}\mathscr{A}}(t_{o1})\ {}^e\mathbf{G}_{\mathscr{F}\mathscr{A}}(t_{s2})\ {}^e\mathbf{G}_{\mathscr{A}\mathscr{F}}(t_{o2}) \cdots {}^e\mathbf{G}_{\mathscr{F}\mathscr{A}}(t_{sn})\ {}^e\mathbf{G}_{\mathscr{A}\mathscr{F}}(t_{on})\mathbf{u}_{\mathscr{F}} \tag{170}$$

In this expression, $\boldsymbol{\phi}_{\mathscr{F}}$ is the initial vector for the first shut time; $\boldsymbol{\phi}_{\mathscr{F}}{}^e\mathbf{G}_{\mathscr{F}\mathscr{A}}(t_{s1})$ then provides the initial vector for the first open time, and so on to the end of the record. The sequence of the openings and shuttings, and all the information on correlations contained in it, is taken into account. The likelihood defined in equation 170 can then be maximized numerically, as described in Section 12.5, to estimate the model parameters.

14. Concluding Remarks

In the first edition of this book many of the basic ideas described in this chapter were already known. The major advances since then have been in (1) the understanding of the importance of information from correlations (Section 10), (2) the development of the theory for nonstationary processes (Section 11), and (3) the development of usable theories for treating the problem of missed events (Section 12), with the concomitant ability to do direct maximum-likelihood fits of a mechanism to observed values simultaneously for several different sets and types of data (Section 13). On the other hand, some important problems, such as the frequent problems in estimating the number of channels in a patch, remain intractable, and there has been little work on the kinetics of mechanisms that involve subconductance levels.

References

Adams, P. R., 1976, Drug blockade of open end-plate channels, *J. Physiol.* **260:**531–552.

Aldrich, R. W., Corey, D. P., and Stevens, C. F., 1983, A reinterpretation of mammalian sodium channel gating based on single channel recording, *Nature* **306:**436–441.

Anderson, C. R., and Stevens, C. F., 1973, Voltage clamp analysis of acetylcholine produced end-plate current fluctuations at frog neuromuscular junction, *J. Physiol.* **235:**655–691.

Armstrong, C. M., 1971, Interaction of tetraethylammonium ion derivatives with the potassium channels of giant axons, *J. Gen. Physiol.* **58:**413–437.

Ball, F. G., and Sansom, M. S. P., 1988a, Single channel autocorrelation functions. The effects of time interval omission, *Biophys. J.* **53:**819–832.

Ball, F. G., and Sansom, M. S. P., 1988b, Aggregated Markov processes incorporating time interval omission, *Adv. Appl. Prob.* **20:**546–572.

Ball, F. G., and Sansom, M. S. P., 1989, Ion-channel mechanisms: Model identification and parameter estimation from single channel recordings, *Proc. R. Soc. Lond.* [*Biol.*] **236:**385–416.

Ball, F. G., and Yeo, G. F., 1994, Numerical evaluation of observed sojourn time distributions for a single ion channel incorporating time interval omission, *Stat. Comput.* **4:**1–12.

Ball, F. G., Kerry, C. J., Ramsey, R. L., Sansom, M. S. P., and Usherwood, P. N. R., 1988, The use of dwell time cross-correlation functions to study single-ion channel gating kinetics, *Biophys. J.* **54:**309–320.

Ball, F. G., McGee, R., and Sansom, M. S. P., 1989, Analysis of post-perturbation gating kinetics of single ion channels, *Proc. R. Soc. Lond.* [*Biol.*] **236:**29–52.

Ball, F. G., Davies, S. S., and Sansom, M. S. P., 1990, Single-channel data and missed events: Analysis of a two-state Markov model, *Proc. R. Soc. Lond.* [*Biol.*] **242:**61–67.

Ball, F. G., Milne, R. K., and Yeo, G. F., 1991, Aggregated semi-Markov processes incorporating time interval omission, *Adv. Appl. Prob.* **23:**772–797.

Ball, F. G., Milne, R. K., and Yeo, G. F., 1993a, On the exact distribution of observed open times in single ion channel models. *J. Appl. Prob.* **30:**522–537.

Ball, F. G., Yeo, G. F., Milne, R. K., Edeson, R. O., Madsen, B. W., and Sansom, M. S. P., 1993b, Single ion channel models incorporating aggregation and time interval omission, *Biophys. J.* **64:**357–374.

Blatz, A. L., and Magelby, K. L., 1986, Correcting single channel data for missed events, *Biophys. J.* **49:**967–980.

Blatz, A. L., and Magleby, K. L., 1989, Adjacent interval analysis distinguishes among gating mechanisms for the fast chloride channel from rat skeletal muscle, *J. Physiol.* **410:**561–585.

Castillo, J. del, and Katz, B., 1957, Interaction at end-plate receptors between different choline derivatives, *Proc. R. Soc. Lond.* [*Biol.*] **146:**369–381.

Colquhoun, D., 1971, *Lectures on Biostatistics,* Clarendon Press, Oxford.

Colquhoun, D., 1981, How fast do drugs work? *Trends Pharmacol. Sci.* **2:**212–217.

Colquhoun, D., and Hawkes, A. G., 1977, Relaxation and fluctuations of membrane currents that flow through drug-operated ion channels, *Proc. R. Soc. Lond.* [*Biol.*] **199:**231–262.

Colquhoun, D., and Hawkes, A. G., 1981, On the stochastic properties of single ion channels, *Proc. R. Soc. Lond.* [*Biol.*] **211:**205–235.

Colquhoun, D., and Hawkes, A. G., 1982, On the stochastic properties of bursts of single ion channel openings and of clusters of bursts, *Phil. Trans. R. Soc. Lond.* [*Biol.*] **300:**1–59.

Colquhoun, D., and Hawkes, A. G., 1987, A note on correlations in single ion channel records, *Proc. R. Soc. Lond.* [*Biol.*] **230:**15–52.

Colquhoun, D., and Hawkes, A. G., 1990, Stochastic properties of ion channel openings and bursts in a membrane patch that contains two channels: Evidence concerning the number of channels present when a record containing only single openings is observed, *Proc. R. Soc. Lond.* [*Biol.*] **240:**453–477.

Colquhoun, D., and Hawkes, A. G., 1994, The interpretation of single channel recordings, in: *Microelectrode Techniques: The Plymouth Workshop Handbook,* 2nd ed. (D. C. Ogden, ed.) pp. 141–188, Company of Biologists, Cambridge.

Colquhoun, D., and Ogden, D. C., 1986, States of the nicotinic acetylcholine receptor: Enumeration, characteristics and structure, in: *Nicotinic Acetylcholine Receptor, NATO ASI Series Cell Biology,* Vol. 3 (A. Maelicke, ed.) pp. 197–218, Springer, Berlin.

Colquhoun, D., and Ogden, D. C., 1988, Activation of ion channels in the frog end-plate by high concentrations of acetylcholine, *J. Physiol.* **395:**131–159.

Colquhoun, D., and Sakmann, B., 1981, Fluctuations in the microsecond time range of the current through single acetylcholine receptor ion channels, *Nature* **294:**464–466.

Colquhoun, D., and Sakmann, B., 1985, Fast events in single-channel currents activated by acetylcholine and its analogues at the frog muscle end-plate, *J. Physiol.* **369:**501–557.

Colquhoun, D., and Sheridan, R. E., 1981, The modes of action of gallamine, *Proc. R. Soc. Lond.* [*Biol.*] **211:**181–203.

Colquhoun, D., Dreyer, F., and Sheridan, R. E., 1979, The actions of tubocurarine at the frog neuromuscular junction, *J. Physiol.* **293:**247–284.

Crouzy, S. C., and Sigworth, F. J., 1990, Yet another approach to the dwell-time omission problem of single-channel analysis, *Biophys. J.* **58:**731–743.

Cull-Candy, S. G., and Usowicz, M. M., 1987, Multiple-conductance channels activated by excitatory amino acids in cerebellar neurons, *Nature* **325:**525–528.

Denbigh, K. G., 1951, *The Thermodynamics of the Steady State,* Methuen, London; John Wiley & Sons, New York.

Draber, S., and Schultze, R., 1994, Correction for missed events based on a realistic model of a detector, *Biophys. J.* **66:**191–201.

Edmonds, B., & Colquhoun, D., 1992, Rapid decay of averaged single-channel NMDA receptor activations recorded at low agonist concentration, *Proc. R. Soc. Lond.* [*Biol.*] **250:**279–286.

Feller, W., 1966, *An Introduction to Probability Theory and its Applications,* 2nd ed., Vol. 2, John Wiley & Sons, New York.

Fredkin, D. R., and Rice, J. A., 1991, *Maximum Likelihood Estimation of Ion Channel Kinetic Parameters Directly from Digitized Data, Technical Report 338,* Department of Statistics, University of California, Berkeley.

Fredkin, D. R., Montal, M., and Rice, J. A., 1985, Identification of aggregated Markovian models: Application to the nicotinic acetylcholine receptor, in: *Procedures of the Berkeley Conference in Honor of Jerzy Neyman and Jack Kiefer,* Vol. I (L. M. Le Cam and R. A. Olshen, eds.), pp. 269–289, Wadsworth Press, Monterey, California.

Gibb, A. J., and Colquhoun, D., 1992, Activation of NMDA receptors by L-glutamate in cells dissociated from adult rat hippocampus, *J. Physiol.* **456:**143–179.

Gration, K. A. F., Lambert, J. J., Ramsey, R. L., Rand, R. P., and Usherwood, P. N. R., 1982, Closure of membrane channels gated by glutamate may be a two-step process, *Nature* **295:**599–601.

Hamill, O. P., and Sakmann, B., 1981, Multiple conductance states of single acetylcholine receptor channels in embryonic muscle cells, *Nature* **294:**462–464.

Hamill, O. P., Marty, A., Neher, E., Sakmann, B., and Sigworth, F. J., 1981, Improved patch-clamp techniques for high-resolution current recording from cells and cell-free membrane patches, *Pflügers Arch.* **391:**85–100.

Hawkes, A. G., Jalali, A., and Colquhoun, D., 1990, The distributions of the apparent open times and shut times in a single channel record when brief events can not be detected, *Phil. Trans. R. Soc.* [*A*] **332:**511–538.

Hawkes, A. G., Jalali, A., and Colquhoun, D., 1992, Asymptotic distributions of apparent open times and shut times in a single channel record allowing for the omission of brief events, *Phil. Trans. R. Soc.* [*B*] **337:**383–404.

Horn, R., 1991, Estimating the number of channels in patch recordings, *Biophys. J.* **60:**433–439.

Horn, R., and Korn, S. J., 1989, Model selection: Reliability and bias, *Biophys. J.* **55:**379–381.

Horn, R., and Lange, K., 1983, Estimating kinetic constants from single channel data, *Biophys. J.* **43:**207–223.

Horn, R., and Vandenberg, C. A., 1984, Statistical properties of single sodium channels, *J. Gen. Physiol.* **84:**505–534.

Howe, J. R., Cull-Candy, S. G., and Colquhoun, D., 1991, Currents through single glutamate-receptor channels in outside-out patches from rat cerebellar granule cells, *J. Physiol* **432:**143–202.

Jackson, M. B., 1985, Stochastic behaviour of a many-channel membrane system, *Biophys. J.* **47:**129–137.

Jackson, M. B., Wong, B. S., Morris, C. E., Lecar, H. and Christian, C. N., 1983, Successive openings of the same acetylcholine receptor channels are correlated in open time, *Biophys. J.* **42:**109–114.

Jalali, A., and Hawkes, A. G., 1992a, The distribution of apparent occupancy times in a two-state Markov process in which brief events can not be detected, *Adv. Appl. Prob.* **24:**288–301.

Jalai, A., and Hawkes, A. G., 1992b, Generalized eigenproblems arising in aggregated Markov processes allowing for time interval omission, *Adv. Appl. Prob.,* **24:**302–321.

Katz, B., and Miledi, R., 1970, Membrane noise produced by acetylcholine, *Nature* **226:**962–963.

Katz, B., and Miledi, R., 1972. The statistical nature of the acetylcholine potential and its molecular components, *J. Physiol.* **224:**665–699.

Kelly, F. P., 1979, *Reversibility and Stochastic Networks,* John Wiley & Sons, Chichester.

Korn, J. S., and Horn, R., 1988, Statistical discrimination of fractal and Markov models of single-channel gating, *Biophys. J.* **54:**871–877.

Labarca, P., Rice J. A., Fredkin, D. R., and Montal, M., 1985, Kinetic analysis of channel gating: application to the cholinergic receptor channel and the chloride channel from *Torpedo californica, Biophys. J.* **47:**469–478.

Läuger, P., 1985, Ionic channels with conformational substates, *Biophys. J.* **47:**581–590.

Läuger, P., 1988, Internal motions in proteins and gating kinetics of ionic channels, *Biophys. J.* **53:**877–884.

Liebovitch, L. S., 1989, Testing fractal and Markov models of ion channel kinetics, *Biophys. J.* **55:**373–377.

Liebovitch, L. S., Fischbarg, J., Koniarek, J. P., Todorova, I., and Wang, M., 1987, Fractal model of ion-channel kinetics, *Biochim. Biophys. Acta* **896:**173–180.

Magleby, K. L., and Song, L., 1992, Dependency plots suggest the kinetic structure of ion channels, *Proc. R. Soc. Lond.* [*Biol.*] **249:**133–142.

Magleby, K. L., and Weiss, D. S., 1990a, Estimating kinetic parameters for single channels with simulation. A general method that resolves the missed event problem and accounts for noise, *Biophys. J.* **58:**1411–1426.

Magleby, K. L., and Weiss, D. S., 1990b, Identifying kinetic gating mechanisms for ion channels by using two-dimensional distributions of simulated dwell times, *Proc. R. Soc. Lond.* [*Biol.*] **241:**220–228.

McManus, O. B., and Magleby, K. L., 1989, Kinetic time constants independent of previous single-channel activity suggest Markov gating for a large conductance Ca-activated K channel, *J. Gen. Physiol.* **94:**1037–1070.

McManus O. B., Blatz, A. L., and Magleby, K. L., 1985, Inverse relationship of the durations of open and shut intervals for Cl and K channels, *Nature* **317:**625–627.

McManus, O. B., Weiss, D. S., Spivak, C. E., Blatz, A. L., and Magleby, K. L., 1988, Fractal models are inadequate for the kinetics of four different ion channels, *Biophys. J.* **54:**859–870.

Millhauser, G. L., Salpeter, E. E., and Oswald, R. E., 1988, Rate–amplitude correlation from single-channel records. A hidden structure in ion channel gating kinetics? *Biophys. J.* **54:**1165–1168.

Milne, R. K., Yeo, G. F., Madsen, B. W., and Edeson, R. O., 1989, Estimation of single channel parameters from data subject to limited time resolution, *Biophys. J.* **55:**673–676.

Mood, A. M., and Graybill, F. A., 1963, *Introduction to the Theory of Statistics,* 2nd ed., McGraw-Hill, New York.

Neher, E., 1983, The charge carried by single-channel currents of rat cultured muscle cells in the presence of local anaesthetics, *J. Physiol.* **339:**663–678.

Neher, E., and Sakmann, B., 1976, Single-channel currents recorded from membrane of denervated frog muscle fibres, *Nature* **260:**799–802.

Neher, E., and Steinbach, J. H., 1978, Local anaesthetics transiently block currents through single acetylcholine-receptor channels, *J. Physiol.* **277:**153–176.

Ogden, D. C. and Colquhoun, D., 1985, Ion channel block by acetylcholine, carbachol and suberyldicholine at the frog neuromuscular junction, *Proc. R. Soc. Lond.* [*Biol.*] **225:**329–355.

Ogden, D. C., Siegelbaum, S. A., and Colquhoun, D., 1981, Block of acetylcholine-activated ion channels by an uncharged local anaesthetic, *Nature,* **289:**596–598.

Olkin, I., Petkau, A. J., and Zidek, J. W., 1981, A comparison of *n* estimators for the binomial distribution, *J. Amer. Stat. Assoc.* **76:**637–642.

Petracchi, D., Barbi, M., Pellegrini, M., Pellegrino, M., and Simoni, A., 1991, Use of conditional distributions in the analysis of ion channel recordings, *Eur. Biophys. J.* **20:**31–39.

Roux, B., and Sauvé, R., 1985, A general solution to the time interval omission problem applied to single channel analysis, *Biophys. J.* **48:**149–158.
Sakmann, B., Patlak, J., and Neher, E., 1980, Single acetylcholine-activated channels show burst-kinetics in presence of desensitizing concentrations of agonist, *Nature* **286:**71–73.
Sansom, M. S. P., Ball, F. G., Kerry, C. J., McGee, R., Ramsey R. L., and Usherwood, P. N. R., 1989, Markov, fractal, diffusion, and related models of ion channel gating, *Biophys. J.* **56:**1229–1243.
Sine, S. M., and Steinbach, J. H., 1986, Activation of acetylcholine receptors on clonal mammalian BC3H-1 cells by low concentrations of agonist, *J. Physiol.* **373:**129–162.
Sine, S. M., and Steinbach, J. H., 1987, Activation of acetylcholine receptors on clonal mammalian BC3H-1 cells by high concentrations of agonist, *J Physiol.* **385:**325–359.
Spiegel, M. R., 1965, *Laplace Transforms,* Schaum's Outline Series, McGraw-Hill, New York.
Srodzinski, K., 1994, *The Joint Distribution of Adjacent Open and Shut Times in Single Ion Channel Recording,* M. Phil. Thesis, University of Wales Swansea.
Stern, P., Béhé, P., Schoepfer, R., and Colquhoun, D., 1992, Single channel properties of NMDA receptors expressed from cloned cDNAs: Comparison with native receptors, *Proc. R. Soc. Lond.* [*Biol.*] **250:**271–277.
Weiss, D. S. and Magleby, K. L., 1989, Gating scheme for single GABA-activated Cl^- channels determined from stability plots, dwell-time distributions and adjacent-interval durations, *J. Neurosci.* **9:**1314–1324.
Yeo, G. F., Milne, R. K., Edeson, R. O., and Madsen, B. W., 1988, Statistical inference from single channel records: Two-state Markov model with limited time resolution, *Proc. R. Soc. Lond.* [*Biol.*] **235:**63–94.
Yeramian, E., Trautmann, A., and Claverie, P., 1986, Acetylcholine receptors are not functionally independent, *Biophys. J.* **50:**253–263.

Chapter 19

Fitting and Statistical Analysis of Single-Channel Records

DAVID COLQUHOUN and F. J. SIGWORTH

1. Introduction

The aims of analysis of single-channel records can be considered in two categories. The first is to allow one to observe results at leisure in order to determine their qualitative features. It may, for example, be found that the single-channel currents were not all of the same amplitude or that they showed obvious grouping into bursts or that artifacts appeared on the record that might be misleading. These effects are often not easy to see on the oscilloscope screen as an experiment proceeds. It is best to have a computer program that allows one, after the experiment, to scroll flexibly through the recorded data and zoom in on portions of the record to observe details at high time resolution.

The second aim is to perform quantitative analyses of measurable variables (e.g., the channel-open durations), in which these quantities are compared with theoretical distributions, and to try to infer a biological mechanism from the result. Although other measurable variables can be studied, in this chapter we consider only the analysis of channel current amplitudes and dwell times. The current through a single channel is assumed to consist of rectangular pulses having one or a few discrete current levels and infinitely short transition times. The analysis procedures we describe involve, first, the estimation of the amplitudes and times of transition in the measured currents and, second, the fitting of distributions to these estimates.

It is undoubtedly true that one of the disadvantages of recording from single ion channels is the length of time that it takes to analyze the results. One reason for this is that the quantities we measure, for example, the length of time for which a channel stays open, are random variables (as discussed in Chapter 18, this volume). In the simplest case of a quantity that has a simple exponential distribution with mean lifetime τ, the standard deviation of an observation should be simply τ (see, for example, Colquhoun, 1971; Chapter 18, this volume). Therefore, the standard deviation of the mean on n observations should be $\tau/\sqrt{n}$. (The usage of the terms standard deviation and standard error is discussed in section 6.7.1.) In order to find the mean lifetime with an accuracy of 10%, it is necessary to measure 100 or so individual lifetimes. In practice, it is advisable to measure many more events than this. The

Note to the reader: At the authors' request this chapter will use the abbreviations ms and μs instead of msec and μsec.

DAVID COLQUHOUN • Department of Pharmacology, University College London, London WC1E 6BT, England. F. J. SIGWORTH • Department of Cellular and Molecular Physiology, Yale University School of Medicine, New Haven, Connecticut 06510.

Single-Channel Recording, Second Edition, edited by Bert Sakmann and Erwin Neher. Plenum Press, New York, 1995.

main reason additional measurements are needed is that one can never be sure in advance of the shape of the distribution. It is very common for the distribution of observations not to be described by a single exponential distribution but by a mixture of two or three or more exponential terms. Indeed, under some circumstances, the distribution need not be described by a mixture of exponentials at all; for example, this is, strictly speaking, the case when the resolution of the observations is limited (see Section 12 of Chapter 18, this volume, and Section 6.11 below). It will rarely be satisfactory to measure fewer than 200 openings, and a few thousand openings will suffice for quite complex distributions *if* the time constants are well separated. For the evaluation of complex models, data sets with millions of events have been acquired and analyzed (e.g., McManus and Magleby, 1988).

2. Acquiring Data

2.1. Pulsed and Continuous Recordings

Some experiments rely on the application of a stimulus to open the channels. A pulse of applied neurotransmitter or a membrane depolarization is given, and the resulting channel currents are measured. In order to obtain a sufficiently large number of events, sometimes hundreds or thousands of pulsed stimuli are presented. Such experiments are best performed using a computer both to control the application of the stimulus and to acquire data directly during an interval (perhaps a few tens or hundreds of milliseconds) surrounding the time of each stimulus. The resulting recorded data then consist of "sweeps" having a precise timing relationship to the stimulus.

In other experiments the activity of channels is observed under steady-state conditions, for example, in the presence of a constant concentration of an agonist or a constant membrane potential. To obtain the maximum information from the experiment the data are best recorded continuously, for example, with an FM tape recorder, on digital audio tape, or with the combination of a PCM adapter and a videotape recorder. The decreasing costs of computer mass storage media (optical disks, digital tape drives) are making it practical to digitize the data and store it directly in the computer. This makes sense, since for analysis the data must be transferred to the computer eventually.

2.2. Filtering the Data

The filtering of the current-monitor signal from a patch-clamp amplifier is both unavoidable and necessary for practical data analysis. The design of the patch-clamp amplifier places a limit on its frequency response (typically up to 100 kHz or so), so that its output signal can be considered a filtered version of the "true" (infinite bandwidth) current signal. Some filtering is also a necessary part of the data-recording process. FM tape recorders use filters to remove the FM carrier frequencies from the output signal. For the analogue-to-digital converters of digital tape recorders and computer data-acquisition systems, the signal must be first be filtered to avoid aliasing; the DAT and PCM systems designed for audio recording typically incorporate sharp-rolloff elliptic filters for this purpose, which strongly attenuate frequency components above 20 kHz. Finally, some filtering is required anyway for data analysis in order to reduce the background noise sufficiently to allow single-channel events to be detected and characterized.

The question of the optimum degree of filtering is discussed below (Section 3.2). The events of interest are rectangular, so it is undesirable to use a filter with a very sharp rolloff, such as a Butterworth or elliptic filter, because this sort of filter distorts a step input to produce an overshoot and "ringing" appearance (although this sort of filter would be appropriate if the single-channel records are to be used for calculation of a noise spectrum). Most commonly, a Bessel filter (four poles or more) is used. On some commercial active filter instruments, this sort of filter characteristic is sometimes referred to as damped mode or low *Q*. The cutoff frequency labeled on the front panel of the active filter is sometimes the frequency at which the high- and low-frequency asymptotes of the log-attenuation versus log-frequency graph intersect. For a Bessel filter, however, the frequency at which the attenuation is −3 dB is about half of that value. This gives rise to an ambiguity in the specification of filtering that is used. It is desirable that the criterion used always be stated, and it is preferable that the cutoff frequency, f_c, always be specified as the −3 dB frequency, as we do in this chapter.

A useful theoretical model for a general-purpose filter is the Gaussian filter, which has a frequency response function $B(f)$ of the form

$$B(f) = e^{-kf^2} \tag{1}$$

where the constant k is chosen to give 3 dB of attenuation at f_c; i.e., $|B(f_c)|^2 = 1/2$, yielding $k = \ln(2)/2f_c^2$.

Some of the useful properties of the Gaussian filter arise from the fact that the Fourier transform of a Gaussian function is itself a Gaussian function. The inverse transform of equation 1 gives the filter's impulse response, which can be written in the same form as a Gaussian probability distribution:

$$h(t) = \frac{1}{(2\pi)^{1/2}\sigma_g} \exp\left(\frac{-t^2}{2\sigma_g^2}\right) \tag{2}$$

where the width of the impulse response is characterized by σ_g, which is analogous to the standard deviation of a probability distribution. Its value is inversely proportional to f_c

$$\sigma_g = \frac{(\ln 2)^{1/2}}{2\pi f_c} \tag{3}$$

Of special interest for single-channel analysis is the property that the frequency response of two Gaussian filters in cascade is itself Gaussian, with the effective cutoff frequency f_c given by

$$\frac{1}{f_c^2} = \frac{1}{f_1^2} + \frac{1}{f_2^2} \tag{4}$$

where f_1 and f_2 are the cutoff frequencies of the two filters. This property allows repeated filtering to be done on the signal with predictable results. Because Gaussian digital filters are simple to program (see Appendix 3), it is possible to refilter data even after it has been digitized and stored in the computer.

The response characteristic of a Bessel filter is well approximated by the Gaussian response, and the two actually become identical as the number of poles in the Bessel filter becomes large. Equation 4 is therefore useful for estimating the final bandwidth of an entire recording system. A typical system might consist of a patch clamp with roughly Bessel response, a DAT recorder with sharp-cutoff elliptic filters in the recording and playback paths, and a Bessel filter to reduce the bandwidth before digitization by the computer. The contribution from the patch clamp and Bessel filter can be combined as in equation 4. To a first approximation, the effect of a sharp-cutoff filter can be neglected, provided its cutoff frequency is at least twice the f_c of the rest of the system.* Thus, for example, a system with a 10-kHz Bessel filter in the patch clamp cascaded with a 5-kHz Bessel filter yields an effective bandwidth of 4.47 kHz; in this situation the presence of a DAT recorder with its sharp-cutoff 20-kHz filter would have essentially no effect on the final response.

For theoretical work, the Gaussian filter is convenient because its impulse response and step response are relatively simple functions of time; the results in Sections 3 and 4 of this chapter have been computed for a Gaussian response for this reason. Some properties of the Gaussian filter can be summarized as follows.

2.2.1. Properties of the Gaussian Filter

The frequency response function of the Gaussian filter is given by equation 1 or, numerically,

$$B(f) = \exp[-0.3466(f/f_c)^2] \tag{5}$$

The impulse response function (equation 2) can be written in terms of the cutoff frequency f_c as

$$h(t) = 3.011\, f_c \exp[-(5.336\, f_c t)^2] \tag{6}$$

The step response is

$$\begin{aligned} H(t) &= \frac{1}{2}\left[1 + \operatorname{erf}\left(\frac{t}{2^{1/2}\,\sigma_g}\right)\right] \\ &= \frac{1}{2}[1 + \operatorname{erf}(5.336\, f_c t)] \end{aligned} \tag{7}$$

In modeling the response to single-channel current pulses, it is useful to know the peak output of the filter in response to a rectangular pulse of length w and unit amplitude, which is

$$y_{\max} = \operatorname{erf}\left(\frac{w}{2^{3/2}\sigma_g}\right) = \operatorname{erf}(2.668\, f_c w) \tag{8}$$

*For Gaussian filters, each term in equation 4 is proportional to the second moment of the impulse response. Thus, the equation follows from the fact that when two functions are convolved, their second moments add. For sharp-cutoff filters, the second moment is approximately zero; indeed, for Butterworth filters, it is exactly zero.

The total noise variance of the output from the Gaussian filter when the input has a (one-sided) spectral density $S(f) = S_0 (1 + f/f_1 + f^2/f_2^2)$ is given by

$$\sigma_n^2 = \int_0^\infty |B(f)^2| S(f) df$$
$$= S_0[a_0 f_c + (a_1/f_1) f_c^2 + (a_2/f_2^2) f_c^3] \quad (9)$$

where $a_0 = 1.0645$, $a_1 = 0.7214$, and $a_2 = 0.7679$.

2.2.2. Risetime of the Filter

A particularly useful descriptive parameter for a filter is the risetime, T_r. Roughly speaking, T_r is the time for the output of a filter to make a transition when a square step is applied to the input. It therefore corresponds to the minimum length of a pulse to which the filter gives a nearly full-amplitude response. One commonly used definition for the risetime is the time between the 10% and 90% amplitude points of the transition in the output of the filter,

$$T_{10\text{–}90} = 2^{3/2}\sigma_g \text{erf}^{-1}(0.8)$$
$$= 0.3396/f_c \quad (10)$$

The definition we use here sets T_r equal to the reciprocal of the slope at the midpoint of the response $H(t)$ to a unit step input,

$$T_r = \left[\frac{dH(t)}{dt}\right]_{t=0}^{-1} \quad (11)$$

which is given by

$$T_r = (2\pi)^{1/2}\sigma_g$$
$$= 0.3321/f_c \quad (12)$$

For a Gaussian filter the two definitions of risetime give essentially identical values. T_r is inversely proportional to f_c and a 1-kHz Bessel or Gaussian filter has a risetime of about 330 μsec. It is often convenient to use T_r rather than f_c to specify the amount of filtering (e.g., one can say that "openings longer than 2 T_r were fitted").

2.3. Digitizing the Data

The data are always acquired, in the first place, in the form of a voltage (analog) signal; they are then converted to digital form for storage on digital tape (DAT or PCM/videotape), or for computer analysis, by an analog-to-digital converter (ADC). The ADC necessarily samples the voltage at discrete times; if the sample rate is too low, information about rapid voltage changes is lost. This loss of information can be described as frequency aliasing, in which high-frequency components of the original signal become converted to lower-frequency ones.

A good criterion for the choice of the sampling frequency is to require that the digitized record, when interpolated by some convenient means, is indistinguishable from the original continuous record. Sampling at the Nyquist rate (i.e., at twice the filter cutoff frequency) is a special case of this criterion, but for our purposes, the Nyquist criterion requires two unreasonable assumptions. First, it requires that the original signal contain no frequency components above a given frequency f_0 to avoid aliasing. This is unreasonable because no practical filter can accomplish this entirely, and Bessel filters are particularly bad in this respect because of their gradual rolloff characteristic. Second, the samples (digitized at the Nyquist rate of $2f_0$) must be interpolated using a very slowly decaying function of the form $\sin(xt)/xt$ in order to reconstruct the original signal properly. This sort of interpolation requires much computation and is not suitable for short records.

Interpolation is important when the original signal is sampled relatively sparsely; it allows one to reconstruct the record to any degree of smoothness for viewing while using a minimum of computer storage for the digitized data. Proper interpolation also reduces errors in certain transition-fitting procedures (see Section 4.1.2). When a cubic spline function is used to interpolate the points, a practical minimum sampling rate for Bessel-filtered data is about five times the -3 dB frequency of the filter, in which case the peak error in the reconstruction is about 2% (Fig. 1). In the cubic spline, cubic polynomials form the interpolation between every two points, with the second derivative being continuous throughout. The

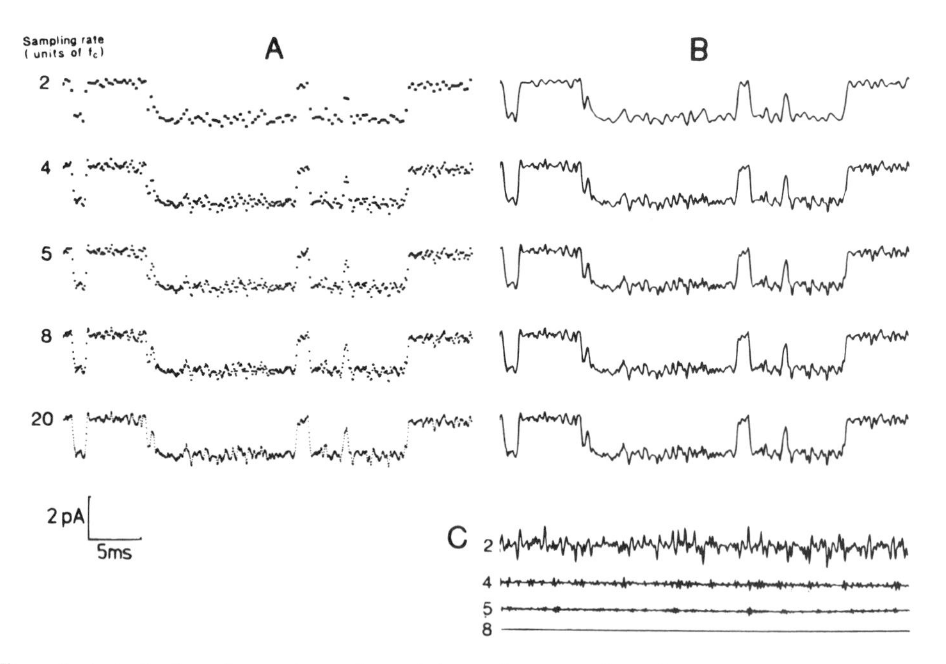

Figure 1. A single-channel current record sampled at various rates. Inward currents through ACh-receptor channels in a rat myoball were recorded cell-attached at 22°C with $V_m = -45$ mV and filtered at $f_c = 2$ kHz with a four-pole Bessel filter. A: Data points as sampled at 2, 4, 5, 8, and 20 times f_c. B: Result of cubic spline interpolation of the sampled data. C: Error traces, computed as the difference between the interpolated traces and the original data sampled at 20 f_c and scaled up by a factor of 4. The single-channel current was -1.5 pA in this recording, and the rms background noise level $\sigma_n = 0.15$ pA.

width of the interpolating function is quite narrow, so that "edge effects" (errors caused by the lack of surrounding data points) persist only about four points in from each edge of a record. A subroutine for spline interpolation is described in Appendix 3. Other interpolation techniques, including simple linear interpolation, can also be used but may require higher sample rates. If interpolation is not used, data sampled at the minimum rate appear sparse and are hard to evaluate by eye; higher rates, such as 10 to 20 times the filter's −3 dB frequency, are needed.

In general, it is best to digitize the entire experimental record. This is a good because it allows all of the data to be inspected directly and because it allows all dwell times, including the longest ones, to be measured directly. Sampling at a rate of 40 kHz (appropriate for a 2 to 4-kHz filter if interpolation is not used) generates 4.8 Mb of data per minute, assuming that data are stored as two-byte integers; thus, only a limited amount of data can be stored in computer memory. In order to digitize a long continuous record without gaps, the computer must have the ability to acquire samples into memory while simultaneously writing the data from memory to disk. This can be done by means of a separate memory buffer incorporated into the ADC system or by using direct memory access (DMA) transfer of data. For high sample rates (say 50 kHz or faster), attention must also be given to the speed at which data can be written to the storage device.

An example of a high-speed continuous acquisition program is the VCatch program for Macintosh computers. It acquires digital samples at a 94-kHz rate directly from the playback of a videotape recording using the VR-10 PCM adapter (Instrutech Corp, Mineola NY) or at sample rates up to 200 kHz using the ITC-16 ADC interface (Instrutech). In each case, the interface hardware includes an internal sample buffer (16k or 32k words of first-in/first-out buffer) that is emptied at regular intervals into a 1 Mb circular buffer in the computer's memory by an asynchronous "timer task" running on the host computer. The main program displays the incoming data and writes blocks of data from this buffer to a large-capacity hard disk. A similar facility is provided by the CED 1401-plus interface (Cambridge Electronic Design, Cambridge, U.K.) for IBM-compatible computers. It uses DMA to transfer ADC samples directly to a 64-kb circular buffer in the computer's memory, allowing analogue voltages to be digitized at rates up to 80 kHz while writing the data continuously to the hard disk. Some commercial interfaces allow continuous sampling and writing to disk only at lower rates than these, e.g., up to 30 kHz. For high-resolution data, this sampling rate may not be sufficient; however, if the original data recording is on FM tape, it is sometimes possible to slow down the tape speed while sampling the data to increase the effective sample rate.

An alternative to digitizing the entire record is to have some sort of automatic detection of the points at which opening transitions occur, and to digitize only the sections that contain openings. In this approach it is necessary that the detection method keep a record of the time intervals between openings, so that the distribution of shut periods can be constructed. This approach is satisfactory only to the extent that the detection system is reliable and the detection parameters have been properly set up before the recording starts. However, the availability of high-capacity disk drives that can store an entire recording makes this approach less attractive than it was in the past.

3. Finding Channel Events

The analysis of single-channel records first involves estimating the time and the amplitude of each transition in the current record. The list of these values is described as an idealized record that approximates the true channel activity and serves as the data set for statistical analysis of the kinetics. In practice, some of the original transitions are missed in the analysis process. To a certain extent, corrections can be made for missing events (see Section 12 of Chapter 18, this volume; Section 6.11 below), but it is important that the idealized record be as complete and unbiased as possible, especially when multistate kinetics are involved.

Finding events and fitting the transitions are considered separately in this section and the next because the two operations are often carried out separately. For example, a simple transition finder can rapidly scan a digitized record for putative channel activity. Once each event is found, it can then be fitted to an idealized time course by a much more time-consuming fitting routine, which may even require the record to be filtered differently. On the other hand, event detection and characterization can be combined in the use of a simple threshold detector, which provides a simple but useful estimator of transition times for event characterization.

3.1. Description of the Problem

The basic problem in identifying channel activity in an experimental record is that short channel openings are indistinguishable from random noise fluctuations about the baseline; similarly, short gaps are indistinguishable from fluctuations away from the open-channel current level. This is because, as a result of filtering, narrow current pulses as well as random noise fluctuations take on roughly the same time course as the recording system's impulse response. Determining whether a particular blip is a channel opening can therefore be done only statistically. In order to estimate the reliability and the limits of detection, we consider a model situation and apply some classical results from communication theory to the problem.

We assume that the channel activity to be detected consists of widely spaced rectangular current pulses of random duration but fixed amplitude A_0. The baseline level is zero. The background noise has a spectral density $S_n(f)$ and is assumed to be Gaussian distributed and independent of the channel activity. (These last two conditions appear to hold in high-quality patch recordings.) The completely unfiltered current signal $x(t)$ (if it could be observed) is represented as the sum of noiseless channel activity $s(t)$ and a noise function $n(t)$, as illustrated in Fig. 2.

The detection strategy is the following: at each time point t we form a linear combination $y(t)$ of signal values according to

$$y(t) = \int_{-\infty}^{\infty} h(t - \tau)x(\tau)d\tau \tag{13}$$

where h is a normalized weighting function that determines, in effect, the amount of time

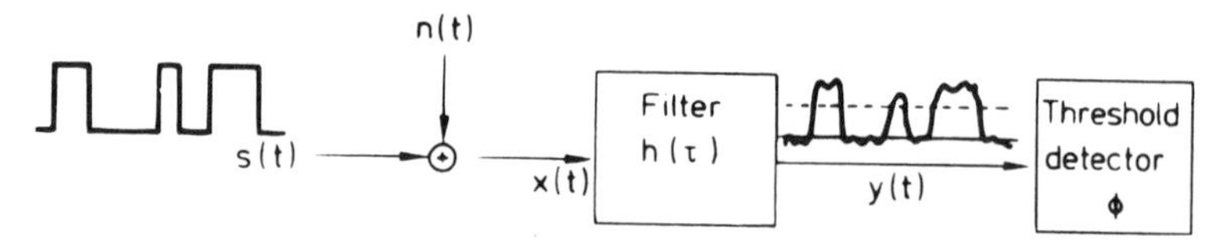

Figure 2. Model of single-channel event detection

averaging that is done in forming y. The value of y is then compared with a threshold ϕ; if $y > \phi$ at some time t, channel activity is said to be detected at t.

This detection scheme is general in the sense that it includes all possible linear signal-processing operations in the specification of the function h. It is also an optimum detection scheme in the sense that, for a signal consisting of pulses of defined shape and size, it can yield the lowest probability of error in detecting these pulses (VanTrees, 1968). We do not know, however, whether it is the optimum scheme for detecting pulses having random widths, as are actually encountered in single-channel records.

The operation described by equation 13 is a filtering operation; in fact, the function $y(t)$ is just what one obtains as the output from a filter with impulse response $h(t)$. Thus, we can represent a linear detection scheme of this kind simply as a filter followed by a threshold detector, as shown in Fig. 2. The filter in this diagram actually represents the transfer function of the entire recording system, including the characteristics of the pipette, patch-clamp amplifier, analog filter, and any computations that are performed on the digital samples. One step in event detection is often performed by a computer program in which y is computed as a weighted sum over discrete sample values rather than as an integral. This is equivalent to operating on the signal by a digital filter, which in turn is equivalent to continuous-time filtering, by the sampling theorem. Regardless of how the filtering is performed, the problem of determining the best way to detect events is reduced to finding a suitable value for the threshold ϕ and a suitable response characteristic for the filter.

3.2. Choosing the Filter Characteristics

3.2.1. Signal-to-Noise Ratio

The filter's cutoff frequency f_c and the form of the filter's frequency response characteristic can be varied to optimize the probability of detection of channel events. One strategy for doing this is to maximize the signal-to-noise ratio (SNR) for the response to a pulse of a given width, w, in the presence of noise. If SNR is defined to be the ratio of the peak amplitude y_{max} of the filtered pulse to the standard deviation of the filtered noise, it can be expressed in terms of the filter transfer function, $B(f)$, and the noise spectrum, $S_n(f)$, as

$$\mathrm{SNR} = \frac{y_{max}}{\sigma_n} = \frac{\left| \int_{-\infty}^{\infty} B(f)X(f)df \right|}{\left[\int_{-\infty}^{\infty} |B(f)|^2 S_n(f)df \right]^{1/2}} \qquad (14)$$

where $X(f)$ is the complex Fourier transform of the original pulse shape. We will see that the choice of the best filter setting depends strongly on the form of S_n. The background noise in the patch clamp should theoretically show flat spectral density at low frequencies (below about 1 kHz) and rise asymptotically as f^2 at high frequencies (see Chapter 4, this volume). In the frequency range between 1 kHz and 10 kHz, the spectral density typically is seen to rise roughly proportionally to f.

Two useful models for background noise are, therefore, the so-called "$1 + f$" spectrum, having the form

$$S_n = S_0\left(1 + \frac{|f|}{f_0}\right)$$

and "$1 + f^2$" noise,

$$S_n = S_0\left(1 + \frac{f^2}{f_0^2}\right)$$

In each case, f_0 is a characteristic "corner" frequency. In order to give numerical values for the results of calculations, we adopt a standard background noise spectrum of the $1 + f$ form with the (one-sided) spectral density $S_0 = 10^{-30}$ A^2/Hz and with $f_0 = 1$ kHz. This is a noise level that can be obtained with present-day amplifiers and pipette technology when some care is exercised.

If we assume a tunable filter with a variable cutoff frequency, f_c, of the form $B(f) = B_0(f/f_c)$, then we can calculate the dependence of σ_n on f_c by evaluating the denominator of equation 14. In the case that $S_n(f)$ is proportional to f_c^a for some exponent a, σ_n will be proportional to $f_c^{(a+1)/2}$.

In the case of a Gaussian filter response, σ_n can be computed directly from equation 9. The dependence of σ_n on f_c for various spectral types (flat, $1 + f$, and $1 + f^2$) is illustrated by the lower curves in Fig. 3.

The numerator of equation 14 is the peak value y_{max} of the filtered pulse. For a rectangular pulse of fixed width, y_{max} is small and proportional to f_c for low f_c values (heavy filtering). For a pulse of amplitude A_0 and width w, the size of the response is related to the filter risetime,

$$y_{max} \approx A_0 \frac{w}{T_r} \qquad w << T_r \tag{15}$$

As f_c is increased, T_r decreases, and y_{max} approaches the original pulse height when $w \geq T_r$. This last condition corresponds to filter bandwidths at which the rectangular shape of the original pulse can be resolved. The relation between y_{max} and f_c is shown by the upper curve in Fig. 3.

The choice of the optimum f_c for the three spectral types is indicated by the dashed lines in Fig. 3. In the case of a flat spectrum, the largest SNR value is obtained for a relatively high value of f_c because σ_n grows only as $f_c^{1/2}$, whereas y_{max} rises more quickly at low f_c values. For $S_n(f)$ rising proportionally to f, the choice of f_c is relatively uncritical, since σ_n and y_{max} rise in parallel. Finally, for $S_n(f)$ rising as f^2, f_c is best chosen to be small, since σ_n is rising relatively steeply, as $f_c^{3/2}$. Figure 3 presents an extreme case in which the pulse width w was chosen to be small (10 μs) compared with the time scale of the corner frequency f_0. As a result, the optimum f_c values differ widely. For longer pulses, the spread in optimal f_c values would be less.

3.2.2. Matched Filter

The exact form of the filter response that maximizes the SNR for a given noise spectrum and pulse shape is the so-called matched filter, which has the transfer function (see, for example, Van Trees, 1968)

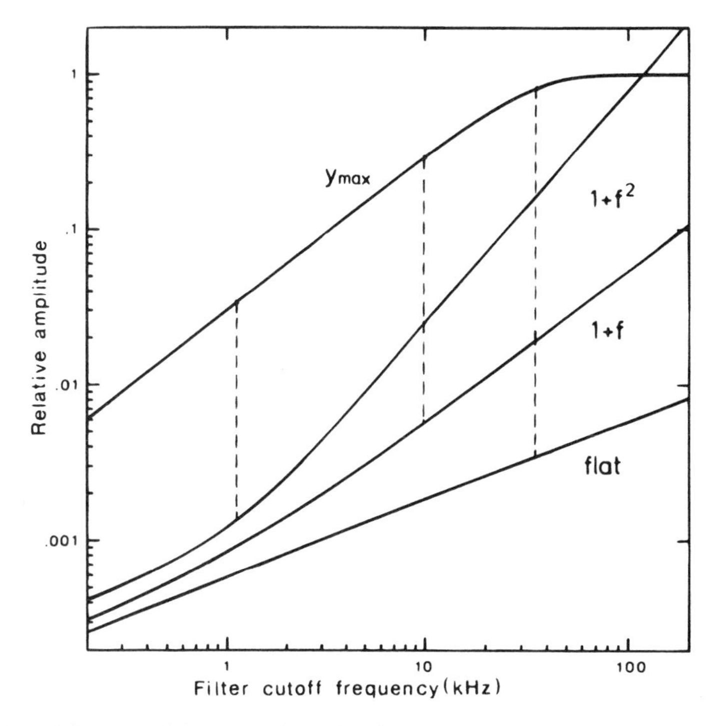

Figure 3. Effect of filter cutoff frequency f_c on signal and noise amplitudes. The upper curve shows the peak amplitude y_{max} of the response of a Gaussian filter to a 10-μs pulse of unit amplitude. Below about 40 kHz, the pulse is appreciably attenuated by the filter. The lower curves show the dependence of the rms noise amplitude σ_n on f_c assuming flat, $1 + f$, and $1 + f^2$ spectral characteristics. (The noise corner frequency was $f_o = 1$ kHz in each case, and S_o values were chosen arbitrarily.) The dashed lines indicate the points of widest separation between y_{max} and σ_n, i.e., the highest signal-to-noise ratios. The f_c values giving the best SNR were 36, 10, and 2 kHz for the three spectral types. In $1 + f^2$ noise, the optimally fitted pulse would be attenuated to only 6% of its original amplitude. The absolute value of σ_n for the "standard" noise spectrum ($S_o = 10^{-30}$ A^2/Hz) can be read directly from the $1 + f$ noise curve if the relative amplitude values are multiplied by 50 pA.

$$B(f) = c\,\frac{X^*(f)}{S_n(f)} \tag{16}$$

where X^* is the complex conjugate of X, and c is an arbitrary gain factor. [The transfer function can be multiplied by an arbitrary delay factor of the form $\exp(-j\ 2\pi\ ft_0)$, but we ignore this.] In the case of a flat noise spectrum, the matched filter's impulse response is a time-reversed copy of the matching signal—in our case, a pulse of width w; the filter is then just a running averager, averaging over a time w. If instead the noise spectrum is not flat, the matched filter has a different form.

It should be noted that the matched filter does not necessarily preserve the shape of the original pulse, since it is optimized only for the peak of the response. In the flat-spectrum case, for example, the response to the matched rectangular pulse is a triangular pulse.

3.2.3. Gaussian Filter

Although matched digital filters are not difficult to program, analog matched filters are difficult to make. Besides, one would prefer to have a general-purpose filter with only one adjustable parameter, say, the cutoff frequency, as opposed to one with the complicated adjustments implied by equation 16. As was mentioned in Section 2.2, the Gaussian filter has various appropriate properties for single-channel analysis. Surprisingly, this filter also gives SNR values nearly as large as those from a matched filter. Figures 4A and D compare SNR values for the matched filter and the Gaussian filter as a function of the pulse width w, assuming noise spectral densities of the $1 + f$ and $1 + f^2$ types, respectively. The SNR values for the Gaussian filter were never less than 0.84 times the matched-filter values and

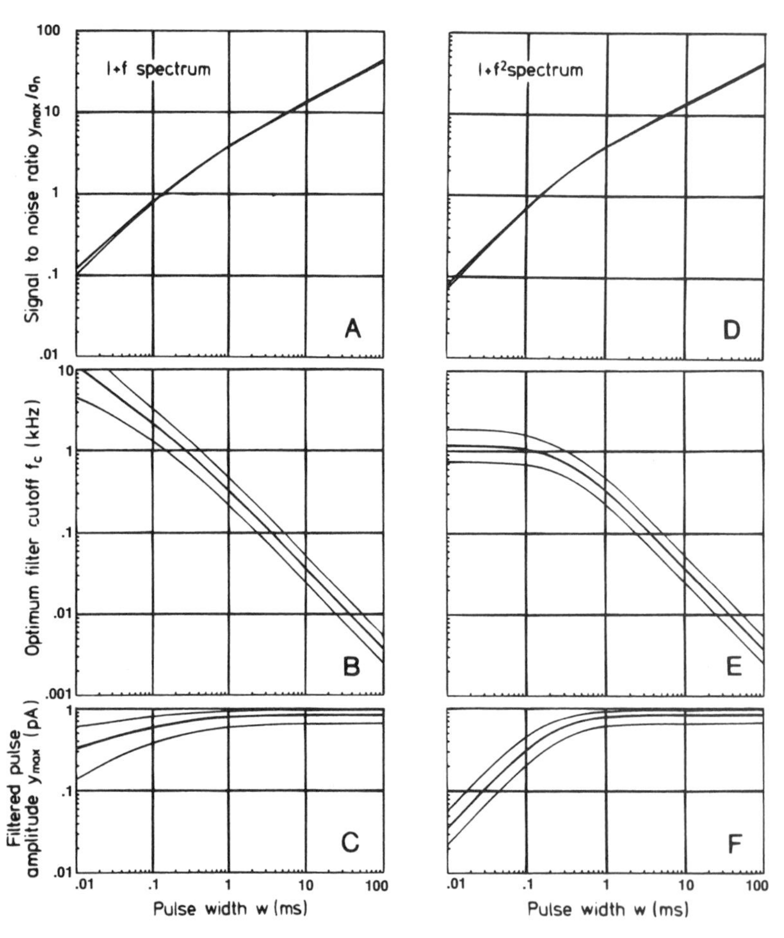

Figure 4. Filtering for optimum signal-to-noise ratios in the presence of background noise with $1 + f$ and $1 + f^2$ spectra. A and D: Ratio of peak signal, y_{max}, to rms noise, σ_n, for the matched filter (heavy curve) and the optimally tuned Gaussian filter (thin curve) as a function of the matched pulse width w. B and E: Gaussian filter cutoff frequency f_c yielding the SNR values plotted above. The choice of f_c is not extremely critical, as indicated by the thin curves, which denote the range of f_c curves giving at least 90% of the maximum SNR. C and F: The corresponding peak signal amplitudes after Gaussian filtering. The thin curves show the range of amplitudes resulting from the range of f_c values in B and E. The noise spectral densities were taken to be one-sided, $S_n = S_o(1 + f/f_o)$ and $S_n = S_o[1 + (f/f_o)^2]$, with $S_o = 10^{-30}$ A²/Hz in each case, and the pulse amplitude $A_o = 1$ pA. The SNR, f_c, and y_{max} values from these curves can be scaled for other values of S_o, f_o, and A_o by forming the ratios $S = S_o/10^{-30}$ A² s, $f = f_o/1$ kHz, and $\hat{A} = A_o/1$ pA. The resulting values SNR′, f_o', and y_o', are given by $\mathrm{SNR}' = [\hat{A}/(Sf)^{1/2}]\mathrm{SNR}(wf)$; $f_c' = ff_c(wf)$; and $y_{max}' = \hat{A}\, y_{max}(wf)$.

were usually much closer to these optimum values. It should be noted, however, that the good performance of the Gaussian filter probably results not so much from its characteristics as from the relatively noncritical nature of the exact filter. Even a simple two-pole low-pass filter can give 0.5 times the optimum SNR.

To construct the curves in Fig. 4 for the Gaussian filter, f_c was allowed to vary to maximize the SNR at each value of *w;* these optimum resulting f_c values are plotted in parts B and D of the figure. As *w* decreases, the optimum f_c increases. In the case of the steeper f^2 spectral asymptote (part E of the figure), the optimum f_c reaches a limiting value of about 1.12 times the corner frequency of the noise spectrum. This behavior can be understood with reference to the $1 + f^2$ curve in Fig. 3: an increase in f_c beyond the limiting value would cause the noise amplitude to grow more quickly than the filtered signal amplitude, even when the original signal is a very narrow pulse.

3.3. Setting the Threshold

Assuming that the filter characteristic has been chosen to be Gaussian, there remain the two parameters, f_c and the threshold ϕ, to be chosen to give the best performance of the event detector. Figure 4 shows how one picks f_c to give the best detection of pulses of a given width. Generally, one does not want to specify a particular value of *w,* however, but instead wants to detect as many events as possible, including the briefest ones. In this situation, the choices of ϕ and f_c are strongly interdependent. It turns out that the simplest procedure, at least in the case of flat and $1 + f$ noise spectra, is to first specify ϕ and then choose f_c.

The threshold needs to be set high enough to avoid counting an excessive number of noise peaks as channel events but low enough to catch as many true events as possible. In background noise having no large periodic components (e.g., containing no contamination at the power-line frequency), the false events appear to be short events occurring at random intervals, roughly like a Poisson process. The average frequency of false events, λ_f, depends strongly on the ratio of the threshold ϕ to the background noise level σ_n; it is also proportional to cutoff frequency f_c of the filter. The probability per unit time of crossing of a threshold by a Gaussian-distributed process is a standard result (see, e.g., Papoulis, 1965); λ_f is half of this rate,

$$\lambda_f = kf_c \exp\left(-\frac{\phi^2}{2\sigma_n^2}\right) \tag{17}$$

where the factor k depends on the filter response characteristic and the form of the spectrum according to

$$k^2 = \frac{1}{f_c^2} \frac{\int_{-\infty}^{\infty} f^2 |B(f)|^2 S_n(f) df}{\int_{-\infty}^{\infty} |B(f)|^2 S_n(f) df} \tag{18}$$

and is of the order of unity. Specifically, for a Gaussian filter, $k = 0.849$ when the noise spectrum is flat, whereas $k = 1.25$ for S_n proportional to f^2; practical recording situations correspond to intermediate values.

The function in equation 17 is plotted in Fig. 5 assuming $f_c = 1$ kHz. The false event rate is seen to be a very steep function of the ratio ϕ/σ_n decreasing from about 10 events/s at $\phi/\sigma_n = 3$ to 0.004 events/s at $\phi/\sigma_n = 5$. What constitutes an acceptable value of λ_f depends on the frequency of true events. For detecting relatively rare channel openings, λ_f should be at least one or two orders of magnitude smaller than the opening rate, which implies a ϕ/σ_n ratio of perhaps 5 or more. On the other hand, in the case that a burst of channel openings has been found, the problem might then be to find all channel-closing events. Since the true events in this case would be much more frequent, λ_f could be larger, and ϕ/σ_n might be chosen to be 3, for example. It is a good idea to be conservative and choose a somewhat larger value for ϕ/σ_n than that given by equation 17 or Fig. 5 to allow for possible errors in the estimation of the baseline level or small changes in the noise level, which could have a large effect on the false-event rate.

The threshold must also be chosen low enough that the desired events will be detected. One strategy for choosing ϕ would be to optimize the detection of the shortest possible events. Let w_{min} be the minimum detectable event width, and y_{max} the peak amplitude of a filtered pulse of this width. If we set $\phi = y_{max}$, approximately half of all such events will be detected, since noise fluctuations will cause some events to cross the threshold and others

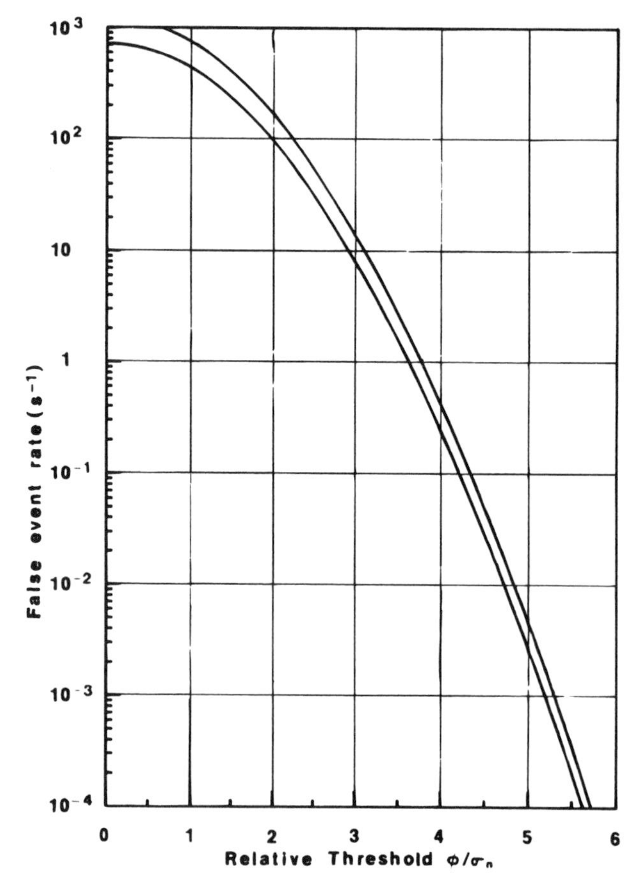

Figure 5. False-event rate, λ_f, as a function of the threshold-to-rms-noise ratio. The curves were calculated according to equation 17 with $f_c = 1$ kHz for the case of $1 + f^2$ (upper curve) and flat spectral densities of background noise. False-event rates corresponding to practical background noise spectra are expected to fall between the curves. Note that λ_f should be scaled proportionally to f_c for other f_c values.

to remain below it. To determine the value of w_{min}, we can use the signal-to-noise ratio curves of Fig. 4. The SNR in this case is just equal to the desired ϕ/σ_n ratio. Given this, the values for w_{min} and f_c can be read from the curves. Unfortunately, this procedure requires that parameters S_0 and f_0 of the noise spectrum be known in order to scale properly the results from Fig. 4.

A simpler approach is suggested by the fact that for the $1 + f$ spectrum, y_{max} varies only weakly with w (Fig. 4C), and for each w, a considerable range of y_{max} values can result in nearly maximum SNR values. Thus, one could pick ϕ equal to a reasonable y_{max} value and then tune the filter while measuring σ_n to give the desired ϕ/σ_n ratio. What is a reasonable y_{max} value? This issue is discussed in Appendix 1; in summary, a good choice of ϕ is 0.7 A_0 in the case of small-amplitude events, which will require heavy filtering ($f_c \leq f_0$), or 0.5 A_0 for larger-amplitude events for which a wider filter bandwidth will be used. This choice of $\phi = 0.5\ A_0$ is of practical interest because it allows simple event characterization as well, as described in Section 4.1.

3.4. Practical Event Detection

3.4.1. Optimal Threshold Detection

A general procedure for setting up the filter and threshold detector can now be summarized as follows: (1) given the channel amplitude A_0, pick a threshold level ϕ, e.g., in the range 0.4 to 0.7 times A_0; (2) adjust the filter's corner frequency to bring the rms noise, σ_n, to the desired fraction, e.g., one-fifth, of ϕ (3) optionally, ϕ can be readjusted slightly in view of the relationship between f_c and the frequency of the corner of the noise spectrum.

In typical patch recordings the background noise spectrum has, up to now, commonly been of the $1 + f$ form, for which the above strategies are appropriate. The final asymptote of the noise spectral density is, however, proportional to f^2, and it is likely that as techniques improve and extraneous noise sources are eliminated, the background noise in practical recordings will more nearly approach this asymptote. Once the noise density is seen to rise more steeply than linearly with frequency, a different strategy for choosing the threshold and filter frequency should be used. Recall that in this case the SNR is not improved when f_c increases beyond a critical value (Fig. 4E); therefore, it would be best in the case of large events to set the filter first to the critical frequency, about 1.2 times f_0. Then, the threshold level can be chosen to be the proper multiple of σ_n to achieve an acceptably low false-event rate.

Some convenient means for measuring σ_n is clearly required in order to set up the filter and threshold in the ways just described. A "true rms" voltmeter can be used to read σ_n directly, provided that sufficiently long event-free stretches are available for the measurement to be made. If the record is digitized, a segment can first be checked visually for the absence of obvious events. A calculation of the standard deviation of all the points in the segment then yields σ_n. A fairly long segment (or collection of segments) is needed for a precise estimate; 1000 points yields a standard deviation for σ of roughly 5%, depending on the spectral type and the relative sampling rate. For example, if the sampling rate is higher than $5f_c$, more points will be required because of the increased correlation between adjacent samples.

Throughout this section, we have assumed that the baseline level is zero. Since in experimental records the baseline current level is nonzero and typically shows a slow drift with time, any event-finding procedure needs to compensate for this. One strategy for automatic compensation is to identify event-free segments of the record and to correct the baseline estimate continuously by a small amount proportional to the difference between the latest segment and the baseline estimate. The estimate is then subtracted to give a zero-baseline record for event detection. This procedure is similar in effect to a first-order high-pass filter and is suitable for records with small drifts and moderate levels of channel activity. Automatic routines can, however, be "confused" by records with high activity (i.e., with little time spent at the baseline level) and by sudden changes in the baseline level. The most reliable technique is probably to fit the baseline, for example, by using a computer display of the data with a superimposed movable baseline cursor. In the method described in Section 4.2, the baseline position is constantly updated by means of a least-squares fit to any section of baseline that is on the screen.

Finally, it should be emphasized that the conditions described in this section for optimum *detection* of channel events are not necessarily the best conditions for *characterizing* channel events. Specifically, the best signal-to-noise ratios for event detection are sometimes obtained with relatively heavy filtering that distorts the shape of brief events. This presents no problem when the goal is to detect short, widely spaced events. However, as is shown in the next section, less filtering is desirable when one wants to discriminate the occurrence of two closely spaced short events from a single longer event or if one wants to determine the amplitude and duration of an event simultaneously.

3.4.2. Alternative Approaches to Event Detection

Sometimes it is not essential to minimize the probability of false events. In the time-course-fitting technique one intentionally places the threshold close to the baseline. Whenever this threshold is crossed, the computer displays the event that has been detected. It is then left to the operator to decide whether to fit the event or not. If an event is obviously false, there is no point in fitting it, but the decision about whether to fit or not is not critical as long as the resolution that is eventually imposed on the data (see Section 5.2) is such as to produce an acceptable false-event rate. The advantage of this approach is that it ensures that all events that are longer than the subsequently-imposed resolution are fitted.

A practical way to check the false-event rate, one that does not require careful measurement of σ_n, the baseline drift, etc., is simply to observe the frequency of detected events having the "wrong" polarity. If, for example, the true channel currents are positive going, any negative-going current pulses are most likely false events.

4. Characterizing Single-Channel Events

Since most single-channel current events appear to be rectangular steps of one or more amplitudes, the crucial step in analyzing a current record containing a single class of channel events is to determine the time of each current transition. These times can then be used for a kinetic analysis of the channel activity. The technical challenge is to characterize as many of the actual channel transitions as possible, including the briefest openings or gaps. In many cases the record can be modeled as a series of brief, widely spaced pulses having a width w that we wish to measure. Depending on the nature of the channel, these pulses could represent either openings or gaps. Special difficulties arise in the fitting process when the pulses are not widely spaced; the interpretation of histograms (see Section 6) is also complicated in this case when the channel openings and gaps are both brief and roughly equal in duration.

It is the rule, rather than the exception, for records to contain more than one open-channel amplitude. This may result from the presence of more than one channel type in the patch and/or from the presence of one sort of channel that can open to more than one level. The question of how constant these levels are is discussed in Section 5.3.1, but regardless of this, the existence of multiple levels causes considerable problems, especially for the more automated methods of analysis. Sometimes amplitude estimates are just averaged together to give the "mean single-channel current" and although this is sometimes a reasonable procedure, it more usually is not. In practice, estimating the amplitude of long events is straightforward, but for short events, the estimation of the amplitude not only is unreliable but also increases the uncertainty in the transition time estimates. The usual practice, therefore, is to fit only the duration of brief events, with the amplitude constrained to some average value.

The methods of channel characterization we consider here are simple ones in which an attempt is made to detect channel-opening and closing events with a minimum of ambiguity. We recommend these methods because the bias and statistical errors in the characterization are relatively well known and because the detection of each event can be readily verified by the user. More sophisticated transition-detection schemes have been applied to single-channel data, including the Hinckley detector and T-test methods (see Chapter 3, this volume); these methods are not much better at characterizing simple isolated channel events but show promise in allowing better characterization of rapid bursts of events and subconductance levels. Still other methods exist that do not rely on the detection of individual events at all but obtain indirect information about dwell times and amplitudes from the statistics of the entire record. Examples of these are power spectra and all-points histograms computed from single-channel records. These provide less information than a full evaluation of closed and open times but can be used to fit simple models and thus estimate dwell times and amplitudes of rapidly switching channels in cases where these parameters cannot otherwise be obtained (see Section 5.3.2; Chapter 3, this volume). A more general technique is the application of "hidden Markov model" signal-processing algorithms (Chung *et al.,* 1990), which allow a complete model of the channel activity to be evaluated from all of the information in the recording. This technique allows the extraction of useful kinetic information from records having a signal-to-noise ratio several times lower than that required for the simple methods discussed here. However, it has not yet been applied widely to real data, or tested directly against alternative methods of analysis; its usefulness as a routine method is, therefore, not yet known.

4.1. Half-Amplitude Threshold Analysis

4.1.1. The Technique

The use of a simple threshold detector is the most widely used method of single-channel analysis and is readily applied to channels having only one nonzero conductance level. An estimate of the channel amplitude A_0 is used to set a threshold level, assumed here to be $A_0/2$. Every crossing of the threshold is interpreted as an opening or closing of the channel, so that the time spent above the threshold, w_t, is taken as an estimate of the channel-open time. As was pointed out by Sachs *et al.* (1982), choosing the threshold to be $A_0/2$ is convenient because w_t is then an unbiased estimate of the true pulse width w_0 for long pulses of either polarity and can therefore be used to estimate both open and closed times. However, for short events with w_0 of the order of the filter risetime T_r, w_t underestimates w_0 (see Fig. 6). Events shorter than a dead time of about $T_r/2$ are missed altogether, because, after filtering they never reach the threshold.

The exact value of the dead time T_d of this detection technique can be either measured experimentally or calculated by finding the pulse width that gives a half-amplitude response from the recording system. If, for example, an analogue filter is used and has its bandwidth set far below that of the other parts of the recording system, it suffices to observe its output while variable-width pulses are applied to the input by a stimulator. In the case of a Gaussian filter, T_d is found (see equation 8) according to

$$\operatorname{erf}(T_d/2^{3/2}\sigma_g) = \frac{1}{2} \tag{19}$$

which yields $T_d = 0.538\ T_r$ or, equivalently, $T_d = 0.179/f_c$. If, for example, a sample rate of $10f_c$ is used (see Section 2.3), T_d is 1.79 sample intervals. Alternatively, a dead time can be imposed retrospectively, as described in Section 5.2 (as long as all events longer than the chosen value have been measured). This method ensures a consistent dead time throughout.

If not only the dead time but also the complete relationship between w_t and w_0 is known, then the distorting effect of the threshold-crossing analysis can be estimated. In terms of the filter step response $H(t)$, which is assumed for simplicity to be symmetrical about $t = 0$, the

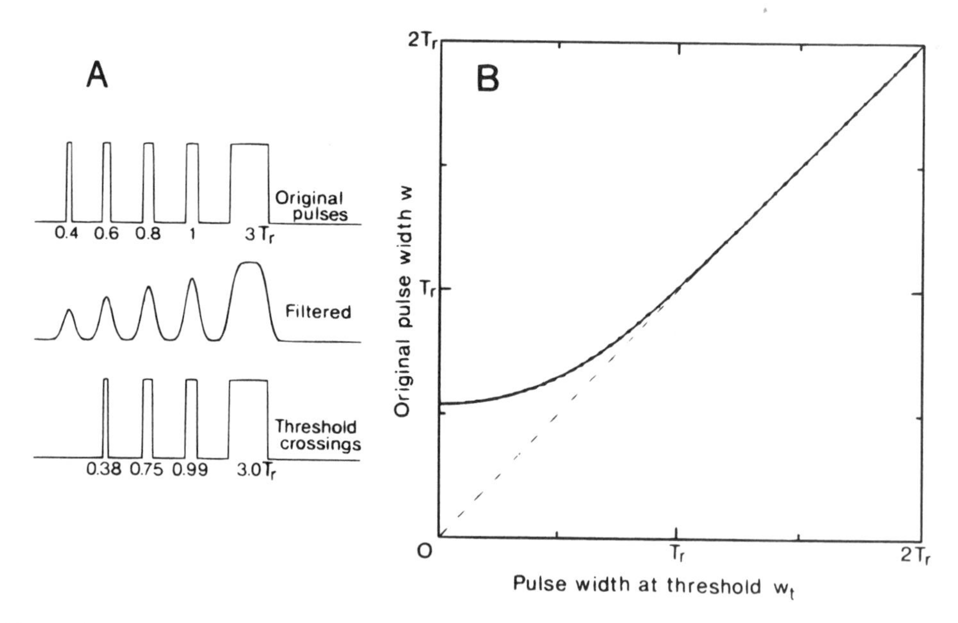

Figure 6. Relationship between true pulses with width w_o and the width w_t at the 50% threshold for Gaussian-filtered pulses. A: Simulated pulses with lengths given in units of T_r. The shortest pulse fails to reach threshold, and the pulses of intermediate width result in low values of the threshold-crossing width, w_t. B: The relationship between w_t and true pulse width in the absence of noise. For w_t equal to T_r or longer, w and w_t are essentially equal (dashed line). The points (barely visible under the curve) are values of the approximation function (equation 21).

relationship is given implicitly by

$$H\left(\frac{w_t + w_0}{2}\right) + H\left(\frac{w_t - w_0}{2}\right) = \frac{1}{2} \tag{20}$$

which must be evaluated numerically. Figure 6B shows this relationship for Gaussian-filtered pulses. A convenient approximation to the relationship, having relative errors less than 10^{-3}, is given by the function

$$\begin{aligned} w_0 &= g(w_t) \\ &= w_t + a_1 \exp(-w_t/a_1 - a_2 w_t^2 - a_3 w_t^3), \qquad w_t > 0 \end{aligned} \tag{21}$$

with $a_1 = 0.5382\, T_r$, $a_2 = 0.837\, T_r^{-2}$, and $a_3 = 1.120\, T_r^{-3}$. These coefficients are alternatively given in terms of the filter cutoff frequency as $a_1 = 0.1787\, /f_c$, $a_2 = 7.58\, f_c^2$, and $a_3 = 30.58\, f_c^3$. The function g can be used directly to convert the observed w_t values to effective w_0 values. Alternatively, the function can be used to predict the probability density function (pdf) of threshold-crossing intervals, $f_t(w_t)$, from the pdf of true durations $f(w_0)$ according to

$$f_t(w_t) = f[g(w_t)]g'(w_t) \tag{22}$$

Thus, in the absence of effects from noise, the distortions of this simple analysis scheme can be compensated by the fitting of a modified distribution to the resulting duration estimates.

4.1.2. Effect of Noise

Noise can be thought of as an instantaneous variation of the threshold level. For relatively long events, the Gaussian-distributed threshold fluctuations cause an approximately Gaussian-distributed random error in the determination of each threshold-crossing time. The standard deviation in the apparent corrected width, w, is approximately

$$\sigma_w = 2^{1/2} \frac{\sigma_n}{A_0} T_r \tag{23}$$

If the duration, w, of short events is corrected, for example, according to equation 21, the error in these estimates for w near T_d is

$$\sigma_w = \frac{\sigma_n}{A_0} T_r \tag{24}$$

and is also approximately Gaussian distributed.

The threshold-crossing technique automatically excludes events with (apparent) w_0 values less than T_d, but because of the effect of noise, some events with true w_0 values less than T_d will be counted, and some larger events will be missed. The general effect of noise is, therefore, a broadening and distortion of the distribution of apparent event durations as a result of randomness in the estimates of w_0. This broadening is most serious when the underlying distribution $f(w_0)$ is rapidly varying. Figure 7 compares theoretical distributions of w_0 with distributions calculated on the basis of threshold-crossing analysis in the presence of a fairly high noise level (with $\phi/\sigma_n = 4$). When the time constant of the underlying w_0 distribution is less than T_r, an exponential fit to the observed, corrected distribution would yield a time constant that is too large (compare curves 1 and 3 in Fig. 7A). A similar effect of noise is to be expected on duration estimates obtained by the time-course-fitting technique, as errors in the estimates cause a "smearing out" of rapidly varying portions of the distribution.

For the best performance of the threshold-crossing analysis, it is generally best to decrease T_r as much as possible (that is, increase the filter cutoff f_c) to reduce the number of missed events. The same consideration applies here as in Section 3.4 above, however, about choosing a sufficiently large ϕ/σ_n ratio to give an acceptable false-event rate. When the threshold-analysis technique is implemented on a computer, an additional problem arises from the nature of digitized records. Because of the finite sample interval, it is possible for a set of digitized current values to lie below a certain threshold even when the original current trace crosses the threshold. This introduces an additional, biased error in the estimates of event durations. It is a good idea to use interpolation in order to minimize this effect, especially when the sample interval is relatively long.

The performance of the threshold-analysis technique has been considered so far only in the case of widely spaced events. A problem with the technique is that it responds poorly to short events that come very closely spaced in time. For example, a brief pulse can be counted as a longer one when it occurs in the vicinity of a second pulse (Fig. 8). This effect becomes significant when both the pulse length and the gap between pulses are roughly T_d or smaller. The systematic errors that are introduced by this failure have not been characterized, but they are probably not serious when either the mean open time or the mean gap time is at least several times T_r.

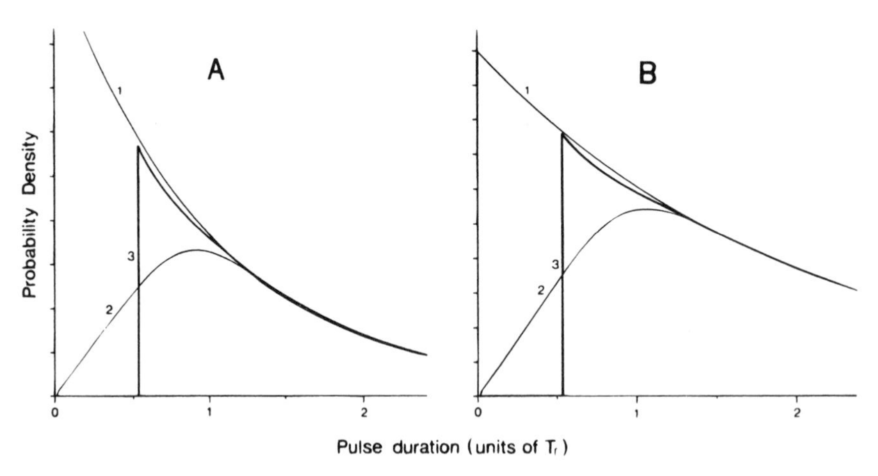

Figure 7. Distortion of pulse duration distributions by the threshold-crossing analysis in the presence of noise. The original distributions of durations w are shown by curve 1 in each part of the figure. The time constants of the distributions were T_r in A and $2T_r$ in B. The distribution of threshold-crossing times, w_t, is shown as curve 2 in each part, and the corrected distribution $g(w_t)$ is shown as curve 3. In the absence of noise, curve 3 would superimpose on the original distribution. A flat background noise spectrum with $\sigma_n = \phi/4$ was assumed.

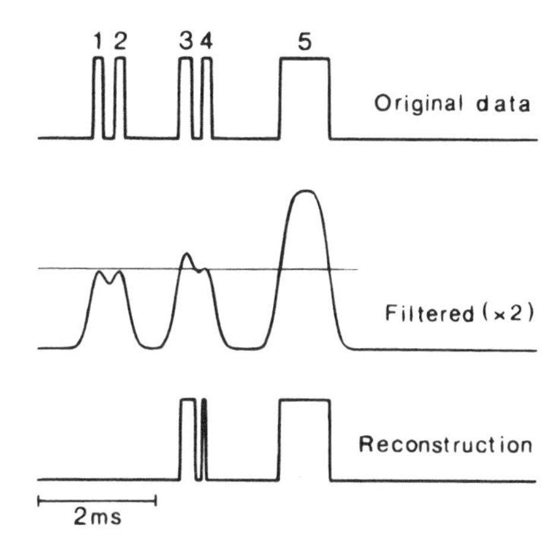

Figure 8. Threshold analysis of closely spaced events. Simulated rectangular events (top trace) were filtered with a 1-kHz Gaussian filter (T_r = 330 μs) and displayed (middle trace) with twice the vertical scaling. The reconstruction obtained from the threshold-crossing analysis is shown in the lower trace. Events 1 and 2 had lengths of $0.5T_r$ and were just below threshold. Event 4 was even shorter ($0.47T_r$) but was detected because it followed event 3 ($0.67T_r$) by a shut interval of only $0.5T_r$. Event 5 had a length of $2.5T_r$.

4.1.3. Estimating the Amplitude

The threshold-crossing technique assumes that the event amplitude is known *a priori,* so that the threshold can be set correctly. In practice, this presents little problem in interactive (as opposed to entirely automatic) fitting programs, since the operator can usually find sufficiently "square" events to provide an initial estimate for the amplitude. An estimate of the amplitude of an individual event, provided it is long enough, can be made by averaging the amplitude of the trace between threshold crossings, excluding the points within a given distance (e.g., 0.7 T_r) of the threshold-crossing points. Because of this exclusion, only events longer than about $2T_r$ can be used for determining the amplitude. As will be shown below, the time-course-fitting technique can give amplitude estimates for events shorter than this, but only at the expense of increased error in the duration estimates. This method suffers from the problem that the amplitude estimates so found will be too low if the region of the trace that is averaged contains brief shuttings that have not been detected because they did not cross the threshold level. If such brief shuttings are at all common (which is often the case), then it is necessary to inspect each amplitude fit to make sure that such bias has not occurred.

4.2. Direct Fitting of the Current Time Course

4.2.1. The Technique

A theoretical time course of the current can be computed on the basis of the step response of the recording system and fitted to the actual record. The step response can be measured by injecting a square-wave signal into the input of the patch-clamp amplifier, for example using a built-in integrator (see Chapter 4, this volume), or by coupling the triangle-wave output of a function generator into the headstage input through a small capacitance (e.g., by simply holding a wire near the headstage). A high-quality triangular wave is needed for this job. The resulting output signal, filtered and digitized in the same way as the data to be analyzed, is stored in a computer file for subsequent use. Usually, a suitable trigger pulse is also recorded, so that several sweeps can be averaged to obtain a smooth output curve. Such a curve is illustrated in Fig. 9A; it is scaled so that it covers the range from 0 to 1.

Once the output of the apparatus to a step is known, it is easy to calculate the output expected for a series of steps such as a channel opening and shutting. The process is illustrated in Fig. 9 for single-channel openings of two different durations, t_0. The response to the opening transition is simply the step response function, which has already been stored. The response to the shutting transition is exactly the same but inverted and displaced to the right by t_0 seconds. If these two curves are added, we obtain the expected output to a rectangular input, as illustrated for two examples in Figs. 9C and F.

This calculated output can be used to fit actual data as follows. The data are displayed on the screen, on which is superimposed the calculated response (output) to a rectangular input, which has been scaled by multiplying it by the amplitude of the opening. The amplitude cannot be measured from the event itself if it is very brief, so the amplitude must then be taken as the mean amplitude of all previous openings that have been fitted or as the amplitude of the last opening fitted. The times of the two transitions are then adjusted until the calculated output superimposes, as well as possible, on the data, as illustrated in Fig. 12. The adjustment of the amplitudes and transition times can be done manually or by means of a least-squares fit, as described below.

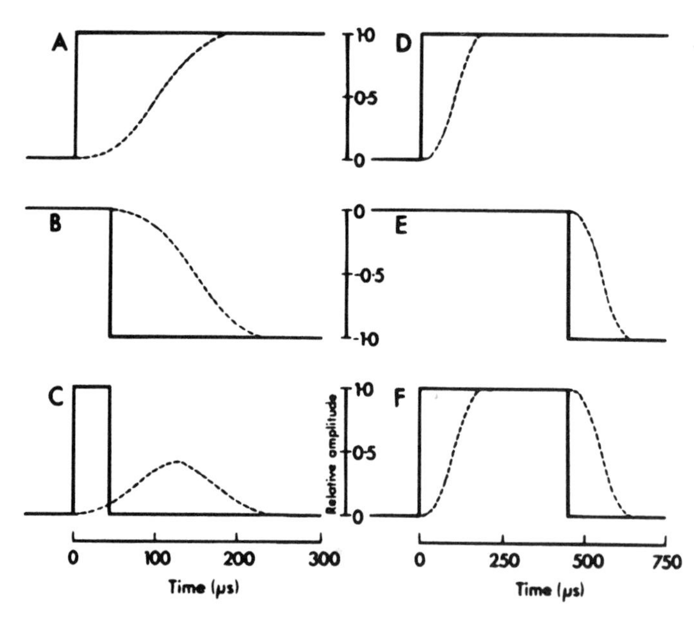

Figure 9. Illustration of the method of calculation of the expected response of the system from the measured response to a step input. The left-hand column illustrates a short (45-μs) pulse, and the right-hand column a longer (450-μs) pulse. The dashed lines in A and D show (on different time scales) the experimentally measured response to a step input, shown schematically as a continuous line, for a system (patch clamp, tape recorder, and filter) for which the final filter (eight-pole Bessel) was set at 3 kHz (−3 dB). A: The response to a unit step at time zero is shown. B shows the same signal but shifted 45 μs to the right and inverted. The sum of the continuous lines in A and B gives the 45 μs unit pulse shown as a continuous line in C. The sum of the dashed lines in A and B is shown as a dashed line in C and is the predicted response of the apparatus to the 45-μs pulse. It reaches about 41% of the maximum amplitude, which is very close to the value of 39% expected for a Gaussian filter (see equation 8). D, E, and F show, except for the time scale, the same as A, B, and C but for a 450-μs pulse, which achieves full amplitude.

4.2.2. Theory

The formal justification of the procedure illustrated in Fig. 9 is as follows. The step input at $t = 0$ is denoted $u(t)$, which is zero for $t < 0$ and unity for $t > 0$. A rectangular pulse input extending from time 0 to time w is therefore

$$s(t) = u(t) - u(t - w) \tag{25}$$

The output expected for this input can then be found (as long as the system behaves linearly) by convolving this input with the impulse response function of the system $h(t)$; i.e.,

$$y(t) = \int_0^t [u(t) - u(t - w)]h(t - \tau)d\tau \tag{26}$$

The system's response to a unit step input $u(t)$ is defined to be the system step response $H(t)$, which is the integral of h. Expressed in terms of $H(t)$, equation 26 simplifies to

$$y(t) = H(t) - H(t - w) \tag{27}$$

This is the calculation illustrated in Fig. 9. When the form of the input is inferred by superimposing this calculated response on the experimental data, we are performing a sort of graphic deconvolution.

This process can be extended to any number of transitions. In Fig. 10, some of the outputs that can result from four transitions (two rectangular pulses) are illustrated. If the transitions are well separated, the output, of course, simply looks like two somewhat rounded rectangular pulses (Fig. 10A). If the middle two transitions are close together, we have an opening with an incompletely resolved short gap (Fig. 10B). If the first three transitions are close together, the response looks like a single opening with an erratic rising phase (Fig. 10C). And if all four transitions are close together, the response looks like a (rather noisy) opening of less than full amplitude (Fig. 10D). If the channel were initially open in Fig. 10D, the response might be mistaken for an incomplete shutting to a conductance sublevel.

Before we go on to discuss the practical aspects of time course fitting, it is appropriate first to discuss the problems that may arise in attempting to fit both duration and amplitude simultaneously.

4.2.3. Simultaneous Determination of Amplitude and Duration

In theory, both the times and amplitudes of transitions in the theoretical trace could be varied to provide a best fit to the time course of the experimental record. The practical difficulty is that for pulse widths, w, shorter than the recording system risetime, T_r, the shape of the observed current pulse is relatively insensitive to w. In Fig. 11A, we compare the time courses of Gaussian-filtered pulses that have widths that differ by a factor of two but equal areas. Even in the absence of noise, the time courses are nearly indistinguishable for w less than about $T_r/2$.

To obtain a quantitative estimate of the errors to be expected in fitting the amplitude and duration simultaneously, the performance of a least-squares fitting routine for fitting the time course was evaluated. Figure 11B shows the behavior of the expected standard deviations,

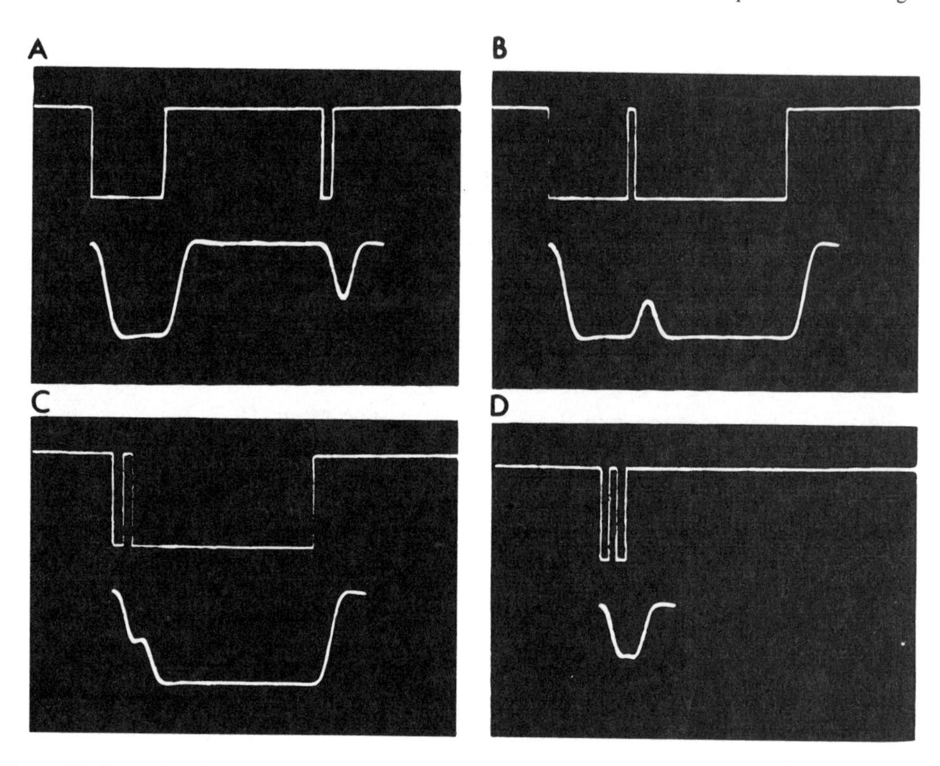

Figure 10. Examples of the calculated output of the apparatus (lower traces) in response to two openings of an ion channel (upper traces). The step response function used to generate the response is that specified in Fig. 9. The curves are generated by a computer subroutine and were photographed on a monitor oscilloscope driven by the digital-to-analogue output of the computer. Openings are shown as downward deflections. A: A fully resolved opening (435 μs) and gap (972 μs) followed by a partially resolved opening (67 μs). B: Two long openings (485 and 937 μs) separated by a partially resolved gap (45.5 μs). C: A brief opening (60.7 μs) and gap (53.1 μs) followed by a long opening (1113 μs); this gives the appearance of a single opening with an erratic opening transition. D: Two short openings (both 58.2 μs) separated by a short gap (48.1 μs); this generates the appearance of a single opening that is only 55% of the real amplitude but appears to have a more-or-less flat top, so it could easily be mistaken for a fully resolved subconductance level.

σ_A and σ_w, for the estimates of the amplitude and width, respectively, that are found using a linearized fitting process. Because the errors are proportional to the background noise standard deviation, σ_n, the values plotted in the figure are normalized with respect to σ_n; i.e., they are σ_A/σ_n and $\sigma_w A_0/\sigma_n T_r$. The behavior of the errors as a function of the original pulse width, w, depends on the form of the background noise spectrum; the two extreme cases of a flat spectrum and an f^2 spectrum are shown.

For long pulses, the error in the estimation of w is constant and is approximately 1.8 and 1.3 times $T_r\sigma_n/A_0$ for the flat and f^2 spectra, respectively. In a typical situation, $A_0/\sigma_n = 10$, which yields σ_w values in the range of 10–20% of T_r. The fact that σ_w is constant at large w can be understood from the way the duration of a long pulse is measured, as the interval between two transitions. If the transitions are far enough apart, the errors caused by noise in the determination of the transition times will be uncorrelated and independent of the time between them. On the other hand, amplitude estimates become more precise for

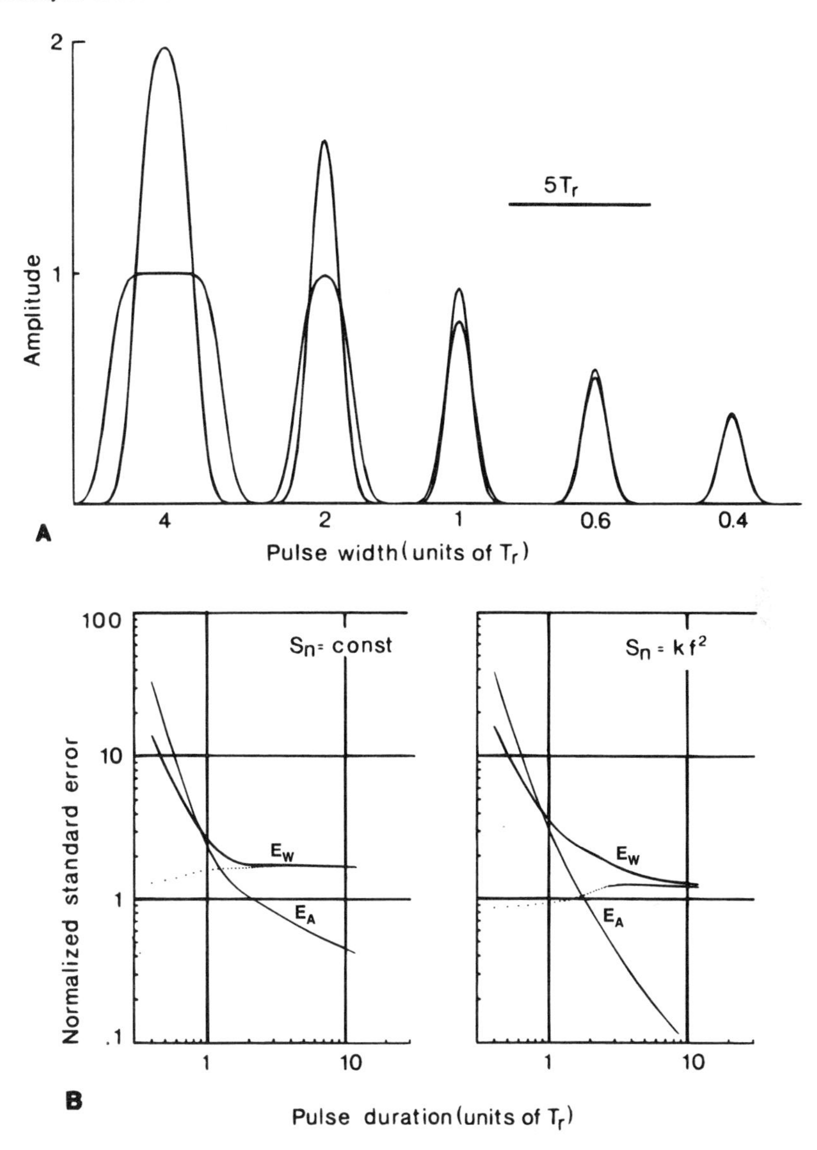

Figure 11. Errors in simultaneous fitting of amplitude and duration. A: Gaussian-filtered pulses of the widths indicated are superimposed with pulses having half the width but twice the amplitude. As the widths become shorter, the time courses become indistinguishable. B: Standard deviations of the estimates of amplitude and duration of Gaussian-filtered pulses in the presence of noise having either a flat or $1 + f^2$ spectrum. Pulse durations are given in units of the filter risetime T_r. The expected errors ϵ_A and ϵ_w are normalized to the background noise σ_n and other parameters according to $\epsilon_A = \sigma_A/\sigma_n$ and $\epsilon_w = \sigma_w A_o/\sigma_n T_r$. The dotted curve gives ϵ_w when the amplitude estimate is constrained to the correct value A_o.

longer pulses, with the error decreasing as $w^{-1/2}$ in the case of a flat background spectrum and large w.

As the pulse width becomes comparable to T_r or shorter, the errors of estimates of both amplitude and width increase sharply, becoming double their asymptotic values at about 0.8 T_r in the flat-spectrum case. This sharp rise does not occur if the amplitude is constrained and the duration alone is fitted, as illustrated by the dotted curves in Fig. 11B. This rise reflects the difficulty of simultaneous fitting. Because it occurs in the vicinity of T_r, it can be seen that a small T_r, i.e., the largest possible filter bandwidth, is best for simultaneous fitting. Of course, in practice the filter bandwidth must be chosen low enough to avoid false events.

If the duration alone is fitted, with the amplitude held fixed, the error in the duration estimate depends only weakly on w and, in fact, decreases slightly as w becomes small, as shown by the dotted curves in Fig. 11B. The absolute size of the error is much smaller, and the criterion for choosing f_c to minimize the error (which is essentially proportional to $T_r\sigma_n$) is similar to that for event detection.

In conclusion, it is possible to obtain some amplitude information from events shorter than the recording system risetime T_r. In practice, this information is difficult to obtain because it is based on fine details of the pulse shape, but it could conceivably be useful for statistically testing hypotheses such as the existence of multiple channel populations. Much more precise estimates for the duration of short channel events can be obtained by fixing the amplitude in the fitting process. For longer events with $w \geq 2\ T_r$ the concurrent estimation of the amplitude has only a small effect on the error of the duration estimates. Estimating the amplitude of these events would then be worthwhile provided that the size of the error in the amplitude estimates (approximately equal in magnitude to σ_n at $w = 2\ T_r$) is acceptable.

4.2.4. Time-Course Fitting in Practice

Personal computers are now fast enough that it has become feasible to fit simultaneously both the duration and amplitude of single-channel openings. With the program SCAN, which is under development at University College London, the fit does not take any noticeable length of time for fitting up to four transitions and is still quite acceptable for fitting say ten or more transitions when run on an 80486 or Pentium processor machine. The data trace is scrolled across the screen until an event is detected, as described in Section 3.4.2. The trace is then expanded, contracted, or shifted as necessary to get a suitable section of data for fitting on the screen. The program then makes initial guesses for the positions of all the transitions and amplitudes, performs a least-squares fit on the basis of these guesses, and displays the fitted curve for acceptance, rejection, or modification by the operator. Three examples of fits done in this way are shown in Fig. 12.

When the channel is shut at each end of the fitted region, as in Fig. 12A and B, fitting n transitions involves estimation of $2n + 1$ parameters (the time at which each transition occurs, the amplitude following each transition, and the amplitude before the first transition). The fit of the amplitude after the last transition is taken as a new estimate of the current baseline position. The program "knows" that the channel is shut at this point, so the next transition must be an opening; the average of a section of trace before the next opening can therefore be taken as a temporary baseline estimate, even if drift has occurred, thereby allowing reasonable initial guesses to be made for the next fitting. In this way it is possible to keep track of the baseline level. Other options in the program allow fits to be done with only one open level for all openings or to be done by specifying the amplitudes in advance and fitting only the transition times (as was always done with earlier programs).

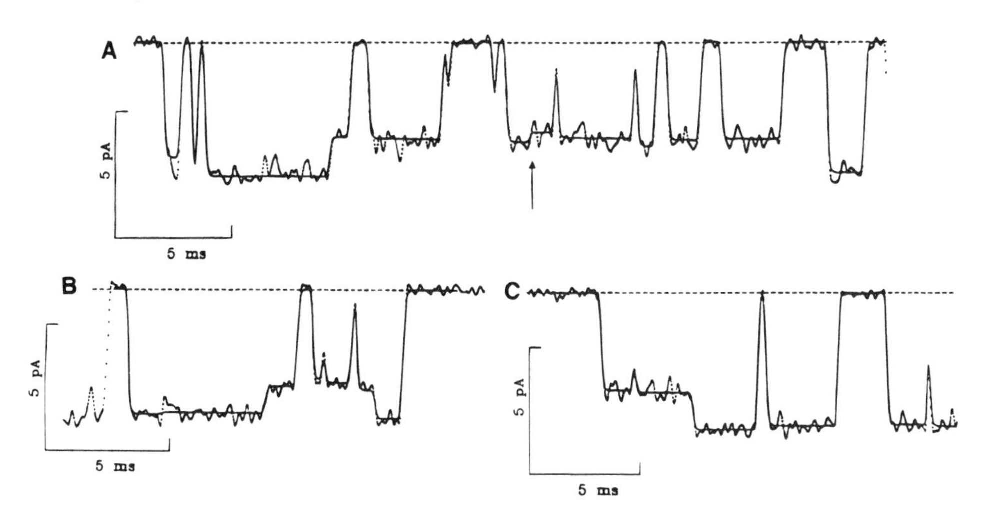

Figure 12. Three examples of fitting with the SCAN program. The record is from an NMDA-type glutamate receptor in a dentate gyrus granule cell (outside-out patch at −100 mV, glutamate 200 nM + glycine 1 μM with 1 mM Ca and no added Mg, eight-pole Bessel filter at 2 kHz, −3 dB, risetime 166 μs; methods as in Gibb and Colquhoun, 1991; data of A. J. Gibb). The dashed line shows the baseline (shut) level. The record was sampled at 50 kHz (though 20 kHz would have been sufficient and more usual). The transition times, and amplitudes (for events that were longer than two risetimes), were fitted simultaneously by least squares. Shut periods shorter than two risetimes had their amplitudes fixed to zero. Open periods shorter than two risetimes had their amplitude constrained to be the same as that of the closest opening that was longer than three risetimes. The fitted curve is the continuous line. A: Two contiguous fittings. The durations and amplitudes in this fit, starting from the first opening, are as follows: 0.707 ms, −4.48 pA; 0.491 ms, 0 pA; 0.248 ms, −5.22 pA; 0.321 ms, 0 pA; 5.33 ms, −5.24 pA; 0.894 ms, −3.69 pA; 0.802 ms, 0 pA; 3.08 ms, −3.75 pA; 0.216 ms, 0 pA; 0.074 ms, −3.73 pA; 1.83 ms, 0 pA; 0.092 ms, −3.91 pA; 0.448 ms, 0 pA; 1.02 ms, −3.91 pA; 1.07 ms, −3.52 pA; 0.131 ms, 0 pA; 3.17 ms, −3.74 pA; 0.156 ms, 0 pA; 0.756 ms, −4.04 pA; 0.511 ms, 0 pA; 1.39 ms, −3.80 pA; 0.865 ms, 0 pA; 2.47 ms, −3.75 pA; 1.92 ms, 0 pA; 1.59 ms, −5.06 pA. Note that the transition from −3.91 pA to −3.52 pA (marked with arrow) is dubious, and this would probably be removed later, at the stage when the resolution is imposed on the data (see text, Section 5.2), when adjacent openings that differ in amplitude by less than some specified amount are concatenated into a single opening (with the average amplitude). B and C: Two more examples. In B there is a very small transition (from −4.98 to −4.90 pA) shortly after the first opening transition; this was triggered by the wobble in the data at this point but would certainly be removed before analysis (see A).

The fitting of amplitudes in this way will be biased if the regions of trace that are fitted contain brief shuttings, as discussed *a propos* threshold-crossing analysis in Section 4.1.3. This problem can be minimized by allowing the program to fit very brief events, even though most of them will be rejected later, when a realistic resolution is imposed (see Section 5.2).

It is, as discussed in Section 4.2.3, not feasible to fit both amplitude and duration to very short openings or shuttings. Shut periods shorter than a specified length (usually two risetimes) have their amplitudes fixed to zero. Open periods shorter than a specified length (also usually two risetimes) have their amplitude constrained to be the same as that of the closest opening that is longer than, say, three risetimes, if such an opening is present in the region of trace being fitted. Otherwise, the amplitude of short openings is fixed at the current mean full amplitude (or some other specified value).

Once a satisfactory fit has been obtained, the data points in the fitted region can be entered into an all-points histogram. Also, those data points that are in regions where the fitted curve is flat can be entered separately into shut-point and open-point histograms, which exclude points that are in the region of transition from one level to another (see Section 5.3.2). This procedure means that these three sorts of histogram can be viewed at any time during the fitting process.

4.2.5. Advantages and Disadvantages of Time-Course Fitting

There are two major advantages in using the time-course-fitting method. The first is that it is the only well-tested method for dealing with records that contain multiple conductances or subconductance states. The second is that the resolution of measurements can be somewhat greater than can be obtained with the threshold-crossing method.

It is quite likely that, during time-course fitting, some of the events fitted will not be real openings or shuttings of the ion channel but merely random noise or small artifacts. This is not really a disadvantage of the method (except insofar as it takes time), because such events should be eliminated at a later stage, when a realistic resolution is imposed on the idealized record (Section 5.2). In fact, it is actually an advantage, because it minimizes the bias in amplitude estimates that result from the presence of brief events that may be detectable but would not normally be fitted.

There will, from time to time, be events on the screen that are ambiguous. It may be impossible to tell whether an event is a genuine channel opening at all, or whether it is some form of interference. And even if the event is "obviously" an opening, it may be impossible to be sure whether it is an opening to a subconductance level or whether it is two or more brief full openings separated by short gap(s) (as illustrated in Figs. 8 and 10). Such events will necessitate a subjective decision by the operator about the most likely interpretation of the data. Magleby (1992) has criticized the method because of the "operator bias" that is introduced into the analysis in this way. However, exactly the same sort of operator bias will occur in any form of threshold-crossing analysis in which the operator inspects and approves or disapproves what the program has done. As mentioned above, it is highly desirable that the operator should know what the program has done. It is equally very desirable that the operator should be aware that the data contain ambiguous events, even if he/she is not sure what to do with them. The only case in which the argument about operator bias seems to be valid is when data are analyzed automatically by the "total simulation" method proposed by Magleby and Weiss (1990). In this case, it is necessary that a completely automatic method of analysis be used because of the immense amount of computation that is involved, and it is necessary that the simulated and experimental records be analyzed by identical methods (including the ambiguous bits). In all other cases, there is little to be gained by sweeping the ambiguities under the carpet.

The question of ambiguous events has been discussed at some length. However, it is probably true, at least for channels that have a reasonably good signal-to-noise ratio, that such events are sufficiently rare that the conclusions from the analysis are unlikely to be much altered by the subjective decisions that must occasionally be made.

4.3. Event Characterization Using a Computer

4.3.1. Data Display

The single most important feature of a computer system for analyzing single-channel data is a responsive and flexible means of displaying the digitized data. Before and during the quantitative event characterization, it is essential that the user be able to examine the recording, millisecond by millisecond if necessary, to be able to judge the quality of the data. Visual inspection can show features that could be missed or misinterpreted by automatic analysis programs, such as the presence of artifacts or superimposed channel events, systematic changes or "rundown" of the channel activity, and subconductance levels.

An example of a suitable display for long, continuous data recordings is that of the DataSelector program shown in Fig. 13. Here the data are shown at three different time resolutions, providing an overview of the entire multimegabyte file (top trace) while also allowing inspection of a selected region at high resolution. One important feature of the program is the ability of the user to select the position and degree of magnification of the data in each trace. As the box in a trace is dragged or resized using the computer's mouse, the trace below it is redrawn to correspond to the region enclosed by the box. Another important feature of the program is the rapid, flicker-free redrawing of the traces as they are

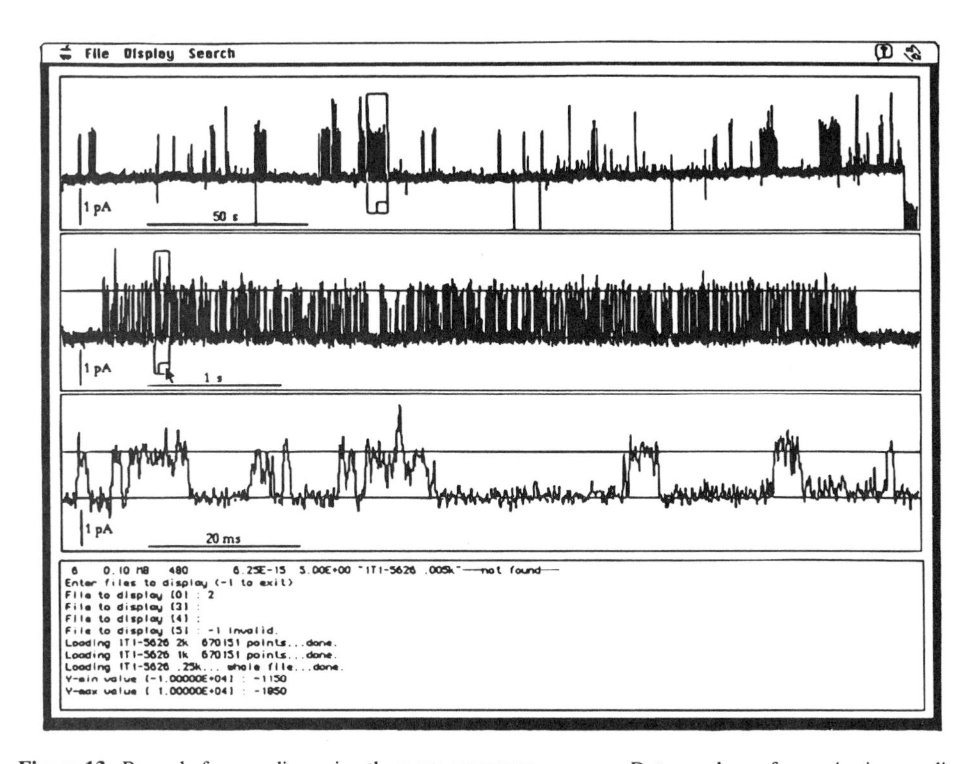

Figure 13. Perusal of a recording using the DATA SELECTOR program. Data are shown from a 4-min recording of potassium-channel currents that includes a slow baseline drift and several spikes from electrical interference. The top trace shows an overview of the entire recording; the region indicated by the box, about 10 s in duration, is expanded as the middle trace. The mouse cursor is positioned to change the size of the box in the middle trace, which selects the 150-ms segment shown in the bottom trace. A brief upward spike that is visible in the upper traces is seen in the bottom trace to be too broad to be a simple noise spike; it also has approximately twice the amplitude of the main channel events, suggesting that it represents an overlapping channel opening. The original recording was obtained with a VR-10 PCM/VCR recording system; the data were transferred directly to the Macintosh computer, creating a 49-MB data file at the 94-kHz sample rate. An off-line Gaussian filtering program, in turn, created synchronized, filtered files with bandwidths of 250 Hz, 1 kHz, and 2kHz. The DATA SELECTOR program reads data from these files as needed to draw and update the display.

rescaled. This is accomplished by first drawing each trace on an off-screen pixel map and then copying it to the screen buffer. The copying operation is very fast, providing an essentially instantaneous update. The drawing operation itself is also fast enough (usually taking less than 100 ms) so that the scrolling and changes of magnification appear smooth and continuous to the user.

For this sort of display, it is important to have fast graphics. The trace-drawing routine used in DataSelector was written in assembly language and is optimized for rapidly graphing arrays of thousands of data points.* It draws directly to the offscreen pixel-map memory rather than making calls to the operating system's graphics routines. Similarly, high-speed displays on IBM-compatible personal computers typically use graphics subroutines that write directly to the video memory rather than using the BIOS interrupts.

For the characterization of events the computer display must also be able to superimpose cursors or reconstructed transitions over the raw data and allow the user to make manual adjustments and corrections. For the 50% threshold analysis, it is sufficient to use the computer's mouse to adjust two variable parameters, the estimated current amplitudes before and after a transition. Time-course fitting requires more adjustable parameters, and for that purpose a set of knobs (i.e., potentiometers that are read by the computer's ADC) can be more flexible than the mouse, though when the method described in Section 4.2.4 works well, the number of manual adjustments that are needed is small, and mouse/keyboard operation is feasible. Use of the numerical keypad, rather than letter keys or mouse, for making menu choices is much more ergonomically satisfactory for operations that are highly repetitive (and single channel analysis is certainly in this category).

4.3.2. Programs

It is still the case, 18 years after the invention of the patch clamp, that no commercial program is available that can perform all of the methods that are described in this chapter. Perhaps the most serious thing that is lacking is a satisfactory program for analyzing records that contain conductance sublevels or multiple conductance levels. At present, if you wish to do things that cannot be done by the commercially available programs, there are two options. You must either write a program yourself or get one from somebody who has done the job you require.

Many programs offer the choice of fully automatic analysis, without any visual inspection of how the program has interpreted your data. Use of such methods is very dangerous unless your data are of high quality and have been subjected to some preliminary check that the baseline stability, conductance sublevels, ambiguous events, and artifacts are all within the range that the program can cope with safely (e.g., see Magleby, 1992). If done thoroughly, such a check may take almost as long as checking individual fits unless your recording is of exceptionally high quality. The speed of automatic methods obviously makes them very

*The drawing algorithm is based on the observation that the display of a trace can be generated by a set of vertical lines, one for each horizontal pixel position in the display. Often there are many more data points to be graphed, say 10^4 or 10^5, than the number of horizontal pixel positions, which might be only 640 or 1024. In simplified form the algorithm can be understood as follows: let n be the number of data points corresponding to a given horizontal pixel position. The endpoints of the vertical line to be drawn at that position are chosen simply to be the minimum and maximum values of $n + 1$ data points (including one from the set of points corresponding to the next horizontal position). Because only vertical lines are to be drawn, the actual drawing routine can be very simple and efficient.

attractive, but the computer maxim "garbage in, garbage out" certainly applies to single-channel analysis, and it may require some investment of time to ensure that you do not get "garbage out."

The earlier forms of time-course fitting were substantially more time consuming than threshold-crossing analysis, even when the fits produced by the latter were inspected. However, the methods described above are faster, and there is now probably not much difference, at least for data that are good enough that initial guesses for transition times and amplitudes are usually satisfactory, so few manual adjustments are needed. As personal computers get faster, so the time taken for least-squares fitting of many parameters will be reduced still further, and the difference between the various methods will become negligible. The speed of the analysis will depend only on the amount of visual checking that is done.

4.3.3. Storing the Idealized Record

The output from these programs is a list of numbers representing the time of each transition in the current record and the amplitude of the transition. This list contains all of the information present in the idealized record that is constructed in the fitting process. Generally, this information is stored in a file by the computer for further processing, such as sorting into histograms or fitting of distributions. Although in principle only two numbers need be stored for each transition in the original record, it is a good idea to include some more information in the file to allow for mistakes that inevitably occur in the analysis process. For example, if the only clue to the number of channels open is the number and polarity of step amplitude values, the corruption of a single entry could cause much confusion. One format for the storage of data, used by the TAC program, which performs threshold-crossing analysis, stores a record containing the following information as an entry for each transition:

1. AbsTime, the time of the transition (LONGREAL in seconds)
2. EventType, the kind of event. This is an enumerated type, having values corresponding to (1) normal transition, (2) interval of data to be ignored, (3) transition between conductance levels, etc.
3. Level, the number of channels open after the transition (INTEGER)
4. PreAmp, the current amplitude before the transition (REAL, in amperes)
5. NumPre, the number of data samples used to estimate the preamplitude (zero if the amplitude was not determined automatically; INTEGER)
6. PostAmp, the current amplitude after the transition (REAL, in amperes)
7. NumPost, the number of data samples used to estimate the postamplitude (INTEGER)

The use of a LONGREAL (64-bit floating-point) value provides sufficient numerical resolution (better than 1 nanosecond in 24 hr) to allow the absolute time of each event to be stored, even when the transition time has been interpolated to a fraction of a sample interval. This greatly simplifies operations in which individual transition records are edited and also allows each event to be synchronized with its position in the raw data file. The current amplitudes are documented by their values as well as the number of points used to estimate them (when automatic amplitude estimation is in effect) so that the reliability of the values can be estimated. This record structure occupies 24 bytes of storage for each event. The resulting event list files are nevertheless much shorter than the raw data files they describe.

The NumPre, NumPost, and EventType indicators allow the subsequent analyses to be carried out with certain values (e.g., ambiguous amplitudes) either included or excluded. This will allow a judgment as to the influence of the ambiguities on the conclusions from the analysis.

5. The Display of Distributions

Analysis of the experimental results by one of the methods described in Section 4 produces an idealized record. This takes the form of an event list that contains the duration of each event and the amplitude of the single-channel current following each transition (or, for some sorts of analysis, only a record of whether the channel was open or shut). We now wish to move on to discuss the ways in which the information in this event list can be viewed and fitted with appropriate curves.

5.1. Histograms and Probability Density Functions

5.1.1. Stability Plots

This section deals mainly with the display of measurements that have been made at equilibrium, so the average properties of the record should not be changing with time. In practice, it is quite common for changes to occur with time, and any such change can easily make the corresponding distribution meaningless. It is, therefore, important to check the data for stability before distributions are constructed or fitted. This can be done by constructing *stability plots* as suggested by Weiss and Magleby (1989). In the case, for example, of measured open times, the approach is to construct a moving average of open times and to plot this average against time or, more commonly, against the interval number (e.g., the number of the interval at the center of the averaged values). A common procedure is to average 50 consecutive open times and then increment the starting point by 25 (i.e., average open times 1 to 50, 26 to 75, 51 to 100, etc). The overlap between samples smoothes the graph (and so also blurs detail). An exactly similar procedure can be followed for shut times and for open probabilities. In the case of open probabilities, a value for P_{open} is calculated for each set of 50 (or whatever number is chosen) open and shut times as total open time over total length. If a shut time is encountered that has been marked as "unusable" during analysis (see Section 4.3.3), then the set must be abandoned and a new set started at the next valid opening.

Figure 14 shows examples of stability plots for amplitudes (in A, C, and E) and for open times, shut times, and P_{open} (in B, D, and F). Graphs for A–D are from experiments with recombinant NMDA receptors. The two amplitude levels are stable throughout the recording for the experiment shown in Fig. 14A and B, though there is a modest tendency in B for shut times to decrease and for P_{open} to increase correspondingly during the experiment. In contrast, Fig. 14C shows a different experiment in which the two amplitude levels both show a sudden decrease after about the 900th interval. Amplitude histograms from such an experiment would show three or four levels but would of course give no hint that there had been a sudden change in the middle of the experiment. The corresponding stability plots for open times, shut times, and P_{open}, shown in Fig. 14D, also show instability; shut times decrease, and P_{open} correspondingly increases, at about the same point in the experiment where the amplitude changes. The open times, however, remain much the same throughout in D, as is also the case for B and F. Figure 14E and F show similar plots from an experiment on adult frog endplate nicotinic receptors, in which all the measured quantities remain stable throughout the recording; data from this experiment were used to construct the shut-time histogram shown in Fig. 15.

Plots of this sort can be used to mark (e.g., by superimposing cursors on the plot)

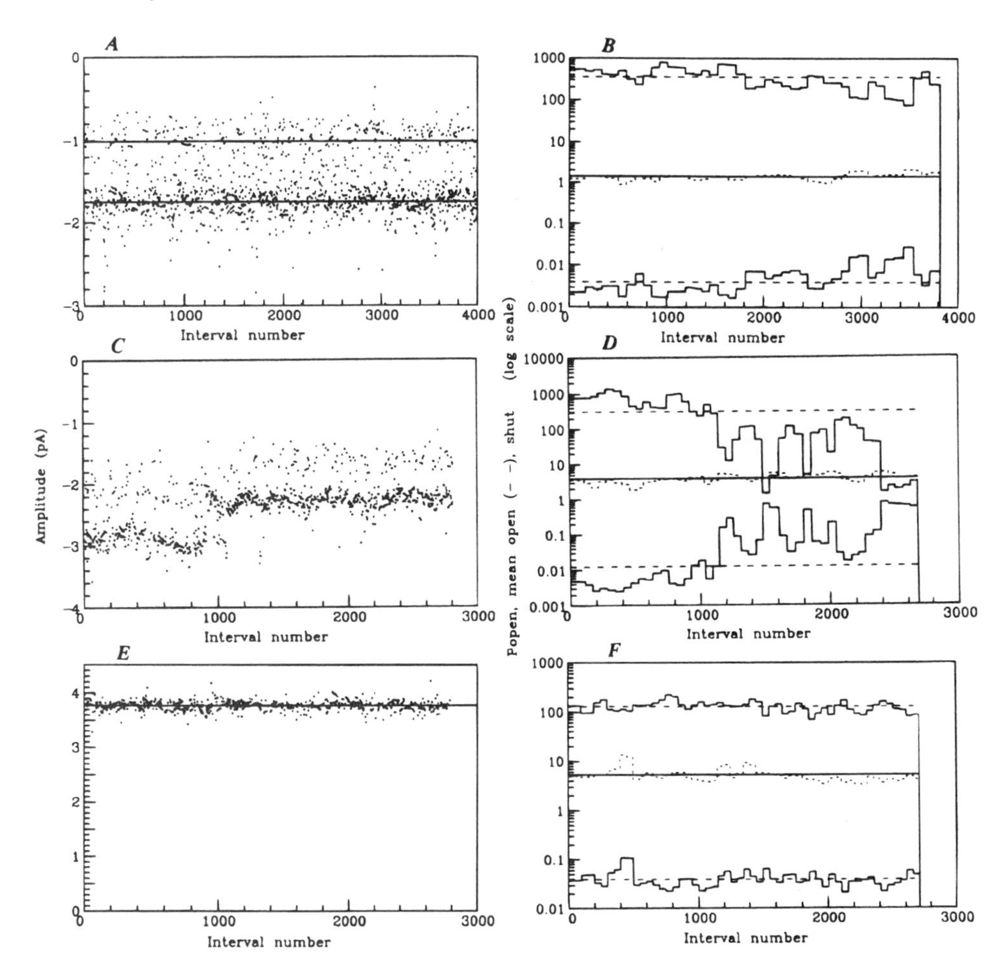

Figure 14. Examples of stability plots. Data for A, B, C, and D are from NMDA-type glutamate receptors expressed in oocytes (unppublished data of P. Stern, P. Béhé, R. Schoepfer, and D. Colquhoun; methods as in Stern *et al.,* 1992). Oocytes were transfected with NR1 + NR2C subunits in A and B (4002 resolved intervals) and with NR1 + NR2A + NR2C subunits in C and D (2810 resolved intervals). A and C show amplitude stability plots; the horizontal lines in A mark the amplitudes that were fitted to the amplitude histogram, −1.01 pA and −1.75 pA. B and D show stability plots for shut time (top), open time (middle), and P_{open} (bottom). Average of 50 values plotted, with increment of 25 intervals. Horizontal lines show the average values for the whole run. E and F show the same two types of stability plot for the same frog endplate nicotinic receptor data that was used to construct the histograms in Fig. 15 (amplitudes are plotted as positive numbers in E).

sections of the data that are to be omitted from the analysis. For example, this approach has been used to inspect, separately, the channel properties when the channel is in a high-P_{open} period and when it is behaving normally.

It should be noted that when the average P_{open} value (the value for the whole of the data) is plotted on the stability plot, it can sometimes appear to be in the wrong position. This may happen when the record contains a very long shut period that reduces the overall P_{open} but affects only one point on the stability plot (which is normally constructed with interval number on the abscissa rather than time).

5.1.2. Probability Density Functions

Most of the data with which we have to deal consist of continuous variables (channel amplitudes, durations of open periods, etc.) rather than discontinuous or integer variables. One exception is the distribution of the number of openings per burst, which is discussed below; this number can, of course, take only integer values. The probability distribution of a continuous variable may be specified as a probability density function, which is a function specified such that the area under the curve represents probability (or frequency). Most commonly, the pdf is an exponential or sum of exponentials (see Chapter 18, this volume). For example, if a time interval has a simple exponential with mean $\tau = 1/\lambda$, its pdf is

$$f(t) = \lambda e^{-\lambda t} \qquad t > 0 \tag{28}$$

which has dimensions of s^{-1}. Alternatively, the exponential density can be written in terms of the time constant, τ, rather than the rate constant, λ. This is preferable for two reasons. First, it is easier to think in terms of time rather that rate or frequency. Second, use of time constants prevents confusion between *observed* rate constants (denoted λ) and the rate constants for transitions between states in the underlying mechanism (see Chapter 18, this volume). Thus, equation 28 will be written in the form

$$f(t) = \tau^{-1}e^{-t/\tau} \tag{29}$$

The area under this curve, as for any pdf, is unity. When there is more than one exponential component, the distribution is referred to as a *mixture of exponential distributions* (or a "sum of exponentials," but the former term is preferred since the total area must be 1). If a_i represents the area of the ith component, and τ_i is its mean, then

$$\begin{aligned} f(t) &= a_1\tau_1^{-1}e^{-t/\tau_1} + a_2\tau_2^{-1}e^{-t/\tau_2} + \cdots \\ &= \Sigma a_i\tau_i^{-1}e^{-t/\tau_i} \end{aligned} \tag{30}$$

The areas add up to unity; i.e.,

$$a_1 + a_2 + \cdots = 1$$

or

$$\Sigma a_i = 1 \tag{31}$$

and they are proportional, roughly speaking, to number of events in each component. The overall mean duration is given by:

$$\text{mean duration} = \Sigma a_i\tau_i \tag{32}$$

In practice, the data consist of an idealized record of time intervals constructed by one of the methods described above (see Section 4). This record may be revised to ensure consistent time resolution (see Section 5.2). The open times, shut times, and other quantities of interest can be obtained from it. For example, the data might consist of a series of n open times $t_1, t_2, \ldots, t_n$. They might be, for example, 1.41, 5.82, 3.91, 10.9 . . . , 6.43 ms. The probability density function is, roughly speaking, proportional to the probability that the observation falls within an infinitesimal interval (from t to $t + dt$; see Chapter 18, this volume). But we have not got an infinite data set, so the pdf of the data looks like a series of delta functions (one at each measured value). This sort of display is not very helpful as it stands, so we smooth it by using a finite binwidth. In other words, we display a histogram as an approximation to the pdf by counting the number of observations that fall in intervals (bins) of specified width. In the example above, we might use 1 ms as the bin width and count the number of observations between 0 and 1 ms, 1 and 2 ms, and so on. These can then be plotted on a histogram as illustrated, for example, in Fig. 15. The histogram is discontinuous, and its ordinate is a dimensionless number. The pdf it approximates is, on the other hand, a continuous variable with dimensions of s^{-1}, so care is needed when both histogram and pdf are plotted on the same graph (see Section 5.1.5).

Figure 15A shows a histogram of shut times, with a time scale running from 0 to 1500 ms, with a bin width of 80 ms. This range includes virtually all the shut times that were

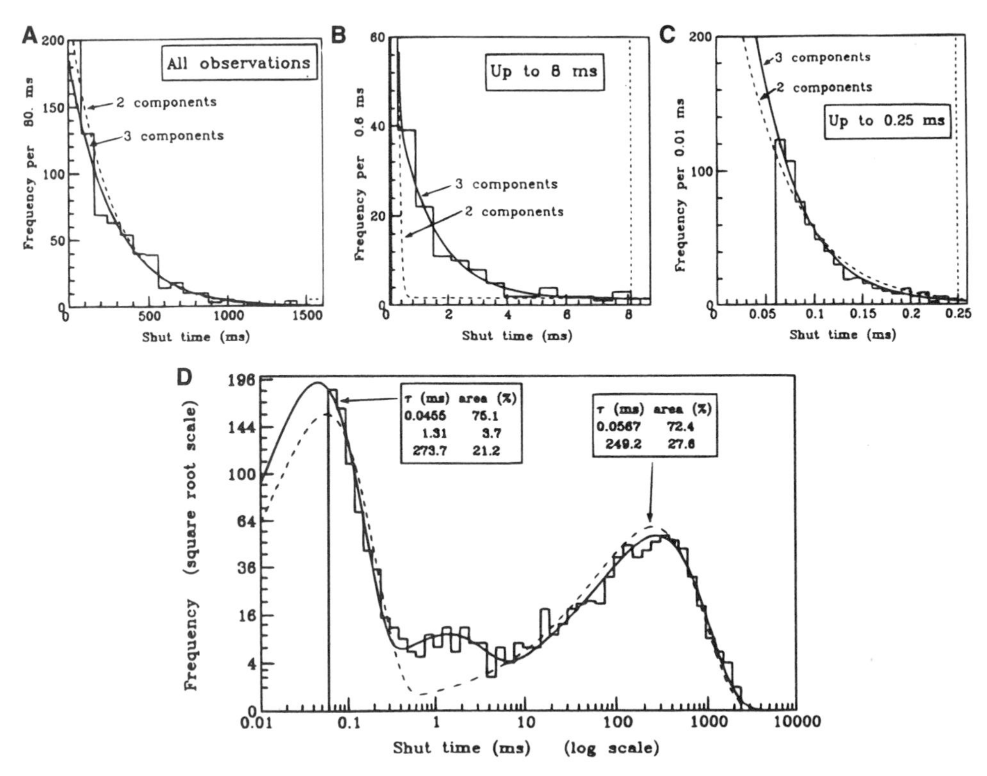

Figure 15. Example of a distribution of shut times. In A, B, and C, the histogram of shut times is shown (on three different time scales), and in D the distribution of log(shut times) for the same data is shown. The data are from nicotinic channels of frog endplate (suberyldicholine 100 nM, −130 mV). Resolutions of 80 μs for open times and 60 μs for shut times were imposed as described in the text; this resulted in 1348 shut times, which were used to construct each of the histograms. The dashed bins (which are off scale in B and C) represent the number of observations above the upper limit. The data were fitted by the method of maximum likelihood with either two exponentials (dashed curve) or three exponentials (continuous curve). The same fits were superimposed on all of the histograms. The estimated parameters are shown in D. (D. Colquhoun and B. Sakmann, unpublished data.)

observed. The first bin actually starts at $t = 60$ μs rather than at $t = 0$ because a resolution of 60 μs was imposed on the data (see Section 5.2 below), so there are no observations shorter than this. All that is visible on this plot is a single slowly decaying component with a mean of about 250 ms, though the first bin, the top of which is cut off on the display, shows that there are many short shut times too. The same data are shown again in Fig. 15C, but only shut times up to 250 μs are shown here, with a bin width of 10 μs; the 60-μs resolution is obvious on this plot. There are many shut times longer than 250 μs of course, and these are pooled in the dashed bin at the right-hand end of the histogram (the top of which is cut off). Again, the histogram looks close to a single exponential, but this time with a mean of about 50 μs. Although it is not obvious from either of these displays, there is in fact a (small) third component in this shut-time distribution. It is visible only in the display of the same data in Fig. 15B, in which all shut times up to 8 ms are shown (with a bin width of 0.6 ms), where an exponential with a mean of about 1 ms is visible. The data were not fitted separately for Figs 15A, B, and C, but one fit was done to all the data (by maximum likelihood—see Section 6) with either two exponential components (dashed line) or 3 exponential components (solid line). This same fit is shown in all four sections of Fig 15. The inadequacy of the two-component fit is obvious only in the display up to 8 ms.

Clearly, the conventional histogram display is inconvenient for intervals that cover such a wide range of values. The logarithmic display described next is preferable.

5.1.3. Logarithmic Display of Time Intervals

It was suggested by McManus *et al.* (1987) and by Sigworth and Sine (1987) that it might be more convenient, when intervals cover a wide range (as in the preceding example), to look at the distribution of the logarithm of the time interval rather than the distribution of the intervals themselves. Note that this is not simply a log transformation of the x axis of the conventional display (which would produce a curve with no peak, and would have bins of variable width on the log scale). Sine and Sigworth suggested, in addition, the use of a square-root transformation of the ordinate in order to keep the errors approximately constant throughout the plot.

The distribution has the following form. If the length of an interval is denoted t, and ln denotes the natural (base e) logarithm, we define

$$x = \ln(t)$$

then we can find the pdf of x, $f_x(x)$, as follows. First we note that if a t is less than some specified value t_1, then it will also be true that $\ln(t)$ is less than $\ln(t_1)$. Thus,

$$\text{Prob}[t < t_1] = \text{Prob}[\ln(t) < \ln(t_1)] = P \tag{33}$$

In other words, the cumulative distributions for t and $\ln(t)$ are the same. Now it is pointed out in Chapter 18 (this volume, Section 3.1) that the pdf can be found by differentiating the cumulative distribution. Thus, denoting the probability defined in equation 33 as P,

$$\begin{aligned} f_x(x) = \frac{dP}{dx} = \frac{dP}{d\ln(t)} &= \frac{dt}{d\ln(t)} \cdot \frac{dP}{dt} \\ &= tf(t) \\ &= \Sigma\, a_i \tau_i^{-1} \exp(x - \tau_i^{-1} e^x) \end{aligned} \tag{34}$$

The second line here follows because dP/dt is simply the original distribution of time intervals, $f(t)$; it shows, oddly, that the distribution of $x = \ln(t)$ can be expressed most simply not in terms of x but in terms of t. When $f(t)$ is multiexponential, as defined in equation 30, and we express $f_x(x)$ in terms of x by substituting $t = e^x$, we obtain the result in equation 34. This function is not exponential in shape but is (for a single exponential component) a negatively skewed bell-shaped curve, the peak of which, very conveniently, occurs at $t = \tau$.

The same data that were displayed in Fig 15A, B, and C are shown in Fig 15D as the distribution of log(shut times). The same fitted curves are also shown (the fitting uses the original intervals, not their logarithms), and the three-component fitted curve shows peaks that occur at the values of the three time constants. It is now clearly visible, from a single graph, that the two-exponential fit is inadequate. (The slow component of the two-exponential fit also illustrates the shape of the distribution for a single exponential because it is so much slower than the fast component that the two components hardly overlap.) This sort of display is now universally used for multicomponent distributions. Its only disadvantage is that it is hard, in the absence of a fitted line, to judge the extent to which the distribution is exponential in shape.

5.1.4. The Cumulative Distribution

The area under the pdf up to any particular value, t, of the time interval is the cumulative form of the distribution, or *distribution function,* namely

$$F(t) = P(\text{time interval} \leq t) = \int_0^t f(t)dt = 1 - e^{-t/\tau} \tag{35}$$

This is a probability and is dimensionless; it increases from 0 to 1 as t increases. Alternatively we may consider the probability that an interval is *longer than t,* which is, for a single exponential,

$$1 - F(t) = P(\text{interval} > t) = \int_t^\infty f(t)dt = e^{-t/\tau}$$

or, for more than one component, the sum of such integrals:

$$1 - F(t) = P(\text{interval} > t) = \Sigma a_i e^{-t/\tau_i} \tag{36}$$

Occasionally, the data histogram is plotted in this cumulative form with the fitted function (36) superimposed on it. This presentation will always look smoother than the usual sort of histogram (the number of values in the early bins is large), but it should *never* be used, because the impression of precision that this display gives is *entirely spurious.* It results from the fact that each bin contains all the observations in all earlier bins, so adjacent bins contain nearly the same data. In other words, successive points on the graph are not independent but are strongly correlated, and this makes the results highly unsuitable for curve fitting.

To make matters worse, it may well not be obvious at first sight that cumulative distributions have been used, because the curve, equation 36, has exactly the same shape as the pdf, equation 30. There are no good reasons to use cumulative distributions to display data; they are highly misleading. In any case, it is much easier to compare results if everyone uses the same form of presentation.

5.1.5. Superimposition of a Probability Density Function on the Histogram

It is helpful to regard the ordinate of the histogram not as a dimensionless number but as a "frequency" or "number per unit time" with dimensions of reciprocal time; the ordinate then becomes directly analogous to probability density. Rather than regarding the height of the histogram block as representing the number of observations between, say, 4 and 6 ms, we regard the *area* of the block as representing this number. The ordinate, the height of the block, will then be the number per 2-ms bin. This is illustrated in Fig. 16 for a hypothetical example of a simple exponential distribution of open time durations with mean $\tau = 10$ ms and rate constant $\lambda = 1/\tau = 100\ \mathrm{s}^{-1}$. The pdf is thus $f(t) = 100e^{-100t}\ \mathrm{s}^{-1}$. It is supposed that there are $N = 494$ observations altogether (including those that might be too short to be seen in practice—see Section 6.1).

The histogram is plotted with a bin width of 2 ms, so the ordinate is number per 2-ms bin. The pdf has, of course, unit area. In order to obtain a curve that can be superimposed on the histogram, we must multiply the pdf by the total number of events and convert its units from s^{-1} to $(2\ \mathrm{ms})^{-1}$ by dividing by 500. The continuous curve is therefore $g(t) = (494/500)f(t) = 98.8\mathrm{e}^{-100t}\ (2\ \mathrm{ms})^{-1}$. The number of observations that are expected between 4 and 6 ms is the area under the continuous curve; i.e., from equation 35 or 36, it is $494(e^{-4/\tau} - e^{-6/\tau}) = 60.6$. This is almost the same as the ordinate of the continuous curve at the midpoint ($t = 5$ ms) of the bin: $g(t) = 98.9\mathrm{e}^{-5/\tau} = 59.9$ (per 2 ms). This approximation will always be good as long as the bin width is much less than τ. Thus, if we actually observed the expected number of observations (60.6) between 4 and 6 ms, the histogram bin would fit the continuous curve closely, as shown in Fig. 16.

Generalizing this argument, the function, $g(t)$, to be plotted on the histogram is

$$g(t) = Nd\,f(t) \tag{37}$$

where $f(t)$ is the probability density function, with units s^{-1} (estimated by fitting the data as described in Section 6), d is the bin width (with units of seconds), and N is the estimated total number of events as calculated by equations 87, 91, or 101, as appropriate. Note that

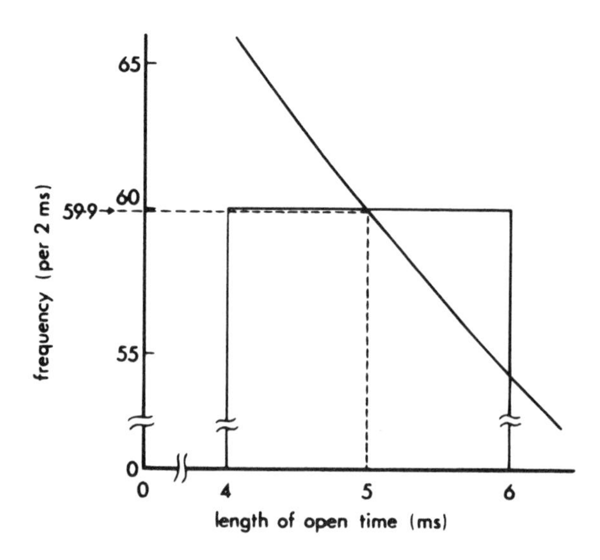

Figure 16. Schematic illustration of the superimposition of a continuous curve (proportional to the fitted theoretical pdf) to a histogram of observed frequencies. The block corresponds to 60 observations between 4 and 6 ms and has an area equal to that under the continuous curve between 4 and 6 ms. The ordinate of the continuous curve at the midpoint of the bin ($t = 5$ ms) is 59.9 $(2\ \mathrm{ms})^{-1}$. See text for further details.

equation 37 is dimensionless, so it is really the pdf that is scaled to the data rather than the other way around.

In the case where the log(interval length) is displayed, as described in section 5.1.3, the probability density function, $f(t)$, would usually be fitted, as described in Section 6, by the method of maximum likelihood applied to the original observations (not to their logarithms). The distribution of $\log_{10}(t)$ is, from equation 34, $2.30259tf(t)$, where the factor 2.30259 [= ln(10)] converts from natural logarithm units to common logarithm units. The curve, $g(t)$, to be plotted on the logarithmic histogram is thus

$$g(t) = Nd' \, 2.30259 \, tf(t) \tag{38}$$

where d' denotes the bin width in $\log_{10}$ units.

5.1.6. Variable Bin Width

The approach discussed above makes it immediately clear how one should construct a histogram with unequal bin widths. It is sometimes useful to use a narrower bin width for shorter intervals than for long ones (there are usually more short intervals, and the pdf changes most rapidly in this region). Thus, if the ordinate is specified as, for example, frequency per 2 ms, then the height of the ordinate for a bin width of 2 ms (say the bin for 6 to 8 ms) is the actual number of observations found to fall within this bin. However, if the shorter intervals are plotted with a bin width of 1 ms rather than 2 ms, then the height of the ordinate for the the 1-ms-wide bins should be twice the number actually observed to fall into the bin. Thus, the area still represents the actual number observed. The plotted function is still as given in equation 37 above, but d is now interpreted as the base width of the bins, i.e., 2 ms in this example, because the ordinate is the frequency per 2 ms bin.

5.1.7. Measurement of P_{open}

One often wishes to measure the probability that a channel is open from a single-channel record. This quantity is usually denoted P_{open} and is sometimes called the "open probability." It is undesirable to refer to P_{open} as the probability of opening, because this sounds like a rate constant (probability of opening in a short time interval; see Chapter 18, this volume), which is not what is intended.

Measurements of P_{open} are useful as an empirical index of the activity in a record, though the overall P_{open} for a whole record will often be so distorted by long sojourns in desensitized or inactivated states as to be uninterpretable. More fundamentally, if it is possible to identify the parts of the record when channels are desensitized, then measurements of P_{open} on the remaining sections provide the best means of constructing equilibrium concentration–response curves (e.g., Colquhoun and Ogden, 1988). Such P_{open} curves have the advantages over other methods that (1) they are corrected for desensitization, (2) they measure response on an *absolute* scale (the maximum possible response is known in advance to be 1), and (3) they allow direct inspection of the channels that underlie the response so there can be little doubt about their identity and homogeneity (see Section 5.9 for tests of homogeneity).

In a record that is in the steady state, P_{open} is simply the average fraction of time spent in the open state. An absolute value for P_{open} can, however, be measured only from a record that contains only one individual channel (or from a section of a record, such as a burst or cluster, where only one channel is active; see Sections 5.6 and 5.9). However, for the purposes of assessment of stability (Section 5.1.1), this is not really important.

When all of the open and shut times have been measured, P_{open} can be calculated as total open time divided by total length of the record. For records where there is essentially only one open level, this is the same thing as the average current level throughout the record, divided by the open-channel current level. In this case, the best method to measure P_{open} is to integrate the record (with an analogue integrator circuit or digitally). This is a good method in principle because the record is filtered, and linear filters do not affect the *area* of the response, only its shape, so integration should be unaffected by the imperfect resolution of open and shut times. Use of digital integration is equivalent to the use of point-amplitude histograms to measure P_{open}, as described in Section 5.3.2. It is important to notice, however, that integration will be satisfactory *only* as long as adequate allowance can be made for the drift in the baseline (shut) level that occurs in most real records.

When the system is not in a steady state, P_{open} will be a function of time and can no longer be defined as the average fraction of time spent in the open state. This is the case, for example, following a voltage or concentration jump or during a synaptic current. In such cases, $P_{\text{open}}(t)$ must be measured by repeating the jump many times and measuring the fraction of occasions when the channel is open at time t.

5.2. Missed Events: Imposition of a Consistent Time Resolution

Unless the mean length of an opening is very long compared with the minimum resolvable duration, it is inevitable that some short openings will remain undetected. Similarly, some short shuttings will also be missed. Methods for making appropriate allowances or corrections for such missed events are considered briefly in Section 6.11 and in rather more detail in Chapter 18 (this volume). In this section we discuss only the aspects of the problem that require action to be taken *before* histograms are constructed.

5.2.1. Definition of Resolution

When the single-channel record is scanned to fit the time of each opening and shutting, as discussed in Sections 3 and 4, the usual procedure would be to fit every detectable opening and gap (shut time). The length of opening (or gap) considered "detectable" will depend on the sort of detection method used. For the threshold-crossing analysis described in Section 4.1, the minimum length is set by T_{d}, although observed durations up to about twice this value are biased and need to be corrected (e.g., with equation 21) before insertion into a histogram. With time-course fitting, the minimum length is not clearly defined and will certainly depend on the details of the method that is used, on who the operator is, and, quite possibly, on how tired he or she is. This will not matter too much as long as care is taken to fit everything that might possibly be an opening or shutting, so that when a realistic resolution is subsequently imposed (Section 5.2.3), it can be said with certainty that events longer than this chosen resolution will not have been omitted during the fitting process.

5.2.2. Effects of Missed Events

Consider, for example, the distribution of the open time when there is a substantial proportion of undetected short gaps; openings will appear to be longer than they actually are, because two (or more) openings separated by an undetected gap will be counted as a single opening (the measured open times are, therefore, more properly referred to as *apparent* open times).

When the histogram of shut or of open times is plotted, the frequency will tend to fall off for very short durations, below which some or all events are too short to be detected. Thus, the distribution may appear to have a peak. One way to deal with this is to look at the histogram and decide on a duration above which it is thought that all openings will be detected and accurately measured; only observations that are longer than this minimum time are used for the fitting process. There is, of course, a large arbitrary element in this decision (and it is also always possible that the open time distribution really does go through a maximum; see Chapter 18, this volume). Nevertheless, if the value chosen is on the safe side, this method may seem to be satisfactory. But it is actually fundamentally inconsistent, as becomes clear when we consider the effect of the open-time resolution on the shut-time distribution.

One way in which inconsistency arises becomes obvious when we consider fitting of *shut* times. If we look at the histogram and see that it has a peak near 100 μs but falls off for shorter shut times, we may decide, quite reasonably, that it is safe to fit (see Section 6) all shut times longer than, say, 140 μs. However, the shut times shorter than 140 μs are still present in the data, and even though they have just been deemed to be too short to be reliable, they will still be regarded as separating two openings, and will therefore, despite their unreliability, shorten the apparent open time to a lower value than it would have if the short gaps had not been detected at all. And, of course, an exactly analogous inconsistency in measurement of apparent shut times can arise when short openings are partially missed.

Another sort of inconsistency will arise if the criterion for the gap length that is detectable does not remain exactly the same throughout the analysis. If it is not constant, the apparent lengths of openings will vary with time, so the distribution of the measured open times will be distorted even if *all* the openings are long compared with the minimum resolvable duration.

A third reason why it is important to know about the resolution is encountered when, for example, measurements of open times are made at different membrane potentials. The resolution for, say, brief shuttings, will be worse when the single-channel currents are smaller (potentials closer to the reversal potential), so more of them will be missed. The apparent open times will therefore appear to be longer at potentials near the reversal potential, even if the true open time does not depend on membrane potential at all.

Finally, the methods that have been developed recently for making corrections for missed brief events almost all require that the resolution of the data be known and consistent. In other words, if the resolution is stated to be 100 μs, then we must be as sure as possible that no events shorter than this are present in the data, and that all events longer than 100 μs have been detected.

For all of these reasons, it is important that the resolution (the shortest event fitted) be stated in published work; without knowing the resolution, it is impossible for other authors to compare their results for quantities as mean "apparent open time" (though this rarely stops them from trying).

5.2.3. Imposition of Resolution

One way to avoid the inconsistencies just described is to impose a resolution on the data retrospectively (Colquhoun and Sakmann, 1985). In the analysis of the original experimental record, every event is fitted even if it is so short that its reality is dubious. While this is done, a judgment is made as to the shortest duration (t_{res} say) that can be trusted (the value of t_{res} may not be the same for open times and for shut times). Again, this is quite subjective; a value on the safe side should be chosen. The most important criterion for the choice of t_{res} is that it should be chosen so that it ensures a sufficiently low false-event rate, e.g., below $10^{-8}\ s^{-1}$ (see Section 3.3).

When the analysis is completed, and the idealized record is stored (see Section 4), the chosen value of t_{res} can be specified and the idealized record revised as follows:

1. All open times that are shorter than t_{res} must be removed. Shut times that are separated by openings shorter than t_{res} are treated as single shut periods. The lengths of all such shut times are concatenated (together with the lengths of intervening short openings) and inserted in the revised data record as single shut times.
2. Similarly, all shut times that are shorter than t_{res} must be removed. If the two openings that are separated by the short gap have both got the same amplitude, then the two open times are concatenated (together with the intervening shut time) and inserted into the revised record as a single opening. If the two openings have different amplitudes, they are inserted into the revised record as two openings with a direct transition from the first open level to the second. This procedure entails deciding exactly what "the same amplitude" means. Some criterion must be specified, which will depend on what amplitude difference is deemed large enough to be detectable; for example, amplitudes that are separated by less than 10% of the full amplitude might be deemed "the same."

In this way a new idealized record, with consistent time resolution throughout, is produced, and it is this that is used for subsequent construction of histograms and fitting. The new record cannot, of course, contain any openings (or gaps) shorter than t_{res}, so the histograms start at this point. As long as the original idealized record is kept, it is easy to repeat the fitting with a different resolution if necessary.

It may be noticed that, for example, imposition of a 50-μs resolution on a perfect record, followed by imposition of 100-μs resolution, will not necessarily give exactly the same result as imposition of 100-μs resolution directly on the perfect record. To the extent that the data we start with are never perfectly resolved, this approach does not give precisely the required results, but it is, nevertheless, the best that can be done.

5.2.4. Resolution, Sublevels, and Fit Range

It must be remembered that events (openings or shuttings) may be *detected* with certainty in the single-channel record even when their duration is shorter than the risetime (T_r) of the recording system. However, their duration must be at least $2T_r$ before their amplitude can be measured accurately (see Section 4). If, for example, it is desired to construct a distribution of the apparent times but to include in the distribution only those open times that are sufficiently long for their amplitudes to be known, then only openings longer than $2T_r$ or $2.5T_r$ can be used. However, this does *not* mean that the resolution of $2T_r$ should be imposed on the data. If this resolution were imposed on the shut times, many brief shuttings, which are nevertheless long enough to be detected with certainty, would be excluded, thus causing the apparent open times to be longer and causing unnecessary error in the estimation of the open time. The resolution that is imposed should depend on what can be *detected* reliably (i.e., distinguished from random noise), but, in the case just described, the range of values that are used for *fitting* should exclude values shorter than $2T_r$. When conditional distributions are used for maximum-likelihood fitting, as described below, there is no problem in fitting only those observations that lie within any specified range.

The distinction between resolution and fit range does not cause too many problems when there are no subconductance levels in the record. In this case, any deflection toward the baseline must represent a complete shutting. But when the channel shows subconductance levels, the problem is more difficult, and it may be desirable to impose different resolutions and/or different filtering for different sorts of analysis.

For example, Howe *et al.* (1991) describe procedures for analysis of NMDA-type glutamate-activated channels that showed conductance levels of 30 pS, 40 pS, and 50 pS. A resolution that produces an acceptable false-event rate for the 50-pS openings may result in an unacceptably high false-event rate for smaller openings in the same record. For analysis of amplitudes, the results were treated as described above; the resolution was set to produce an acceptable false-event rate for 50-pS events, but events shorter than $2.5T_r$ were excluded from fitting. For distributions of shut times the open-time resolution was set to give an acceptable false-event rate for 50-pS openings, but the shut-time resolution was set to ensure that events described as shuttings were unlikely to be transitions from 50 pS to 40 or 30 pS or transitions from 40 pS to 30 pS. To achieve this, the resolution was set to duration w, such that events are counted as shuttings only if they are seen to reach a level safely (say 2 standard deviations) below the 30-pS level. This can be achieved by solving for w (e.g., by bisection)

$$\frac{A_{50} - (A_{30} - 2s_{30})}{A_{50}} = \mathrm{erf}(0.886w/T_r)$$

The right-hand side of this equation gives the fraction of its maximum amplitude attained by a rectangular pulse of length w (see equations 8 and 12). On the left hand side, A_{50} and A_{30} are the absolute current amplitudes for the 50-pS and 30-pS openings, and s_{30} is the standard deviation of the 30-pS currents (these values being obtained from fitting of amplitude histograms). For further details, see Howe *et al.* (1991).

5.3. The Amplitude Distribution

Single-channel current amplitudes are interesting for two main reasons. First, the ways the amplitude varies with ionic composition of the bathing medium and with membrane potential are important for the study of ion permeation mechanisms. Second, amplitude measurements are often a useful way to characterize channel types, e.g., types with different subunit compositions or with mutations.

It has been shown in Section 4.2.3 that the amplitude of a channel opening can be measured accurately only if the duration of the opening is at least twice the risetime (T_r) of the recording system. Amplitude measurements should, therefore, be included in the amplitude histogram only when the opening is longer than some specified length such as $2T_r$ or $2.5T_r$. This can, of course, be done properly only if the amplitude is estimated separately for every opening, and there are, unfortunately, still many analysis programs in use that cannot do this.

Often there will be more than one channel in the patch of membrane from which the recording is made, and in this case, more than one channel may be open at the same time, so that current amplitudes that are integer multiples of the single-channel current are seen. This question is discussed further in Section 5.4.

5.3.1. How Variable Are Single-Channel Amplitudes?

The amplitudes of single-channel currents are, in some cases at least, very consistent. For example, Fig. 17 shows a distribution of amplitudes measured from adult rat endplate nicotinic acetylcholine receptors. It has been fitted (arbitrarily) with a Gaussian curve and shows a mean of 6.62 pA and a standard deviation of 0.12 pA (i.e., 1.8% of the mean). The variability from one opening to the next of the same ion channel or of different channels in the patch seems to be very small, possibly no greater than the error in the fitting of the amplitude. In this case the amplitude is not an inherently random variable like the open time but is, for practical purposes, a more or less fixed quantity.

However, channel amplitudes more commonly are not exactly constant. It seems that just about every sort of channel shows extra open-channel noise; i.e., the current record is somewhat noisier when the channel is open than when it is shut (e.g., Sigworth, 1985, 1986). If it is assumed that the excess open-channel noise is independent of the baseline noise, so their variances are additive, the root mean square excess noise, s_{excess}, can be estimated as

$$s_{\text{excess}} = (s_{\text{open}}^2 - s_{\text{shut}}^2)^{0.5} \tag{39}$$

The extent of this excess open-channel noise varies greatly from one sort of channel to another; it is very small for adult frog muscle nicotinic receptors (D. Colquhoun, unpublished

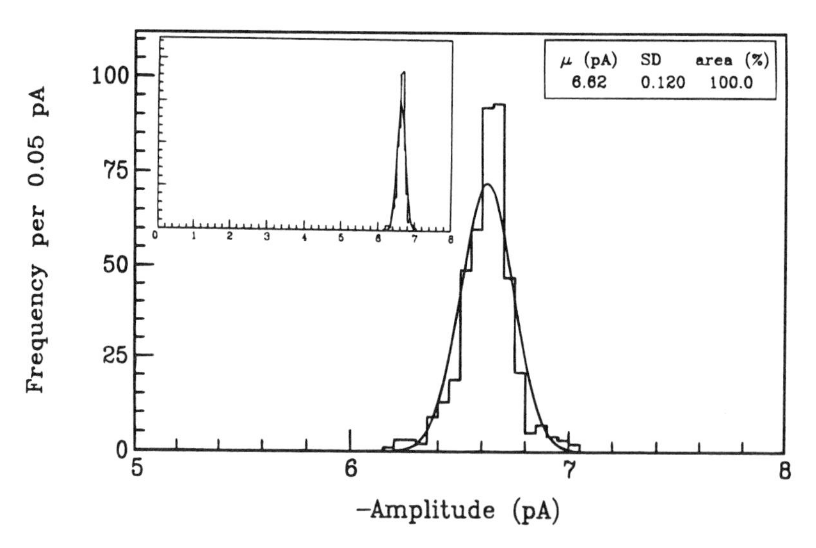

Figure 17. An example of the distribution of the fitted amplitudes of single-channel currents; amplitudes were defined by eye, by means of a cursor on the computer screen. Unpublished data of D. C. Ogden and N. K. Mulrine; channel openings elicited by 100 nM acetylcholine in cell-attached patch on adult (200-g) rat endplate in extracellular solution (with 20 mM K^+ and 1 mM Ca^{2+}), at resting potential −80 mV. After a resolution of 50 μs was imposed for both openings and shuttings, there were 1100 resolved intervals, and the histogram was constructed from 433 amplitudes of openings that were longer than 2 risetimes. The continuous curve is a Gaussian distribution, which was fitted to the data by the method of maximum likelihood; it has a mean of 6.62 pA and a standard deviation of 0.12 pA. The main display covers only the range from 5 pA to 8 pA in order to show clearly that the observed distribution has a sharper peak and broader tails than the Gaussian curve, as predicted in Section 5.3.3 and Appendix 2. The inset shows the same distribution plotted over the range 0–8 pA to show that there is only one narrow peak in the distribution.

data) but very large for some neuronal nicotinic receptors (e.g., Mathie *et al.,* 1991). It is likely that the phenomenon is intrinsic to the receptor protein; it appears in recombinant receptors and can be strongly influenced by small mutations. It seems likely that it can be regarded as resulting from fluctuations in channel structure that produce small changes in conductance or from entry into subconductance states that are short-lived and/or differ only slightly from the main conductance level. The appearance of extra open-channel noise can also be mimicked by frequent and brief channel blockages (e.g., Ogden and Colquhoun, 1985).

It is common for more than one conductance level to appear in single-channel recordings. One, probably quite common, reason for this is heterogeneity of the channels in the membrane patch. In addition, though, it has become apparent that most types of ion channel have more than one conductance level. For some types these conductance sublevels are rare, but for others they are quite common. For example, the NMDA-type glutamate receptors all show this phenomenon clearly, as illustrated in Figs. 12 and 18. These channels have a 50-pS main level and a briefer 40-pS sublevel. It is not known whether such sublevels have any functional importance (though it seems unlikely), but they are certainly useful for characterizing subunit combinations (Stern *et al.,* 1992).

In this case of NMDA receptors, the 50-pS and 40-pS peaks are quite clear and reproducible from experiment to experiment. There is, however, some question as to whether all "50-pS" openings have exactly the same conductance (apart from random measurement errors). There is some reason to suspect that they may not.

The various methods that are used for investigation of amplitudes are discussed next.

5.3.2. Point-Amplitude Histograms

The simplest procedure is to make a histogram of the values of the individual digitized data points (after subtracting the baseline value, so the shut channel appears with zero amplitude). This is often known as a *point-amplitude histogram* to distinguish it from histograms formed from fitted amplitudes (see Section 5.3.3). There will be a lot of points in such histograms, but the points are not independent, so the large number of points does not necessarily imply high precision. In order for the sample points in filtered data to be approximately independent, they would need to be about one risetime (T_r) apart, but the sample rate is normally a good deal higher than this. For statistical purposes, the "effective number of points" could taken roughly as (sample duration)/T_r.

The relative areas of the peaks in a point-amplitude histogram represent the number of data points, i.e., the length of time spent, at each amplitude level (cf. next section). The areas of the peaks can therefore be used to estimate the fraction of time for which the channel is open, i.e., the probability of being open (P_{open}), as long as all data points are included (see also Section 5.1.7).

5.3.2a. ***The All-Point-Amplitude Histogram.*** The crudest method is simply to make the histogram directly from all points in the data record. In fact, this is the *only* method that is available in many commercial programs. The main problem is that the method obviously depends on the baseline remaining exactly constant throughout the record. This is rarely true, so in practice it is possible to use the method only on relatively short stretches of data for which the baseline can be checked carefully. Alternatively, if both baseline and open levels have been fitted, as illustrated in Fig. 12, all the data points in the region that has been fitted (and approved) can be entered into the histogram; this provides excellent compensation for baseline changes, but the results cannot be used to estimate P_{open} because many shut points are omitted.

An example of an all-point histogram constructed in the latter way is shown in Fig. 18B and C. The peaks for the shut level and for the main (about 5-pA) open level are obvious. However, there is a smear of points between the two (the data points that lie in the transition regions between open and shut), and this partially obscures the small peak that corresponds to the sublevel at about 4 pA; this is shown on an enlarged scale in Fig. 18C. This smearing can be reduced as follows.

5.3.2b. *Open-Point and Shut-Point Amplitude Histograms.* Once transitions have been located by one of the methods described in Section 4, then it becomes possible to exclude data points that lie on the transitions from one conductance level to another. Knowledge of the step-response function of the recording system allows the transition period to be defined accurately. An example is shown in Fig. 12; only those data points that correspond to the flat sections of the fitted curve (i.e., areas where no transitions were detected) are entered into the histogram. The open-point amplitude histogram in Fig. 18E was constructed in this way. Most of the smearing has gone, and the rather small 4-pA component is more clearly defined than in the all-point histogram, as shown on an enlarged scale in Fig. 18F. And, since the baseline adjacent to the openings is fitted along with the openings, there should be no distortion caused by baseline drift.

The data points that correspond to shut periods are entered into a separate histogram, as for the open points. A shut-point histogram is shown in Fig. 18D; it is usually found, as in this case, that the shut-point histogram is fitted very well by a simple Gaussian curve (i.e., the baseline noise is Gaussian). Open-point histograms, on the other hand, may not be perfectly Gaussian because of such effects as undetected sublevel transitions or brief closures.

5.3.2c. *Analysis of Flickery Block.* The asymmetry in point-amplitude histograms contains information about the nature of open-channel noise. This information can be interpreted by use of either noise analysis (Sigworth, 1985, 1986; Ogden and Colquhoun, 1985; Heinemann and Sigworth, 1990) or the amplitude histogram itself (Yellen, 1984; Heinemann and Sigworth, 1991). These methods have been used, for example, to analyze rapid channel block. High concentrations of a low-affinity channel-blocking agent produce so-called "flickery noise." Because of the high concentration, blockages are frequent, and openings are short, and when blockages are so brief that they cannot be resolved easily in the single-channel record, the open channel appears to be very noisy and to have a reduced amplitude (see Chapter 18, this volume). Such flickery noise, when it happens to be in the right frequency range, will produce a characteristically shaped smear in the all-point amplitude histogram. If the blocking process is approximated as a two-state process, and we look at the channel only while it is open or blocked, the mechanism can be written thus

$$\text{open} \underset{k_{-\mathrm{B}}}{\overset{k_{+\mathrm{B}}x_{\mathrm{B}}}{\rightleftharpoons}} \text{blocked,} \tag{40}$$

where $k_{-\mathrm{B}}$ is the dissociation rate constant for the blocker, $k_{+\mathrm{B}}$ is the association rate constant, and x_{B} is the blocker concentration, so $k_{+\mathrm{B}}x_{\mathrm{B}}$ is the transition rate from open to blocked state (per unit open time).

One approach, which works best for events that are close to being resolvable (mean duration comparable to the filter risetime), is based on the work of Fitzhugh (1983). This theory showed that, for data that have been filtered through a simple RC filter with time constant $\tau_{\mathrm{f}} = RC$, the point-amplitude histogram should be described by the beta distribution. The beta distribution was used to analyse fast block by Yellen (1984). If we denote as y the

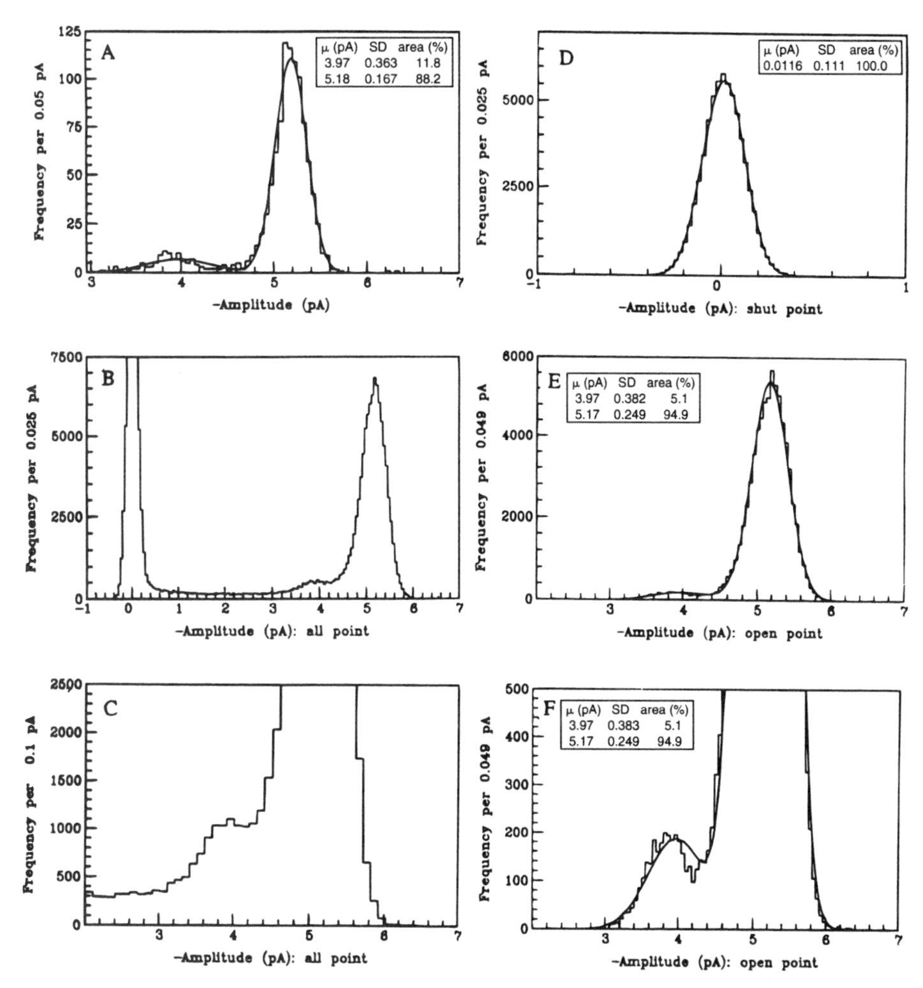

Figure 18. Examples of various sorts of amplitude histograms. Data were from a 10-s recording of NMDA channels (same data as were illustrated in Fig. 12, where details are given). Resolution was set to 30 μs for shuttings and 40 μs for openings, with concatenation of contiguous open levels that differed by less than 0.5 pA. A: Distribution of fitted amplitudes (of the type listed in legend of Fig. 12). Openings with a duration of less than two risetimes (332 μs) were excluded, which left 1049 amplitudes to be fitted (between 3.4 and 6.0 pA) with a mixture of two Gaussian distributions by maximum-likelihood method using the original values. The components had means of 3.97 pA and 5.18 pA (the usual "40-pS" and "50-pS" components seen in 1 mM Ca). The areas of the components were 11.8% and 88.2%, and the standard deviations were 0.36 pA and 0.17 pA, respectively. B and C: All-points amplitude histogram. This histogram shows the amplitude of all data points within the fitted range (solid line in Fig. 12). This ensures freedom from the effects of baseline drift but means that the relative area occupied by the shut points is arbitrary. The small "40-pS" component is shown on an enlarged scale in C; this also makes more obvious the smearing that is inevitable in an all-points histogram. D, E, and F: Separate open-point and shut-point histograms. The data points that correspond to the regions where the fitted curve (see Fig. 12) was flat were collected separately for regions where the channel was shut and where it was open. This eliminates the smeared points during the transition from shut to open. The shut-point histogram in D is well fitted with a single Gaussian (standard deviation 0.12 pA). The open-point histogram in E (and, on an enlarged scale, in F) shows much clearer demarcation of the subconductance level than the all-points histogram. The fit with two Gaussian components is not perfect, though the fitted means, 3.97 pA and 5.17 pA, are almost identical to those found from the fitted amplitudes in A.

current amplitude, normalised to lie between the values of 0 and 1, the probability density function for the beta distribution can be written as

$$f(y) = \frac{\Gamma(a + b)}{\Gamma(a)\Gamma(b)} y^{(a-1)}(1 - y)^{(b-1)} \quad (41)$$

where

$$a = k_{-B}\tau_f \quad \text{and} \quad b = k_{+B}x_B\tau_f.$$

In this result, $\Gamma(x)$ denotes the gamma function, which is a continuous version of the factorial function, such that $\Gamma(x + 1) = x!$ when x is an integer. The gamma function is tabulated by Abramovitz and Stegun (1965) or can be computed as described by Press *et al.* (1992). This result does not include the background noise in the recording, but by first convolving it with the baseline noise distribution and then fitting the result to the point-amplitude histogram, one can obtain estimates of the blocking and unblocking rate constants.

The beta distribution method makes some undesirable assumptions. For example, a simple *RC* filter is never used in practice. For any more realistic filters, the theory becomes a great deal more complicated (A. Jalali and A. G. Hawkes, unpublished results), though Yellen (1984) showed that use of the beta distribution could work reasonably well with Bessel-filtered data. Second, it makes no allowance for spontaneous shuttings or for the excess open-channel noise (Sigworth, 1985, 1986) that exist in the absence of blocker.

For these reasons, Ogden and Colquhoun (1985) preferred to use the noise spectrum (spectral density function) of the open-channel noise. Use of noise analysis allows approximate corrections to be made for excess open-channel noise, and it allows use of a realistic filter; expressions are given by Colquhoun and Ogden for the variance of Gaussian-filtered, or Butterworth-filtered Lorentzian noise. They were able to estimate the mean duration of blockage of a nicotinic channel by carbachol as about 9 μs (which is similar to that measured by direct time-course fitting). Heinemann and Sigworth (1990) used a noise analysis to estimate the mean duration of block of gramicidin channels by Cs^+ as about 1 μs. This latter value was confirmed by Heinemann and Sigworth (1991) by inspection of the cumulants of the point-amplitude distribution. The cumulant method provides a method of analysis of point-amplitude histograms that is complementary to the beta-function approach. The beta function works best with events that are comparable with the filter risetime, but the cumulant method is better for events that are much shorter than the risetime (but are widely spaced).

5.3.3. Amplitude Histograms from Fitted Amplitudes

The other main method for display of single-channel amplitudes is to measure the amplitude of each opening separately and to make a histogram of the results. The amplitudes can be measured by placing a cursor on the data on the computer screen, by eye, or by using a least-squares fit to the data as illustrated in Fig. 12. In either case, the amplitude can be measured only for events that are longer than about two risetimes, as explained in Section 4.2.3. And in either case, the estimates are susceptible to bias resulting from undetected brief closures. The latter problem can be minimized by fitting all possible closures, even if they are so short that they will eventually be eliminated when a safe resolution is imposed (see Section 5.2.3). The histogram has only one value for each opening, so if more than one open level is present, the relative areas of the components will represent the relative *frequencies* with which the levels occur (rather than the relative time spent at each level).

An example of an amplitude histogram constructed in this way is shown in Fig 18A, for the same data that were used to illustrate the point-amplitude histograms. In principle, this histogram should show components even more clearly than the open-point-amplitude histogram, because smearing between transitions from one open level to another is avoided, as is smearing between open and shut levels. It is, on the other hand, somewhat less objective than the open-point-amplitude histogram.

It is usual to fit (as described in Section 6.8.3) a Gaussian or a mixture of Gaussians both to amplitude histograms and to point-amplitude histograms in an attempt to resolve different conductance levels. In fact, this may not be appropriate in either case (as discussed in the preceding section for point-amplitude histograms). In the case of fitted amplitudes, the distribution often shows a sharper peak and broader tails than is expected for a Gaussian, as illustrated in Fig. 18A or, particularly clearly, in Fig. 17.

A distribution of this sort is to be expected because the amplitude values are obtained from events of variable duration. The long events give the most precise estimates and cluster around the true value to give the sharp peak. Short events give values with more scatter and contribute to the tails. The distribution that would be expected is derived in Appendix 2. This result, although preferable to Gaussian fits, has not yet been used much in practice, probably because of the inconvenience involved in determining the background noise spectrum.

5.3.4. Mean Low-Variance Amplitude Histograms

Patlak (1988) suggested a method for detection of peaks in amplitude histograms by searching the digitized data record for sections where the channel is open and the record is "flat." This is done by looking at sections of the data of fixed length (e.g., ten points). The mean and standard deviation of each such section is calculated, and this process is repeated after advancing the start of the data section by, for example, one point, until the end of the data is reached. A data section is deemed to be flat and therefore to represent a well-defined conductance level if its standard deviation is less then some specified multiple (e.g., 0.5 to 2) of the standard deviation of the baseline noise. All sections that have a larger standard deviation than this are rejected, and a histogram is constructed of the mean amplitudes of the remaining sections. Three different values have to be specified to construct the histogram (the section length, the number of points to advance, and the threshold standard deviation), so a variety of histograms can be produced. This method may work well on some sorts of data, especially if the conductance levels are reasonably long-lived. However, use of sufficient filtering will make long-lived conductance levels obvious by any method of analysis. It is much harder to distinguish subconductance states that are short-lived. If the data sections are made shorter, their increased scatter means the histogram is more scattered, so peaks become hard to distinguish. And if the threshold standard deviation for inclusion is made lower, the number of points in the histogram is reduced, so again peaks become hard to distinguish with certainty.

5.3.5. Subconductance Transition Frequencies

When there is more than one open-channel conductance level, it may be of interest to measure the frequency of transitions from one open level to another (and from each open level to the shut level). This provides another way to characterize quantitatively different receptors or subunit combinations (Howe *et al.*, 1991; Stern *et al.*, 1992). It can also provide useful information about reaction mechanisms, and it allows a test of the principle of microscopic reversibility (see Chapter 18, this volume). For example, in the data shown in Fig. 18, there are components with means of about 4 pA and 5 pA (i.e., conductances of about 40 pS and 50 pS). When amplitudes have been fitted to each opening (Section 5.3.3), it is simple to categorize each transition in the idealized record as 0 → 50 pS, 50 → 0 pS, 0 → 40 pS, 40 → 0 pS, 40 → 50 pS or 50 → 40 pS. This cannot, of course, be done with programs that produce only point-amplitude histograms.

Calculation of a Critical Amplitude

The amplitude components almost always overlap to some extent, so the classification of openings (into 40-pS and 50-pS classes in the above example) will not be entirely unambiguous. A critical amplitude, A_c, that minimizes the total number of amplitudes misclassified was used by Howe *et al.* (1991). This number is proportional to

$$\begin{aligned} n_{\text{mis}} &= a_1 \int_{A_c}^{\infty} f_1(A)dA + a_2 \int_{0}^{A_c} f_2(A)dA \\ &= 0.5\{a_1[1 - \text{erf}(u_1/\sqrt{2})] + a_2[1 - \text{erf}(u_2/\sqrt{2})]\} \end{aligned} \tag{42}$$

where f_1 and f_2 are the Gaussian densities for the components with smaller and larger means, respectively; a_1 and a_2 are proportional to the areas of these components; erf represents the error function (see Appendix 3.3); and u_1 and u_2 are standard normal deviates; i.e., $u_1 = |(A - \mu_1)/\sigma_1|$, $u_2 = |(A - \mu_2)/\sigma_2|$, where μ_1, σ_1 and μ_2, σ_2 are the means and standard deviations of the components. This is at a minimum when

$$(a_1/\sigma_1)e^{-u_1^2/2} = (a_2/\sigma_2)e^{-u_2^2/2}$$

Thus, A_c may be found by solving the quadratic equation

$$aA_c^2 + bA_c + c = 0 \tag{43}$$

where the coefficients are defined as

$$\begin{aligned} a &= (1/\sigma_2^2) - (1/\sigma_1^2), \\ b &= 2[(\mu_1/\sigma_1^2) - (\mu_2/\sigma_2^2)], \\ c &= (\mu_2^2/\sigma_2^2) - (\mu_1^2/\sigma_1^2) - 2\ln[(a_2/\sigma_2)/(a_1/\sigma_1)]. \end{aligned} \tag{44}$$

5.4. The Open and Shut Lifetime Distributions

There are only two directly observable types of distribution, the distribution of open times and the distribution of shut times or gaps (i.e., of the durations of the intervals between openings). Although the open times are an obvious focus of attention, the shut times are equally if not more informative (see Chapter 18, this volume). Usually it is sensible to look at the shut-time distribution first, because it is this that dictates whether or not it is feasible to divide the openings into bursts.

It is preferable to refer to these distributions as those of *apparent* open times and *apparent* shut times because the effects of undetected shuttings and openings, respectively, mean that the results will rarely be accurate (see Sections 5.2 and 6.11 and Chapter 18, this volume). For example, if some shut times are too short to be resolved, then the measured openings will be too long, because some actually consist of two or more openings separated by unresolved gaps. The shut times may also be too long if they contain brief undetected openings. However the word "apparent" will, for brevity, be dropped when the intention is clear from the context.

Both distributions are usually fitted by mixtures of exponentials, as in equation 30. The number of components in the open-time distribution should be equal to the number of open states, and the number of components in the shut-time distribution should be equal to the number of shut states. It is, of course, always possible that some of the components will be too small or too fast to be detected, so the distributions can provide only a lower bound for the numbers of states. Although these distributions are much more susceptible to errors resulting from missed events than are distributions such as that of the total open time per burst (see below), it is remarkable that such errors should not much affect the *number* of components that are found, even when the time constants of the components are quite wrong (see Section 12 of Chapter 18, this volume, Hawkes *et al.* 1992).

When the patch contains more than one channel, even when no multiple openings are seen, there is no way to be sure whether or not a particular opening originates from the same channel as the preceding opening. This complicates the interpretation of the results (see Chapter 18, this volume). In cases in which the openings are observed to occur in bursts, there is often reason to think that all of the openings in one burst may originate from the same channel, even if the next burst originates from a different channel, so the gaps within bursts may be easier to interpret. It is therefore usually interesting to analyze the characteristics of bursts of openings when it is possible to do so. Distributions that are relevant to this case are considered in Section 5.5.

5.4.1. Multiple Openings

If the experimental record has periods when more than one channel is open, measurement of apparent open lifetimes becomes more difficult. Such records are useful for averaging to simulate a relaxation or for calculation of the noise from the patch recording. They may also be useful for estimating the number of active channels in the patch (see Chapter 18, this volume) and for testing for the mutual independence of channels. In general, however, records with multiple openings are unsuitable for looking at distributions of open times and shut times because, if two channels are open, there is no way of telling, when one of them closes, whether the one that closes is that which opened first or that which opened second.

Although it is possible to recover open- and shut-time distributions from records with multiple openings (Jackson, 1985), it is generally desirable to use records that have only one channel open at a time or only very few multiple openings. In order to use the method of Jackson (1985), the number of active channels must be known, and in most cases this is difficult to estimate accurately (see Section 8 of Chapter 18, this volume), and this method cannot cope with subconductance levels, which almost all channels show to some extent.

When there are only a few multiple openings in a record, one way to deal with them is to omit all the openings in the group where multiple openings occur and to measure the lifetimes of only the single openings before and after this group. The time between these openings is not a valid shut time and must be marked as "unusable" in the idealized list of shut times so that it can be excluded from the shut-time distribution. This procedure tends to select against long openings, so the open times thus measured will be slightly too short on average. An alternative procedure would be to take the length of the group of multiple openings as a single open time, which would make the open times too long on average. If there are enough multiple openings in the record that the bias could be substantial, then both of these methods could be used; if the two methods give results that disagree by enough to matter, then the number of multiple openings is too large to allow any simple analysis.

5.4.2. Distributions of Open Times Conditional on Amplitude

When there is more than one conductance level, it will usually be interesting to look separately at open times for each level. For example, in the data shown in Fig. 18 there are components with means of about 4 pA and 5 pA (i.e., conductances of about 40 pS and 50 pS). When amplitudes have been fitted to each opening (Section 5.3.3), it is simple to go through each opening and select the openings whose amplitudes lie in a specified range. The histogram is then plotted using the durations of these openings. A method for calculating an optimum critical amplitude that minimizes the number of misclassified amplitudes has been given above, in Section 5.3.5.

It is, of course, necessary to exclude openings that are too short for their amplitudes to be well defined. This is done by excluding from *fitting* (see Section 6.8.1) all values below $t_{min} = 2T_r$ or $2.5T_r$, rather than by imposing a low resolution on the data, as described in Section 5.2.4. Such analyses obviously can not be done with computer programs that do not fit an amplitude to every opening but rely only on all-point amplitude histograms.

5.5. Burst Distributions

5.5.1 Definition of Bursts

In extreme cases, it will be obvious to a casual observer that openings are occurring in groups, separated by long silent periods, rather than at random (exponentially distributed) intervals. For example, Colquhoun and Sakmann (1985) observed groups of channel openings separated by very short shut periods of average duration around 40 μs, even though the agonist concentration was so low that these groups occurred, on average, at intervals of the order of 500 ms (i.e., 10^4 times longer). Empirically speaking, openings will appear to be grouped into bursts whenever the distribution of all shut times requires two (or more) exponentials to fit it. If the time constants for the exponentials are very different, as in the above example, the bursts will be very obvious, and it will usually be quite clear whether any particular shut period should be classified as being within a burst or between bursts. If the time constants differ by less than a factor of 100 or so, the distinction becomes progressively more ambiguous.

Burst characteristics can be rigorously defined in at least two different ways. These two definitions will be, for practical purposes, equivalent in cases (such as the example given above) in which the bursts are very obvious, but in general, they are different. The definitions are as follows.

1. A burst of openings can be defined empirically as any series of openings separated

by gaps that are all less than a specified length (t_{crit}, say). In the example given above, we might take $t_{crit} = 0.4$ ms; the probability that a gap with a mean duration of 50 μs will be longer than 0.4 ms is about 0.3 per 1000, and the probability that a gap with a mean duration of 500 ms is less than 0.4 ms is about 0.8 per 1000. Thus, there is little chance that a gap would be wrongly classified in this case. A suitable value for t_{crit} must be chosen by inspection of the distribution of all shut periods before burst analysis is attempted.

2. A gap within a burst can be defined, for a particular mechanism, as a sojourn in a particular (short-lived) state (or set of states), for example, the blocked state in the case of a simple ion channel-blocker mechanism (see Section 4 of Chapter 18, this volume). Gaps between bursts are then similarly defined as sojourns in a different (long-lived) state or set of states. This definition was adopted by Colquhoun and Hawkes (1982; see Chapters 18 and 20, this volume). Unlike the first definition, it depends on an interpretation of the observations in terms of mechanism. Conversely, though, it allows inferences about mechanism from the observations; it connects the theory with the observations. On the other hand, unlike the first definition, it is not an algorithm that can be automatically and empirically applied to a set of data regardless of subsequent interpretation.

Choice of the Critical Shut Time for Definition of Bursts

There is no unique criterion for the optimum way to divide an experimental record into bursts. At least three methods have been proposed.

Suppose that we wish to find a value of t_{crit} that lies between two components of the shut-time distribution. The slower component has, say, an area a_s and mean t_s, and the faster component is specified by a_f and τ_f (see equation 30).

Jackson *et al.* (1983) proposed that t_{crit} should be defined as the shut-time duration that minimizes the *total number* of misclassified intervals. This criterion involves solving for t_{crit} the equation

$$\frac{a_f}{\tau_f} e^{-t_{crit}/\tau_f} = \frac{a_s}{\tau_s} e^{-t_{crit}/\tau_s} \tag{45}$$

The criterion proposed by Magleby and Pallotta (1983) and by Clapham and Neher (1984) is to choose t_{crit} so that *equal numbers* of short and long intervals are misclassified. This involves solving for t_{crit} the equation

$$a_f e^{-t_{crit}/\tau_f} = a_s(1 - e^{-t_{crit}/\tau_s}). \tag{46}$$

A third approach is to choose t_{crit} so that *equal proportions* of short and long intervals are misclassified (Colquhoun and Sakmann, 1985). In this case, t_{crit} is given by solving

$$e^{-t_{crit}/\tau_f} = 1 - e^{-t_{crit}/\tau_s} \tag{47}$$

None of these three equations can be solved explicitly, but the value of t_{crit} can be found easily by numerical solution by, for example, the bisection method (Press *et al.,* 1992) with τ_f and τ_s as the initial guesses between which t_{crit} must lie.

The three methods defined by equations 45–47 all give different values for t_{crit}, though 46 and 47 will be the same when $a_s = a_f$. When the areas for short and long intervals differ greatly, the first two methods (especially the first) may result in misclassification of a large proportion of the rarer type of interval, and so it may sometimes be felt to be more appropriate to use the third method, despite the fact that it does not minimize the total number of misclassifications.

When the time constants, τ_f and τ_s are very different, as in the example above, it will make very little difference which of the methods is used. But the difference that is needed is often underestimated. If the record contains N shut times (and N open times) the number of bursts that are found will be N times the probability that a shut time is greater than t_{crit}. The latter probability is, from equation 36,

$$P(\text{shut time} > t_{crit}) = \Sigma a_i e^{-t_{crit}/\tau_i} \tag{48}$$

In the case of the two-component shut-time distribution,

$$P(\text{shut time} > t_{crit}) = a_f e^{-t_{crit}/\tau_f} + a_s e^{-t_{crit}/\tau_s}$$

so, if $t_{crit} >> \tau_f$, the first term will be very small, and if $t_{crit} << \tau_s$, then the second term will be approximately a_s, so the number of bursts located will be Na_s for any value of t_{crit} that satisfies these criteria. Nevertheless, equation 48 shows that the number of bursts found decreases monotonically as t_{crit} is increased. There is no genuine plateau where it becomes independent of t_{crit}.

Consider, for example, the case where τ_s is 100 times longer than τ_f; e.g., $\tau_f = 1$ ms, and $\tau_s = 100$ ms. When $a_f = a_s = 0.5$, the three methods in equations 45–47 give, respectively, $t_{crit} = 4.65$ ms, 3.40 ms, and 3.40 ms. The total number misclassified per 100 openings is, respectively, 2.75, 3.34, and 3.34, but the first method misclassifies 4.5% of long openings and 0.95% of short openings, whereas the last misclassifies 3.34% of both. When there are more short openings than long (i.e., many openings per burst), say $a_f = 0.9$, $a_s = 0.1$, the results are the same for the last method, but 45 and 46 give $t_{crit} = 6.87$ ms and 5.18 ms respectively, and equation 45 gives the total number misclassified per 100 as only 0.757, though 6.6% of long openings and 0.10% of short are misclassified.

If, however, τ_s is only 10 times longer than τ_f, e.g., $\tau_f = 1$ ms and $\tau_s = 10$ ms, then, when $a_f = a_s = 0.5$, the three methods give $t_{crit} = 1.80$ ms, 1.80 ms, and 2.56 ms, respectively, but even equation 45 misclassifies 15.2 shut times per 100, with 22.6% of long shut times being misclassified. Clearly, a factor of 10 is not big enough. This is apparent immediately from the fact that 16.5% of intervals with a mean length of 1 ms are greater than $t_{crit} = 1.80$ ms, and 16.5% of intervals with a mean length of 10 ms are shorter than 1.80 ms.

Once bursts have been defined, many sorts of distribution can be constructed from the idealized record, some of which are now listed.

5.5.2. The Distribution of the Number of Openings per Burst

We simply count the number of apparent openings (r, say) in each burst. Unlike the other distributions to be considered, this is a discontinuous variable; it can take only the integer values 1,2, . . . , ∞. This number will, of course, be underestimated if some gaps are too short to be resolved (see Sections 5.2 and 6.11 and Chapter 18, this volume). If there is only one sort of open state, the number of openings per burst is expected to follow a geometric distribution, i.e.,

$$P(r) = (1 - \rho)\rho^{r-1} \tag{49}$$

which decreases with r (because $\rho < 1$). The mean number of openings per burst is

$$\mu = 1/(1 - \rho) \tag{50}$$

Further details are given in Sections 6.1 and 6.8. Notice that $P(r)$ decreases by a constant factor (ρ) each time r is incremented by 1. This property is characteristic of exponential curves, and the geometric distribution is in fact the discrete equivalent of the exponential distribution encountered elsewhere. When the mean becomes large, the distribution approximates the exponential distribution with mean μ, namely, $\mu^{-1}e^{-r/\mu}$.

In general, the distribution will be a mixture of several such geometric terms; the number of terms will often be equal to the number of open states but may be fewer in principle (apart from the problem that not all components may be detectable). The question of the expected number of components is quite complex and is discussed in Section 13.4 of Chapter 18 (this volume).

5.5.3. The Distribution of Burst Length

This is the length of time from the beginning of the first opening of a burst to the end of the last opening. Clearly, it will be relatively unaffected by the presence of short unresolved gaps, compared with the distributions of open times and of number of openings per burst. The distribution should be described by a mixture of exponentials, as in equation 30, under the usual assumptions. The number of exponential components is, in principle, quite large, being equal to the number of open states plus the number of short-lived shut states (see Chapter 18, this volume; Colquhoun and Hawkes, 1982). In practice it is unlikely that all components will be resolved, and under some circumstances the burst length distribution may be well-approximated by a single exponential, as described in Section 5.3 of Chapter 18 (this volume).

5.5.4. The Distribution of the Total Open Time per Burst

This is the total length of all the openings in each burst. It is also relatively insensitive to undetected brief openings or shuttings (shuttings that are brief enough to be missed will cause only a small error in measuring the total open time). This distribution should also be described by a mixture of exponentials, as in equation 30. It is, in principle, simpler than the distribution of burst length, because the number of components is expected to be equal to the number of open states (Chapter 18, this volume; Colquhoun and Hawkes, 1982). This, together with the fact that it is less sensitive to missed events than the distribution of apparent open times, makes it the best distribution to look at in order to make inferences about the (minimum) number of open states. The distribution of the total open time per burst is also of interest because it is predicted, surprisingly, that it will not be affected by a simple channel blocker (see Chapter 18, this volume). This prediction provides a useful way of investigating blocking mechanisms (Neher and Steinbach, 1978; Neher, 1983; Colquhoun and Ogden, 1985; Johnson and Ascher, 1990).

The distribution of the total shut time per burst may also be of interest for some sorts of interpretation (Colquhoun and Hawkes, 1982).

5.6. Cluster Distributions

Sakmann *et al.* (1980) observed that bursts of openings could themselves be grouped together into clusters of bursts with long gaps between clusters. They were looking at nicotinic channels with high agonist concentrations, and the long silent periods between clusters occurred when all the ion channels in the patch were in long-lived desensitized states. In records of this sort it is often possible to say, with a high degree of certainty, that all of the openings in one cluster originate from the same individual ion channel. All of the shut times within a cluster can therefore be interpreted in terms of mechanism, even when the number of channels in the patch is not known (see Section 8 of Chapter 18, this volume). Such clusters are also useful for measurement of the probability that a channel is open (P_{open}), as described in Section 5.1.7.

Another case in which clusters of bursts (and superclusters of clusters) have been observed is the NMDA-type glutamate receptor (Gibb and Colquhoun, 1991, 1992). Measurements at very low agonist concentrations allow resolution of this unusually complex structure if the individual channel activations and subdivision of the record into bursts of openings and into clusters of bursts should aid in the interpretation of such records. The relevant theory has been given by Colquhoun and Hawkes (1982). This can, of course, be done only when the time constants of the shut-time distribution are sufficiently well separated (see Section 5.1). The mean gap between clusters should preferably be at least 100 times greater than the mean gap between bursts (within a cluster); and the latter should preferably be 100 times greater than the shut times within a burst.

Of course, we are quite free to treat the whole cluster as a long burst by an appropriate choice of t_{crit} (see Section 5.5.1); these bursts can then be analyzed like any other (they will have a rather complex distribution of gaps within bursts). Equally, we can ignore the clustering and analyze the individual bursts as above (the distribution of gaps between bursts would then be rather complex).

When the record is divided into clusters of bursts, a large number of different sorts of distributions can then be constructed, for example, the length of the kth burst in a cluster and the distribution of gaps between bursts within clusters; further details are given by Colquhoun and Hawkes (1982).

5.7. Measurement and Display of Correlations

Certain types of mechanism can give rise to correlations between the length of one opening and the next or between the length of an opening and that of the following shut time. When this happens, the correlation will gradually die out over successive openings: there will be a smaller correlation between the length of an opening and the length of the next but one opening (described as a correlation with lag = 2), and so on for increasing lags. Such correlations have been observed for both nicotinic and NMDA receptors. Measurements of correlation can give information about mechanisms, in particular information about how states are connected, that cannot be found in any other way. The origin and interpretation of correlations are discussed in Section 10 of Chapter 18 (this volume), where appropriate references will be found. We shall discuss here the ways in which correlations may be measured and displayed.

5.7.1. Correlation Coefficients and Runs Test

Perhaps the simplest way to test for correlations is to use a *runs test,* as employed by Colquhoun and Sakmann (1985). To do this, open times (or shut times or burst lengths, etc.) that are shorter than some specified length (e.g., 1 ms) are represented as 0, and values longer than this length are represented as 1. We then ask whether runs of consecutive 0 values (or of consecutive 1 values) occur with the frequency expected for independent events. If, for example, long openings tend to occur together, this will produce long runs of 1 values. Say there are n_0 zero values, n_1 unity values, and $n = n_0 + n_1$ values altogether. The number of runs, N_r say, in the data is then counted, a run being defined as a contiguous section of the series that consists entirely of (one or more) 0 values or entirely of 1 values (thus 110001 has three runs). If the series is random, then the mean and variance of N_r will be

$$E(N_r) = \frac{2n_0n_1}{n} + 1 \qquad \text{var}(N_r) = \frac{2n_0n_1(2n_0n_1 - n)}{n^2(n-1)} \tag{51}$$

The test statistic

$$z = \frac{N_r - E(N_r)}{[\text{var}(N_r)]^{1/2}} \tag{52}$$

will have an approximately Gaussian distribution with zero mean and unit standard deviation, so a value of $|z|$ larger than about 2 is unlikely to occur by chance.

The extent of correlation for any specified lag m can be calculated as a correlation coefficient, r_m. If the observations (e.g., open times, shut times, burst lengths, etc.) are denoted $t_1, t_2, \ldots, t_n$, with mean $\bar{t}$, then the correlation coefficient is calculated as

$$r_m = \frac{\sum_{i=1}^{i=n-m} (t_i - \bar{t})(t_{i+m} - \bar{t})}{\sum_{i=1}^{i=n} (t_i - \bar{t})^2} \tag{53}$$

5.7.2. Distributions Conditional on Length of Adjacent Event

The calculations in the last section give no visual impression of the strength of correlations, but various graphical displays that do so can be made. For example, the distribution of the length of openings conditional on the length of adjacent shut time can be constructed. Examples of such conditional distributions are shown in Chapter 18, this volume, (Section 10, Fig. 13). If, as in these examples, short openings tend to occur next to long shuttings, then the distribution of open times, conditional on the open time being next to a long shutting, will show an excess of short openings (relative to the overall open-time distribution). In order to construct such a conditional distribution from experimental data, it is necessary to specify a range of shut times rather than a single value. For example, to construct a distribution of open times conditional on the adjacent shut time being between 0.05 and 0.3 ms, simply locate all the open times that are adjacent to shut times that fall in this range and plot the histogram of these openings.

A more synoptic view can be obtained by restricting attention to the mean open times rather than looking at their distribution. Define several shut-time ranges and then plot the mean open time (for openings that are adjacent to shut times in each range) against the midpoint of the range. It will generally be best to center these shut-time ranges around the time constants of the shut-time distribution. The mean open time may also be plotted against the *mean* of the shut times in the range rather than against the midpoint of the range. An example is shown in Chapter 18 (this volume, Fig. 12). The mean open time decreases as the adjacent shut time increases.

A third way to display correlation information is to construct a two-dimensional dependency plot (Magleby and Song, 1992). This plot is explained and illustrated in Chapter 18 (this volume, Section 10).

5.7.3. Distribution of Open Time Conditional on Position within the Burst

The distributions of quantities such as (1) the length of the kth opening in a burst or (2) the length of the kth opening in a burst for bursts that have exactly r openings are potentially informative when there are correlations in the data. If these distributions differ for different values of k (or of r), these variations can be tested against the predictions of specific mechanisms, which can be calculated as described by Colquhoun and Hawkes, 1982; Chapter 20, this volume). Such distributions are, however, likely to be rather sensitive to undetected brief events (see Section 6.11 below; Section 12 of Chapter 18, this volume). Their potential has yet to be exploited.

5.8. Distributions following a Jump: Open Times, Shut Times, and Bursts

It is often of interest to measure single-channel currents following a rapid (step-like) change of membrane potential or of ligand concentration (a *voltage jump* or *concentration jump*). The principles underlying such measurements are discussed and exemplified in Section 11 of Chapter 18 (this volume).

Notice that application of a rectangular pulse (of membrane potential or of ligand concentration) is actually *two* concentration jumps. In terms of macroscopic current, the first step is sometimes referred to as the "on-relaxation," and the second, when the stimulus is returned (usually) to the prejump condition, is referred to as the "off-relaxation." In the context of voltage-activated channels (but, for no particular reason, not for agonist-activated channels), the off-relaxation is often referred to as a "tail current"; it is probably rather unhelpful, though harmless, to use a separate term for an off-jump, since it does not differ in principle from an on-jump. Sometimes attention is focused mainly on the on-relaxation (e.g., when a step depolarization opens a voltage-activated channel); sometimes the main focus is more on the off-relaxation (e.g., the events following a brief pulse of agonist applied to an agonist-activated channel).

The distribution of the latency until the first opening occurs is of crucial importance for understanding topics such as the shape of synaptic currents or the mechanism of inactivation of sodium channels (see Chapter 18, this volume). In principle it is easy to measure it from experimental records. The main problem in practice is that it cannot be interpreted unless there is only one channel in the patch (or at least a known number of channels). This is often hard to achieve.

Even when the channel shows no correlations, the distribution of first latencies is expected to differ from that of other shut times (see Chapter 18, this volume), though in this case the distributions of all subsequent shut and open times should be the same as those at equilibrium. When the channel shows correlations, the distributions (and hence means) of the first, second, . . . open time, and shut time, after the jump may differ from their equilibrium values. If the channel also shows correlations between burst lengths, then the distributions of the first, second, etc. burst length following the jump will also differ. After a sufficient number of openings has occurred, the equilibrium distributions will eventually be attained. Further details and examples can be found in Chapter 18 (this volume, Sections 10 and 11).

5.8.1 Delays in the Recording System

When first latencies are being measured, it is obviously very important that we know precisely when the step was applied (i.e., where $t = 0$ lies on the experimental record).

Voltage Jumps

In the case of voltage jumps, this problem has been discussed in detail by Sigworth and Zhou (1992). It is important to compensate properly for the large capacitative current artifact that accompanies a voltage jump applied with the patch clamp. Methods for doing this are discussed in Chapter 7 (this volume) and by Sigworth and Zhou (1992). The voltage jump may not be applied to the patch at the precise moment that the command pulse is applied. This can happen because vagaries of the relative timing of DAC outputs and ADC inputs: these depend on the characteristics of the computer's real-time interface and on precisely how it is programmed. Delays may also occur when the command pulse is filtered (to reduce its maximum rate of rise). The true $t = 0$ point on the record can be estimated by measuring the time from when the command pulse starts to the midpoint of the instantaneous current (the current that flows "instantly" through channels that are already open when the potential changes). Alternatively, the capacity compensation can be slightly misadjusted, and then one can measure the time to the peak of the resulting capacitative current. These procedures are illustrated by Sigworth and Zhou (1992).

Concentration Jumps

In the case of concentration jumps, delays may be much greater than for voltage jumps. Typically, a jump is applied to an outside-out patch by moving (by means of a piezoelectric device) a theta glass pipette from which two solutions flow, so the interface between the solutions moves across the patch. Delays arise primarily because of the time taken for the command pulse to be translated into movement of the piezo and the time taken for the solution leaving the theta glass to reach the patch. The delay can be measured as follows. Break the patch at the end of the experiment and flow a hypotonic solution through one side of the application pipette; then measure the time from application of a command pulse to the piezo to the appearance of a junction response. It is obviously important that the relative position of patch and application pipette remains the same throughout. It is still better if the measurement of delay can be made with the patch intact, as it is during the experiment proper. This may be possible, for example, by applying a step change in potassium concentration while a potassium-permeable channel is open (the channel opening itself can be used to trigger the command pulse to the piezo). This method was used by Colquhoun *et al.* (1992) to estimate the rate at which concentration changes at the patch surface; it is also an ideal method to measure delay (as long as an appropriate channel can be found).

There will also be a delay in the current-measurement pathway, essentially all of which is caused by filtering. An eight-pole Bessel filter introduces a delay (in seconds) of $0.51/f_c$, where f_c is the −3 dB frequency in Hertz. For example, a 1-kHz filter introduces a delay of 510 μs (Sigworth and Zhou, 1992).

The fitting of the results of jump experiments is considered later, in Section 6.13.

5.9. Tests for Heterogeneity

It is, unfortunately, quite common for more than one sort of channel to be in the patch of membrane from which a recording is made. This may be the case not only with native receptors but also with recombinant channels expressed in oocytes (e.g., Gibb *et al.,* 1990); injection of a defined set of subunit RNAs does not necessarily guarantee that a single well-defined sort of channel will be produced (see also Edmonds *et al.,* 1995a,b). This sort of heterogeneity will make distributions confusing and serious kinetic analysis impossible. It is, therefore, important to know when it is present.

One criterion that has been used for agonist-activated channels is based on P_{open} measurements (see Section 5.1.7). At high agonist concentrations, when the probability of the channel being open is high, openings appear in long clusters separated by even longer desensitized periods (Sakmann *et al.,* 1980; see also Section 5.6). Because all of the openings in one cluster are likely to arise from the same individual channel, a value of P_{open} can be measured from each cluster (by integration or by measuring individual open and shut times; see Colquhoun and Ogden, 1988, for example). The next cluster may arise from a different channel, but it should give, within sampling error, the same value for P_{open} if all the channels in the patch are identical.

An excellent method for assessing whether the P_{open} values (or open times or shut times, etc.) vary to a greater extent than expected from sampling error was proposed by Patlak *et al.* (1986). They used a randomization test (an elementary account of the principles of randomization tests is given by Colquhoun, 1971). This method has been used, for example, by Mathie *et al.* (1991) and by Newland *et al.* (1991). Suppose that measurements are made on N clusters of openings, and n_i is the number of openings in the ith cluster. The observed scatter of the measurements, S_{obs}, can be measured as

$$S_{obs} = \sum_{i=1}^{N} n_i(y_i - \bar{y})^2 \tag{54}$$

where y_i represents the measurement of interest (e.g., P_{open} or mean open time or mean shut time) for the ith cluster. The probability of observing a value of S_{obs} (or larger) on the null hypothesis that the clusters are homogeneous, can then be found as follows. Take all the measured open and shut times from all the clusters as a single group and select at random from them N groups of n_i values. Then calculate the scatter from these artificially generated clusters, using equation 54 in exactly the same way as was done for the real measurements; this will produce a value that may be denoted S_{ran}. This randomization procedure is then repeated many times (e.g., 1000 or more). A histogram can be constructed from the values of S_{ran} so generated. The fraction of cases in which S_{ran} exceeds the observed value, S_{obs}, is the required probability. If it is very small, then it is unlikely that the null hypothesis was correct, and it must be supposed that the channels are heterogeneous.

6. The Fitting of Distributions

6.1. The Nature of the Problem

The term *fitting* means the process of finding the values of the constants in some specified equation that produce the best fit of that equation to the experimental data. This definition implies that one must (1) decide on an appropriate equation to fit to the data, (2) decide what the term "best" means, and (3) find an algorithm that can then find the best fit. It is perhaps worth noting that the process of fitting involves thinking in a somewhat inverted way. Normally, one thinks of the data as being variable and the parameters in an equation (e.g., the slope and intercept of a straight line) as being constants. During the process of fitting, though, the data are constant (whatever we happened to observe), but the parameters ("constants") are varied to make the equation fit the observations.

There are two quite different approaches to fitting, which may be called (1) empirical fitting and (2) fitting a mechanism directly. In the former case, exponentials (or geometrics) are fitted without necessarily specifying any particular mechanism; the parameters are the time constants and areas of the exponential components. In the latter approach, the parameters to be fitted are not the time constants of the exponentials but the underlying rate constants in a specified mechanism. The former approach is by far the most common, and it will be discussed next. The direct fitting of mechanisms requires consideration of missed events and will be discussed later, in Section 6.12.

6.1.1. Empirical Fitting of Exponentials

In practice, this usually means fitting a mixture of exponential distributions to data that consist of a list of time intervals (e.g., a list of apparent open times, shut times, or burst lengths, found as described earlier). Similarly, a mixture of geometric distributions may be used to fit the number of openings per burst, etc. This process is not entirely empirical, however, because there is good reason to expect that these may be appropriate equations. Any "memoryless" reaction mechanism is expected to result in observations that can be described by exponentials (or geometrics), as described in Chapter 18 (this volume), so they are obviously sensible things to fit. There is, of course, no guarantee that they will fit adequately. For example, (1) the effect of limited time resolution will, in principle, result in nonexponential distributions (e.g., Section 6.11, Chapter 18, this volume; Hawkes *et al.*, 1990, 1992), or (2) the transition rates may not be constant, e.g., because membrane potential or ligand concentration are not constant, or (3) the mechanism may be genuinely non-Markovian. These topics are discussed at greater length in Chapter 18 (this volume).

The general form for a mixture of exponential densities has already been given in equation 30. If a_i represents the area of the ith component, and τ_i is its mean or time constant, then, when there are k components,

$$f(t) = a_1\tau_1^{-1}e^{-t/\tau} + a_2\tau_2^{-1}e^{-t/\tau_2} + \cdots$$
$$= \sum_{i=1}^{k} a_i\tau_i^{-1}e^{-t/\tau_i} \tag{55}$$

The areas add up to unity, i.e., $\Sigma a_i = 1$, and the overall mean duration is $\Sigma a_i\tau_i$. Although it was stated above that the areas are proportional, roughly speaking, to the number of events in each component, it must be emphasized that, in general, the areas and time constants (means) of the components have no separate physical significance. An approximate physical interpretation of the components may be possible in particular cases (some examples are given in Chapter 18, this volume), but they must be demonstrated separately in each case. The density is sometimes written in the alternative form

$$f(t) = \sum_{i=1}^{k} w_i e^{-t/\tau_i} \tag{56}$$

where the coefficients w_i are the amplitudes (dimensions s^{-1}) of the components at $t = 0$. Clearly, they are related to the areas thus:

$$w_i = a_i/\tau_i. \tag{57}$$

The cumulative exponential distributions have already been given in equations 35 and 36.

6.1.2. Empirical Fitting of Geometrics

The general form for a mixture of geometric distributions (see Section 5.5.2) with k components, is

$$P(r) = \sum_{i=1}^{k} a_i(1 - \rho_i)\rho_i^{r-1}, \qquad r = 1, 2, \ldots, \infty \tag{58}$$

where a_i is the area of the ith component, and ρ_i is a dimensionless parameter ($\rho_i < 1$) (see Chapters 18 and 20, this volume). Alternatively, this can be written as

$$P(r) = \sum_{i=1}^{k} w_i\rho_i^{r-1} \tag{59}$$

where the coefficients w_i are the relative amplitudes, at $r = 1$, of the components. The area and amplitudes are related by

$$a_i = w_i/(1 - \rho_i) \tag{60}$$

The "means", μ_i, of the individual components (which, as for exponentials, will not generally have any separate physical significance) are

$$\mu_i = 1/(1 - \rho_i) \tag{61}$$

and the overall mean is

$$\mu = \sum_{i=1}^{k} a_i \mu_i \tag{62}$$

Thus, yet another general form for a mixture of geometric distributions is

$$P(r) = \sum_{i=1}^{k} a_i \mu_i^{-1}(1 - \mu_i^{-1})^{r-1} \tag{63}$$

Under certain circumstances (see Section 13.4 of Chapter 18, this volume), it is predicted that there will be a component with $\rho = 0$, i.e., from equation 62, a component with a mean of exactly one opening per burst. Such a component will contribute to $P(1)$ only.

The cumulative form of the geometric distribution, i.e., the probability that we observe n or more (e.g., the probability of observing n or more openings per burst) is

$$P(r \geq n) = \sum_{i=1}^{k} a_i \rho_i^{n-1} \tag{64}$$

6.1.3. The Number of Parameters to Be Estimated

In the cases of both exponential and geometric distributions there are $2k - 1$ parameters, the values of which must be estimated from the data by the fitting process. For exponentials there are k different time constants, τ_i (or rate constants, $\lambda_i = 1/\tau_i$) and $k - 1$ values for the areas, a_i (the areas must sum to 1, so estimation of $k - 1$ values defines the kth value). For geometrics, the parameters could be k values of μ_i (or of ρ_i), plus $k - 1$ values for the areas, a_i. Sometimes it may be desirable not to estimate all of these parameters from the data but to fix the values of one or more of them (they might, for example, be fixed at values that have already been determined from earlier experiments). This should improve the precision of the remaining parameters that are estimated from the data.

It will always be sensible to constrain the values of the time constants, τ_i, to be positive when fitting exponentials. Negative values are obviously impossible, so the fitting routine should be prevented from trying negative values. Similarly, in fitting geometrics, the values of μ_i should be constrained to be not less than 1 (or, if fitting ρ_i, the ρ_i values should be constrained to lie between 0 and 1). When fitting steady-state distributions, the areas, a_i, of the components are expected to be nonnegative too, so it may help the fitting process if they too are constrained. However, some sorts of distribution (for example that of the shut time preceding the first opening after a jump) may well have one or more negative areas; in such cases it is important that the program *not* constrain areas to be positive (see Sections 7 and 11 of Chapter 18, this volume).

6.2. Criteria for the Best Fit

The usual approach is to define a measure of the goodness of fit (or of the badness of fit) of the fitted distribution to the experimental observations. The parameter values are then chosen to maximize the goodness of fit (or to minimize the badness of fit). Different measures of goodness of fit will give different estimates of the parameters from the same data.

For conventional curve fitting (e.g., to ordinary graphs or to macroscopic currents), the weighted least-squares criterion is usually supposed (and in some cases has been shown) to be the best method. In such cases the distribution of the observations is almost always unknown. In the single-channel context, though, the problem is rather different. The distribution of the observations *is* known—it is what is being fitted. It is, therefore, possible to do better by using the maximum-likelihood approach.

The likelihood function provides a measure of goodness of fit and is discussed in Sections 6.5–6.9. Other, less good, methods appear in the literature, e.g., the χ^2 statistic (which provides one measure of badness of fit and is discussed in Section 6.4). Still worse, one can find some wholly inappropriate use of least-squares criteria, or even "curve stripping" on semilogarithmic plots, but they are not worth discussion here.

All fitting methods will give much the same results if the amount of data is very large and the fit very close, but this is rarely the case in the real world. The maximum-likelihood method is undoubtedly preferable to any other for the purposes of fitting distributions, and the speed of computers is now such that there is no reason to use any other method.

6.2.1. How Many Components Should Be Fitted?

If a specified mechanism is being fitted (see Section 6.12), the mechanism dictates the number of exponential components, so there is no problem. But when exponentials are being fitted without reference to a mechanism, it is often difficult to decide how many components should be fitted to the observations. For example, in Fig. 15 the shut-time data are shown fitted with both a two-exponential fit and a three-exponential fit. The fastest and slowest components are obvious, but the intermediate component (with $\tau = 1.31$ ms) has only 3.7% of the area and could easily be missed, especially if the log display were not used and the histogram that reveals this component most clearly (Fig. 15*B*) were not inspected. It could also be missed easily if the amount of data were smaller.

There are three ways in which to judge the number of components that are needed: (1) visual inspection of the histograms—e.g., in Fig. 15 the need for the third component is pretty convincing when the appropriate display is inspected. (2) By checking the reproducibility of the time constants and areas from one experiment to another (if they are not reasonably reproducible you are probably trying to fit too many components). And (3) statistical tests - the question can be asked 'is there a statistically-significant improvement in the fit when an extra component is added?' The second of these methods is by far the most reliable.

The Statistical Approach

The statistical approach is easy to apply when the fitting is done by the method of maximum likelihood (see below). This and related questions are discussed by Horn (1987). Denote as L the maximum value for the log(likelihood), i.e., the value evaluated with the best-fit parameters, $L(\hat{\theta})$ (see Sections 6.5–6.8). Suppose that the same data are fitted twice. First we fit k_1 components (and hence $n_1 = 2k_1 - 1$ parameters), yielding a maximum value for the log-likelihood of L_1. Next we fit the same data again, but with more components (say k_2 components and hence $n_2 = 2k_2 - 1$ parameters); this time the maximum value for the log-likelihood is L_2. With more free parameters to adjust, the second fit is bound to be better (so $L_2 > L_1$), but is it *significantly* better or not? The extent of the improvement in fit can be measured by

$$R = L_2 - L_2$$

where R is the *log likelihood ratio,* i.e., the logarithm of the ratio of the two likelihoods, defined such that the ratio is greater than 1 ($R > 0$). It can be shown (e.g., Rao, 1973) that, if the correct number of components were k_1, then $2R$ would have (for large samples) a χ^2 distribution with $n = n_2 - n_1$ degrees of freedom. Thus, by obtaining the probability corresponding to $2R$ (by computation or from a χ^2 table), it is possible to judge whether the second fit is significantly better. The P value so found is the (approximate) probability that fitting with k_2 components would produce, by chance, an improvement in fit equal to (or greater than) that observed, if in fact the fit with k_1 components were correct. If P is sufficiently small, it would be concluded that chance alone is unlikely to account for the observed improvement, so the larger number of components is justified.

The Criterion of Reproducibility

The problems with the statistical approach are, as for all significance tests, of two sorts. First, a nonsignificant difference does not mean that there is no difference, merely that a difference could not be detected (possibly because it was not a good experiment). Second, the test copes only with random errors and cannot allow for the systematic errors that are so common in real experiments. Nevertheless, if the experiment cannot be repeated, this is probably the best approach.

Normally, though, a distribution (such as that in Fig. 15) is not determined only once but many (or at least several) times in separate experiments. The question then arises about what should be done if some experiments appear to be fitted well by two components but others require three. This question shows the inadequacy of the statistical approach. The number of components that are required is dictated by the mechanism involved and does not change from one experiment to another (as long as they are all done with the same channel type and are not invalidated by heterogeneity of the channels). However, the amount, and quality, of the data, and hence one's ability to *detect* components, may vary considerably from experiment to experiment. This is illustrated nicely by the history of the data shown in Fig. 15. At first distributions of this sort were usually fitted with two components. However, it became apparent that quite often the data needed three components, as in the case shown. Once this had become quite convincing, the earlier data sets were all refitted with three components, *whether or not this produced a significant improvement* in any individual experiment. The results showed that, within reasonable limits, the time constant and area of all three components were reproducible from one experiment to another. This is the strongest sort of evidence for the need for three components, and it is the procedure that should be adopted whenever possible.

6.3. Optimizing Methods

In order to begin, one should obtain a good optimizing computer subroutine or procedure. These are general-purpose programs that are designed to find (given initial guesses for them) the parameter values that minimize (or maximize) any specified function and so can be used to maximize the likelihood (or, equivalently, to minimize the negative likelihood—most routines are designed to mimimize).

The user has to supply only a subroutine or procedure that, when supplied with values for the parameters, will calculate a value for the quantity to be minimized (e.g., a value for the minus log-likelihood; see below). The minimization subroutine then adjusts the values for the parameters, and for each set of parameters it calls the user's routine to see how well the parameters fit the data. It is not expected that the user's routine will itself change the parameter values (though it can be useful to have it do so, as described below).

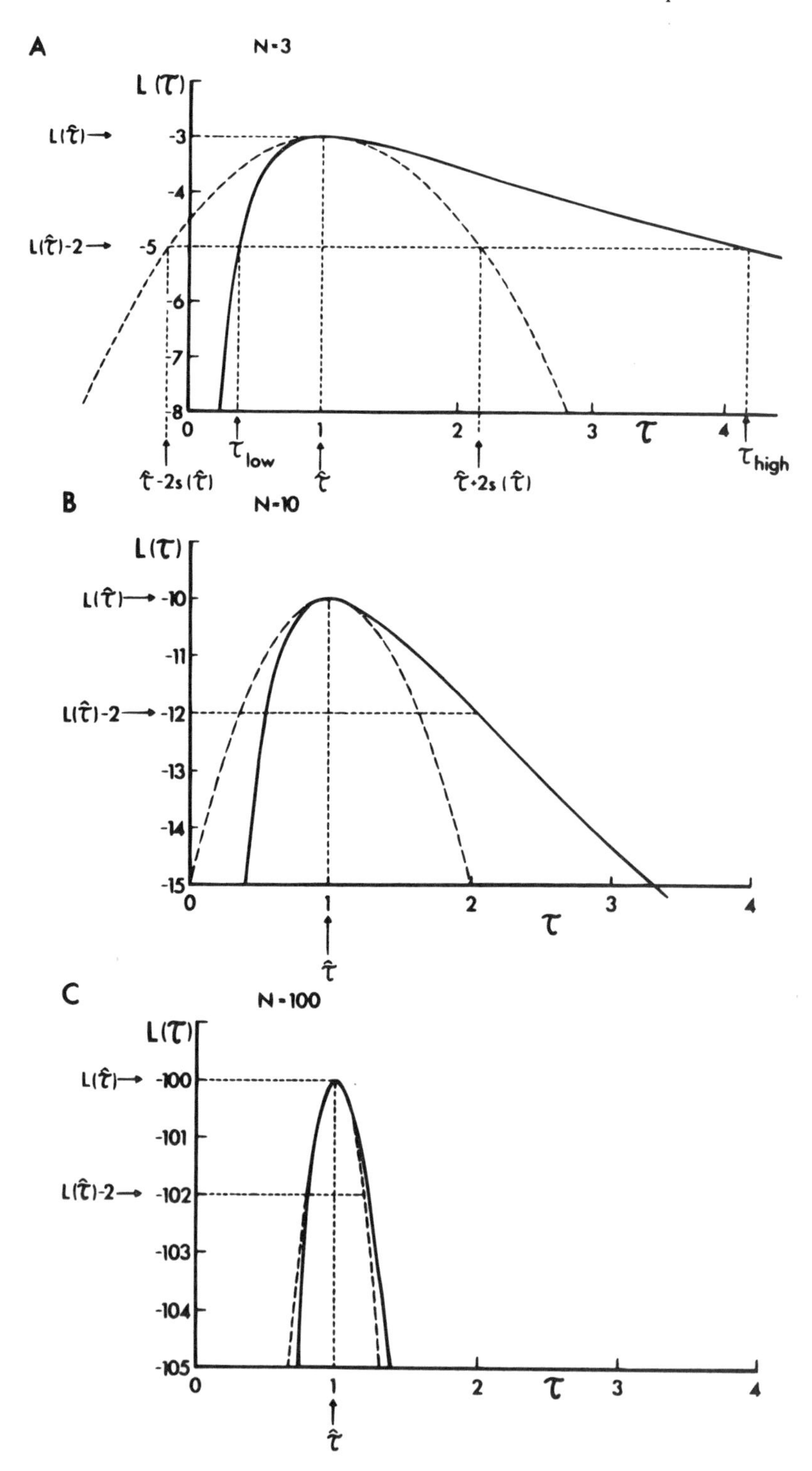
A
N=3
L(τ)
L(τ̂)→
-3
-4
L(τ̂)-2→
-5
-6
-7
-8
0
1
2
3
4
τ
τ_low
τ_high
τ̂-2s(τ̂)
τ̂
τ̂+2s(τ̂)
B
N=10
L(τ)
L(τ̂)→ -10
-11
L(τ̂)-2→ -12
-13
-14
-15
0
1
2
3
4
τ
τ̂
C
N=100
L(τ)
L(τ̂)→ -100
-101
L(τ̂)-2→ -102
-103
-104
-105
0
1
2
τ
3
4
τ̂

The fitting process is shown graphically in Fig. 19 in the case where there is only one parameter to be estimated. We simply find the value of the parameter that corresponds to the maximum log-likelihood. In the case where two parameters are to be fitted, we would need a three-dimensional version of this graph, with the possible values of the two parameters on the x and y axes and the value of the likelihood that corresponds to each pair of parameter values sticking out of the paper on the z axis. This sort of graph is often shown as a contour map in two dimensions, with parameter values on x and y axes and the corresponding likelihood values marked on contours. The contour map portrays, in geographic terms, a hill, and the problem is now simply to find the top of the hill; this is the maximum likelihood, and the pair of parameters that are the coordinates of the maximum are the *maximum-likelihood estimates* of the parameters. Since optimizing subroutines usually minimize functions, the more common geographic analogy is that we are searching for the bottommost point in a valley. These graphical analogies are usually an excellent way to picture what is happening (though in ill-behaved cases it is possible for contour lines to cross each other, which is not allowed in ordinary maps; e.g., see Colquhoun, 1971, Fig. 12.8.2).

There are very many programs available. They may be found in many standard libraries (such as NAG or IMSL) or by inquiry from your local computer center. The main choice lies between simple search methods and more complicated gradient methods. A much faster, noniterative method for fitting macroscopic exponential curves by use of Chebyshev polynomials is available, but it is inappropriate for fitting distributions. Even for macroscopic exponentials, it cannot be recommended until such time as the properties of the estimates it provides (in comparison with least-squares estimates) have been fully explored.

6.3.1. Simple Search Methods

The simple search methods look for the bottom of a valley by trying various sets of parameter values and simply noting whether one set of values is better than the previous set. "Better" means "further down the valley"; i.e., the user-defined subroutine produces a smaller value of the quantity to be minimized.

An advantage of search methods is that they usually converge (approach the bottom of the valley) reliably, even with poor initial guesses or when the function is ill behaved. These properties can be quite important. It is a considerable advantage in practice to be able to give rather rough initial guesses (it takes time to find good guesses). Even more important is the ability of these methods to cope with any sort of constraint on the values being fitted. For example, in fitting the time course of single-channel openings, as described in Section

---

Figure 19. The log-likelihood, $L(\tau)$, of a particular value of the time constant, τ, plotted against τ (continuous line) for the case of a simple exponential distribution (from equation 71). Graphs are given for samples of size $n = 3$ (A), $n = 10$ (B), and $n = 100$ (C). The dashed line shows the quadratic curve that has the same curvature at the maximum as $L(\tau)$, namely $Q(\tau) = Q(\hat{\tau}) - (\tau - \hat{\tau})^2/2s^2$ where $s = \hat{\tau}/\sqrt{n}$ (see equation 80). The curves have been drawn for the case $\hat{\tau} = 1$, and the abscissa can be interpreted as $\tau/\hat{\tau}$. In graph A, the definition of standard deviations and likelihood intervals is illustrated for the case of $m = 2$ unit likelihood intervals and the corresponding ± 2-standard-deviation intervals (see Table I and Section 6.7.2). A horizontal dashed line is drawn two units below the maximum, i.e., at $L(\hat{\tau}) - 2$. The points at which this intersects the continuous line give the lower and upper limits (τ_{low} and τ_{high}) for $\hat{\tau}$. The points of intersection with the dashed line give the 2-standard-deviation limits, $\hat{\tau} \pm 2s(\hat{\tau})$. For large samples, the dashed and continuous lines become similar, so the two approaches to error specification give similar results (see also Table I).

4.2, it is desirable to constrain the amplitude of a short opening to be the same as that of the nearest opening that is long enough to have a well-defined amplitude. But as the parameter estimates are adjusted, what is considered "short" and "long" may change, so the function being fitted changes as the fitting progresses, and this function may itself change the parameter values. A similar sort of thing occurs in fitting distributions. If the $k - 1$ areas being fitted are adjusted by the minimization routine so that they add up to 1 or more, and it is desired to prevent the kth area being negative (this is not *always* desirable—see Section 6.1.3), then the function that is being minimized can scale all the area values down so they add up to, say, 0.99, and return the altered values to the minimization routine. Such tricks are very useful, but gradient methods tend to take grave exception to them, whereas search methods, which care only about whether the function is reduced or not, carry on quite happily. Search methods also take little computer memory (though this is rarely critical with modern computers). On the other hand, search methods are usually rather slow, especially in the later stages of convergence when high precision is demanded.

Simple search methods include *patternsearch* (see Colquhoun, 1971), and the *simplex* method (Nelder and Mead, 1965; O'Neill, 1971; Hill, 1978; Press *et al.,* 1992). Care is needed because there are many versions of *simplex* in circulation, some of which are not very good. The version given by Press *et al.* (1992), called AMOEBA, appears to be quite satisfactory; the version they give is somewhat inconvenient to use as it stands, so an example is given, in Appendix 3 (Section A3.4), of a small subroutine that may conveniently be used to call AMOEBA. The program as it stands is rather minimal; it can be improved, for example, by adding code (1) to print out the progress of the iterations, (2) to abort the program from the keyboard if it appears to be stuck, (3) to test the convergence by the parameter step size rather than by the reduction in the function, (4) to keep track of the absolute minimum encountered (which may sometimes be better than the final result), and (5) to restart the minimization if a local search after convergence suggests that further improvement is possible. A particularly valuable addition is code to allow the values of specified parameters to be fixed (e.g., at values determined from other experiments) rather than estimated. This can be achieved by defining the parameter array (*theta,* in Section A3.4) to contain all of the parameters (so it can be used for calls to the function or for printing the current parameter values), but defining a second array from which the fixed values are omitted for use by *simplex* when it is adjusting the parameter values.

6.3.2. Gradient Methods

There are many types of gradient methods (see Press *et al.,* 1992, for a brief survey). They have in common the characteristic that, given a set of parameter values that define a point on the surface of the value, they calculate the slope of the surface at that point and use this value to work out the next set of parameter values to try. For example, they may work out the direction of steepest descent and follow this path in the hope that it is the fastest way to the bottom of the valley.

Gradient methods fall into two main categories, as far as the user is concerned. One category requires only that the user supply a subroutine to calculate, for a specified set of parameter values, the value of the function to be minimized, exactly as for search methods. The other category requires that, in addition, the user supply a subroutine to calculate the first derivatives of the function to be minimized. The latter type allows gradients to be calculated faster, but is much less flexible for the user, because for each function that is to be fitted the user must differentiate it algebraically and write a subroutine to evaluate these derivatives, which may be quite complicated.

The gradient methods usually take fewer iterations to converge and so may be much faster than search methods. On the other hand, they often converge less reliably and require better initial guesses, and it may be difficult or impossible to impose the required constraints on the fit with this sort of method, as exemplified above.

6.4. The Minimum-χ^2 Method

This method is really obsolete, but it will be described here because it has been quite widely used in the past and will give satisfactory results if the data are good enough. In order for this method to be used, the observations must first be grouped into a histogram. The data for the fitting are the frequencies of the observations in each bin. Thus, the parameter values will depend, to some extent, on the bin widths that are chosen to display the histogram (this is not the case with the maximum-likelihood method). The observed number of values in the jth bin will be denoted f_j^{obs}. The χ^2 statistic is a measure of the deviation of this observed value from the fitted (or calculated or expected) frequency. The value of the expected frequency depends, of course, on the values of the parameters (time constants, etc.) that are chosen, so we shall denote it $f_j(\theta)$ where θ represents the values of all the parameters. For example, when fitting two exponentials, the parameters could be τ_1, τ_2, and a_1, so $\theta = [\tau_1 \ \tau_2 \ a_1]$ (this is, in the notation of the appendix to Chapter 20, this volume, a *vector*, but it can be read here as a set of parameter values). The expected frequency is calculated from the equation for the distribution (e.g., equation 55), which, as discussed in Section 5.1.5, is approximately proportional to the frequencies if the bin width is not too wide. The values of the parameters are adjusted (by the optimizing program) to minimize χ^2, i.e., to minimize the badness of fit.

The χ^2 statistic is defined as

$$\chi^2 = \sum_{j=1}^{n_{\text{bin}}} \frac{[f_j^{\text{obs}} - f_j(\theta)]^2}{f_j(\theta)} \tag{65}$$

where n_{bin} is the number of bins in the histogram. Notice that, as in any fitting procedure, the data are treated as constants, and the parameters are treated as variables.

This method can be regarded as a sort of weighted least-squares approach; the denominator would be an estimate of the variance (reciprocal weight) of the numerator for a Poisson-distributed variable (the observed frequency in a given bin should be multinomially distributed, and this may approximate a Poisson distribution).

The fact that the denominator depends on the values of the parameters slows down the fitting procedure, and sometimes a modified χ^2 method has been used in which the observed, rather than the expected, values are used in the denominator. In other words, the parameters are chosen to minimize

$$\sum_{j=1}^{n_{\text{bin}}} \frac{[f_j^{\text{obs}} - f_j(\theta)]^2}{f_j^{\text{obs}}}. \tag{66}$$

If an observed value, f_j^{obs}, is zero, it must be replaced by unity (or by an average value over nearby bins) to avoid division by zero.

The χ^2 criterion, though reasonable, is arbitrary. It is also clear that, in principle, some information must be lost when the original time intervals are grouped into bins. For example, observations of 1.1 msec and 1.9 msec are treated as though they were both 1.5 msec if they are pooled into a bin from 1 to 2 msec. There is another, more natural way to fit the results that does not involve these disadvantages, namely, the method of maximum likelihood. This method also allows sensible estimates of error for the fitted parameters and is described next.

6.5. The Method of Maximum Likelihood: Background

When we have done an experiment and wish to choose the best values of the parameters, it seems sensible to ask what values of the parameters are, in the light of our data, the most probable. Although this may appear an innocent enough question, it has given rise to fierce debate for over three centuries. The debate still continues. The essential argument is about whether it is proper to talk about the probability of a hypothesis at all. If we measure durations of ion channel openings, we imagine that there is some real true value of the mean open time. Suppose our observed mean is 8 ms, and the true mean (which is never known of course) is 10 ms. The probability of the hypothesis that the true mean is 10 ms is unity; the probability that it is anything else (including 8 ms) is zero. Therefore, one cannot speak of the probability that the parameters have particular values (not, at least, if we wish to retain the familiar frequency interpretation of probability). Most people now think that the best way to circumvent this problem is to speak not of the probability of a hypothesis (given some data) but of the probability of getting the data (given an hypothesis). This latter probability was first used to measure the plausibility of hypotheses by Bernoulli in 1777; it was greatly developed and popularized by R. A. Fisher, who termed it *likelihood* from 1921 onwards.

The probability of observing the data, given a hypothesis, is just an ordinary probability distribution if the hypothesis is regarded as fixed and the data as varying. However, when we regard the data as fixed (as they are when we wish to analyze a particular experiment) and the hypothesis as varying, then this quantity no longer behaves like a probability, and we term it *the likelihood of the hypothesis*. In summary, denoting likelihood by Lik,

$$\text{Prob}[\text{data}\,|\,\text{hypothesis}] \equiv \text{Lik}[\text{hypothesis}\,|\,\text{data}] \tag{67}$$

In this expression, the vertical bar stands for "given" (see Section 2 of Chapter 18, this volume).

The method of maximum likelihood consists of varying the values of the parameters (the hypothesis) so as to maximize expression 67. Thus, we choose the parameter values that maximize the probability of observing our data.

This approach can be justified in two ways. The likelihood advocate would simply say that if you wish to use parameter values (such as minimum χ^2 values) that make the data less probable than his, then it is for you to justify your apparently perverse decision (see Edwards, 1972). Another approach is to examine closely the statistical properties of the method, which are, in most cases, at least as good as those of any other approach (see Rao, 1973).

Of course, in order to calculate the probability of getting the data, given some hypothetical parameter values, we need to know what probability distribution the observations follow. In most experimental work, this is not known with any certainty, so, although the method of least squares is actually the same as the method of maximum likelihood if errors follow a Gaussian distribution, the former term is usually used because knowledge of the distribution is uncertain. However, with data of the sort we are discussing here, we do know about the distribution. It is the very thing that we look at and wish to fit; this is why maximum likelihood is a natural procedure to adopt.

6.6. Maximum Likelihood for a Simple Exponential Distribution

These ideas can most easily be made clear by discussing data that follow a simple exponential distribution before going on to more general cases.

The data consist, say, of n time intervals, which we can denote $t_1, t_2 \ldots, t_n$. These are fixed, and this list provides the data on which fitting is based. Histogram frequencies are not used, and the values obtained are quite independent of the bin width(s) that are chosen for the histogram. It is still necessary to construct a histogram in order to display the final results of the fit, and the appearence of the histogram will vary to some extent according to the bin width(s) that are chosen, but the fitted line will not.

What, given some hypothetical value of the time constant τ, is the probability of making these observations; in other words, what is the likelihood of this value of τ? The time values are (in principle) continuous variables, so we must use probability densities rather than probabilities (but this does not matter much because we only need something that is directly proportional to the likelihood). The simple exponential distribution can be written

$$f(t) = \tau^{-1}e^{-t/\tau} \qquad 0 < t < \infty, \tag{68}$$

so the probability (density) of making the first observation t_1 is

$$f(t_1) = \tau^{-1}e^{-t_1/\tau} \tag{69}$$

The probability of making all the observations (t_1 *and* t_2 *and . . . and* t_n) is, if the observations are independent, simply proportional to the product of the separate probability densities, and this is the likelihood of the specified value of τ. Thus,

$$\mathrm{Lik}(\tau) = f(t_1)f(t_2)\cdots f(t_n). \tag{70}$$

It is more convenient to work with the logarithm of this quantity (so we get sums rather than products), and this log-likelihood is denoted $L(\tau)$. From equations 69 and 70, it is simply

$$L(\tau) = \sum_{i=1}^{n} \ln f(t_i) = n\ln(\tau^{-1}) - \tau^{-1}\sum_{i=1}^{n} t_i \tag{71}$$

This log-likelihood must, of course, reach its maximum at the same value of τ as does the likelihood (equation 70) itself. When $L(\tau)$ is plotted against various possible values of τ, it produces a curve like those shown in Fig. 19 (continuous lines). This curve summarizes all of the information that the data contain about τ. The curve goes through a maximum and the value of τ at the maximum is the value that makes the data most probable. It is the maximum-likelihood estimate (denoted $\hat{\tau}$) of the unknown true value of τ, i.e., of the true mean lifetime.

The position of the maximum can easily be found analytically in this simple case by differentiating equation 71 with respect to τ and equating the result to zero. This gives

$$\hat{\tau} = \sum_{i=1}^{n} t_i/n = \bar{t} \tag{72}$$

Not surprisingly, the estimate is simply the arithmetic mean of the observations.

With the help of equation 72, $L(\tau)$ from equation 71 can be written in a form that shows that the decline of the graph on either side of the maximum depends only on the ratio, $\tau/\hat{\tau}$, namely,

$$L(\tau) = L(\hat{\tau}) - n[\ln(\tau/\hat{\tau}) + (\tau/\hat{\tau})^{-1} - 1] \tag{73}$$

Consider next the (usual) case in which resolution is limited. Suppose that it is impossible to measure reliably any intervals (t_i values) less than some specified amount, $t_{\min}$ (see Sections 5.2, 6.8, and 6.11). Our observations are restricted to the range $t_{\min}$ to infinity. Therefore, rather than the simple exponential pdf in equation 68, we need the conditional pdf for t given that it is greater than $t_{\min}$. To obtain this, we divide by the probability that an observation is greater than $t_{\min}$ (see Section 2 in Chapter 18, this volume), which, from equation 36, is simply $\exp(-t_{\min}/\tau)$. This gives

$$f(t) = \frac{\tau^{-1}e^{-t/\tau}}{P(t > t_{\min})} = \frac{\tau^{-1}e^{-t/\tau}}{e^{-t_{\min}/\tau}} = \tau^{-1}e^{-(t-t_{\min})/\tau} \tag{74}$$

The log-likelihood is therefore

$$L(\tau) = \sum_{i=1}^{n} \ln f(t_i) = n\ln(\tau^{-1}) - \tau^{-1}\sum_{i=1}^{n}(t_i - t_{\min}) \tag{75}$$

Differentiating and equating to zero gives the maximum-likelihood estimate of the mean lifetime as

$$\hat{\tau} = \bar{t} - t_{\min} \tag{76}$$

i.e., we subtract the lower limit $t_{\min}$ from the mean of the observations. This relationship was used by Neher and Steinbach (1978); it is generalized in Section 6.8. The same relationship can be obtained by noting that for an exponentially distributed variable with mean τ, the mean of all observations longer than $t_{\min}$ is simply $\tau + t_{\min}$ (see the more general result following equation 85).

Once estimates of the parameters of the pdf have been found, we can estimate the true number (N) of observations, which includes those that have been missed because they are less than $t_{\min}$. This is is done simply by dividing the observed number, n, by the probability that an observation is greater than $t_{\min}$, i.e.,

$$N = \frac{n}{e^{-t_{\min}/\tau}} \tag{77}$$

This is generalized below, in equation 87. The expected frequency in a bin between t and $t + \Delta t$ is then simply

$$N(e^{-t/\hat{\tau}} - e^{-(t+\Delta t)/\hat{\tau}}) \tag{78}$$

This can be compared directly with the observed frequency (see Section 5.1).

In these cases there was no need for iterative computer optimization because the maximum-likelihood estimates could easily be calculated explicitly from equation 72 or 76. This cannot be done in general (see Section 6.8).

Non-independent Observations

The multiplication in equation 70 is correct only if the observations are independent. This is not always true. It is quite common, for example, for open times to be correlated; in the case of the muscle nicotinic receptor a long opening tends to be followed by another long opening. The question of correlations is discussed in more detail in Sections 5.7 and 5.8 and particularly in Chapter 18 (this volume, Sections 10, 11, and 13). When such correlations are present, the estimates obtained by the methods described here will not be genuine maximum-likelihood estimates, and errors calculated for the estimates will, to some extent, be erroneous. The effect of correlations on the fitting process has never been investigated in detail. It seems unlikely that the effects will be serious, and the bias of the estimates is unlikely to be worse than that of genuine maximum-likelihood estimates.

6.7. Errors of Estimates: The Simple Exponential Case

Once an estimate ($\hat{\tau}$) of the mean lifetime is obtained, it is natural to ask how accurate this estimate is likely to be. Estimates of error calculated from within a single experiment are notoriously unreliable and overoptimistic. The only reasonable estimate of error is found by repeating the whole experiment several times. Nevertheless, internal error estimates may be useful as a warning when an attempt is made to extract more information than the data contain, or in cases where repetition of the experiment is impossible. Two ways of estimating errors follow naturally from the maximum-likelihood approach. They are discussed next for the simple exponential case and generalized below. (It should be noted that these are not the only ways in which errors can be assessed; there is no general agreement about how this should be done in nonlinear problems.)

6.7.1. Approximate Standard Deviations

The first approach is to attach some sort of standard deviation to the estimate, $\hat{\tau}$, that has been found. A standard approach is to calculate the observed information by differentiating $-L(\tau)$ twice and then substituting $\hat{\tau}$ for τ. From equation 71 or 75 we obtain

$$-\left[\frac{\partial^2 L(\tau)}{\partial \tau^2}\right]_{\tau=\hat{\tau}} = \frac{n}{\hat{\tau}^2} \approx \frac{1}{\mathrm{var}(\hat{\tau})}. \tag{79}$$

The quantity in equation 79 has a simple interpretation. The second derivative measures the curvature of the graph (e.g., Fig. 19) near the maximum. If it is small, the graph is flat; i.e., the likelihood is rather insensitive to the exact value of τ; therefore, $\hat{\tau}$ is rather ill defined and has a large standard deviation. The reciprocal of expression 79 provides an estimate of the variance of $\hat{\tau}$, and its square root is an estimate of the standard deviation of $\hat{\tau}$, denoted $s(\hat{\tau})$. Thus, we obtain

$$s(\hat{\tau}) \simeq \hat{\tau}/\sqrt{n} \tag{80}$$

It should be noted that the validity of this estimate of error depends entirely on the assumption that the observations really do come from a population described by a single exponential pdf, so that we are fitting the right thing. Insofar as this will never be exactly true, the estimate is optimistic (or even meaningless).

The standard deviation for $\hat{\tau}$ found above, as for any standard deviation, can be interpreted in terms of a confidence interval only if we know the distribution of $\hat{\tau}$ (i.e., what the distribution of $\hat{\tau}$ values would be if we had many such estimates). If we suppose that $\hat{\tau}$ has a Gaussian distribution, which, from the central limit theorem, will be approximately true when the number of observations is large, then an approximate 95% confidence interval for $\hat{\tau}$ might be calculated as $\hat{\tau} \pm 2$ standard deviations; i.e.,

$$\hat{\tau} \pm 2s(\hat{\tau}). \tag{81}$$

The imperfection of this approach can easily be illustrated by an extreme example. Suppose we have only three observations, and their mean indicates that $\hat{\tau} = 2$ ms. Then the standard deviation of the mean is estimated as $2/\sqrt{3} = 1.15$ ms. Now calculate a confidence interval for $\hat{\tau}$ by taking two standard deviations on either side of $\hat{\tau}$, i.e., 2 ± 2.3 ms or -0.3 ms to $+4.3$ ms. According to this calculation, a value of $\tau = -0.3$ ms for the true mean lifetime is compatible with the observations, although it is obvious that all negative values are actually quite impossible. One way of looking at the reason for this silly result is that intervals calculated in this way are necessarily symmetrical (the Gaussian distribution is symmetrical), but more realistic error limits, such as those described in the next section, will not generally be symmetrical.

This example may be thought not to matter much because we never use such small numbers of observations. However, in some cases, we do wish to calculate the mean of quite small numbers. Consider, for example, the "intermediate shut times" (with $\tau = 1.31$ ms) in Fig. 15. Their mean length is of interest, but even in a long experiment, not many values can be observed, so absurdities like that just illustrated can easily occur in practice. They can be avoided by the method described in Section 6.7.2.

Standard Deviations and Standard Errors

Since the time intervals, t_i, follow a simple exponential distribution in this case, the standard deviation of the individual observations should, on average, be equal to the mean lifetime (e.g., Colquhoun, 1971); i.e., $s(t_i) = \bar{t}$. The standard deviation of the mean of n lifetimes, often known as the standard error of the mean, is calculated as $s(t_i)/\sqrt{n}$, which, since $\hat{\tau} = \bar{t}$ in this case, is just the result obtained in equation 80, but here it was obtained *via* the rather general method of equation 79. When quantities like that in equation 80 are obtained, it is often asked whether they are standard deviations or standard errors. This question is based on a common misunderstanding, because these are not two separate things. In fact, there is only one sort of measure of variability involved, and that is the standard deviation. This measure can be applied to any sort of variable quantity, as an index of how variable it is. It can be applied, for example, to a set of measured time intervals, t_i, and it will measure how much they vary for one interval to another. It can equally be applied to the mean of n lifetimes to measure how much repeated measurements of such means vary.

Or it can be applied to a time constant of a distribution (a τ value, equation 30), as a measure of how much repeated measurements of that τ value will vary. The standard error, a term that perhaps causes more misunderstanding than any other in elementary statistics, is not a separate sort of thing but is merely a piece of jargon standing for "the standard deviation of the mean of n observations" or, more generally, for "the (predicted) standard deviation of any quantity derived from the raw observations." The term standard error of the mean is still worse—it is not only misleading but also tautologous. The valid distinction is not between standard deviation and standard error but between (1) standard deviations that are estimated directly from a set of replicate observations (e.g., a set of measurements of individual lifetimes), the scatter of which can be directly observed, and (2) standard deviations that are calculated indirectly (e.g., standard deviation of the mean, or the standard deviation of a τ value) when we have actually got only one value (for the mean or for τ). In order to understand what the standard deviation means in the latter cases, we need to consider the standard deviation as a measure of how scattered the values would be *if* the quantity in question (the mean, or the τ value) were repeatedly estimated under identical conditions.

6.7.2. Likelihood Intervals

The second approach to estimation of errors, the calculation of likelihood intervals, overcomes these problems. This is quite easy in the case of simple exponentials (but uses quite a lot of computer time in more complex cases; see Section 6.9). The method is simply illustrated by the graph of the log-likelihood function, $L(\tau)$, against τ shown in Fig. 19. The maximum on the graph is at $\tau = \hat{\tau}$, so it is $L(\hat{\tau})$. If a horizontal line is drawn at a fixed distance, m $\log_e$ units, below the maximum, it intersects the graph at two points, one below $\hat{\tau}$ and one above $\hat{\tau}$.

The values of τ at these intersection points, τ_{low} and τ_{high} say, are, more formally, the (two) solutions of

$$L(\tau) = L(\hat{\tau}) - m \tag{82}$$

The values of τ_{low} and τ_{high} are clearly both less likely than $\hat{\tau}$ to the same extent (m $\log_e$-likelihood units), so it seems that they are good candidates to provide limits for the uncertainty in $\hat{\tau}$. They are called m-unit likelihood intervals or support intervals (see Edwards, 1972).

Conventional confidence intervals have an exact probability associated with them, but this is generally not possible in nonlinear problems of the sort that we have. Consider, however, a Gaussian variable with mean μ. In this case, the curve $L(\mu)$ has a simple quadratic form with constant curvature, from equation 79, and $\hat{\mu}$ is simply the arithmetic mean. In this case, the m-unit likelihood interval is just the conventional confidence interval defined as μ plus or minus $(2m)^{1/2}$ standard deviations, i.e.,

$$\hat{\mu} \pm (2m)^{1/2} s(\hat{\mu}) \tag{83}$$

Thus, there is a correspondence between $m = 0.5$ limits and one-standard-deviation limits; similarly, there is a correspondence between $m = 2$ limits and two-standard-deviation limits, and between $m = 4.5$ limits and three-standard-deviation limits.

The likelihood curves for a simple exponential distribution from equation 71 are plotted as continuous lines in Fig. 19 for samples of size $n = 3$, $n = 10$, and $n = 100$. The dashed curves in Fig. 19 show the corresponding quadratic curves that are implicitly assumed in the calculation of the approximate standard deviations. The values for error limits are tabulated in Table I (which is, like Fig. 19, normalized to unit value of $\hat{\tau}$). Thus, to return to the example that follows equation 81 with $\hat{\tau} = 2$ ms and $n = 3$, the two-unit likelihood interval, from Table I, is seen to be 2×0.379 to 2×4.16, i.e., 0.758 ms to 8.32 ms. These limits are unsymmetrical (from $\hat{\tau} - 1.242$ ms to $\hat{\tau} + 6.32$ ms), and are far more realistic limits than $\hat{\tau} - 2.3$ ms to $\hat{\tau} + 2.3$ ms, which were found from the "approximate standard deviation" approach.

It is clear from Fig. 19 and Table I that in the simple exponential case, approximate limits from equation 81 are quite satisfactory for samples of 100 or more, for which $\hat{\tau}$ has a nearly Gaussian distribution.

6.8. Maximum-Likelihood Estimates: The General Case

The case of a single exponential distribution has been discussed in Sections 6.6 and 6.7. The results given there generalize easily to any number of components.

In general terms, we denote the values of the parameters to be estimated (θ_1, θ_2, . . .) by the symbol θ and denote the jth observation as y_j, so the n data values are y_1, y_2, . . . , y_n. The probability (density) of a particular observation, y_1 say, given some trial values of the parameters, θ, is denoted $f(y_1|\theta)$. The probability of observing all of our particular data values is, for the specified θ, proportional to the product of all the individual probabilities (densities). This is, by definition, the likelihood of θ for our particular data. As before, we prefer to work with the logarithm of this quantity, which is

$$L(\theta) = \sum_{j=1}^{n} \ln f(y_j|\theta) \tag{84}$$

This can be calculated as soon as we specify the distribution explicitly. An optimizing computer routine can then find the values of the parameters that maximize $L(\theta)$; these are the maximum-likelihood estimates, and they are collectively denoted $\hat{\theta}$.

Table I. Likelihood Intervals[a] and Standard Deviations

	Sample size (n)			
$s(\hat{\tau}) = 1/\sqrt{n}$	3 0.577	10 0.316	100 0.100	Approximate probability[b]
$m = 0.5$	0.591, 1.89	0.741, 1.40	0.906, 1.11	
$\hat{\tau} \pm s(\hat{\tau})$	0.423, 1.58	0.684, 1.32	0.900, 1.10	0.68
$m = 2$	0.379, 4.16	0.564, 2.03	0.824, 1.23	
$\hat{\tau} \pm 2s(\hat{\tau})$	−0.155, 2.16	0.368, 1.63	0.800, 1.20	0.95
$m = 4.5$	0.260, 11.1	0.441, 3.08	0.751, 1.37	
$\hat{\tau} \pm 3s(\hat{\tau})$	−0.732, 2.73	0.051, 1.95	0.700, 1.30	0.997

[a]Comparison of m-unit likelihood intervals (from equation 82) and corresponding intervals calculated from approximate standard deviations (equation 81) for three sample sizes. The numbers in the table can be obtained from the graphs in Fig. 19-19, or by solving the equations. The numbers given are the limits on either side of $\hat{\tau} = 1.0$; they should be multiplied by the observed value of $\hat{\tau}$.

[b]This probability is based on the normal deviate by which the standard deviations are multiplied; use of Student's t statistic would give a better approximation.

This procedure can be made clearer if it is illustrated by the three most common sorts of distribution.

6.8.1. Mixtures of Exponentials

Distributions that have the form of a mixture of a number (k) of exponential densities are the most common; they have already been defined in Section 6.1. The parameters in this case are the time constants, $\tau_1, \tau_2, \ldots, \tau_k$, and the relative areas, $a_1, a_2, \ldots, a_{k-1}$. Alternatively, we could estimate the τ_i and the amplitudes w_i, or we could estimate the rate constants λ_i and the areas, a_i. It makes no difference which of these ways we choose, because, for example, $\hat{\tau}_i = 1/\hat{\lambda}_i$, so we get the same result whether the distribution is written in terms of rate constants or of time constants. However, the areas (a_i) are likely to be more nearly independent of the time constants than are the amplitudes (see also Section 6.10), so convergence may be easier if areas are estimated.

The distribution can be written, if we choose the time constants and areas as parameters, as in equation 55. Notice again that there are not $2k$ parameters but $2k - 1$, because the areas must add up to unity, as in equation 31.

Limiting the Fitted Range

In practice, limited frequency resolution means that nothing shorter than t_{min} can be measured; this limitation can be incorporated into the fitting procedure, as described in Section 6.6. Sometimes we may wish to exclude values below some t_{min} value that is greater than the resolution. We may also sometimes wish to exclude from the fit all values that are longer than some specified length t_{max} (e.g., to exclude a small number of exceptionally large values). Therefore, we need, in general, the conditional pdf, given that all the observations are between t_{min} and t_{max}. This is given by

$$f(t) = \frac{\sum_{i=1}^{k} a_i \tau_i^{-1} e^{-t/\tau_i}}{P(t_{min} < t < t_{max})} \qquad (t_{min} < t < t_{max}), \tag{85}$$

which is a generalization of equation 74. The mean value of such censored observations is

$$E(t) = \frac{\sum_{i=1}^{k} a_i [(t_{min} + \tau_i) e^{-t_{min}/\tau_i} - (t_{max} + \tau_i) e^{-t_{max}/\tau_i}]}{P(t_{min} < t < t_{max})}$$

The denominator in these results is simply the probability that an observation with the distribution in equation 55 lies between t_{min} and t_{max}, namely, from equation 36,

$$P(t_{min} < t < t_{max}) = \sum_{i=1}^{k} a_i (e^{-t_{min}/\tau_i} - e^{-t_{max}/\tau_i}) \tag{86}$$

The observations consist of n measured time intervals $t_1, t_2, \ldots, t_n$. Equation 85 can be evaluated for each of these in turn using some particular trial values (θ) of the parameters.

The logarithms of these values are added to give, from equation 84, the value of $L(\theta)$. The optimizing program then adjusts the parameter values so as to maximize $L(\theta)$. The values of parameters that do this are the maximum-likelihood estimates $\hat{\tau}_1, \hat{\tau}_2, \ldots, \hat{a}_1, \hat{a}_2, \ldots$ An estimate of the true number of observations, N (including those shorter than t_{min} or longer than t_{max}), can then be obtained from the observed number, n, as in equation 77:

$$N = \frac{n}{P(t_{min} < t < t_{max})} \tag{87}$$

where the denominator is as given by equation 86, with $\hat{\tau}_i$, $\hat{a}_i$ substituted for τ_i, a_i. A numerical example is considered in Section 6.10.

6.8.2. Mixtures of Geometric Distributions

In general, the distribution of the number of openings per burst, and similar quantities, is expected to be a mixture of one or more (k, say) geometric distributions of the sort defined already in equation 58 (see also Chapter 18, this volume). The distribution gives the probability of observing r (openings per burst, for example), and it can be written in a number of different ways. Alternative forms are given in equations 58, 59, and 63. In general, we may wish to include in the fitting process only those observed values that are between r_{min} and r_{max} inclusive. Thus, as in the exponential case, we need the conditional distribution, which, from equation 58, is:

$$P(r) = \frac{\sum_{i=1}^{k} a_i(1 - \rho_i)\rho_i^{r-1}}{P(r_{min} \leq r \leq r_{max})}, \qquad (r_{min} \leq r \leq r_{max}) \tag{88}$$

From equation 64, the denominator is given by

$$P(r_{min} \leq r \leq r_{max}) = \sum_{i=1}^{k} a_i(\rho_i^{r_{min}-1} - \rho_i^{r_{max}}) \tag{89}$$

The data consist of a series of n observations of the variable r, which we can denote $r_1, r_2, \ldots, r_n$. These might be, for example, the number of openings observed in n different bursts. The probability of observing all of these values is given by the product of the $P(r_j)$ values, so the log-likelihood is

$$L(\theta) = \sum_{j=1}^{n} \ln P(r_j) \tag{90}$$

where $P(r_j)$ is calculated from equation 88 for particular values of the parameters (a_i and p_i), which are collectively denoted θ. The optimizing program adjusts the values of these parameters until $L(\theta)$ is maximized, as usual. If there is only a single component, there is only one parameter, and if all observations are included ($r_{min} = 1$, $r_{max} = \infty$), then $L(\theta)$ can be maximized analytically in this case. This gives the maximum-likelihood estimate of the mean, $\hat{\mu}$, simply as $\bar{r}$, the arithmetic mean of the observations, and hence, from equation 61, $\hat{\rho} = 1 - (1/\hat{\mu}) = 1 - (1/\bar{r})$.

An estimate of the true number of observations, N (including those below r_{min} or greater than r_{max}), can then be obtained from the observed number, n, thus:

$$N = \frac{n}{P(r_{min} \leq r \leq r_{max})} \tag{91}$$

where the denominator is given by equation 89, with $\hat{\rho}_i$, $\hat{a}_i$ substituted for ρ_i, a_i.

6.8.3. Mixtures of Gaussian Distributions

The principles are exactly the same as in the other cases. Suppose that the variable y (usually a single-channel amplitude measurement in the present context) has a Gaussian distribution with mean μ and standard deviation σ. Its probability density function is

$$f(y) = \frac{1}{\sigma(2\pi)^{1/2}} e^{-u^2/2} \tag{92}$$

where

$$u = \frac{(y - \mu)}{\sigma} \tag{93}$$

is the "standard Gaussian deviate."

A mixture of k Gaussians is, therefore,

$$f(y) = \sum_{i=1}^{k} a_i f_i(y) \tag{94}$$

where $f_i(y)$ represents the Gaussian in equation 92 with mean μ_i and standard deviation σ_i, and a_i are the relative areas of the components.

The cumulative form of the Gaussian distribution, the probability that y is less than some specified value, y_1 say, is the integral of $f(y)$,

$$P(y \leq y_1) = \int_{y=-\infty}^{y_1} f(y)dy \tag{95}$$

Unlike the other cumulative distributions given above, this one cannot be written in an explicit form. However, it is easy to calculate values for it in a computer program, since all mathematical function libraries contain routines to calculate values of the error function, erf(x) (see also Appendix 3). The cumulative Gaussian distribution is simply related to the error function, thus:

$$P(y \leq y_1) = 0.5[1 + \text{erf}(u_1/\sqrt{2})] \tag{96}$$

where $u_1 = (y_1 - \mu)/\sigma$.

We shall often want to fit constants over a restricted range of values, excluding values below y_{min} and values greater than y_{max}. Again, we need the distribution of y conditional on y being between y_{min} and y_{max}. This is given by dividing $f(y)$, from equations 92 and 94, by $P(y_{min} < y < y_{max})$, which, from equation 96, can be calculated as

$$P(y_{min} < y < y_{max}) = 0.5 \sum_{i=1}^{k} a_i [\mathrm{erf}(u_i^{max}/\sqrt{2}) - \mathrm{erf}(u_i^{min}/\sqrt{2})] \tag{97}$$

where

$$u_i^{max} = \frac{(y_{max} - \mu_i)}{\sigma_i} \quad \text{and} \quad u_i^{min} = \frac{(y_{min} - \mu_i)}{\sigma_i} \tag{98}$$

The distribution of y, conditional on y being between y_{min} and y_{max}, is therefore

$$f(y|y_{min} < y < y_{max}) = \frac{f(y)}{P(y_{min} < y < y_{max})} \tag{99}$$

where $f(y)$ is given by equation 94 and the denominator is given by equation 97.

The data consist of a series of n observations of the variable y (e.g., channel amplitudes), which we can denote $y_1, y_2, \ldots, y_n$. The probability of observing all of these values is given by the product of the $f(y_j)$ values, so the log-likelihood is

$$L(\theta) = \sum_{j=1}^{n} \ln f(y_j) \tag{100}$$

where $f(y_j)$ is calculated from equation 99 for particular values of the parameters (a_i, μ_i, and σ_i), which are collectively denoted θ. In the case of Gaussian fits, there are $3k - 1$ parameters to be estimated. In cases where components overlap too much for all of these parameters to be estimated successfully, it may be helpful to constrain the standard deviation to be the same for all k components. In this case, there will be $2k$ parameters to be estimated, namely, k values of the means (μ_i), $k - 1$ values for the areas (a_i), and one value of σ. The optimizing program adjusts the values of these parameters until $L(\theta)$ is maximized, as usual.

An estimate of the true number of observations, N (including those below y_{min} or greater than y_{max}), can then be obtained from the observed number, n, as before, from

$$N = \frac{n}{P(y_{min} < y < y_{max})} \tag{101}$$

where the denominator is given by equation 97 with the maximum-likelihood values substituted for the parameters.

6.8.4. Binned Maximum-Likelihood Fits

The full maximum-likelihood fitting method is quite fast enough for it to be feasible, on a fast PC, to fit up to, say, five exponential components to several thousand intervals.

With more components or more data (or a slow computer), the full fit may become inconveniently slow. If a faster method is really necessary, the binned maximum likelihood method (Sigworth and Sine, 1987) should be used. In this method we use, to calculate the likelihood, not the probability (density) of observing a particular interval (given a set of parameter values) but, rather, the probability that our particular bin frequencies will be as observed. The values for the fitted parameters will, therefore, no longer be independent of how the bin boundaries are chosen. However, it has been shown, for logarithmically binned data (see Section 5.1.3), that the results are likely to be close to those from the full maximum-likelihood fit if at least 8–16 bins per decade are used (Sigworth and Sine, 1987).

The quantity to be maximized, the "binned log-likelihood," can be written in the form

$$L(\theta) = \sum_{j=1}^{n_{\text{bin}}} n_j \ln\left[\frac{F(t_{j+1}) - F(t_j)}{P(t_{\text{min}} < t < t_{\text{max}})}\right] \tag{102}$$

where the number of terms summed is now the number of bins, n_{bin} (rather than the total number of intervals), n_j is the number of observations in the jth bin, and t_j is the lower boundary of the jth bin. The numerator of this expression uses the cumulative distribution, $F(t)$, as given in equation 35 or 36 to calculate the probability, for the specified parameter values, θ, that an observation lies in the jth bin. The denominator, which was defined in equation 86, gives the probability that an observation is within the fitted range, t_{min} to t_{max}, the values of which must, in this case, correspond to bin boundaries.

6.9. Errors of Estimation in the General Case

The treatment in Section 6.7 can be generalized with the help of matrix notation, so that the two sorts of error calculation can be calculated for distributions with any number of parameters. An introduction to this notation is given in Chapter 20 (this volume). Further details can be found in Box and Coutie (1956), Beale (1960), Bliss and James (1966), Edwards (1972), and Colquhoun (1979). The following procedures are reasonable approaches to the specification of errors, but they are not unique.

6.9.1. Approximate Standard Deviations

Denote the parameters, ν in number, as $\theta = (\theta_1, \theta_2, \ldots, \theta_\nu)$. The analogue of equation 79 is the observed information matrix, $\mathbf{I}(\theta)$, which is a $\nu \times \nu$ matrix with elements

$$-\left(\frac{\partial^2 L(\theta)}{\partial\theta_i \partial\theta_j}\right)_{\theta=\hat{\theta}} \tag{103}$$

This form is known as a Hessian matrix. The inverse of this matrix gives the covariance matrix, $\mathbf{C}(\theta)$, of the parameter estimates, so

$$\mathbf{C}(\theta) \simeq \mathbf{I}(\theta)^{-1} \tag{104}$$

The elements of this matrix may be denoted $\text{cov}(\theta_i,\theta_j)$. The diagonal elements of $\mathbf{C}(\theta)$ give estimates of the variances of the parameter estimates, θ_i. Thus,

$$\mathrm{var}(\theta_i) = \mathrm{cov}(\theta_i,\theta_i) \tag{105}$$

The square root of this gives the approximate standard deviation of the parameter estimate, θ_i. The off-diagonal elements ($i = j$) give the covariances of these estimates. These measure the tendency of the estimate of θ_i to be large if the estimate of θ_j happens to be large (see Section 6.10 for examples). This tendency is more conveniently expressed as a correlation coefficient, r_{ij}, between the two estimates; this can be calculated as

$$r_{ij} = \frac{\mathrm{cov}(\theta_i,\theta_j)}{[\mathrm{var}(\theta_i)\mathrm{var}(\theta_j)]^{1/2}} \tag{106}$$

It has been noted that if we fit the sum of k exponentials, only $k - 1$ areas ($a_1, \ldots, a_{k-1}$, say) are estimated. The area, a_k, for the kth component follows immediately from the fact that the total area for the pdf is unity:

$$a_k = 1 - \sum_{i=1}^{k-1} a_i . \tag{107}$$

A standard deviation can be attached to a_k by the relationship

$$\mathrm{var}(a_k) = \sum_{i=1}^{k-1} \mathrm{var}(a_i) + 2 \sum_{i=1}^{k-1} \sum_{\substack{j=1 \\ j<i}}^{k-1} \mathrm{cov}(a_i,a_j) \tag{108}$$

The right-hand side of this equation is simply the sum of all the elements in those rows and columns of $\mathbf{C}(\theta)$ that refer to the $k - 1$ estimates of areas. If there are only two components, it reduces to $\mathrm{var}(a_2) = \mathrm{var}(a_1)$. For three components it reduces to $\mathrm{var}(a_3) = \mathrm{var}(a_1) + \mathrm{var}(a_2) + 2\mathrm{cov}(a_1,a_2)$.

Explicit algebraic derivation of equation 103 or 104 would be a formidable task in all but the simplest cases, but, fortunately, it is not necessary. The second derivatives in equation 103 can be estimated by standard numerical methods as long as we have a subroutine to calculate $L(\theta)$ for specified values of the parameters. The Hessian so found can be inverted numerically by means of a matrix-inversion routine (see Chapter 20, this volume) to give the covariance matrix according to equation 104.

6.9.2. Likelihood Intervals and Likelihood Regions

Likelihood intervals can also be calculated in the general case, and this is probably one of the best ways of expressing errors for the parameters taken one at a time. In principle, it would be better to calculate a joint likelihood region for all k parameters, but such a k-dimensional region cannot be simply represented when $k > 3$. An example of a joint likelihood for the case where two parameters are estimated is shown in Fig. 20 (see also Colquhoun, 1979). The graph shows a contour for $L(\theta) = L(\hat{\theta}) - 2$, so any pair of parameter values, θ_1 and θ_2, that lie on this contour are 2 log-likelihood units less likely than the best estimates, $\hat{\theta}_1$ and $\hat{\theta}_2$. The obliqueness of the contour shows that the estimates of θ_1 and θ_2 are positively correlated in this case; i.e., if both θ_1 and θ_2 were decreased, or both were increased, the fit would be little worse; i.e., $L(\theta)$ would be reduced only slightly. This may be compared with the effect of increasing θ_1 and decreasing θ_2 (or vice versa); this would cause the fit to become much worse. The tangents to the contour are also shown in Fig. 20; they define (see text) 2-unit limits for $\hat{\theta}_1$ and $\hat{\theta}_2$ *separately*. When the parameter estimates are correlated, as in this example, these limits for the individual parameters are, in a sense, pessimistic: if, for example, the true value of θ_1 were actually near $\hat{\theta}_1^{\text{low}}$, the correlation makes it improbable that the true value of θ_2 would be near $\hat{\theta}_2^{\text{high}}$. In order properly to take into account the correlation between the parameter estimates, a joint likelihood region (the contour in Fig. 20) is preferable. Points outside this region define pairs of θ_1, θ_2 values that are unlikely.

The numerical calculations that are needed to calculate likelihood regions or intervals take a good deal longer than those for the approximate standard deviations but are perfectly feasible on fast personal computers. The principle is very simple. The *m*-unit likelihood limits (see Section 6.7 for explanation of this term) for a particular parameter, θ_1, say, are defined as the values of θ_1 such that, if θ_1 is held constant at that value, and the likelihood $L(\theta)$ is maximized again, allowing all the parameters except θ_1 to vary freely, then the maximum value of $L(\theta)$ that can be attained is $L(\hat{\theta}) - m$; i.e., it is m units less than the true maximum, $L(\hat{\theta})$, which is attained when all of the parameters are allowed to vary.

In order to calculate the lower or upper limit for θ_1, iterative procedures are used. An initial guess is made, and the minimization is performed with θ_1 fixed at this value; if the maximum attained is not $L(\hat{\theta}) - m$, then the whole process is repeated by any standard iterative method (e.g., bisection or Newton–Raphson). This process is illustrated graphically

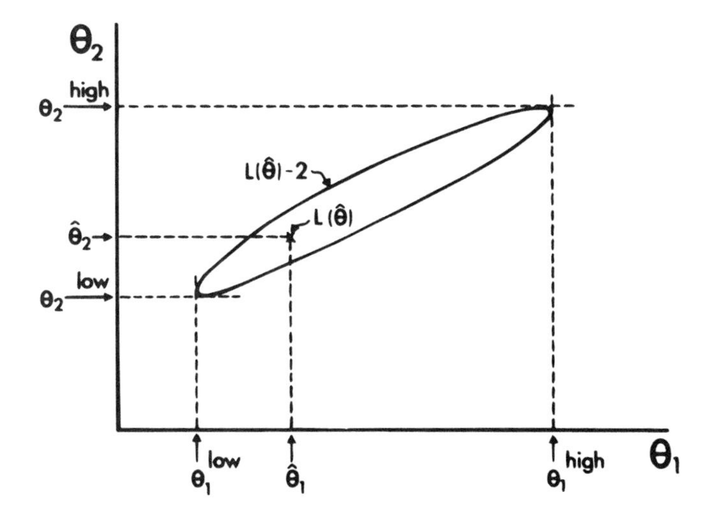

Figure 20. Schematic illustration of a joint likelihood region and of likelihood limits for the separate parameters in a case in which there are two parameters, θ_1 and θ_2, so $\theta = (\theta_1\ \theta_2)$. The graph shows a contour map of $L(\theta)$ with the peak of the hill, $L(\hat{\theta})$, marked with a cross; this corresponds with the maximum-likelihood estimates, $\hat{\theta}_1$ and $\hat{\theta}_2$, of the two parameters, as shown. The contour for $L(\hat{\theta}) - 2$ is shown. Further explanation is given in the text.

for the case when there are two parameters in Fig. 20. A numerical example is illustrated in Fig. 21.

If there are two components, the likelihood limits for a_2 are simply unity minus the limits for a_1. If there are more than two components (cf. equations 107 and 108 above), then in order to find limits for a_k it is necessary to refit the whole curve, so that a_k becomes one of the parameters that is estimated rather than the one inferred from the fact that the total area is unity.

6.10. Numerical Example of Fitting of Exponentials

The simultaneous fit of a triple-exponential pdf can be illustrated by data on shut times that were obtained with a low concentration (100 nM) of suberyldicholine (*R. temporaria*, cutaneous pectoris endplate $E_m = -123$ mV, 10°C). The results are similar to those shown in Fig. 15. The total number of openings fitted was 1021, but after imposition (see Section 5.2) of a minimum resolvable time of 50 μs (for both openings and gaps) and elimination of a few shut times that were unusable (because, for example, they contained ambiguous openings or simultaneous openings of more than one channel), the number of shut times to be fitted was 934. It was decided (see Section 5.2) to fit all durations between $t_{min} = 50$ μs and $t_{max} = 2000$ s, a total of 931 shut times.

The estimates of the time constants for the three components, found by maximizing $L(\theta)$ from equation 84 with $f(t)$ given by equation 85 were $\hat{\tau}_1 = 45.2$ μs $\hat{\tau}_2 = 1.28$ ms and $\hat{\tau}_3 = 440$ ms. The areas under the pdf accounted for by these components were, respectively, 74.0% (i.e. $\hat{a}_1 = 0.740$), 2.3% ($\hat{a}_2 = 0.023$), and 23.7% ($\hat{a}_3 = 0.237$). The maximum value of $L(\theta)$ attained was $L(\hat{\theta}) = -2899.33$. The fitted curve resembles that shown in Fig. 15. This fit implies, from equations 86 and 87, that the true number of shut times is $N = 1860.0$, of which 931 are in the observed range (the data), 922.3 are shorter than 50 μs, and 4.7 are above 2000 s.

The component with intermediate rate ($\hat{\tau}_2 = 1.28$ ms) is quite small and, as expected, has the largest relative errors. Nevertheless, the error calculations below give no real reason to doubt its reality; and, far more important, the need for this component is visible to the eye when the data are displayed appropriately (e.g., as in Fig. 15B or D), and it is reproducible from experiment to experiment.

In general, of course, it is quite improper to speak of short gaps, intermediate gaps, and long gaps on the basis of this fit; there is one pdf (which happens to be described by the sum of three exponentials), not three simple exponential pdfs. At least, it is improper unless we *define* the term "short gaps" in the manner suggested below, in which case the problem arises only when we wish to interpret the gaps so defined in terms of dwell times in particular states or sets of states. In some cases this convenient terminology can be justified, but only insofar as separate physical meanings can be attached, as an approximation, to the three components (see, for example, Colquhoun and Hawkes 1982; Chapter 18, this volume). Insofar as such an interpretation is valid, the data suggest that there are $N_f = \hat{a}_1 N = 1376$ "short gaps," $N_m = \hat{a}_2 N = 42.8$ "intermediate gaps," and $N_S = \hat{a}_3 N = 440.8$ "long gaps." Of the "short gaps," only $N_f e^{-50/45.2} = 455$ would be above 50 μs and therefore detectable.

First consider the errors for these estimates found by the approximate standard deviation approach (see Section 6.9.1). The second derivatives in equation 103 were estimated numerically; reasonable numerical accuracy is obtained by incrementing the parameters by ±10% from the maximum-likelihood values given above or by incrementing each parameter by enough to decrease $L(\theta)$ by 0.1. This provides an estimate of the observed information matrix.

Table II. Analysis of the Triple-Exponential Fit to Shut Time Duration[a]

Parameter	ML estimate $\hat{\theta}$	Approx SD $s(\hat{\theta})$	Likelihood intervals $m = 0.5$	Likelihood intervals $m = 2$	$2s(\hat{\theta})$
τ_1 (μs)	45.2	2.4	42.9–47.7 (−2.3 to +2.5)	40.6–50.5 (−4.6 to +5.3)	4.8
$100a_1$ (%)	74.0	1.6	72.2–75.5 (−1.8 to +1.5)	70.5–77.1 (−3.5 to +3.1)	3.1
τ_2 (ms)	1.28	0.42	0.90–1.76 (−0.38 to +0.48)	0.67–2.45 (−0.61 to +1.17)	0.84
$100a_2$ (%)	2.29	0.43	1.93–2.74 (−0.36 to +0.45)	1.55–3.32 (−0.74 to +1.03)	0.86
τ_3 (ms)	440.0	24.0	418–466 (−22 to +26)	396–494 (−44 to +54)	48.0
$100a_2$ (5)	23.7	1.5	22.3–25.4 (−1.4 to +1.7)	20.8–27.0 (−2.9 to +3.3)	3.0

[a]The maximum likelihood estimate, $\hat{\theta}$, of each parameter is given, with its approximate standard deviation, $s(\hat{\theta})$. Likelihood intervals are given in the form of intervals, and also, in parentheses, in the form of the deviation from $\hat{\theta}$. This deviation may be compared with $s(\hat{\theta})$ for the $m = 0.5$ unit intervals, and with $2s(\hat{\theta})$ (which is listed in the last column) for the $m = 2$ unit intervals.

This is then inverted numerically by means of any standard matrix-inversion subroutine to give the covariance matrix (equation 104) as follows (it is symmetric, so only the lower part is given):

$$\operatorname{cov}(\theta) \simeq \begin{bmatrix} 5.73 \times 10^{-6} & & & & \\ -2.20 \times 10^{-5} & 2.44 \times 10^{-4} & & & \\ 2.51 \times 10^{-4} & -1.97 \times 10^{-4} & 0.18 & & \\ 9.07 \times 10^{-8} & -1.88 \times 10^{-5} & -8.40 \times 10^{-5} & 1.81 \times 10^{-5} & \\ 1.50 \times 10^{-3} & -1.21 \times 10^{-2} & 1.28 & 3.40 \times 10^{-3} & 578.3 \end{bmatrix} \begin{matrix} \tau_1 \\ a_1 \\ \tau_2 \\ a_2 \\ \tau_3 \end{matrix} \tag{109}$$

(columns, left to right: τ_1, a_1, τ_2, a_2, τ_3)

The diagonal elements of this give the approximate variances of the parameter estimates (the order of the parameters is shown above, and to the right of, the matrix). The square roots of these variances are the standard deviations of the estimates and are shown in Table II. For example, for $\hat{\tau}_2$ the standard deviation is $s(\hat{\tau}_2) = (0.18)^{1/2} = 0.42$. The standard deviation for the area of the slowest component ($\hat{a}_3$) is obtained from equation 108 as $\operatorname{var}(\hat{a}_3) = 2.44 \times 10^{-4} + 1.88 \times 10^{-5} + 2(-1.88 \times 10^{-5}) = 2.25 \times 10^{-4}$, so the standard deviation for $\hat{a}_3$ is $(2.25 \times 10^{-4})^{1/2} = 1.5 \times 10^{-2}$, or 1.5%, as shown in Table II.

The correlation matrix is found from equation 109 by means of equation 106. It is

$$\begin{bmatrix} — & & & & \\ -0.59 & — & & & \\ 0.25 & -0.03 & — & & \\ 0.009 & -0.28 & -0.05 & — & \\ 0.03 & -0.03 & 0.13 & 0.03 & — \end{bmatrix} \begin{matrix} \tau_1 \\ a_1 \\ \tau_2 \\ a_2 \\ \tau_3 \end{matrix} \tag{110}$$

(columns, left to right: τ_1, a_1, τ_2, a_2, τ_3)

The correlation coefficient, for example, between the estimates of τ_1 and a_1 is $r_{21} = -2.20\times10^{-5}/[5.73\times10^{-6})(2.44\times10^{-4})]^{1/2} = -0.59$. This modest correlation is the strongest found; it reflects the intuitively obvious fact that the fit would be almost as good if τ_1 were decreased and a_1 increased, or *vice versa*. In other words, a faster time constant for the fast component would not reduce the goodness of fit much if the area allocated to this component were simultaneously increased (this implies a considerable increase in the amplitude of the fast component, $w_1 = a_1/\tau_1$: see equation (57). This correlation is aggravated by the lack of observations below 50 μs. There is also a small negative correlation (−0.28) between $\hat{a}_1$ and $\hat{a}_2$ and a small positive correlation (+0.25) between $\hat{\tau}_1$ and $\hat{\tau}_2$. Apart from these, the estimates are virtually independent. The fact that the slow component is well separated from, and nearly independent of, the other components means that a rough estimate of the standard deviation of its time constant can be calculated (compare equation 79) as $\hat{\tau}/\sqrt{N_S} = 21$ ms, which is not far from the value of 24 ms given by the full calculation (see Table II). For the small intermediate component, this approximation is, however, very poor; it gives $s(\hat{\tau}_2) = 0.19$ ms, compared with 0.42 ms from the full calculation.

The fact that only modest correlations are found for this fit is a good sign; it implies that the parameters are well-defined. If, for example, a strong positive correlation were found between two parameters, this would mean that if both were increased the quality of the fit would be little affected. In other words, the ratio of the two parameters is well defined, but their separate values are dubious.

The likelihood intervals for $m = 0.5$ and $m = 2.0$ (see Sections 6.7 and 6.9) are given for each parameter in Table II. It can be seen that the former are not far from what is expected from the approximate standard deviations, even for the small intermediate component, in this example (which has quite a large number of observations). The difference between the two approaches is larger in the case of the two-unit intervals, especially for the small component; for example, $\hat{\tau}_2 = 1.28$ ms, and $\hat{\tau}_2 \pm 2s(\hat{\tau}_2)$ implies an interval about $\hat{\tau}_2$ of $\hat{\tau}_2 - 0.84$ to $\hat{\tau}_2 + 0.84$ ms, whereas the two-unit likelihood interval gives $\hat{\tau}_2 - 0.61$ to $\hat{\tau}_2 + 1.17$ ms. The estimation of the limits for $\hat{\tau}_2$ is illustrated in Fig. 21 (see also Sections 6.7 and 6.9).

6.11. Effects of Limited Time Resolution

Virtually all experimental records contain intervals that are too short to be detected or measured, and this can cause serious distortion of distributions of open times, shut times, and number of openings per burst. The effect of missing brief events will be much less on distributions such as those of the burst length or the total open time per burst, so one way of dealing with the problem is to present only these distributions.

The practical aspects of this problem have already been described in Section 5.2.

The question of making corrections for missed events can be dealt with in two ways. The first, and most common, case occurs when no specific mechanism is being postulated for the channel under investigation. In this case it may be possible to make approximate corrections for missed events retrospectively. This can be done only in the case that *either* short openings *or* short gaps, but not both, are missed to any substantial extent. Such approximate corrections can also be done only when the kinetics of the observations are relatively simple. For example, if the distribution of (apparent) open times has more than one exponential component, then such corrections become difficult (though not necessarily impossible). Methods for making this sort of approximate correction are discussed, for

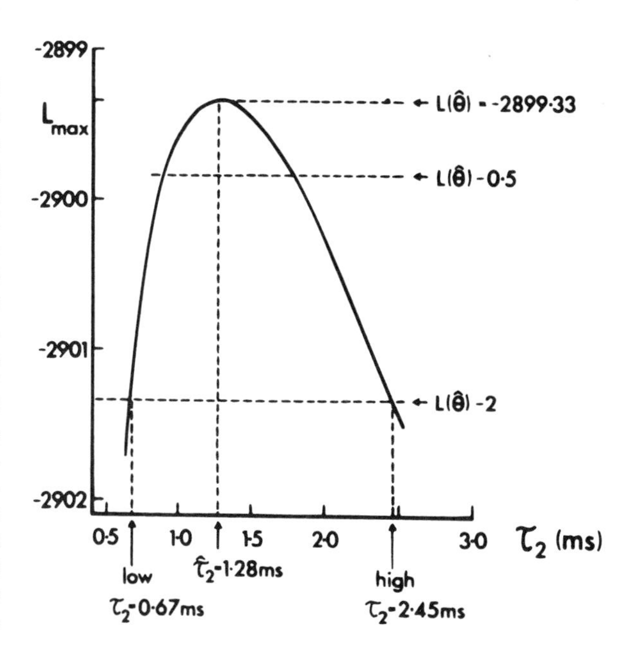

Figure 21. Estimation of likelihood intervals for τ_2 in the numerical example given in Section 6.10 (see Table II). The procedure is a generalization (to more than two parameters) of that illustrated in Fig. 20. The graph shows L_{max} plotted against τ_2, where L_{max} was found by holding τ_2 constant at the value shown on the abscissa and maximizing $L(\theta)$ with respect to the other four parameters (τ_1, a_1, a_2, and τ_3). The peak of the curve is, therefore, the overall maximum $L(\hat{\theta}) = -2899.33$ and corresponds to $\hat{\tau}_2 = 1.28$ ms. The values of τ_2 corresponding to $L_{max} = L(\hat{\theta}) - 2 = -2901.33$ are the 2-unit limits: 0.67 and 2.45 ms. The 0.5-unit limits can similarly be read off at $L(\hat{\theta}) - 0.5 = -2899.83$. In practice, it would be uneconomical to calculate this whole graph; the required points are found numerically by iteration (see text).

example, by Colquhoun and Sakmann (1985), and in Section 12 of Chapter 18 (this volume), and they are justified in more detail by Hawkes *et al.* (1992).

Exact corrections for missed events are possible only when a specific mechanism for channel operation is postulated. The methods that are available for doing exact corrections are discussed in Sections 12 and 13.7 of Chapter 18 (this volume). A particular benefit of these methods is that they have made it possible to fit reaction mechanisms directly to idealized data, as discussed next.

6.12. Direct Fitting of Mechanisms

The discussion so far has concerned the empirical fitting of exponentials (or geometrics) without specifying any particular reaction mechanism; the parameters to be fitted are the time constants and areas of the exponential components. Most investigations of reaction mechanisms have used such fits as the basis for a *post hoc* attempt to infer a mechanism. This procedure is less than ideal. One problem with it is that the information obtained from one sort of distribution may overlap strongly with that from another sort. For example, the distributions of burst length and of total open time per burst will be similar if the gaps within bursts are short (or rare). No method is known for combining the information from different sorts of fit in an optimal way to obtain the best idea about how well a specified mechanism fits the observations. Likewise, this approach makes it hard to compare two different putative mechanisms. Another problem with the *post hoc* approach is that, since each sort of distribution is fitted separately, the constraints on the relationship between them, which are implicit in the mechanism, are not taken into account.

Clearly, as mentioned in Section 6.1, it would be preferable to fit, as the adjustable parameters, not the time constants of the exponentials but the underlying rate constants in a specified mechanism (e.g., the values of k_{-1}, α_2, etc. in the mechanism specified in equation

110 of Chapter 18). Furthermore, since one set of values for these rate constants should be able to predict *all* the results from any sort of experiment, it is obviously preferable to do one simultaneous fit of all the observations that have been made. For example, it is desirable to fit, simultaneously, observations on steady-state records at several different agonist concentrations or membrane potentials, observations on channel openings following jumps under various conditions, and any other data that may have been obtained. Furthermore, it is undesirable to fit open times and shut times separately, because this procedure cannot take advantage of the information available from the sequence in which they occur (i.e., information from correlations—see Sections 5.7 and 5.8 above and Sections 10–13 of Chapter 18, this volume).

The sort of optimum approach to direct fitting just described was already well understood at the time of the first edition of this book (see Section 6.1.2 of Chapter 11 of the first edition), and attempts to implement direct fits had already been made (Horn and Lange, 1983). The problems were that the observations in the idealized record that are to be fitted suffer from omission of brief events and that retrospective corrections for missed events are not useful if a direct fit is to be attempted. Nothing very effective could be done until methods were devised to predict the distributions of what is *actually* observed rather than what would have been observed if time resolution had been perfect. Such methods now exist and are summarized in Sections 12 and 13.7 of Chapter 18 (this volume). They are now coming into use (e.g., Sine *et al.,* 1990). The approach is to calculate one value of the total likelihood from all the sets of data that are being fitted and to find the parameters that maximize this likelihood. The likelihood is calculated from the sequence of open and shut times rather than separately from each, so information from correlations is included in the fitting process. An example is given in Section 12.5 of Chapter 18 (this volume), and the general theory is summarized in Section 13.7 of Chapter 18 (this volume).

6.13. Fitting the Results after a Jump

The first problem is to get the results. Apart from the problem of estimating the number of channels, it is also the case that only one first latency can be measured for each jump, and it may be hard to get enough values in one experiment to make a decent-looking distribution. There will also be only one value per jump of each subsequent open and shut time if the first, second, etc. values differ (and this will not be known until their distributions have been looked at separately). It is perhaps for this reason that first latencies have often been displayed as cumulative distributions; the spurious appearance of precision that characterizes this sort of display (see Section 5.1.4) makes them look better than they are; this is highly undesirable.

Channel openings can be fitted by one of the methods already described, and a defined resolution can be imposed as described in Section 5.2 (this is especially desirable if the results are to be fitted with allowance for missed events). First latencies would then be corrected for recording delays (see preceding section). If the shut-time components are sufficiently well separated, it may be possible to define bursts of openings in the record. The theoretical distributions describing openings, shuttings, and bursts after a jump are given by Colquhoun and Hawkes (1987) in the case of a single channel and no missed events (see also Chapter 18, this volume). It is also possible to fit a mechanism directly, with allowance for missed events, as described for stationary records in Chapter 18 (this volume) and Section 6.11 (A. G. Hawkes, A. Jalali and D. Colquhoun, unpublished data).

When empirical mixtures of exponentials are being fitted to the first-latency distribution, it should be remembered that the areas of some components may be negative, as explained and illustrated in Chapter 18 (this volume Section 11). It is therefore important to be sure that your fitting program does not constrain all the areas to be positive (see Section 6.1.3).

6.13.1. Latencies with *N* Channels

If more than one channel is present, the first latencies will, of course, appear to be shorter than they really are. In the case of the first-latency distribution (but not any of the others), it is relatively simple to correct the observations *if* the number of channels is known. When N independent channels are present, the observed first latency will be greater than t if the first latencies for all N individual channels are greater than t. Thus, from equation 36,

$$P(\text{all } N \text{ latencies} > t) = 1 - F_N(t) = [1 - F_1(t)]^N$$

where $F_1(t)$ is the probability, for one channel, that the latency is equal to or less than t (see Section 5.1.4), and the observed cumulative distribution provides an estimate of $F_N(t)$ (Aldrich *et al.*, 1983). The pdf of the first latency is the first derivative of $F_1(t)$, so if we denote the pdf for N channels as $f_N(t)$ we get (Colquhoun and Hawkes, 1987)

$$f_1(t) = \frac{f_N(t)}{N[1 - F(t)]^{N-1}}$$

6.13.2. Effect of Finite Sample Length

The rectangular pulse of voltage, or ligand concentration, will be of fixed finite length, and the length of the data record collected after the end of the pulse will usually also be of fixed length. There will, therefore, always be an incomplete interval at the end of each record; if the channel was shut at the end of the record, the length of the shutting is not known because the next opening has not been recorded, and conversely, if the channel is open at the end of the record, the length of this last opening is not known. But we do know, in either case, that the interval was *at least* as long as the bit of it that was observed. It is easy to take into account this information when doing maximum-likelihood fitting (with or without allowance for missed events). For all complete intervals of length t_i, the log-likelihood is found as $L = \Sigma \ln f(t_i)$ (see equation 84); the probability of observing an interval of length *at least* t is $1 - F(t)$, so a separate term, $\Sigma \ln[1 - F(t_i)]$, can be added to the log-likelihood for the incomplete intervals (of length t_i). We then maximize the sum of these two terms, which is the overall log-likelihood (Hoshi and Aldrich, 1988).

Appendix 1. Choice of the Threshold for Event Detection

The choice of the threshold setting that allows the detection of the briefest events was considered in Section 3.3. However, optimizing the detection of the shortest pulses is not necessarily the best strategy for detection of single-channel events, because one is interested in counting events of all widths. The ideal event detector would have a sharp transition at some width, w_{min}, such that events narrower than this would be missed but essentially all longer events would be counted. In practice, the transition, as seen in a graph of the probability of detection as a function of w, is not necessarily sharpest when ϕ and f_c are chosen as described in Section 3.3.

Figure A1 demonstrates this property for pulses in the presence of $1 + f$ noise. The probability of detection p_{det} depends not only on w but also on ϕ, f_c, the channel amplitude, and the spectral characteristics. Part A of Fig. A1 corresponds to the case of a low channel amplitude (specifically, $A_0 = 0.22$ pA when the standard noise spectrum is assumed) in which events of width $w_{min} = 3$ ms or longer could be resolved. To construct each curve in the figure, a value of ϕ was first selected, and f_c was then chosen to give $\phi/\sigma_n = 5$. On the basis of these parameters, $p_{det}(w)$ was then estimated. The value of ϕ giving the best detection

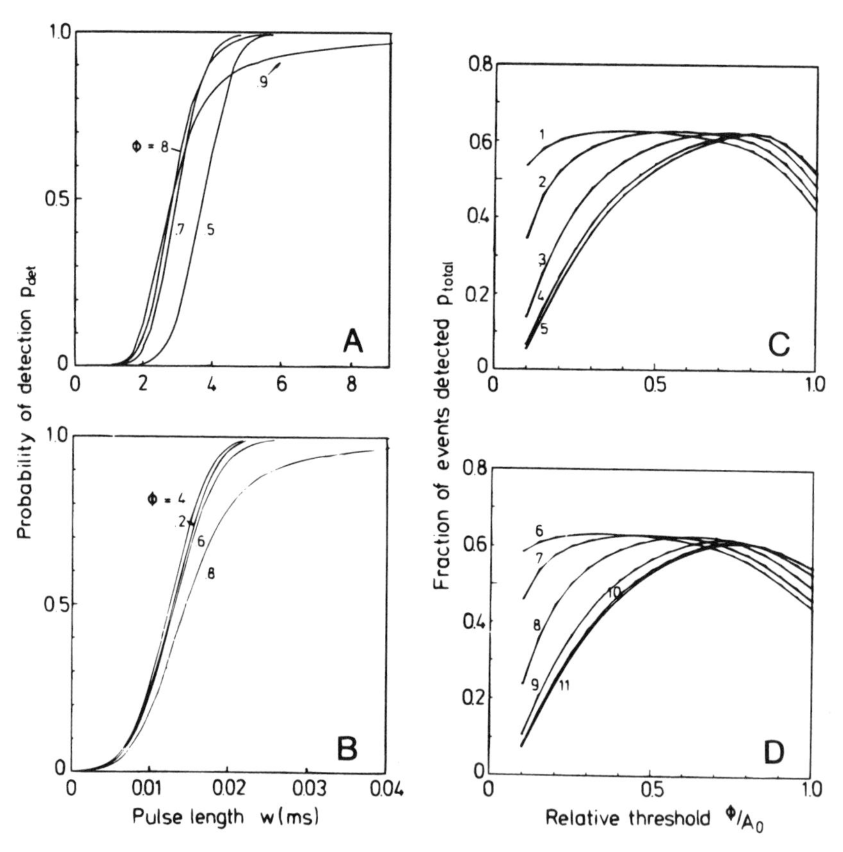

Figure A1. Performance of the event detector at various settings of the threshold ϕ. A and B: The probability of detection of isolated pulses of unit amplitude as a function of pulse width w. The parameters in A correspond to very small currents ($A_0 = 0.22$ pA in $S_o = 10^{-30}$ A²/Hz noise), giving $w_{min} \approx 3$ ms. At each value of ϕ, f_c was adjusted to give $\sigma_n = \phi/5$ to keep the false-event rate approximately constant. In B, the parameters correspond to relatively large currents ($A_o = 7.1$ pA in the same noise); much shorter events ($w_{min} \approx 13$ μs) can be detected. In this case, f_c was adjusted to keep $\phi/\sigma_n = 3$, corresponding to a higher false-event rate. C and D show the overall fraction, p_{total}, of pulses detected, given exponential distributions of pulse widths (equation A1). Each curve represents a different effective event amplitude, with the lowest-numbered curves corresponding to the largest amplitudes. Values for the amplitudes, time constants of the distribution, and other parameters are given in Table AI. In C, ϕ/σ_n was fixed at 5, whereas in D, $\phi/\sigma_n = 3$. The larger σ_n values in D cause the curves to be broadened and the optimum ϕ values to be slightly lower. Curves 4 and 6 were computed for the same conditions as in parts A and B, respectively.

of the shortest pulses was about 0.8 times A_0; when ϕ was reduced to 0.7 A_0, the transition moved to a slightly higher value of w but was steeper. On the other hand, increasing to 0.9 A_0 broadened the transition, so that whereas the very briefest pulses could be detected with higher probability, pulses even twice as long as w_{min} would be detected with only 80% probability. This sort of broadening of the transition region is very undesirable because it biases the selection of events in a way that can cause distortion of experimental lifetime distributions.

The broadening of the transition region is most severe when ϕ approaches A_0. This can be understood intuitively from the fact that if ϕ is near the full event amplitude, even moderately long events may fail to exceed ϕ when noise fluctuations are present. When ϕ is set lower (with f_c concurrently set lower), the longer events will have relatively larger peak amplitudes and will have a better chance of exceeding the threshold.

Figure A1B shows $p_{det}(w)$ curves for pulses of relatively large amplitude (A_0 = 7.1 pA in the standard noise spectrum; $\phi/\sigma_n = 3$). The minimum pulse width is about 13 μs in this case, and the optimum ϕ for detection of short pulses is much smaller, approximately 0.36 A_0. As the figure shows, the position and shape of the p_{det} curve depend only weakly on ϕ in the range 0.2 to 0.5 times A_0.

Parts C and D of Fig. A1 give a summary of the performance of an event detector in situations with various ratios of channel amplitude to background noise level. The quantity that is plotted here is the total fraction of events detected, p_{total}, out of an ensemble of pulse-shaped events having a probability density function $f(w)$ of widths,

$$p_{total} = \int_0^{\infty} p_{det}(w) f(w) dw \tag{A1}$$

where $f(w)$ was chosen to be exponential, $f(w) = (1/\tau)\exp(-w/\tau)$. For each curve, τ was fixed at the value $2w_{min}$; the actual values used are given in Table AI. The maximum values

Table A1. Parameters for the Curves in Fig. A1C,D[a]

					$\phi = 0.5\ A_0$		$\phi = 0.7\ A_0$	
Curve	ϕ/σ_n	$A_0^2/S_0 f_0$	A_0 (pA)	w_{min} (msec)	f_c (kHz)	p'_{total}	f_c (kHz)	p'_{total}
1	5	5000	7.1	0.023	7.62	1.00	10.94	0.96
2	5	500	2.2	0.089	2.00	1.00	3.02	0.99
3	5	50	0.71	0.43	0.375	0.95	0.641	1.00
4	5	5	0.22	3.84	0.046	0.87	0.086	0.98
5	5	0.5	0.07	38.4	0.0047	0.85	0.0092	0.98
6	3	5000	7.1	0.013	13.16	0.99	18.7	0.94
7	3	500	2.2	0.048	3.71	1.00	5.45	0.97
8	3	50	0.71	0.206	0.834	0.98	1.34	0.99
9	3	5	0.22	1.21	0.121	0.91	0.222	1.00
10	3	0.5	0.07	10.4	0.0129	0.88	0.025	1.00
11	3	0.05	0.02	104.0	0.0013	0.88	0.0025	1.00

[a]Part C was computed with $\phi/\sigma_n = 5$ (low false-event rate; curves 1–5), and D with $\phi/\sigma_n = 3$ (curves 6–11). Each curve represents a different value of the signal-to-noise parameter $A_0^2/S_0 f_0$, which corresponds to the given A_0 value in the standard case ($S_0 = 10^{-30}$ A^2/Hz, f_0 = 1 kHz, 1 + f spectrum). The w_{min} values give the effective minimum detectable pulse width. The distribution of pulse widths for calculating p_{total} was chosen to be exponential in each case, with the time constants $\tau = 2w_{min}$. For ϕ = 0.5 and 0.7, the corresponding f_c values and the relative detection efficiency $p'_{total} = p_{total}(\phi)/p_{total}(\max)$ are given. The maximum value $p_{total}(\max)$ was always within a few percent of $\exp(-w_{min}/\tau) = \exp(-1/2)$, the probability expected if only those events shorter than w_{min} were not detected.

of p_{total} computed in this way were near 0.6, which is to be expected since, if $p_{det}(w)$ were zero for $w < w_{min}$ and unity for all larger w, p_{total} would equal $\exp(-w_{min}/\tau) = 0.61$.

A comparison of the A_0 and w_{min} columns of Table AI shows the approximate limits of pulse detection, and the f_c columns show typical corresponding filter bandwidths. The choice of the ϕ/σ_n ratio equal to 3 instead of 5 allows pulses shorter by a factor of 2–3 to be detected, but at the cost of higher false-event rates. For large pulses ($A_0 > 1$ pA in this case), w_{min} decreases as $1/A_0$, whereas for smaller pulses, w_{min} varies as $1/A_0^2$. The A_0 values given correspond to the standard noise spectrum; for other $1 + f$ spectra, the dimensionless parameter $A_0^2/(S_0 f_0)$ is the appropriate measure for the signal-to-noise relationship, and w_{min} values should be scaled as $1/f_0$ for f_0 differing from 1 kHz.

Although this analysis has been quite complicated, the practical conclusions can be stated simply. First, for detecting channels of relatively low amplitude, implying that f_c must be set to be below f_0 (1 kHz in this example) to obtain a suitable background noise level, a good choice for ϕ is about $0.7A_0$. This is near the peaks of the corresponding p_{total} curves but is low enough to insure a sharp transition in the $p_{det}(w)$ curves. Second, for detecting larger channel events, for which f_c can be larger than f_0, the exact choice of ϕ is less critical, with the range 0.4 to $0.5A_0$ generally being best. The special case $\phi = 0.5\,A_0$ is of interest for event characterization. It can be seen from Fig. A1C and D that p_{total} is always at least 85% of its peak value when $\phi = 0.5\,A_0$ is chosen.

Appendix 2. The Expected Distribution of Fitted Amplitudes

We derive here the distribution of channel amplitudes that would be expected when amplitudes are estimated by averaging. Points are averaged over an interval w_a that lies within the "flat-top" portion of an event. This estimate, A, has an expected value (long-term average) equal to the true channel amplitude, A_0.

We assume that the background noise spectrum is flat and that the noise does not change appreciably when a channel opens. In this case, A has a variance that depends on w_a according to

$$\sigma_A^2(w_a) \approx S_0/w_a \tag{A2}$$

where S_0 is the (one-sided) spectral density. Strict equality holds in the limit when w_a is very large compared with the recording system risetime T_r, but the approximation is actually very good for all $w_a \geq T_r$. It is also a good approximation to the error in least-square fitting of the time course (Fig. 11B).

In practice, the background noise spectrum rises with frequency, but it is usually flat below 1 kHz. Since the frequencies that predominantly contribute to σ_A^2 are below $f = 1/2w_a$, for w_a on the order of 1 ms or larger the flat-spectrum assumption is usually justified, with S_0 being taken as the low-frequency spectral density.

Assuming that the baseline level is known exactly, σ_A^2 is the entire variance of the channel amplitude estimate. If we assume that the background noise is Gaussian distributed, the probability density of values of A for a given w_a is also Gaussian:

$$g_w(A;w_a) = \frac{1}{(2\pi)^{1/2}\sigma_A(w_a)} \exp\left[\frac{-(A - A_0)^2}{2\sigma_A^2(w_a)}\right] \tag{A3}$$

In practice, one does not want to hold the averaging interval constant but instead allows it to vary with the channel-open time, t_0. We assume the relationship

$$w_a = t_0 - t_m \qquad (t_0 \geq t_m) \tag{A4}$$

where t_m is the (fixed) length of an event that is "masked off" before averaging; this would typically be chosen to be between 1 and 2 risetimes in length to avoid any bias toward lower estimates as a result of the rising and falling edges of the pulse. Finally, we wish to ignore amplitude estimates from the briefest events by setting a lower limit w_{min} for averaging widths. The resulting pdf for the amplitude from an ensemble of events having random widths is then given by

$$g(A) = \int_{w_{min}}^{\infty} g_w(A;w_a)f(w_a)dw_a \tag{A5}$$

where $f(w_a)$ is the pdf of averaging widths. If t_0 is distributed according to a mixture of exponential densities, as in equation 30, then $f(w_a)$ is also multiexponential,

$$f(w_a) = \Sigma a_i \tau_i^{-1} e^{-w_a/\tau_i} \qquad w_a > 0 \tag{A6}$$

Substituting equations A6 and A3 into the integral A5 yields

$$g(A) = \frac{1}{(2\pi S_0)^{1/2}} \int_{w_{min}}^{\infty} w_a^{1/2} \exp\left[\frac{w_a(A - A_0)^2}{2S_0}\right][\Sigma a_i \tau_i e^{-w_a/\tau_i}]dw_a \tag{A7}$$

It is helpful to change the variable of integration to $x_i = (w_a/\tau_i)^{1/2}$ and to introduce the definitions

$$\begin{aligned} x_{0i} &= (w_{min}/\tau_i)^{1/2} \\ \sigma_{0i} &= (S_0/\tau_i)^{1/2} \end{aligned} \tag{A8}$$

where x_{0i} is dimensionless and gives a measure of the spread of the distribution of w_a values, and σ_{0i} is the standard deviation of an amplitude estimate when $w_a = \tau_i$. Finally, we set $\delta_i = (A - A_0)/2^{1/2}\,\sigma_{0i}$ so that δ_i are the normalized deviations of A from its expected value. The integral can then be evaluated to yield

$$\begin{aligned} g(A) = \frac{1}{(2\pi)^{1/2}} \sum_{i=1}^{k} \frac{a_i}{\sigma_{0i}(1 + \delta_i^2)} \\ \times \left\{x_{0i} \exp[-x_{0i}^2(1 + \delta_i^2)] + \frac{\pi^{1/2}}{2(1 + \delta_i^2)^{1/2}} \operatorname{erfc}[x_{0i}(1 + \delta_i^2)^{1/2}]\right\} \end{aligned} \tag{A9}$$

where erfc is the complementary error function. (A formula for numerically evaluating this function is given in Appendix 3).

Figure A2 shows plots of this distribution for various values of x_0 in the case where the open time has a simple exponential distribution with mean τ. Since σ_0 is kept constant, the figure demonstrates the effect of changing the duration limit $w_{\min}$ on the shape of the amplitude distribution obtained from a given set of single-channel events. When x_0 is larger than unity, the first term of equation A9 predominates, so that the distribution is essentially Gaussian in shape and has a standard deviation $\sigma_a \approx \sigma_0/x_0 = (S_0/w_{\min})^{1/2}$. Large x_0 corresponds to the case in which $w_{\min}$ is large compared to τ, so that the distribution of w values dies off quickly beyond $w_{\min}$. A nearly Gaussian amplitude distribution is therefore to be expected from the tightly clustered w_a values.

As x_0 decreases, the tails of the distribution become wider, and the distribution becomes distinctly non-Gaussian, but it remains symmetrical. To obtain the sharpest distribution, it is best to choose $w_{\min}$ (and therefore x_0) to be large. However, a high $w_{\min}$ value implies that fewer events will be counted in the amplitude histogram. A good compromise is to choose $w_{\min} = \tau/2$, yielding $x_0^2 = 0.5$. This allows the fraction $\exp(-1/2) \simeq 0.6$ of the maximum number of events to be counted while yielding a distribution that is nearly indistinguishable from a Gaussian having the standard deviation $\sigma = 1.24\ \sigma_0$ (Fig. A2B).

Rather than computing the background noise power spectrum to determine S_0, it may be more convenient in practice to estimate σ_0^2 directly. This can be done by forming the averages of a large number of successive stretches, of length τ, of the background trace. The variance of these values can then be used directly as an estimate of σ_0^2.

Appendix 3. Numerical Techniques for Single-Channel Analysis

A3.1. A Digital Gaussian Filter

This digital filter forms output values y_i from input values x_i by forming a weighted sum

$$y_i = \sum_{j=-n}^{n} a_j x_{i-j} \tag{A10}$$

where the a_j are coefficients that sum to unity.

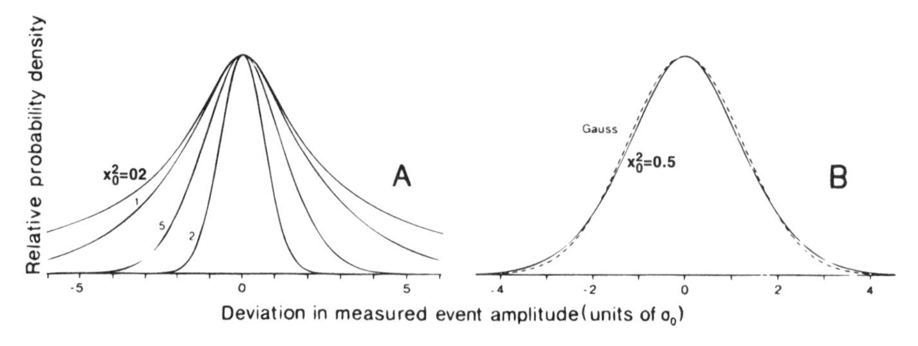

Figure A2. A: Plots of the function in equation A9 for various values of the parameter x_0^2, in the case where the open time has a simple exponential distribution. The plots were scaled to superimpose the peak values. B: Comparison of equation A9 with a Gaussian function. The parameter x_0 was chosen to be $1/\sqrt{2}$; the Gaussian function (dotted curve) was fitted by eye and had a standard deviation equal to $1.24\sigma_0$.

A continuous-time Gaussian filter is characterized by the width parameter or "standard deviation" σ_g of its impulse response, which is related to the cutoff frequency f_c according to (see equation 2)

$$\sigma_g = 0.1325/f_c \tag{A11}$$

Similarly, for a discrete filter, σ_g can be defined in units of sample intervals, in which case equation A11 holds if f_c is expressed in units of the sampling rate.

For a discrete Gaussian filter having width σ_g, the coefficients have the form

$$a_j = \frac{1}{\sqrt{2\pi}\sigma_g} \exp\left(\frac{-j^2}{2\sigma_g^2}\right) \tag{A12}$$

and the number of terms, n, is chosen so that the missing terms are negligible in size; in the implementation described here, n is chosen to be $4\sigma_g$.

If σ_g is relatively small, coefficients of the form of equation A12 sum to less than unity and yield a filter with wider bandwidth than f_c; these errors exceed 1% when σ_g is less than about 0.6. Since small σ_g corresponds to relatively light filtering, a suitable choice for the coefficients in this case is

$$\begin{aligned} a_1 &= \sigma_g^2/2 \\ a_0 &= 1 - 2a_1 \\ a_{-1} &= a_1 \end{aligned} \tag{A13}$$

so that each output value of the filter depends only on the corresponding input value and its two neighboring points. This simple filter function causes no problems with aliasing, provided the original data points are sampled at a sufficient rate, e.g., five times the cutoff frequency of Bessel-response prefiltering.

Filter procedures are presented in Fig. A3 for FORTRAN and in Fig. A4 for MODULA-2. The FORTRAN implementation operates on an array of integer input values and produces integer output; intermediate computations are however, performed in floating point. Note that because the number of coefficients n (this value is called NC in the FORTRAN subroutine, NumCoeffs in the MODULA-2 version) increases inversely as f_c, sufficient room in the coefficient array A should be provided for the smallest expected f_c value. For example, n = 53 for $f_c = 0.01$, but $n = 5$ for $f_c = 0.1$. The MODULA-2 implementation consists of two procedures, one to compute the coefficients and the other to perform the actual filtering. The latter, DoFilter, operates on real (floating-point) values and is capable of decimating the data, i.e., producing fewer output points than input points.

As an example of the use of these subroutines, suppose that we have a digitized record that was filtered with a Bessel filter at 2 kHz and sampled at a 10-kHz rate. To reduce the effective bandwidth to 1 kHz, the second filtering operation should have a cutoff frequency (see equation 4), of $(1 - 1/4)^{-1/2} = 1.15$ kHz. In calling the filter routine, the FC or Frequency variable should therefore be set to 0.115.

In both of the implementations shown, the evaluation of the sum (equation A10) is done only after checking that the input array bounds will not be exceeded; the result is that the values of the input points before the beginning and after the end of the input array are in effect assumed to be zero. Although the points in the middle of a long data array will not

```
      SUBROUTINE FILTER (IN, OUT, NP, FC)
C
C     Gaussian filter subroutine.  Accepts integer
C     data from the array IN, filters it with a -3db
C     frequency FC (in units of sampling frequency)
C     and returns the integer results in the OUT array.
C
      INTEGER IN(NP), OUT(NP)
      REAL A(54)
C          (Coefficient array.  54 terms are sufficient
C           for FC >= .01
C
C     -----First, calculate the coefficients-----
      SIGMA = 0.132505 / FC
      IF (SIGMA.LT. 0.62) GOTO 10
C
C        Standard gaussian coefficients.
C        NC is the number of coefficients not counting
C        the central one A(0).
      NC = INT( 4.0 * SIGMA )
      IF (NC .GT. 53) NC = 53
      B = -0.5 / ( SIGMA * SIGMA )
      A(1) = 1.0
      SUM = 0.5
C
      DO 5, I = 1, NC
      TEMP = EXP( (I*I) * B )
      A(I+1) = TEMP
      SUM = SUM + TEMP
 5    CONTINUE
C        Normalize the coefficients
      SUM = SUM * 2.0
      DO 7, I = 1, NC + 1
      A(I) = A(I) / SUM
 7    CONTINUE
      GOTO 20
C
C        Alternate routine for narrow impulse
         response.  Only three terms are used.
10    A(2) = SIGMA * SIGMA / 2.0
      A(1) = 1.0 - 2.0 * A(2)
      NC = 1
C
C     -----Actual filtering is done here-----
20    DO 40, I = 1, NP
      JL = I - NC
      IF (JL .LT. 1) JL = 1
      JU = I + NC
      IF (JU .GT. NP) JU = NP
      SUM = 0.0
C
      DO 30, J = JL, JU
      K = IABS(J-I) + 1
      SUM = SUM + IN(J) * A(K)
30    CONTINUE
C
      OUT(I) = SUM
40    CONTINUE
      RETURN
      END
```

Figure A3

```
IMPLEMENTATION MODULE FilterReal;

FROM SYSTEM IMPORT ETOX; (* exp function *)
FROM InOut IMPORT WriteString, WriteInt, WriteLn;

CONST
  MaxFilterCoeffs = 220;

 (* Module global variables *)
VAR
  NumCoeffs : INTEGER;
  Coeffs    : ARRAY[0..MaxFilterCoeffs] OF REAL;


PROCEDURE SetGaussFilter ( Frequency: REAL );
 (* Load the filter coefficient values according to the cutoff
    frequency (in units of the sample frequency) given.
 *)

VAR
  sigma, b, sum   : REAL;
  i               : INTEGER;

BEGIN
  sigma:=0.132505/Frequency;
  IF sigma < 0.62 THEN   (* light filtering *)

    Coeffs[1] := sigma*sigma*0.5;
    Coeffs[0] := 1.0 - sigma*sigma;
    NumCoeffs:=1;

  ELSE  (* normal filtering *)

    NumCoeffs:= TRUNC(4.0 * sigma);
    IF NumCoeffs > MaxFilterCoeffs THEN
      WriteString ("FilterReal.SetGaussFilter: Too many coefficients:");
      WriteInt( NumCoeffs, 4 ); WriteLn;
      NumCoeffs:= MaxFilterCoeffs;
    END;
    b:= -1.0/(2.0*sigma*sigma);

      (* First make the sum for normalization *)
    sum:= 0.5;
    FOR i:=1 TO NumCoeffs DO
      sum:= sum + ETOX( b * FLOAT(i*i) );
    END;
    sum:= sum * 2.0;

      (* now compute the actual coefficients *)
    Coeffs[0]:= 1.0 / sum;
    FOR i:=1 TO NumCoeffs DO
      Coeffs[i]:= ETOX( FLOAT(i*i) * b ) / sum;
    END;
  END;
END SetGaussFilter;
```

Figure A4

be affected by this, the first and last n output values are reduced in magnitude by this truncation of the sum. This becomes an important issue when one wishes to filter a long digitized recording that does not fit into a single array of length N. The way to avoid the "edge effects" is to read overlapping segments of data into the input array and then to write out only the central $N - 2n$ points of the output array each time (with the exception of the first and last segments, where the initial and final "edges" should be written to preserve the total number of points).

```
PROCEDURE DoFilter( VAR Input, Output : ARRAY OF REAL;
                    NumInputPoints    : INTEGER;
                    Compression       : INTEGER);
  (* From the Input array, create a filtered Output that is
     decimated by Compression.  Thus the number of output points
     is equal to NumInputPoints DIV Compression.  SetGaussFilter
     must be called before this procedure to set up the filter coefficients.
  *)

VAR
  i0, i, j       : INTEGER;
  jmax, jmin     : INTEGER;
  sum            : REAL;

BEGIN
  FOR i0 := 0 TO (NumInputPoints DIV Compression) - 1 DO
    i := i0 * Compression;

     (* Make sure we stay within bounds of the Input array *)
    jmax := NumCoeffs;
    jmin := NumCoeffs;
    IF jmin > i THEN jmin := i END;
    IF jmax >= NumInputPoints - i THEN jmax := NumInputPoints - i - 1 END;

    sum := Coeffs[0] * Input[i];      (* Central point *)

    FOR j := 1 TO jmin DO                    (* Early points *)
      sum := sum + Coeffs[j] * Input[i-j];
    END;

    FOR j := 1 TO jmax DO                    (* Late points *)
      sum := sum + Coeffs[j] * Input[i+j];
    END;

     (* Assign the output value *)
    Output[i0] := sum;

  END;  (* FOR i0 *)
END DoFilter;

END FilterReal.
```

Figure A4. *Continued.*

A3.2. Cubic Spline Interpolation

A very useful interpolation technique for single-channel recording is the cubic spline, in which a cubic polynomial spans the interval between each pair of data points. A different polynomial is used for each interval, with coefficients chosen to match the function values as well as the first and second derivatives at the sample points. An introduction to the theory can be found in Hamming (1975). Briefly, we wish to find an interpolating function f whose values $f(1), f(2) \ldots$ match the data values $y_1, y_2, \ldots$ obtained at equally spaced sample times. Intermediate values $f(k + \theta)$ for θ between 0 and 1 are given by

$$f(k + \theta) = y_k\rho + y_{k+1}\theta + a_k(\rho^3 - \rho) + a_{k+1}(\theta^3 - \theta) \quad \text{(A14)}$$

where $\rho = 1 - \theta$. Before the interpolation is done, the coefficients a_k must be computed. They are specified by the system of equations

$$a_{k-1} + 4a_k + a_{k+1} = y_{k-1} - 2y_k + y_{k+1} \quad \text{(A15)}$$

which can be solved by Gaussian elimination.

```
      SUBROUTINE SPLINE (IN, OUT, A, N, NOUT)
C
C     This subroutine accepts N integer values from array
C     IN, interpolates them by the factor NE = NOUT/N
C     and returns the NOUT-NE+1 output points in the
C     array OUT.  The array A is used internally for
C     coefficients of the cubic term of the interpolating
C     polynomial.
C
      INTEGER IN(N), OUT(NOUT)
      REAL A(N)
C
      B = -1.0 / (2.0 + SQRT( 3.0 ))
      NE = NOUT / N
      NE1 = NE - 1
      E = NE
C
C        Form the coefficient array
C
      A(1) = 0.0
      A(N) = 0.0
      DO 10, I=2, N-1
        TEMP = 2 * IN(I) - IN(I-1) - IN(I+1)
        A(I) = B * (TEMP + A(I-1))
 10   CONTINUE
C
      DO 20, I=1, N-1
        J = N-I
        A(J) = A(J) + B * A(J+1)
 20   CONTINUE
C
C        Insert the original points into OUT
C
      DO 30, I=1, N
        K = NE*I - NE1
        OUT(K) = IN(I)
 30   CONTINUE
C
C        Handle the intermediate points
C
      DO 40, J=1, NE1
        P = J/E
        Q = 1.0 - P
        P3 = P * (P * P - 1.0)
        Q3 = Q * (Q * Q - 1.0)
        DO 40, I=1, N-1
          I1 = I+1
          K = NE * I + J - NE1
          OUT(K) = Q*IN(I) + P*IN(I1) + Q3*A(I) + P3*A(I1)
 40   CONTINUE
      RETURN
      END
```

Figure A5

```
IMPLEMENTATION MODULE SplineReal;

PROCEDURE Spline (VAR In, Work, Out: ARRAY OF REAL;
                      InNumber     : INTEGER;
                      Expansion    : INTEGER);

  (* From the InNumber input points, make (InNumber-1) * Expansion - 1
     output points, using cubic spline interpolation.  The output
     points Out[ Expansion * i] are equal to the corresponding input
     points In[i].
     The Work array must have at least InNumber elements.
  *)
  CONST
    c = 0.2674919;  (* equals 1 / ( 2 + sqr(3) ) *)

  VAR
    p, q,
    p3, q3        : REAL;
    i,j,k,ini     : INTEGER;

  BEGIN
     (* Compute coefficients: forward calculation *)
    Work[0] := 0.0;
    FOR i := 1 TO InNumber-2 DO
      Work[i]:= c * ( In[i-1] - 2.0 * In[i] + In[i+1] - Work[i-1] );
    END;

     (* Back-substitution *)
    Work[InNumber-1] := -c * Work[InNumber-2];
    FOR i := InNumber-1 TO 1 BY -1 DO
      Work[i-1] := Work[i-1] - c * Work[i];
    END;

     (* Copy the original points *)
    j := 0;               (* j is the output pointer *)
    FOR i:=0 TO InNumber-1 DO
      Out[j]:=In[i];
      INC(j, Expansion);  (* increment j by Expansion *)
    END;

      (* Compute the interpolated points *)
    FOR k:=1 TO Expansion-1 DO
      p:= FLOAT(k) / FLOAT(Expansion);
      q:= 1.0 - p;
      p3 := p * ( p * p - 1.0);
      q3 := q * ( q * q - 1.0);
      j  := k;
      ini:= 0;
      FOR i:=0 TO InNumber-2 DO
        Out[j]:= q * In[ini] + p * In[ini+1]
               + q3 * Work[i] + p3 * Work[i+1];
        INC(ini);
        INC(j, Expansion);
      END;
    END;
  END Spline;

END SplineReal.
```

Figure A6

```
        subroutine AMOCALL(npar,nvert,simp,theta,stepfac,functol,funk,
     &     fmin,niter)
c Subroutine to simplify call of AMOEBA.FOR from Press et al. (1992)
c This subroutine uses the input values (see below) to:
c (1) set up the starting simplex in simp(21,20)
c (2) set the corresponding function values in fval(21)
c
c SIMP should be declared in calling program, e.g. as simp(21,20).
c      (SIMP is defined here, but because of problems in passing values
c      in 2-dimensional arrays with variable dimensions, it is simpler
c      to declare SIMP in the calling program)
c
c INPUT:
c     npar = number of parameters
c     nvert = npar+1
c     theta(npar) = initial guesses for parameters
c     stepfac = value to control initial step size, e.g. stepfac=0.1
c        starts with step size=0.1*initial guess.
c     functol = tolerance for convergence (should be set to machine
c        precision, or a bit larger -see Press et al.)
c     funk = name of subroutine that calculates the value to be
c        minimized
c
c OUTPUT:
c     theta = final values of parameters (in this version, set to the
c        parameters corresponding to the best vertex of final simplex).
c     fmin = corresponding minimum value for funk(theta)
c     niter = number of function evaluations done
c
      real simp(nvert,npar),fval(21),theta(npar),step(20)
      EXTERNAL funk
c
c     nvert=npar+1             ! # of vertices in simplex
```

Figure A7

The FORTRAN subroutine SPLINE (Fig. A5) accepts an integer array of N data values and fills a second integer array with the original points and interpolated values. The subroutine first computes the coefficients in an efficient manner that is equivalent to Gaussian elimination and backsubstitution. The N coefficients are kept in a real array A for further use if desired. The subroutine forces the second derivative of f to be zero at the first and last data points. This means that if a long record is to be interpolated in shorter segments, the segments

```
      do j=1,npar
        step(j)=stepfac*theta(j)
      enddo
      do j=1,npar
        simp(1,j)=theta(j)              !!start values=vertex #1
      enddo
      fval(1)=funk(theta)      !function value for these
      fac=(sqrt(float(nvert))-1.)/(float(npar)*sqrt(2.))
      do i=2,nvert
        do j=1,npar
          simp(i,j)=simp (1,j) + step(j)*fac
        enddo
        simp(i,i-1)=simp(1,i-1) + step(i-1)*(fac+1./sqrt(2.))
        do j=1,npar
          theta(j)=simp(i,j)       !copy paramters into theta (for funk)
        enddo
        fval(i)=funk(theta)            !function value for these
      enddo
c
      call AMOEBA(simp,fval,nvert,npar,npar,functol,funk,niter)
c
c     Return the best vertex
      fmin=fval(1)
      do i=2,nvert
        if(fval(i).lt.fmin) then
          fmin=fval(i)
          do j=1,npar
           theta(j)=simp(i,j)
          enddo
        endif
      enddo
c
      RETURN
      end
```

Figure A7 *Continued.*

should have some overlap (e.g., ten data points) to allow smooth "splicing" of the interpolated segments.

A MODULA-2 implementation of the same algorithm is given in Fig. A6. Here the coefficients a_k are stored in the Work array while the input and output data are assumed to be in arrays of real values.

A3.3. Error Function Evaluation

For computations involving the step response of Gaussian filters, a numerical approximation for the error function is required. One of the simplest approximations for the complementary error function is

$$\mathrm{erfc}(x) = (a_1 t + a_2 t^2 + a_3 t^3)\exp(-x^2) \tag{A16}$$

where $t = 1/(1 + px)$; $p = 0.47047$; $a_1 = 0.3480242$; $a_2 = -0.0958798$; $a_3 = 0.7478556$; and where x is restricted to positive values. The error in this approximation is less than 2.5×10^{-5}.

The error function itself can be evaluated as

$$\mathrm{erf}(x) = 1 - \mathrm{erfc}(x)$$

and for negative values of x,

$$\mathrm{erf}(x) = -\mathrm{erf}(-x)$$

The formula in equation A16 is from Hastings (1955), which also contains more exact formulas. These formulas can also be found in Abramovitz and Stegun (1964), p. 299.

A3.4. A Calling Routine for AMOEBA

The subroutine (in FORTRAN) designed to simplify calling of the simplex minimization routine by Press *et al.* (1992) is given in Fig. A7. This was discussed in Section 6.3.

References

Abramovitz, M., and Stegun, I. A., 1965, *Handbook of Mathematical Functions,* Dover Publications, New York.

Aldrich, R. W., Corey, D. P., and Stevens, C. F., 1983, A reinterpretation of mammalian sodium channel gating based on single channel recording, *Nature* **306:**436–441.

Beale, E. M. L., 1960, Confidence regions in non-linear estimation, *J. R. Statist. Soc.* **B22:**41–76.

Bliss, C. I., and James, A. T., 1966, Fitting the rectangular hyperbola, *Biometrics* **22:**573–602.

Box, G. E. P., and Coutie, G. A., 1956, Application of digital computers in the exploration of functional relationships, *Proc. IEEE* **103**(Part B, Suppl. 1)**:**100–107.

Clapham, D. E., and Neher, E., 1984, Substance P reduces acetylcholine-induced currents in isolated bovine chromaffin cells, *J. Physiol.* **347:**255–277.

Colquhoun, D., 1971, *Lectures on Biostatistics,* Clarendon Press, Oxford.

Colquhoun, D., 1979, Critical analysis of numerical biological data, in: *Proceedings of the Sixth International CODATA Conference* (B. Dreyfus, ed.), pp. 113–120, Pergamon Press, Oxford.

Colquhoun, D., and Hawkes, A. G., 1982, On the stochastic properties of bursts of single ion channel openings and of clusters of bursts. *Phil. Trans. R. Soc. Lond.* [*Biol.*] **300:**1–59.

Colquhoun, D., and Hawkes, A. G., 1987, A note on correlations in single ion channel records, *Proc. R. Soc. Lond.* [*Biol.*] **230:**15–52.

Colquhoun, D., and Ogden, D. C., 1988, Activation of ion channels in the frog endplate by high concentrations of acetylcholine, *J. Physiol.* **395:**131–159.

Colquhoun, D., and Sakmann, B., 1985, Fast events in single-channel currents activated by acetylcholine and its analogues at the frog muscle end-plate. *J. Physiol.* **369:**501–557.

Colquhoun, D., Jonas, P., and Sakmann, B., 1992, Action of brief pulses of glutamate on AMPA/kainate receptors in patches from different neurones of rat hippocampal slices, *J. Physiol.* **458:**261–287.

Chung, S. H., Moore, J. B., Xia, L., Premkumar, L. S., and Gage, P. W., 1990, Characterization of single channel currents using digital signal processing techniques based on hidden Markov models, *Phil. Trans. R. Soc. Lond. [Biol.]* **329:**265–285.

Edmonds, B., Gibb, A. J., and Colquhoun, D. 1995a, Mechanisms of activation of glutamate receptors and the time course of excitatory synaptic currents, *Ann. Rev. Physiol.,* in press.

Edmonds, B., Gibb, A. J., Colquohoun, D., 1995b, Mechanisms of activation of muscle nicotinic acetylcholine receptors, and the time course of endplate currents, *Ann. Rev. Physiol.,* in press.

Edwards, A. W. F., 1972, *Likelihood,* Cambridge University Press, Cambridge.

Fitzhugh, R., 1983, Statistical properties of the asymmetric random telegraph signal, with applications to single-channel analysis, *Math. Biosci.* **64:**75–89.

Gibb, A. J., and Colquhoun, D., 1991, Glutamate activation of a single NMDA receptor-channel produces a cluster of channel openings, *Proc. R. Soc. Lond. [Biol.]* **243:**39–45.

Gibb, A. J., and Colquhoun, D., 1992, Activation of NMDA receptors by L-glutamate in cells dissociated from adult rat hippocampus, *J. Physiol.* **456:**143–179.

Gibb, A. J., Kojima, H., Carr, J. A., and Colquhoun, D., 1990, Expression of cloned receptor subunits produces multiple receptors, *Proc. R. Soc. Lond. [Biol.]* **242:**108–112.

Hawkes, A. G., Jalali, A., and Colquhoun, D., 1990, The distributions of the apparent open times and shut times in a single channel record when brief events can not be detected, *Phil. Trans. R. Soc. Lond. [A]* **332:**511–538.

Hawkes, A. G., Jalali, A., and Colquhoun, D., 1992, Asymptotic distributions of apparent open times and shut times in a single channel record allowing for the omission of brief events, *Phil. Trans. R. Soc. Lond. [Biol.]* **337:**383–404.

Heinemann, S. H., and Sigworth, F. J., 1990, Open channel noise. V. Fluctuating barriers to ion entry in gramicidin A channels, *Biophys. J.* **57:**499–514.

Heinemann, S. H., and Sigworth, F. J., 1991, Open channel noise. VI. Analysis of amplitude histograms to determine rapid kinetic parameters, *Biophys. J.* **60:**577–587.

Hill, I. D., 1978, A remark on algorithm AS47: Function minimization using a simplex procedure, *Appl. Statist.* **27:**280–382.

Horn, R., 1987, Statistical methods for model discrimination. Applications to gating kinetics and permeation of the acetylcholine receptor channel, *Biophys. J.* **51:**255–263.

Horn, R., and Lange, K., 1983, Estimating kinetic constants from single channel data, *Biophys. J.* **43:**207–223.

Hoshi, T., and Aldrich, R. W., 1988, Gating kinetics of four classes of voltage-dependent K^+ channels in pheochromocytoma cells, *J. Gen. Physiol.* **91:**107–131.

Howe, J. R., Cull-Candy, S. G., and Colquhoun, D., 1991, Currents through single glutamate-receptor channels in outside-out patches from rat cerebellar granule cells, *J. Physiol.* **432:**143–202.

Jackson, M. B., 1985, Stochastic behaviour of a many-channel membrane system, *Biophys. J.* **47:**129–137.

Jackson, M. B., Wong, B. S., Morris, C. E., Lecar, H., and Christian, C. N., 1983, Successive openings of the same acetylcholine receptor channel are correlated in open time, *Biophys. J.* **42:**109–114.

Johnson, J. W., and Ascher, P., 1990, Voltage-dependent block by intracellular Mg^{2+} of *N*-methyl-D-aspartate-activated channels, *Biophys. J.* **57:**1085–1090.

Magleby, K. L., 1992, Preventing artifacts and reducing errors in single-channel analysis, *Methods Enzymol.* **207:**763–791.

Magleby, K. L., and Pallotta, B. S., 1983, Burst kinetics of single calcium-activated potassium channels in cultured rat muscle, *J. Physiol.* **344:**605–623.

Magleby, K. L., and Song, L., 1992, Dependency plots suggest the kinetic structure of ion channels, *Proc. R. Soc. Lond. [Biol.]* **249:**133–142.

Magleby, K. L., and Weiss, D. S., 1990, Estimating kinetic parameters for single channels with simulation, *Biophys. J.* **58:**1411–1426.

Mathie, A., Colquhoun, D., and Cull-Candy, S. G., 1991, Conductance and kinetic properties of single channel currents through nicotinic acetylcholine receptor channels in rat sympathetic ganglion neurones, *J. Physiol.* **439:**717–750.

McManus, O. B., and Magleby, K. L., 1988, Kinetic states and modes of single large-conductance calcium-activated potassium channels in cultured rat skeletal muscle, *J. Physiol.* **402:**79–120.

McManus, O. B., Blatz, A. L., and Magleby, K. L., 1987, Sampling, log-binning, fitting and plotting durations of open and shut intervals from single channels, and the effects of noise, *Pflügers Arch.* **410:**530–553.

Neher, E., 1983, The charge carried by single-channel currents of rat cultured muscle cells in the presence of local anaesthetics, *J. Physiol.* **339:**663–678.

Neher, E., and Steinbach, J. H., 1978, Local anaesthetics transiently block currents through single acetylcholine-receptor channels, *J. Physiol.* **277:**153–176.

Nelder, J. A., and Mead, R., 1965, A simplex method for function minimization, *Comput. J.* **7:**308–313.

Newland, C. F., Colquhoun, D., and Cull-Candy, S. G., 1991, Single channels activated by high concentrations of GABA in superior cervical ganglion neurones of the rat, *J. Physiol.* **432:**203–233.

Ogden, D. C., and Colquhoun, D., 1985, Ion channel block by acetylcholine, carbachol and suberyldicholine at the frog neuromuscular junction, *Proc. R. Soc. Lond.* [*Biol.*] **225:**329–355.

O'Neill, R., 1971, Algorithm AS47, function minimization using a simplex procedure, *Appl. Statist.* **20:**338–345.

Papoulis, A., 1965, *Probability, Random Variables and Stochastic Processes,* McGraw-Hill, New York.

Patlak, J. B., 1988, Sodium channel subconductance levels measured with a new variance-mean analysis, *J. Gen. Physiol.* **92:**413–430.

Patlak, J. B., Ortiz, M., and Horn, R., 1986, Opentime heterogeneity during bursting of sodium channels in frog skeletal muscle, *Biophys. J.* **49:**773–777.

Press, W. H., Teukolsky, S. A., Vetterling, W. T., and Flannery, B. P., 1992, *Numerical Recipes,* 2nd ed., Cambridge University Press, Cambridge.

Rao, C. R., 1973, *Linear Statistical Inference and Its Applications,* 2nd ed., John Wiley & Sons, New York.

Sachs, F., Neil, J., and Barkakati, N., 1982, The automated analysis of data from single ionic channels, *Pflügers Arch.* **395:**331.

Sakmann, B., Patlak, J., and Neher, E., 1980, Single acetylcholine-activated channels show burst-kinetics in presence of desensitizing concentrations of agonist, *Nature* **286:**71–73.

Sigworth, F. J., 1985, Open channel noise: I. Noise in acetylcholine receptor currents suggests conformational fluctuations, *Biophys. J.* **47:**709–720.

Sigworth, F. J., 1986, Open channel noise. II. A test for coupling between current fluctuations and conformational transitions in the acetylcholine receptor, *Biophys. J.* **49:**1041–1046.

Sigworth, F. J., and Sine, S. M., 1987, Data transformations for improved display and fitting of single-channel dwell time histograms, *Biophys. J.* **52:**1047–1054.

Sigworth, F. J., and Zhou, J., 1992, Analysis of nonstationary single-channel currents, *Methods Enzymol.* **207:**746–762.

Sine, S. M., Claudio, T., and Sigworth, F. J., 1990, Activation of *Torpedo* acetylcholine receptors expressed in mouse fibroblasts: Single channel current kinetics reveal distinct agonist binding affinities, *J. Gen. Physiol.* **96:**395–437.

Stern, P., Béhé, P., Schoepfer, R., and Colquhoun, D., 1992, Single channel properties of NMDA receptors expressed from cloned cDNAs: Comparison with native receptors, *Proc. R. Soc. Lond.* [*Biol*] **250:**271–277.

Van Trees, H. L., 1968, *Detection, Estimation and Modulation Theory,* Part 1, New York, John Wiley & Sons.

Weiss, D. S., and Magleby, K. L., 1989, Gating scheme for single GABA-activated Cl-channels determined from stability plots, dwell-time distributions, and adjacent-interval durations, *J. Neurosci.* **9:**1314–1324.

Yellen, G., 1984, Ionic permeation and blockade in Ca-activated K channels of bovine chromaffin cells, *J. Gen. Physiol.* **84:**157–186.

Chapter 20

A Q-Matrix Cookbook

How to Write Only One Program to Calculate the Single-Channel and Macroscopic Predictions for Any Kinetic Mechanism

DAVID COLQUHOUN and ALAN G. HAWKES

1. Introduction

It is clear from the examples in Chapter 18 (this volume) that the algebra involved in kinetic arguments can be quite lengthy, even for simple mechanisms with only three states. For more complex mechanisms it becomes rapidly worse. Furthermore, this complicated algebra would have to be carried out separately for every kinetic mechanism that was of interest. On the other hand, the use of matrix notation allows perfectly general solutions to be written down. Not only are the results general, but they are also compact and simple-looking. They do not result in pages of complicated-looking algebra. For example, once a solution has been obtained for a quantity such as the distribution of the burst length, this result can be applied to *any* kinetic mechanism that is postulated. There is no need for further algebra (or for further programming) when a new mechanism is considered. With a general computer program, all that is needed is to supply the program with a definition of the states and the values of the transition rates for the mechanism you wish to study.

Chapter 18 (this volume) includes a brief introduction to matrix-based theory. More comprehensive treatments are given, for example, by Colquhoun and Hawkes (1982). The paper by Colquhoun and Hawkes (1981) contains some explicit algebraic examples for common mechanisms, but this paper should not be consulted for the underlying theory, which is dealt with more elegantly in the 1982 paper.

The purpose of this chapter is to provide a guide to programming the sort of general matrix results needed, so that you can obtain numerical predictions from them. It is intended as a practical 'cookbook' guide rather than an explanation of the underlying theory. We consider a number of examples based on one particular mechanism. Although these do not exhaust all the possible things you can do, they exhibit the usefulness of the approach, and the computational elements needed are readily adapted to other situations.

Note to the reader: At the author's request this chapter will use British spelling and the abbreviations ms and μs instead of msec and μsec.

DAVID COLQUHOUN • Department of Pharmacology, University College London, London WC1E 6BT, England. ALAN G. HAWKES • European Business Management School, University of Wales Swansea, Swansea SA2 8PP, Wales.

Single-Channel Recording, Second Edition, edited by Bert Sakmann and Erwin Neher. Plenum Press, New York, 1995.

Within the body of this chapter we assume that the reader is familiar with the basic notation and operations of matrix algebra such as addition, subtraction, and multiplication of matrices, and the determinant of a square matrix; the fourth basic operation, division, is known as *matrix inversion,* the inverse (if it exists) of a matrix $\mathbf{A}$ being denoted by $\mathbf{A}^{-1}$. These basic operations are described briefly in the Appendix. *If you are not familiar with the basic operations, please read the Appendix before proceeding further.*

Modern computer libraries have excellent routines for matrix manipulation, so it is not too hard to write an entirely general program to produce numerical results for *any* specified mechanism. For example there are the widely available NAG routines in FORTRAN and PASCAL, similar routines in the packages LINPACK and EISPACK, and very good matrix algebra facilities within the languages/packages such as APL, GENSTAT, SPLUS, and MATLAB and the computer algebra packages such as MAPLE, MATHEMATICA, MACSYMA, REDUCE, and DERIVE. General computational guidance and some code can be found in the invaluable book, *Numerical Recipes,* by Press *et al.* (1993).

Because of all this variety, we do not, in general, give detailed computer code in any language, except a few simple examples in languages that are particularly appropriate for this kind of work but that may be less well known. We concentrate instead on discussing the principles involved in one or two of the trickier parts and give numerical examples against which others can check the results obtained from their own programs.

Numerical Solutions and Explicit Solutions

One virtue of writing out the algebra the hard way is that one can see every term in the equations, and therefore one may be able (if the results are not too complex) to get a feeling for how the equations work, e.g., which are the important bits of the equation, and which bits can be neglected. On the other hand, if it gets complex, one may get easily confused and make mistakes.

At the other extreme, it is sometimes possible to solve the equations *numerically* without solving them *algebraically* at all. For example, the macroscopic behaviour of a kinetic system is described by a set of linear first-order differential equations, and standard algorithms exist in all computer libraries (e.g., the Runge–Kutta method) for producing the numerical solution of such sets of equations, given values for the rate constants, etc. The solution comes out as a set of numbers, e.g., the fraction of open channels at each of a set of specified times. Not much work is needed to get the results, but, on the other hand, the results have no generality. Thus, if the value of a rate constant or concentration is changed, the whole calculation must be done again.

In some cases, this sort of numerical solution is all that is possible. For example, if the ligand concentration is varying with time (as during a synaptic current), then the coefficients of the differential equations are not constant, and explicit solutions are usually not possible. If, however, the coefficients *are* constant (e.g., concentrations and membrane potential are constant throughout), then solutions to most problems can be found in the form of a sum of exponential terms. Once this has been done, values can be calculated at any time point with little further effort, and, moreover, the time constants involved in these expressions may be interpreted physically. The problem is that the algebraic expressions for these time constants, and for the coefficients of the exponential terms, are very complicated except in the simplest cases. In mechanisms with three distinct states, for example, the time constants must be found as the solution of a quadratic equation (illustrated in Chapter 18). With four states the time constants are the solutions of a cubic equation. Although a cubic equation can, like a quadratic, be solved explicitly, the results are even more untidy.

With more states than four, the higher-degree polynomials that must be solved for the time constants are not generally solvable in an explicit form. This fact blurs the distinction between numerical and explicit results. The use of matrix notation allows explicit solutions, in the form of sums of exponential terms, to be written compactly and elegantly. But, in order to calculate numbers from these 'explicit' results, it will usually be most convenient (though not necessary) to evaluate the time constants of the exponentials. This process involves solving a polynomial equation for the time constants, as just described, a process that is known, in matrix jargon, as finding the *eigenvalues* of a matrix, as explained below. This polynomial, if higher order than a cubic, will have to be solved by some sort of numerical method (routines for doing this are available in all computer matrix libraries, so you do not have to do this yourself). This does mean, though, that even explicit solutions will involve numerical steps when one wishes to calculate values from them.

2. Basic Notation and a Particular Mechanism

In this section we introduce the basic notation we will use and exemplify it by a particular mechanism that we will use throughout the chapter to give some numerical examples (see also the final section in Chapter 18, this volume). Further notation will be introduced as needed for calculating particular quantities. In general, an ion channel can be considered at any time to be in one of k physical states, which we will label by the integers 1 to k, and the dynamics are governed by the transition rates q_{ij}, which here are interpreted as giving rise to probabilities as follows: if the channel is in state i at time t then, for a small time increment δt,

$$P[\text{channel moves to state } j \text{ during the time interval } (t, t+\delta t)] \simeq q_{ij}\delta t \qquad (1)$$

where $\simeq$ means 'approximately equal to'. More precisely, q_{ij} is the limit as δt tends to zero of the ratio of the above probability and the increment δt (see Chapter 18, this volume, Section 1.2). Clearly, this definition makes sense only when i and j are different (i.e., $i \neq j$). The matrix $\mathbf{Q}$ is a square array of values with k rows and k columns. The entry in the ith row and jth column of this array is q_{ij}, and the value (for $i \neq j$) is usually simply the rate constant for transitions between states (these transition rates all have dimensions of reciprocal time, so association rate constants must be multiplied by the ligand concentration, as discussed in Chapter 18 (this volume), to obtain the transition rate). Thus equation 1 is the microscopic probabilistic manifestation of the law of mass action. This defines all the entries in $\mathbf{Q}$ except for those along its diagonal (i.e., the values for $i = j$). These diagonal elements q_{ii} are then chosen so that the sum of the values in each row is zero; thus, q_{ii} is negative, and $-q_{ii}$ represents the total rate at which the channel leaves state i, or $-1/q_{ii}$ is the average duration of a sojourn in state i.

In order to relate the algebra to the experimentally observed channel behaviour, the next step is to classify the k states according to their observable characteristics. The simplest classification is into open states and shut states, and we will use the symbol $\mathcal{A}$ to denote the index set of the open states and $\mathcal{F}$ to denote the set of shut states (examples are given below). If we are interested in burst characteristics, it will be useful to subclassify the shut states as being 'brief' or 'long'. The brief shut states are those that have short lifetimes on average and can be associated with short shut times within a burst; the index set of these will be denoted by $\mathcal{B}$. The remaining 'long' shut states, associated with long gaps between bursts, will be denoted by $\mathscr{C}$. Then $\mathscr{F} = \mathscr{B} \cup \mathscr{C}$, the union of the two sets; i.e., all shut states are classified as either brief or long. It is convenient to number the states so that $\mathscr{A}$ contains the smallest index numbers, $\mathscr{C}$ the largest and $\mathscr{B}$ the intermediate values. Let the numbers of states in each group be denoted $k_{\mathscr{A}}$, $k_{\mathscr{B}}$, and $k_{\mathscr{C}}$; then $k_{\mathscr{A}} + k_{\mathscr{B}} + k_{\mathscr{C}} = k$.

2.1. A Five-State Mechanism

All the illustrations in this chapter are done with respect to the five-state mechanism shown in equation 110 of Chapter 18 (this volume) and reproduced in equation 2. There are $k_{\mathscr{A}} = 2$ open states, which are numbered 1 and 2 in equation 2, so the set of open states is $\mathscr{A} = \{1,2\}$.

$$
\begin{array}{ccccc}
\text{State number} & & & & \text{State number}\\
5 & \mathrm{R} & & & \\
 & k_{-1}\Big\uparrow\Big\downarrow 2k_{+1} & & & \\
4 & \mathrm{AR} & \underset{\alpha_1}{\overset{\beta_1}{\rightleftharpoons}} & \mathrm{AR}^* & 1 \\
 & 2k_{-2}\Big\uparrow\Big\downarrow k_{+2} & & 2k^*_{-2}\Big\uparrow\Big\downarrow k^*_{+2} & \\
3 & \mathrm{A_2R} & \underset{\alpha_2}{\overset{\beta_2}{\rightleftharpoons}} & \mathrm{A_2R}^* & 2
\end{array} \qquad (2)
$$

We suppose that the two states for which agonist molecules are bound, but the channel is shut, are both short-lived, so we number them as states 3 and 4 and classify them as $\mathscr{B}$ states; i.e., we take $\mathscr{B} = \{3,4\}$, with $k_{\mathscr{B}} = 2$. It should be noticed at this point that, since we wish to identify brief observed shut times with sojourns in this set of states, it is actually not good enough to say that the lifetimes of states 3 and 4 are both (on average) brief; we actually require that a sojourn within subset $\mathscr{B}$ should be (on average) brief. The latter does not necessarily follow from the former because, if the rate constants were such that there were many $3 \leftrightarrow 4$ transitions before $\mathscr{B}$ was left, it is possible that a long time could be spent within $\mathscr{B}$ even though states 3 and 4 were both short-lived. Finally the set $\mathscr{C} = \{5\}$ consists of the single shut state, $k_{\mathscr{C}} = 1$, in which no agonist molecules are bound to the channel receptors. This is supposed to have a long lifetime; this will, of course, be true only when the agonist concentration is low, so the calculations refer only to this condition.

The transition rate from state 3 to state 4 is labeled, on diagram 2, as $2k_{-2}$ rather than k_{-2}. This is because two agonist molecules are bound to state 3, and one *or* the other must dissociate to make a transition to state 4. If the dissociation rate constant for a *single* agonist–receptor complex is k_{-2}, and if the two bound molecules both behave in the same way, the fact that *either* may dissociate makes the total transition rate $2k_{-2}$. The rate for a single site, k_{-2}, is known as a *microscopic rate constant,* whereas the net transition rate, $2k_{-2}$, is a *macroscopic rate constant.* Similar remarks apply to the $2 \rightarrow 1$ transition and to the $5 \rightarrow 4$ transition. We shall use microscopic rate constants here because they are the most fundamental physical quantities (but remember the implicit assumption of equivalent binding sites).

2.2. The Q Matrix

The general Q matrix for this mechanism is shown in equation 127 of Chapter 18 (this volume). As in Colquhoun and Hawkes (1982), we take for our calculations the particular values $\beta_1 = 15\ s^{-1}$, $\beta_2 = 15{,}000\ s^{-1}$, $\alpha_1 = 3000\ s^{-1}$, $\alpha_2 = 500\ s^{-1}$, $k_{-1} = k_{-2} = 2000\ s^{-1}$, $k_{+1} = 5 \times 10^7\ M^{-1}\ s^{-1}$, $k_{+2} = k^*_{+2} = 5 \times 10^8\ M^{-1}s^{-1}$. Then the principle of microscopic reversibility implies that we must also have $k^*_{-2} = (1/3)s^{-1}$ (see Chapter 18, this volume, Section 7). Finally, we take the agonist concentration to be $x_A = 100$ nM; thus, for example, $q_{54} = 2k_{+1}x_A = 2 \times (5 \times 10^7\ M^{-1}s^{-1}) \times (100 \times 10^{-9}\ M) = 10\ s^{-1}$, as shown in row 5, column 4 of **Q,** below. When these values are substituted into the general forms given in equation 127 of Chapter 18 (this volume), we get the Q matrix:

$$\mathbf{Q} = \left[\begin{array}{cc|ccc} -3050 & 50 & 0 & 3000 & 0 \\ 0.666667 & -500.666667 & 500 & 0 & 0 \\ \hline 0 & 15000 & -19000 & 4000 & 0 \\ 15 & 0 & 50 & -2065 & 2000 \\ 0 & 0 & 0 & 10 & -10 \end{array}\right] \tag{3}$$

Alternatively, it may be more convenient to work on a millisecond time scale, in which case we multiply the above transition rates by 10^{-3} giving, in ms^{-1},

$$\mathbf{Q} = \left[\begin{array}{cc|cc|c} -3.050 & 0.05 & 0 & 3 & 0 \\ 0.000666667 & -0.500666667 & 0.5 & 0 & 0 \\ \hline 0 & 15 & -19 & 4 & 0 \\ 0.015 & 0 & 0.05 & -2.065 & 2 \\ \hline 0 & 0 & 0 & 0.01 & -0.01 \end{array}\right] \tag{4}$$

This has also an advantage in having entries closer in value to 1 than the previous version of **Q,** as some matrix routines can become numerically unstable if the entries are too large or too small. This is the example we will work with. The element q_{21} is actually $(2/3) \times 10^{-3}$; again, because of numerical sensitivity of some matrix operations, one should make this reasonably accurate, and we have expressed it as a decimal to six significant figures.

The matrix in equation 3 has been partitioned into blocks, indicated by the horizontal and vertical hairlines; these divisions correspond to the division of states into the open states (set $\mathscr{A}$) and the shut states (set $\mathscr{F}$). The matrix in equation 4 has also been partitioned into blocks, the divisions in this case correspond to the division of states into the sets $\mathscr{A}$, $\mathscr{B}$, and $\mathscr{C}$. This is an example of the general partition of **Q** into the forms

$$\mathbf{Q} = \begin{bmatrix} \mathbf{Q}_{\mathscr{A}\mathscr{A}} & \mathbf{Q}_{\mathscr{A}\mathscr{F}} \\ \mathbf{Q}_{\mathscr{F}\mathscr{A}} & \mathbf{Q}_{\mathscr{F}\mathscr{F}} \end{bmatrix} = \begin{bmatrix} \mathbf{Q}_{\mathscr{A}\mathscr{A}} & \mathbf{Q}_{\mathscr{A}\mathscr{B}} & \mathbf{Q}_{\mathscr{A}\mathscr{C}} \\ \mathbf{Q}_{\mathscr{B}\mathscr{A}} & \mathbf{Q}_{\mathscr{B}\mathscr{B}} & \mathbf{Q}_{\mathscr{B}\mathscr{C}} \\ \mathbf{Q}_{\mathscr{C}\mathscr{A}} & \mathbf{Q}_{\mathscr{C}\mathscr{B}} & \mathbf{Q}_{\mathscr{C}\mathscr{C}} \end{bmatrix}, \tag{5}$$

where, for example, $\mathbf{Q}_{\mathscr{A}\mathscr{B}}$ indicates a submatrix of the matrix **Q** obtained by selecting the rows belonging to the index set $\mathscr{A}$ (in this case 1,2) and the columns belonging to the index set $\mathscr{B}$ (in this case 3,4, so $\mathbf{Q}_{\mathscr{A}\mathscr{B}}$ has two rows and two columns). Thus, from equation 4 we have in the present case

$$\mathbf{Q}_{\mathscr{A}\mathscr{B}} = \begin{bmatrix} 0 & 3 \\ 0.5 & 0 \end{bmatrix}. \tag{6}$$

The use of submatrices of this sort is very common in all calculations that concern single-channel properties. In contrast, calculations concerning the average properties of a large number of channels (i.e., relaxation or noise analysis) use the whole **Q** matrix as it stands, as illustrated in Section 4 below. The fact that single-channel observations allow use of a smaller matrix is one way of viewing the reason single channels can provide simpler and more direct inferences than macroscopic observations.

In computer programs for the evaluation of single-channel models, it is, therefore, very commonly required to find a submatrix of **Q,** i.e., in the example above, to move the values in rows 1,2 and columns 3,4 as in equation 4 into rows 1,2 and columns 1,2, as in equation 6. It is easy to write a subroutine in any language that will move the elements in any specified rows and columns into the top left-hand corner in this way. Some languages make this very simple. For example, the computer language APL has very good array-handling properties, and one example is that if one assigns, say, A←1 2 and B←3 4, then the expression Q[A;B] is equivalent to $\mathbf{Q}_{\mathscr{A}\mathscr{B}}$.

3. Equilibrium State Occupancies

The fraction of molecules in each state at equilibrium can be obtained from explicit algebraic expressions, which are not difficult to obtain even for complex mechanisms. However, it is very convenient to have them evaluated by the same computer program that does the subsequent, more complex, kinetic calculations. Evaluation of equilibrium occupancies from the **Q** matrix directly raises some problems that may not be obvious, so this problem will be discussed next.

We shall denote the occupancy of state i at time t as $p_i(t)$. Let $\mathbf{p}(t)$ be a row vector with k elements (a $1 \times k$ matrix; see Appendix) that contains these occupancies for each of the k, states.

$$\mathbf{p}(t) = [p_1(t) \quad p_2(t) \quad \cdots \quad p_k(t)] \tag{7}$$

where $p_i(t)$ is the probability that the channel is in state i at time t. The corresponding vector of derivatives is

$$\frac{\mathrm{d}\mathbf{p}(t)}{\mathrm{d}t} = \begin{bmatrix} \frac{\mathrm{d}p_1(t)}{\mathrm{d}t} & \frac{\mathrm{d}p_2(t)}{\mathrm{d}t} & \cdots & \frac{\mathrm{d}p_k(t)}{\mathrm{d}t} \end{bmatrix} \tag{8}$$

With this notation, the kinetic equations that describe the system can be written as

$$\frac{\mathrm{d}\mathbf{p}(t)}{\mathrm{d}t} = \mathbf{p}(t)\mathbf{Q} \tag{9}$$

This result follows directly from the law of mass action, which describes the rate of change of the concentration of each reactant (or, equivalently, of the fraction of the system in each state). The result can easily be verified for particular examples by multiplying out the right-hand side. Note, though, that matrix notation has allowed us to write an equation, equation 9, that is correct for *any* mechanism, however complex and with any number of states.

The term *steady state* means that these occupancies (probabilities) do not change with time, and so the derivatives are zero. In the absence of an energy input, the existence of steady state means that the system is at equilibrium (see Chapter 18, this volume, Section 7). After sufficient time ($t \to \infty$), equilibrium will be reached, and the equilibrium occupancy of state i is denoted $p_i(\infty)$. These equilibrium values are in the equilibrium vector of probabilities $\mathbf{p}(\infty)$, which, from equation 9, satisfies the equation

$$\mathbf{0} = \mathbf{p}(\infty)\mathbf{Q} \tag{10}$$

subject to $\Sigma p_i(\infty) = 1$, because the total probability (or total occupancy) must be 1. Here, the symbol $\mathbf{0}$ represents a matrix with elements that are all zeros. [Mathematical note: for *ergodic* mechanisms, and this will include all reversible mechanisms with at most one closed set of states (*i.e.,* a set that the channel cannot get out of), $\mathbf{p}(t)$ will tend to a unique limit vector $\mathbf{p}(\infty)$ as $t \to \infty$, independently of initial conditions, and this will be the same as the equilibrium vector $\mathbf{p}$ above. In these circumstances the above equation will have a unique solution.]

Several systematic methods for obtaining the steady-state distribution for any mechanism are available (all, of course, are just standard methods for solving simultaneous equations).

3.1. The Determinant Method

The best-known method is, perhaps, that based on the use of determinants (e.g., Huang, 1979; Colquhoun and Hawkes, 1987). The Appendix gives a brief definition of a determinant, and all matrix program libraries contain routines for calculation of determinants, so it will never be necessary to program this oneself. The procedure is as follows. To find the value for $p_i(\infty)$, cross out the ith row and the ith column of $\mathbf{Q}$. Then calculate the determinant of the matrix that remains (which now has $k - 1$ rows and $k - 1$ columns). Call this determinant d_i. Then $p_i(\infty)$ can be calculated as d_i divided by the sum of all k values of d. In general

$$p_i(\infty) = d_i / \sum_{j=1}^{k} d_j \tag{11}$$

3.2. The Matrix Method

In order to obtain numerical results from a computer program, it is generally more convenient to use matrix methods than to use determinants. This is, however, not quite as straightforward as it seems, essentially because the number of unknowns is one less than the number of equations, so one equation is superfluous. This is reflected in the fact that $\mathbf{Q}$ has a determinant of zero (because the rows each add up to 0) and so cannot be inverted (see Appendix). There are two ways around this problem. One is to define a modified (*reduced*) **Q** matrix that has only $k - 1$ rows and columns rather than k (where k is the number of states in which the system can exist). This gets rid of the superfluity and thus allows a straightforward solution. The other, and perhaps more straightforward, solution is to use a trick (described below) to solve the equations directly. Both methods will be described.

3.2.1. The Reduced Q-Matrix Method

The procedure is to subtract the elements in the bottom row of **Q** from each of the other rows and to omit the last column. Thus, if we denote the reduced **Q** matrix as **R,** with elements r_{ij}, then

$$r_{ij} = q_{ij} - q_{kj} \quad \text{for} \quad 1 \leq i, \;\; j \leq k - 1 \tag{12}$$

Thus, for the **Q** matrix in equation 4, the reduced version would be

$$\mathbf{R} = \begin{bmatrix} -3.050 & 0.05 & 0 & 2.99 \\ 0.000666667 & -0.500666667 & 0.5 & -0.01 \\ 0 & 15 & -19 & 3.99 \\ 0.015 & 0 & 0.05 & -2.075 \end{bmatrix} \tag{13}$$

Define also a reduced version of the row vector that contains the equilibrium probabilities, with the last value, $p_k(\infty)$, omitted. The last value can be found at the end from the fact that the probabilities must add to 1. Call this vector

$$\mathbf{p}(\infty)' = [p_1(\infty) \quad p_2(\infty) \quad \cdots \quad p_{k-1}(\infty)]$$

Finally, define a row vector, $\mathbf{r} = [q_{k1} \quad q_{k2} \ldots q_{k,k-1}]$, that contains the first $k - 1$ elements of the bottom row of **Q.** In this example, from equation 4, we have

$$\mathbf{r} = [0 \quad 0 \quad 0 \quad 0.01] \tag{14}$$

The equations $\mathbf{p}(\infty)\mathbf{Q} = \mathbf{0}$ imply $\mathbf{p}(\infty)'\mathbf{R} + \mathbf{r} = \mathbf{0}$. Because **R,** unlike **Q,** can be inverted (it is not singular), the solution can be written as

$$\mathbf{p}(\infty)' = -\mathbf{r}\mathbf{R}^{-1} \tag{15}$$

All computer matrix libraries contain procedures for matrix inversion so it is not necessary to program this operation oneself (indeed is is not desirable, because enormous specialist effort has gone into writing algorithms that will give numerically accurate results).

3.2.2. Solution of $\mathbf{p}(\infty)\mathbf{Q} = \mathbf{0}$ Directly

This can be done by adding a unit column (all values are 1) on to the right-hand end of **Q** to produce a matrix with k rows and $k + 1$ columns. Call this matrix **S.** Also define a row vector, $\mathbf{u}$ say, containing k values all equal to 1. For example, from the $\mathbf{Q}$ matrix in equation 4, let $\mathbf{u} = [1\ 1\ 1\ 1\ 1]$ and

$$\mathbf{S} = \begin{bmatrix} -3.050 & 0.05 & 0 & 3 & 0 & 1 \\ 0.000666667 & -0.500666667 & 0.5 & 0 & 0 & 1 \\ 0 & 15 & -19 & 4 & 0 & 1 \\ 0.015 & 0 & 0.05 & -2.065 & 2 & 1 \\ 0 & 0 & 0 & 0.01 & -0.01 & 1 \end{bmatrix} \tag{16}$$

The solution for the equilibrium occupancies can then be found, for any mechanism, as

$$\mathbf{p}(\infty) = \mathbf{u}(\mathbf{SS}^{\mathrm{T}})^{-1} \tag{17}$$

In this result, $\mathbf{S}^{\mathrm{T}}$ represents the transpose of $\mathbf{S}$ (see Appendix), so $\mathbf{SS}^{\mathrm{T}}$ is a matrix with k rows and k columns, but, unlike $\mathbf{Q}$, it is not singular so it can be inverted to get the required solution. Hawkes and Sykes (1990) discuss this solution and show that in APL it is computed remarkably simply (see Appendix 1).

With any of these three methods, the solution for our example matrix (equation 4) is, to four significant figures,

$$\mathbf{p}(\infty) = (0.00002483 \quad 0.001862 \quad 0.00006207 \quad 0.004965 \quad 0.9931). \tag{18}$$

Under these conditions the channel is shut most of the time; it is open only 0.189% of the time, and for 99.3% of the time it is in the unoccupied state (state 5).

4. Relaxation to Equilibrium

4.1. General Solutions for the Rate of Approach to Equilibrium

In any problem that involves the average behaviour of a large number of channels, the problem is to find how occupancies change with time. This description encompasses all macroscopic voltage-jump and concentration-jump experiments, for example. The problem, then, is to solve equation 9 for $\mathbf{p}(t)$; this vector contains the occupancy of each state at time t. We expect, under the conditions mentioned at the start, that the time course of these occupancies will be described by the sum of $k - 1$ exponential terms (Colquhoun and Hawkes, 1977).

4.1.1. Initial Occupancies at Equilibrium

The aim, in a jump experiment, is to keep the membrane potential and ligand concentrations constant at all times (except for the actual moment of the jump). The approach to equilibrium after a jump should therefore be dictated by the $\mathbf{Q}$ matrix calculated for the conditions (the potential and concentrations) that exist *after* the jump. In order to calculate this time course, we need to know the occupancy of each state at the moment ($t = 0$, say) when the jump was applied, i.e., the initial occupancies, $\mathbf{p}(0)$. If the system was at equilibrium before the jump was imposed, then these initial occupancies will simply be the equilibrium occupancies, calculated as described in Section 3 from the **Q** matrix that describes the conditions *before* the jump.

4.1.2. Initial Occupancies Not at Equilibrium

In a more complex case we might wish to calculate the time course of the response to a brief pulse of membrane potential or concentration. This will involve two separate calculations, one for the onset of the response from the moment ($t = 0$) that the pulse starts and another for the 'offset' of the response after the pulse ends. In this case it is reasonable to suppose that the system has equilibrated before the pulse is applied, so the initial occupancies for calculating the onset of the response can be found as above. However, if the pulse is brief (duration t_p say), there will not be time for the system to come to equilibrium before the end of the pulse. The initial occupancies for calculating the 'offset' time course will simply be the occupancies, $\mathbf{p}(t_p)$, that were found from the onset calculation to obtain at the moment $t = t_p$ when the pulse ends, and the calculation can then be completed using the **Q** matrix appropriate to the conditions after the end of the pulse.

Formally, the solution of differential equation 9 is just

$$\mathbf{p}(t) = \mathbf{p}(0)e^{\mathbf{Q}t} \tag{19}$$

where $\mathbf{p}(0)$ contains the occupancy probabilities at $t = 0$. This result, despite being completely general for any mechanism, looks no more complicated than its scalar (nonmatrix) equivalent, which would describe only the simplest two-state shut ↔ open reaction. This astonishingly simple result is, in a sense, all that there is to be said about calculating the time course on the basis of some specified reaction scheme. Needless to say, though, there is a bit more to be said. In particular, it may not be at all obvious what $e^{\mathbf{Q}t}$ means. In this expression, e represents the usual (scalar) constant (2.71828 . . . , the base of natural logarithms), but the exponent $\mathbf{Q}t$ is a matrix! The exponential of a matrix is an unfamiliar object even to many mathematicians. It is to some extent reassuring to find that it is defined by the usual power series

$$e^{\mathbf{Q}t} = \mathbf{I} + \mathbf{Q}t + \frac{(\mathbf{Q}t)^2}{2!} + \frac{(\mathbf{Q}t)^3}{3!} + \frac{(\mathbf{Q}t)^4}{4!} + \cdots \tag{20}$$

where **I** is the identity matrix (see Appendix), with matrices and their powers replacing the usual scalar terms with which we are familiar. It is immediately clear from this that, since **Q** is a $k \times k$ matrix, $e^{\mathbf{Q}t}$ is itself also a $k \times k$ matrix. The (infinite) series in equation 20 involves nothing more complex than multiplying and adding matrices, so it can easily be evaluated, stopping after a finite number of terms of course. However, use of this expansion, though sometimes satisfactory, is not generally the fastest or the most accurate way of evaluating the exponential numerically. Furthermore, its use does not generate the $k - 1$ exponential components that an experienced experimenter expects to see.

4.2. Evaluation of p(*t*) as a Sum of Exponential Components

The trick needed to accomplish this is a beautifully elegant technique called the *spectral expansion* of a matrix. This is critical to most of the results in this chapter, and the problem is therefore looked at in more detail in Section 9. For the moment, we merely state that $e^{\mathbf{Q}t}$ can be written in the form

$$e^{\mathbf{Q}t} = \sum_{i=1}^{k} \mathbf{A}_i \exp(-\lambda_i t), \tag{21}$$

where λ_i are *eigenvalues* of the matrix $-\mathbf{Q}$, k in number (equal to the number of states), and the $\mathbf{A}_i$ are a set of square matrices derived from $\mathbf{Q}$ and known as the spectral matrices of $\mathbf{Q}$. How these quantities are obtained from $\mathbf{Q}$ will be discussed in Section 9, below. Given this relationship, our solution can be written in the form

$$\mathbf{p}(t) = \mathbf{p}(0)e^{\mathbf{Q}t} = \mathbf{p}(0) \sum_{i=1}^{k} \mathbf{A}_i \exp(-\lambda_i t) \tag{22}$$

We now have the exponentials in the familiar scalar form. However there are k terms, but we are expecting only $k - 1$ exponentials. The explanation of this is that $\mathbf{Q}$ is singular, so that one of the eigenvalues, say λ_1, is zero; thus, $\exp(-\lambda_1 t) = \exp(0) = 1$, regardless of t. The other eigenvalues can all be shown to be real and positive, so that $\exp(-\lambda_i t)$ tends to zero as $t \to \infty$ for all eigenvalues except λ_1. Thus, the limit, letting $t \to \infty$, of the above equation shows that

$$\mathbf{p}(\infty) = \mathbf{p}(0)\mathbf{A}_1 \tag{23}$$

and so

$$\mathbf{p}(t) = \mathbf{p}(\infty) + \mathbf{p}(0) \sum_{i=2}^{k} \mathbf{A}_i \exp(-\lambda_i t)$$

or, in terms of time constants,

$$\mathbf{p}(t) = \mathbf{p}(\infty) + \mathbf{p}(0) \sum_{i=2}^{k} \mathbf{A}_i \exp(-t/\tau_i) \tag{24}$$

where $\tau_i = 1/\lambda_i$ is the time constant of the ith component. The approach to the equilibrium is thus a mixture of $k - 1$ exponential components. Notice that the *same* set of $k - 1$ time constants describe the time course of change for all the states; all that differs from one state to another is the amplitude of the components, i.e., the size (and sign) of the coefficients that multiply each exponential term.

4.3. Expressing the Coefficients as Scalars

In the result in equation 24, the exponential terms $\exp(-t/\tau_i)$ are ordinary scalars, but the coefficients of these exponential terms are still in matrix form. Both sides of equation 24 are $1 \times k$ matrices (vectors), the ith entry being $p_i(t)$. The last term, with the summation sign, involves the products of $\mathbf{p}(0)$, which is $1 \times k$, with $\mathbf{A}_i$, which is $k \times k$, and this product is $1 \times k$. If the matrices on the right-hand side are multiplied out, the result can be put into an entirely scalar form. Thus, the time course of the occupancy of the jth state, $p_j(t)$, can be written as

$$p_j(t) = p_j(\infty) + w_{2j}e^{-t/\tau_2} + w_{3j}e^{-t/\tau_3} + \cdots + w_{kj}e^{-t/\tau_k} \tag{25}$$

where w_{ij} ($i = 2, 3, \ldots, k$) are the $k - 1$ coefficients that define the amplitudes of the components for the jth state (w_{ij} is the coefficient for the component with time constant τ_i). The result in equation 25 can be written more compactly as

$$p_j(t) = p_j(\infty) + \sum_{i=2}^{i=k} w_{ij}e^{-t/\tau_i} \tag{26}$$

These coefficients are given by

$$w_{ij} = \sum_{r=1}^{r=k} p_r(0)a_{rj}^{(i)} \tag{27}$$

where $a_{rf}^{(i)}$ denotes the value in the rth row and jth column of $\mathbf{A}_i$. Thus, once the λ_i and the $\mathbf{A}_i$ have been found (see below), the occupancy of any state at any time can easily be calculated.

4.4. The Current through a Channel

Calculations of the sort outlined above will most commonly be aimed at calculating an observable quantity such as the time course of the current through ion channels. There may, in general, be more than one open state through which current can pass: in the notation introduced earlier there are $k_{\mathcal{A}}$ open states. These states may not all have the same conductance. Thus, if γ_i is the conductance of the ith open state, the expected current at time t, for one channel, will be

$$(V - V_{\text{rev}})[\gamma_1 p_1(t) + \gamma_2 p_2(t) + \cdots + \gamma_{k_{\mathcal{A}}} p_{k_{\mathcal{A}}}(t)]$$

i.e., just add up the probabilities of all the open states, each multiplied by the appropriate conductance, and then multiply that sum by the effective voltage across the membrane (the difference between the membrane potential, V, and the reversal potential, V_{rev}, at which no current flows). In macroscopic studies there are typically many channels, N say, so the observed current should be N times the above expected current. Using equation 24 we get, in scalar form, the current as the sum of $k - 1$ exponential terms,

$$I(t) = I(\infty) + \sum_{i=2}^{k} b_i \exp(-t/\tau_i), \tag{28}$$

where the (scalar) coefficients of each exponential are

$$b_i = N(V - V_{\text{rev}})\mathbf{p}(0)\mathbf{A}_i\mathbf{v} \tag{29}$$

In this result the column vector **v** has k elements of which the first $k_{\mathcal{A}}$ are the conductances γ_l, and the rest are zeroes (the conductance of the shut states). Multiplying by this is equivalent to adding the probabilities as described above. Multiplying out the matrices in equation 29 shows that the coefficients can be expressed entirely in terms of scalar quantities, thus:

$$b_i = N(V - V_{\text{rev}}) \sum_{r=1}^{r=k} \sum_{j=1}^{j=k_{\mathcal{A}}} p_r(0)\gamma_j a_{rj}^{(i)} \tag{30}$$

Without the factor N, equations 28–30 also give the time course of the current you would expect to see if you averaged several repetitions of the step experiment with a single channel (see Chapter 18, this volume).

4.5. Numerical Results

Using the methods discussed in Section 9, we find the eigenvalues of $-\mathbf{Q}$ are

$$\begin{cases} \lambda_1 = 0 \text{ ms}^{-1} \\ \lambda_2 = 0.1018 \text{ ms}^{-1} \\ \lambda_3 = 2.022 \text{ ms}^{-1} \\ \lambda_4 = 3.094 \text{ ms}^{-1} \\ \lambda_5 = 19.41 \text{ ms}^{-1} \end{cases} \tag{31}$$

and these correspond to time constants $\tau_i = 1/\lambda_i$ (apart from the case $\lambda_1 = 0$) of

$$\begin{cases} \tau_2 = 9.821 \text{ ms} \\ \tau_3 = 0.4945 \text{ ms} \\ \tau_4 = 0.3233 \text{ ms} \\ \tau_5 = 51.52\ \mu\text{s} \end{cases} \tag{32}$$

The k spectral matrices are as follows:

$$\mathbf{A}_1 = \begin{bmatrix} 0.00002483 & 0.001862 & 0.00006207 & 0.004965 & 0.9931 \\ 0.00002483 & 0.001862 & 0.00006207 & 0.004965 & 0.9931 \\ 0.00002483 & 0.001862 & 0.00006207 & 0.004965 & 0.9931 \\ 0.00002483 & 0.001862 & 0.00006207 & 0.004965 & 0.9931 \\ 0.00002483 & 0.001862 & 0.00006207 & 0.004965 & 0.9931 \end{bmatrix}$$

$$\mathbf{A}_2 = \begin{bmatrix} 1.670\text{E-}5 & 0.03497 & 0.0009297 & 0.001728 & -0.03764 \\ 4.662\text{E-}4 & 0.9763 & 0.02596 & 0.04825 & -1.051 \\ 3.719\text{E-}4 & 0.7787 & 0.0207 & 0.03848 & -0.8383 \\ 8.641\text{E-}6 & 0.01809 & 0.0004811 & 0.0008941 & -0.01948 \\ -9.411\text{E-}7 & -0.001971 & -0.00005239 & -0.00009738 & 0.002121 \end{bmatrix}$$

$$
\mathbf{A}_3 = \begin{bmatrix} 0.04051 & -0.06356 & 0.006321 & 2.779 & -2.762 \\ -0.0008475 & 0.00133 & -0.000132 & -0.05813 & 0.05778 \\ 0.002525 & -0.003961 & 0.0003934 & 0.1732 & -0.1721 \\ 0.01389 & -0.0218 & 0.002165 & 0.9531 & -0.9473 \\ -0.00006905 & 0.0001083 & -0.00001076 & -0.004737 & 0.004708 \end{bmatrix}
$$

$$
\mathbf{A}_4 = \begin{bmatrix} 0.9594 & 0.0272 & -7.9\text{E-}3 & -2.785 & 1.807 \\ 0.0003627 & 0.00001028 & -2.987\text{E-}6 & -0.001053 & 0.000683 \\ -0.00316 & -0.0000896 & 2.602\text{E-}5 & 0.0009174 & -0.00595 \\ -0.01393 & -0.0003949 & 1.147\text{E-}4 & 0.04043 & -0.02622 \\ 0.00004517 & 0.000001281 & -3.719\text{E-}7 & -0.0001311 & 0.00008504 \end{bmatrix}
$$

$$
\mathbf{A}_5 = \begin{bmatrix} 1.455\text{E-}7 & -0.0004734 & 0.0005968 & -1.377\text{E-}4 & 1.419\text{E-}5 \\ -6.312\text{E-}6 & 0.02053 & -0.02588 & 5.971\text{E-}3 & -6.157\text{E-}4 \\ 2.387\text{E-}4 & -0.7765 & 0.9788 & -2.258\text{E-}1 & 2.328\text{E-}2 \\ -6.883\text{E-}7 & 0.002239 & -0.002823 & 6.512\text{E-}4 & -6.714\text{E-}5 \\ 3.548\text{E-}10 & -0.000001154 & 0.000001455 & -3.357\text{E-}7 & 3.461\text{E-}8 \end{bmatrix} \tag{33}
$$

Note that every row of the matrix $\mathbf{A}_1$ is the same as the equilibrium vector shown in equation 18, so its columns consist of identical numbers. This is why we get this limit as $t \to \infty$ regardless of the initial probability vector $\mathbf{p}(0)$.

There are just two open states, with conductances γ_1 and γ_2 say, so we have

$$
\mathbf{v} = \begin{bmatrix} \gamma_1 \\ \gamma_2 \\ 0 \\ 0 \\ 0 \end{bmatrix} \tag{34}
$$

4.6. Example of a Concentration Jump

Suppose the membrane potential is $V = -100$ mV, with a reversal potential $V_{\text{rev}} = 0$ mV, and that the conductances of the two open states are $\gamma_1 = 40$ pS and $\gamma_2 = 50$ pS. There was zero drug concentration before time $t = 0$, so, at equilibrium, all channels are in the unoccupied shut state (state 5). The initial vector is therefore

$$
\mathbf{p}(0) = (0 \quad 0 \quad 0 \quad 0 \quad 1)
$$

At $t = 0$, the concentration is suddenly increased to 100 nM (the concentration used to calculate the $\mathbf{Q}$ matrix in equation 4 and its eigenvalues and spectral matrices in equations

31–33. The current will then rise from $I(0) = 0$ at $t = 0$ toward its equilibrium value for an agonist concentration of 100 nM. For $N = 1$ channel, this is

$$I(\infty) = (V - V_{\mathrm{rev}})[\gamma_1 p_1(\infty) + \gamma_2 p_2(\infty)]$$

$$= -9.4095 \times 10^{-3}\ \mathrm{pA}$$

The current will follow a time course described by the sum of four exponential terms, as in equation 28, with time constants as specified in equation 32. The coefficients, b_i, for each of these terms can be calculated by equation 29 or 30, and the values (in picoamperes) are $b_2 = 9.8563 \times 10^{-3}$ pA, $b_3 = -0.2655 \times 10^{-3}$ pA, $b_4 = -0.1871 \times 10^{-3}$ pA, and $b_5 = 0.005770 \times 10^{-3}$ pA. By far the largest component is the second, that with $\tau_2 = 9.821$ ms (its amplitude, b_2, is 37 times greater than that of the next largest component), so the relaxation is quite close to being a single exponential with this time constant. Notice that the sum of the four b_i values comes to $+9.409 \times 10^{-3}$ pA, thus ensuring that the current at $t = 0$ is indeed zero.

5. Distribution of Open Times and Shut Times

The theory enabling prediction of the distributions of open times and shut times in a single channel record is described in some detail in Colquhoun and Hawkes (1982). We will not go into the theory here but merely quote results (the form CH82 followed by a number will be used to refer to equations from that paper) and comment on computational aspects.

5.1. Distribution of Open Times

The distribution of all open times is given by (CH82-3.64) as

$$f(t) = \boldsymbol{\phi}_{\mathrm{o}} \exp(\mathbf{Q}_{\mathcal{A}\mathcal{A}} t)(-\mathbf{Q}_{\mathcal{A}\mathcal{A}})\mathbf{u}_{\mathcal{A}} \tag{35}$$

The beauty of this result is that, despite being quite general, it looks (apart from an initial and final vector) very much like the simple exponential distribution, $f(t) = \lambda \exp(-\lambda t)$, with $-\mathbf{Q}_{\mathcal{A}\mathcal{A}}$ in place of λ. In equation 35, $\boldsymbol{\phi}_{\mathrm{o}}$ is a row vector ($1 \times k_{\mathcal{A}}$) containing the probabilities of starting an open time in each of the $k_{\mathcal{A}}$ open states; $\mathbf{Q}_{\mathcal{A}\mathcal{A}}$ is a $k_{\mathcal{A}} \times k_{\mathcal{A}}$ matrix, the subsection of the $\mathbf{Q}$ matrix relating to the open states only (see Section 2), and $\mathbf{u}_{\mathcal{A}}$ is a column vector ($k_{\mathcal{A}} \times 1$) whose elements are all 1 (this has the effect of summing over the $\mathcal{A}$ states—see Appendix 1, equation A4). Thus, the result in equation 35 is scalar. To evaluate it we need only have routines to multiply matrices and a way of evaluating $\exp(\mathbf{Q}_{\mathcal{A}\mathcal{A}} t)$. The latter can be found in exactly the same way as used for finding $\exp(\mathbf{Q}t)$, as outlined in Section 4 and specified in detail in Section 9. The only differences are that (1) this time we have a smaller matrix; i.e., from equation 4,

$$\mathbf{Q}_{\mathcal{A}\mathcal{A}} = \begin{bmatrix} -3.050 & 0.05 \\ 0.000666667 & -0.500666667 \end{bmatrix} \tag{36}$$

and (2) unlike $\mathbf{Q}$, this time the matrix is unlikely to be singular, so none of the eigenvalues will be zero. In this example, the eigenvalues of $-\mathbf{Q}_{\mathscr{A}\mathscr{A}}$ are

$$\lambda_1 = 0.500654 \text{ ms}^{-1}, \qquad \lambda_2 = 3.05001 \text{ ms}^{-1} \tag{37}$$

so the time constants are

$$\tau_1 = 1.99739 \text{ ms}, \qquad \tau_2 = 0.327867 \text{ ms}$$

The distribution of open times will have two ($k_{\mathscr{A}}$) exponential components with these time constants. To get the distribution in the form of a sum of (scalar) exponentials, we again, as in Section 4, use the spectral expansion trick. By direct analogy with equation 21, we can write

$$\exp(\mathbf{Q}_{\mathscr{A}\mathscr{A}}t) = \sum_{i=1}^{k_{\mathscr{A}}} \mathbf{A}_i \exp(-\lambda_i t) \tag{38}$$

where $\mathbf{A}_i$ now represents the $k_{\mathscr{A}}$ spectral matrices (each $k_{\mathscr{A}} \times k_{\mathscr{A}}$) of $-\mathbf{Q}_{\mathscr{A}\mathscr{A}}$, and λ_i are the eigenvalues of $-\mathbf{Q}_{\mathscr{A}\mathscr{A}}$, already given in equation 37. In the present example, the spectral matrices are

$$\mathbf{A}_1 = \begin{bmatrix} 5.1288 \times 10^{-6} & 0.019613 \\ 2.6150 \times 10^{-4} & 0.999995 \end{bmatrix}$$

$$\mathbf{A}_2 = \begin{bmatrix} 0.999995 & -0.019613 \\ -2.6150 \times 10^{-4} & 5.1288 \times 10^{-6} \end{bmatrix}. \tag{39}$$

To complete the calculation, we need to find the relative areas of these two components. In general, a mixture of exponential densities can be represented as

$$f(t) = \Sigma a_i (1/\tau_i) \exp(-t/\tau_i) \tag{40}$$

the sum running over the number of components (here $i = 1$ to $k_{\mathscr{A}}$). Here a_i represents the area of the ith component (the total area being 1). From equations 35 and 38, the areas are given by

$$a_i = -\tau_i \boldsymbol{\phi}_o \mathbf{A}_i \mathbf{Q}_{\mathscr{A}\mathscr{A}} \mathbf{u}_{\mathscr{A}} \tag{41}$$

We now have everything needed to evaluate these areas, apart from the initial vector, $\boldsymbol{\phi}_o$. This is given (CH82-3.63) as

$$\boldsymbol{\phi}_o = \mathbf{p}_{\mathscr{F}}(\infty)\mathbf{Q}_{\mathscr{F}\mathscr{A}} / \mathbf{p}_{\mathscr{F}}(\infty)\mathbf{Q}_{\mathscr{F}\mathscr{A}}\mathbf{u}_{\mathscr{A}} \tag{42}$$

In this expression, $\mathbf{p}_{\mathscr{F}}(\infty)$ represents the part of the equilibrium occupancies found in equation 18 for the shut states only. The equilibrium vector was

$$\mathbf{p}(\infty) = (0.00002483 \quad 0.001862 \mid 0.00006207 \quad 0.004965 \quad 0.9931) \tag{43}$$

but a vertical line has been added that partitions $\mathbf{p}(\infty)$ into $\mathbf{p}_{\mathcal{A}}(\infty)$ (the first two elements, for the open states) and $\mathbf{p}_{\mathcal{F}}(\infty)$ (the last three elements for the shut states). Thus

$$\mathbf{p}_{\mathcal{F}}(\infty) = (0.00006207 \quad 0.004965 \quad 0.9931) \tag{44}$$

and when this is postmultiplied by $\mathbf{Q}_{\mathcal{F}\mathcal{A}}$, which, from equation 4, is the $k_{\mathcal{F}} \times k_{\mathcal{A}}$ matrix

$$\mathbf{Q}_{\mathcal{F}\mathcal{A}} = \begin{bmatrix} 0 & 15 \\ 0.015 & 0 \\ 0 & 0 \end{bmatrix} \tag{45}$$

the result is a $1 \times k_{\mathcal{A}}$ vector that forms the numerator of $\boldsymbol{\phi}_o$ in equation 42. The denominator, $\mathbf{p}_{\mathcal{F}}(\infty)\mathbf{Q}_{\mathcal{F}\mathcal{A}}\mathbf{u}_{\mathcal{A}}$, is a simple scalar, the sum of the elements in the numerator, which ensures that the elements of $\boldsymbol{\phi}_o$ add up to 1. When these are multiplied out, the result is

$$\boldsymbol{\phi}_o = (0.07407 \quad 0.92593) \tag{46}$$

Thus, any individual opening has a 7.4% chance of starting in open state 1 and a 92.6% chance of starting in open state 2. The areas can now be found from equation 41 and come to $a_1 = 0.9276$ and $a_2 = 0.07238$. The final probability density function for open times, in the form given in equation 40, is therefore

$$f(t) = 0.9276(1/1.997)e^{-t/1.997} + 0.07238(1/0.3279)e^{-t/0.3279} \tag{47}$$

The slower component, $\tau_1 = 1.99739$ ms, predominates, having 92.8% of the area.

When channel openings can be divided into bursts, there are many other open-time distributions of potential interest, e.g., the distribution of the first opening of a burst of openings or of all openings in bursts with one opening, etc. All of them have the basic form for the probability density:

$$f(t) = \boldsymbol{\phi} \exp(\mathbf{Q}_{\mathcal{A}\mathcal{A}}t)\mathbf{c} \tag{48}$$

where $\boldsymbol{\phi}$ is a suitable row vector containing the probabilities of starting an open time in each of the $k_{\mathcal{A}}$ open states, and $\mathbf{c}$ is some appropriate column vector (expressions are given by Colquhoun and Hawkes, 1982, for various cases). They all involve $\exp(\mathbf{Q}_{\mathcal{A}\mathcal{A}}t)$, so they all have $k_{\mathcal{A}}$ components with the same time constants as before. Only $\boldsymbol{\phi}$ and $\mathbf{c}$, and hence the relative areas, differ from one sort of distribution to another.

5.2. Distribution of Shut Times

The distribution of all shut times can be found in exactly the same way as just described for open times. In fact, all that has to be done is to interchange $\mathcal{A}$ and $\mathcal{F}$ in the equations already given. The distribution is thus

$$f(t) = \boldsymbol{\phi}_s \exp(\mathbf{Q}_{\mathcal{FF}}t)(-\mathbf{Q}_{\mathcal{FF}})\mathbf{u}_{\mathcal{F}} \tag{49}$$

where the initial vector, which gives the probabilities that a shut period starts in each of the $k_{\mathcal{F}}$ shut states, is

$$\boldsymbol{\phi}_s = \mathbf{p}_{\mathcal{A}}(\infty)\mathbf{Q}_{\mathcal{AF}}/\mathbf{p}_{\mathcal{A}}(\infty)\mathbf{Q}_{\mathcal{AF}}\mathbf{u}_{\mathcal{F}} \tag{50}$$

To complete the calculations, we need the 'shut' portion of the $\mathbf{Q}$ matrix; i.e., from equation 4,

$$\mathbf{Q}_{\mathcal{FF}} = \begin{bmatrix} -19 & 4 & 0 \\ 0.05 & -2.065 & 2 \\ 0 & 0.01 & -0.01 \end{bmatrix} \tag{51}$$

and

$$\mathbf{Q}_{\mathcal{AF}} = \begin{bmatrix} 0 & 3 & 0 \\ 0.5 & 0 & 0 \end{bmatrix} \tag{52}$$

with, from equation 43,

$$\mathbf{p}_{\mathcal{A}}(\infty) = (0.00002483 \quad 0.001862) \tag{53}$$

The calculations proceed in exactly the same way as for open times. We find the $k_{\mathcal{F}}$ (=3 in this case) eigenvalues and spectral matrices for $\mathbf{Q}_{\mathcal{FF}}$ and use them to express the shut-time distribution in the form of $k_{\mathcal{F}}$ exponential components, as in equation 40. The relative areas of the components are (cf. equation 41)

$$a_i = -\tau_i \boldsymbol{\phi}_s \mathbf{A}_i \mathbf{Q}_{\mathcal{FF}}\mathbf{u}_{\mathcal{F}} \tag{54}$$

where the time constants, τ_i, are now those for the shut-time distribution. The results are as follows. The eigenvalues of $-\mathbf{Q}_{\mathcal{FF}}$ are

$$\lambda_1 = 0.263895 \times 10^{-3}\ \text{ms}^{-1}, \qquad \lambda_2 = 2.06293\ \text{ms}^{-1}, \qquad \text{and} \quad \lambda_3 = 19.0118\ \text{ms}^{-1};$$

the corresponding time constants, $\tau_i = 1/\lambda_i$, are

$$\tau_1 = 3789.4\ \text{ms}; \qquad \tau_2 = 0.484747\ \text{ms}; \qquad \tau_3 = 52.5989\ \mu\text{s}. \tag{55}$$

The relative areas of these components, from equation 54, are

$$a_1 = 0.261946; \qquad a_2 = 0.00836704; \qquad a_3 = 0.729687 \tag{56}$$

Thus, most shut times are either very short, 52.6 μs (73%), or very long, 3789 ms (26%).

The intermediate quantities needed to obtain the above areas are $\boldsymbol{\phi}_s$, given by equation 50 and the spectral matrices $\mathbf{A}_i$. The results are

$$\boldsymbol{\phi}_s = [0.92593 \quad 0.07407 \quad 0] \tag{57}$$

$$\mathbf{A}_1 = \begin{bmatrix} 2.613\text{E-}6 & 9.931\text{E-}4 & 0.2040 \\ 1.241\text{E-}5 & 4.717\text{E-}3 & 0.9690 \\ 1.275\text{E-}5 & 4.845\text{E-}3 & 0.9953 \end{bmatrix}$$

$$\mathbf{A}_2 = \begin{bmatrix} 6.934\text{E-}4 & 0.2349 & -0.2288 \\ 2.936\text{E-}3 & 0.9946 & -0.9689 \\ -1.430\text{E-}5 & -4.845\text{E-}3 & 4.720\text{E-}3 \end{bmatrix}$$

$$\mathbf{A}_3 = \begin{bmatrix} 0.9993 & -0.2359 & 2.483\text{E-}2 \\ -2.949\text{E-}3 & 6.960\text{E-}4 & -7.326\text{E-}5 \\ 1.552\text{E-}6 & -3.663\text{E-}7 & 3.855\text{E-}8 \end{bmatrix} \tag{58}$$

6. Distribution of the Number of Openings per Burst

It is a very common observation that channel openings occur in *bursts* of several openings in quick succession rather than singly. This will be the case when, as in our example, the shut-time distribution contains some components that are very brief (short shuttings within a burst) and some that are very long (shut times between bursts). This is the case in our numerical example, as found in equations 55 and 56.

A burst of openings must obviously contain at least one opening. In general, it may contain r openings separated by $r - 1$ brief shuttings, where r is random. For the present purposes we define a 'shut time within a burst' as a shut time spent entirely within the short-lived shut states, set $\mathcal{B}$. There are, in this case, $k_{\mathcal{B}} = 2$ such states, states 3 and 4 (see equations 4–6). An entry into the long-lived shut state (state 5) will produce a 'shut time between bursts'. There is only one way out of state 5, with rate $q_{54} = 0.01\ \text{ms}^{-1}$, so the mean lifetime of a single sojourn in state 5 is 1/0.01 = 100 ms. Notice that this is much shorter than the 3789-ms component of shut times; this is because the channel will oscillate several times between the shut states 5, 4, and 3 between one burst and the next and is likely to visit state 5 several times before the next opening occurs (i.e., before the next burst starts). A channel in state 4 is much more likely to return to state 5 (rate 2 ms^{-1}) than either to proceed to state 3 (rate = 0.05 ms^{-1}) or to open to state 1 (rate = 0.015 ms^{-1}).

In order to analyze the burst structure we need two new matrices, denoted $\mathbf{G}_{\mathcal{A}\mathcal{B}}$ and $\mathbf{G}_{\mathcal{B}\mathcal{A}}$, which are defined as

$$\mathbf{G}_{\mathcal{A}\mathcal{B}} = -\mathbf{Q}_{\mathcal{A}\mathcal{A}}^{-1}\mathbf{Q}_{\mathcal{A}\mathcal{B}} \qquad \mathbf{G}_{\mathcal{B}\mathcal{A}} = -\mathbf{Q}_{\mathcal{B}\mathcal{B}}^{-1}\mathbf{Q}_{\mathcal{B}\mathcal{A}} \tag{59}$$

The interpretation of $\mathbf{G}_{\mathcal{A}\mathcal{B}}$ is as follows. It is a $k_{\mathcal{A}} \times k_{\mathcal{B}}$ matrix, the i,j element (that in row i, column j) of which gives the probability that an open channel, initially in state i (one of the $\mathscr{A}$ states), will eventually (after any number of transitions among the open states) arrive in state j, one of the short-lived shut states (set $\mathscr{B}$). $\mathbf{G}_{\mathscr{B}\mathscr{A}}$ has an exactly analogous interpretation. The product $\mathbf{G}_{\mathscr{A}\mathscr{B}}\mathbf{G}_{\mathscr{B}\mathscr{A}}$, which is a $k_{\mathscr{A}} \times k_{\mathscr{A}}$ matrix, thus describes routes from open states to brief shut states and back to open. This is just what happens during a burst of openings, so it is intuitively reasonable that the distribution of the number of openings per burst depends on this product. The distribution is, of course, discontinuous: the number of openings per burst (r) can take only the values 1, 2, 3, ... The probability, $P(r)$, of observing r openings per burst is (CH82-3.5):

$$P(r) = \boldsymbol{\phi}_{\mathrm{b}}(\mathbf{G}_{\mathscr{A}\mathscr{B}}\mathbf{G}_{\mathscr{B}\mathscr{A}})^{r-1}(\mathbf{I} - \mathbf{G}_{\mathscr{A}\mathscr{B}}\mathbf{G}_{\mathscr{B}\mathscr{A}})\mathbf{u}_{\mathscr{A}} \tag{60}$$

This is a *geometric distribution* (see also Chapters 18 and 19, this volume). As for the open time distribution, it starts with a $1 \times k_{\mathscr{A}}$ vector, $\boldsymbol{\phi}_{\mathrm{b}}$, which contains the $k_{\mathscr{A}}$ probabilities that the first opening in a burst starts in each of the $k_{\mathscr{A}}$ open states. These are not, in general, the same as the probabilities (in $\boldsymbol{\phi}_{\mathrm{o}}$, see equations 42 and 46) for *any* opening. This 'start-of-burst' vector is (CH82-3.2)

$$\boldsymbol{\phi}_{\mathrm{b}} = \frac{\mathbf{p}_{\mathscr{C}}(\infty)(\mathbf{Q}_{\mathscr{C}\mathscr{B}}\mathbf{G}_{\mathscr{B}\mathscr{A}} + \mathbf{Q}_{\mathscr{C}\mathscr{A}})}{\mathbf{p}_{\mathscr{C}}(\infty)(\mathbf{Q}_{\mathscr{C}\mathscr{B}}\mathbf{G}_{\mathscr{B}\mathscr{A}} + \mathbf{Q}_{\mathscr{C}\mathscr{A}})\mathbf{u}_{\mathscr{A}}} \tag{61}$$

Taking $\boldsymbol{\phi}_{\mathrm{b}}$ first, we have in our example, from equation 43,

$$\mathbf{p}_{\mathscr{C}}(\infty) = 0.9931 \tag{62}$$

(there is only one long-lived shut state, $k_{\mathscr{C}} = 1$, so this is a simple scalar, the equilibrium occupancy of state 5), and from equation 4 we have the various submatrices of $\mathbf{Q}$ needed to evaluate equations 59 and 61 as

$$\mathbf{Q}_{\mathscr{A}\mathscr{B}} = \begin{bmatrix} 0 & 3 \\ 0.5 & 0 \end{bmatrix}, \quad \mathbf{Q}_{\mathscr{B}\mathscr{B}} = \begin{bmatrix} -19 & 4 \\ 0.05 & -2.065 \end{bmatrix}, \quad \mathbf{Q}_{\mathscr{B}\mathscr{A}} = \begin{bmatrix} 0 & 15 \\ 0.015 & 0 \end{bmatrix},$$

$$\mathbf{Q}_{\mathscr{C}\mathscr{B}} = [0 \quad 0.01], \qquad \mathbf{Q}_{\mathscr{C}\mathscr{A}} = [0 \quad 0]$$

and $\mathbf{Q}_{\mathscr{A}\mathscr{A}}$, which has already been given in equation 36. Multiplying the matrices gives

$$\boldsymbol{\phi}_{\mathrm{b}} = (0.275362 \quad 0.724638) \tag{63}$$

Comparing this with $\boldsymbol{\phi}_{\mathrm{o}}$, evaluated in equation 46, shows that the first opening of a burst has a greater probability (0.275) of starting in open state 1 than is the case for *all* openings (0.0741). This is clearly a result of the fact that, before the burst, the channel must have been in state 5, from which it must move to state 4, which communicates directly with state 1. Conversely, once the burst has started, much of the time is spent oscillating between shut state 3 and open state 2, so many of these openings start in open state 2.

Next, calculate $\mathbf{G}_{\mathscr{A}\mathscr{B}}$ and $\mathbf{G}_{\mathscr{B}\mathscr{A}}$ from equation 59. Inversion of $\mathbf{Q}_{\mathscr{A}\mathscr{A}}$ gives

$$\mathbf{Q}_{\mathscr{A}\mathscr{A}}^{-1} = \begin{bmatrix} -0.327876 & -0.0327439 \\ -4.36586 \times 10^{-4} & -1.99738 \end{bmatrix} \tag{64}$$

so we find

$$\mathbf{G}_{\mathscr{AB}} = \begin{bmatrix} 0.016372 & 0.983628 \\ 0.998690 & 0.0013098 \end{bmatrix} \qquad \mathbf{G}_{\mathscr{BA}} = \begin{bmatrix} 0.0015371 & 0.793519 \\ 0.0073011 & 0.019214 \end{bmatrix}. \tag{65}$$

Multiplication of these gives

$$\mathbf{G}_{\mathscr{AB}}\mathbf{G}_{\mathscr{BA}} = \begin{bmatrix} 0.0072068 & 0.031890 \\ 0.0015446 & 0.792504 \end{bmatrix} \tag{66}$$

We can now evaluate the distribution in equation 60 for any value of r. The value of $(\mathbf{G}_{\mathscr{AB}}\mathbf{G}_{\mathscr{BA}})$ raised to the power $r - 1$ can be calculated by repeatedly multiplying $\mathbf{G}_{\mathscr{AB}}\mathbf{G}_{\mathscr{BA}}$ by itself the necessary number of times.

However, once again, the spectral resolution trick proves useful. First, it provides a much quicker way of raising a matrix to a power than the obvious method of repeated multiplication. Second, it allows the distribution to be put in the scalar form of a mixture of simple geometric distributions (directly analogous with the exponentials in equations 40 and 47). The matrix $(\mathbf{G}_{\mathscr{AB}}\mathbf{G}_{\mathscr{BA}})^n$ can be expressed (see Appendix 1) in the form

$$(\mathbf{G}_{\mathscr{AB}}\mathbf{G}_{\mathscr{BA}})^n = \sum_{i=1}^{i=k_{\mathscr{A}}} \mathbf{A}_i \rho_i^n \tag{67}$$

where ρ_i are the eigenvalues of $\mathbf{G}_{\mathscr{AB}}\mathbf{G}_{\mathscr{BA}}$ (they are not denoted λ here because they are dimensionless rather than being rates), and $\mathbf{A}_i$ are the spectral matrices of $\mathbf{G}_{\mathscr{AB}}\mathbf{G}_{\mathscr{BA}}$, found as before (see Section 9). This enables the distribution in equation 60 to be put into the entirely scalar form of a mixture of simple geometric distributions:

$$P(r) = \sum_{i=1}^{i=k_{\mathscr{A}}} a_i(1 - \rho_i)\rho_i^{r-1} \qquad (r = 1, 2, \ldots) \tag{68}$$

where a_i represents the area of each component. It can be seen that the ith component distribution has the form of a geometrically decaying series, the value of ρ_i (which is less than 1) being the factor by which P is reduced each time r is increased by 1. This is a discrete analogue of exponential decay, and the 'mean number of openings per burst' for each component (analogous with the time constant for exponentials) is

$$\mu_i = 1/(1 - \rho_i). \tag{69}$$

These results show that the areas of the components can be calculated as

$$a_i = \frac{\boldsymbol{\phi}_{\mathrm{b}}\mathbf{A}_i(\mathbf{I} - \mathbf{G}_{\mathscr{AB}}\mathbf{G}_{\mathscr{BA}})\mathbf{u}_{\mathscr{A}}}{(1 - \rho_i)} \tag{70}$$

In our example, we find the eigenvaues of $\mathbf{G}_{\mathcal{AB}}\mathbf{G}_{\mathcal{BA}}$ to be

$$\rho_1 = 0.0071441 \qquad \rho_2 = 0.792567 \tag{71}$$

and these, from equation 69, correspond to components with

$$\mu_1 = 1.0072 \qquad \mu_2 = 4.8208 \tag{72}$$

The spectral matrices of $\mathbf{G}_{\mathcal{AB}}\mathbf{G}_{\mathcal{BA}}$ are

$$\mathbf{A}_1 = \begin{bmatrix} 0.999920 & -0.040603 \\ -1.9666 \times 10^{-3} & 7.9857 \times 10^{-5} \end{bmatrix} \qquad \mathbf{A}_2 = \begin{bmatrix} 7.9857 \times 10^{-5} & 0.040603 \\ 1.9666 \times 10^{-3} & 0.999920 \end{bmatrix} \tag{73}$$

Thus, from equation 70, the areas of the two components are

$$a_1 = 0.262793 \qquad a_2 = 0.737207 \tag{74}$$

In words, 73.7% of the area is accounted for by a component with a 'mean' of 4.82 openings per burst, but 26.3% of the area corresponds to a 'mean' of 1.007 openings per burst. There are more bursts that have only one opening than would be predicted from the former component alone.

7. Distribution of Burst Length

The total duration of a burst of openings is the sum of the durations of the r open times and $r - 1$ shut times that constitute the burst. The probability density of these burst lengths is given (CH82-3.17, 3.4) as

$$f(t) = \boldsymbol{\phi}_\mathbf{b}[\exp(\mathbf{Q}_{\mathcal{EE}}t)]_{\mathcal{AA}}(-\mathbf{Q}_{\mathcal{AA}})(\mathbf{I} - \mathbf{G}_{\mathcal{AB}}\mathbf{G}_{\mathcal{BA}})\mathbf{u}_{\mathcal{A}} \tag{75}$$

(see also Section 13.4 of Chapter 18, this volume).

Once again, we need only to do some matrix multiplication, apart from finding the exponential of yet another matrix. This time we need to find $\exp(\mathbf{Q}_{\mathcal{EE}}t)$, where $\mathcal{E}$ is the set of 'burst states' consisting of the open states, $\mathcal{A}$, and the short shut states, $\mathcal{B}$. Thus, $\exp(\mathbf{Q}_{\mathcal{EE}}t)$ is a $k_{\mathcal{E}} \times k_{\mathcal{E}}$ matrix, where $k_{\mathcal{E}} = k_{\mathcal{A}} + k_{\mathcal{B}}$ (= 4 in our example), and the distribution will have $k_{\mathcal{E}}$ time constants $\tau_i = 1/\lambda_i$ where λ_i are now the eigenvalues of $-\mathbf{Q}_{\mathcal{EE}}$. As before, we shall also need the $k_{\mathcal{E}}$ spectral matrices, $\mathbf{A}_i$, of $\mathbf{Q}_{\mathcal{EE}}$, and these will each be of dimension $k_{\mathcal{E}} \times k_{\mathcal{E}}$. However, equation 75 contains a bit of notation that has not occurred before. Although $\exp(\mathbf{Q}_{\mathcal{EE}}t)$ is a $k_{\mathcal{E}} \times k_{\mathcal{E}}$ matrix (4 × 4 in our case), the notation $[\exp(\mathbf{Q}_{\mathcal{EE}}t)]_{\mathcal{AA}}$ means that we need only take the upper left $k_{\mathcal{A}} \times k_{\mathcal{A}}$ portion of it (the upper left 2 × 2 block in our case). This corresponds elegantly to the fact that a burst must both start and end in an open state and allows a general expression for a complicated quantity like the burst length to be written, in equation 75, in a form that is little more complicated than the distribution of open times.

As usual, the spectral expansion trick allows the distribution to be written as a scalar mixture of $k_{\mathcal{E}}$ exponential components, as in equation 40. In this case, the areas of each component are given by

$$a_i = -\tau_i \boldsymbol{\phi}_{\mathbf{b}}[\mathbf{A}_i]_{\mathcal{A}\mathcal{A}}\mathbf{Q}_{\mathcal{A}\mathcal{A}}(\mathbf{I} - \mathbf{G}_{\mathcal{A}\mathcal{B}}\mathbf{G}_{\mathcal{B}\mathcal{A}})u_{\mathcal{A}} \tag{76}$$

The quantities $\boldsymbol{\phi}_{\mathbf{b}}$, $\mathbf{G}_{\mathcal{A}\mathcal{B}}$, $\mathbf{G}_{\mathcal{B}\mathcal{A}}$, etc. are all found as before. In our example, we have

$$\mathbf{Q}_{\mathcal{E}\mathcal{E}} = \begin{bmatrix} -3.050 & 0.05 & 0 & 3 \\ 0.000666667 & -0.500666667 & 0.5 & 0 \\ 0 & 15 & -19 & 4 \\ 0.015 & 0 & 0.05 & -2.065 \end{bmatrix} \tag{77}$$

The eigenvalues of this are $\lambda_1 = 0.10160 \text{ ms}^{-1}$, $\lambda_2 = 2.01260 \text{ ms}^{-1}$, $\lambda_3 = 3.0933 \text{ ms}^{-1}$, and $\lambda_4 = 19.408 \text{ ms}^{-1}$, and the corresponding time constants are

$$\tau_1 = 9.84244 \text{ ms} \qquad \tau_2 = 0.49687 \text{ ms} \qquad \tau_3 = 0.323283 \text{ ms} \qquad \tau_4 = 51.5246 \ \mu\text{s} \tag{78}$$

This time, we shall not list all the spectral matrices, but the first one is

$$\mathbf{A}_1 = \left[\begin{array}{cc|cc} 1.8765 \times 10^{-5} & 0.037102 & 9.8702 \times 10^{-4} & 2.03952 \times 10^{-3} \\ 4.94699 \times 10^{-4} & 0.978107 & 0.026020 & 0.053767 \\ \hline 3.9481 \times 10^{-4} & 0.780608 & 0.020766 & 0.042910 \\ 1.0198 \times 10^{-5} & 0.020162 & 5.3638 \times 10^{-4} & 1.1083 \times 10^{-3} \end{array}\right] \tag{79}$$

In this matrix, the upper left hand $k_{\mathcal{A}} \times k_{\mathcal{A}}$ part has been marked by thin lines, so that part of $\mathbf{A}_1$ to be used to calculate the areas in equation 76 is

$$(\mathbf{A}_1)_{\mathcal{A}\mathcal{A}} = \begin{bmatrix} 1.8765 \times 10^{-5} & 0.037102 \\ 4.94699 \times 10^{-4} & 0.978107 \end{bmatrix} \tag{80}$$

The areas, from equation 76, come out to be

$$a_1 = 0.73561 \qquad a_2 = 0.01424 \qquad a_3 = 0.25007 \qquad a_4 = 0.0000772 \tag{81}$$

Only the first and third components have sufficiently large areas to be detectable experimentally.

7.1. Comparison with Relaxation

The time constants for this burst length distribution (rather than those for open times) are very similar to those found in Section 4 (equation 32) for the macroscopic relaxation (at the same agonist concentration), the time course of which thus depends largely on the burst length. The predominant component in both has a time constant of 9.82 ms. All the other components were small (and so not likely to be detectable in an experiment) in the case of the relaxation, and the same is true for components 2 and 4 in the burst length distribution, which have very small areas. However, the burst-length distribution gives information that could not have been found from the relaxation, because the component with $\tau_3 = 0.323283$ ms has 25% of the area and therefore should be easily measurable in experiments.

7.2. Total Open Time per Burst

The distribution of the total open time per burst, expressions for which are given by Colquhoun and Hawkes (1982), is simpler than that for the burst length. Like the simple open time distribution, it has only $k_{\mathcal{A}}$ components (two in our case), and these are quite similar to the two components with non-negligible area in the burst-length distribution just discussed. The similarity of the distributions is, of course, a result of the fact that most shut times within a burst are very short in this example, so their omission makes little difference.

8. Channel Openings after a Jump

The whole discussion so far has concerned channels at equilibrium. It is also possible to study channel openings following application of some perturbation such as voltage or concentration jump. Following the jump there will be a period, before a new equilibrium is established, when openings can be observed in nonequilibrium conditions. The necessary theory for calculating distributions from the **Q** matrix under such conditions has been given by Colquhoun and Hawkes (1987). In general, the *time constants* are still calculated from the subsections of **Q**, so they are the same as found at equilibrium. What differs are the relative areas associated with each time constant.

A typical calculation of this sort would be to predict the probability distribution of the time to the first opening (the *first latency*) after the jump. In practice this can be a little tricky because, depending on the state of the system before and after the jump, some of the relevant submatrices become singular if the agonist concentration is zero. Furthermore, there may then be a nonzero chance of the channel failing ever to open at all. We will not, therefore give general results here.

We give only a simple case for our example mechanism when there is no agonist present initially, so that the channel must start in state 5. If there is an instantaneous jump in agonist concentration to level x_A, giving the general **Q** matrix of equation 4, then the time to first opening is simply a shut time starting from state 5. It therefore has a probability distribution that is found exactly as described in Section 5 for shut times, except that in this case the initial vector is taken as $\boldsymbol{\phi}_s = (0 \quad 0 \quad 1)$, rather than using the equilibrium initial vector given in equation 50 and numerically in equation 57. The time constants are therefore exactly the same as in equation 55, the $\mathbf{A}_i$ are as given in equation 58, but with this new $\boldsymbol{\phi}_s$ substituted into equation 54, we now have areas

$$a_1 = 1.000138 \qquad a_2 = -0.0001392 \qquad a_3 = 1.224 \times 10^{-6} \tag{82}$$

Note that although these areas sum to 1 as usual, the second one is negative. It is easily verified that, with these values substituted into equation 40, the corresponding probability density function is zero at $t = 0$, rises to a maximum, and then decays back to zero. This occurs because the system starts in state 5 and must make a transition into state 4 before there is any chance at all of an opening taking place. The density is depicted in Fig. 1a; on this time scale it is not easy to see exactly what is happening near $t = 0$, and the distribution is very nearly just a single exponential with mean 3789 ms but with just a little something funny at the origin. Figure 1b shows more clearly the behaviour at the origin; the density rises smoothly over the first 2 ms and is then almost flat up to 5 ms before decaying as in Fig. 1a.

9. Calculating the Exponential of a Matrix

In the preceding sections we have seen that all that is needed to calculate a wide range of theoretical distributions for any mechanism is some straightforward matrix algebra and the ability to find the exponential of a matrix. In most cases one would want to do this via the spectral expansion, giving rise to mixed exponential (or geometric) functions, but we also discuss briefly a method that does not require this.

9.1. Functions of a Matrix

The central, and beautiful, result that underlies the tricks described above can be put in the following form. For any analytic function, f, and any $n \times n$ matrix $\mathbf{M}$ whose eigenvalues are distinct, the corresponding function of the matrix $\mathbf{M}$, denoted $f(\mathbf{M})$, can be written in the form

$$f(\mathbf{M}) = \sum_{i=1}^{i=n} \mathbf{A}_i f(\lambda_i) \tag{83}$$

where the λ_i are the eigenvalues of $\mathbf{M}$ and the $\mathbf{A}_i$ are its spectral matrices, the calculation of which is described below. The beauty of this result is that when we have some function of a matrix such as $\exp(\mathbf{Q})$, the meaning of which is not immediately apparent, it is changed into a function of the λ values, $\exp(\lambda)$, and, since the λ values are ordinary scalar numbers, the problem disappears.

For example, equation 83 shows immediately, for $f(\mathbf{M}) = \mathbf{M}^0$, that

$$\sum_{i=1}^{i=n} \mathbf{A}_i = \mathbf{I} \tag{84}$$

and for $f(\mathbf{M}) = \mathbf{M}$, we also see immediately that

$$\sum_{i=1}^{i=n} \mathbf{A}_i \lambda_i = \mathbf{M} \tag{85}$$

These two results are useful in checking calculations of the spectral matrices, $\mathbf{A}_i$.

It may also be mentioned that the spectral matrices have the interesting property that multiplying together any pair of them results in a matrix that is all zeroes; i.e., $\mathbf{A}_i \mathbf{A}_j = \mathbf{0}$ $(i \neq j)$. In fact, this property is responsible for many of their useful characteristics.

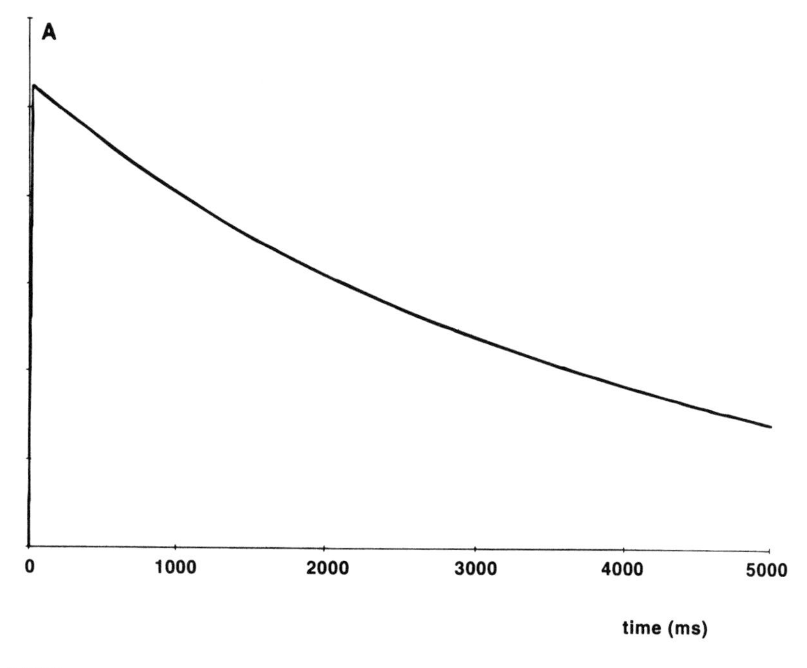

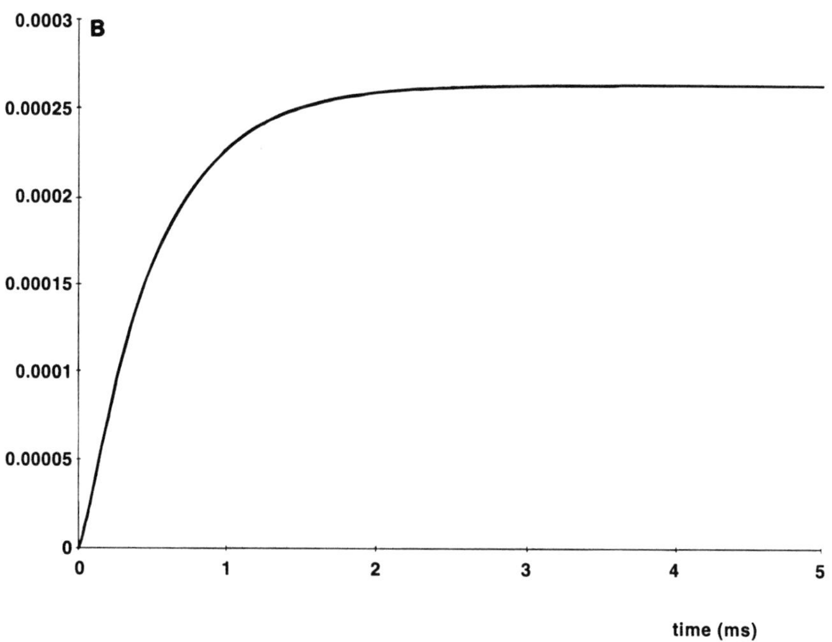

Figure 1. Probability density of first latency (a) over 5 sec and (b) near the origin.

The two particular examples of the application that we have used in the foregoing sections are, first, to calculate powers of a matrix in Section 6 (equation 67), i.e., taking $f(\mathbf{M}) = \mathbf{M}^r$

$$\mathbf{M}^r = \sum_{i=1}^{i=n} \mathbf{A}_i \lambda_i^r \tag{86}$$

and, second, the ubiquitous exponential of a matrix, $f(\mathbf{M}) = \exp(\mathbf{M})$,

$$\exp(\mathbf{M}) = \sum_{i=1}^{i=n} \mathbf{A}_i \exp(\lambda_i). \tag{87}$$

In most of our examples the matrix $\mathbf{M}$ represents $\mathbf{Q}t$ or some subsection of $\mathbf{Q}$ multiplied by t.

9.2. Calculation of the Spectral Matrices

Now we come, at last, to describing how actually to calculate the spectral matrices, the step that underlies almost everything described in the preceding sections. It is not hard because library routines are available to do all the difficult bits.

First we will just describe the bare bones of the algorithm that enables computation of the quantities $\mathbf{A}_i$ and λ_i that were first introduced in equation 21. Later we will discuss the mathematics of it a little, but that can be skipped by those who do not care to know it.

Any good computer library contains routines that do a trick called 'finding the eigenvalues of a general square matrix'. The term *eigenvalue* is discussed briefly in Appendix 1, but all we need to know here is that the required values of λ_i are nothing other than the eigenvalues of the matrix $-\mathbf{Q}$, and the time constants are the reciprocals of these values, $\tau_i = 1/\lambda_i$. We therefore simply put the values for the transition rates into $\mathbf{Q}$, change the signs of all the elements to get $-\mathbf{Q}$, and use a standard routine to calculate the eigenvalues and hence the time constants (alternatively, find the eigenvalues of $\mathbf{Q}$, and change their sign). A matrix of size $k \times k$, such as $\mathbf{Q}$, will generally have k eigenvalues. The matrix $\mathbf{Q}$ (and $-\mathbf{Q}$) is singular, which means that one of the eigenvalues will be zero (and we do not divide by zero to get a time constant!); it is usually convenient to arrange for this to be the first one, so we will assume $\lambda_1 = 0$.

Standard computer library routines will also find the *eigenvectors* of the matrix $-\mathbf{Q}$ (usually these will be calculated in the same routine as the eigenvalues). The routine will often produce the k eigenvectors, each a column vector $\mathbf{x}_i$ associated with the value λ_i, already in the form of the columns of a matrix $\mathbf{X}$ (if it does not supply them in this form, then use the $\mathbf{x}_i$ to create a matrix $\mathbf{X}$). Thus, $\mathbf{X}$ can be thought of as a set of columns

$$\mathbf{X} = [\mathbf{x}_1 \quad \mathbf{x}_2 \quad \cdots \quad \mathbf{x}_k] \tag{88}$$

This is now inverted to produce a matrix which we shall call $\mathbf{Y} = \mathbf{X}^{-1}$; again, a standard routine will do that. We shall denote the ith row of $\mathbf{Y}$ as $\mathbf{y}_i$ (a $1 \times k$ matrix). We are now in a position to calculate the k *spectral matrices*, denoted $\mathbf{A}_i$, that are needed to complete the trick. These are calculated as

$$\mathbf{A}_i = \mathbf{x}_i \mathbf{y}_i, \qquad i = 1 \text{ to } k \tag{89}$$

These are each square $k \times k$ matrices, as they arise from a column postmultiplied by a row (see example A5 in the Appendix). You may not easily find this final part as a ready-made program, but it is very simple to program, just a few things to multiply together. Then $e^{\mathbf{Q}t}$ is represented in the form of equation 21, which leads, as we have seen in several examples, to various functions of interest: all have the form of a mixture of exponentials with rate constants τ_i and weights that are computed by pre- and postmultiplying the matrices $\mathbf{A}_i$ by various appropriate vectors.

9.3. Other Exponentials

We have described the spectral representation of $e^{\mathbf{Q}t}$. Exactly the same procedure applies if you want to use $\mathbf{Q}_{\mathcal{A}\mathcal{A}}$, $\mathbf{Q}_{\mathcal{B}\mathcal{B}}$, etc. instead of $\mathbf{Q}$. The only differences are that, instead of k, you will get $k_{\mathcal{A}}$, $k_{\mathcal{B}}$, etc. eigenvalues and $\mathbf{A}_i$ matrices, and usually none of the eigenvalues will take the value zero.

9.4. Calculation of Spectral Matrices Using the NAG Library

As a particular example of the sort of code needed, the following is a subroutine (in FORTRAN) that calls the NAG library subroutine F02AGF, which returns the eigenvalues and eigenvectors of a general square matrix. If, for example, we want the spectral matrices and eigenvalues of $\mathbf{Q}_{\mathcal{F}\mathcal{F}}$, we put the appropriate elements of $\mathbf{Q}$ into an array QFF and call QMAT thus:

```
call QMAT(QFF, Amat, kF, eigenval, ifail)
```

This subroutine might be made more complete by allowing adjustable array dimensions, by adding a routine to sort the eigenvalues (along with their associated eigenvectors) into ascending or descending order, and by adding checks on the accuracy of matrix inversion (e.g., by checking that $\mathbf{XY} = \mathbf{I}$ to acceptable accuracy), but the version shown works fine.

```
      subroutine QMAT(Q,Amat,k,eigenval,ifail)
c   Calculation of eigenvalues and spectral matrices using the NAG library
c   routine F02AGF
c   INPUT:
c      Q=double precision matrix (k×k)
c      k=size of Q
c   OUTPUT:
c      eigenval(i)=eigenvalues of Q (i=1,...,k)
c      A(m)=mth spectral matrix of Q (m=1,...,k). Each A(m) is k×k in
c         size and Amat(i,j,m) is value in ith row and jth column of A(m)
c      ifail=0 if there are no errors in the NAG routine, F02AGF
```

```
c
      IMPLICIT double precision (A-H,O-Z)
      double precision Q(10,10),Amat(10,10,10),eigenval(10)
      double precision X(10,10),Y(10,10)
c   next 2 lines for F02AGF
      double precision QD(10,10),eigimag(10),Ximag(10,10)
      integer*4 iwork(10)
c
c   Define QD = -Q; (this also preserves the input value of Q)
        do i = 1,k
            do j = 1,k
                QD(i,j) = -Q(i,j)
            enddo
        enddo
c
        km = 10       !maximum k, defined by declarations
        ifail = 0
        call F02AGF(QD,km,k,eigenval,eigimag,X,km,Ximag,km,iwork,ifail)
c   X now has columns that are the real parts of the k column eigenvectors
c   of Q. Now invert X using a matrix inversion subroutine; put result in Y.
        call MATINV(X,k,km,Y,km)
c   Calculate the spectral matrices, A(m), from X and Y.
        do m=1, k
            do i=1, k
                do j=1, k
                    Amat(i, j, m)=X(i, m)*Y(m, j)
                enddo
            enddo
        enddo
        RETURN
        end
```

Another generally useful bit of code is the following fragment that will calculate scalars, w_m, by premultiplying $\mathbf{A}_m$ by a row vector, defined as an array *row*(*i*), and postmultiplying it by a column vector defined as an array *col*(*i*). The distributions described in earlier sections mostly end up in this form. The subscript m is used here for $\mathbf{A}_m$ so we can keep to the usual notation of using *i,j* for rows and columns.

```
        do m = 1,k
            w(m) = 0.0
            do i = 1,k
                do j = 1,k
                    w(m) = w(m) + row(i) * Amat(i,j,m) * col(j)
                enddo
            enddo
        enddo
```

9.5. Calculation of Matrix Exponentials Using MAPLE

In recent years a number of programs have been developed that are able to carry out symbolic manipulation, such as factorization, differentiation, integration, and many other mathematical operations as well as doing numerical calculations. We give here a simple example in one of these programs, called MAPLE.

Let us take the case of the shut-time distributions, for which we need the exponential of $\mathbf{Q}_{\mathcal{FF}}$. First we must start the linear algebra package and then create the matrix

with(linalg):

Qff:=matrix(3, 3, [−19, 4, 0, 0.05, −2.065, 2, 0, 0.01 ,−0.01]); (90)

Then $e^{\mathbf{Q}_{\mathcal{FF}}t}$ is evaluated at $t = 2$, say, simply be typing

exponential(*Qff**2); (91)

giving the result

$$\begin{bmatrix} 1.381 \times 10^{-5} & 0.004786 & 0.2002 \\ 5.982 \times 10^{-5} & 0.02078 & 0.9529 \\ 1.251 \times 10^{-5} & 0.004764 & 0.9948 \end{bmatrix}. \tag{92}$$

If instead we want the spectral expansion, the eigenvalues are returned by

eigenvals(−*Qff*); (93)

giving

$$[19.01, 2.063, 0.0002639], \tag{94}$$

and

eigenvects(−*Qff*); (95)

gives the above eigenvalues together with the eigenvectors, which we report below as columns of a matrix $\mathbf{X}$ in the same order as the corresponding eigenvalues in equation 94:

$$\mathbf{X} = \begin{bmatrix} 9.996 & 0.2358 & 0.02042 \\ -0.02949 & 0.9986 & 0.09702 \\ 1.55 \times 10^{-5} & -0.004864 & 0.09965 \end{bmatrix} \tag{96}$$

From these one can obtain the spectral matrices as described at the beginning of this section. Notice, however, that the eigenvalues in equation 94 have come out in the reverse order from those above equation 55, so the A_i calculated from $\mathbf{X}$ in equation 96 will also come out in reverse order compared to those in equation 58; none of this matters as long as you keep track of what goes with what. As discussed in the Appendix, the eigenvectors are not unique, as they can each be multiplied by arbitrary nonzero constants (so do not worry too much if your program gives different results), but the resulting $\mathbf{A}_i$ should be the same.

9.6. Calculation of Spectral Matrices and Matrix Exponentials Using MATHEMATICA

Another powerful modern computer algebra package, having broadly similar features to MAPLE, is MATHEMATICA. We illustrate the use of this also with some simple examples. As in the previous section we will look at shut times.

$$\text{Qff} = \{\{-19, 4, 0\}, \{0.05, -2.065, 2\}, \{0, 0.01, -0.01\}\}; \tag{97}$$

is equivalent to the MAPLE statement 90, and MatrixForm[Qff] will print it so that it looks like the result in equation 51.

$$\text{MatrixExp[Qff 2];} \tag{98}$$

is the equivalent of statement 91 and gives the same result. If we want to use the spectral expansion explicitly, we can use a series of statements as follows.

The eigenvalues and eigenvectors can be found from

$$\{\text{vals,vects}\} = \text{Eigensystem}[-\text{Qff}]; \tag{99}$$

The eigenvectors are stored as rows, and we use columns, so

$$\text{X} = \text{Transpose[vects]}; \qquad \text{Y} = \text{Inverse[X]}; \tag{100}$$

Now we can define a function, let us call it expqt to calculate $e^{\mathbf{Q}t}$ for any t using equation 107, below. Thus,

$$\text{expqt[t\_]:= X.DiagonalMatrix[E\^{}(-vals t)].Y;} \tag{101}$$

We can now use this function for any *t;* for example,

```
expqt[2]
```

now gives the same result as equation 98.

Notice that in these examples a space indicates multiplication by a scalar, and a dot indicates matrix multiplication.

To extend the example, we can do the calculations for the probability density of first latency discussed in Section 8. We need to set some vectors

```
phis = {0,0,1};     uf = {1,1,1};
```

then define a new function

$$\text{latencypdf[t\_]:= -phis.expqrt[t].Qff.uf;} \tag{102}$$

which is a representation of equation 49. Then we can use this for any t; for example, latencypdf[0] returns the value 0, confirming the discussion in Section 8, where it was said that the density starts at 0 when $t = 0$. A plot of this density over the interval $0 \leq t \leq 5$ ms can be had simply by issuing the command

```
Plot[latencypdf[t],{t,0,5}]
```

You need a few extra options to label it and make it pretty.

The spectral matrices have, in effect, been built into equation 101. If you want to find them explicitly, they can be found as

$$\text{A1 = Outer[Times,vects[[1]],Y[[1]]]} \tag{103}$$

This does the calculation of equation 89 for the first eigenvector and the first row of the matrix **Y**. The second and third spectral matrices can be found similarly, simply replacing the number 1 wherever it occurs in equation 103 by 2 or 3, as appropriate. The results should be the same as equation 58, although not necessarily in the same order (see discussion above concerning MAPLE).

9.7. Another Way to Calculate the Exponential of a Matrix

Most people will want to use the spectral representation method because they feel that they can interpret the time constants that it yields. Sometimes, however, you may not need that, and an alternative would do. A number of ways are discussed by Moler and van Loan (1978). We discuss here one method that we have found robust; an APL program that implements it appears in Hawkes (1984).

Let $\mathbf{M} = \mathbf{Q}t$, for some fixed t. Then we want to calculate $e^{\mathbf{M}}$. There are two essential parts to this method.

9.7.1. Core Method

If **M** is in some sense 'small', then it is easy and quite accurate to use a truncated series calculation similar to equation 20; i.e.,

$$e^{\mathbf{M}} = \mathbf{I} + \mathbf{M} + \frac{\mathbf{M}^2}{2!} + \frac{\mathbf{M}^3}{3!} + \cdots \frac{\mathbf{M}^N}{N!} \tag{104}$$

We have found stopping after $N = 13$ terms is sufficient. What is meant by small? Take the sum of the modulus of each element of **M** in the ith row; do this for each row and let Δ be the biggest of these sums. The core method works well if $\Delta < 1/2$.

9.7.2. Squaring Method

If $\Delta > 1/2$, then find an integer r such that $\Delta/R < 1/2$, where $R = 2^r$. The essence of the method is that

$$(e^{\mathbf{M}/\mathbf{R}})^R = e^{\mathbf{M}} \tag{105}$$

so calculate $\mathbf{T} = e^{\mathbf{M}/R}$ using the core method, and then find $\mathbf{T}^R$ by matrix multiplication. This is quite cunning, but the choice of R to have the form 2^r is also very cunning because it means you only need to do r matrix multiplications instead of $R - 1$, by calculating successive squares. For example, if $r = 4$ so $R = 16$, then $\mathbf{T}^2 = \mathbf{T} \times \mathbf{T}$; $\mathbf{T}^4 = \mathbf{T}^2 \times \mathbf{T}^2$; $\mathbf{T}^8 = \mathbf{T}^4 \times \mathbf{T}^4$; $\mathbf{T}^{16} = \mathbf{T}^8 \times \mathbf{T}^8$; and you have calculated $\mathbf{T}^{16}$ with a series of just four multiplications instead of 15 if you did it the hard way: $\mathbf{T}^{16} = \mathbf{T} \times \mathbf{T} \times \mathbf{T} \times \mathbf{T} \times \mathbf{T} \ldots \times \mathbf{T}$.

9.8. Further Mathematical Notes

In this section we discuss briefly the mathematical justification of the spectral expansion. It is not essential reading for those who merely want to do the calculations but serves to satisfy the mathematically curious.

In the Appendix we describe how a square matrix, $\mathbf{M}$, can, in some circumstances, be represented in the form $\mathbf{M} = \mathbf{X}\mathbf{\Lambda}\mathbf{X}^{-1}$, where $\mathbf{\Lambda}$ is a diagonal matrix containing the eigenvalues of $\mathbf{M}$ and the columns of the matrix $\mathbf{X}$ consist of the corresponding eigenvectors of $\mathbf{M}$. It can be shown that this is always possible if $\mathbf{M}$ is the $\mathbf{Q}$ matrix of a reversible Markov process and that the eigenvalues are real and negative (apart from the one zero value $\lambda_1 = 0$), and the eigenvectors have real elements. The eigenvalues, λ_i, of the matrix $-\mathbf{Q}$ are simply minus the eigenvalues of $\mathbf{Q}$ and are therefore nonnegative; the eigenvectors of $\mathbf{Q}$ and $-\mathbf{Q}$ can be taken as the same. This is also true if $\mathbf{Q}$ is replaced by any of the submatrices $\mathbf{Q}_{\mathscr{A}\mathscr{A}}$, $\mathbf{Q}_{\mathscr{B}\mathscr{B}}$, $\mathbf{Q}_{\mathscr{F}\mathscr{F}}$, $\mathbf{Q}_{\mathscr{E}\mathscr{E}}$ with dimensions reduced from k to $k_{\mathscr{A}}$ or $k_{\mathscr{B}}$, etc., and in those cases none of the eigenvalues are zero.

Equation (A15) of the Appendix then implies that

$$\mathbf{Q}^r = \mathbf{X}(-\mathbf{\Lambda})^r\mathbf{X}^{-1} \tag{106}$$

Now from equation 20 we find that

$$e^{\mathbf{Q}t} = \mathbf{I} + \mathbf{Q}t + \frac{(\mathbf{Q}t)^2}{2!} + \frac{(\mathbf{Q}t)^3}{3!} + \frac{(\mathbf{Q}t)^4}{4!} + \cdots$$

$$= \mathbf{I} + \mathbf{X}(-\mathbf{\Lambda}t)\mathbf{X}^{-1} + \frac{\mathbf{X}(-\mathbf{\Lambda}t)^2\mathbf{X}^{-1}}{2!} + \frac{\mathbf{X}(-\mathbf{\Lambda}t)^3\mathbf{X}^{-1}}{3!} + \frac{\mathbf{X}(-\mathbf{\Lambda}t)^4\mathbf{X}^{-1}}{4!} + \cdots$$

$$= \mathbf{X}\left[\mathbf{I} + (-\mathbf{\Lambda}t) + \frac{(-\mathbf{\Lambda}t)^2}{2!} + \frac{(-\mathbf{\Lambda}t)^3}{3!} + \frac{(-\mathbf{\Lambda}t)^4}{4!} + \cdots\right]\mathbf{X}^{-1}$$

The expression in brackets is obviously $e^{-\Lambda t}$, but if we look at it in detail we see that it is a sum of diagonal matrices, so it must again be diagonal. Then

$$e^{\mathbf{Q}t} = \mathbf{X}e^{-\Lambda t}\mathbf{X}^{-1} \tag{107}$$

Furthermore, the ith element on the diagonal of $e^{-\Lambda t}$ is just

$$1 + (-\lambda_i t) + \frac{(-\lambda_i t)^2}{2!} + \frac{(-\lambda_i t)^3}{3!} + \frac{(-\lambda_i t)^4}{4!} + \cdots = \exp(-\lambda_i t)$$

The key result (equation 21) therefore follows immediately from equation A20 of the Appendix. We have thus justified the general statement (equation 83) for the particular case of the exponential function (remembering that, for convenience, we choose to work with the eigenvalues of $-\mathbf{Q}$ instead of $\mathbf{Q}$).

10. Time Interval Omission

All of the preceding distributions are derived under ideal assumptions. It has long been recognised, however, that the limitations on observational resolution caused by noise and filtering by the recording equipment can distort the observation of open times and shut times through failure to observe very short intervals. Some discussion of this phenomenon is given in Chapter 18 (this volume). We are unable to go into details here but note that, although it is naturally more complicated, and the nice mixture-of-exponentials feature now only appears as a very good asymptotic approximation, the kinds of matrix operations involved are mostly no more difficult than those discussed here. The main complication arises from the need to find some sort of generalised eigenvalues, which is equivalent to finding at what values of some parameter a certain determinant of a matrix vanishes: that is not all that difficult. Some details are given by Hawkes *et al.* (1992).

11. Concluding Remarks

The results of the classical theory are quite easy in terms of matrix algebra provided one can calculate the exponential of certain matrices. Computer software is readily available in many languages or packages that will carry out all of the necessary operations, including (sometimes with a little bit of effort) those exponentials. Thus, it should not be too difficult for anyone to put together a program in his or her own favourite system. We hope this chapter helps to clarify what is needed.

The necessary tools can be summmarised as follows. It is important to note that all calculations should be done using double-precision arithmetic. You will need:

1. Subroutines/procedures to add, subtract, multiply, and invert matrices.
2. A routine to extract a submatrix (consisting of specified rows and columns) from a larger matrix (unless the other routines are capable of operating directly on subsections of a matrix).

3. A routine to find the eigenvalues and eigenvectors of a general square matrix, preferably sorted into ascending or descending order of the eigenvalues, and to calculate the spectral matrices from them.

The numerical examples provide a benchmark against which you can test the results from your own programs. Good luck!

Appendix 1. A Brief Introduction to Matrix Notation

This account is a brief synopsis. For further details see, for example, Stephenson (1965) or Mirsky (1982).

A1.1. Elements of a Matrix

A matrix is a table and is usually denoted by a bold type symbol. It is rather like a spreadsheet, the entries in the table being defined by the row and column in which they occur. The entry in the ith row and the jth column of the matrix **A**, an *element* of **A**, is usually denoted in lower case italic as a_{ij}. A matrix with n rows and m columns is said to be an $n \times m$ matrix, or to have *shape* $n \times m$. If $n = m$, we have a square matrix. Thus, for example, 2×2 and 2×3 matrices can be written as

$$\mathbf{A} = \begin{bmatrix} a_{11} & a_{12} \\ a_{21} & a_{22} \end{bmatrix} \qquad \mathbf{C} = \begin{bmatrix} c_{11} & c_{12} & c_{13} \\ c_{21} & c_{22} & c_{23} \end{bmatrix}. \tag{A1}$$

Clearly, a 1×1 matrix has just one element and can, for most purposes, be treated as an ordinary number (called a *scalar,* which has no shape), though strictly speaking it is not.

The elements for which $i = j$ are called *diagonal elements* (e.g., a_{11} and a_{22}, in this example), and the rest ($i \neq j$) are called *off-diagonal elements.*

A1.2. Vectors

Vectors are nothing new. In the context of matrix algebra they are simply matrices that happen to have only one row (a *row vector* or $1 \times n$ matrix) or only one column (a *column vector* or $n \times 1$ matrix). They are manipulated just like any other matrix.

A1.3. Equality of Matrices

Two matrices are said to be *equal* if all the corresponding elements of each are equal. Clearly, in that case, the two matrices must both be the same shape. For example $\mathbf{A} = \mathbf{B}$ means, in the 2×2 case,

$$\begin{bmatrix} a_{11} & a_{12} \\ a_{21} & a_{22} \end{bmatrix} = \begin{bmatrix} b_{11} & b_{12} \\ b_{21} & b_{22} \end{bmatrix}$$

which is a shorthand way of writing the four separate relationships: $a_{11} = b_{11}$, $a_{12} = b_{12}$, $a_{21} = b_{21}$, and $a_{22} = b_{22}$.

A1.4. Addition and Subtraction of Matrices

This is very easy. You just add (or subtract) the corresponding elements in each of them (clearly, the two matrices must be the same shape). Thus, for example,

$$\mathbf{C} = \mathbf{A} + \mathbf{B} = \begin{bmatrix} a_{11} & a_{12} \\ a_{21} & a_{22} \end{bmatrix} + \begin{bmatrix} b_{11} & b_{12} \\ b_{21} & b_{22} \end{bmatrix} = \begin{bmatrix} (a_{11} + b_{11}) & (a_{12} + b_{12}) \\ (a_{21} + b_{21}) & (a_{22} + b_{22}) \end{bmatrix}$$

shape: 2×2 2×2 2×2

or, more briefly, $c_{ij} = a_{ij} + b_{ij}$ for all i and j. Subtraction, $\mathbf{A} - \mathbf{B}$, is defined in the obvious equivalent manner.

A1.5. Multiplication of Matrices

The definition of the product of two matrices may, at first sight, seem a bit perverse, but it turns out to be exactly what is needed for the convenient representation of, for example, simultaneous equations or for representation of all the possible routes from one state to another [see, for example, section 2 of Colquhoun and Hawkes (1982)]. Multiplication goes 'row into column.' For example:

$$\mathbf{C} = \mathbf{AB} = \begin{bmatrix} a_{11} & a_{12} \\ a_{21} & a_{22} \end{bmatrix} \begin{bmatrix} b_{11} & b_{12} \\ b_{21} & b_{22} \end{bmatrix} = \begin{bmatrix} (a_{11}b_{11} + a_{12}b_{12}) & (a_{11}b_{12} + a_{12}b_{22}) \\ (a_{21}b_{11} + a_{22}b_{21}) & (a_{21}b_{12} + a_{22}b_{22}) \end{bmatrix} \quad \text{(A2)}$$

shape: 2×2 2×2 2×2

Thus, the element in the ith row and the jth column of the product, **C**, is obtained by taking the ith row of **A** and the jth column of **B**, multiplying the corresponding elements, and adding all these products. Clearly, the number of columns in **A** must be the same as the number of rows in **B**. If **A** is an $n \times m$ matrix, and **B** is $m \times k$, then the product, $\mathbf{C} = \mathbf{AB}$ will be an $n \times k$ matrix. More formally, the element c_{ij} is obtained from the sum of products

$$c_{ij} = \sum_{r=1}^{m} a_{ir} b_{rj}$$

Some computer languages allow matrices to be added or multiplied symbolically, e.g., by simply writing $\mathbf{A} = \mathbf{B} * \mathbf{C}$; in others a subroutine or procedure must be called to do this. In the APL language, matrix multiplication is a special case of a powerful idea called the *inner product* operator. The APL expression $C \leftarrow A + . \times B$, although not standard mathematical notation, reflects the fact that the result is obtained by a combination of adding and multiplying (guess what the result of $A \times . + B$ is).

It is important to note that, contrary to the case with ordinary (scalar) numbers, it is *not* generally true that **AB** and **BA** are the same (multiplication of matrices is *not* necessarily *commutative*). Indeed, both products will exist only if **A** has shape $n \times m$ and **B** has shape $m \times n$: then the product **AB** has shape $n \times n$, whereas **BA** has shape $m \times m$. The two products will only have the same shape if **A** and **B** are square matrices of the same shape, and even then the products are not necessarily the same. We must therefore distinguish between *premultiplication* and *postmultiplication*. In the product **AB**, it is said that **A** premultiplies **B** (or that **B** postmultiplies **A**).

If a matrix is multiplied by an ordinary (scalar) number, x say, this means simply that every element of the matrix is multiplied by x. Thus $x\mathbf{A}$, which is equal to $\mathbf{A}x$, is a matrix with elements xa_{ij}.

It can be shown that matrix multiplication is *associative;* i.e., parentheses are not needed because, for example, **A**(**BC**) = (**AB**)**C**, which can therefore be written unambiguously as the triple product **ABC**.

A1.6. Some More Examples of Matrix Multiplication

A few more examples may help to clarify the rules given above. If we postmultiply a row vector (a $1 \times n$ matrix) by a column vector ($n \times 1$ matrix), we get an ordinary (scalar) number (a 1×1 matrix). For example,

$$\mathbf{ab} = [a_1 \quad a_2 \quad a_3]\begin{bmatrix} b_1 \\ b_2 \\ b_3 \end{bmatrix} = (a_1b_1 + a_2b_2 + a_3b_3) \tag{A3}$$

shape: 1×3 3×1 1×1

When looking at a matrix expression, it is always useful to note the size of each array in the expression, as written below the results above. This makes it instantly obvious, for example, that the result in equation A3 is scalar. Likewise, the expression **aXb** where **a** is a $1 \times n$ row vector, **X** is an $n \times m$ matrix, and **b** is an $m \times 1$ column vector, is clearly also scalar. Note, however, that we sometimes do need to distinguish between a scalar and a 1×1 matrix; for example, if **C** is a 3×3 matrix and **a** and **b** are as above, then (**ab**)**C** makes sense if **ab** is regarded as a scalar, but it is not equal to **a**(**bC**) because the multiplication **bC** is not possible (you cannot multiply a 3×1 and a 3×3 matrix), and you have broken the associative law mentioned above. If you consider **ab** as a 1×1 matrix, however, you cannot multiply that by **C** either, so that is all right. Users of logical software, such as APL or MATHEMATICA may need to be careful of this distinction.

Note that it follows from equation A3 that $\mathbf{aa}^{\mathrm{T}}$, where $\mathbf{a}^{\mathrm{T}}$ is the transpose of **a** (see below), is a sum of squares, thus:

$$\mathbf{aa}^{\mathrm{T}} = [a_1 \quad a_2 \quad a_3]\begin{bmatrix} a_1 \\ a_2 \\ a_3 \end{bmatrix} = \sum a_i^2$$

shape: 1×3 3×1 1×1

Similarly, if **u** is a unit column vector (all elements $u_i = 1$), then premultiplying it by a row vector simply sums the elements of the latter (this is a common feature in the distributions described in this chapter), thus:

$$\mathbf{au} = [a_1 \quad a_2 \quad a_3]\begin{bmatrix}1\\1\\1\end{bmatrix} = \sum a_i \tag{A4}$$

If the multiplication is done the other way around, **ba** instead of **ab,** we are multiplying an $n \times 1$ matrix by a $1 \times n$ matrix, so the result has shape $n \times n$, thus:

$$\mathbf{ba} = \begin{bmatrix}b_1\\b_2\\b_3\end{bmatrix}[a_1 \quad a_2 \quad a_3] = \begin{bmatrix}b_1a_1 & b_1a_2 & b_1a_3\\b_2a_1 & b_2a_2 & b_2a_3\\b_3a_1 & b_3a_2 & b_3a_3\end{bmatrix} \tag{A5}$$

shape: $3\times1 \qquad 1\times3 \qquad 3\times3$

A1.7. The Identity Matrix

The matrix equivalent of the number 1 is the identity matrix, denoted **I.** This is a square matrix for which all the diagonal elements are 1, and all others are zero. Actually there are lots of different identity matrices, one for each possible shape, 1×1, 2×2, 3×3, etc.; often we do not bother to indicate the shape because it is usually obvious from the context. It has the property that multiplication of any matrix by **I** of the appropriate shape does not change the matrix. For the 2×2 and 3×3 cases, we have, respectively,

$$\mathbf{I} = \begin{bmatrix}1 & 0\\0 & 1\end{bmatrix} \quad \text{and} \quad \mathbf{I} = \begin{bmatrix}1 & 0 & 0\\0 & 1 & 0\\0 & 0 & 1\end{bmatrix} \tag{A6}$$

Thus, for example, using the matrices defined in equation A1,

$$\mathbf{AI} = \mathbf{IA} = \mathbf{A}$$

$$\mathbf{CI} = \mathbf{IC} = \mathbf{C}$$

Note that the **I** matrix that postmultiplies **C** must have shape 3×3, whereas all the others must be 2×2.

A1.8. Determinants

A determinant is a *number* (not a table) that can be calculated from the elements of a square matrix. The determinant is usually written just like the matrix, except that it is enclosed by vertical lines rather than brackets (it therefore looks rather similar to the matrix, and it is important to remember that the whole symbol represents a single number). For example, the determinant of the 2×2 matrix in (A1), denoted det(**A**), is

$$\det(\mathbf{A}) \equiv |\mathbf{A}| \equiv \begin{vmatrix} a_{11} & a_{12} \\ a_{21} & a_{22} \end{vmatrix} = a_{11}a_{22} - a_{12}a_{21} \tag{A7}$$

The right-hand side of this shows how the number is calculated from the elements of **A.** A matrix that has a determinant of zero, det(**A**) = 0, is said to be a *singular* matrix; such a matrix has a linear relationship between its rows or between its columns and cannot be inverted (see below).

For larger matrices the definition gets a bit more complicated. One does not have to worry too much about the details, as most scientific software has functions to calculate a determinant. The next paragraph may therefore be safely skipped.

For an $n \times n$ matrix **A,** the determinant is defined as

$$\det(\mathbf{A}) \equiv |\mathbf{A}| = \sum (-1)^{\sigma(\pi)} a_{1\pi(1)} a_{2\pi(2)} \cdots a_{n\pi(n)}$$

where the summation is over all permutations π of the integers 1 to n, and $\sigma(\pi)$ equals 1 if π is an 'even' permutation and equals -1 if π is on 'odd' permutation. For example, if **A** is a 3×3 matrix, then

$$|\mathbf{A}| = a_{11}a_{22}a_{33} + a_{12}a_{23}a_{31} + a_{13}a_{21}a_{32} - a_{13}a_{22}a_{31} - a_{11}a_{23}a_{32} - a_{12}a_{21}a_{33}$$

For more detail see any standard text, such as one of those referenced above.

A1.9. 'Division' of Matrices

If **C** = **AB,** then what is **B**? In ordinary scalar arithmetic we would simply divide both sides by **A** to get the answer, provided **A** was not zero. When **A** and **B** are matrices, the method is analogous. We require some matrix analogue of what, for ordinary numbers, would be called the reciprocal of a square matrix **A.** Suppose that some matrix exists, which we shall denote $\mathbf{A}^{-1}$, that behaves like the reciprocal of **A** in the sense that, by analogy with ordinary numbers,

$$\mathbf{A}\mathbf{A}^{-1} = \mathbf{I} = \mathbf{A}^{-1}\mathbf{A}$$

where **I** is the identity matrix. The matrix $\mathbf{A}^{-1}$ is called the *inverse* of **A.** Thus, the solution to the problem posed initially can be found by premultiplying both sides of **C** = **AB** by $\mathbf{A}^{-1}$ to give the result as $\mathbf{B} = \mathbf{A}^{-1}\mathbf{C}$. In the case of a 2×2 matrix the inverse can be written explicitly as

$$\mathbf{A}^{-1} = \begin{bmatrix} a_{22}/\det(\mathbf{A}) & -a_{12}/\det(\mathbf{A}) \\ -a_{21}/\det(\mathbf{A}) & a_{11}/\det(\mathbf{A}) \end{bmatrix}$$

where det(**A**) is the number defined above. It can easily be checked that this result is correct by multiplying it by **A**: the result is the identity matrix, **I**. It is apparent from this that the inverse cannot be calculated if det(**A**) = 0 because this would involve division by zero. Thus, singular matrices (see above) cannot be inverted.

The inverse of a product can be found as

$$(\mathbf{AB})^{-1} = \mathbf{B}^{-1}\mathbf{A}^{-1}$$

provided **A** and **B** are both invertible.

We will not define here how to write down explicitly the elements of the inverse of a matrix for larger matrices. The explicit form rapidly becomes very cumbersome, and accurate numerical calculation of the inverse of large matrices requires special techniques that are available in any computer library.

A1.10. Differentiation of a Matrix

This involves no new ideas. You just differentiate each element of the matrix. Thus, the expression $d\mathbf{A}/dt$ simply means, in the 2×2 case,

$$\frac{d\mathbf{A}}{dt} = \begin{bmatrix} da_{11}/dt & da_{12}/dt \\ da_{21}/dt & da_{22}/dt \end{bmatrix} \tag{A8}$$

A1.11. Transpose of a Matrix

Swapping the rows and columns of a matrix is referred to as transposition. If $\mathbf{A} = [a_{ij}]$, then its transpose, denoted $\mathbf{A}^T$, is $\mathbf{A}^T = [a_{ji}]$. In the case of the matrices defined in equation A1, we have, therefore,

$$\mathbf{A}^T = \begin{bmatrix} a_{11} & a_{21} \\ a_{12} & a_{22} \end{bmatrix} \qquad \mathbf{C}^T = \begin{bmatrix} c_{11} & c_{21} \\ c_{12} & c_{22} \\ c_{13} & c_{23} \end{bmatrix}$$

The transpose of a product is

$$(\mathbf{AB})^T = \mathbf{B}^T\mathbf{A}^T$$

A1.12. Eigenvalues and Eigenvectors of a Matrix

Any square matrix can be represented in a simple form that has nice properties that depend on its eigenvalues and eigenvectors. If **M** is an $n \times n$ matrix, then a nonzero ($n \times 1$) column vector **x** is said to be an *eigenvector* of **M** if there is a scalar, λ, such that

$$\mathbf{Mx} = \lambda\mathbf{x} \tag{A9}$$

Then λ is called the *eigenvalue* corresponding to $\mathbf{x}$. Such an eigenvector is not unique because you can multiply $\mathbf{x}$ by any nonzero scalar and the equation above is still satisfied, with the same value of λ.

Because it makes no difference if any matrix or vector is multiplied by an identity matrix, the above equation can be written as $\mathbf{Mx} = \lambda\mathbf{Ix}$ or

$$(\mathbf{M} - \lambda\mathbf{I})\mathbf{x} = \mathbf{0} \tag{A10}$$

Now, any square matrix, when multiplied by a nonzero vector, can only yield a zero result if it is *singular,* i.e., if its determinant is zero. Thus,

$$|\mathbf{M} - \lambda\mathbf{I}| = 0 \tag{A11}$$

The eigenvalues are the solutions to this equation. When the equation is evaluated algebraically, it turns out to be a polynomial in λ of degree n and so, by a well-known theorem in algebra, it has n solutions, $\lambda_1, \lambda_2, \ldots \lambda_n$. For example, in the case $n = 2$, using equation A7,

$$0 = |\mathbf{M} - \lambda\mathbf{I}| = \begin{vmatrix} m_{11} - \lambda & m_{12} \\ m_{21} & m_{22} - \lambda \end{vmatrix} = (m_{11} - \lambda)(m_{22} - \lambda) - m_{12}m_{21}$$

or

$$\lambda^2 - \lambda(m_{11} + m_{22}) + m_{11}m_{22} - m_{12}m_{21} = 0$$

This is a quadratic equation in λ having two roots λ_1 and λ_2. Note that, from school algebra, we have that the sum of the roots, $\lambda_1 + \lambda_2$, is equal to minus the coefficient of λ in the equation, namely $m_{11} + m_{22}$ and the product of the roots, $\lambda_1\lambda_2$, is equal to the constant term $m_{11}m_{22} - m_{12}m_{21}$, which we recognize as being the determinant det($\mathbf{M}$). These are examples of two quite general results:

1. The sum of the eigenvalues of a square matrix equals the sum of the its diagonal elements, which is known as the *trace* of the matrix.
2. The product of the eigenvalues equals the determinant of the matrix.

Although there are always n roots of an $n \times n$ matrix, they are not necessarily distinct. For example, if $\mathbf{M}$ is the 2×2 identity matrix, the above equations become $(1 - \lambda)^2 = 0$, so both roots are 1; i.e., $\lambda = 1$ is a *repeated* root. Finding roots of polynomials is not always easy, but there are standard computer programs widely available to find these eigenvalues and the eigenvectors $\mathbf{x}_i$ that go with them. For given λ_i, the corresponding eigenvector satisfies equation A9 or, equivalently, A10. Note that equation A10 is very much like the transpose of equation 10 for finding an equilibrium vector (if you identify $\mathbf{M} - \lambda_i\mathbf{I}$ with $\mathbf{Q}$), so it is not too difficult. Remember that the scaling of an eigenvector is arbitrary: any constant times the eigenvector is also an eigenvector, but the arbitrary scaling factors cancel out during subsequent calculations so they are not important for the applications discussed here.

The set of equations

$$\mathbf{Mx}_i = \lambda_i\mathbf{x}_i, \qquad i = 1 \text{ to } n$$

can be written as a single matrix equation, $\mathbf{MX} = \mathbf{X\Lambda}$, or, simply interchanging the two sides of the equation,

$$\mathbf{X\Lambda} = \mathbf{MX} \tag{A12}$$

where $\mathbf{X}$ is a matrix whose columns are the eigenvectors $\mathbf{x}_i$. Thus, $\mathbf{X}$ can be written as

$$\mathbf{X} = [\mathbf{x}_1 \quad \mathbf{x}_2 \quad \cdots \quad \mathbf{x}_n] \tag{A13}$$

There is one eigenvector for each eigenvalue, so it does not matter which order we put the eigenvalues in, as long as the eigenvectors are kept in the corresponding order. $\mathbf{\Lambda}$ is an $n \times n$ *diagonal* matrix with the eigenvalues λ_i down the diagonal and zeroes everywhere else.

In general, matters can get complicated from here on if some of the eigenvalues are repeated roots; in that case we need something called the Jordan canonical form, which can be found in advanced textbooks on algebra or, in the Markov process case, Cox and Miller (1965, Chapter 3). Fortunately, in the case of reversible Markov processes, these complications do not generally arise. So we will assume, sufficiently for our purpose, that the eigenvalues are all distinct, and then it can be shown that the matrix $\mathbf{X}$ is invertible, and so we can postmultiply equation A12 to obtain

$$\mathbf{X\Lambda X}^{-1} = \mathbf{MXX}^{-1} = \mathbf{MI} = \mathbf{M} \tag{A14}$$

The value of all this comes when we want to raise the matrix $\mathbf{M}$ to some power. For example,

$$\mathbf{M}^2 = \mathbf{X\Lambda X}^{-1}\mathbf{X\Lambda X}^{-1} = \mathbf{X\Lambda I\Lambda X}^{-1} = \mathbf{X\Lambda}^2\mathbf{X}^{-1}$$

It is easy to see that we can keep on doing this, so that

$$\mathbf{M}^r = \mathbf{X\Lambda}^r\mathbf{X}^{-1} \tag{A15}$$

The important thing about this is that, although, in general, raising the matrix $\mathbf{M}$ to the power r is quite difficult, it is very easy for the diagonal matrix $\mathbf{\Lambda}$: the matrix $\mathbf{\Lambda}^r$ is simply another diagonal matrix whose diagonal elements are simply powers of the eigenvalues λ_i^r so you only have to calculate powers of scalars.

It should be noted that, in general, the eigenvalues and the elements of the eigenvectors of $\mathbf{M}$ may be complex numbers. Fortunately, this is not the case in ion-channel models.

There is another way in which we can represent equation A15. To generalize a little, let $\mathbf{D}$ be any $n \times n$ diagonal matrix with diagonal elements d_i, and let $\mathbf{Y}$ denote the inverse $\mathbf{X}^{-1}$, which we now consider as a set of *rows*

$$\mathbf{Y} = \begin{bmatrix} \mathbf{y}_1 \\ \mathbf{y}_2 \\ \vdots \\ \mathbf{y}_i \\ \vdots \\ \mathbf{y}_n \end{bmatrix} \tag{A16}$$

Then it is easy to see that the matrix product $\mathbf{DY}$ can be represented as a similar set of rows

$$\mathbf{DY} = \begin{bmatrix} \mathbf{y}_1 d_1 \\ \mathbf{y}_2 d_2 \\ \vdots \\ \mathbf{y}_i d_i \\ \vdots \\ \mathbf{y}_n d_n \end{bmatrix} \tag{A17}$$

Note that, because the d_i are scalars, it does not matter if we write them before the $\mathbf{y}_i$ or after. Then

$$\mathbf{XDY} = [\mathbf{x}_1 \quad \mathbf{x}_2 \quad \cdots \quad \mathbf{x}_n] \begin{bmatrix} \mathbf{y}_1 d_1 \\ \mathbf{y}_2 d_2 \\ \vdots \\ \mathbf{y}_i d_i \\ \vdots \\ \mathbf{y}_n d_n \end{bmatrix}$$

Now this looks just like a row times a column, similar to example A3, despite the fact that the elements here are vectors rather than scalars. But the nice thing about *partitioned matrices* is that, provided everything is the right shape, they behave formally just like ordinary matrices. Thus, in this case,

$$\mathbf{XDY} = \sum_{i=1}^{n} \mathbf{x}_i \mathbf{y}_i d_i \tag{A18}$$

But $\mathbf{x}_i \mathbf{y}_i$ is a column times a row, similar to example A4, and so the result is an $n \times n$ matrix

$$\mathbf{A}_i = \mathbf{x}_i \mathbf{y}_i \qquad (i = 1 \text{ to } n) \tag{A19}$$

These matrices $\mathbf{A}_i$ are called the *spectral matrices* of the matrix **M.** Then equation A18 can be written as

$$\mathbf{XDX}^{-1} = \mathbf{XDY} = \sum_{i=1}^{n} \mathbf{A}_i d_i \tag{A20}$$

As a particular example, equation A15 can now be written as

$$\mathbf{M}^r = \mathbf{X}\mathbf{\Lambda}^r\mathbf{X}^{-1} = \sum_{i=1}^{n} \mathbf{A}_i \lambda_i^r \tag{A21}$$

Thus, once we have calculated the eigenvalues and eigenvectors, and hence the spectral matrices, of the matrix **M,** we have only to calculate the powers of the (scalar) eigenvalues, multiply them by the constant matrices $\mathbf{A}_i$, and add them instead of doing a lot of matrix multiplications to get $\mathbf{M}^r$.

Appendix 2. Some APL Code

APL is a powerful computer language that uses an "executable notation," which differs from normal mathematical notation but is far more logically consistent. It has the advantage that, when you get used to the notation, the instructions you give the computer are essentially the same as you would write on paper. We illustrate some of the calculations for Sections 5 and 6 to show what is possible. The version we have used is DYALOG APL.

The hard part with any system is finding the eigenvalues and eigenvectors, and any code is too complex to show here, so let us assume there is a function EIGEN that results in a vector of eigenvalues and an **X** matrix of eigenvectors when supplied with a square matrix as argument. Suppose the transition rates shown in equation 4 are already stored in the matrix **Q.**

Identify the index sets of subclasses and create a vector **U** containing five 1's

```
A←1 2 ◇ B←3 4 ◇ C←,4 ◇ F←B∪C ◇ E←A∪B ◇ U←5⍴1        (A22)
```

Note that the comma in the above line is important: it makes sure that C, consisting of a single number, is a vector, not a scalar; the ◇ character simply allows several statements on one line. Next

```
y←0 0 0 0 0 1 ◇ PINF←y⌹⍉Q, 1        (A23)
```

finds the equilibrium distribution, **p**(∞), using the method of equation 17 (note that **u** in that equation is **S***y* while **S** is **Q**,1), with the numerical result of equation 18. The remaining numerical results arising from the APL statements shown below are reported in Section 5.

```
PHIS←PINF[A]+.×Q[A;F]÷PINF[A]+.×Q[A;F]+.×U[F]        (A24)
```

finds the initial vector $\boldsymbol{\phi}_s$ (see equation 50 and the numerical result of equation 57).

Now let

```
L X←EIGEN Q[F;F] ◇ TAU←÷L ◇ SM←(↓⍉X)∘.×¨↓⌹X        (A25)
```

After that L contains the eigenvalues of $\mathbf{Q}_{\mathcal{FF}}$, **X** the **X** matrix of eigenvectors, while **A** contains the set of all three matrices $\mathbf{A}_1$ $\mathbf{A}_2$ $\mathbf{A}_3$ by doing each of the vector products in equation 89; the symbol ¨ is the "each operator" in APL and ⌹ is the matrix inversion function. The result is shown in equation 56. The coefficients a_i are then formed into the vector **a,** with three elements in this case, by a version of equation 52.

```
a← −TAU×(⊂PHIS)+.×¨A+.×¨⊂Q[F;F]+.×U[F]        (A26)
```

with the numerical results shown in equation 56.

To get the distribution of the number of openings per burst, evaluate equation 59 as

```
GAB← −(⌹Q[A;A])+.×Q[A;B] ◇ GBA← −(⌹Q[B;B])+.×G[B;A]        (A27)
```

with results shown in equation 65.

The initial vector given by equation 61 is evaluated as

PHIB←*num*÷(*num*←PINF[C]+.×Q[C;A]+Q[C;B]+.×GBA)+.×U[A] (A28)

where the numerator is stored in *num* to save calculating the same thing twice.

The result is shown in equation 63.

This time the spectral expansion needed is given by

RHO X←EIGEN GAB+.×GBA ◇ MU←÷1−RHO ◇ SM←(↓⍉X)∘.×¨↓⌹X (A29)

with numerical results for RHO, MU, and SM shown in equations 71–73, respectively. The calculation for MU follows from equation 69.

Now form an identity matrix of shape $k_{\mathcal{A}} \times k_{\mathcal{A}}$

I← DIAG (ρA)ρ1

The formula for the areas is then given by equation 70, and the numerical results by equation 74:

a← MU×(⊂PHIB)+.×¨SM+.×¨⊂(I−GAB+.×GBA)+.×U[A] (A30)

Comparison of these equations with the corresponding ones in Sections 5 and 6 shows how easy it is (with a few little tricks you have to get used to) to translate the usual mathematical formulas into an equivalent executable notation. Results from the other sections can be obtained in much the same manner.

References

Colquhoun, D., and Hawkes, A. G., 1977, Relaxation and fluctuations of membrane currents that flow through drug-operated channels, *Proc. R. Soc. Lond.* [*B*]**199:**231–262.

Colquhoun, D., and Hawkes, A. G., 1981, On the stochastic properties of ion channels, *Proc. R. Soc. Lond.* [B]**211:**205–235.

Colquhoun, D., and Hawkes, A. G., 1982, On the stochastic properties of bursts of single ion channel openings and of clusters of bursts, *Phil. Trans. R. Soc. Lond.* [*B*]**300:**1–59.

Colquhoun, D., and Hawkes, A. G., 1987, A note on correlations in single ion channel records, *Proc. R. Soc. Lond.* [*B*]**230:**15–52.

Cox, D. R., and Miller, H. D., 1965, *The Theory of Stochastic Processes,* Chapman and Hall, London.

Hawkes, A. G., 1984, Complex numbers in APL, *Vector J. Br. APL Assoc.* **1:**107–120.

Hawkes, A. G., and Sykes, A. M., 1990, Equilibrium distributions of finite-state Markov processes, *IEEE Trans. Reliability* **29**(5)**:**592–595.

Hawkes, A. G., Jalali, A., and Colquhoun, D., 1992, Asymptotic distributions of apparent open times and shut times in a single channel record allowing for the omission of brief events, *Phil. Trans. R. Soc. Lond.* [*B*]**337:**383–404.

Huang, C. Y., 1979, Derivation of initial velocity and isotopic exchange rate equations, *Methods Enzymol.* **63:**54–85.

Mirsky, L., 1982, *An Introduction to Linear Algebra,* Dover, New York.

Moler, C., and van Loan, C., 1978, Nineteen dubious ways to compute the exponential of a matrix, *SIAM Rev.* **20**(4)**:**801–836.

Press, W. H., Flannery, B. P., Teukolsky, S. A., and Vetterling, W. T., 1993, *Numerical Recipes: The Art of Scientific Computing,* 2nd ed., Cambridge University Press, Cambridge.

Stephenson, G., 1965, *An Introduction to Matrices, Sets and Groups,* Longman, London.

Part IV

"CLASSICS"

Chapter 21

Geometric Parameters of Pipettes and Membrane Patches

BERT SAKMANN and ERWIN NEHER

1. Introduction

The object of the patch-current measurement is a small membrane patch and a pipette tip of comparable dimensions. Both are at the limit of resolution of the optical microscope. Only rarely is it possible to "see" the membrane patch during the experiment, and even then quantitative evaluation of its size is impossible or difficult. The purpose of this chapter is to provide some hints toward estimation of the tip geometry of patch pipettes and of patch sizes by a combination of light and electron microscopy and electrical measurements. It should be emphasized, however, that there are wide variations in tip geometry and patch sizes, even if all the parameters that are under control of the experimenter are kept constant. The numbers given should therefore only be considered as order-of-magnitude estimates.

2. Geometry of Patch Pipettes

The purpose of this section is to relate the dimensions of "typical" patch pipettes to their conductance properties. The conductance measured before the tip touches the cell membrane can provide an estimate of the pipette tip geometry. For this end, the appearance in the scanning electron microscope of several types of patch pipettes commonly used for cell-attached, cell-free, and whole-cell recording is described. To relate the tip geometry to the pipette conductance, we have measured in specified conditions [150 mM KCl as pipette-filling solution, pipette tips immersed in a standard test (150 mM NaCl) solution] the conductance and the electron microscopic appearance of the same pipette. Alternatively, we have measured the geometry and the conductance of pairs of pipettes fabricated under the same conditions. We used one of the pipettes for scanning electron microscopy and the other for the electrical measurements.

2.1. Tip Shape of Soft Glass Pipettes

Pipettes fabricated from commercially available soft glass (CEE BEE® capillaries, 1.6-mm outer diameter and 0.3-mm wall thickness) were pulled by the two-stage procedure

BERT SAKMANN • Department of Cell Physiology, Max-Planck-Institute for Medical Research, D-69120 Heidelberg, Germany. ERWIN NEHER • Department of Membrane Biophysics, Max-Planck-Institute for Biophysical Chemistry, Am Fassberg, D-37077 Göttingen, Germany.
Single-Channel Recording, Second Edition, edited by Bert Sakmann and Erwin Neher. Plenum Press, New York, 1995.

described by Hamill *et al.* (1981). These pipettes have DC resistances ranging from 1 MΩ to 5 MΩ when filled with 150 mM KCl; the exact value is determined by the coil heat setting of the final pull. Figure 1 shows the appearance in the scanning electron microscope of the tip of a pipette having a resistance of 2.5 MΩ, a "typical" value for pipettes used for cell-attached recordings. The average diameter of the tip opening measured from the "tip-on" view of such pipettes (Fig. 1A) was 1.13 ± 0.29 μm (mean ± S.D., $n = 14$). This corresponds to a tip opening area of 1.0 μm^2. The width of the rim, estimated from the shadowed ring surrounding the tip opening, is 0.19 ± 0.04 μm. The taper of the tip and the thickness of the glass wall were determined from side-on views (Fig. 1B) and from pipettes with partially broken-off tips. The dimensions of the tip of a typical soft glass pipette are shown in Fig. 1C. It represents a longitudinal section through a pipette of 2–2.5 MΩ resistance.

The tip shape is approximately conical, with an angle ϕ of 24° for the average type of pipette. When the pipette is modeled as having an approximately cylindrical shank and a conical tip, the total resistance of the pipette is given by the sum of the tip and the shank resistances (Snell, 1969).

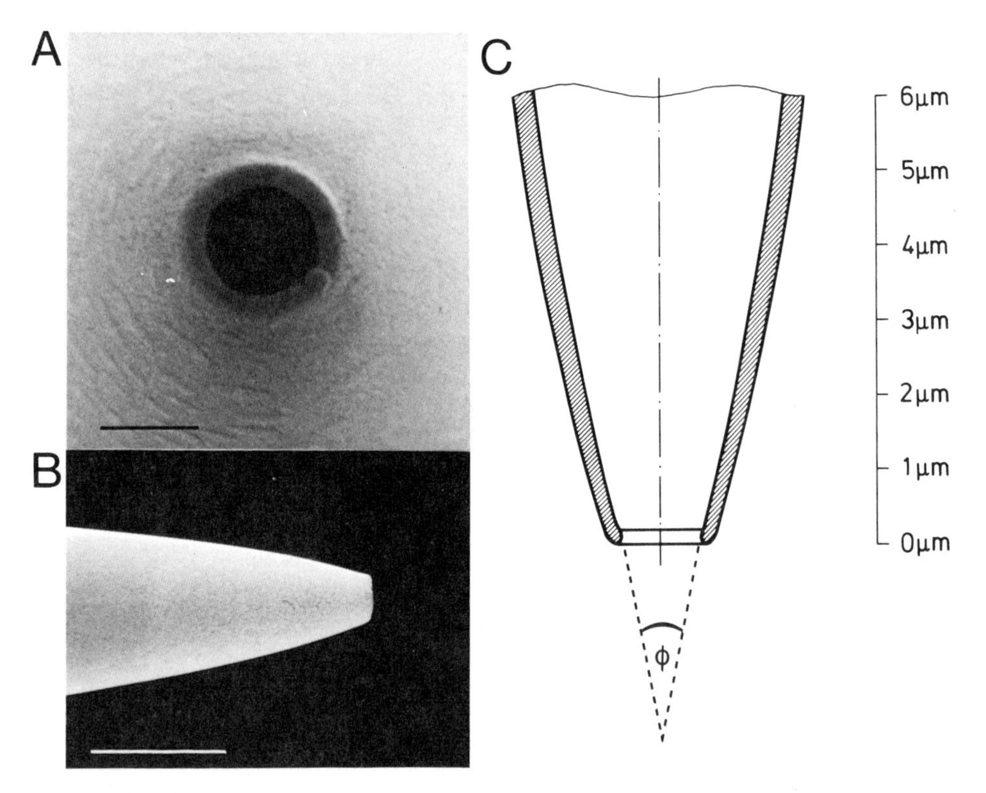

Figure 1. Tip geometry of a soft glass patch pipette. A: Scanning electron micrograph of a tip-on view of the pipette opening. The resistance of this type of pipette is between 2 and 2.5 MΩ. The darker ring represents the rim of the pipette tip. The tip opening diameter is 1.1 μm. The width of the rim is 0.2 μm. Scale bar represents 1 μm. B: Scanning electron micrograph of a side-on view of the same pipette tip. Scale bar represents 4 μm. For scanning electron microscopy, the pipette tips were sputtered with gold, resulting in a gold layer of ≅200 Å thickness. C: Reconstruction of a longitudinal section through the pipette tip of a 2- to 2.5-MΩ pipette. The wall thickness (shaded) was determined from pipettes broken at various distances from the tip opening. The tip cone angle ϕ is 24°. It is measured as indicated in the figure.

$$R = \frac{\rho l}{\pi \cdot r_s^2} + \frac{\rho \cot(\phi/2)}{\pi}\left(\frac{1}{r_t} - \frac{1}{r_s}\right) \qquad (1)$$

Since the radius of the cylindrical shank r_s is much larger (>50 μm) than that of the radius of the tip opening r_t, the resistance of the tip dominates. The pipette tip resistance is 1.2 MΩ, assuming ρ = 51 Ω · cm for the specific resistivity of 150 mM KCl (at 25°C), ϕ = 24° for the angle of the tip cone, and a tip opening radius r of 0.6 μm. The measured resistance is, however, 2–2.5 MΩ. This reflects the fact that the resistance of the pipette shank is not negligible. Breaking off the tip leaves one with pipettes of 1 MΩ resistance, which is an estimate of the shank resistance.

The tip opening areas of pipettes with various conductances are plotted in Fig. 2. According to equation 1, one would expect the tip opening area to increase with the square of the pipette conductance for pipettes in which the resistance of the tip dominates. The data points in Fig. 2 can be approximated by such a square relationship for pipettes with resistances >2.5 MΩ. For practical purposes, however, i.e., for the whole range of pipette resistances of 1–5 MΩ, an approximately linear relationship is seen between pipette conductance and tip opening area. The broken line in Fig. 2 has a slope of 6.9 μm^2 MΩ.

2.2. Tip Shape of Hard Glass Pipettes

Patch pipettes can also be fabricated from hard glass, i.e., borosilicate glass. The advantage of hard glass pipettes is that the specific resistivity of this glass is much higher than

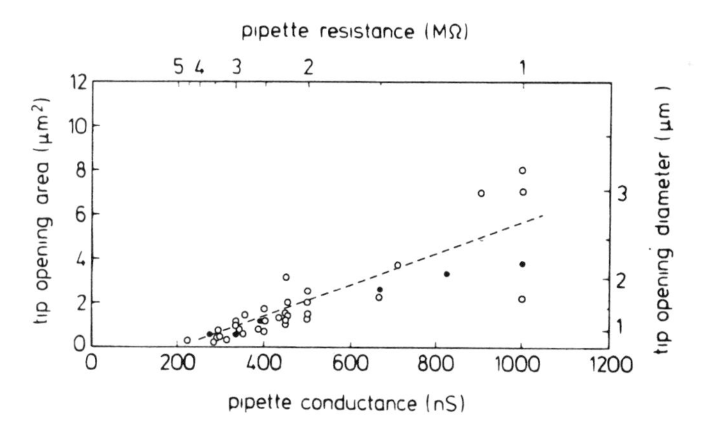

Figure 2. Relationship between tip opening area and pipette conductance (pipettes were filled with 150 mM KCl) for pipettes fabricated from thin-walled capillaries. The tip opening area is plotted for pipettes of various resistances. To obtain this plot, the pipette resistance and the tip opening area were measured on the same pipette (closed symbols); or the opening area was measured on one of a pair of pipettes and the resistance was measured on the other one (open symbols). Data from both heat-polished and unpolished pipettes are included. Round symbols, soft glass pipettes; square symbols, borosilicate glass pipettes; triangular symbols, aluminosilicate glass pipettes. The broken line is the regression line to all data points and has a slope of 6.9 μm^2/1000 nS. To obtain measurements of conductance and tip geometry of the same pipette, the KCl solution in the pipette was replaced first by distilled water, then by methanol. Subsequently, the methanol was allowed to dry out by keeping the pipette at +60°C for 10 min and maintaining a constant negative pressure. Then, the pipette tip was prepared for scanning electron microscopy.

that of soft glass. As a consequence, the background noise contribution from the capacitance across the glass wall is smaller, and the quality of coating of the pipette shank and tip is less critical. It is sufficient to coat the pipette only up to a distance about 200 μm from the tip opening in order to obtain pipettes with noise properties similar to those of "heavily coated" soft glass pipettes. Hard glass pipettes are pulled from capillaries of similar dimensions to the soft glass tubing described above. For example, when borosilicate capillaries from Kimax® with a wall thickness of 0.2 mm are used, patch pipettes can be fabricated with a tip geometry very similar to that of the soft-glass pipette shown in Fig. 1 (see Table I for details). The relationship between tip opening area and pipette conductance is also very similar to that given in Fig. 2.

The specific conductance of glass containing aluminum, i.e., aluminosilicate glass, is even lower than that of the borosilicate hard glass described above. We have only limited experience with pipettes made from this type of glass. It is possible to pull pipettes that have resistances of 1–5 MΩ and a tip geometry similar to that of soft glass shown in Fig. 1C provided one uses thin-walled capillaries. For example, commercially available aluminosilicate glass from A–M Systems (outer diameter, 1.5 mm; wall thickness, 0.3 mm) can be used. The main difference in the pulling procedure is that with this glass the length of the first pull should be only about 5 mm, and the recentering distance should be 3 mm. The heat settings must be increased for both the first and the final pull. The relationship between tip opening area and the pipette conductance is comparable to that for soft glass (Fig. 2).

2.3. Tip Shape of Thick-Walled Pipettes

When pipettes are pulled from thick-walled glass capillaries, e.g., with a wall thickness of 0.5 mm or larger, the width of the rim and the wall thickness of the pipette tip are about

Table I. Geometric Parameters of Tips of Pipettes Fabricated from Various Types of Glass Capillaries[a]

Pipettes fabricated from	Tip opening area (μm²)	Rim area (μm²)	Rim width (μm)	Cone angle (deg)
Thin-walled capillaries				
CEE BEE (n = 14) (soft glass)	1.0	0.79	0.19	24
Kimax (n = 2) (hard borosilicate glass)	1.2	0.82	0.2	20
Standard aluminum glass (n = 2) (hard aluminosilicate glass)	1.0	0.9	0.22	25
Thick-walled capillaries				
Jencons (n = 8) (hard borosilicate glass)	1.01	1.71	0.39	10

[a]Thin-walled pipettes have resistances of 2–2.5 MΩ, thick walled of 8–10 MΩ, when filled with 150 m² · KCl. All numbers are mean values; the number of pipettes is given in parentheses. Pipettes of the tip geometry listed above are obtained by two-stage pulling. Pulling was done by gravity with a weight of 500 g. The length of the first pull has to be shorter for hard glass pipettes than for soft glass pipettes. Concomitantly, the recentering distance is shorter. The setting for the heating coil for the second pull is varied to obtain pipettes of the geometry listed in the table. The length of the first pull and the length of recentering are as follows for the various glass capillaries: CEE BEE, 9 mm and 6 mm; Kimax, 6 mm and 4 mm; A–M glass, 5 mm and 3 mm; Jencons, 9 mm and 6 mm. The glass capillaries are commercially available from the following manufacturers: CEE BEE, Type 101-PS plain, from C. Bardram, Braunstien 4, 3460-Birkorod, Denmark; Kimax: Article No. 34500, from Kimble Products, Vineland, NJ, U.S.A.; A–M Glass, Cat. No. G.CASS-150-4, from A–M Systems, Inc., Everett, WA 98024, U.S.A; Jencons, Cat. No. H15/10, from Jencons (Scientific) Ltd., Leighton Buzzard, England.

twice as large as those of pipettes pulled from thin-walled (≤0.3 mm) capillaries. The advantages of thick-walled hard glass pipettes are, first, that the greater wall thickness further reduces the shunt resistance across the glass wall, and second, that in a number of preparations with these pipettes, we have found that gigohm seals of greater stability are obtained, and the success rate of obtaining gigohm seals is significantly larger than with conventional thin-walled pipettes.

Thick-walled hard glass pipettes can easily be fabricated from commercially available hard glass (Jencons Scientific, Catalogue No. H 10/15) of 1.8-mm outer diameter and 0.5-mm wall thickness using the same two-stage pulling procedure as described for fabrication of thin-walled pipettes (for details, see Table I). Polishing of the tip is, however, done without cooling of the tip by an air jet. The shrinkage of the tip during polishing is observed as a slight increase in the wall thickness of the initial 2–3 μm of the pipette tip. The appearance in the scanning electron microscope of the tips of thick-walled hard glass pipettes, which readily form gigohm seals on both enzyme-treated cells (heart myocytes, chromaffin cells, skeletal muscle fibers) and tissue-cultured cells (spinal cord neurons), is shown in Fig. 3. Pipettes with tip dimensions shown in Fig. 3 have resistances of 8–11 MΩ when filled with 150 mM KCl. The size of the tip opening area is ≅1 μm^2, i.e., comparable to that of soft

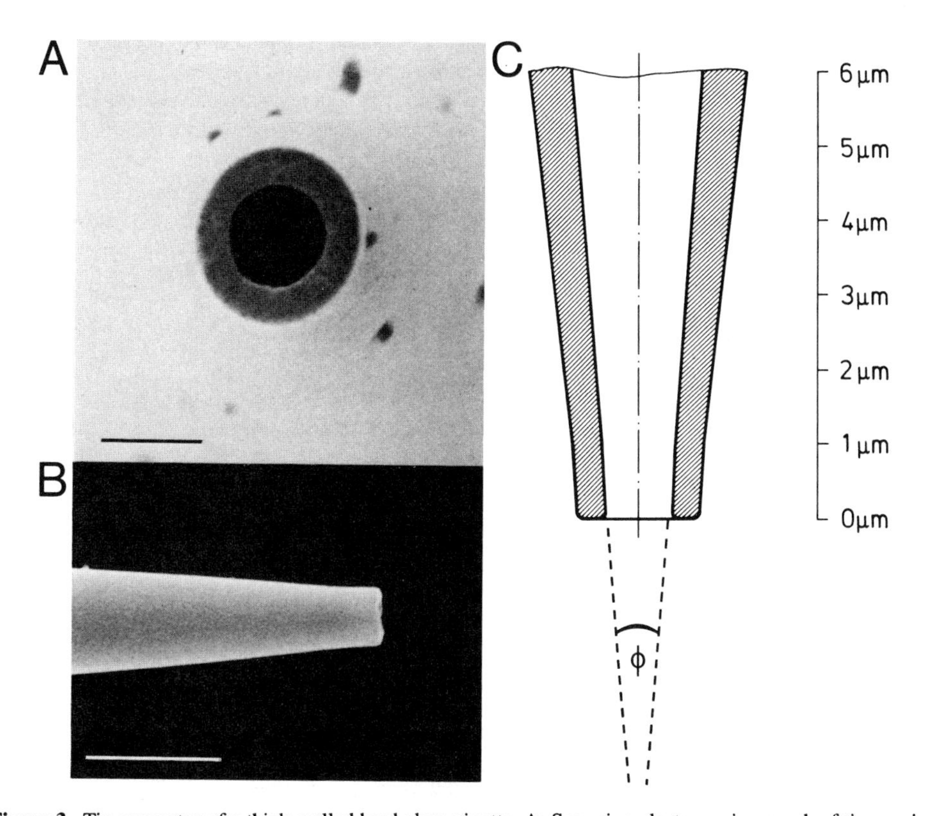

Figure 3. Tip geometry of a thick-walled hard glass pipette. A: Scanning electron micrograph of tip-on view of 10-MΩ hard glass pipette. The diameter of the tip opening is 0.98 μm. The width of the glass rim is 0.4 μm. The scale bar represents 1 μm. B: Scanning electron micrograph of side-on view of the same pipette. Scale bar represents 4 μm. C: Reconstruction of a longitudinal section through the tip of a thick-walled hard glass pipette. Reconstruction of the outer dimensions is based on tip-on and side-on views of pipettes as shown in A and B. The dimensions of the wall are based on measurements of pipettes with broken-off tips.

glass pipettes of 2.5–3 MΩ resistance. However, the width of the rim is 0.4–0.5 μm, which is about twice that of thin-walled pipettes. It is also obvious that the rim is not always completely smooth even after extensive polishing.

In comparison to pipettes fabricated from thin-walled glass, the pipette tip is much slimmer. This is shown by the side-on view of the pipette in Fig. 3B. The geometry of "typical" thick-walled hard glass patch pipettes used for cell-attached recordings is shown in Fig. 3C, which represents a longitudinal section through a tip of a typical pipette of 8–10 MΩ resistance. In comparison with thin-walled pipettes, the tip cone is much sharper. The cone angle is 8–12°. The sharper tip can partly account for the larger resistance of thick-walled pipettes having a tip opening area similar to that of low-resistance pipettes with thinner walls. The pipette solution in the cone-shaped tip has a resistance of about 3 MΩ (using a tip opening diameter of 0.9 μm and a cone angle $\phi = 12°$). The measured resistance is 8–10 MΩ. This means that the larger part of the pipette resistance resides in the pipette shank. It is consistent with the observation that breaking the tip of these pipettes reduces the pipette resistance to 4–5 MΩ. Thus, for thick-walled hard glass pipettes, the resistance is a much less reliable indicator of the tip opening area than for thin-walled pipettes.

Pipettes that have resistances of 8–11 MΩ are very reproducible in terms of their tip geometry. This is illustrated in Fig. 4, where the tip opening area of thick-walled pipettes is shown for pipettes with different resistances. It is seen that pipettes with resistances in the range of 8–10 MΩ have a tip opening area that varies only about twofold, i.e., between 0.6 μm^2 and 1.2 μm^2, whereas for pipettes with lower resistances, large variations in the tip opening area are found. The variability in resistance of thick-walled pipettes with values <7 MΩ most likely reflects changes both in the diameter and length of the pipette shank and in the shape of the tip cone rather than changes in the tip opening diameter.

Thick-walled hard glass pipettes for whole-cell recording, i.e., with a much lower pipette resistance (2–5 MΩ), are made from thick-walled capillaries by reducing the length of the first pull to 5 mm and recentering the pipette by 3 mm. Pipettes fabricated by this procedure

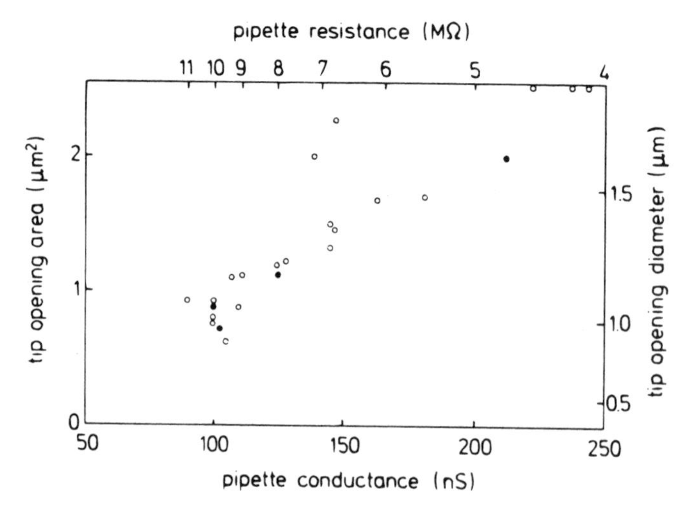

Figure 4. Relationship between tip opening area and pipette conductance for thick-walled hard glass pipettes. Pipettes are filled with 150 mM KCl solution. Filled symbols represent measurements of conductance and tip opening area on the same pipette. The other data are based on experiments on pairs of pipettes. Note large scatter of values for pipettes of less than 7 MΩ resistance.

have tip diameters of 2–3 μm and resistances of 3–4 MΩ. The advantage of this type of pipette is the better stability of the seal in the whole-cell recording mode.

3. Geometry of Membrane Patches

3.1. Patch Area by Observation in the Light Microscope

In some tissues, such as enzyme-dispersed heart myocytes or enzyme-treated skeletal muscle fibers, gigohm seals sometimes form spontaneously; i.e., upon touching the cell surface with the pipette tip, a tight seal forms without release of the slight pressure and without application of suction. In these cases, visual observation of the pipette tip and the cell surface membrane at high (×1200) magnification does not show any deformation of the cell membrane. The area of the electrically insulated membrane patch under the pipette tip presumably is very similar to that of the tip opening area. Most likely, the glass–membrane seal forms on the rim of the pipette tip. The area of the membrane patch can be estimated from the pipette resistance measured before seal formation has occurred (Fig. 2). More frequently, however, a seal forms only after slight suction has been applied to the pipette interior. During suction, part of the cell membrane is pulled into the opening of the pipette tip, forming an omega-shaped deformation of the cell membrane. The development of membrane deformation during establishment of a gigohm seal can be observed in the light microscope, and estimates of the area of the membrane patch in the pipette tip can be obtained.

As illustrated in Fig. 5, we have estimated the area of membrane patches pulled into the pipette tip under two experimental conditions. First, a patch of membrane was drawn into the pipette tip until a gigohm seal had formed. Membrane deformation was observed at ×500 or ×1250 magnification using either a ×40 water-immersion objective in a normal microscope or a ×100 oil-immersion objective in an inverted microscope. The axis of the pipette was nearly parallel (<10° deviation) to the plane of focus to give a side-on view. The membrane patch in the tip was isolated mechanically from the rest of the cell surface membrane by retracting the tip 20 to 50 μm from the cell surface. At this stage, the pipette tip was photographed (Fig. 5A). When further suction was applied to the pipette interior, the membrane patch lost its contact with the glass wall. It then adopted the shape of a spherical vesicle, which bounced up and down in the tip of the pipette. On application of slight pressure (<5 cm H_2O) to the pipette interior, it was caged in the tip and was photographed (Fig. 5B).

From photomicrographs such as those shown in Fig. 5A,B, the area of the membrane patch drawn into the pipette tip during gigohm seal formation was estimated. We have measured the area in the omega-shaped configuration, i.e., when the patch adhered to the pipette wall, and after formation of a membrane vesicle. The pipettes used for these experiments had resistances of 2 MΩ, with a tip geometry very similar to that illustrated in Fig. 1.

In the omega-shaped configuration, the membrane area can be estimated by modeling the patch as a cone, representing the "sealed" membrane area in contact with the glass wall, and a hemispherical area, which represents the "free" unsealed membrane facing the pipette solution. In the vesicular configuration, the patch area was determined from its apparent diameter. The values of membrane areas estimated in this way are given in Table II. In the omega-shaped configuration, the ratio of the "free" to the "sealed" membrane area was 0.5 to 0.6. This ratio was determined for relatively large patches of membrane when a visual discrimination of "free" and "sealed" portion of the patch was possible. In each experiment,

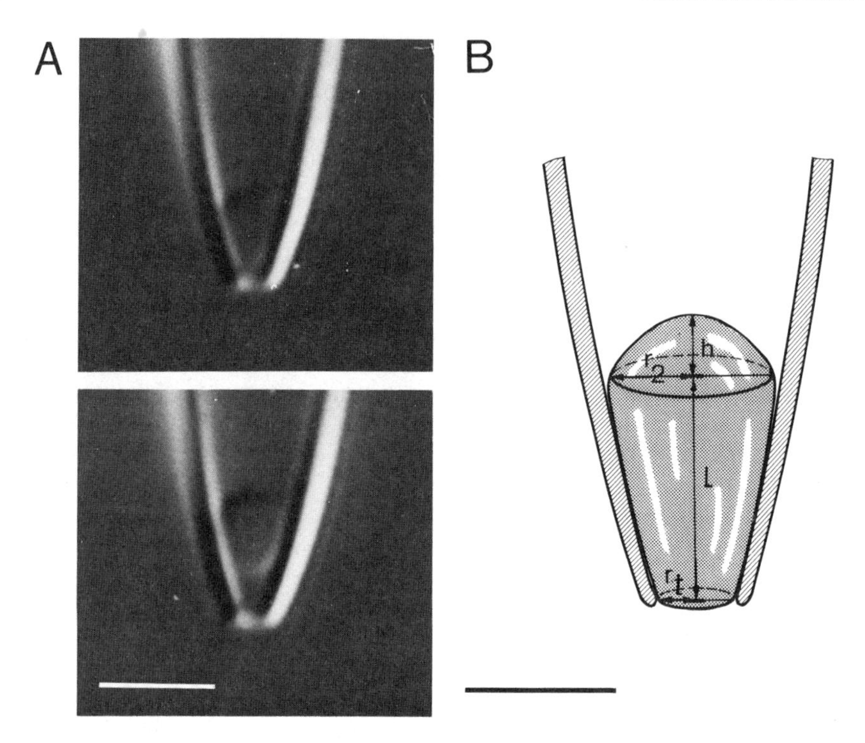

Figure 5. Geometry of mechanically isolated membrane patches in the tip of a soft-glass pipette of 2-MΩ resistance. A: Photomicrograph of a pipette tip after isolation of a membrane patch from a rat myoball. A × 100 oil-immersion objective in an inverted microscope (Zeiss Invertoskop) equipped with Normarski interference contrast was used. The membrane patch forms an omega-shaped vesicle that "seals" the tip opening. Pipette access resistance was 2 GΩ. Part of the patch is sealed to the pipette wall, and part faces the pipette solution. B: Photomicrograph of pipette tip after the mobilization of the membrane patch by suction applied to the pipette interior. Same membrane patch as shown in A. The pipette access resistance has dropped to 2 MΩ. The isolated membrane patch has the shape of a vesicle. The apparent diameter is 2.75 μm; the membrane area is 23 μm^2. C: Schematic diagram of the geometry of an omega-shaped vesicle as shown in A. The measured distance from the pipette tip opening to the top of the vesicle ($l + h$ as indicated in the figure) is 3.75 μm. The distance from the tip opening to the base of the spherical part is 3 μm; the height h of the spherical part is 0.75 μm. The radius r_2 of the pipette 3 μm from the tip opening is 1.25 μm. The tip opening radius r_1 is 0.5 μm. From these values, one obtains for the area of the "free" membrane portion $A_f = \pi(h^2 + 2r_2^2) = 11\ \mu m^2$. The area of the "sealed" membrane portion is $A_s = \pi/(r_2 + r^1) = 17\ \mu m^2$, the total membrane area is 29 μm^2, assuming that the tip opening is also spanned by the membrane. From the area of the sealed membrane, 17 μm^2, and the measured seal resistance, 2 GΩ in this example, one can estimate the dimensions of contact between cell membrane and pipette wall. Assuming a specific resistivity of 51 Ω · cm for 150 mM KCl solution, and assuming free mobility of K^+ and Cl^-, one obtains a value <1 Å for the distance between membrane and glass.

the total area of the same membrane patch estimated in the two configurations, omega-shaped and vesicle-shaped, was nearly identical. Since the estimation of membrane area in the vesicle configuration is more straightforward, most area measurements were made from vesicles. The average of seven membrane patches isolated with pipettes of 2.5 MΩ resistance and measured in the vesicle configuration was 14 ± 5 μm^2. Assuming that the geometry of these patches in the omega shape is as illustrated in Fig. 5C, i.e., that the ratio of the "free" to the "sealed" portion is 0.5, the area of the "free" membrane is of the order of 5 μm^2.

Table II. Comparison of the Area of the Same Membrane Patch Determined from Photomicrographs of Omega-Shaped Vesicles and Spherical Vesicles[a]

Pipette	Omega-shaped vesicle "Sealed" area	"Free" area	Total area	Spherical vesicle
1	24 μm^2	12 μm^2	37 μm^2	32 μm^2
2	14 μm^2	9 μm^2	24 μm^2	19 μm^2
3	14 μm^2	8 μm^2	23 μm^2	19 μm^2
4	17 μm^2	11 μm^2	23 μm^2	23 μm^2

[a]All experiments were done with pipettes fabricated from CEE BEB® capillaries with 2–2.5 MΩ resistance filled with 150 mM KCl. See Fig. 5 for determination of "sealed" and "free" area. It has been assumed that the membrane spanning the tip opening has an area of 1 μm^2.

A few additional observations were made in these experiments. (1) When suction is applied in the cell-attached configuration to destroy the membrane patch in the pipette for whole-cell recording, initially the omega shape in the pipette tip increases in size. The membrane then forms a vesicle, which is sucked into the pipette shank. At this stage, the access resistance of the pipette dropped by orders of magnitude, indicating a low-resistance access to the cell's interior. (2) Once a vesicle has formed in the tip, it could be easily expelled into the bath solution through the pipette tip by application of a slight pressure to the pipette interior. (3) When omega-shaped patches were pulled into the tip opening without formation of a gigohm seal, they always spontaneously formed vesicles.

3.2. Patch Area as Measured by Patch Capacitance

The area of patches in the pipette tip can be estimated from patch capacitance if the specific capacitance of the membrane is known. Biological membranes quite generally have specific capacitance values close to 1 μF/cm^2. This value was confirmed by Fenwick *et al.* (1982) for bovine chromaffin cells, and it was adopted for the measurements described below.

3.2.1. Measurement of Patch Capacitance

The capacitance of the pipette–patch assembly can be measured quite accurately by various techniques (see, for instance, Neher and Marty, 1982). The main problem is to determine which part of the measured capacitance is that of the patch and which is caused by the pipette. We accomplished this distinction by observing capacitance changes during manipulations with the pipette, as illustrated schematically in Fig. 6. These included patch formation (cell attached), withdrawal of the pipette, and pressing of the pipette against a nearby Sylgard® microsphere. Withdrawal was either done after rupture of the patch membrane (which resulted in an outside-out patch) or with an intact patch, which resulted in a vesicle. In the latter case, an inside-out patch could be formed by briefly ($\simeq$1 sec) touching the Sylgard® microsphere. Pressing the pipette against Sylgard® completely closed off the tip ($R_{seal} > 100$ GΩ). The capacitance reading at this stage was taken as the baseline. The capacitance of the different patch configurations are given with respect to this baseline.

Several tests were performed to show that this procedure yields valid estimates of patch

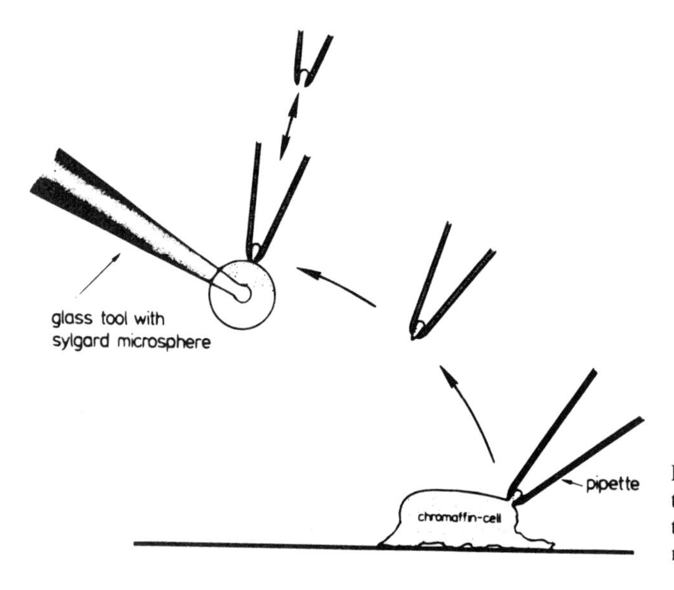

Figure 6. Schematic diagram of the manipulations performed with the pipette for determination of membrane patch capacitance.

capacitance or else that the capacitance of the Sylgard®-sealed tip actually represents that part of the capacitance of a pipette–patch assembly that has to be considered background in a measurement on an intact patch. Two possible sources of error were considered. (1) The depth of immersion of the pipette may change during withdrawal of the pipette from the cell or during any other manipulation. Changes in the depth of immersion would certainly change pipette capacitance, and this would be indistinguishable from a change in patch capacitance. (2) Pressing the pipette onto Sylgard® may cover some of the tip region of the pipette and thereby change its capacitance.

The tests were performed with pipettes the tips of which had been sealed completely by slightly more heat polishing than is usually done on patch pipettes. They were Sylgard®-coated, filled, and handled otherwise identically to normal patch pipettes.

Figure 7A shows a protocol of such a test. It is a penwriter record of the output of a lock-in amplifier set up in connection with a patch clamp for capacitance measurement as described by Neher and Marty (1982). The record starts with a 100 fF calibration signal produced by manually changing the capacitance neutralization setting on the patch clamp. At that point, the pipette was already immersed in solution. The pipette was then advanced two times by 100-μm steps (arrows), each of which resulted in a capacitance increase of approximately 20 fF. Then the pipette was withdrawn by 100 μm (simulating vesicle formation), and it is seen that the capacitance changes only very little (circle).

This experiment was done repeatedly with the same result. It can be explained by the observation that on first immersion, the meniscus in the air–water–Sylgard® contact zone advances steadily. However, it comes to a standstill on withdrawal in a hysteresis-like fashion. It reverses only after 200 to 300 μm of withdrawal. In the experiment shown in Fig. 7A, the pipette was then moved sideways by 200 μm (triangle) and subsequently advanced by 50 μm (arrow) without changes in the capacitance reading. At that point, it touched the Sylgard® microsphere. Then, the pipette was advanced in 5-μm steps (small arrows). These movements produced only very little increase in capacitance. In some experiments, a decrease in capacitance of 10–20 fF was observed at that point. This occurred only when the pipette tip cut deeply (>10 μm) into the Sylgard® sphere.

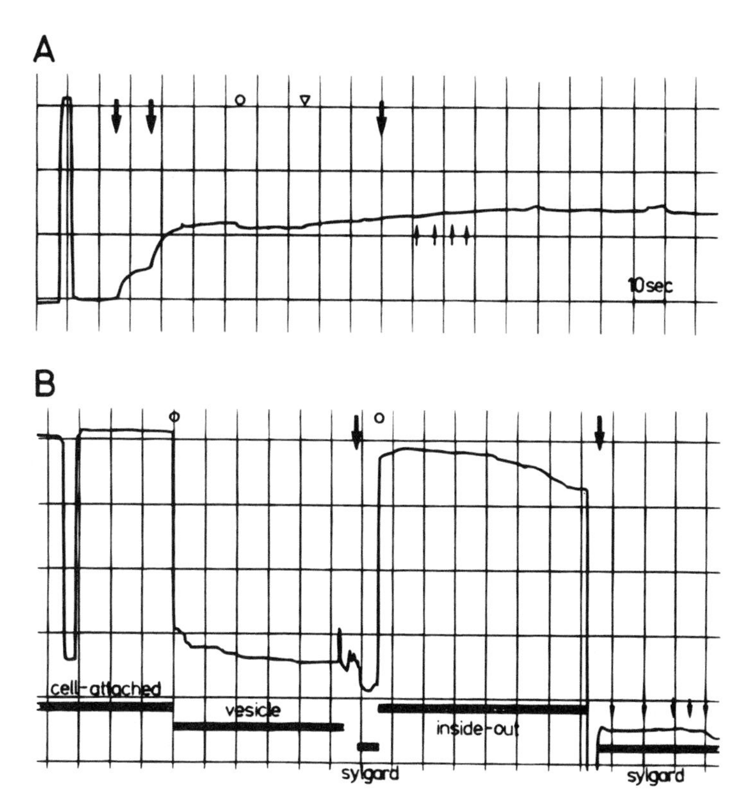

Figure 7. Recording of capacitance. Both records are tracings of the output signal of a lock-in amplifier operated at 3.2 kHz, 18 mV rms. Part A is a measurement on a pipette with a permanently sealed tip. The changes in capacitance reading mainly reflect changes in immersion depth during various manipulations performed with the pipette as explained in the text. The deflection at the beginning of the record is a 100-fF calibration mark. Large arrows, 100-μm advancements or advancement towards Sylgard® microsphere; circle, 100-μm withdrawal; triangle, sideways movement; small arrows, 5-μm forward movement pushing into Sylgard® (see text). Part B is a measurement with a saline-filled pipette of 1.8 MΩ resistance. The record starts in the cell-attached configuration. First, a negative-going (capacitance decrease) calibration signal of 100 fF was applied. Then the pipette was withdrawn by 56 μm, which resulted in vesicle formation and a concomitant capacitance decrease of approximately 100 fF. Subsequently, the pipette was moved toward the Sylgard® surface. When the Sylgard® surface occluded the tip, the capacitance reading decreased by a further 10–20 fF. The level obtained can be considered the "baseline" of the capacitance measurement (see text). Further on, the pipette was withdrawn, which resulted in an increase of capacitance reading almost back to the level of the cell-attached patch. In other experiments, it was established (e.g., by observing Na^+ channels) that at this stage an intact inside-out patch was present. The patch broke after 1 min. This resulted in a large (offscale) deflection. Subsequent movement of the patch toward the Sylgard® reestablished the baseline. It is not clear why there is a small difference in baseline between readings before and after the inside-out configuration. This difference was not consistently found and could be a slow continuous change in the geometry of the meniscus surrounding the pipette. A similar slow drift of opposite polarity can be seen in part A. At the end of the experiment, the pipette was advanced in several 5-μm steps (small arrows). Only the last step, which caused the pipette to cut visibly into the Sylgard®, led to a noticeable capacitance decrease. Symbols same as in part A.

From these and similar experiments it can be concluded that the experimental protocol does allow establishment of a stable baseline for measurement of patch capacitances provided four criteria are met: (1) the pipette must be heavily Sylgard® coated; (2) the movements of the pipette should be limited to approximately 100 μm in the longitudinal direction; (3) the pipette should not be pushed into Sylgard® by more than 10 μm; (4) the point of deepest immersion should be reached at the beginning of the measurement, i.e., in the cell-attached configuration, in order to exploit the hysteresis properties of the meniscus.

In some experiments, small spontaneous changes in capacitance, presumably because of spontaneous changes in the meniscus, did occur. However, these were usually smaller than 10 to 20 fF. This quantity should therefore be considered as the limit of resolution for the measurements.

3.2.2. Range of Capacitance Values

Capacitance measurements either were taken as described above (see also Fig. 7B) or else were read directly from the calibrated dial setting of the capacitance compensation network built into the patch-clamp amplifier (see Chapter 4, this volume). Bovine chromaffin cells as well as neuroblastoma cells were used for the measurements.

Capacitance values of patches ranged from <10 fF to 250 fF, corresponding to patch areas <1 μm^2 to 25 μm^2. There was no obvious difference between the different patch configurations or between different preparations. However, a definite correlation between patch size and pipette resistance could be observed (Fig. 8). Much of the scatter in Fig. 8 is probably caused by differences in the strength and duration of the suction that was applied before gigaseal formation.

Some additional observations made during these experiments were the following. (1) Very often, but not always, the capacitance in the cell-attached configuration was exactly the same as that of the inside-out patch. This is consistent with the view that it is basically the same patch of membrane that is being measured on in both cases. (2) When a pipette with an outside-out patch touched the Sylgard®, the patch usually broke. In some cases, however, it survived, although at much decreased capacitance. Also, a sudden decrease in

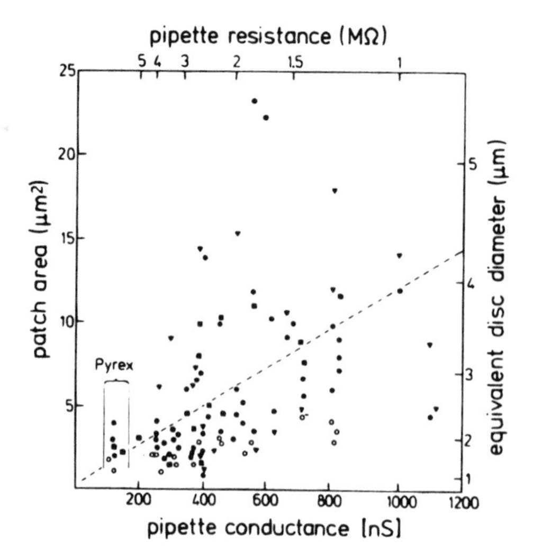

Figure 8. Patch area as determined from capacitance measurement plotted versus pipette conductance (resistance). A specific capacitance of 1 $\mu F/cm^2$ was adopted for converting capacitance into area. Each symbol reflects a capacitance reading with respect to a baseline as defined in the text and in Fig. 7. ●, cell-attached configuration; ■ inside-out patch; ▼, outside-out patch; ○, vesicle. Pooled data from chromaffin cells and neuroblastoma cells (14 measurements) recorded with either saline-filled or KCl-filled pipettes. No striking differences were observed between the types of cells and filling solutions. The straight line is a linear regression to the data points obtained in the cell-attached configuration, excluding the values obtained with thick-walled pipettes (marked Pyrex®), according to $a = 12.6(1/R + 0.018)$, where a is area (μm^2) and R is resistance (MΩ).

capacitance could sometimes be observed as it approached the Sylgard® surface, before the pipette visually touched the Sylgard®. The existence of intact outside-out patches in these situations could be confirmed by observing single Na^+-channel currents with normal properties. (3) Sometimes patches changed size spontaneously. These changes were higher than the spontaneous changes in background value mentioned above. (4) Vesicles always had lower capacitance values than the patches before or after vesicle formation. This is consistent with the view that the vesicle represents two membranes in series. (5) In a few cases in which the pipette had been withdrawn by one to two cell diameters, an outside-out patch formed by visual and electrical criteria. Still, sideways movement of the pipette resulted in a large change in patch size. It seemed that the patch had still been connected to the cell by submicroscopic cytoplasmic bridges (see also Fig. 5A). (6) In one case, a cell-attached patch was so large that its dimensions could be judged visually by means of an eyepiece micrometer. Comparison of electrically measured patch size with the optical determination suggested that only the hemispherical top of the omega-like patch contributed to the electrical measurement.

4. Conclusions

From the observations reported above, the following conclusions can be drawn.For pipettes fabricated from soft and hard glass and using glass capillaries with ≤0.3-mm wall thickness, the size of the tip opening area can be varied easily between 1 μm^2 and 5–8 μm^2 by varying the heat setting of the final pull. In this range, the pipette conductance is mostly governed by the conductance of the cone-shaped tip. When pipettes are filled with 150 mM KCl, the pipette resistance varies between 1 MΩ and 5 MΩ in an approximately linear fashion with the tip opening area. The thickness of the tip rim and the glass wall is about 0.2 μm.

Pipettes fabricated from thick-walled borosilicate glass capillaries (wall thickness 0.5 mm) have excellent sealing properties when this tip opening area is <2 μm^2. When filled with 150 mM KCl, these pipettes have resistances of 8–11 MΩ. The main advantage of thick-walled hard glass pipettes is the fact that the Sylgard® coat must not extend very close to the tip opening (i.e., less than 200 μm). On most cell types the pipette–membrane seal is more stable with thick-walled pipettes than with thin-walled ones.

Visual observation of the pipette tip during formation of gigohm seals when suction is applied to the pipette interior shows that the cell surface membrane is deformed. An omega-shaped membrane patch is pulled about 2–3 μm into the tip opening. During spontaneous seal formation, this deformation is not visible. When the patch is mechanically isolated by retracting the pipette tip from the cell surface, the omega shape of the patch does not change measurably. On further suction, a round vesicle forms. The area of the membrane patch can be determined in both configurations. The average area of membrane patches isolated with "typical" pipettes of 2–3 MΩ resistance is 5–20 μm^2.

A method for determining the membrane capacitance of sealed membrane patches is described. Assuming a specific membrane capacitance of 1 $\mu F/cm^2$, the capacitance value can be interpreted in terms of membrane area. Such capacitance measurements show that an approximately linear relationship exists between the area of the membrane patch and the pipette conductance. The membrane area contributing to the patch capacitance varies between 2 μm^2 and 25 μm^2. The area of the membrane patch is always larger than the tip opening area. This is consistent with the visual observation of membrane patches pulled 2–3 μm into the pipette tip.

A comparison of the patch area determined from measurements of vesicle size in the light microscope and the patch area determined from capacitance measurements suggests that about 30% of the membrane patch in the tip is "free" and contributes to the capacitance, whereas about 70% is sealed to the glass wall. It suggests that the high-resistance membrane–glass contact extends over a relatively long distance (1–2 μm) from the tip opening.

ACKNOWLEDGMENT. We would like to thank Mrs. I. Kraeft and Mr. J. Winkler for help with scanning electron microscopy.

References

Fenwick, E., Marty, A., and Neher, E., 1982, A patch-clamp study of bovine chromaffin cells and of their sensitivity to acetylcholine, *J. Physiol* **331**:577–597.

Hamill, O. P., Marty, A., Neher, E., Sakmann, B., and Sigworth, F. J., 1981, Improved patch-clamp techniques for high-resolution current recording from cells and cell-free membrane patches, *Pflügers Arch.* **391**:85–100.

Neher, E., and Marty, A., 1982, Discrete changes of cell membrane capacitance observed under conditions of enhanced secretion in bovine adrenal chromaffin cells, *Proc. Natl. Acad. Sci. U.S.A.* **79**:6712–6716.

Snell, I. M., 1969, Some electrical properties of fine-tipped pipette microelectrodes, in: *Glass Microelectrodes* (M. Lavallee, O. F. Schanne, and M. C. Hebert, eds.), pp. 111–123, John Wiley & Sons, New York.

Chapter 22

Conformational Transitions of Ionic Channels

P. LÄUGER

1. Introduction

Ion transport through a channel may be described as a series of thermally activated processes in which the ion moves from a binding site over an energy barrier to an adjacent site. The "binding sites" are the minima in the potential-energy profile that result from interactions of the ion with ligand groups of the channel. In the traditional treatment of ionic channels, the energy levels of wells and barriers are considered to be fixed, i.e., independent of time and not influenced by the movement of the ion. This description, which corresponds to an essentially static picture of protein structure, represents a useful approximation in certain cases. Recent findings on the dynamics of proteins, however, suggest a more general concept of barrier structure.

A protein molecule in thermal equilibrium may assume a large number of conformational states and may rapidly move from one state to another (Frauenfelder *et al.*, 1979; Karplus, 1982). Evidence for fluctuations of protein structure comes from X-ray diffraction and Mössbauer studies (Huber *et al.*, 1976; Parak *et al.*, 1981), optical experiments (Lakowicz and Weber, 1980), and the kinetic analysis of ligand rebinding to myoglobin after photodissociation (Austin *et al.*, 1975). These and other studies have shown that internal motions in proteins occur in the time range from picoseconds to seconds.

Long-lived conformational substates of ionic channels may be directly observed in single-channel records (Hamill and Sakmann, 1981). The detection of fast transitions between substates is limited, however, by the finite bandwidth of the measurement. Apart from instrumental limitations, an inherent restriction is imposed by Nyquist's theorem, which sets a lower limit of $\tau \simeq kT\Lambda/(V\Delta\Lambda)^2$ for the lifetime of a substate that can be detected from a single-channel record (Λ is the conductance of the channel, $\Delta\Lambda$ the conductance increment of the substate, and V the driving force of ion flow expressed as a voltage). With Λ = 10 pS, $\Delta\Lambda$ = 1 pS, and V = 100 mV, the theoretical limit of detection would be $\tau \simeq 4$ μsec. In practice, however, the resolution is still lower. This means that the observed single-channel current is likely to represent an average over many unresolved conductance states. As is discussed below, the existence of such "hidden" substates may strongly influence the observable properties of (average) single-channel conductance such as dependence on ion concentration and voltage.

Of particular interest is the possibility that transitions between conformational states of the channel protein may be coupled to the translocation of the ion within the channel (Frehland, 1979). In this case, the permeability of the channel depends explicitly on the rate

P. LÄUGER • Department of Biology, University of Konstanz, D-78434 Konstanz, Germany. Dr. Läuger is deceased. Correspondence to H.-J. Appel at this address.
Single-Channel Recording, Second Edition, edited by Bert Sakmann and Erwin Neher. Plenum Press, New York, 1995.

constants of conformational transitions. An extreme situation arises when a channel can assume two conformations, one with the binding site accessible only from the left and the other with the binding site accessible only from the right. In this case, where ion translocation through the channel is limited by the rate of interconversions of the two states, the channel exhibits a carrier-like behavior. Another interesting consequence of coupling between ion translocation and conformational transitions is the following. If ions are driven through the channel by an external force (a difference of electrochemical potential), a nonequilibrium distribution of conformational states is created. As discussed below, this may result in an apparent violation of microscopic reversibility, i.e., in a situation in which the frequency of transitions from state A to B is no longer equal to the transition frequency from B to A.

2. Two-State Channel with a Single Binding Site

We consider a channel that (in the conducting state) fluctuates between two conformations A and B. We assume that the rate of ion flow through the channel is limited by two (main) barriers on either side of a single (main) binding site (Fig. 1). In series with the rate-limiting barriers, smaller barriers may be present along the pathway of the ion. This model corresponds to a channel consisting of a wide, water-filled pore and a narrow part, acting as a selectivity filter, in which the ion interacts with ligand groups (Hille, 1971). Since the binding site may be empty or occupied, the channel may exist in four substates (Fig. 2): A^0, conformation A, empty; A^*, conformation A, occupied; B^0, conformation B, empty; B^*, conformation B, occupied. The rate constants for transitions between A and B depend, in general, on whether the binding site is empty or occupied (i.e., $k_{AB}^0 \neq k_{AB}^*$ and $k_{BA}^0 \neq k_{BA}^*$).

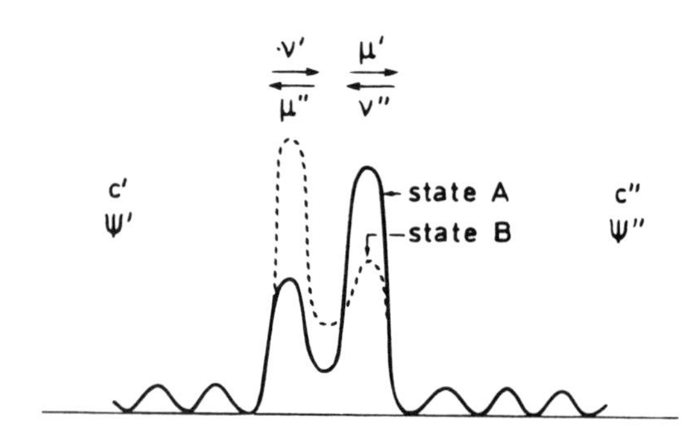

Figure 1. Energy profile of a channel with two conformational states: ν' and ν'' are the frequencies of jumps from the solutions into the empty site; μ' and μ'' are the jumping frequencies from the occupied site into the solutions; c', c'' and ψ', ψ'' are the ion concentrations and the electrical potentials in the left and right aqueous solutions.

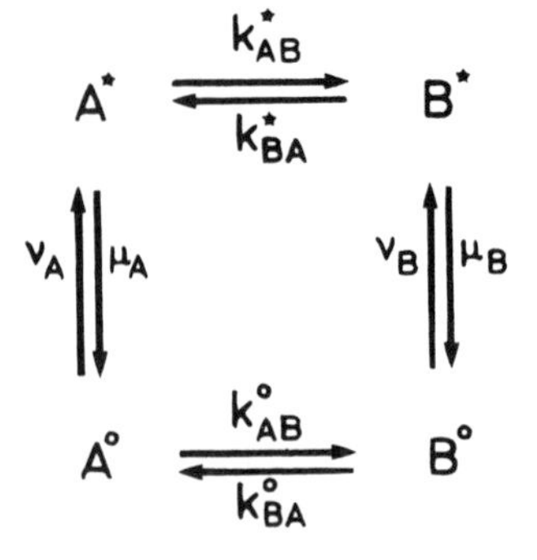

Figure 2. Transitions among four substates of a channel with one binding site (A^0, conformation A, empty; A^*, conformation A, occupied; B^0, conformation B, empty; B^*, conformation B, occupied).

The coulombic field around the ion tends to polarize the neighborhood by reorienting dipolar groups of the protein. In this way, the probability of a given transition may be strongly affected by the presence of the ion in the binding site.

Transitions between empty and occupied states occur by exchange of an ion between the binding site and the left or right aqueous phase (Fig. 1):

$$\nu_A = \nu'_A + \nu''_A = c'\rho'_A + c''\rho''_A \tag{1}$$

$$\nu_B = \nu'_B + \nu''_B = c'\rho'_B + c''\rho''_B \tag{2}$$

$$\mu_A = \mu'_A + \mu''_A \qquad \mu_B = \mu'_B + \mu''_B \tag{3}$$

In equations 1 and 2, it is assumed that ions in the energy wells outside the rate-limiting barriers are always in equilibrium with the corresponding aqueous phase. The jumping frequencies ν'_A, ν''_A, ν'_B, ν''_B into the empty side are then proportional to the aqueous ion concentrations c' and c'', whereas the rate constants μ'_A, μ''_A, μ'_B, and μ''_B for leaving the site are independent of c' and c''.

The principle of microscopic reversibility requires that the rate constants obey the following relationship (Läuger *et al.*, 1980):

$$\frac{\nu'_A\mu'_A}{\nu''_A\mu''_A} = \frac{\nu'_B\mu'_B}{\nu''_B\mu''_B} = \frac{\nu'_A\mu'_B k^*_{AB} k^0_{AB}}{\nu''_B\mu''_A k^0_{AB} k^*_{BA}} = \exp[z(u - u_o)] \tag{4}$$

In this, z is the valence of the ion; u is the voltage across the channel, and u_0 the equilibrium voltage of the ion, both expressed in units of kT/e_0 (k, Boltzmann's constant; T, absolute temperature; e_0, elementary charge):

$$u = \frac{\psi' - \psi''}{kT/e_0} \tag{5}$$

$$zu_0 = \ln(c''/c') \tag{6}$$

(compare Fig. 1). Introducing the equilibrium constants k^0 and k^* between the conformational states,

$$\kappa^0 = k^0_{AB}/k^0_{BA} \qquad \kappa^* = k^*_{AB}/k^*_{BA}. \tag{7}$$

The stationary ion flux Φ through the channel from solution 1 to solution 2 is obtained as (Läuger *et al.*, 1980):

$$\Phi = (1/\sigma)[1 - \exp(zu_0 - zu)][\nu'_A\mu'_A(1 + \nu_B/k^0_{BA} + \mu_B/k^*_{BA}) + \nu'_B\mu'_B(\kappa^0\kappa^* + \kappa^*\nu_A/k^0_{BA} + \kappa^0\mu_A/k^*_{BA}) + \kappa^*\nu'_A\mu'_B + \kappa^0\nu'_B\mu'_A] \tag{8}$$

$$\sigma \equiv (1 + \kappa^0)(\mu_A + \kappa^*\mu_B + \mu_A\mu_B/k^*_{BA}) + (1 + \kappa^*)(\nu_A + \kappa^0\nu_B + \nu_A\nu_B/k^0_{BA}) + \nu_A\mu_B(\kappa^*/k^0_{BA} + 1/k^*_{BA}) + \nu_B\mu_A(\kappa^0/k^*_{BA} + 1/k^0_{BA}) \tag{9}$$

It is seen from equations 8 and 9 that the ion flux Φ explicitly depends on the rate constants k^0_{AB}, k^*_{AB}, k^0_{BA}, and k^*_{BA}. This is an expression of the phenomenon of coupling between ion translocation and conformational transitions. Similarly, it is found that the equation for the probability of a given conformation state not only contains the equilibrium constants κ^0, κ^*, ν_A / μ_A, and ν_B / μ_B but depends explicitly on the translocation rate constants μ'_A, μ''_A, μ'_B, and μ''_B.

An essential condition for the occurrence of coupling is the assumption that transitions between the two conformations can take place both in the empty and in the occupied state of the binding site. If transitions can start only from one of the states ($k^0_{AB} \approx k^0_{BA} \approx 0$ or $k^*_{AB} \approx k^*_{BA} \approx 0$), then the dependence of Φ on the rate constants of conformational transitions is lost.

2.1. Concentration Dependence of Conductance

The ohmic conductance Λ of the channel under the condition $c' = c'' = c$ is obtained from equations 8 and 9 in the form:

$$\Lambda(c) = \frac{z^2 e_0^2}{kT} \cdot \frac{c(\alpha + \beta c)}{\gamma + \delta c + \epsilon c^2} \tag{10}$$

The parameters α, β, γ, δ, and ϵ are concentration-independent combinations of the rate constants:

$$\alpha \equiv \rho'_A \mu'_A (1 + \mu_B / k^*_{BA}) + \rho'_B \mu'_B \kappa^0 (\kappa^* + \mu_A / k^*_{BA}) + \rho'_A \mu'_B \kappa^* + \rho'_B \mu'_A \kappa^0 \tag{11}$$

$$\beta \equiv (\rho_B \rho'_A \mu'_A + \rho_A \rho'_B \mu'_B \kappa^*) / k^0_{BA} \tag{12}$$

$$\gamma \equiv (1 + \kappa^0)(\mu_A + \mu_B \kappa^* + \mu_A \mu_B / k^*_{BA}) \tag{13}$$

$$\delta \equiv \rho_A \mu_B (1/k^*_{BA} + \kappa^* / k^0_{BA}) + \rho_B \mu_A (1/k^0_{BA} + \kappa^0 / k^0_{BA}) + (1 + \kappa^*)(\rho_A + \rho_B \kappa^0) \tag{14}$$

$$\epsilon \equiv \rho_A \rho_B (1 + \kappa^*) / k^0_{BA} \tag{15}$$

$$\rho_A \equiv \rho'_A + \rho''_A \qquad \rho_B \equiv \rho'_B + \rho''_B \tag{16}$$

It is seen from equation 10 that $\Lambda(c)$ is a nonlinear function of ion-concentration-containing terms that are quadratic in c. This behavior may be compared with the properties of a one-site channel with fixed barrier structure, which always exhibits a simple saturation characteristic of the form

$$\Lambda(c) = \frac{z^2 e_0^2}{kT} \cdot \frac{\rho c}{\mu + \rho c} \cdot \frac{\mu' \mu''}{\mu} \tag{17}$$

It can easily be shown that for certain combinations of rate constants, $\Lambda(c)$ goes through a maximum with increasing ion concentration. Such a nonlinear concentration dependence of conductance is usually taken as evidence for ion–ion interaction in the channel or for the existence of regulatory binding sites. In the channel model discussed here, the nonlinearity of $\Lambda(c)$ is a direct consequence of the coupling between ion flow and conformational transitions.

For further discussion of equation 10, it is useful to consider two limiting cases in which conformational transitions are either much slower or much faster than ion translocation.

2.1.1. Case 1: $k^0_{AB}, k^*_{AB}, k^0_{BA}, k^*_{BA}, \ll \nu_A, \nu_7, \mu_A, \mu_B$

Under this condition, the mean lifetime of a given state is much longer than the average time an ion spends in the energy well, which may be as short as 10^{-11} sec. Since many ions pass through the channel during the lifetime of the individual state, a well-defined conductance can be assigned to each state. On the other hand, the frequency of transitions between states A and B may still be much too high to be resolved in a single-channel record. The observed current is then averaged over the two rapidly interconverting conductance states (Fig. 3). Under the conditions given above, equation 10 reduces to

$$\Lambda(c) = p_A \Lambda_A + (1 - p_A)\Lambda_B \tag{18}$$

where Λ_A and Λ_B are the conductances of the channel in states A and B, respectively, which have the form of equation 17; p_A is the probability of finding the channel in state A (A^0 or A^*), which, in the vicinity of equilibrium, is given by (with $K_A \equiv \rho_A/\mu_A$; $K_B \equiv \rho_B/\mu_B$; $\kappa^* K_A = \kappa^0 K_B$):

$$p_A = \frac{1 + cK_A}{1 + \kappa^0 + (1 + \kappa^*)cK_A} \tag{19}$$

According to equation 18, Λ is equal to the weighted average of the conductances in states A and B. Since not only Λ_A and Λ_B but also p_A contains the ion concentration c, the concentration dependence of Λ is different from the simple saturation characteristic given by equation 17.

If the frequency of conformational transitions is so low that discrete conductance states can be observed in a single-channel current record, the mean lifetimes τ_A and τ_B of the two conductance states can be determined. If $\tilde{p}^0_A = \tilde{p}^0_A/p_A$ is the conditional probability that the channel is in state A^0 (given that it is in state A^0 or A^*), the transition frequency $1/\tau_A$ is equal to $\tilde{p}^0_A k^0_{AB} + (1 - \tilde{p}^0_A)k^*_{AB}$. This yields, in the limit of slow conformational transitions,

$$\tau_A = \frac{\mu_A + \nu_A}{\mu_A k^0_{AB} + \nu_A k^*_{AB}} \qquad \tau_B = \frac{\mu_B + \nu_B}{\mu_B k^0_{BA} + \nu_B k^*_{BA}} \tag{20}$$

Thus, the mean lifetimes depend on ion concentration (through ν_A and ν_B). Only when the transition frequencies are unaffected by the presence of the ion in the binding site ($k^0_{AB} = k^*_{AB}$ $k^0_{BA} = k^*_{BA}$) are the lifetimes given by the usual concentration-independent relationships $\tau_A = 1/k^0_{AB}$ and $\tau_B = 1/k^0_{BA}$.

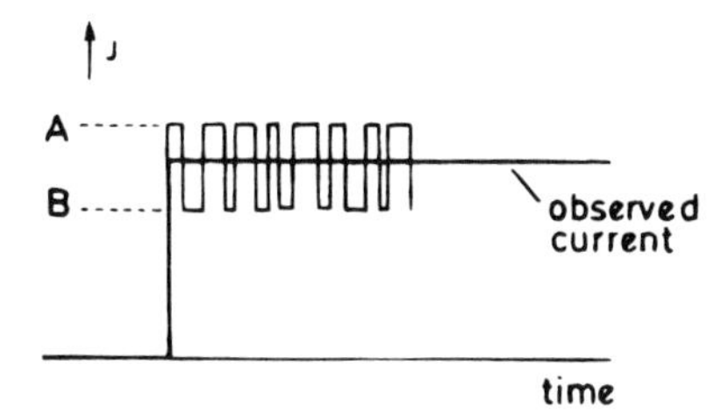

Figure 3. If the mean lifetimes of conformations A and B are much longer than the average dwelling time of an ion in the binding site, the channel fluctuates between discrete conductance states. The frequency of transitions between A and B, however, may be much higher than the bandwidth of the measurement; in this case, only an average current is observed.

2.1.2. Case 2: $k_{AB}^{0}, k_{AB}^{*}, k_{BA}^{0}, k_{BA}^{*} \gg \nu_A, \nu_B, \mu_A, \mu_B$

When interconversion of states A and B is much faster than ion transfer between binding site and water, coupling between ion translocation and conformational transitions is lost, since states A and B are always in equilibrium with each other, even for nonzero ion flow through the channel. This equilibrium may be described by introducing the probability $\bar{p}_A^0$ that an empty channel is in state A and the probability $\bar{p}_A^*$ that an occupied channel is in state A:

$$\bar{p}_A^0 = \frac{1}{1 + \kappa^0} \qquad \bar{p}_A^* = \frac{1}{1 + \kappa^*} \tag{21}$$

Under the condition of fast conformational transitions, equation 10 reduces to the simple form of equation 17 when the following substitutions are introduced:

$$\rho = \bar{p}_A^0 \rho_A + (1 - \bar{p}_A^0)\rho_B \tag{22}$$

$$\mu' = \bar{p}_A^* \mu_A' + (1 - \bar{p}_A^*)\mu_B' \tag{23}$$

$$\mu'' = \bar{p}_A^* \mu_A'' + (1 - \bar{p}_A^*)\mu_B'' \tag{24}$$

This means that in the limit of fast interconversion of states, the equation for Λ becomes formally identical to the corresponding equation derived for a channel with fixed barrier structure, provided that the rate constants are replaced by weighted averages of the rate constants in the two states. This result [which can be generalized to multisite channels with more than two conformational states (Läuger *et al.*, 1980)] has to be expected, since under the above conditions the lifetime of a given conformation is much shorter than the time an ion spends in the binding site, and therefore, the ion "sees" an average barrier structure. Despite the formal identity of the conductance equation with equation 17, the interpretation of the transport process in a channel with variable barrier structure is different, because an ion will preferably jump over the barrier when the barrier is low; this means that the jump rate largely depends on the frequency with which conformational states with low barrier heights are assumed.

2.2. Carrier-like Behavior of Channels

A special situation (with strong coupling) occurs when in state A the barrier to the right is very high (binding site mainly accessible from the left) and in state B the barrier to the left is very high (binding site mainly accessible from the right). In this case, neither state is ion conducting, but ions may pass through the channel by a cyclic process in which binding of an ion in state A from the left is followed by a transition from A to B and release of the ion to the right (Fig. 4). In other words, the channel approaches the kinetic behavior of a carrier. (A carrier is defined as a transport system with a binding site that is exposed alternately to the left and to the right external phase.) Indeed, in the limit $\mu'_A \approx 0$, $\rho''_A \approx 0$, $\mu''_B \approx 0$, $\rho'_B \approx 0$, equation 10 reduces to the expression for the conductance of a carrier with a single binding site (Läuger, 1980). This means that channel and carrier mechanisms are not mutually exclusive possibilities; rather, a carrier may be considered as a limiting case of a channel with multiple conformational states.

2.3. Single-Channel Currents with Rectifying Behavior

An example of an ionic transport system that exhibits a strongly asymmetric current–voltage characteristic in symmetrical electrolyte solutions is the so-called inwardly rectifying potassium channel, which has been found in a number of biological membranes (for a review, see Thompson and Aldrich, 1980). It has recently been demonstrated that the open-channel current itself shows a rectifying characteristic. A possible explanation of this finding is the assumption that the channel fluctuates between different conductance states with voltage-dependent transition frequencies that are too high to be resolved in a single-channel record.

In order to illustrate the possibility that hidden conformational states give rise to rectifying single-channel currents, we assume that in the channel model described above, the ion translocation rates in conformation B vanish ($\nu_B = \mu_B = 0$), so that this state becomes nonconducting. Furthermore, the mean lifetime of state A is assumed to be much longer than the mean dwell time of an ion in the binding site. The average single-channel current I is then given by

$$I = p_A I_A \tag{25}$$

where p_A is the probability of finding the channel in conformation A:

$$p_A = \frac{\mu_A + \nu_A}{\mu_A(1 + \kappa^0) + \nu_A(1 + \kappa^*)} \tag{26}$$

and I_A is the current through the channel in state A,

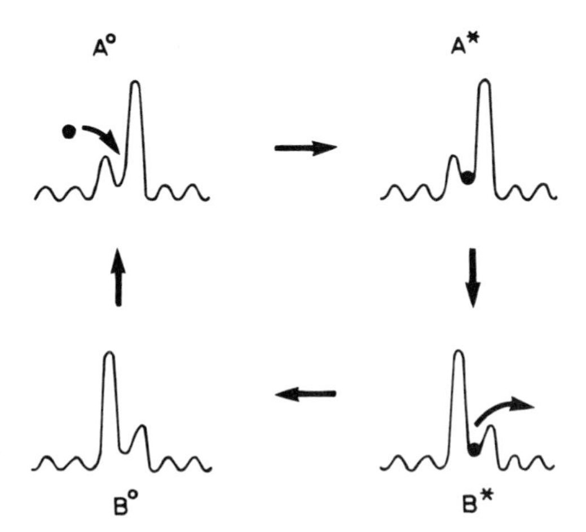

Figure 4. Carrier-like behavior of an ion channel results from transitions between state A with a high barrier to the right and state B with a high barrier to the left. During the cycle $A^0 \rightarrow A^* \rightarrow B^* \rightarrow B^0 \rightarrow A^0$, an ion is translocated from left to right.

$$I_A = ze_0[1 - \exp(zu_0 - zu)] \frac{\nu'_A \mu'_A}{\nu_A + \mu_A} \tag{27}$$

In order to simplify the analysis further, we assume that the equilibrium constants κ^0 and κ^* for the conformational states are identical ($\kappa^0 = \kappa^* \equiv \kappa$) so that $p_A = 1/(1 + \kappa)$. The voltage dependence of the transitions between conformations A and B may be formally described by introducing a gating charge of magnitude $a'e_0$, so that

$$p_A = \frac{1}{1 + \bar{\kappa} \cdot \exp(au)} \tag{28}$$

where $\bar{\kappa}$ is the value of κ for $u = 0$. Thus, for negative voltages u, the probability p_A approaches unity but vanishes for increasing positive values of u. This means that the single-channel current I (equation 25) exhibits a strongly rectifying behavior for sufficiently large values of a.

3. Nonequilibrium Distribution of Long-Lived Channel States

In this section we consider a channel with a single binding site (Fig. 1) and three conformational states A, B, and C. The lifetimes of the states are assumed to be sufficiently long that transitions may be directly observed in current records (Fig. 5). In this case, again, many ions enter and leave the binding site during the lifetime of a given conductance state. A macroscopically observable transition, say from A to B, can result, at the microscopic level, from a transition $A^0 \rightarrow B^0$ (binding site empty) or from a transition $A^* \rightarrow B^*$ (binding site occupied). The distinction between these two elementary processes is meaningful as long as the actual duration of a conformational transition is shorter than the mean lifetimes of the empty and occupied states of the binding site. Accordingly, the microscopic description of the transition frequencies may be based on the scheme shown in Fig. 6.

Transition frequencies f_{XY} from state X to state Y (X,Y = A,B,C) may be obtained from single-channel current records such as those shown in Fig. 5 (Hamill and Sakmann, 1981). For a cyclic interconversion of three states:

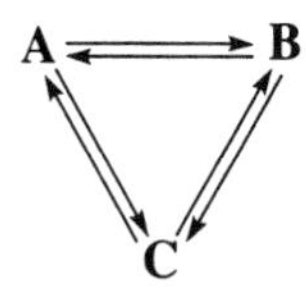

the principle of microscopic reversibility requires that under equilibrium conditions the transition frequencies in both directions be the same ($f_{XY} = f_{YX}$). If, however, transitions

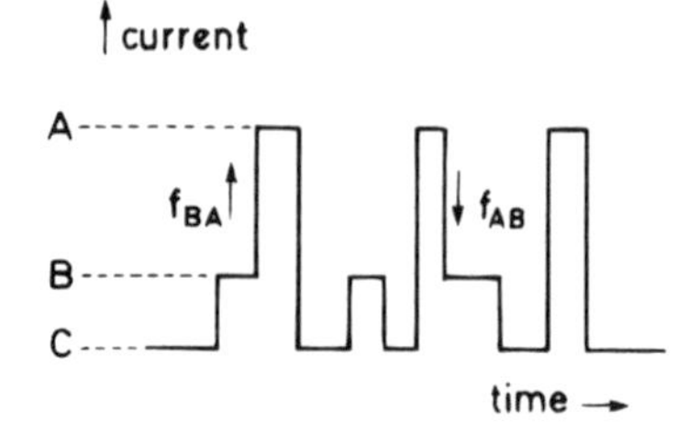

Figure 5. Single-channel record showing transitions among three different conformational states of the channel: f_{AB} and f_{BA} are the observed frequencies of transitions $A \rightarrow B$ and $B \rightarrow A$, respectively.

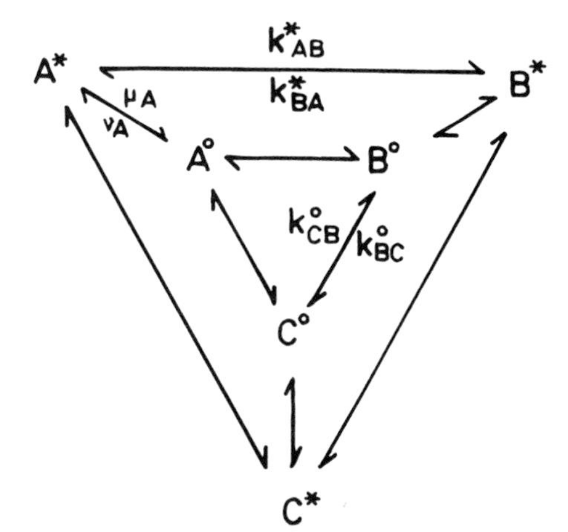

Figure 6. Transitions among three conformational states A, B, and C, of a channel. A^0, B^0, C^0: binding site empty; A^*, B^*, C^*: binding site occupied.

between conformational states are coupled to ion translocation, one may expect that the symmetry relationship $f_{XY} = f_{YX}$ no longer holds when ions are driven through the channel by an external force. Deviations from the symmetry relationship have been observed in studies of acetylcholine-activated channels (Hamill and Sakmann, 1981).

The expectation that f_{XY} and f_{YX} may become unequal in the presence of a driving force for ion flow is borne out by an analysis of the reaction scheme of Fig. 6. In order to simplify the formal treatment, we assume that in states A and C the binding site is always in equilibrium with the left-hand solution (solution′) and in state B with the right-hand solution (solution″). This means that even for nonzero ion flow, the ratio of the probabilities of occupied and empty states is given by the corresponding equilibrium constant:

$$p^*_A / p^0_A = \nu'_A / \mu''_A \equiv Q_A \tag{29}$$

$$p^*_B / p^0_B = \nu''_B / \mu'_B \equiv Q_B \tag{30}$$

$$p^*_C / p^0_C = \nu'_C / \mu''_C \equiv Q_C \tag{31}$$

Microscopic reversibility requires that the following relationships hold:

$$\frac{\nu'_X \mu'_X}{\nu''_X \mu''_X} = \frac{\nu'_X \mu'_Y k^0_{YX} k^*_{XY}}{\nu''_Y \mu''_X k^0_{XY} k^*_{YX}} = \exp[z(u - u_0)] \tag{32}$$

$$k^0_{AB} k^0_{BC} k^0_{CA} = k^0_{BA} k^0_{AC} k^0_{CB} \qquad k^*_{AB} k^*_{BC} k^*_{CA} = k^*_{BA} k^*_{AC} k^*_{CB} \tag{33}$$

The total transition frequency f_{XY} of a single channel results from transitions $X^0 \rightarrow Y^0$ and $X^* \rightarrow Y^*$. Thus,

$$f_{XY} = p^0_X k^0_{XY} + p^*_X k^*_{XY} \tag{34}$$

The asymmetry in the transition frequencies may be expressed by the quantity ρ_{XY}:

$$\rho_{XY} \equiv \frac{f_{XY} - f_{YX}}{f_{XY} + f_{YX}} \tag{35}$$

ρ_{XY} is obtained by calculating the probabilities p^0_X and p^*_X in the stationary state. The result reads:

$$\rho_{AB} = \frac{Q_A}{\epsilon_{AB}}[1 - \exp(zu - zu_0)](k^0_{AB}k^*_{BA}k_{CA} - k^0_{CB}k^*_{BA}k_{AC}) \tag{36}$$

$$\rho_{AC} = \frac{Q_B}{\epsilon_{AC}}[1 - \exp(zu - zu_0)](k^0_{CB}k^*_{BA}k_{AC} - k^0_{AB}k^*_{BC}k_{CA}) \tag{37}$$

$$\epsilon_{AB} \equiv 2k_{AB}k_{BA}(k_{CA} + k_{CB}) + k_{AB}k_{BC}k_{CA} + k_{BA}k_{AC}k_{CB} \tag{38}$$

$$\epsilon_{AC} \equiv 2k_{AC}k_{CA}(k_{BA} + k_{BC}) + k_{AB}k_{BC}k_{CA} + k_{BA}k_{AC}k_{CB} \tag{39}$$

$$k_{XY} \equiv k^0_{XY} + Q_X k^*_{XY} \tag{40}$$

An analogous expression for $\rho_{CB} = -\rho_{BC}$ is obtained from equation 36 by interchanging the subscripts A and C. The result contained in equations 36–40 may be summarized in the following way. The transition frequencies are asymmetric ($\rho_{XY} = 0$) as long as a driving force for ion flow is present. On the other hand, the asymmetry disappears ($\rho_{XY} = 0$) at equilibrium, where $\exp(zu - zu_0)$ becomes equal to unity. This has to be expected, since the asymmetry of transition frequencies is a manifestation of a nonequilibrium distribution of conformational states created by ion flow through the channel.

It may also be shown using equations 32 and 33 that ρ_{AB} and ρ_{AC} vanish when the transition rate constants for empty and occupied binding sites are the same ($k^0_{XY} - k_{XY}$). In general, however, $k^0{}_{XY}$ and k^0_{XY} are different since the presence of a charge in the binding site changes the electrostatic interaction of the channel with the external field.

Another limiting case in which the transition frequencies become symmetrical is given by the condition that conformational transitions can occur only in the empty state of the channel ($k^*_{XY} = k^*_{YX} = 0$) (Marchais and Marty, 1979).

4. Current Noise in Open Channels

Acetylcholine-activated channels have been shown to exhibit random current fluctuations in the open-channel state 19, this volume. A common source of current noise in open channels is given by the statistical nature of ion translocation over barriers (Stevens, 1972; Läuger, 1978; Frehland, 1980). It is clear, however, that this "transport noise," which is frequency independent up to very high frequencies, represents only a minor noise component in the acetylcholine-receptor channel. Evidence for a second noise source comes from the large amplitude and the dispersion in the millisecond range of the current fluctuations. A likely explanation for the predominant component of the observed noise is the assumption that conductance fluctuations are induced by thermal fluctuations of channel structure (C. F. Stevens, personal communication). Indeed, a protein channel that can assume many conformational states will exhibit conductance fluctuations with a frequency spectrum that is determined by the relaxation times of the protein molecule. The spectral intensity $S_I(f)$ of current noise may be expected to be of the form (f is the frequency):

$$S_I(f) = S_I(\infty) + \sum_i \frac{A_i}{1 + (2\pi f \tau_i)^2} \tag{41}$$

The time constants τ_i and the amplitudes A_i are functions of the transition rate constants. Even for the two-state channel with a single binding site discussed above, the expressions for the τ_i and A_i are rather complex (Frehland, 1979). In fact, $S_I(f)$ contains as limiting cases the frequency spectra of ion carriers and of channels with simple open–closed kinetics, which are entirely different in shape (Neher and Stevens, 1977; Kolb and Läuger, 1978).

For a two-state channel with simple open–closed behavior ($k^0_{AB} = k^*_{AB} = 1/\tau_A$, $k^0_{BA} = k^*_{BA} = 1/\tau_B$), the spectral intensity is given by

$$S_I(f) = (I_A - I_B)^2 \frac{4\tau^2}{\tau_A + \tau_B} \cdot \frac{1}{1 + (2\pi f \tau)^2} \tag{42}$$

$$\tau = \frac{\tau_A \tau_B}{\tau_A + \tau_B} \tag{43}$$

where I_A and I_B are the single-channel currents in states A and B, and τ_A and τ_B are the mean lifetimes. If the experimentally accessible frequencies f are much smaller than $1/\tau$, only white noise of intensity $S_I(0)$ can be observed. According to equation 42, $S_I(0)$ vanishes when the transitions become very fast (τ_A, $\tau_B \rightarrow 0$).

When a channel carries out random transitions among many conformational substates, and when the single transitions cause only minor conductance changes, then a pseudocontinuous behavior results ("channel breathing"). The analysis of such open-channel current noise may be expected to yield important information on the dynamics of channel proteins.

5. Conclusion

Recent studies on the dynamics of proteins suggest that ionic channels can assume a large number of conformational states. Although many of these substates will have lifetimes too short to be detected in single-channel current records, they nevertheless may influence the observable properties of the channel such as concentration or voltage dependence of conductance. Of particular interest is the possibility that conformational transitions of the channel protein are coupled to ion translocation between binding sites. Such coupling occurs when conformational transitions can take place both in the empty and in the occupied state of a binding site. Coupling between conformational transitions and ion translocation may lead to a nonmonotonic concentration dependence of conductance and may result in carrier-like behavior of the channel in which the rate of ion flow is limited by the rate of conformational transitions. Furthermore, ion flow through the channel driven by an external force may create a nonequilibrium distribution of conformational states, resulting in observable asymmetries in the transition frequencies among the states.

References

Austin, R. M., Beeson, K. W., Eisenstein, L., Frauenfelder, H., and Gunsalus, I. C., 1975, Dynamics of ligand binding to myoglobin, *Biochemistry* **14:**5355–5373.

Frauenfelder, H., Petsko, G. A., and Tsernoglu, D., 1979, Temperature-dependent X-ray diffraction as a probe of protein structural dynamics, *Nature* **280:**558–563.

Frehland, E., 1979, Theory of transport noise in membrane channels with open–closed kinetics, *Biophys. Struct. Mechanism* **5:**91–106.

Frehland, E., 1980, Nonequilibrium ion transport through pores. The influence of barrier structures on current fluctuations, transient phenomena and admittance, *Biophys. Struct. Mechanism* **7:**1–16.

Hamill, O. P., and Sakmann, B., 1981, Multiple conductance states of single acetylcholine receptor channels in embryonic muscle cells, *Nature* **294:**462–464.

Hille, B., 1971, The permeability of the sodium channel to organic cations in myelinated nerve, *J. Gen. Physiol.* **58:**599–619.

Huber, R., Deisenhofer, J., Colman, P. M., Matshushima, M., and Palm, W., 1976, Crystallographic structure studies of an IgG molecule and an Fc fragment, *Nature* **264:** 415–420.

Karplus, M., 1982, Dynamics of proteins, *Ber. Bunsenges. Phys. Chem.* **86:**386–395.

Kolb, H.-A., and Läuger, P., 1978, Spectral analysis of current noise generated by carrier-mediated ion transport, *J. Membr. Biol.* **41:**167–187.

Lakowicz, J. R., and Weber, G., 1980, Nanosecond segmental mobilities of tryptophan residues in proteins observed by lifetime-resolved fluorescence anisotropy, *Biophys. J.* **32:**591–600.

Läuger, P., 1978, Transport noise in membranes. Current and voltage fluctuations at equilibrium, *Biochim. Biophys. Acta* **507:**337–349.

Läuger, P., 1980, Kinetic properties of ion carriers and channels, *J. Membr. Biol.* **57:**63–178.

Läuger, P., Stephan, W., and Frehland, E., 1980, Fluctuations of barrier structure in ionic channels, *Biochim. Biophys. Acta* **602:**167–180.

Marchais, D., and Marty, A., 1979, Interaction of permanent ions with channels activated by acetylcholine in *Aplysia* neurones, *J. Physiol.* **297:**9–45.

Neher, E., and Stevens, C. F., 1977, Conductance fluctuations and ionic pores in membranes, *Annu. Rev. Biophys. Bioeng.* **6:**345–381.

Parak, F., Frolov, E. N., Mössbauer, R. L., and Goldanskii, V. I., 1981, Dynamics of metmyoglobin crystals investigated by nuclear gamma resonance absorption, *J. Mol. Biol.* **145:**825–833.

Stevens, C. F., 1972, Inferences about membrane properties from electrical noise measurements, *Biophys. J.* **12:**1028–1047.

Thompson, S. H., and Aldrich, R. W., 1980, Membrane potassium channels, in: *The Cell Surface and Neuronal Function: Cell Surfaces Reviews,* Vol. 6 (C. W. Cotman, G. Poste, and G. L. Nicolson, eds.), pp. 49–85, North-Holland, Amsterdam.

Appendix

Improved Patch-Clamp Techniques for High-Resolution Current Recording from Cells and Cell-Free Membrane Patches

O. P. Hamill, A. Marty, E. Neher, B. Sakmann, and F. J. Sigworth

Max-Planck-Institut für biophysikalische Chemie, Postfach 968, Am Fassberg, D-3400 Göttingen, Federal Republic of Germany

Abstract. 1. The extracellular patch clamp method, which first allowed the detection of single channel currents in biological membranes, has been further refined to enable higher current resolution, direct membrane patch potential control, and physical isolation of membrane patches.

2. A description of a convenient method for the fabrication of patch recording pipettes is given together with procedures followed to achieve giga-seals i.e. pipette-membrane seals with resistances of $10^9 - 10^{11}$ Ω.

3. The basic patch clamp recording circuit, and designs for improved frequency response are described along with the present limitations in recording the currents from single channels.

4. Procedures for preparation and recording from three representative cell types are given. Some properties of single acetylcholine-activated channels in muscle membrane are described to illustrate the improved current and time resolution achieved with giga-seals.

5. A description is given of the various ways that patches of membrane can be physically isolated from cells. This isolation enables the recording of single channel currents with well-defined solutions on both sides of the membrane. Two types of isolated cell-free patch configurations can be formed: an inside-out patch with its cytoplasmic membrane face exposed to the bath solution, and an outside-out patch with its extracellular membrane face exposed to the bath solution.

6. The application of the method for the recording of ionic currents and internal dialysis of small cells is considered. Single channel resolution can be achieved when recording from whole cells, if the cell diameter is small (< 20 μm).

7. The wide range of cell types amenable to giga-seal formation is discussed.

Key words: Voltage-clamp – Membrane currents – Single channel recording – Ionic channels

Introduction

The extracellular patch clamp technique has allowed, for the first time, the currents in single ionic channels to be observed (Neher and Sakmann 1976). In this technique a small heat-polished glass pipette is pressed against the cell membrane, forming an electrical seal with a resistance of the order of 50 MΩ (Neher et al. 1978). The high resistance of this seal ensures that most of the currents originating in a small patch of membrane flow into the pipette, and from there into current-measurement circuitry. The resistance of the seal is important also because it determines the level of background noise in the recordings.

Recently it was observed that tight pipette-membrane seals, with resistances of 10 – 100 GΩ, can be obtained when precautions are taken to keep the pipette surface clean, and when suction is applied to the pipette interior (Neher 1981). We will call these seals "giga-seals" to distinguish them from the conventional, megaohm seals. The high resistance of a "giga-seal" reduces the background noise of the recording by an order of magnitude, and allows a patch of membrane to be voltage-clamped without the use of microelectrodes (Sigworth and Neher 1980).

Giga-seals are also mechanically stable. Following withdrawal from the cell membrane a membrane vesicle forms occluding the pipette tip (Hamill and Sakmann 1981; Neher 1981). The vesicle can be partly disrupted without destroying the giga-seal, leaving a cell-free membrane patch that spans the opening of the pipette tip. This allows single channel current recordings from isolated membrane patches in defined media, as well as solution changes during the measurements (Horn and Patlak 1980; Hamill and Sakmann 1981). Alternatively, after giga-seal formation, the membrane patch can be disrupted keeping the pipette cell-attached. This provides a direct low resistance access to the cell interior which allows potential recording and voltage clamping of small cells.

These improvements of the patch clamp technique make it applicable to a wide variety of electrophysiological problems. We have obtained giga-seals on nearly every cell type we have tried. It should be noted, however, that enzymatic treatment of the cell surface is required in many cases, either as part of the plating procedure for cultured cells, or as part of the preparation of single cells from adult tissues.

In this paper we describe the special equipment, the fabrication of pipettes, and the various cell-attached and cell-free recording configurations we have used. To illustrate the capabilities of the techniques we show recordings of AChR-channel currents in frog muscle fibres and rat myoballs, as well as Na currents and ACh-induced currents in bovine chromaffin cells.

Part I

Techniques and Preparation

Giga-seals are obtained most easily if particular types of pipettes are used and if certain measures of cleanliness are

Send offprint requests to B. Sakmann at the above address

Reprinted from *Pflügers Arch.* **391:**85–100. Copyright Springer-Verlag 1981.

taken. The improved resolution requires a more careful design of the electronic apparatus for lowest possible background noise. These experimental details will be described in this section.

1. Pipette Fabrication and Mechanical Setup

Pipette Fabrication.* Patch pipettes are made in a three-stage process: pulling a pipette, coating of its shank with Sylgard, and the final heat polishing of the pipette tip.

First step-pulling: Patch pipettes can be pulled from flint glass or borosilicate glass. Flint glass has a lower melting point, is easier to handle, and forms more stable seals than borosilicate glass, which however has better electrical properties (see below). We routinely use commercially available flint capillaries made for hemocytometric purposes (Cee-Bee hemostat capillaries), or melting point determination capillaries. The borosilicate (Pyrex) glass is in the form of standard microelectrode capillaries (Jencons, H 15/10). The pipettes are pulled in two stages using a vertical microelectrode puller (David Kopf Instruments, Tujunga, CA, USA, Model 700C) and standard Nichrome heating coils supplied with it. In the first (pre-)pull the capillary is thinned over a length of 7 – 10 mm to obtain a minimum diameter of 200 μm. The capillary is then recentered with respect to the heating coil and in the second pull the thinned part breaks, producing two pipettes. To obtain large numbers of pipettes of similar properties it is advisable to use a fixed pulling length and fixed settings for the two stages. For example with Cee-Bee capillaries and the David Kopf puller we use the following settings. The prepull is made at 19 A with a pulling length of 8 mm. The thinned part of the capillary is then recentered by a shift of approximately 5.5 mm. The final pull is made at a critical heat setting around 12 A. Slight variations of the heat setting around this value produce tip openings between fractions of a μm and several μm. We aim at openings between 1 and 2 μm. These pipettes, then, have steep tapers at the very tip (see for example Fig. 10C). The Pyrex capillaries require higher heat settings of 24 and 15 A for the two stages; the resulting pipettes have thicker walls at the tip, and often the tips break unevenly in pulling.

Second step-coating: In order to reduce the pipette-bath capacitance and to form a hydrophobic surface, pipette shanks are coated with Sylgard to within about 50 μm from the tip. Already-mixed Sylgard can be stored for several weeks at $-20°$ C. It is applied to the pipette using a small glass hook taking care that the very tip remains uncoated. We apply the Sylgard while the pipette is mounted in a microforge and cure it by bringing the heated filament close to the pipette for a few seconds. The Sylgard coating is not required for giga-seal formation; it only serves to improve background noise.

Third step – heat polishing: Polishing of the glass wall at the pipette tip is done on a microforge shortly after Sylgard coating. We observe this step at 16 × 35 magnification using a compound microscope with a long-distance objective. The heat is supplied by a V-shaped platinum-iridium filament bearing a glass ball of ≃0.5 mm diameter. The filament is heated to a dull red glow and a stream of air is directed towards the glass ball, restricting the heat to its immediate vicinity. The tip of the pipette is brought to within 10 – 20 μm of the ball for a few seconds; darkening of the tip walls indicates polishing of the tip rim. If the pipettes are coated with Sylgard, it is preferable to heat-polish them within an

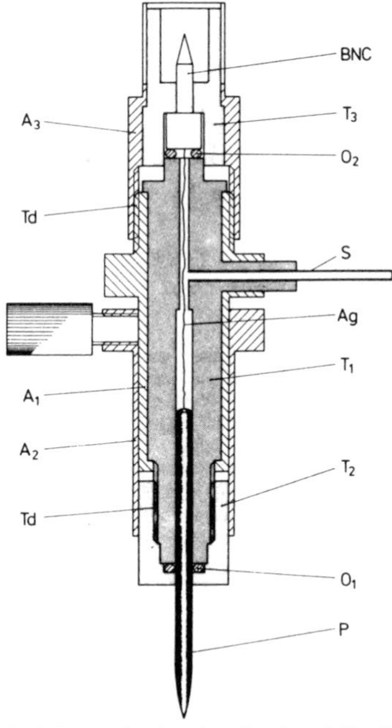

Fig. 1. Cross section through suction pipette holder. The holder serves two basic functions, firstly to provide electrical connection between the patch pipette solution and the pin of a BNC connector, and secondly to allow suction or pressure to be applied to the pipette interior. The holder has a Teflon body T_1 with a central bore for tight fitting of a patch pipette P and a chlorided silver wire Ag which is soldered to the pin of a BNC connector. The BNC pin is held by Teflon piece T_3. The pipette is tightened by a screw cap T_2. Outlet S connects to Silicone rubber tubing for application of suction or pressure to the inner compartment, which is made airtight by the O-rings O_1 and O_2. A_1 and A_3 are aluminium shields to the body; A_2 is a sliding shield to the pipette. Td indicates screw threads. The unit (without pipette) is 55 mm long

hour after coating; after this time, it is difficult to obtain a steep taper at the pipette tip. When pipettes have to be stored more than a few hours they should be cleaned before use by immersion in methanol while a positive pressure is applied to their interior.

Sylgard-coated patch pipettes usually do not fill by capillary forces when their tip is immersed into solution. They can be filled quickly by first sucking in a small amount of pipette solution and then back-filling. All the solutions used for filling should be filtered using effective pore sizes smaller than 0.5 μm. We use pipettes with resistance values in the range 2 – 5 MΩ. These have opening diameters between 0.5 and 1 μm.

Mechanical Setup. The patch pipettes are mounted on a suction pipette holder shown schematically in Fig. 1. It consists of inner parts made of Dynal or Teflon T_1, T_2, T_3) and is shielded by metal caps (A_1, A_2, A_3). The outlet S is connected to silicone rubber tubing through which suction is applied, usually by mouth. It is critical that the O-rings, O_1 and O_2 fit tightly. Otherwise the pipette tip can move slightly

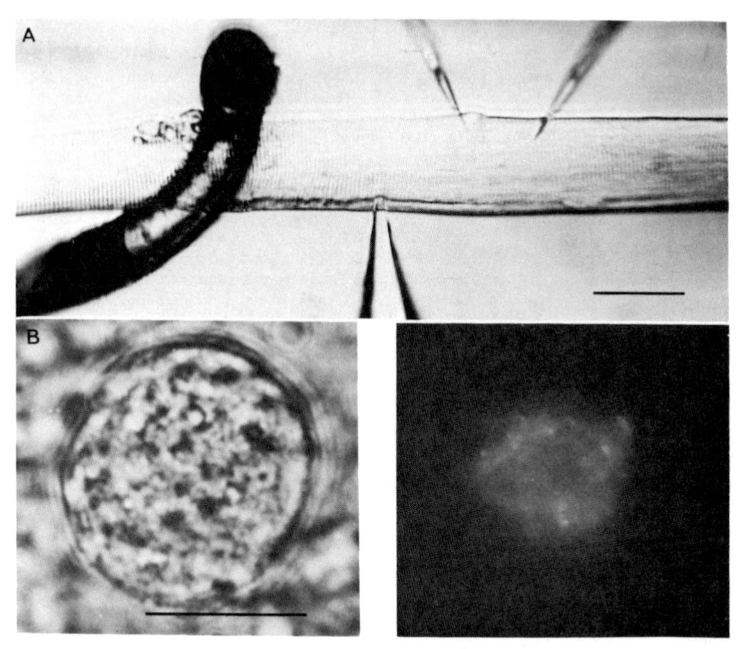

Fig. 2 A and B. Single cell preparations used for demonstration of improved patch clamp techniques. (**A**) Enzyme treated frog (*R. temporaria*) cutaneous pectoris muscle fibre. The end-plate region of this fibre is viewed by Normarski optics. The fibre is supported by a glass hook. The fibre is stripped of its nerve terminal. The patch pipette is seen in contact with the synaptic trough. Two intracellular glass microelectrodes are used here to voltage clamp the fibre locally. Alternatively, the measurement can be performed at the natural resting potential without intracellular electrodes. (**B**) Primary culture of rat myoball. The same myoball is viewed in bright field optics on the left side and, using fluorescence microscopy, on the right side after labelling with fluorescent Rhodamine-conjugated α-BuTX. The fluorescence pattern illustrates the "patchy"/distribution of AChR's in this preparation. Calibration bars: 50 μm (upper), and 25 μm (lower)

during suction, tearing off a membrane patch from the cell. The pipette holder connects to a BNC connector of the amplifier head stage which is mounted on a Narashige MO-103 hydraulic micromanipulator. This, in turn, is mounted onto another manipulator for coarse movements (Narishige MM 33). The pipette holder should be repeatedly cleaned by methanol and a jet of nitrogen.

2. Preparations

The development of giga-seals requires a "clean" plasma membrane; that is, no sign of a surface coat should be detectable in conventionally-stained EM-sections. This requirement is met in many tissue-cultured cells, for example myotubes, spinal cord cells and dorsal root ganglion cells. In adult tissue however individual cells are covered with surface coats and enzymatic cleaning of the cell surface must precede the experiment. The exact protocol of enzymatic cleaning varies from tissue to tissue (see Neher 1981). Here we describe a treatment procedure adequate for frog skeletal muscle fibres. We also briefly describe the preparation of rat myoballs. These cells, as well as the chromaffin cells, require no enzyme treatment before use.

a) End-Plate Region of Frog Muscle Fibres. From innervated muscle a useable preparation can be obtained within 2 – 3 h using the following procedure. The whole cutaneous pectoris muscle is bathed for an hour at room temperature in normal frog Ringer solution containing 1 mg/ml collagenase (Sigma type I). At this point overlying fibre layers can be easily cut away, such that a monolayer of fibres remains. The muscle endplate region is subsequently superfused with Ringer solution containing 0.07 mg/ml Protease (Sigma, type VII) for 20 – 40 min. The tendinous insertions of the muscle fibres are protected by small 3 – 7 mm guides made from Perspex to restrict the flow of protease containing solution to the endplate region of the muscle (Neher et al. 1978). This procedure results in a preparation of ≈ 20 fibres with ends firmly attached to skin and sternum. When a single fibre is viewed the bare synaptic trough can be easily seen with a × 16 objective (Zeiss 0.32) and × 16 eyepieces using Nomarski interference contrast optics (Fig. 2A). Although currents can be recorded from the synaptic area, the perisynaptic AChR density within 10 – 50 μm of the synaptic trough is high enough in most preparations to allow recording of ACh activated single channel currents at low ACh concentrations (< 1 μM). Preparations kept in phosphate-buffered Ringer solution remain viable and can be used for up to 48 h when kept at < 10° C. All bath solutions contain 10^{-8} M Tetrodotoxin to avoid muscle contraction during the dissection.

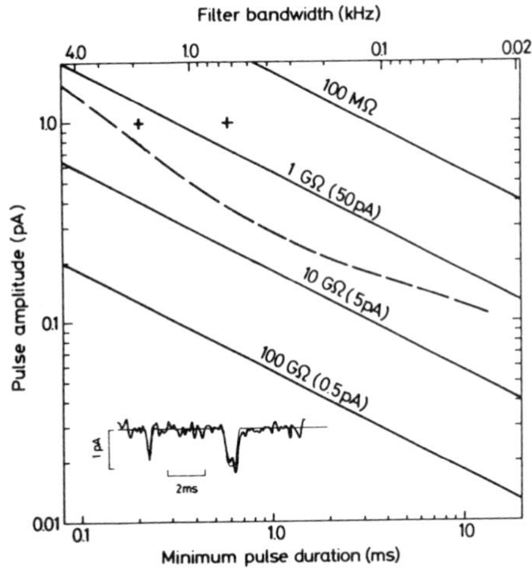

Fig. 3. Limits of pulse detection due to background noise. The relationship between filter settings (*top scale*) or minimum detectable rectangular pulse durations (bottom scale) and the pulse amplitudes (*vertical scale*) is shown for various background noise levels. The solid lines represent theoretical limits imposed by Johnson noise and shot noise sources and are discussed below. The dashed curve represents the background noise in an actual recording situation (50 GΩ seal on a myoball) and was computed from the spectrum in Fig. 5. Using this curve and the top scale, an appropriate low-pass filter setting can be found for observing single-channel currents whose amplitude is given by the vertical scale. The filter bandwidths (−3dB frequency, Gaussian or Bessel response) are chosen to make the standard deviation of the background noise $^1/_8$ of the given amplitude. When the bottom scale is used the curve shows whether or not a current pulse of given amplitude and duration can be recognized in the presence of the background noise. Each combination of amplitude and duration corresponds to a point in the figure. All points lying above the curve represent pulses that can be detected reliably. For example, 1 pA channel currents can be detected if the duration is at least 0.15 ms; pulses of 0.2 pA amplitude can be detected only if the duration is greater than 2.2 ms. Comparison of the top and bottom scales in the figure then gives the filter setting yielding the best signal-to-noise ratio (peak signal to rms noise) for a pulse of the given minimum duration. However, at this filter setting, the time-course of such a minimum-width pulse is distorted. This is illustrated in the *inset*, which shows the simulated response of our recording system at 2 kHz to two rectangular pulses, both with original amplitudes of 1 pA but with durations of 0.2 and 0.6 ms. The thin trace shows the response in the absence of noise; in the thicker trace a recording of background noise (same data as in Fig. 5) has been added. The parameters of the two pulses are indicated as crosses in the figure. The shorter pulse represents the movement of 1250 elementary charges; it is clearly detectable, but the *a. posteriori*, simultaneous estimation of pulse amplitude and duration is impossible. The rectangular form of the longer pulse is just recognizable, however, allowing these parameters to be estimated. It can be seen from this example that for kinetic analysis of channel gating, the shortest mean event duration should be considerably longer (preferably by an order of magnitude or more) than the minimum given in the figure. The solid lines show the ultimate theoretical detection limits imposed by Johnson noise in the seal and membrane resistance, computed from Eq. (1). Since seal resistances above 100 GΩ are often observed, it should in principle be possible to resolve much smaller pulses than is presently possible. These lines also represent the resolution limits that would be imposed by shot noise (Eq. 2) in channels carrying the indicated current levels in the case that shot noise is the predominant noise source. For the calculation of the detection limits in this figure a filter with Gaussian frequency response was assumed; the filter bandwidth was chosen to give a risetime t_r equal to 0.9 times the minimum pulse duration. The minimum detectable pulse amplitude was taken to be 8 times the standard deviation σ of the background noise. A minimum-width pulse is attenuated to 6.7 σ by the filter; with a detection threshold of 4.7 σ the probability of missing an event is less than 0.02, while the probability per unit time of a background fluctuation being mistaken for a pulse is less than $3 \times 10^{-6}/t_r$

b) Myoballs from Embryonic Rat Muscle. The procedure to obtain spherical "myoballs" is essentially the same as that used by other laboratories (Horn and Brodwick 1980). The growth medium (DMEM + 10% fetal calf serum) is changed on day 3 and on day 6 after plating of the cells on 18 mm cover slips placed into culture dishes. For 2 days starting on day 8, medium containing 10^{-7} M colchicine is used. Thereafter normal growth medium is used again and changed every third day. This procedure results in ≃ 100 spherical myoballs of 30–80 μm diameter (Fig. 2B) per cover slip. A single cover slip can be cracked with a scapel blade into 8–10 small pieces which can be transferred individually into the experimental chamber. The culture medium is exchanged for normal bath solution before the experiment. This solution has the following composition (in mM): 150 NaCl, 3 KCl, 1 $MgCl_2$, 1 $CaCl_2$, 10 HEPES, pH adjusted to 7.2 by NaOH.

As visualized with fluorescent α-bungarotoxin, the ACh receptors are unevenly distributed in myoballs (Fig. 2B); however in virtually all patches single channel currents could be recorded with low ACh concentrations (< 1 μM). For experiments with ACh-activated channels it is advisable to work within 2–5 days following colchicine treatment. During a later period 5–10 days following colchicine treatment myoballs are a suitable preparation for investigating properties of electrically excitable Na and K channels as well as Ca-dependent K channels.

c) Chromaffin Cells. As an example of cells obtained by enzymatic dispersion* of an adult organ we use bovine chromaffin cells. These cells are dispersed by perfusion with collagenase of the adrenal gland (Fenwick et al. 1978), and are subsequently kept in short term culture for up to 8 days (Medium 199, supplemented with 10% fetal calf serum and 1 mg/ml BSA).

3. Background Noise and Design of Recording Electronics†

One of the main advantages of the giga-seal recording technique is the improvement, by roughly an order of magnitude, in the resolution of current recordings. The resolution is limited by background noise from the membrane, pipette and recording electronics.

Theoretical Limits. Apart from noise sources in the instrumentation there are inherent limits on the resolution of the patch clamp due to the conductances of the patch membrane and the seal. One noise source is the Johnson noise of the membrane-seal combination, which has a one-sided current spectral density

$$S_I(f) = 4\,kT\,Re\,\{Y(f)\} \qquad (1)$$

where $4\,kT = 1.6 \times 10^{-20}$ Joule at room temperature, and $Re\,\{Y(f)\}$ is the real part of the admittance, which depends, in general, on the frequency f. If the membrane-seal parallel combination is modelled as a simple parallel $R-C$ circuit, then $Re\,\{Y(f)\} = 1/R$. Integrating the resulting (in this case constant) spectral density over the frequency range of interest gives the noise variance, which decreases with increasing patch resistance. From the variance we have calculated the size of the smallest detectable current pulses, which is plotted in Fig. 3 against the minimum pulse durations for various values of R.

Another background noise source is the "shot noise" expected from ions crossing the membrane, for example through leakage channels or pumps. Although the size and spectrum of this noise depends on details in the ion translocation process, a rough estimate of the spectral density can be made assuming that an ion crosses the membrane rapidly (Stevens 1972; Läuger 1975),

$$S_I = 2\,Iq \tag{2}$$

where q is the effective charge of the current carrier (we assume a unit charge $q_e = 1.6 \times 10^{-19}$ Coulomb) and I is the unidirectional current. The shot noise at $I = 0.5$ pA is nearly the same as the Johnson noise with $R = 100\,\mathrm{G\Omega}$. If R is determined mainly by "leakage channels" in the membrane patch, the shot noise may be comparable to the Johnson noise in size.

Intrinsic Noise in the Pipette. As can be seen in Fig. 3, the background noise in our present recording system is several times larger than the limit imposed by the patch resistance. The excess results from roughly equal contributions from noise sources in the pipette and sources in the current-to-voltage converter. We are aware of three main sources of Johnson noise in the pipette, each of which can be roughly modelled by a series $R-C$ circuit. The current noise spectral density in such a circuit is given by (1) with

$$Re\,\{Y\,(f)\} = \frac{\alpha^2}{R(I+\alpha^2)}, \tag{3}$$

where $\alpha = 2\,\pi f RC$. In the high frequency limit (α large) this approaches $1/R$; in the low frequency limit $Re\,\{Y\} = (2\,\pi f C)^2 R$, which increases with frequency.

The potentially most serious noise source arises from a thin film of solution that creeps up the outer wall of an uncoated pipette. Evidence for the presence of this film is that, when a small voltage step is applied to the pipette, a slow capacitive transient is observed whose size and time constant are influenced by air currents near the pipette. The film apparently has a distributed resistance R of the order of 100 MΩ, and a distributed wall capacitance $C \approx 3$ pF. In the high frequency regime the noise (like that in a 100 MΩ resistor) is very large. A Sylgard coating applied to the pipette reduces the noise considerably: the hydrophobic surface prevents the formation of a film, and the thickness of the coating reduces C.

Secondly, we find that the bulk conductivity of the pipette glass can be significant. The Cee-Bee capillaries, for example, show substantial conductivity above 100 Hz, as evidenced by capacitance transients and noise spectra from pipettes with closed tips. Coating the pipette helps, but even in a Sylgard-coated pipette the effective values of R and C are roughly 2 GΩ and 2 pF. Pyrex electrode glass (Jencons H 15/10) has at least an order of magnitude lower conductivity. However, it is more difficult to make pipettes with this hard glass because of its higher melting point.

Finally, the pipette access resistance R_{acc} (in the range 2 – 5 MΩ) and the capacitance of the tip of the pipette C_{tip} (of the order of 0.3 pF) constitute a noise source. Since the time constant is short, the low-frequency limit of (3) holds. The resulting spectral density increases as f^2, becoming comparable to the 1 GΩ noise level around 10 kHz. This noise could be reduced in pipettes having steeper tapers near the tip, reducing R_{acc}, or having the coating extend closer to the tip, reducing C_{tip}.

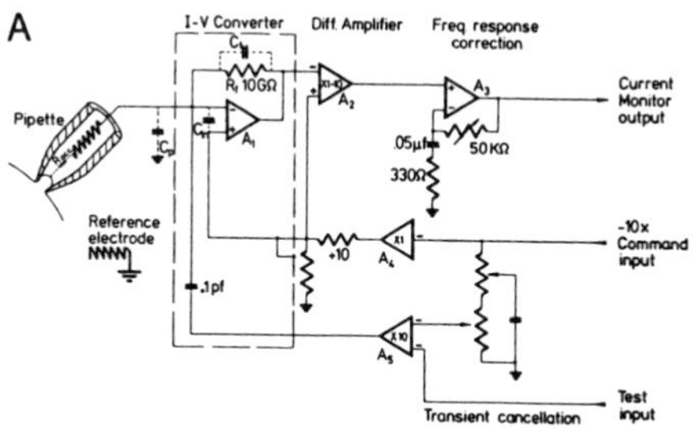

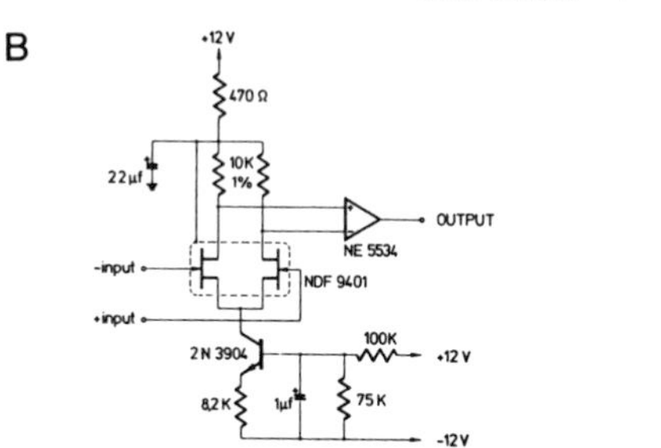

Fig. 4A and B. Circuit diagrams. **(A)** A simplified diagram of the recording system. The current-to-voltage converter is mounted on a micromanipulator, and the pipette holder (Fig. 1) plugs directly into it. Important stray capacitances (indicated by dotted lines) are the feedback capacitance $C_f \approx 0.1$ pf and the total pipette and holder capacitance $C_p = 4-7$ pF. C_{in} represents the input capacitance of amplifier A_1, which is either a Burr Brown 3523J or the circuit shown in **B**. With the values shown, the frequency response correction circuit compensates for time constants $R_f C_f$ up to 2.5 ms and extends the bandwidth to 10 kHz. The transient-cancellation amplifier A_5 sums two filtered signals with time constants variable in the ranges 0.5 – 10 µs and 0.1 – 5 ms; only one filter network is shown here. The test input allows the transient response of the system to be tested: a triangle wave applied to this input should result in a square wave at the output. Amplifiers A_2, A_4 and A_5 are operational amplifiers with associated resistor networks. The op amps for A_2 and A_4 (NE 5534, Signetics or LF 356, National Semiconductor) are chosen for low voltage noise, especially above 1 kHz; more critical for A_3 and A_5 (LF 357 and LF 356) are slew rate and bandwidth. For potential recording from whole cells (see part IV), a feedback amplifier is introduced between current monitor output and the voltage command input. **(B)** Circuit of a low-noise operational amplifier for the $I-V$ converter with a selected NDF 9401, dual FET (National Semiconductor) and the following approximate parameters: Input bias current, 0.3 pA; input capacitance, 8 pF; voltage noise density at 3 kHz, 5×10^{-17} V²/Hz; and gain-bandwidth product 20 MHz. The corresponding values for the 3523J are 0.01 pA, 4 pF, 4×10^{-16} V²/Hz and 0.6 MHz. The lower voltage noise of this amplifier is apparent in the $I-V$ converter's background noise above 500 Hz. The high gain-bandwidth product of the amplifier results in a loop bandwidth of ≈ 300 kHz in the $I-V$ converter, so that the frequency response in the 5 – 10 kHz region is negligibly affected by changes in C_p. The loop bandwidth with the 3523 is about 5 kHz

Noise in the Current-to-Voltage Converter. Figure 4A shows a simplified diagram of the recording electronics. The pipette current is measured as the voltage drop across the high-valued

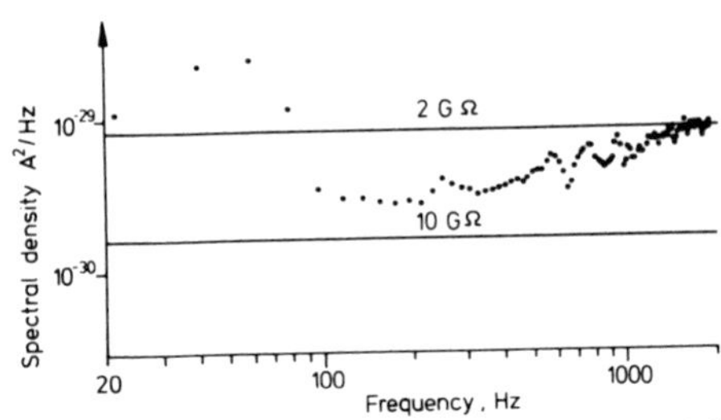

Fig. 5. Power spectrum of the total background noise from a rat myoball membrane patch at resting potential (dots). The amplifier of Fig. 4B was used, with a coated, hard-glass pipette. The patch resistance was 50 GΩ. Lines indicate the lower limit of the noise imposed by the 10 GΩ feedback resistor, and for comparison, the Johnson noise in a 2 GΩ resistor

resistor R_f; the Johnson noise in this resistor is the predominant noise source in the $I-V$ converter below a few hundred Hertz. The substantial shunt capacitance $C_f \approx 0.1$ pF across this resistor affects the frequency response of the $I-V$ converter but makes no contribution to $Re\{Y\}$, and therefore to the noise current, assuming that it is a pure capacitance. This assumption appears to hold for the colloidfilm resistors (Type CX65, Electronic GmbH, Unterhaching/Munich, FRG) we use, since, after correction of the frequency response, we found that the $I-V$ converter's noise spectrum was unchanged when we substituted a home-made tin-oxide resistor having $C_f < 0.01$ pF for the commercial resistor. Other resistor types, including the conductive-glass chip resistors we have previously used (Neher et al. 1978) and colloid-film resistors with higher values do not show the transient response characteristic of a simple $R-C$ combination and therefore probably have a frequency-dependence of $Re\{Y\}$. Correcting the frequency response of these resistors is also more complicated; this is the primary reason why we have not yet used values for R_f above 10 GΩ, even though this might improve the low-frequency noise level.

The other main noise source in the $I-V$ converter is the operational amplifier itself. With both of the amplifiers we use (Burr Brown 3523J, and the circuit of Fig. 4B) the low-frequency (4 – 100 Hz) spectral density is essentially equal to the value expected from R_f, suggesting that the amplifier current noise is negligible. At higher frequencies however the amplifier voltage noise becomes the dominant noise source. This voltage noise is imposed by the feedback loop on the pipette and the input of the amplifier, causing a fluctuating current to flow through R_f to charge C_p and C_{in} (see Fig. 4). The resulting contribution to the current fluctuations has the spectrum

$$S_I(f) = [2\pi f(C_p + C_{in})]^2 \, S_{V(A)}(f), \tag{4}$$

where $S_{V(A)}$ is the amplifier voltage noise spectral density. The f^2-dependence dominates over the constant or $1/f$ behavior of $S_{V(A)}$ giving an increase of S_I with frequency. This noise source can be reduced by minimizing C_p by using a low solution level and avoiding unnecessary shielding of the pipette and holder. It can also be reduced by choosing an operational amplifier having low values for C_{in} and $S_{V(A)}$; the amplifier of Fig. 4B was designed for these criteria.

Figure 5 shows the spectrum of the background noise during an actual experiment. Because the noise variance is the integral of the spectrum the high-frequency components have much greater importance than is suggested by this logarithmic plot. Below 90 Hz the excess fluctuations mainly come from 50 Hz pickup. In the range 100 – 500 Hz the spectral density is near the level set by the 10 GΩ feedback resistor. Above 500 Hz, electrode noise sources and the amplifier voltage noise contribute about equally to the rising spectral density.

Capacitance Transient Cancellation. For studying voltage-activated channels voltage jumps can be applied to the pipette. However, a step change in the pipette potential can result in a very large capacitive charging current. For example, charging 5 pF of capacitance to 100 mV in 5 µs requires 100 nA of current, which is 4 – 5 orders of magnitude larger than typical single channel currents. We use three strategies to reduce this transient to manageable sizes. First, we round the command signal (e.g. with a single time constant of 20 µs) to reduce the peak current in the transient. Second, we try to reduce the capacitance to be charged as much as possible. Metal surfaces near the pipette and holder (excepting the ground electrode in the bath!) are driven with the command signal; this includes the microscope and stage, and the enclosure for the $I-V$ converter. (Alternatively, an inverted command signal could be applied only to the bath electrode.) This measure reduces the capacitance to be charged to 1 – 2 pF when coated pipettes and low solution levels are used. Notice, however, that while the capacitance to be charged by an imposed voltage change is reduced, C_p is unchanged for the purpose of the noise calculation (Eq. 4).

Third, we use a transient cancellation circuit which injects the proper amount of charge directly into the pipette, so that the $I-V$ converter is required to supply only a small error current during the voltage step. The charge is injected through a small, air-dielectric capacitor (see Fig. 4A) which is driven with an amplified and shaped version of the command voltage. The same capacitor can be used to inject currents for test purposes. With these three measures the transient from a 100 mV step can be reduced to below 10 pA (at 2 kHz bandwidth), which is small enough to allow computer subtraction of the remainder.

Part II

Patch Current Recording with Giga-Seals

1. Development of the Giga-Seal

In the past, seal resistances as high as 200 MΩ could be obtained by pressing a pipette tip against a cell membrane and applying suction. This same procedure can also lead to the formation of a seal in the gigaohm range; the only difference is that precautions must be taken to ensure the cleanliness of the pipette tip (Neher 1981). The main precautions are (1) the use of filtered solutions in the bath as well as in the pipette, and (2) using a fresh pipette for each seal. Further precautions are listed below.

The formation of a giga-seal is a sudden, all-or-nothing increase in seal resistance by as much as 3 orders of magnitude. Figure 6 shows the time-course of the development of a 60 GΩ seal in the perisynaptic region of a frog muscle fibre. When the tip of the pipette was pressed against the enzymatically cleaned muscle surface the seal resistance was 150 MΩ. (The resistance was measured by applying a 0.1 mV voltage pulse in the pipette and monitoring the

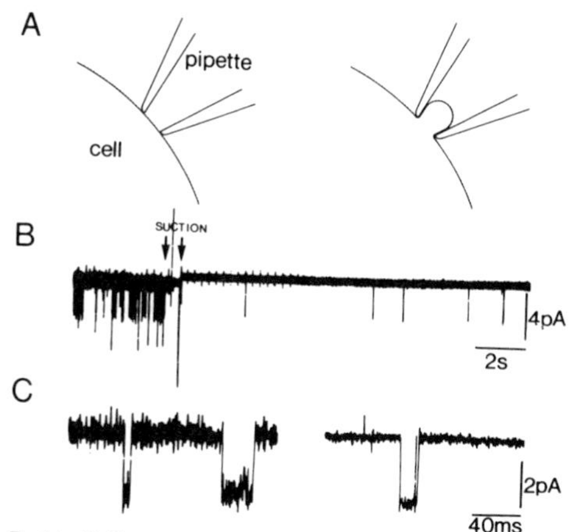

Fig. 6A–C. Giga-seal formation between pipette tip and sarcolemma of frog muscle. (**A**) Schematic diagrams showing a pipette pressed against the cell membrane when the pipette-membrane seal resistance is of the order of 50–100 Megohms (*left*), and after formation of a gigaseal when a small patch of membrane is drawn into the pipette tip (*right*). (**B**) The *upper trace* is a continuous current record before, during and after application of suction. In this experiment a pipette-membrane seal resistance of 150 MΩ was achieved by pressing the pipette against the membrane. Single suberyldicholine-induced channel currents are apparent. During the time indicated by the two arrows slight suction was applied to the pipette interior resulting in the formation of a giga-seal of 60 GΩ resistance. Note reduction in background noise level. The decrease in channel opening frequency presumably resulted from depletion of agonist in the pipette tip during suction. It increased again during the minute following giga-seal formation. The two large current deflections represent artifacts. The *lower traces* show single channel currents at higher resolution before (*left*) and after (*right*) formation of a giga-seal. The single channel current pulse on the right is preceded by capacitive artifacts from a calibration pulse. All records were made at the cell's resting potential of −92 mV and at 11 °C. They were low pass filtered at 1 kHz (*upper trace*) or 3 kHz (*lower traces*)

resulting current flow). When a slight negative pressure of 20–30 cm H_2O was applied (arrows) the resistance increased within a few seconds to 60 GΩ. The development of giga-seals usually occurs within several seconds when a negative pressure is applied; seals always remain intact when the suction is subsequently released. In some cases giga-seals develop spontaneously without suction. In other cases suction has to be applied for periods of 10–20 s, or a seal may develop only after suction has been released.

It was previously suggested that upon suction the membrane at the pipette tip is distored and forms and Ω-shaped protrusion (Neher 1981). This is indeed supported by measurements of patch capacitance after giga-seal formation (Sigworth and Neher 1980). The increase in area of glass-membrane contact, going along with such a distortion, probably explains the gradual, 2–4-fold increase of seal resistance which is usually observed during suction shortly before giga-seal formation, and is also seen in cases when giga-seals do not develop. Giga-seal formation, however, is unlikely to be explained solely by such an area increase. Crude estimates of the thickness of a water layer interposed between membrane and glass give values in the range 20–50 Å for values of the seal resistance between 50 and 200 MΩ. These distances are characteristic for equilibrium separations between hydrophilic surfaces in salt solutions (Parsegian et al. 1979; Nir and Bentz 1978).

A seal resistance larger than 10 GΩ, however, is consistent only with glass-membrane separations of the order of 1 Å, i.e. within the distance of chemical bonds. The abrupt change in distance involved may therefore represent the establishment of direct contact between the surfaces, as occurs during transfer of insoluble surface monolayers on to glass substrates (Langmuir 1938; Petrov et al. 1980).

Also in favor of a tight membrane-glass contact is evidence that small molecules do not diffuse through the seal area. After establishment of a giga-seal the application of high ACh concentrations (in the range of 5–10 μM) outside the pipette does not activate single channel currents in the patch, even though the rest of the cell is depolarized by 20–50 mV.

The high-resistance contact area between glass and membrane also seems to be well delineated. We conclude this from the observation that the so-called "rim-channel" currents, which are quite common when using thick-walled pipette tips, are rarely observed after formation of a giga-seal (see below).

Reproducibility of Giga-Seals. The success rate for the establishment of giga-seals varies for different batches of patch pipettes. This variability probably results from a combination of several factors. The following general rules have been found helpful so far.

1. To avoid dirt on the pipette tip, pipettes should always be moved through the air-water interface with a slight positive pressure (10 cm H_2O). Even the first pipette-cell contact should be made with pipette solution streaming outwards. When the pressure is released while the pipette touches the cell the pipette membrane seal resistance should increase by a factor >2.

2. Each pipette should be used only once after positive pressure has been relieved.

3. Following enzyme treatment of muscle preparations the surface of the bathing solution is frequently covered with debris which readily adheres to the pipette tip, preventing giga-seal formation. The water surface can be cleaned by wiping with lens paper or by aspiration.

4. HEPES-buffered pipette solutions should be used when Ca^{2+} is present in the pipette solution. In phosphate buffer small crystals often form at the pipette tip by precipitation.

5. When slightly (10%) hypoosmolar pipette solutions are used the giga-seals develop more frequently. With these precautions, about 80% of all pipettes will develop giga-seals on healthy preparations. However, even after giga-seal formation, irregular bursts of fast current transients are observed on some patches. We interpret these as artifacts due to membrane damage or leakage through the seal.

2. Improved Current Recording After Giga-Seal Formation

When a pipette is sealed tightly onto a cell it separates the total cell surface membrane into two parts: the area covered by the pipette (the patch area) and the rest of the cell. Current entering the cell in the patch area has to leave it somewhere else. Thus, the equivalent circuit of the whole system consists of two membranes arranged in series. This, and the resulting complications will be discussed in a later section (see also Fig. 10). Here we will focus on the simple case that the total cell membrane area is very large with respect to the patch area.

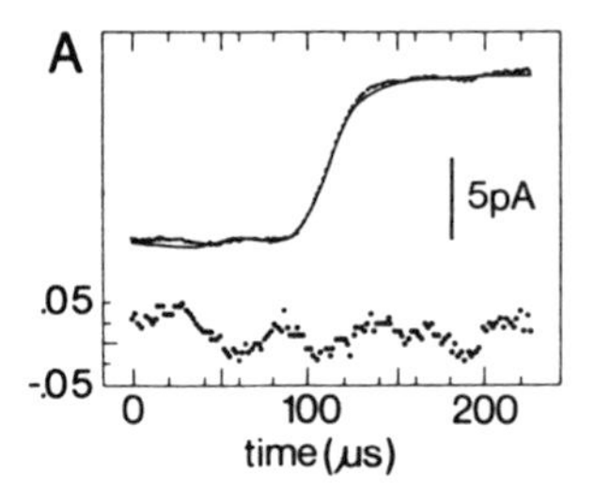

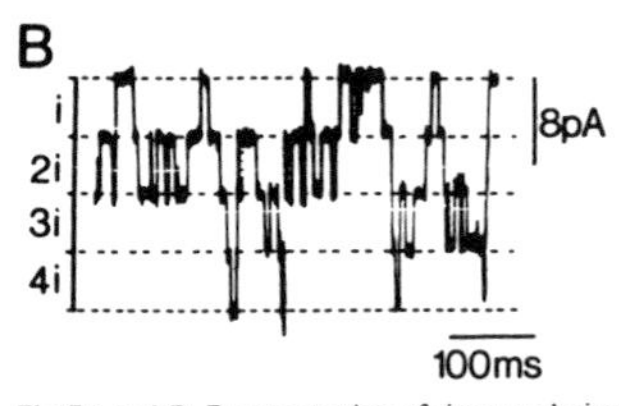

Fig. 7A and B. Demonstration of time resolution and of uniformity in step sizes. (**A**) A digitized record (1.87 μs sample interval) of the opening time course of channels activated by 100 nM SubCh on the perisynaptic region of an adult frog muscle fiber (11° C). The patch was hyperpolarized to approximately − 160 mV and a solution containing 100 mM CsCl was used. The single channel currents were − 10.5 pA in amplitude. Six individual channel opening events were averaged by superposition after alignment with respect to the midpoint of the transition. The same procedure was followed to obtain the instrument's step-response, using records from capacitively-injected current steps. The step response (continuous line) is superimposed on the channel's opening time course after amplitude scaling. The relative difference between the two curves is plotted below. The amplitude of the fluctuations in the difference record is the same during the transition and during the rest of the record. Based on the size of these fluctuations, an upper limit was estimated for the transition time between the closed and open states of the channel by the following procedure: The transfer function of the electronic apparatus was calculated from the known step response by Fourier transform methods. Then, theoretical responses to open-close transitions of various shapes were calculated and compared to the experimental step response. It was found that the predicted deviations between the two curves were significantly larger than the observed ones only when the open-close transitions were spread out in time over 10 μs or more. (**B**) A current record from a myoball under the following conditions: 1 μM ACh; 18° C; − 140 mV holding potential. Individual single channel currents superimpose to form regularly spaced amplitude levels

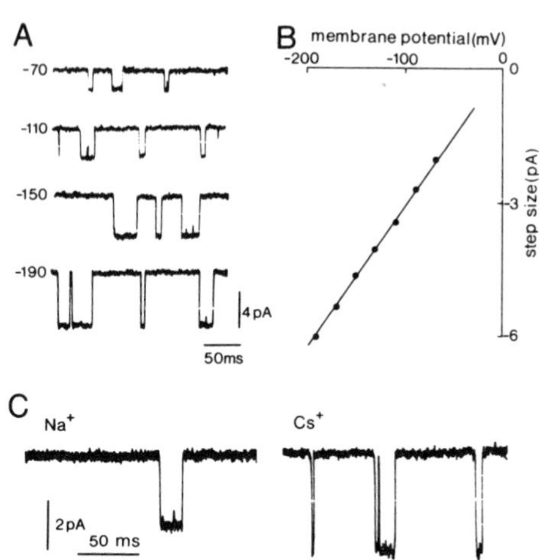

Fig. 8A–C. Control of voltage and of ionic environment in the pipette tip after formation of a giga-seal. (**A**) Single channel current recordings at 50 nM SubCh in the perisynaptic membrane of cutaneous pectoris muscle fibre; 11° C. The membrane potential of this fibre was − 89 mV measured by an intracellular microelectrode. The pipette potential was shifted by different amounts to obtain the membrane potential indicated on the left of each trace (in mV). The pipette was filled with Ringer solution in which NaCl concentration was reduced to 100 mM to improve gigaseal formation. (**B**) Current-voltage relationship of channel currents, derived from the experiment shown in **A**. Each point represents the mean current amplitude of ten individual current events. The straight line is drawn by eye and represents a single channel (chord) conductance of 32 pS. (**C**) Single channel current recordings under the conditions of part A at − 90 mV membrane potential. The main salt in the pipette solution was 100 mM CsCl (*right*) and 100 mM NaCl (*left*). The average single channel current amplitudes were 2.8 pA and 3.8 pA in Na^+ and Cs^+ solutions respectively

Then, the small patch currents will not noticeably alter the cell's resting potential. For instance, a myoball with 50 MΩ input resistance will be polarized less than 0.5 mV by a patch current of 10 pA. Thus, a patch can be considered "voltage clamped" even without the use of intracellular electrodes. The clamp potential is equal to the difference between the cell potential and the potential in the pipette. Some properties of current recordings done under this type of "voltage clamp" are illustrated here.

a) Increased Amplitude and Time Resolution; the Time Course of Channel Opening is Fast. Due to low background noise and an improved electronic circuit the time course of channel opening and closing can be observed at 5 – 10 kHz resolution. Figure 7A illustrates the average opening time-course of suberyldicholine-activated channels at the frog endplate region. The large single channel currents and the low background noise level allowed a much higher time resolution to be obtained than in previous recordings. Still, the rising phase of the conductance (dotted line) is seen to be indistinguishable from the step response of the measuring system (continous line). From a comparison of the two time courses it can be estimated that the actual channel opening occurs within a time interval smaller than 10 μs (see legend, Fig. 7).

b) Lack of Rim-Channel Currents. A major problem of the extracellular patch clamp technique has been the occurrence of currents from channels in the membrane area under the rim of the pipette. These currents are not uniform in size; the resulting skewed step-size histograms complicate the estimation of the single channel current amplitudes (Neher et al. 1978). Current recordings with giga-seals, however, show amplitude distributions that are nearly as narrow as expected from noise in the baseline. The improvement reflects a more sharply delineated seal region. A recording is shown in Fig. 7B which demonstrates the regularity in current amplitudes.

The regularity also allows (i) the unequivocal discrimination between channel types of slightly different conductance, e.g. synaptic and extrasynaptic ACh receptor channels, and (ii) analysis of channel activity when the currents of several channels overlap.

c) Voltage Control of the Membrane Patch. Previously the potential inside the pipette had to be balanced to within less than 1 mV of the bath potential; otherwise large, noisy leakage currents flowed through the seal conductance. The high seal resistance now allows the patch membrane potential to be changed. For example, a 100 mV change in pipette potential will drive only 5 pA of current through a 20 GΩ seal. This leakage is comparable in size to a single channel current and is easily manageable by leakage subtraction procedures. The ability to impose changes in membrane potential has been used to activate Na channels in myoballs (Sigworth and Neher 1980). It also allows measurement of single channel currents in adult muscle fibres over a wide range of potential that is not accessible with the conventional two-microelectrode voltage clamp because of local contractions. Figure 8A and B shows representative traces and the current-voltage relationship of single channel currents recorded from a muscle fibre at various patch membrane potentials ranging from −70 to −190 mV.

d) Control of Extracellular Ion Composition. Giga-seals form a lateral diffusion barrier for ions (see above, p. 91). Here we show that the ionic composition on the external side of the patch membrane is that of the pipette solution. Figure 8C illustrates ACh-activated single channel currents from a frog muscle fibre when the major salt in the pipette was CsCl (100 mM) while the bath contained normal Ringer. The open channel conductance was seen to be 1.3 times larger than the conductance in standard Ringer solution, consistent with the larger estimates of conductance that have been made from fluctuation analysis (Gage and Van Helden 1979).

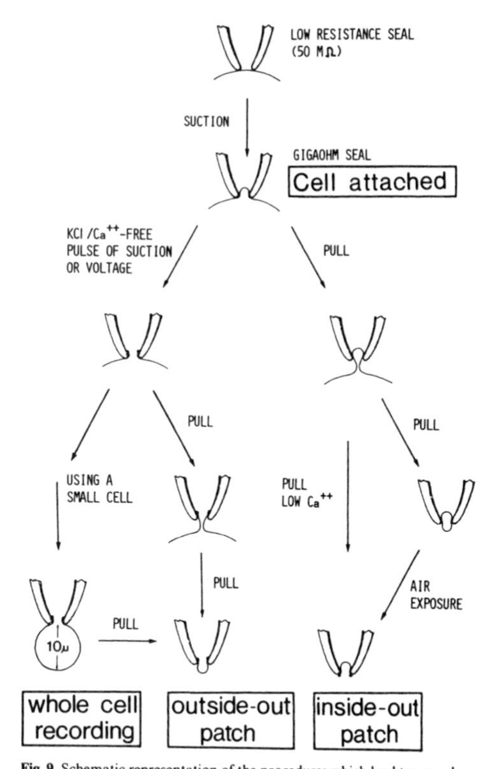

Fig. 9. Schematic representation of the procedures which lead to recording configurations. The four recording configurations, described in this paper are: "cell-attached", "whole-cell recording", "outside-out patch", and "inside-out patch". The upper most frame is the configuration of a pipette in simple mechanical contact with a cell, as has been used in the past for single channel recording ((Neher et al. 1978). Upon slight suction the seal between membrane and pipette increases in resistance by 2 to 3 orders of magnitude, forming what we call a cell-attached patch. This configuration is described in part II of this article. The improved seal allows a 10-fold reduction in background noise. This stage is the starting point for manipulations to isolate membrane patches which lead to two different cell-free recording configurations (the outside-out and inside-out patches described in part III). Alternatively, voltage clamp currents from whole cells can be recorded after disruption of the patch membrane if cells of sufficiently small diameter are used (see part IV of this article). The manipulations include withdrawal of the pipette from the cell (pull), short exposure of the pipette tip to air and short pulses of suction or voltage applied to the pipette interior while cell-attached

Part III

Single Channel Current Recording from "Cell-Free" Membrane Patches

Apparently the contact between cell membrane and glass pipette after formation of a giga-seal is not only electrically tight, but also mechanically very stable. The pipette tip can be drawn away from the cell surface without a decrease in the seal resistance (Hamill and Sakmann 1981; Neher 1981). As will be shown below a tight vesicle sealing off the tip forms, when this is done in normal Ca^{2+}-containing bath solution. Procedures are described by which the resistance of either the inner or the outer part of the vesicle can be made low (< 100 MΩ) without damaging the giga-seal. The remaining intact membrane can then be studied as before. Either "inside-out" or "outside-out" patches can be isolated in this way (see Fig. 9). By varying the composition of the bath solution, the effect of drugs or ion concentration changes on single channel currents can be studied at either the cytoplasmic or the extracellular face of the membrane.

1. Vesicle Formation at the Pipette Tip

The top trace in Fig. 10A shows single channel currents recorded in a frog muscle fibre at its resting potential of −90 mV. The patch pipette was filled with standard Ringer solution plus 50 nM suberyldicholine (SubCh) and was sealed against the surface membrane in the perisynaptic region. When the pipette tip was slowly withdrawn a few µm from the cell surface the shape of single channel currents became rounded and they decreased in size (middle trace). Often a fine cytoplasmic bridge could then be observed between the cell surface and the pipette as shown in Fig. 10D. Upon further removal the cytoplasmic bridge tied off but left the giga-seal intact. Single channel currents further decreased in size and finally disappeared in the background noise within the next 1−2 min. The pipette input resistance remained high (110 GΩ, see bottom trace).

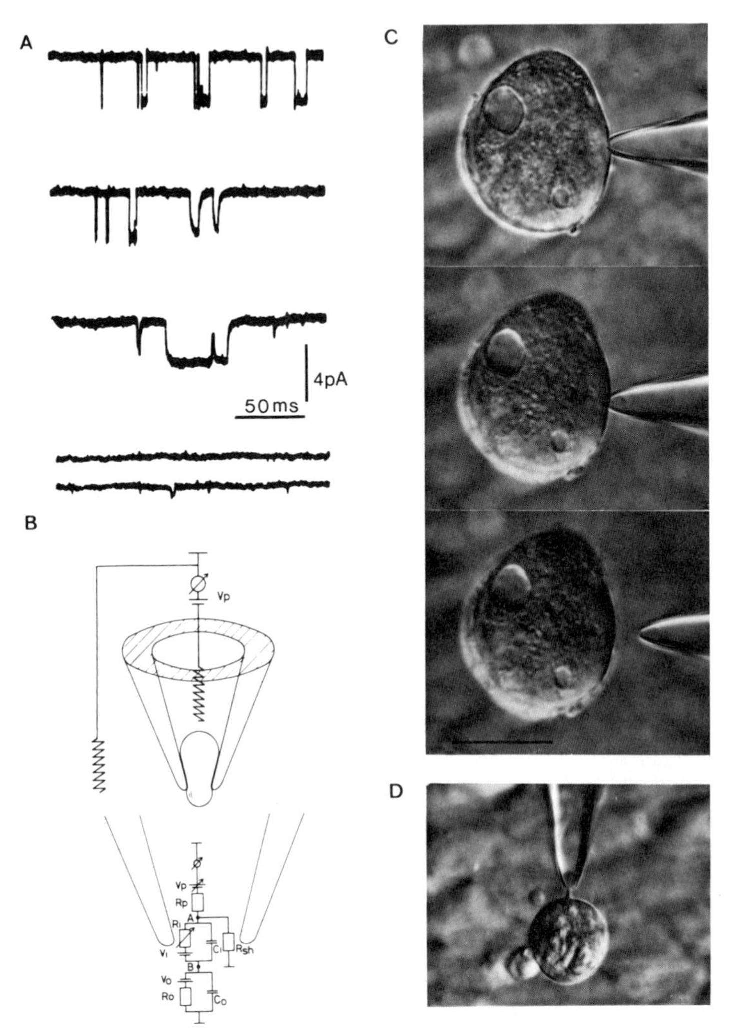

Fig. 10A–D. Formation of a membrane vesicle at the pipette tip. (A) Patch current recording during withdrawal of the pipette tip from muscle sacrolemma. The *upper trace* illustrates single channel currents after formation of a gigaseal under conditions similar to those of Fig. 8. During withdrawal of the pipette single channel currents suddenly appeared distorted in shape; currents showed rounded rising and falling time courses and progressively lower amplitudes. In this experiment single channel currents became undetectable within 30 s following the appearance of rounded current events. The pipette-membrane seal resistance remained high, (> 30 GΩ) and current pulses remained undetectable even after hyperpolarizing the membrane by 90 mV. This suggests that the pipette opening became occluded by a membraneous structure, which most probably was a closed vesicle as illustrated in part **B**. **(B)** Schematic diagram of recording situation following formation of a membrane vesicle. The equivalent circuit is modelled by parallel current pathways through the shunt resistance R_{sh} and through the vesicle. The two halves of the vesicle are represented by $R-C$ combinations in series. The inner membrane, exposed to the pipette solution, has resistance R_i and capacitance C_i. The outer membrane, represented by R_o and C_o, is exposed to the bath solution. Point **B** represents the interior of the vesicle. V_o and V_i are the electromotive forces of the outer and inner membranes. R_p is the access resistance of the unsealed pipette. Opening of a channel in the presence of ACh in the pipette solution decreases the resistance of the inner membrane, R_i. Therefore, R_i which in fact is a parallel combination of membrane resistance and open channel resistance R_c is modelled as a variable resistor. **(C)** Visualisation of membrane patch isolation from a rat myoball. The upper micrograph shows the tip of a patch pipette in contact with a myoball. When suction is applied, an Ω-shaped membrane vesicle is pulled into the pipette tip (*middle micrograph*). Following slight withdrawal of the pipette tip from the myoball, a cytoplasmic bridge of sarcolemma between pipette tip and myoball surface was observed (not shown). At this stage of membrane isolation distorted single channel current pulses are recorded like those shown in **A**. Upon further withdrawal the cytoplasmic bridge ruptures leaving a small vesicle protruding from the pipette tip (*lower micrograph*). Note "healing" of the sarcolemmal membrane of the myoball. **(D)** Visualisation of cytoplasmic bridge between myoball and vesicle in the pipette tip in another experiment. Calibration bar: 20 μm. (Normarski interference optics, × 40 water immersion objective)

The most likely explanation for this sequence of events is that during rupture of the cytoplasmic bridge between pipette tip and cell surface a closed vesicle formed at the tip of the pipette. The formation of a vesicle is illustrated in Fig. 10C on a myoball. The vesicle is partly exposed to the bath solution ("outer membrane") and partly to the pipette solution ("inner membrane") in the way shown schematically in Fig. 10B.

Three kinds of observations give indications on the electrical properties of the vesicle. (i) When ACh is added at a low concentration to the bath solution, current pulses of reduced size are sometimes recorded at an applied pipette potential V_p of +90 mV or more. These current pulses reflect an outward current flow through AChR channels on the outer membrane of the vesicle. (ii) Destruction of the barrier properties of the outer membrane (see the next section) results in the reappearance of single channel currents through the inner membrane. (iii) In some experiments single channel currents of decreased amplitude and distored shape could be recorded when the pipette potential was increased to $> +70$ mV even several minutes following pipette withdrawal. Examples of such distorted current pulses recorded from patches and vesicles are illustrated in Figs. 10A and 11. Single channel currents were of rectangular shape as long as the membrane patch was attached to the cell.

Equivalent circuit. The electrical signals after vesicle formation can be explained by an equivalent circuit shown in Fig. 10B. The two halves of the vesicle form a series combination of two RC-circuits, shunted by the leakage resistance R_{sh}. Assuming a specific conductance of the membrane of 10^{-4} $\mathrm{Scm^{-2}}$ and an area of a semi-vesicle in the range of several μm^2, values for the membrane resistances R_i and R_o should be in the range of several hundred GΩ, much larger than the resistance of a single open AChR-channel (about 30 GΩ). Local damage or partial breakdown of the membrane could bring this value down into the range of single open channels. Then, opening of a single channel in either of the two membranes would result in an attenuated current flow through the series combination. Consider the simple case that R_i and R_o are the same and are identical to the resistance of a single channel R_c. Then, upon opening of a single channel the apparent conductance at steady state is attenuated by a factor of 6 with respect to the true conductance.

More generally, the apparent conductance G_{app} of the channel (assumed to open on the inner side of the vesicle) will be

$$G_{app} = R_i^2/((R_i+R_o)\,(R_cR_o+R_cR_i+R_iR_o)).$$

The time course will be governed by the time course of the voltage at point B in the equivalent circuit. For a series combination of two RC-elements and a step-like perturbation this will be a single exponential (neglecting R_p of Fig. 10B). The time constant τ of the exponential is equal to

$$\tau = R_c \| R_i \| R_o \cdot (C_i+C_o),$$

where $R_c \| R_i \| R_o$ is the parallel combination of $R_c R_i$ and R_o (see Fig. 10B). For the simple case above $\tau_1 = R_c\,(C_i+C_o)/3$ while the channel is open and $\tau_2 = R_c\,(C_i+C_o)/2$ with the channel closed. The measurement of a time constant, therefore, gives the total membrane capacitance of the vesicle. More generally, determination of G_{app}, τ_1, and τ_2 allows calculation of the three unknowns R_i, R_o and (C_i+C_o) if the conductance of the single channel $(1/R_c)$ is known. This analysis neglects the effects of R_{sh}, which however adds only a constant offset in current (at V_p = constant and $R_p \ll R_i, R_o$).

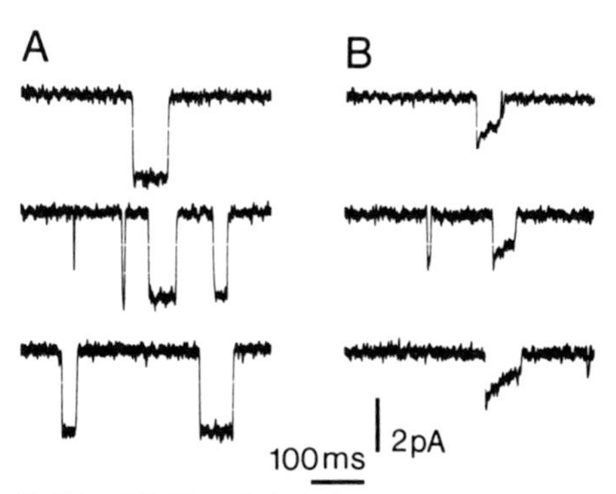

Fig. 11A and B. Distorted single channel current shapes following isolation of a membrane vesicle from myoball sarcolemma. (**A**) Single channel currents recorded from a cell-attached membrane patch. In this experiment the Na^+ in the pipette solution was reduced to 50 mM. Pipette potential was +80 mV. Assuming a resting potential of −70 mV the membrane potential across the patch was −150 mV. The slope conductance was 21 pS. (**B**) Single channel currents after physical isolation of the membrane patch from the myoball. Pipette potential was +80 mV. Single channel currents could be observed at the pipette zero potential indicating that the vesicle had a membrane potential. The apparent slope conductance of the initial peak amplitude of these current events was 12 pS. All records filtered at 0.4 kHz, low pass; 18°C. The duration of the current events is not representative since long duration currents were selected to illustrate differences in the time course in cell-attached and cell-free configurations. Also, the time constant of decay was unusually long in this experiment. In some experiments single channel currents similar to those shown here but of very small amplitude could be observed also without application of a pipette potential. This indicates that the vesicle can have a resting potential

Experimental data gave estimates for C_i+C_o in the range 0.03 − 0.3 pF. Assuming a specific capacitance of the vesicle membrane of $1\,\mu F/cm^2$ this results in a vesicle area of $3-30\,\mu m^2$. The waveform of individual channel responses (Fig. 11) can display either rising or decaying relaxations. The form depends on the specific relation between R_o, R_i, C_o and C_i.

Although closed vesicles form regularly in standard bath solution 1 mM Ca^{2+} and 1 M Mg^{2+}, formation of tight vesicles is infrequently observed when divalent metal cations are left out of the bath solution or if they are chelated by EGTA. Horn and Patlak (1980) used F^--containing bath solution to prevent formation of closed vesicles.

Simultaneous intracellular recording of the membrane potential of the cell under study shows that isolating a membrane vesicle from the cell surface membrane does not damage the cell. In a few cases even the opposite was observed. Cells that depolarized partially while the seal was being formed returned to the normal resting potential after pipette withdrawal.

2. Formation of "Cell-Free" Membrane Patches at the Pipette Tip

Vesicle formation at the tip opening offers the possibility of measuring single channel currents in cell-free patches by selectively disrupting either the inner or the outer membrane of the vesicle. Figure 9 illustrates schematically how this can be done. When the outer membrane of the vesicle is disrupted the cytoplasmic face of the inner membrane is exposed to the bath solution. Its extracellular face is exposed to the pipette solution. We call this an "inside-out" membrane patch. When, on the other hand, the inner membrane of the vesicle is disrupted the cytoplasmic face of the outer membrane of the vesicle is exposed to the pipette solution, its extracellular face to the bath solution. This will be called an "outside-out" patch. The next two sections describe formation of these two configurations of membrane patches.

The "Inside-Out" Membrane Patch. To obtain a membrane whose cytoplasmic face is exposed to the bath solution, either formation of the outer vesicle membrane has to be prevented (Horn and Patlak 1980) or it has to be disrupted once it has been formed (Hamill and Sakmann 1981). This can be done either mechanically or chemically.

A prerequisite for isolated patch formation is a seal of $>20\ G\Omega$. Disruption of the outer vesicle membrane is done by passing the pipette tip briefly through the air-water interface of the bath or by touching an air bubble held by a nearby pipette. Brief contact with a drop of hexadecane will sometimes disrupt the vesicle. Figure 12 illustrates the disruption of the barrier properties of the outer vesicle membrane. Initially the pipette tip was sealed against a myoball membrane. The pipette solution contained 0.5 μM ACh. Single channel currents were recorded at the cell's resting potential as shown in the uppermost trace. After withdrawal of the pipette tip from the cell surface single channel currents disappeared. Upon increasing the pipette potential to +70 mV the trace became noisier but single currents were not resolved. By briefly (1 – 2 s) passing the tip through the air-water interface the outer membrane of the vesicle was disrupted. Upon reimmersion of the tip into the bath solution single channel currents of the expected size were recorded (Fig. 12). Single channel currents, which in most membrane patches of myoballs fall into two classes (Hamill and Sakmann 1981) were similar in their respective amplitudes in the cell-attached and cell-free configuration as shown in Fig. 12B on another patch at –100 mV membrane potential.

The outer membrane of the vesicle can also be made leaky by exposing it to a Ca-free, 150 mM KCl bath solution during isolation, as was originally used by Kostyuk et al. (1976) to disrupt neuronal membranes for internal dialysis. A vesicle tends to reform with this procedure, and mechanical disruption is usually required in addition to open the vesicle completely. The stability of inside-out patches is greatly improved when most of the Cl^- in the bath solution is replaced by SO_4^{2-}. For membrane patches isolated from myoballs an anion mixture of 4 mM Cl^- and 75 mM SO_4^{2-} was found to yield stable recordings for up to several hours.

The "Outside-Out" Membrane Patch. In order to work with the outer membrane of the vesicle, the inner membrane can be made leaky or can be disrupted in very much the same way as described for the case above, i.e. by exposing the inner vesicle membrane to a pipette solution containing 150 mM KCl and only a low ($<10^{-6}$ M) concentration of Ca. Alternatively it can be opened by mechanical rupture of the patch preceding vesicle formation. Isolation of a membrane patch in the outside-out configuration is illustrated in Fig. 13. A pipette containing 150 mM KCl and 3 mM HEPES buffer was used. A few minutes following the establishment of a giga-seal, the background noise increased progressively by several orders of magnitude. This was accompanied by a decrease of the

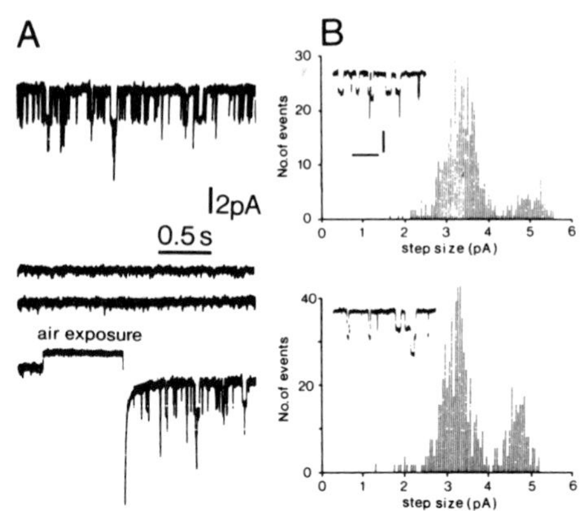

Fig. 12A and B. Isolation of an "inside-out" membrane patch from rat myoball sarcolemma. (**A**) The upper three traces were recorded during the process of vesicle formation and are analogous to those shown in Fig. 10A. Standard solutions were used. The pipette contained 0.5 μM ACh. Single channel currents (*first trace*, recorded at the cell's resting potential of ~ –70 mV) became undetectable following withdrawal of the pipette tip. Upon shifting the pipette potential from 0 (*second trace*) to +70 mV (*third trace*) the current trace changed by only 0.8 pA, indicating formation of a closed vesicle where $R_o > R_i$ (see Fig. 10B). The pipette tip was then passed briefly (1 – 2 s) through the air-water interface which disrupted the outer membrane of the vesicle. Upon reimmersion single channel currents were recorded at a pipette potential of +70 mV (*fourth trace*), which were similar to those of the cell-attached case (shown in the first trace). (**B**) Step size distribution of single channel currents recorded from the same patch of membrane in the cell-attached configuration (*upper graph*) and in the cell-free "inside-out" configuration (*lower graph*), both at a membrane potential of –100 mV. The pipette solution contained 0.5 μM ACh. The distribution of step sizes is doubly peaked. It indicates that in this patch two types of AChR channels, junctional and extrajunctional, with slightly different open channel conductances were activated. The larger spread of step size distributions in the cell-attached recording configuration is due to the larger background noise which in this experiment was mostly caused by mechanical instabilities and which disappears following isolation of the patch. In some experiments the average opening frequency of AChR-channels decreased (up to 30%) following isolation of the membrane patch. This is probably due to a decrease of the inner membrane area exposed to the pipette solution. Insets show examples of single channel current events (Calibration bars: 4 pA and 50 ms). A downward deflection of the current trace in this and all other figures indicates cation transfer from the compartment facing the extracellular membrane side to the compartment facing the cytoplasmic side. This is from the pipette to the bath solution for an inside-out patch

pipette-bath resistance to less than 1 GΩ and by the development of a large inward current. Upon withdrawal of the pipette tip from the cell surface the background noise decreased within 1 – 2 s to the initial low level and the pipette-bath resistance simultaneously increased to a value larger than 10 GΩ.

We attribute the initial decrease of the pipette input resistance to the disruption of the membrane patch and not to an increased leakage of the pipette-membrane seal. This follows from the observation that the inward current inverts at a rather large negative pipette potential presumably equal to the cell resting potential. Also the pipette input capacitance shows an increase corresponding to the cell capacitance (see below, Part IV).

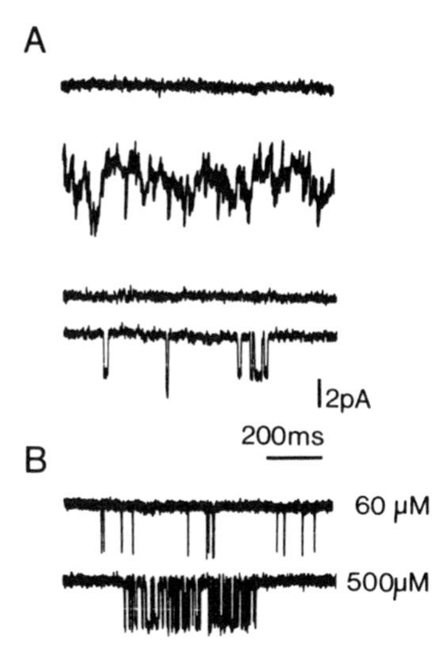

Fig. 13A and B. Isolation of an "outside-out" membrane patch from rat myoball sarcolemma. (**A**) After formation of a "gigaseal" using a pipette containing 150 mM KCl and 3 mM HEPES at pH 7.2 the background noise increased within 2 – 3 min. The upper two traces represent recordings immediately following gigaseal formation and 3 min later. At this stage a leakage current of > lnA developed which eventually drove the feedback amplifier into saturation. The leakage currents could be reduced or inverted when the pipette potential was shifted to – 50 to – 70 mV. The pipette access resistance decreased to values < 100 MΩ. Upon withdrawal of the pipette tip the pipette access resistance increased again into the GΩ range, and the background noise decreased (third trace). Addition of 0.5 – 1 μM ACh to the bath solution induced single channel currents of 2.5 pA amplitude at – 70 mV membrane potential (fourth trace). All records are filtered at 0.5 kHz; temp. 18° C. (**B**) Demonstration of equilibration of bath-applied agonists at the extracellular face of an outside-out patch. Single channel currents were recorded at – 70 mV membrane potential from the same membrane patch when either 60 μM carbachol (*above*) or 500 μM carbachol (*below*) was added to the bath solution. At the lower concentration single channel current events appeared at random, at the higher concentration current pulses appeared in "bursts". Addition of 10^{-6} M α-BuTx irreversibly blocked agonist activated currents (not shown)

The subsequent sealing observed when pulling away the pipette apparently results from the formation of a new membrane bilayer at the pipette tip, with its external side facing the bath solution (outside-out patch, see Fig. 9). Several observations lead to this conclusion: (i) Air exposure of the pipette tip as described in the previous section results in a decrease of the pipette-bath resistance to values < 100 MΩ. (ii) Addition of low ACh concentrations to the bath solution activates single channel current pulses. At a pipette potential of – 70 mV they are similar in their amplitude and average duration to those observed on cell-attached membrane patches (Fig. 13A). (iii) In experiments where only the pipette contained ACh, no single channel currents were recorded at this stage.

The breakdown of the initial membrane patch, which was spontaneous in the experiment illustrated in Fig. 13A, can be accelerated or initiated by applying brief voltage (of up to 200 mV) or negative pressure pulses (matching the pressure of a 100 cm H_2O column) to the pipette interior. In this way, the access resistance can be lowered to a value near the pipette resistance.

The extracellular face of outside-out membrane patches equilibrates rapidly (< 1 s) and reversibly with ACh added to the bath solution by perfusion of the experimental chamber or applied by flow of ACh-containing solution from a nearby pipette of 20 – 50 μm tip diameter. Figure 13B shows a recording of single channel currents from an outside-out patch when carbachol is applied at 60 μM and 500 μM. At the low carbachol concentration current pulses appear at random intervals, whereas bursts of current pulses are recorded at high agonist concentration as previously reported for cell-attached membrane patches (Sakmann et al. 1980).

3. Equilibration of the Cytoplasmic Face of Cell-Free Patches with Bath or Pipette Solutions

In the previous two sections it was shown that by suitable manipulations either the inner membrane or the outer membrane of the vesicle can be disrupted such that it represents a low series resistance. In order to check whether, using either of the two configurations of cell-free membrane patches, the cytoplasmic face of the patch equilibrates with the experimental solution we have measured I – V relations of ACh-activated channels under various ionic conditions in both inside-out and outside-out configurations (Fig. 14A). The *I – V* relations show the following features which are expected for ionic equilibration between the cytoplasmic membrane face and the bath solutions or pipette solutions: (i) when Na^+ concentration is reduced on one side of the membrane, one branch of the *I – V* relation changes strongly whereas the other branch is minimally affected in its extremes (ii) the two recording configurations result in overlapping *I – V* relationships when ionic compositions on both sides of the membrane are the same (iii) changes in the shape of the *I – V* relations can be reversed and are reproducible from one patch to the next.

It is well established that the disrupted membrane of the vesicle does not represent an appreciable electrical series resistance (see for instance Fig. 12). However it might well contribute to changes in the *I – V* relation if its conductance were high, but ion-selective. In such a case a potential would develop across the disrupted membrane which would produce a parallel shift in the *I – V* relation along the voltage axis. This, however, is contrary to the results shown in Fig. 14A where changes occur in curvature, and neighbouring curves approach each other asymptotically.

Further evidence for ionic equilibration is provided by kinetic studies. It was found that after successful patch isolation the *I – V* relations did not change their properties with time if solutions in the pipette and in the bath were kept constant. Furthermore steady state properties were obtained instantaneously upon a change of environment. This is shown in Fig. 14B.

Both the changes in curvature and the fact that the changes occur instantaneously point towards very efficient exchange of ionic contents between the interior of a vesicle and the compartment neighbouring its disrupted membrane. In addition, ionic exchange across a disrupted membrane between the pipette interior and the cell interior was observed in the whole cell recording configuration (see part IV). These findings, together, make it very likely that true equilibration

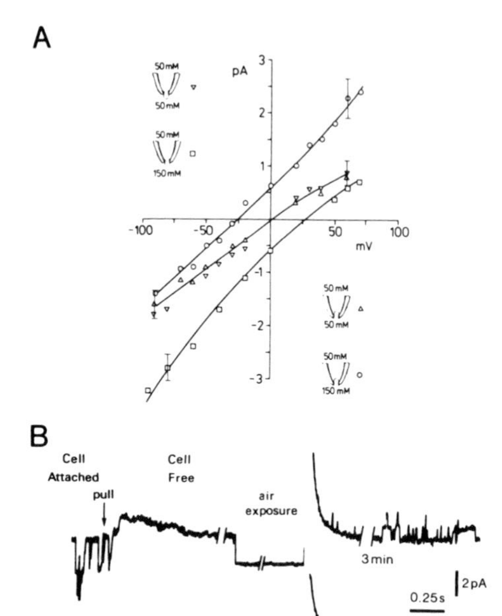

Fig. 14A and B. Equilibration of the cytoplasmic face of cell-free membrane patches with bath or pipette solutions. (**A**) $I-V$ relationships for acetylcholine-activated channels in inside-out and outside-out membrane patches measured under different ionic conditions. For all measurements the pipette solution contained 50 mM NaCl as the predominant salt. The bath solution contained initially 150 mM NaCl, which was changed to a solution containing 50 mM after patch isolation. Note that under symmetrical conditions $I-V$ relationships in both patch configurations overlapped, passed through the pipette zero potential, and showed slight rectification. Under asymmetrical conditions the reversal potentials were shifted by 25 mV. The $I-V$ relationships showed the expected curvature, and, in one branch each, approached the neighbouring symmetrical $I-V$. The symbols ○, △, ▽, □ represent the mean channel currents determined from 16, 3, 2, and 3 separate experiments. SEM's are shown for some averages. All measurements were made at 18 C. Changes in junction potentials caused by solution changes were measured independently, and were corrected for. (**B**) Current records before, during, and after formation of an inside-out patch at zero pipette potential. For both upper and lower traces the pipette was withdrawn from the cell while in normal bath solution. The pipette solutions contained 50 mM NaCl. During withdrawal of the pipette from the myoball, ACh-activated currents, evident while cell-attached, disappeared. For the upper trace the pipette tip was briefly exposed to air, and then returned to the normal bath solution. Inverted currents immediately appeared consistent with the vesicle being opened to the bath solution. The currents recorded did not change their properties over a three-min period. They displayed an $I-V$ relation similar to the inside-out asymmetrical case described in part **A**. For the *lower trace*, after pipette withdrawal from the cell, the bath solution was changed to one containing 50 mM NaCl and the pipette tip was then exposed shortly to air. No ACh-induced currents were evident at the pipette zero potential. They appeared upon polarization and displayed a similar $I-V$ relationship as described for symmetrical cases in part **A**

between the cytoplasmic membrane face and adjacent bulk solutions takes place.

Part IV

*Recording of Whole-Cell Voltage Clamp Currents**

It was pointed out above that the membrane patch which separates the pipette from the cell interior can be broken without damaging the seal between the pipette rim and the cell membrane. This is the situation occurring at the initial stage during the formation of outside-out patches (see p. 96). Here, we demonstrate the suitability of this configuration for studying the total ionic currents in small cells. The technique to be described can be viewed as a microversion of the internal dialysis techniques originally developed for molluscan giant neurons (Krishtal and Pidoplichko 1975; Kostyuk and Krishtal 1977; Lee et al. 1978) and recently applied to mammalian neurones (Krishtal and Pidoplichko 1980). As it is appropriate only for cells of less than 30 μm in diameter, we take as an example bovine chromaffin cells in short-term tissue culture. These cells have a diameter of 10–20 μm. Their single channel properties will be detailed elsewhere (Fenwick, Marty, and Neher, manuscript in preparation).

The pipette was filled with a solution mimicking the ionic environment of the cell interior (Ca-EGTA buffer, high K^+). After establishment of a giga-seal, the patch membrane was disrupted, usually by suction, as previously described (see above). The measured zero-current potential was typically −50 to −70 mV. This corresponds to the cell resting potential (Brandt et al. 1976). Applying small voltage jumps from this potential revealed a resistance value in the range of 10 GΩ. This resistance is mainly due to the cell membrane since markedly larger resistances (20–50 GΩ) were obtained when using a CsCl solution in the pipette interior and tetrodotoxin in the bath. Small voltage jumps also showed that the disruption of the intial patch is accompanied by a large increase of the input capacitance (Fig. 15). The additional capacitance was about 5 pF, in good agreement with the value expected from the estimated cell surface, assuming a unit capacity of 1 $\mu F/cm^2$. The time constant of the capacity current was of the order of 100 μs, which shows that the series resistance due to the pipette tip is no more than 20 MΩ. However, larger time constants were occasionally observed, indicating an incomplete disruption of the initial membrane patch.

Depolarizing voltage commands elicited Na and K currents which could be well resolved after compensation of the cell capacitance current (Fig. 16). The cell can be considered under excellent voltage clamp since, (i) at the peak inward current, the voltage drop across the series resistance is small (less than 2 mV in the experiment of Fig. 16, assuming a 20 MΩ series resistance), and (ii) the clamp settles within 100 μs as indicated above. The background noise was somewhat larger than that of a patch recording due to the conductance and capacitance of the cell. However, for small cells, resolution was still good enough to record large single channel responses. This is illustrated in Fig. 17 which shows individual ACh-activated channel currents in a chromaffin cell.

The pipette provides a low-resistance access to the cell interior which we used to measure intracellular potentials under current clamp conditions. The recordings showed spontaneous action potentials resembling those published by

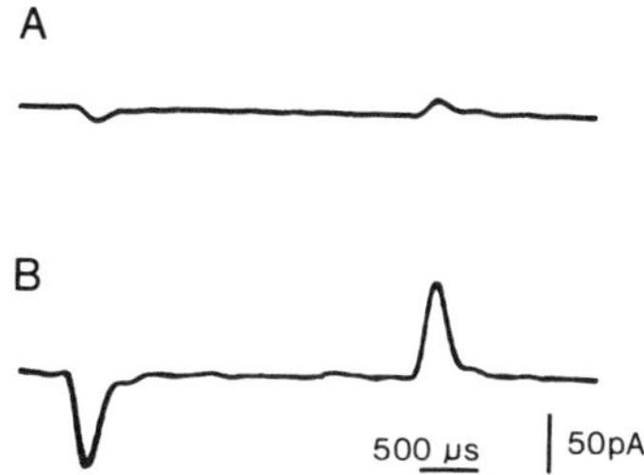

Fig. 15A and B. Capacitive current of a chromaffin cell. (**A**) After establishing a gigaseal on a chromaffin cell, a 20 mV pulse was applied to the pipette interior starting from a holding potential equal to the bath potential. The capacitance of the pipette and of the patch was almost completely compensated (see part I), resulting in very small capacitive artifacts. The pipette input resistance was 20 GΩ. (**B**) After disruption of the patch the response to a 3 mV pulse at a holding potential of −54 mV was measured. The capacitive current was much larger than in **A**, as the cell membrane capacitance had to be charged (compare the amplitudes of the voltage steps). From the integral of the capacitive current a cell membrane capacitance of 5 pF is calculated. The cell had a diameter of 13 µm. Assuming a spherical shape, one obtains a unit capacitance of about 1 µF/cm². The DC current was smaller than 2 pA, indicating a cell membrane resistance of several GΩ. The time constant of the capacitive current was less than 0.1 ms, which, together with the value of the cell capacitance, indicates a series resistance smaller than 20 MΩ

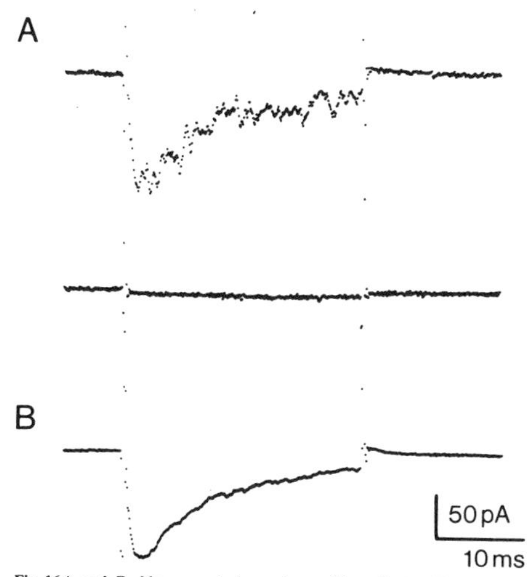

Fig. 16A and B. Na currents in a chromaffin cell. (**A**) Single sweep responses to 34 mV depolarizing (*above*) and hyperpolarizing (*below*) pulses starting from a holding potential of −56 mV. (**B**) Average responses to 25 depolarizing voltage commands, as above. Room temperature; 1 kHz low pass

Brandt et al. (1976). The resting potential, averaging around −60 mV, displayed large fluctuations presumably due to spontaneous opening and closing of ionic channels. Similarly, action potentials and EPSPs could be recorded from small cultured spinal cord neurons.

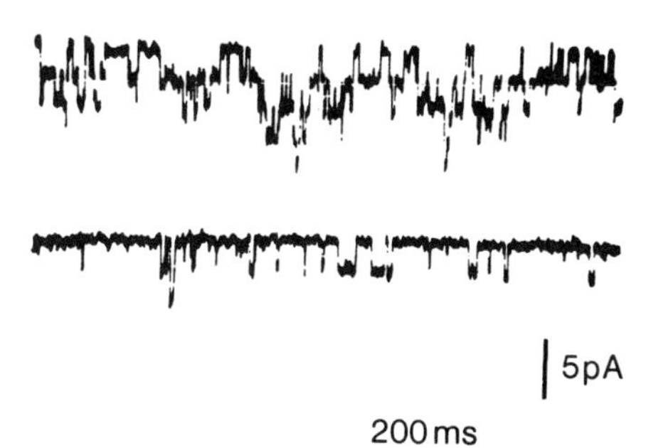

Fig. 17. Single channel records from a whole chromaffin cell. Normal bath solution with 100 µM ACh added to the chamber. Single ACh-induced currents appeared at varying frequency due to slow changes in ACh concentration and due to desensitization. Two examples at different mean frequency are given. Holding potential was −65 mV; the amplitude of single channel currents was 2.3 pA and their mean open time approximately 30 ms. Filter bandwith 200 Hz

When the pipette was withdrawn from the cell it sealed again, forming an outside-out patch (see Part III,2) of small dimensions. The pipette input capacitance dropped back to a value close to that observed during establishment of the initial giga-seal. Concomitantly background noise was reduced and small channel currents, like those of individual Na channels, could be observed.

In some experiments, the pipette was filled with a Cs-rich solution. The resting potential dropped from a normal value to zero within 10 s after disruption of the initial patch. This observation suggests that the intracellular solution exchanges quickly with the pipette interior. Thus, the method described in this section may be applied not only to record membrane currents, but also to alter the cell's ionic contents. Compared to the nystatin method (Cass and Dalmark 1973) it has the advantage to allow the exchange of divalent ions and of macromolecules.

The present method offers several advantages over the usual recording techniques using glass microelectrodes. It avoids the leakage due to cell penetration with the microelectrode, it allows reliable voltage clamp of small cells, and it offers the possibility of studying macroscopic currents and single channel currents in the same cell. It also allows at least partial control of the ionic milieu of the cell interior.

Conclusions

The methods described here provide several options for voltage- or current-clamp recording on cells or cell-free membrane patches. The size of the cells is not a restriction to the applicability of at least two of these methods. The only requirement is a freely accessible cell surface. This requirement is naturally fulfilled for a number of preparations. It is also fulfilled for most other preparations after enzymatic cleaning. The variety of cell types on which giga-seal formation has been successful is illustrated in Table 1.

The manipulations described here provide free access to either face of the membrane for control of the ionic environment. The giga-seal allows ionic gradients across the membrane to be maintained and high resolution measurements of current through the membrane to be performed.

Table 1. A listing of preparations on which giga-seals have been obtained. Only a fraction of the preparations have been investigated in detail. Bovine chromaffin cells and guinea pig liver cells were plated and kept in short term tissue culture. We acknowledge receiving cells from J. Bormann, T. Jovin, W. D. Krenz, E.-M. Neher, L. Piper, and G. Shaw, Göttingen, FRG; I. Schulz, Frankfurt, FRG; G. Trube, Homburg, FRG, and I. Spector, Bethesda, MD, USA

Cell lines in tissue culture:	Single cells:
Mouse neuroblastoma	Human and avian erythrocytes
Rat basophilic leukaemia cells	Mouse activated macrophages
Primary tissue culture:	Enzymatically dispersed cells:
Rat myotubes and myoballs	Bovine chromaffin cells
Mouse and rabbit spinal cord cells	Guinea pig heart myocytes
Rat cerebellar cells	Guinea pig liver cells
Rat dorsal root ganglion cells	Mouse pancreatic cells
Torpedo electrocytes	Enzyme treated cells:
Rat fibroblasts	Frog skeletal muscle fibres
	Rat skeletal muscle fibres
	Snail ganglion cells

The manipulations are simple and, with some practice, appear to be performed more easily than standard voltage clamp experiments. We expect that they will help to clarify physiological mechanisms in a number of preparations which, so far, have not been amenable to electrophysiological techniques.

Acknowledgements. We thank E. Fenwick for providing dispersed chromaffin cells. We also thank H. Karsten for culturing myoballs and Z. Vogel for a gift of Rhodamine-labelled α-bungarotoxin. O. P. Hamill and F. J. Sigworth were supported by grants from the Humboldt foundation. E. Neher and A. Marty were partially supported by the Deutsche Forschungsgemeinschaft.

References

Brandt BL, Hagiwara S, Kidokoro Y, Miyazaki S. (1976) Action potentials in the rat chromaffin cell and effects of Acetylcholine. J Physiol (Lond) 263:417–439

Cass A, Dalmark M (1973) Equilibrium dialysis of ions in nystatin-treated red cells. Nature (New Biol) 244:47–49

Fenwick EM, Fajdiga PB, Howe NBS, Livett BG (1978) Functional and morphological characterization of isolated bovine adrenal medullary cells. J Cell Biol 76:12–30

Gage PW, Van Helden D (1979) Effects of permeant monovalent cations on end-plate channels. J Physiol (Lond) 288:509–528

Hamill OP, Sakmann B (1981) A cell-free method for recording single channel currents from biological membranes. J Physiol (Lond) 312:41–42P

Horn R, Brodwick MS (1980) Acetylcholine-induced current in perfused rat myoballs. J Gen Physiol 75:297–321

Horn R, Patlak JB (1980) Single channel currents from excised patches of muscle membrane. Proc Natl Acad Sci USA 77:6930–6934

Kostyuk PG, Krishtal OA (1977) Separation of sodium and calcium currents in the somatic membrane of mollusc neurones. J Physiol (Lond) 270:545–568

Kostyuk PG, Krishtal OA, Pidoplichko VI (1976) Effect of internal fluoride and phosphate on membrane currents during intracellular dialysis of nerve cells. Nature 257:691–693

Krishtal OA, Pidoplichko VI (1975) Intracellular perfusion of Helix neurons. Neurophysiol (Kiev) 7:258–259

Krishtal OA, Pidoplichko VI (1980) A receptor for protons in the nerve cell membrane. Neuroscience 5:2325–2327

Läuger P (1975) Shot noise in ion channels. Biochim Biophys Acta 413:1–10

Langmuir I (1938) Overturning and anchoring of monolayers. Science 87:493–500

Lee KS, Akaike N, Brown AM (1978) Properties of internally perfused, voltage clamped, isolated nerve cell bodies. J Gen Physiol 71:489–508

Neher E (1981) Unit conductance studies in biological membranes. In: Baker PF (ed), Techniques in cellular physiology. Elsevier/North-Holland, Amsterdam

Neher E, Sakmann B (1976) Single channel currents recorded from membrane of denervated frog muscle fibres. Nature 260:799–802

Neher E, Sakmann B, Steinbach JH (1978) The extracellular patch clamp: A method for resolving currents through individual open channels in biological membranes. Pflügers Arch 375:219–228

Nir S, Bentz J (1978) On the forces between phospholipid bilayers. J Colloid Interface Sci 65:399–412

Parsegian VA, Fuller N, Rand RP (1979) Measured work of deformation and repulsion of lecithin bilayers. Proc Natl Acad Sci USA 76: 2750–2754

Petrov JG, Kuhn H, Möbius D (1980) Three-Phase Contact Line Motion in the deposition of spread monolayers. J Colloid Interface Sci 73:66–75

Sakmann B, Patlak J, Neher E (1980) Single acetylcholine-activated channels show burst-kinetics in presence of desensitizing concentrations of agonist. Nature 286:71–73

Sigworth FJ, Neher E (1980) Single Na^+ channel currents observed in cultured rat muscle cells. Nature 287:447–449

Stevens CF (1972) Inferences about membrane properties from electrical noise measurements. Biophys J 12:1028–1047

Received March 11 / Accepted May 27, 1981

Index

f = figures; *t* = tables.